Operator Theory: Advances and Applications
Vol. 127

Recent Advances in Operator Theory and Related Topics

The Béla Szökefalvi-Nagy Memorial Volume

László Kérchy
Ciprian Foias
Israel Gohberg
Heinz Langer
Editors

Springer Basel AG

Editors:

László Kérchy
Bolyai Institute
University of Szeged
Aradi Vértanuúk Tere 1
6720 Szeged
Hungary

Ciprian I. Foias
Department of Mathematics
Indiana University
Bloomington, IN 47405-4301
USA

Prof. I. Gohberg
School of Mathematical Sciences
Raymond and Beverly Sackler
Faculty of Exact Sciences
Tel Aviv University
Ramat Aviv 69978
Israel

Prof. H. Langer
Mathematik
Technische Universität Wien
Wiedner Hauptstrasse 8–10/1411
1040 Wien
Austria

2000 Mathematics Subject Classification 47-06

A CIP catalogue record for this book is available from the Library of Congress, Washington D.C., USA

Deutsche Bibliothek Cataloging-in-Publication Data

Recent advances in operatory theory and related topics : the Béla Szökefalvi-Nagy memorial volume / Lászlo Kérchy ... ed.. - Basel ; Boston ; Berlin : Birkhäuser, 2001
(Operator theory ; Vol. 127)
ISBN 3-7643-6607-9

ISBN 978-3-0348-9539-2 ISBN 978-3-0348-8374-0 (eBook)
DOI 10.1007/978-3-0348-8374-0

Member of the BertelsmannSpringer Publishing Group
Printed on acid-free paper produced from chlorine-free pulp. TCF ∞
Cover design: Heinz Hiltbrunner, Basel

ISBN 3-7643-6607-9

www.birkhäuser-science.com

9 8 7 6 5 4 3 2 1

Contents

Preface

Béla Szőkefalvi-Nagy, one of the founders of modern operator theory, and one of its major contributors, passed away on December 21, 1998. To honour him, a *Memorial Conference for Béla Szőkefalvi-Nagy* was held on August 2–6, 1999, in Szeged, Hungary, in which 91 mathematicians from all over the world took part. There were 19 plenary lectures in the morning and 63 talks in two parallel sessions in the afternoon.

The present volume contains proceedings and other research papers of the conference participants, and also 6 articles of distinguished experts who were unable to attend. These 35 articles present original recent results in various areas of operator theory and connected fields, many of them strongly related to contributions of Béla Sz.-Nagy. As usual, all the papers in this volume were refereed.

The camera ready copy of the volume, made by the Fyx Bt. (Szeged), was financed by the Bolyai Institute of the University of Szeged and by the Hungarian NFS Research Grant T 035123. Special thanks go to Erzsébet Szőkefalvi-Nagy for providing the photos from the family album.

Béla Szőkefalvi-Nagy, 1980

Farewell speech

Ciprian Foias

On August 3, 1999, at the grave site in Szeged
to the participants of the Memorial Conference

Dear Colleagues,

We have come here to bring our hommage to Béla Szőkefalvi-Nagy, a man of high distinction, an affectionate and effective parent and a great mathematician. Although a direct descendent on the male line from a military man ennobled for his valor in battles against the Turks, Béla did not grow up as the scion of an opulent family. His family had to quit their town, where Béla was born, on a matter of principle. They came to Szeged, where Béla's father, an accomplished mathematician, found adoquate position. Still, Béla during his college years had to privately tutor other students. Béla was a student of F. Riesz and A. Haar, but not in the sense in which we understand it today, neither in the sense in which I was Béla's de facto student in Operator Theory. In that time Béla had to bootstrap himself up to eventually become one of F. Riesz's peers in the branch of mathematics which is now Operator Theory. In fact, while still in his twenties, Béla wrote the first monograph in pure Operator Theory. It was written in French, but when van der Waerden received a copy of the manuscript he immediately accepted it for publication in the prestigious "Ergebnisse" series, but in a German translation. Almost all leading mathematicians in Béla's generation learned Spectral Theory from Béla's short but dense monograph. The scientific collaboration between Riesz and Sz.-Nagy effectively took off only after Béla has proven himself as an exceptional mathematician. The culminating result of that collaboration was the Riesz–Sz.-Nagy book on Functional Analysis, one of the most read and qouted mathematical treatises in the second half of this century. Among the other outstanding works of Béla in that time let me mention the following two: First Béla's discovery that the Banach generalized limit (which until then was only a mathematical curiosity) can be made a nonconstructive but very effective tool in Operator Theory. A lot of mathematics came out from that discovery. Béla had the magic wand that changed any mathematical object touched into either a mine of mathematical gold or into a mean for finding such a mine. The second work is Béla's "Appendix", dedicated to F. Riesz, which he inserted in a later edition of his great book with F. Riesz. This is one of the most impressive synthesis work in modern mathematics. It contains the most general and powerful dilation theorem in Operator Theory, the potential of which is still far from being fully exploited. This is not suprising if one is aware of how rich in consequences a particular case of Béla's general dilation theorem in the "Appendix", namely the famous Béla's power dilation theorem, turned out to be. One of the great chances in my life was to become Béla's collaborator in trying to understand the reaches of that particular dilation theorem. It took us many years and many joined us in that quest which is still going on. All these years I learned from Béla. It took me almost 10 years to understand his views and his approaches to mathematics and I still try to assimilate many of them. In particular whenever I tried or try to contribute to a question in Operator Theory I ask myself: How would Béla solve this problem and what kind of solution or what kind of proof would find he acceptable? And my students do the same and so will do their students.

We are here today, many linked to Béla in many ways. In what concerns me, Béla's departure was also as that of a close older relative, a dear uncle. And I probably am expressing the feelings of many present here. But cheer up, a man like Béla does not die. Only his physical presence is fading away in the mist of time. Béla continues to live in many ways. First he lives genetically in the large and successful clan he left behind. By their accomplishments and their physical appearance, Béla's heirs show how much he lives through them. For all who have known Béla personally, whom he befriended, helped or led, Béla is still alive in their hearts as long as those hearts throb. Finally, a great mathematician like Béla survives much, much longer. Béla Szőkefalvi-Nagy will live in the minds of generation after generation of mathematicians for all the future mankind has.

Reminiscences of Béla Szőkefalvi-Nagy

Israel Gohberg

Béla Szőkefalvi-Nagy was one of my important teachers. I never was a formal student of his, but I studied systematically his book as well as the books he co-authored with F. Riesz and with C. Foias; they were always on my table. His papers and his results impressed me very much; they played a crucial role in my mathematical education. The papers of B. Sz.-Nagy also served as examples for me of how to write mathematics. M. G. Krein told me about the work and personality of B. Sz.-Nagy, so even before our first meeting I was very well informed.

I met B. Sz.-Nagy for the first time in the Moscow University at the Conference of Functional Analysis in January 1956. These were the first years of my career and I knew about the importance of his contributions from M. G. Krein, and I was very happy to be introduced to him. We also had common interests, and I tried to have a talk with him. Unfortunately, I did not speak English and my German was not very good, but we soon discovered that we could converse in Rumanian, and this is how we communicated until I learned English. I very much liked B. Sz.-Nagy's papers, and on a few occasions I found in his work answers to questions which were bothering me. The meeting with B. Sz.-Nagy impressed me, and soon after the conference I received from him a wonderful present — a large set of his reprints accompanied by a warm presentation. Later I followed his joint papers with C. Foias with great interest.

I met B. Sz.-Nagy a few times in Moscow. Once he related to me the following story. During one of his first visits to Moscow he tried to buy a map of the city in order to be more independent. He quickly discovered that such maps are not available for sale. He then asked his guides (who were officially provided by the Soviet Academy) to make a plan for him of that part of Moscow which included the Moscow University, the administrative offices of the Academy, the metro and other sites. Even such a map he could not obtain, and only many years later did he receive a map of Moscow. He was not always able to use this map because the distances were wrongly shown.

B. Sz.-Nagy and M. G. Krein had a high regard for each other. They met frequently at conferences and congresses in the Soviet Union, and at least once B.

Sz.-Nagy visited M. G. Krein in Odessa. In 1968 B. Sz.-Nagy visited Moscow together with A. Rényi. The aim of their visit was to improve cooperation between the Academy of Sciences of Hungary and the Soviet Academy of Sciences in Mathematics. At the request of B. Sz.-Nagy, M. G. Krein was invited by the Soviet Academy to Moscow. By chance I was in Moscow during these days to receive an exit visa for a visit to Hungary at the invitation of B. Sz.-Nagy. From our conversations I understood that B. Sz.-Nagy was considering cooperation with M. G. Krein and his school as an important part of his plans. Very soon after the formal meetings B. Sz.-Nagy understood that the administration of Steklov Institut was against such plans and the cooperation was restricted to a formal cooperation with the Steklov Institute. At the same time B. Sz.-Nagy continued to discuss the matter of cooperation with M. G. Krein and he continued to get advice from him.

B. Sz.-Nagy was planning that after Moscow he would visit Chisinau (Kishinev), where he was invited by the Academy. He understood that in this case I would probably have to return my exit visa for my visit to Hungary, and it was questionable whether I would be able to receive it again. He therefore sacrificed his visit and said that he would visit Chisinau another time. Unfortunately there never was an opportunity for another visit. B. Sz.-Nagy visited me in Tel Aviv where he presented a very nice Toeplitz Lecture. We also met in Amsterdam at the invitation of our friend, Rien Kaashoek. B. Sz.-Nagy was very interested in hearing about how I succeeded in emigrating from the Soviet Union, and about life in Israel.

B. Sz.-Nagy and his family were very religious Catholics. During my first visit to Szeged in 1968 he took me on a walking tour through the town, and the first place he showed me was the great synagogue in Szeged. I was very impressed by it, but I was also afraid that my visit to the synagogue could lead to problems in regard to other trips abroad. Many years later I learned that the synagogue was built by the grandfather of my friend Terry Horvath (the wife of my colleague and friend, John Horvath from College Park, Maryland).

In 1970 B. Sz.-Nagy organized a conference on functional analysis in Tihany on the Balaton Lake in Hungary. As with everything organized by him, it was perfect from every point of view. The timing and the selection of participants were very good; it came after the International Mathematical Congress in Nice and many westerners came, as well as a large group of mathematicans from the East. I participated in this conference and enjoyed it very much. Moreover this conference proved to be very important for me later after emigration because here I met for the first time the majority of my western colleagues.

M. G. Krein never traveled abroad. The reason for this was that he had never been granted an exit visa. Only once he was given permission to travel abroad.

That was in 1970 to attend the conference in Tihany. It probably worked out this time because he used a private invitation which did not have to go through the high official channels. But this time he could not use the visa because precisely at that time there was an epidemic of cholera in Odessa and no one was allowed to leave Odessa. At the request of M. G. I gave Professor B. Sz.-Nagy regards from M. G. and told him the reason why M. G. could not come. Professor B. Sz.-Nagy smiled and answered, "So it's now called cholera, is it?". In the West people were already used to the various reasons that were invented to justify the absence of M. G. Krein. This was the only time that the reason given was the true reason, but no one believed it already.

The following event took place during the conference in Tihany. Lewis Coburn presented a nice talk; this talk was held before a long break. During the talk he used up all the available chalk. Béla — the organizer of the conference — immediately passed to him a piece of soft yellow chalk. Lewis finished his talk with this piece of chalk which colored his hands yellow. During the break that followed the talk Lewis looked for somewhere to wash his hands. On the way he showed everyone his yellow hands and explained what was the cause. After a while he dropped his left hand and showed only his right hand. In this way he approached also Béla who was involved in a discussion with a participant. Being busy Béla did not really take in what Lewis was saying to him, and he also did not notice the colour of the hand which was extended to him. He had the impression that Lewis just wanted to shake hands with him, so he grasped Lewis' yellow hand in a handshake. At this point Lewis started to jump around and yell. Béla immediately understood what had happened — he was under the impression that Lewis wanted to pay him back for giving him the yellow chalk. With a smile he started to stroke Lewis on his head and face, saying, "You are a good boy, a good boy," thereby turning Lewis' head and face yellow. All those who witnessed this incident started laughing. This laugh resounded throughout the courtyard.

The last time Béla Sz.-Nagy and I met was in the summer of 1993 during the conference in Szeged dedicated to his eightieth birthday. His wife was already very sick and this affected him strongly. Nevertheless he made an effort to attend all the talks given at the conference, and to be active during them. Together with C. Foias we discussed the organization of the next edition of the Sz.-Nagy and Foias book. I understood that he had detailed plans for it and he very much wanted to see this new edition. There was an atmosphere of sadness at this meeting, and I felt that this was probably the last time we would meet. Unfortunately, this sad prediction turned out to be correct.

Béla died in 1998, but I and my colleagues always feel his presence among us. We see him smiling at us through his books and papers, he is with us at our desks, in our seminar rooms and classrooms. His influence will be felt for a very long time.

B. Sz.-Nagy in the second row, on the ceremony of receiving the Gy. Kőnig Award in Budapest, 1942. Front row from the left: L. Fejér, F. Riesz and Gy. Sz.-Nagy

B. Sz.-Nagy and M. G. Krein in Moscow, 1956

Á. Korányi, B. Sz.-Nagy, B. Pukánszky and I. Kovács in Szeged, 1956

B. Sz.-Nagy and M. A. Naimark in Balatonföldvár, Hungary, 1964

V. Ptak and B. Sz.-Nagy in Herculane, Romania, 1981

B. Sz.-Nagy and C. Foias in Oberwolfach, 1986

B. Sz.-Nagy and P. R. Halmos in Wabash, Indiana, 1982

I. Gohberg, M. A. Kaashoek, B. Sz.-Nagy's wife and B. Sz.-Nagy in Amsterdam, 1985

Participants of the Conference in Balatonföldvár, Hungary, 1964

Participants of the 80th Anniversary Conference in Szeged, 1993

Béla Szőkefalvi-Nagy
1913–1998

László Kérchy and Heinz Langer

I. Biographical sketch

Béla Szőkefalvi-Nagy was born on July 29, 1913, in the city of Kolozsvár in Transylvania, which was in the Austro-Hungarian Empire at that time. His name refers to noble origin of the family; actually, Szőkefalva is a small village near the town Erzsébetváros, where his father was born. Since most of his scientific papers were signed as B. Sz.-Nagy, we shall also use this shortened form of his name.

Béla's father, Gyula, was also a mathematician. He taught in the 'Marianum', which was a monastic school with great historical tradition in Kolozsvár. Later he had the chair of the Department of Geometry at the University of Szeged. Béla's mother, Jolán Bölöni, was a secondary school teacher, specializing in mathematics, physics and natural history. In an interview, Béla Sz.-Nagy remembered his childhood as a wonderful period of his life. The little boy showed much interest in almost everything. He was especially attracted by the beauties and secrets of nature, and he liked gardening very much. Later he found great pleasure in the study of the grammatical structure of languages. Besides his native language, Hungarian, he was fluent in Romanian, French, German and later in English; at school he also learnt Greek. Signs of his mathematical talent were observed early. Already at the age of six, Béla surprised his parents by his cleverness in solving puzzles.

World War I changed various borders. In 1929 Béla's father, having lost his job in Kolozsvár (Cluj Napoca), Romania, was invited to the Mathematics Department of the Teachers' Training College in Szeged. The family moved to Hungary, and two years later Béla Sz.-Nagy started his university studies in mathematics and physics at the University of Szeged. In those years modern physics was undergoing a revolutionary development. Already as a secondary school student, Béla had chosen J. von Neumann's book on the mathematical foundations of quantum physics, and B. L. van der Waerden's book on the relation between quantum theory and the theory of groups, as his favorite readings. He was so delighted by these texts, that at the sections he found especially exciting he could not remain seated, but jumped up and ran to his mother in the kitchen to tell her what miraculouos things he had read. At the university he greatly enjoyed the courses of Zoltán Bay in physics. At the same time, he was deeply influenced by his professors of mathematics, Frigyes Riesz, Alfréd Haar and Béla Kerékjártó. From von Neumann's book he already knew the importance of Hilbert space operators in the study of quantum physics, and at the university he could learn operator theory from an original source: from the lectures of F. Riesz, one of the founders of this theory.

In an interview Sz.-Nagy remembered the time when he changed from being a student to a colleague of F. Riesz in the following way.

"I thought that I found an essentially simpler approach to a result of Riesz. (That was about a new proof for Stone's representation of one-parameter semigroups of unitary operators.) I showed him. In two days he told me that there was a false step in my work. I did not get discouraged. I felt that I started in the right direction, so I undertook to correct the mistake. And that happened in the following way: I laid down on the divan and started to think about the problem. If one is unable to cross the river at some point, then one must look for another, more suitable place. In a couple of days I found the correct step in the proof. It was a warm summer day. Riesz was resting on the bank of the river Tisza. Though I knew that maybe it was not proper to disturb the relaxation of such a highly respected professor, nevertheless I visited him at the boathouse. He told me to come back the next day. And by next day he was changed. Already when he was shaking my hands in a friendly way, I felt that by that gesture he had adopted his student as colleague. Of course, for me he still remained a master."

Béla Sz.-Nagy wrote his PhD thesis on isomorphic function systems, a topic related to the research of Alfréd Haar. In 1937–38 he spent eight months in Leipzig, where van der Waerden and Heisenberg were working at that time. In the first semester of 1939 he continued his studies at the Universities of Grenoble and Paris, where — among others — he met Hadamard and Denjoy. In 1939, following his father in this position, he was appointed to the Department of Mathematics of the Teachers' Training College in Szeged. In 1940 he became Privatdozent at the University of Szeged, and then in 1948 a full professor. (He was proud of the letter of recommendation written by von Neumann, whom he met only once, in Budapest.) First, he was the head of the Department of Descriptive Geometry, then he directed the Department of Analysis until his retirement in 1983.

His treatise 'Spektraldarstellung linearer Transformationen des Hilbertschen Raumes', published in 1942 in the Ergebnisse series of Springer Verlag, made him world famous. Generations learnt spectral theory of normal operators from this concise masterpiece. The monograph 'Leçons d'analyse fonctionelle', written jointly with Frigyes Riesz, was published in 1952. It became a fundamental textbook in, and reference for functional analysis, and was translated into six languages.

Frigyes Riesz died in 1956 and in the same year Sz.-Nagy met the gifted young Romanian mathematician, Ciprian Foias. Once Sz.-Nagy compared this event in his life with the apperance of the new Dalai Lama in Tibet. For the purposes of quantum physics the spectral theory of normal operators was of primary interest. However, other applied areas of mathematics (e.g. scattering theory or the theory of electrical networks) required the study of nonnormal operators. Starting from Sz.-Nagy's Dilation Theorem, Sz.-Nagy and Foias developed a new branch of the theory of nonnormal operators. They summarized their results in the monograph 'Analyse harmonique des opérateurs de l'espace de Hilbert', published in 1967, and translated later into English and Russian, as well.

II. Scientific results

In the sequel we illustrate Sz.-Nagy's scientific activity by some examples; because of his diverse mathematical interests and results our discussion cannot embrace all of his works.

First papers. His first scientific articles were algebraic-analytic studies connected with group theory and the theory of orthogonal functions. They were still strongly related to the research of his professors Frigyes Riesz and Alfréd Haar. Riesz's interest was particularly aroused by the work [3]*, where Sz.-Nagy at the age of 22 gave, among other results, a new proof for M. H. Stone's theorem on the spectral representation

$$U(t) = \int_{-\infty}^{\infty} e^{i\lambda t}\, dE_\lambda$$

of a strongly continuous semigroup $(U(t))_{t\geq 0}$ of unitary operators, and at the same time he extended the result for semigroups of normal operators (see also [18]). We mention that Stone's theorem plays a fundamental role in theoretical physics, in the theory of stochastic processes, and in other applications.

The results of Sz.-Nagy's doctoral thesis are contained in the papers [6–9]. Here, following the work of Haar, he shows among other results the following. Suppose the multiplication tables of the orthogonal function systems (ψ_j) and $(\hat{\psi}_j)$, defined on the subsets M and $\hat{M}$ of the real line $\mathbf{R}$, respectively, coincide. In other words

* References are to 'Research papers' in the 'Publications of Béla Szőkefalvi-Nagy'.

suppose the relations

$$\int_{\hat{M}} \hat{\psi}_j \overline{\hat{\psi}_k}\, ds = \int_M \psi_j \overline{\psi_k}\, ds, \qquad \int_{\hat{M}} \hat{\psi}_j \hat{\psi}_k \overline{\hat{\psi}_l}\, ds = \int_M \psi_j \psi_k \overline{\psi_l}\, ds$$

hold for all indices j, k, l. Then the two systems can be transformed into each other by a one-to-one, measure preserving mapping of the measurable subsets of M into the measurable subsets of $\hat{M}$.

Fourier series and approximation. Béla Sz.-Nagy contributed to the theory of Fourier series and to approximation theory in a comprehensive way. In 1937, in collaboration with his untimely departed colleague Antal Sólyi (Strausz), they gave a new proof in [17] of the Bohr inequality

$$|F(x)| \leq \frac{\pi}{2U} \sup |f(x)|$$

(see also [57]), following previous research by H. Bohr (brother of the physicist N. Bohr) and J. Favard. Here f is a trigonometric polynomial

$$f(x) = \sum_k (a_k \cos u_k x + b_k \sin u_k x),$$

where $u_k \geq U > 0$, and F is the corresponding integral function

$$F(x) = \sum_k \frac{1}{u_k}(a_k \sin u_k x - b_k \cos u_k x).$$

In the papers [19, 20] he continued the study of related questions in a more general form (see also [14, 15]). In the sequel we outline one of his results, specifying it for periodic functions.

Let K_m denote the set of measurable functions f of period 2π such that $|f(x)| \leq 1$ and the corresponding Fourier series begins with terms containing $\cos mx$ and $\sin mx$ (m being a given nonnegative integer), so it has the form

$$\sum_{k=m}^{\infty} (a_k \cos kx + b_k \sin kx). \tag{1}$$

Furthermore, given a real sequence $\lambda = (\lambda(k))_{k=m}^{\infty}$ and a real number δ, we associate with (1) the transformed series

$$\sum_{k=m}^{\infty} \lambda(k) \left(a_k \cos\left(kx + \frac{1}{2}\delta\pi\right) + b_k \sin\left(kx + \frac{1}{2}\delta\pi\right)\right). \tag{2}$$

The problem is to find conditions which ensure that the transformed series (2) is a Fourier series of a continuous function. Let $T^m_{\lambda\delta}f$ denote this continuous function (if it exists). Now the question can be put as follows. Is there a constant $M^m_{\lambda\delta}$, depending on the quantities m, λ, δ but independent of the choice of the function $f \in K_m$, such that $|(T^m_{\lambda\delta}f)(x)| \leq M^m_{\lambda\delta}$ holds for every x? In particular, one may ask for the smallest possible constant of this type and those 'extremal functions' f_0 for which this smallest constant is attained. Sz.-Nagy showed in [19] that these problems can be solved exactly, provided the sequence λ has certain monotonicity properties. For example, if $m \geq 0, \delta = 0$ and λ is a threefold monotonic zero sequence (that is λ is a zero sequence, and the first, second and third difference-sequences of λ are all nonnegative), then for any function $f \in K_m$ there exists such a continuous function $T^m_{\lambda 0}f$, the value of $M^m_{\lambda 0}$ can be given explicitly as

$$M^m_{\lambda 0} = \frac{4}{\pi}\sum_{k=0}^{\infty}(-1)^k\frac{\lambda((2k+1)m)}{2k+1},$$

and the unique (up to translations) extremal function is $f_0(x) = \operatorname{sign}\cos mx$. A similar statement is true for $\delta = 1$ and a twofold monotonic zero sequence λ satisfying the condition $\sum \lambda(k)/k < \infty$; in this case

$$M^m_{\lambda 1} = \frac{4}{\pi}\sum_{k=0}^{\infty}\frac{\lambda((2k+1)m)}{2k+1}.$$

From these comprehensive results one can deduce generalizations of theorems on harmonic functions, due to H. A. Schwarz and P. Koebe. Furthermore, inequalities concerning integral functions of entire or nonentire order corresponding to (1) can be derived, including certain statements of H. Bohr, S. N. Bernstein, J. Favard, N. I. Akhieser and M. G. Krein. Finally, some corollaries connected with the approximation of certain continuous functions by trigonometric polynomials can be obtained, and they are also extensions of theorems by the latter three authors. We note that these investigations have been frequently quoted and developed even in recent times, mainly by Russian mathematicians.

Béla Sz.-Nagy's achievements in the theory of Fourier series are reviewed in several monographs in this field. The paper [39] provides necessary and sufficient conditions for a decreasing, Lebesgue integrable function g, defined on $[0, \pi]$ and bounded from below, to have Fourier coefficients

$$a_n = \frac{2}{\pi}\int_0^{\pi} g(x)\cos nx\, dx, \quad b_n = \frac{2}{\pi}\int_0^{\pi} g(x)\sin nx\, dx$$

satisfying the conditions

$$\sum_{n=1}^{\infty} \frac{|a_n|}{n^\gamma} < \infty, \quad \sum_{n=1}^{\infty} \frac{|b_n|}{n^\gamma} < \infty.$$

These results were also the starting point for further studies carried out by Hungarian and foreign mathematicians as well; see, for example, R. P. Boas' book 'Integrability theorems for trigonometric transforms' in the series 'Ergebnisse der Mathematik und ihrer Grenzgebiete'.

A central question in the theory of Fourier series is to describe the effectiveness of an approximation of a function f by partial sums of its Fourier series or by means of the partial sums determined by a summability method. In [37] Sz.-Nagy discusses a very general class of summability methods. He gives estimates for the 'Lebesgue constants'

$$\sigma_n = \sup_{\|f\|_\infty \leq 1} \|\sigma_n(f)\|_\infty,$$

where $\sigma_n(f)$ is the nth approximation formed by the method investigated, and for the approximation constants

$$\rho_n = \sup_f \|f - \sigma_n(f)\|_\infty,$$

assuming that the functions f used in the definition of ρ_n, or some of their derivatives, satisfy a Hölder condition with constant 1. These studies continued the work of Lipót Fejér, S. N. Bernstein, D. Jackson, Ch. de la Vallée Poussin, S. M. Nikolskii, György Alexits and others.

Considering the expansion of a given function f with respect to an arbitrary orthogonal system, let us denote by $s_n(f)$ the nth partial sum or nth Fejér mean of this expansion. Extending results by W. Rudin, who dealt with the case $p = 2$, Béla Sz.-Nagy gives in [61] a lower estimate for the quantity

$$\sup \frac{\|f - s_n(f)\|_p}{N(f)}.$$

Here the supremum is taken over the elements of a certain class of functions, for example, a Lipschitz class; $N(f)$ is a certain functional corresponding to the class, for example, the smallest Lipschitz constant; and finally $\|f\|_p$ is the norm in the space L^p. It turns out that, in a certain sense, the trigonometric system and the Haar system possess the best approximation properties in the class of all orthonormal systems.

Geometry. In the papers [25, 38, 66, 81] Béla Sz.-Nagy made excursions into the area of geometry. These works also stimulated other mathematicians and, for example in connection with control theory, the paper [25] is still of great interest today, even after so many years. Here we give a short outline of the results of [81].

Consider an arbitrary nonempty plane set G whose closure is not the whole plane. Let G_t be the parallel set of G, corresponding to distance $t > 0$, i.e., the union of all closed discs with center in G and of radius t. If G is convex and bounded, then with $A(G_t)$ and $L(G_t)$ as the area of G_t and the length of the boundary of G_t, respectively, the Steiner formulas

$$A(G_t) = A(G_0) + L(G_0)t + \pi t^2, \quad L(G_t) = L(G_0) + 2\pi t$$

are valid for every $t > 0$. A nonempty, closed subset of the plane is called of type (n, ν) ($n \geq 0$ is an integer, $\nu = 0, 1$), if it consists of n bounded and ν unbounded components; the unbounded component is assumed to contain the exterior of a disc. Extending results achieved by Endre Makai, Béla Sz.-Nagy shows in [81] that for a set G of type (n, ν) the function

$$t \mapsto A(G_t \backslash G_0) - \pi(n - \nu)t^2$$

is continuous and concave on an interval $0 \leq t < \rho^*$ with a suitable $\rho^* > 0$. It follows that, for $0 \leq t < \rho^*$, the one sided derivatives

$$L_\pm(G_t) = \lim_{h \to 0\pm} h^{-1} A(G_{t+h} \backslash G_t)$$

exist, they are monotone functions of the variable t, and the inequality $L_-(G_t) \leq L_+(G_t)$ holds. For points t where $L_+(G_t) = L_-(G_t) =: L(G_t)$ is true (all but countably many t have this property), the estimate

$$L(G_t) \leq L(G_0) + 2\pi t \quad (t > 0)$$

is obtained. Before, this inequality was proved by Endre Makai in a different way for more specific domains and assuming the existence of the length of the boundary.

The function space L^2. In the papers [12, 13, 40] Sz.-Nagy gave internal characterizations of the set of nonnegative real functions and the set of characteristic functions in the function space L^2, defined with respect to an appropriately chosen positive Borel measure μ on the real line. For example in [13] he proved the following statements.

Let $\mathcal{P}$ be a subset of the (separable) Hilbert space $\mathcal{H}$. There exists a linear, isometric mapping of $\mathcal{H}$ onto the space L^2, which transforms the set $\mathcal{P}$ onto the set of nonnegative real functions in L^2 if and only if the following conditions are fulfilled:

(A) an element u in $\mathcal{H}$ belongs to $\mathcal{P}$ if and only if $\langle u, v \rangle \geq 0$ holds for every element v in $\mathcal{P}$;

(B) if $u_1 + u_2 = v_1 + v_2$ is true for elements of $\mathcal{P}$, then there exist elements $w_{11}, w_{12}, w_{21}, w_{22}$ in $\mathcal{P}$ such that $u_i = \sum_k w_{ik}$, $v_k = \sum_i w_{ik}$.

If the condition (A) is weakened to

(A′) $\mathcal{P}$ is a closed cone in $\mathcal{H}$ with vertex 0, and is generating (that is $\mathcal{P} - \mathcal{P} = \mathcal{H}$) and normal (that is $\|u\| + \|v\| \leq K\|u + v\|$ is true for every $u, v \in \mathcal{P}$ with some constant K),

then (A′) and (B) are necessary and sufficient for the existence of an 'affinity' (i.e., an invertible continuous linear transformation, which is not necessarily an isometry) of $\mathcal{H}$ onto an L^2-space, mapping $\mathcal{P}$ onto the set of nonnegative real functions in L^2 [40].

Linear operators. A central topic of interest in Béla Sz.-Nagy's mathematical activity is the theory of linear operators of Hilbert spaces and its applications in different areas of mathematical analysis. We select some of his many interesting and important results. By now the following statement of [33] has become a classic and frequently referenced theorem. If S is a linear operator on a Hilbert space $\mathcal{H}$ with an (everywhere defined) inverse S^{-1} such that all powers of S with positive and negative exponents remain below a common bound, that is

$$\|S^n\| \leq K, \quad n = 0, \pm 1, \pm 2, \dots \tag{3}$$

is true with an appropriate constant $K (< \infty)$, then S is similar to a unitary operator, which means that there exists a bounded and boundedly invertible operator A such that the operator $U = ASA^{-1}$ is unitary. An analogous theorem holds for one-parameter operator groups (S_t), $-\infty < t < \infty$, with the boundedness property $\|S_t\| \leq K < \infty$ $(-\infty < t < \infty)$.

These results were generalized by several mathematicians, in particular by J. Dixmier and M. M. Day. In these extensions the additive group of integers or real numbers is replaced by an arbitrary 'amenable' group. However, up to now it is an open question whether the statement is true for an arbitrary group Γ, that is whether any family $(S(\gamma))_{\gamma \in \Gamma}$ of bounded invertible operators acting on the space $\mathcal{H}$ and satisfying the conditions

$$S(\gamma^{-1}\delta) = S(\gamma)^{-1}S(\delta), \quad \|S(\gamma)\| \leq K \quad (\gamma, \delta \in \Gamma)$$

is similar to a corresponding family of unitary operators.

The statement above was supplemented in the paper [83], where Béla Sz.-Nagy proved that every compact operator S, which satisfies the inequalities (3) only for nonnegative exponents n, is similar to a contraction, that is to an operator with norm not greater than 1. Later R. S. Fougel showed by an example that for an arbitrary operator this statement is false.

Perturbation theory for an isolated eigenvalue of a selfadjoint operator arose already in J. W. Rayleigh's work, and in E. Schrödinger's papers it became a general method in quantum theory. The problem is the following. Consider a family of operators $A(\varepsilon)$, $|\varepsilon| < \delta$ (ε is a real or complex parameter), where $A(\varepsilon)$ depends continuously or analytically on ε, and assume that the operator $A(0)$ has an isolated eigenvalue $\lambda(0)$. What is the spectrum of $A(\varepsilon)$ in the neighborhood of $\lambda(0)$, if ε is sufficiently small? In particular, if $A(\varepsilon)$ depends analytically on ε, can we choose analytic functions $\lambda_j(\varepsilon)$, such that $\lambda_j(0) = \lambda(0)$ and in a neighborhood of $\lambda(0)$ the spectrum of $A(\varepsilon)$ consists exactly of the points $\lambda_j(\varepsilon)$ ($j = 1, 2, \ldots$), if ε is small? Similar questions can be asked for other parts of the spectrum, and also for nonisolated eigenvalues. In the papers [29, 30, 34, 50] Béla Sz.-Nagy, applying the Riesz functional calculus and developing a method based essentially on a new induction argument, sharpened and extended relevant results of F. Rellich. Namely, he found better estimates for the domains of convergence of the power series arising, and he generalized a number of statements to closed operators acting in Banach spaces.

The results of [52] also belong to the perturbation theory of linear operators. There Sz.-Nagy proved for the first time that the index of a closed, not necessarily bounded, operator is stable under a compact perturbation. It is well known that statements concerning the index play an important role, e.g., in the theory of singular integral equations.

In his papers [71, 73, 80], together with his former student Ádám Korányi, Béla Sz.-Nagy established a strong relationship between the Nevanlinna–Pick problem and similar questions in the theory of analytic functions on one hand and the generalized resolvents of isometric and hermitian Hilbert space operators on the other hand. The well-known Nevanlinna–Pick problem is the following. A function f, given on a subset S of the open unit disc $\mathbf{D}$, should be extended to a function $\widetilde{f}$ on the whole disc $\mathbf{D}$ such that $\widetilde{f}$ is analytic on $\mathbf{D}$ and its real part is nonnegative there. Such an extension $\widetilde{f}$ exists if and only if the kernel function

$$k(s,t) = \frac{f(s) + \overline{f(t)}}{1 - s\bar{t}}$$

is positive definite on the set $S \times S$. The connection with generalized resolvents of isometries is the following: The relation

$$(4) \qquad \widetilde{f}(s) = ib + \langle (I + sU)(I - sU)^{-1}v, v\rangle \quad (s \in S)$$

establishes a bijection between the set of all such extensions $\widetilde{f}$ of f and all unitary extensions U of an isometry in a suitable Hilbert space $\mathcal{H}$; v is a vector in $\mathcal{H}$ and b is a real constant. This method, which in the cited papers was also used for the study of analogous problems for operator valued functions, has been exploited by many authors since.

Hilbert space contractions. In the last 30 years of Béla Sz.-Nagy's scientific activity, Hilbert space contractions of general type (that is not necessarily selfadjoint, unitary or normal) were in the centre of his research. Besides the monograph, which was mentioned above, he published more than 75 papers in this topic, among them about 50 jointly with Ciprian Foias. In the history of mathematics, only a few collaborations were so longstanding and intensive, and, what is more, led to such an elaborated theory. Starting from the fifties, the theory of contractions was one of the three main research directions in the study of nonselfadjoint Hilbert space operators; the other two directions were determined by N. Dunford and J. T. Schwartz (the theory of spectral operators) and by M. G. Krein and his school. On the other hand, the theory of Sz.-Nagy and Foias was strongly related to the Soviet school, in particular, to the research of M. S. Livsic and M. S. Brodskii. It is not accidental that the deep and beautiful results of Sz.-Nagy and Foias inspired mathematicians of many countries to study and to develop the new theory further. So by now an extremely extensive literature exists on Hilbert space contractions and the end of the development is not yet in sight.

The main goal of this research is to find the structure of a general bounded linear operator T in a Hilbert space $\mathcal{H}$. It can be assumed that T is a contraction, that is $\|T\| \le 1$, since an arbitrary bounded linear operator satisfies this condition after multiplying it by an appropriate positive constant. The starting point of these investigations of Sz.-Nagy and Foias was a result of Béla Sz.-Nagy ([62], see also [64, 68, 110] for further proofs), which states that any contraction T in a Hilbert space $\mathcal{H}$ has a unitary dilation. This means that there exists a unitary operator in a larger Hilbert space $\mathcal{K}$ such that

$$(5) \qquad T^{[n]}f = P_{\mathcal{H}}U^n f, \quad f \in \mathcal{H}$$

holds for $n = 0, \pm 1, \pm 2, \ldots$; here $P_{\mathcal{H}}$ is the orthogonal projection of the space $\mathcal{K}$ onto $\mathcal{H}$ and

$$T^{[n]} = \begin{cases} T^n & \text{if} \quad n = 0, 1, \ldots \\ T^{*|n|} & \text{if} \quad n = -1, -2, \ldots . \end{cases}$$

Prior results related to this general theorem are Paul Halmos' statement about the existence of a unitary operator U_1 satisfying (5) for $n = 1$, that is having the property $Tf = P_{\mathcal{H}} U_1 f$ $(f \in \mathcal{H})$, and M. A. Naimark's theorem on the representation of a semispectral measure as the orthogonal projection of a spectral measure. Since the structure of unitary operators is rather well-known, relation (5) makes it possible to gain insight into the structure of T through a study of the connection between T and U.

To illustrate the relationship between T and U in (5), we present here the matrix construction of U due to J. J. Schäffer, as one possible way for the introduction of U. Given a contraction T on the space $\mathcal{H}$, consider the operator U defined by the doubly infinite matrix

$$U = \begin{bmatrix} & \cdot & \cdot & \cdot & \cdot & \\ & \cdot & \cdot & \cdot & \cdot & \\ & \cdot & \cdot & \cdot & \cdot & \\ \cdots & I & 0 & 0 & 0 & \cdots \\ \cdots & 0 & D_{T^*} & T & 0 & \cdots \\ \cdots & 0 & -T^* & D_T & 0 & \cdots \\ \cdots & 0 & 0 & 0 & I & \cdots \\ & \cdot & \cdot & \cdot & \cdot & \\ & \cdot & \cdot & \cdot & \cdot & \\ & \cdot & \cdot & \cdot & \cdot & \end{bmatrix}$$

on the space

$$\mathcal{K} = \sum_{n=-\infty}^{\infty} \oplus \mathcal{H}_n \quad (\mathcal{H}_n = \mathcal{H}; n = 0, \pm 1, \pm 2, \ldots),$$

where $D_T := (I - T^*T)^{1/2}$, $D_{T^*} := (I - TT^*)^{1/2}$. Here T is the entry in the matrix with index $0, 0$. The remaining entries on the diagonal below the main diagonal are equal to the identity mapping I, and all the other entries are equal to the zero operator. The upper left and the lower right parts of the matrix represent shift operators, while the T-dependent part of the matrix is formed by the block

$$\begin{bmatrix} U_{0,-1} & U_{0,0} \\ U_{1,-1} & U_{1,0} \end{bmatrix} = \begin{bmatrix} D_{T^*} & T \\ -T^* & D_T \end{bmatrix}.$$

Therefore, the structure of the operator U corresponds to a principle which is applied, for example, in the theory of electrical networks. It states that in certain systems the output can be produced from the input by a simple shift, and the process which takes place inside the system can be described by a 'black box' located in the center.

The unitary dilation U in (5) can be chosen minimal in the sense that the vectors $U^n f$, $f \in \mathcal{H}$, $n = 0, \pm 1, \pm 2, \ldots$ span the space $\mathcal{K}$; then U is uniquely determined up to an isomorphism. The contraction T is called completely nonunitary if $\mathcal{H}$ does not contain a nonzero subspace where T induces a unitary operator. It turns out that for a completely nonunitary contraction the spectral measure of the minimal unitary dilation is absolutely continuous with respect to Lebesgue measure. Consequently, the functional calculus can be extended from polynomials to all elements of the function class H^∞ (that is to bounded, analytic functions on the open unit disc), and also to those possibly unbounded functions on the unit disc which can be written as a quotient uv^{-1} of a function $u \in H^\infty$ and $v \in K_T^\infty$. Here K_T^∞ stands for the set of functions $v \in H^\infty$ such that the operator $v(T)$ has a densely defined (but not necessarily bounded) inverse. For an arbitrary completely nonunitary contraction T, the class K_T^∞ contains among others the 'outer' functions of H^∞ (in the sense of Beurling). These results of Sz.-Nagy and Foias from the early sixties were very stimulating for the further study of contractions.

It is known that, for any operator T on a finite dimensional space, there exists a polynomial $u(\not\equiv 0)$ such that $u(T) = 0$. In generalizing this property Sz.-Nagy and Foias introduced the class C_0 of those completely nonunitary contractions T for which there exists a function $w \in H^\infty$ $(w \not\equiv 0)$ such that $w(T) = 0$. For any $T \in C_0$, there also exists a 'minimal' function m_T with the prescribed property; more precisely, there exists an 'inner function' m_T (i.e. $|m_T(e^{it})| = 1$ holds almost everywhere for the boundary values of m_T) such that $m_T(T) = 0$ and m_T divides in H^∞ every function w satisfying the condition $w(T) = 0$. To illustrate the interesting properties of contractions T belonging to the class C_0, we list the following statements:

(i) for every vector $f \in \mathcal{H}$, $T^n f \to 0$ and $T^{*n} f \to 0$ as $n \to \infty$;

(ii) the spectrum $\sigma(T)$ consists of the zeros of m_T inside the unit disc and those points on the unit circle through which m_T can not be continued analytically to the exterior of the disc;

(iii) T has a nontrivial invariant subspace;

(iv) the algebra generated by T and I, which is closed in the strong operator topology, coincides with the bicommutant $(T)''$ of T, and with the set of operators X of the form $X = v(T)^{-1}u(T)$, $u \in H^\infty, v \in K_T^\infty$;

(v) T has a cyclic vector x_0 (which means that the whole space is spanned by the vectors $T^n x_0$, $n = 0, 1, 2, \ldots$) if and only if the commutant $(T)'$ of T is commutative.

We remind the reader that the commutant $(T)'$ of an operator T is the set of operators S which commute with T (that is $TS = ST$); and, correspondingly, the bicommutant $(T)''$ of T consists of those operators which commute with every element of $(T)'$. The properties (i)–(v) of the class C_0 are generalizations of similar properties of finite dimensional operators. However, the proofs in the general case are deep and sophisticated.

Unitary-equivalence model. One of the crucial results of the theory of Sz.-Nagy and Foias is the construction of a unitary-equivalence model for a completely nonunitary contraction. To describe this model, for any Hilbert space $\mathcal{L}$, let $L^2_{\mathcal{L}}$ denote the space of measurable, square-integrable functions u, taking values in $\mathcal{L}$ and with the norm

$$\|u\| = \left(\frac{1}{2\pi}\int_0^{2\pi} \|u(e^{it})\|^2\, dt\right)^{1/2}.$$

The Hardy space $H^2_{\mathcal{L}}$ consists of those functions $u \in L^2_{\mathcal{L}}$, with a Fourier series of the form $u(e^{it}) = \sum_{n\geq 0} u_n e^{int}$ $(u_n \in \mathcal{L})$. Furthermore, in the sequel let Θ be a function, which is analytic on the open unit disc, takes values in the set of contractions from a Hilbert space $\mathcal{L}$ into a Hilbert space $\mathcal{L}_*$, and satisfies the condition $\|\Theta(0)x\| < \|x\|$ $(x \in \mathcal{L}, x \neq 0)$. Such a function Θ is called an $(\mathcal{L}, \mathcal{L}_*)$-contractive analytic function. If, additionally, the limiting values $\Theta(e^{it})$ are isometries almost everywhere, then Θ is said to be an inner function.

First we construct the model in the special case when the contraction T satisfies the condition

$$T^{*n}h \to 0 \quad (h \in \mathcal{H} \text{ and } n \to \infty). \tag{6}$$

Let Θ be an $(\mathcal{L}, \mathcal{L}_*)$-contractive analytic inner function. Consider the Hilbert space

$$\mathcal{H}_0(\Theta) := H^2_{\mathcal{L}_*} \ominus \Theta H^2_{\mathcal{L}},$$

and define the operator

$$S_0(\Theta)u := P_{\mathcal{H}_0(\Theta)}(\chi u) \quad (u \in \mathcal{H}_0(\Theta)),$$

where $P_{\mathcal{H}_0(\Theta)}$ is the orthogonal projection of the space $L^2_{\mathcal{L}_*}$ onto $\mathcal{H}_0(\Theta)$, and $\chi(\lambda) = \lambda$. Then the operator $T := S_0(\Theta)$ is a contraction with property (6). Conversely, given a contraction T satisfying (6), construct the so-called characteristic function

$$\Theta_T(\lambda) := [-T + \lambda D_{T^*}(I - \lambda T^*)^{-1} D_T] | \overline{D_T \mathcal{H}} \quad (|\lambda| < 1), \tag{7}$$

which is $(\mathcal{L}, \mathcal{L}_*)$-contractive, inner and analytic, with $\mathcal{L} := \overline{D_T \mathcal{H}}$ and $\mathcal{L}_* = \overline{D_{T^*}\mathcal{H}}$. Then T is unitarily equivalent to the operator $S_0(\Theta_T)$. In the special case (6) considered above, this model was obtained by different methods by G. C. Rota, J. Rovnyak and H. Helson too, though without the explicit form (7) of the function Θ_T.

Now the general model of Sz.-Nagy and Foias for an arbitrary completely nonunitary contraction T can be described in the following way. For an arbitrary $(\mathcal{L}, \mathcal{L}_*)$-contractive analytic function Θ, define the function

$$\Delta(e^{it}) := [I_{\mathcal{L}} - \Theta(e^{it})^* \Theta(e^{it})]^{1/2},$$

the Hilbert space

$$\mathcal{H}(\Theta) := \left(H^2_{\mathcal{L}_*} \oplus \overline{\Delta L^2_{\mathcal{L}}}\right) \ominus \{\Theta w \oplus \Delta w : w \in H^2_{\mathcal{L}}\}, \tag{8}$$

where $\overline{\Delta L^2_{\mathcal{L}}}$ means the closure in $L^2_{\mathcal{L}}$ of the set $\{\Delta v : v \in L^2_{\mathcal{L}}\}$ in $L^2_{\mathcal{L}}$, and define the operator

$$S(\Theta)(u \oplus v) := P_{\mathcal{H}(\Theta)}(\chi u \oplus \chi v) \tag{9}$$

on the space $\mathcal{H}(\Theta)$. Then $S(\Theta)$ is a completely nonunitary contraction. Conversely, given any completely nonunitary contraction T, the operator $S(\Theta_T)$ defined by (9) is unitarily equivalent to T, provided Θ_T is constructed by (7) with $\mathcal{L} := \overline{D_T \mathcal{H}}$ and $\mathcal{L}_* := \overline{D_{T^*} \mathcal{H}}$.

Nowadays, the Sz.-Nagy–Foias model of contractions and the related characteristic function play an important role in the theory of linear operators in Hilbert space for many theoretical and practical questions. For example, by means of this model theory, necessary and sufficient conditions can be given for a contraction to be similar to a unitary operator, and the existence of a nontrivial invariant subspace for certain classes of contractions can be shown. In fact, these subspaces are strongly related to factorizations of the characteristic function. Furthermore, characteristic functions have found substantial applications in the theory of electrical networks and in other fields of practical importance. Finally, we note that, in connection with their studies in scattering theory, Peter Lax and Ralph S. Phillips constructed a model for certain classes of operators, which — by a suitable transformation — can be related to the Sz.-Nagy–Foias model if assumption (6) is satisfied.

Lifting Theorem. In order to formulate another important relevant result of Sz.-Nagy and Foias, let us recall the definition of a minimal isometric dilation of a contraction T in a Hilbert space $\mathcal{H}$. This is an isometric operator V in a Hilbert space $\mathcal{K}_+ \supset \mathcal{H}$ such that

$$T^n f = P_{\mathcal{H}} V^n f \quad \text{for } n = 0, 1, 2, \dots \text{ and } f \in \mathcal{H},$$

$\mathcal{K}_+$ being the closed linear span of the vectors $V^n f$ $(n = 0, 1, 2, \dots; f \in \mathcal{H})$. A minimal isometric dilation of T always exists, and is uniquely determined (up to an isomorphism). In the model (8)–(9) of T we have $\mathcal{K}_+ = H^2_{\mathcal{L}_*} \oplus \overline{\Delta L^2_{\mathcal{L}}}$ and $V(u \oplus v) = \chi u \oplus \chi v$.

Now the so-called 'Lifting Theorem' of Sz.-Nagy and Foias [118, 119] can be formulated as follows. For $i = 1, 2$, let T_i be a contraction on the Hilbert space $\mathcal{H}_i$ with minimal isometric dilation V_i acting on the space $\mathcal{K}_i$, and let us assume that X is a bounded linear transformation from the space $\mathcal{H}_1$ into $\mathcal{H}_2$ such that $T_2X = XT_1$. Then there exists a bounded linear transformation Y from the space $\mathcal{K}_1$ into $\mathcal{K}_2$ satisfying the conditions $V_2Y = YV_1$, $P_2Y(I - P_1) = 0$, $P_2Y|\mathcal{H}_1 = X$, and the relation $\|Y\| = \|X\|$. This theorem considerably generalizes a former result of D. Sarason, and it has many important applications. Thus, for example, T. Ando's result about the existence of a unitary dilation of a pair of commuting contractions can be deduced from it. It was also applied in the study of the operator equation $S^*XT = X$ with given contractions S and T, in connection with extremal problems concerning Hankel matrices, in the study of the structure of contractions belonging to the previously mentioned class C_0, and in other areas. The book 'The Commutant Lifting Approach to Interpolation Problems' by A. E. Frazho and C. Foias gives an account of more recent progress achieved by exploitation of this theorem.

Quasisimilarity models. Besides the models which are unitarily equivalent to the given operator (called unitary-equivalence models), similarity or quasisimilarity models are often of interest. For example, one of the results of Béla Sz.-Nagy, mentioned above, can be formulated also as follows. For an operator S with property (3) the corresponding unitary operator U is a similarity model. Now, Sz.-Nagy and Foias call the operators T on the space $\mathcal{H}$, and T' on the space $\mathcal{H}'$, quasisimilar if there exist one-to-one bounded linear transformations

$$X: \mathcal{H} \to \mathcal{H}', \quad Y: \mathcal{H}' \to \mathcal{H}$$

with dense ranges such that

$$T'X = XT, \quad TY = YT'.$$

In a finite dimensional space similarity and quasisimilarity obviously coincide.

By a well-known result of linear algebra, every operator in a finite dimensional space is similar to a Jordan operator, that is if an appropriate basis is chosen then the operator can be represented by a block diagonal matrix with the following

'Jordan blocks' on the diagonal:

$$\begin{bmatrix} \lambda & 1 & 0 & \cdot & \cdot & 0 \\ 0 & \lambda & 1 & \cdot & \cdot & 0 \\ \cdot & \cdot & \cdot & \cdot & \cdot & \cdot \\ \cdot & \cdot & \cdot & \cdot & \cdot & \cdot \\ \cdot & \cdot & \cdot & \cdot & \cdot & \cdot \\ 0 & 0 & 0 & \cdot & \cdot & 1 \\ 0 & 0 & 0 & \cdot & \cdot & \lambda \end{bmatrix}.$$

Within the framework of their theory of Hilbert space contractions, Sz.-Nagy and Foias succeeded in describing a class of operators, the so-called Jordan operators, which on the one hand generalize the Jordan normal form of a matrix, and on the other hand are quasisimilarity models for a large class of contractions. These Jordan operators can be characterized in the following way. Let $u_1, u_2, \ldots, u_k$ be nonconstant inner functions in H^∞ such that u_{j+1} divides u_j in H^∞, $j = 1, 2, \ldots, k-1$. Form the orthogonal sum

$$S(u_1) \oplus S(u_2) \oplus \ldots \oplus S(u_k) =: S(u_1, u_2, \ldots, u_k), \tag{10}$$

where the definition of the operators $S(u_j)$ is the same as that of $S(\Theta)$ before. The operators of the form (10) are completely nonunitary contractions. Moreover, writing S for $S(u_1, u_2, \ldots, u_k)$, we have $u_1(S) = 0$. Two such operators $S(u_1, u_2, \ldots, u_k)$ and $S(v_1, v_2, \ldots, v_l)$ are quasisimilar if and only if $k = l$ and $u_j = v_j$, $j = 1, 2, \ldots, k$. In this context, one of the main results of Sz.-Nagy and Foias is the following [125, 126]. The class of Jordan operators of the form (10) serves as a quasisimilarity model for those contractions $T \in C_0$ which admit a finite generating system. The latter means that there exist vectors $h_1, h_2, \ldots, h_k$ in $\mathcal{H}$ such that $\mathcal{H}$ is the closed linear span of the vectors $T^n h_j$ $(n = 0, 1, 2, \ldots; j = 1, 2, \ldots, k)$; the minimal value of k is equal to the corresponding index k occurring in the Jordan operator (10).

In their subsequent works Sz.-Nagy and Foias, in collaboration with Hari Bercovici, completed and generalized the previous result providing quasisimilarity model for all C_0-contractions, but we shall not go into details here. We note only that their studies connected with Jordan models of operators were the starting point of a new quasiequivalence theory of (finite or infinite) matrices over the algebra H^∞ of bounded analytic functions on the open unit disc, elaborated by E. A. Nordgren (finite case) and by Béla Sz.-Nagy (infinite case) [146]. Bercovici's book 'Operator Theory and Arithmetic in H^∞' gives a good summary of the results achieved up to 1988 on C_0-contractions

III. Awards, social activity

Béla Sz.-Nagy's scientific accomplishments, realized in 167 articles and 3 monographs, were appropriately acknowledged. He was only in his early thirties when he was elected as corresponding member of the Hungarian Academy of Sciences in 1945. He became an ordinary member in 1956. He was elected honorary member of the Soviet (1971), Irish (1973) and Finnish (1976) Academies. He received the Lomonosov Gold Medal of the Soviet (1980) and the Gold Medal of the Hungarian (1987) Academies. He was presented with the title of Doctor Honoris Causa by the Universities of Dresden (1965), Turku (1970), Bordeaux (1987) and Szeged (1988).

Sz.-Nagy was an outstanding lecturer, with a characteristic transparent, elegant style. He wrote textbooks on complex and on real analysis; the second one also has an English translation. Generations learnt from Béla Sz.-Nagy not only the subject itself, but also to experience the pleasure of disciplined, logical thinking. He was rigorous at the exams, he deemed that the student who could not express himself clearly did not truly understand the subject in question. After passing an exam with Sz.-Nagy, many students had the feeling that no more insurmountable obstacle could be encountered in their life. It was an honour and a real challenge to lecture in his weekly seminar, which was also frequently attended by many foreign visitors. Severe criticism characterized the atmosphere of these meetings; the lecturer could not pass over critical points which were only superficially understood by him. On the other hand, his recognition and encouragement inspired and lent wings to younger colleagues.

Acta Scientiarum Mathematicarum was launched by Alfréd Haar and Frigyes Riesz in 1922, after the university was moved from Kolozsvár to Szeged. Thanks to their dedication, it soon became a popular periodical with readers from all around the world. It was Béla Sz.-Nagy who, from 1946, continued the work of the founding editors. He was Editor-in-Chief until 1982, and after that an Honorary Editor-in-Chief until his death. Many young authors learnt from him, how to write a mathematical article correctly, on a high level, via his editorial remarks and suggestions. He was also Editor-in-Chief of the periodical *Analysis Mathematica* launched in 1975 jointly by the Hungarian and Soviet Academies, and a member of the editorial boards of numerous other prominent international journals and book series.

Béla Sz.-Nagy served in many positions at the University of Szeged and at the Hungarian Academy of Sciences. He directed the Mathematical Committee of the Hungarian Academy for a long period of time, from 1953 to 1990. He was a member of the presidential committee for 8 years, and from 1970 to 1985 he was the head of the Szeged Branch of the Hungarian Academy. He was dean of the Faculty of Sciences at the University of Szeged for two periods (1951/52 and 1963/66). He accomplished his service as a leader with caution and with a sense of excellent diplomacy. It was characteristic of the respect in which he was held that he was elected as president of the Revolutionary Committee of the University in 1956. Thanks to his international scientific contacts (which included leading Soviet mathematicians), his 'only' punishement for that activity was not getting permission to attend the World Congress of Mathematics in Edinburgh. In his absence his lecture was presented by Paul Halmos and it was a resounding success. His scientific and public activity was rewarded by the highest state decorations: he received the Kossuth Prize in 1950 and 1953, the State Prize in 1978, the Decoration with Flag of the Hungarian People's Republic in 1983, and the Middle Cross of the Order of the Hungarian Republic in 1991.

Béla Sz.-Nagy was also actively involved in public life in Szeged. Among other things, he played a vital role in the establishment of the Game Reserve in Szeged. He was awarded the Grand Prize of the Pro Szeged Foundation in 1990, and in 1991 he became an Honorary Citizen of Szeged.

He was a religious man, who loved his family and raised six children with his wife, Jolán, who was a history teacher and a talented singer, and whom he married in 1941. All of their children had succesful careers. Proceeding in the order of birth: Katalin is a gifted artist of singing, their only son, Zoltán, is a Doctor of Physical Sciences, Mária graduated as a physicist and works as a computer scientist, Erzsébet graduated in the faculty of arts and is a leading librarian, Ágnes is a physicist, and Zsuzsanna is an economist.

When Béla Szőkefalvi-Nagy passed away on December 21, 1998, the world lost an excellent mathematician and a warm-hearted human being of exceptional character.

Publications of Béla Szőkefalvi-Nagy

I. Books

[1] B. Sz.-Nagy, *Spektraldarstellung linearer Transformationen des Hilbertschen Raumes,* (Ergebnisse der Mathematik und ihrer Grenzgebiete, V/5), Springer Verlag, Berlin, 1942, IV + 80 pp. — New edition in the USA made by photographic way: 1947; — Second, revised edition: Springer Verlag, Berlin – Heidelberg – New York, 1967, VI + 81 pp.

[2a] F. Riesz & B. Sz.-Nagy, *Leçons d'analyse fonctionnelle,* Akadémiai Kiadó, Budapest. First edition: 1952, VIII + 449 pp. — Second edition: 1953, VIII + 455 pp. — Third edition: 1955, VIII + 488 pp. — Fourth, revised edition: 1965, VIII + 490 pp. — Fifth, unchanged edition: 1968. (Published jointly with Gauthier–Villars beginning with the third edition.)

[2b] ———, *Functional analysis,* Frederick Ungar Publishing Co., New York, 1955, XII + 468 pp. (English translation of the first edition of [2a].)

[2c] ———, *Lekcii po funcionalnomu analizu,* First edition: Foreign Literary Publ. Co., Moscow, 1954, 499 pp. Second edition, revised and supplemented by S. A. Teljakovskii: "Mir", Moscow, 1979, 589 pp. (Russian translation of the second edition of [2a].)

[2d] ———, *Vorlesungen über Funktionalanalysis,* (Hochschulbücher für Mathematik, Bd. 27), VEB Deutscher Verlag der Wissenschaften, Berlin. First edition: 1956, XI + 482 pp. — Second edition: 1968. — Third edition: 1973. — Fourth edition: 1982, 518 pp. (German translation of the third edition of [2a], including the appendices.)

[2e] ———, *Leçons d'analyse fonctionnelle* (in Japanese), Tokyo, 1973, Vol. 1: XII + 282 pp. Vol. 2: XII + 320 pp. (Japanese translation of the fifth edition of [2a], including the appendices.)

[2f] ———, *Funkcionálanalízis,* Tankönyvkiadó, Budapest, 1988, 534 pp. (Hungarian translation of the fourth edition of [2a], including the appendices.)

[2g] ———, *Functional Analysis,* (Dover Books on Advanced Mathematics), Dover Publications, Inc., New York, 1990, XII + 504 pp. (English translation of the second French edition.)

[3a] B. Sz.-Nagy, *Prolongements des transformations linéaires de l'espace de Hilbert qui sortent de cet espace,* Akadémiai Kiadó, Budapest, 1955, 36 pp. (Published separately and as an appendix in the third edition of [2a].)

[3b] ———, *Extensions of linear transformations in Hilbert space which extend beyond this space,* Frederick Ungar Publishing Co., New York, 1960, 37 pp. (English translation of [3a].)

[3c] ———, Prodolženija operatorov v gilbertovom prostranstve s vyhodom iz etogo prostranstva, *Matematika,* **9:6** (1965), 109-144. (Russian translation of [3a].)

[4a] ———, *Valós függvények és függvénysorok,* (University textbook), Tankönyvkiadó, Budapest, 1954, 307 pp. — Second, expanded edition: 1961, 370 pp. — Seventh reprinting: 1981.

[4b] ———, *Introduction to real functions and orthogonal expansions,* Akadémiai Kiadó – Oxford University Press, Budapest – New York, 1964, XI + 447 pp. (English translation of the second edition of [4a].)

[5] ———, *Haar Alfréd összegyűjtött munkái – Alfred Haar Gesammelte Arbeiten,* Akadémiai Kiadó, Budapest, 1959.

[6a] B. Sz.-Nagy & C. Foias, *Analyse harmonique des opérateurs de l'espace de Hilbert,* Akadémiai Kiadó – Masson et Cie, Budapest – Paris, 1967, XI + 373 pp.

[6b] ———, *Harmonic analysis of operators on Hilbert space,* North-Holland Publishing Co., Amsterdam – London; American Elsevier Publishing Co., Inc., New York; Akadémiai Kiadó Budapest; 1970, XIII + 389 pp. (Revised and expanded edition of [6a] in English.)

[6c] ———, *Garmoničeskij analiz operatorov v gilbertovom prostranstve,* Izdat. "Mir", Moscow, 1970, 431 pp. (Revised and expanded edition of [6a] in Russian, with a foreword by M. G. Krein.)

[7] B. Sz.-Nagy, *Unitary dilations of Hilbert space operators and related topics,* (Expository lectures from the CBMS Regional Conference held at the University of New Hampshire, June 7–11, 1971), American Mathematical Society, Providence R. I., 1974, VIII + 54 pp.

II. Research papers

[1] B. Sz.-Nagy, Ein Verfahren zur Gewinnung von Atomformfaktoren, *Zeitschrift f. Phys.*, **91** (1934), 105–110.

[2] ———, Berechnung einiger neuen Atomfaktoren, *Zeitschrift f. Phys.*, **94** (1935), 229–230.

[3] ———, Über messbare Darstellungen Liescher Gruppen, *Math. Annalen*, **112** (1936), 286–296.

[4] ———, Sur la mesure invariante dans des groupes topologiques, *C. R. Acad. Sci. Paris*, **202** (1936), 1248–1250.

[5] ———, Über eine Frage aus der Theorie der orthogonalen Funktionensysteme, *Math. Zeitschrift*, **41** (1936), 541–544.

[6] ———, Izomorf függvényrendszerekről, *Mat. Term. Tud. Értesítő*, **54** (1936), 712–735.

[7] ———, Über isomorphe vollständige Funktionensysteme, *Math. Zeitschrift*, **43** (1937), 1–16.

[8] ———, Über in sich abgeschlossene Funktionensysteme, *Math. Zeitschrift*, **43** (1937), 17–31.

[9] ———, Önmagában zárt ortogonális függvényrendszer szorzótáblázatáról, *Mat. Term. Tud. Értesítő*, **53** (1937), 574–591.

[10] ———, Bedingungen für die Multiplikationstabelle eines in sich abgeschlossenen orthogonalen Funktionensystems, *Annali di Pisa*, **6** (1937), 211–224.

[11] ———, Zur Theorie der Charaktere Abelscher Gruppen, *Math. Annalen*, **114** (1937), 373–384.

[12] ———, Über die Gesamtheit der charakteristischen Funktionen im Hilbertschen Funktionenraum, *Acta Sci. Math. (Szeged)*, **8** (1937), 166–176.

[13] ———, On the set of positive functions in L_2, *Annals of Math.*, **39** (1938), 1–13.

[14] ———, Propriétés extrémales des séries de Fourier transformées par des suites absolutment monotones, *C. R. Acad. Sci. Paris*, **206** (1938), 808–811.

[15] ———, Sur des suites de facteurs multiplement monotones, *C. R. Acad. Sci. Paris*, **206** (1938), 1342–1344.

[16] ———, Projektív sokszögekről és sokoldalakról, *Mat. Term. Tud. Értsítő*, **57** (1938), 105–120.

[17] A. Starusz & B. Sz.-Nagy, Egy Bohr-féle tételről, *Mat. Term. Tud. Értesítő*, **57** (1938), 121–135.

[18] B. Sz.-Nagy, On semigroups of selfadjoint transformations in Hilbert space, *Proceedings National Acad. USA*, **24** (1938), 559–560.

[19] ______, Über gewisse Extremalfragen bei transformierten trigonometrischen Entwicklungen. I. Periodischer Fall, *Berichte Akad. Wiss. Leipzig*, **90** (1938), 103–134.

[20] ______, Über gewisse Extremalfragen bei transformierten trigonometrischen Entwicklungen. II. Nichtperiodischer Fall, *Berichte Akad. Wiss. Leipzig*, **91** (1939), 3–24.

[21] ______, Sur un problème d'extrémum pour les fonctions définies sur tout l'axe réel, *C. R. Acad. Sci. Paris*, **208** (1939), 1865–1867.

[22] ______, Über ein geometrisches Extremalproblem, *Acta Sci. Math. (Szeged)*, **9** (1940), 253–257.

[23] ______, Über Integralungleichungen zwischen einer Funktion und ihrer Ableitung, *Acta Sci. Math. (Szeged)*, **10** (1941), 64–74.

[24] ______, Egy Carlson-féle és néhány azzal rokon egyenlőtlenségről, *Mat. Fiz. Lapok*, **48** (1941), 162–175.

[25] ______, Sur un problème pour les polyèdres convexes dans l'espace n-dimensionnel, *Bulletin Soc. Math. de France*, **69** (1941), 3–4.

[26] ______, Függvények megközelítése Fourier-sorok számtani közepeivel, *Mat. Fiz. Lapok*, **49** (1942), 122–138.

[27] F. Riesz & B. Sz.-Nagy, Über Kontraktionen des Hilbertschen Raumes, *Acta Sci. Math. (Szeged)*, **10** (1943), 202–205.

[28] B. Sz.-Nagy, A Hilbert-féle tér normális átalakításainak félcsoportjairól, *Szent István Akadémia Értesítője* **28** (1943), 87–96.

[29] ______, Perturbációk a Hilbert-féle térben. I, *Mat. Term. Tud. Értesítő*, **61** (1942), 755–775.

[30] ______, Perturbációk a Hilbert-féle térben. II, *Mat. Term. Tud. Értesítő*, **62** (1943), 63–79.

[31] ______, Sur les lattis linéaires de dimension finie, *Commentarii Math. Helvetici*, **17** (1944), 209–213.

[32] ______, Approximation der Funktionen durch die arithmetischen Mittel ihrer Fourierschen Reihen, *Acta Sci. Math. (Szeged)*, **11** (1946), 71–84.

[33] ______, On uniformly bounded linear transformations in Hilbert space, *Acta Sci. Math. (Szeged)*, **11** (1947), 152–157.

[34] ______, Perturbations des transformations autoadjointes dans l'espace de Hilbert, *Commentarii Math. Helvetici*, **19** (1947), 347–366.

[35] ______, Vibrations d'une corde non homogène, *Bulletin Soc. Math. France*, **75** (1947), 193–208.

[36] ______, Expansion theorems of Paley–Wiener type, *Duke Math. J.*, **14** (1947), 975–978.

[37] ———, Sur une classe générale de procédés de sommation pour les séries de Fourier, *Hungarica Acta Math.*, **1** (1948), 14–52.

[38] L. Rédei & B. Sz.-Nagy, Eine Verallgemeinerung der Inhaltsformel von Heron, *Publ. Math. Debrecen*, **1** (1949), 42–50.

[39] B. Sz.-Nagy, Séries et intégrales de Fourier des fonctions monotones non bornées, *Acta Sci. Math. (Szeged)*, **13** (1949), 118–135.

[40] ———, Une caractérisation affine de l'ensemble des fonctions positives dans l'espace L^2, *Acta Sci. Math. (Szeged)*, **12 A** (1950), 228–239.

[41] ———, Méthodes de sommation des séries de Fourier. I, *Acta Sci. Math. (Szeged)*, **12 B** (1950), 204–210.

[42] ———, Riesz Frigyes tudományos munkásságának ismertetése, *Mat. Lapok*, **1** (1950), 170–181.

[43] ———, Méthodes de sommation des séries de Fourier. II, *Časopis Pest. Mat. Fys.*, **74** (1949), 210–219.

[44] ———, Méthodes de sommation des séries de Fourier. III, *Acta Sci. Math. (Szeged)*, **13** (1950), 247–251.

[45] ———, Über die Konvergenz von Reihen orthogonaler Polynome, *Math. Nachr.*, **4** (1951), 50–55.

[46] ———, Sur l'ordre de l'approximation d'une fonction par son intégrale de Poisson, *Acta Math. Acad. Sci. Hungar.*, **1** (1950), 183–187.

[47] ———, Szovjet eredmények a funckionális analízis terén, *Mat. Lapok*, **2** (1951), 5–53.

[48] ———, Ortogonális polinomsorok konvergenciájáról, *Az Első Magyar Matematikai Kongresszus Közleményei*, Augusztus 27–Szeptember 2, 1950, Akadémiai Kiadó, Budapest, 1952, 249–258.

[49] ———, Sajátértékfeladatok perturbációszámítása, *Magyar Tud. Akad. III. Oszt. Közl.*, **1** (1951), 288–293.

[50] ———, Perturbations des transformations linéaires fermées, *Acta Sci. Math. (Szeged)*, **14** (1951), 123–137.

[51] ———, Eredmények az analízis területén, *Magyar Tud. Akad. III. Oszt. Közl.*, **2** (1952), 59–71.

[52] ———, On the stability of the index of unbounded linear transformations, *Acta Math. Acad. Sci. Hungar.*, **3** (1952), 49–52.

[53] ———, On a spectral problem of Atkinson, *Acta Math. Acad. Sci. Hungar.*, **3** (1952), 61–66.

[54] ———, Magyar matematikusok hozzájárulása a spektrálelmélethez, *Magyar Tud. Akad. III. Oszt. Közl.*, **3** (1953), 85–100.

[55] ———, Pozitív polinomok. I, *Mat. Lapok*, **3** (1952), 140–147.

[56] ———, Pozitív polinomok. II, *Mat. Lapok*, **4** (1953), 13–17.

[57] ______, Über die Ungleichung von H. Bohr, *Math. Nachr.*, **9** (1953), 255–259.

[58] ______, A moment problem for selfadjoint operators, *Acta Math. Acad. Sci. Hungar.*, **3** (1952), 285–293.

[59] ______, Momentumprobléma önadjungált operátorokra, *Magyar Tud. Akad. III. Oszt. Közl.*, **4** (1954), 163–171.

[60] ______, Az 1952. évi Schweitzer Miklós matematikai emlékverseny, *Mat. Lapok*, **4** (1953), 126–155.

[61] ______, Approximation properties of orthogonal expansions, *Acta Sci. Math.* (*Szeged*), **15** (1953), 31–37.

[62] ______, Sur les contractions de l'espace de Hilbert, *Acta Sci. Math.* (*Szeged*), **15** (1953), 87–92.

[63] ______, O sopražennyh konusah v gilbertovom prostranstve, *Uspehi Matem. Nauk. III*, **5 (57)** (1953), 167–168.

[64] ______, Transformations de l'espace de Hilbert, fonctions de type positif sur un groupe, *Acta Sci. Math.* (*Szeged*), **15** (1954), 104–114.

[65] ______, Kontrakciók és pozitív definit operátorfüggvények a Hilbert-térben, *Magyar Tud. Akad. III. Oszt. Közl.*, **4** (1954), 189–204.

[66] ______, Ein Satz über die Parallelverschiebung konvexer Körper, *Acta Sci. Math.* (*Szeged*), **15** (1954), 169–177.

[67] ______, Riesz Frigyes 1880–1956, *Magyar Tud. Akad. III. Oszt. Közl.*, **6** (1956), 143–156.

[68] ______, Forsetzungen linearer Transformationen des Hilbertschen Raumes mit Austritt aus dem Raum, *Schr. Forschungsinst. Math.*, **1** (1957), 289–302.

[69] ______, Remark on S. N. Roy's paper "A useful theorem in matrix theory", *Proc. Amer. Math. Soc.*, **7** (1956), 1.

[70] ______, Contributions en Hongrie à la théorie spectrale des transformations linéaires, *Czechoslovak Math. J.*, **6(81)** (1956), 166–176.

[71] ______, Remarks to the preceding paper of Á. Korányi, *Acta Sci. Math.* (*Szeged*), **17** (1956), 71–75.

[72] ______, Preobrazovanija gilbertova prostranstva, položiteljno opredelennye funkcii na polugruppe, *Uspehi Matem. Nauk.*, **11** (1956), 173–182.

[73] B. Sz.-Nagy & Á. Korányi, Relations d'un problème de Nevanlinna et Pick avec la théorie des opérateurs de l'espace hilbertien, *Acta Math. Acad. Sci. Hungar.*, **7** (1956), 295–303.

[74] B. Sz.-Nagy, Sur les contractions de l'espace de Hilbert. II, *Acta Sci. Math.* (*Szeged*), **18** (1957), 1–14.

[75] ———, A Hilbert-tér normális transzformációinak gyengén konvergens sorozatairól, *Magyar Tud. Akad. III. Oszt. Közl.*, **7** (1957), 295–303.

[76] ———, Suites faiblement convergentes de transformations normales de l'espace hilbertien, *Acta Math. Acad. Sci. Hungar.*, **8** (1957), 295–302.

[77] ———, Note on sums of almost orthogonal operators, *Acta Sci. Math. (Szeged)*, **18** (1957), 189–191.

[78] ———, Neumann János munkássága az operátorelmélet területén, *Mat. Lapok*, **8** (1957), 185–210.

[79] ———, Sur les contractions de l'espace de Hilbert. III, *Acta Sci. Math. (Szeged)*, **19** (1958), 26–45.

[80] B. Sz.-Nagy & Á. Korányi, Operatortheoretische Behandlung und Verallgemeinerung eines Problemkreises in der komplexen Funktionentheorie, *Acta Math.*, **100** (1958), 171–202.

[81] B. Sz.-Nagy, Über Parallelmengen nichtkonvexer ebener Bereiche, *Acta Sci. Math. (Szeged)*, **20** (1959), 36–47.

[82] B. Sz.-Nagy & C. Foias, Une relation parmi les vecteurs propres d'un opérateur de l'espace de Hilbert et de l'opérateur adjoint, *Acta Sci. Math. (Szeged)*, **20** (1959), 91–96.

[83] B. Sz.-Nagy, Completely continuous operators with uniformly bounded iterates, *Magyar Tud. Akad. Mat. Kutató Int. Közl.*, **4** (1959), 89–92.

[84] C. Foias, L. Gehér & B. Sz.-Nagy, On the permutability condition of quantum mechanics, *Acta Sci. Math. (Szeged)*, **21** (1960), 78–79.

[85] B. Sz.-Nagy, Spectral sets and normal dilations of operators, *Proc. Internat. Congress Math.*, Edinburgh, 1958, Cambridge Univ. Press, New York, 1960, 412–422.

[86] B. Sz.-Nagy & C. Foias, Sur les contractions de l'espace de Hilbert. IV, *Acta Sci. Math. (Szeged)*, **21** (1960), 251–259.

[87] B. Sz.-Nagy, Bemerkungen zur vorstehenden Arbeit des Herrn G. Brehmer, *Acta Sci. Math. (Szeged)*, **22** (1961), 112–114.

[88] ———, On Schäffer's construction of unitary dilations, *Ann. Univ. Sci. Budapest Eötvös Sect. Math.*, **3–4** (1960/61), 343–346.

[89] B. Sz.-Nagy & C. Foias, Sur les contractions de l'espace de Hilbert. V. Translations bilatérales, *Acta Sci. Math. (Szeged)*, **23** (1962), 106–109.

[90] ———, Sur les contractions de l'espace de Hilbert. VI. Calcul fonctionnel, *Acta Sci. Math. (Szeged)*, **23** (1962), 130–167.

[91] ———, Remark to the preceding paper of J. Feldman, *Acta Sci. Math. (Szeged)*, **23** (1962), 272–273.

[92] B. Sz.-Nagy, Hilbert Dávid, *Magyar Tud. Akad. III. Oszt. Közl.*, **12** (1962), 203–216.

[93] ______, The "outer functions" and their role in functional calculus, *Proc. Internat. Congress Math.*, Stockholm, 1962, Inst. Mittag-Leffler, Djurshalm, 1963, 421–425.

[94] ______, Un calcul fonctionnel pour les opérateurs linéaires de l'espace hilbertien et certaines de ses applications, *Studia Math.*, **1** (1963), 119–127.

[95] B. Sz.-Nagy & C. Foias, Modèles fonctionnels des contractions de l'espace de Hilbert. La fonction caractéristique, *C. R. Acad. Sci. Paris*, **256** (1963), 3236–3238.

[96] ______, Propriétés des fonctions caractéristiques, modèles triangulaires et une classification des contractions de l'espace de Hilbert, *C. R. Acad. Sci. Paris*, **256** (1963), 3413–3415.

[97] B. Sz.-Nagy, Isometric flows in Hilbert space, *Proc. Cambridge Philos. Soc.*, **60** (1964), 45–49.

[98] B. Sz.-Nagy & C. Foias, Sur les contractions de l'espace de Hilbert. VII. Triangulations canoniques. Fonction minimum, *Acta Sci Math. (Szeged)*, **25** (1964), 12–37.

[99] ______, Sur les contractions de l'espace de Hilbert. VIII. Fonctions caractéristiques. Modéles fonctionnels, *Acta Sci. Math. (Szeged)*, **25** (1964), 38–71.

[100] ______, Une caracterisation des sous-espaces invariants pour une contraction de l'espace de Hilbert, *C. R. Acad. Sci. Paris,* Groupe 1, **258** (1964), 3426–3429.

[101] ______, Sur les contractions de l'espace de Hilbert. IX. Factorisations de la fonction caractéristique. Sous-espaces invariants, *Acta Sci. Math. (Szeged)*, **25** (1964), 283–316.

[102] B. Sz.-Nagy, Un calcul fonctionnel pour les contractions, *Seminari dell'Instituto Nazionale di Alta Matematica,* 1962–63, Ediz. Cremonese, Rome, 1965, 525–528.

[103] ______, Sur la structure des dilatations unitaires des opérateurs de l'espace de Hilbert, *Seminari dell'Instituto Nazionale di Alta Mathematica,* 1962–63, Ediz. Cremonese, Rome, 1965, 529–554.

[104] B. Sz.-Nagy & C. Foias, Sur les contractions de l'espace de Hilbert. X. Contractions similaires à des transformations unitaires, *Acta Sci. Math. (Szeged)*, **26** (1965), 79–91.

[105] B. Sz.-Nagy, Positive definite kernels generated by operator-valued analytic functions, *Acta Sci. Math. (Szeged)*, **26** (1965), 191–192.

[106] B. Sz.-Nagy & C. Foias, Corrections et compléments aux contractions. IX, *Acta Sci. Math. (Szeged)*, **26** (1965), 193–196.

[107] ———, Sur les contractions de l'espace de Hilbert. XI. Transformations unicellulaires, *Acta Sci. Math. (Szeged)*, **26** (1965), 301–324.

[108] ———, Quasi-similitude des opérateurs et sous-espaces invariants, *C. R. Acad. Sci. Paris*, **261** (1965), 3938–3940.

[109] ———, Décomposition spectrale des contractions presque unitaires, *C. R. Acad. Sci. Paris*, **262** (1966), 440–442.

[110] B. Sz.-Nagy, Positiv-definite, durch Operatoren erzeugte Funktionen, *Wiss. Z. Techn. Univ. Dresden*, **15** (1966), 219–222.

[111] B. Sz.-Nagy & C. Foias, On certain classes of power-bounded operators in Hilbert space, *Acta Sci. Math. (Szeged)*, **27** (1966), 17–25.

[112] ———, Sur les contractions de l'espace de Hilbert. XII. Fonctions intérieures admettant des facteurs extérieurs, *Acta Sci. Math. (Szeged)*, **27** (1966), 27–33.

[113] ———, Correction: "Sur les contractions de l'espace de Hilbert. XI. Transformations unicellulaires", *Acta Sci. Math. (Szeged)*, **27** (1966), 265.

[114] ———, Forme triangulaire d'une contraction et factorisation de la fonction caractéristique, *Acta Sci. Math. (Szeged)*, **28** (1967), 201–212.

[115] ———, Echelles continues de sous-espaces invariants, *Acta Sci. Math. (Szeged)*, **28** (1967), 213–220.

[116] ———, Similitude des opérateurs de classe C_ρ à des contractions, *C. R. Acad. Sci. Paris Série A*, **264** (1967), 1063–1065.

[117] B. Sz.-Nagy, Szovjet-magyar matematikai kapcsolatok a szegedi Acta Scientiarum Mathematicarum tükrében, *Ünnepi Acta*, Szeged, 1967, 45–57.

[118] B. Sz.-Nagy & C. Foias, Dilatation des commutants d'opérateurs, *C. R. Acad. Sci. Paris Série A*, **266** (1968), 493–495.

[119] ———, Commutants de certains opérateurs, *Acta Sci. Math. (Szeged)*, **29** (1968), 1–17.

[120] B. Sz.-Nagy, Products of operators of classes C_ρ, *Rev. Roumaine Math. Pures Appl.*, **13** (1968), 897–899.

[121] B. Sz.-Nagy & C. Foias, Vecteurs cycliques et quasi-affinités, *Studia Math.*, **31** (1968), 35–42.

[122] ———, Opérateurs sans multiplicité, *Acta Sci. Math. (Szeged)*, **30** (1969), 1–18.

[123] B. Sz.-Nagy, Sur la norme des fonctions de certains opérateurs, *Acta Math. Acad. Sci. Hungar.*, **20** (1969), 331–334.

[124] ———, Hilbertraum-Operatoren der Klasse C_0, *Abstract Spaces and Approximation* (Proc. Conf., Oberwolfach, 1968), Birkhäuser, Basel, 1969, 72–81.

[125] B. Sz.-Nagy & C. Foias, Modèle de Jordan pour une classe d'opérateurs de l'espace de Hilbert, *Acta Sci. Math. (Szeged)*, **31** (1970), 91–115.

[126] ——, Compléments à l'étude des opérateurs de classe C_0, *Acta Sci. Math. (Szeged)*, **31** (1970), 287–296.

[127] B. Sz.-Nagy, Matematika, *Magyar Tudomány* (1970), 269–283.

[128] B. Sz.-Nagy & C. Foias, The "Lifting Theorem" for intertwining operators and some applications, (Proc. Internat. Symposium on Operator Theory, Indiana Univ. Bloomington, 1970), *Indiana Univ. Math. J.*, **20** (1971), 901–904.

[129] ——, Local characterization of operators of class C_0, *J. Funct. Anal.*, **8** (1971), 76–81.

[130] B. Sz.-Nagy, Vecteurs cycliques et commutativité des commutants, *Acta Sci. Math. (Szeged)*, **32** (1971), 177–183.

[131] ——, Sous-espaces invariants d'un opérateur et factorisations de sa fonctions caractéristique, *Actes du Congrès International des Mathématiciens*, Nice, Septembre 1970, Gauthiers-Villars, Paris, **2** (1971), 459–465.

[132] ——, Quasi-similarity of operators of class C_0, *Hilbert Space Operators and Operator Algebras* (Proc. Internat. Conf., Tihany, 1970), North-Holland, Amsterdam, 513–517.

[133] B. Sz.-Nagy & C. Foias, Compléments à l'étude des opérateurs de classe C_0. II, *Acta Sci. Math. (Szeged)* **33** (1972), 113–116.

[134] ——, Accretive operators: Corrections, *Acta Sci. Math. (Szeged)*, **33** (1972), 117–118.

[135] ——, Echelles continues de sous-espaces invariants. II, *Acta Sci. Math. (Szeged)*, **33** (1972), 355–356.

[136] B. Sz.-Nagy, Cyclic vectors and commutants, *Linear Operators and Approximation* (Proc. Conf., Oberwolfach, 1971), Birkhäuser, Basel, 1972, 62–67.

[137] B. Sz.-Nagy & C. Foias, On the structure of intertwining operators, *Acta Sci. Math. (Szeged)*, **35** (1973), 225–254.

[138] ——, Regular factorizations of contractions, *Proc. Amer. Math. Soc.*, **43** (1974), 91–93.

[139] ——, Injection of shifts into strict contractions, *Linear Operators and Approximation. II* (Proc. Conf. Math. Res. Inst., Oberwolfach, 1974), Birkhäuser, Basel, 1975, 29–37.

[140] ——, Jordan model for contractions of class $C_{\cdot 0}$, *Acta Sci. Math. (Szeged)*, **36** (1974), 305–322.

[141] B. Sz.-Nagy, On a property of operators of class C_0, *Acta Sci. Math. (Szeged)*, **36** (1974), 219–220.

[142] ———, Models of Hilbert space operators, (Spectral Theory Symposium, Trinity College, Dublin, 1974), *Proc. Royal Irish Acad. Sec. A*, **74** (1974), 263–270.

[143] ———, A general view on unitary dilations, (Internat. Conf., Madras, 1973; dedicated to Alladi Ramakrishnan), *Lecture Notes in Math.*, **399** (1974), 382–395.

[144] B. Sz.-Nagy & C. Foias, An application of dilation theory to hyponormal operators, *Acta Sci. Math. (Szeged)*, **37** (1975), 155–159.

[145] H. Bercovici, C. Foias & B. Sz.-Nagy, Compléments à l'étude des opérateurs de classe C_0. III, *Acta Sci. Math. (Szeged)*, **37** (1975), 313–322.

[146] B. Sz.-Nagy, Digonalization of matrices over H^∞, *Acta Sci. Math. (Szeged)*, **38** (1976), 223–238.

[147] B. Sz.-Nagy & C. Foias, Commutants and bicommutants of operators of class C_0, *Acta Sci. Math. (Szeged)*, **38** (1976), 311–315.

[148] ———, On contractions similar to isometries and Toeplitz operators, *Annales Acad. Sci. Fennicae, Series A I*, **2** (1976), 553–564.

[149] ———, Vecteurs cycliques et commutativité des commutants. II, *Acta Sci. Math. (Szeged)*, **39** (1977), 169–174.

[150] B. Sz.-Nagy, Quasi-similarity of Hilbert-space operators, *Proc. Internat. Conf. on Differential Equations,* Uppsala, 1977, Almqvist & Wiksell, Stockholm, 1977, 179–188.

[151] B. Sz.-Nagy & C. Foias, On injections intertwining operators of class C_0, *Acta Sci. Math. (Szeged)*, **40** (1978), 163–167.

[152] B. Sz.-Nagy, Nevanlinna class functions of operators, *Proceedings of the Rolf Nevanlinna Symposium on Complex Analysis,* Silivri, 1976, Univ. Istanbul, Istanbul, 1978.

[153] ———, Diagonalization of matrices over H^∞, *Linear Spaces and Approximation* (Proc. Conf., Math. Res. Inst., Oberwolfach, 1977), Birkhäuser, Basel, 1978, 37–46.

[154] H. Bercovici, C. Foias, L. Kérchy & B. Sz.-Nagy, Compléments à l'étude des opérateurs de classe C_0. IV, *Acta Sci. Math. (Szeged)*, **41** (1979), 29–31.

[155] B. Sz.-Nagy & C. Foias, The function model of a contraction and the space L^1/H_0^1, *Acta Sci. Math. (Szeged)*, **41** (1979), 403–410.

[156] B. Sz.-Nagy, A brief review of my work in mathematics (Russian), *Vestnik Akad. Nauk SSSR*, **6** (1980), 50–56.

[157] C. Foias, C. Pearcy & B. Sz.-Nagy, The functional model of a contraction and the space L^1, *Acta Sci. Math. (Szeged)*, **42** (1980), 201–204.

[158] H. BERCOVICI, C. FOIAS & B. SZ.-NAGY, Reflexive and hyper-reflexive operators of class C_0, *Acta Sci. Math. (Szeged)*, **43** (1981), 5–13.

[159] C. FOIAS, C. PEARCY & B. SZ.-NAGY, Contractions with spectral radius one and invariant subspaces, *Acta Sci. Math. (Szeged)*, **43** (1981), 273–280.

[160] H. BERCOVICI, C. FOIAS, C. M. PEARCY & B. SZ.-NAGY, Functional models and extended spectral dominance, *Acta Sci. Math. (Szeged)*, **43** (1981), 243–254.

[161] B. SZ.-NAGY, Some lattice properties of the space L^2, *From A to Z,* Leiden, 1982, Math. Centrum, Amsterdam, 1982, 101–112.

[162] B. SZ.-NAGY & C. FOIAS, Toeplitz type operators and hyponormality, (Dilation theory, Toeplitz operators, and other topics; Timisoara/Herculane, 1982), *Operator Theory: Adv. Appl.,* Birkhäuser, Basel – Boston, **11** (1983), 371–388.

[163] E. DURSZT & B. SZ.-NAGY, Remark to a paper: "Models for noncommuting operators" by A. E. Frazho, *J. Funct. Anal.*, **52** (1983), 146–147.

[164] B. SZ.-NAGY & C. FOIAS, Contractions without cylic vectors, *Proc. Amer. Math. Soc.*, **87** (1983), 671–674

[165] H. BERCOVICI, C. FOIAS, C. PEARCY & B. SZ.-NAGY, Factoring compact operator-valued functions, *Acta Sci. Math. (Szeged)*, **48** (1985), 25–36.

[166] B. SZ.-NAGY, Sets similar to the positive cone in $L^2(m)$, *Operator Theory: Adv. Appl.,* Birkhäuser, Basel, **24** (1987), 313–320.

[167] ———, Bohr inequality and an operator equation, *Operator Theory: Adv. Appl.,* Birkhäuser, Basel, **24** (1987), 321–327.

Operator Theory: Advances and Applications, Vol. 127, 1–27

Inverse problems associated to a canonical differential system

DANIEL ALPAY and ISRAEL GOHBERG

To the memory of Béla Szőkeflavi-Nagy, with respect and affection

We solve the inverse problems associated with a differential expression of the form (1.1) for the Weyl coefficient function and the reflection coefficient function in the case of potentials $k(t)$ of the special form (1.2).

1. Introduction

In the present paper we pursue our study of canonical differential expressions of the form

$$-iJ\frac{\mathrm{d}f}{\mathrm{d}t}(t,\lambda) = \lambda f(t,\lambda) + V(t)f(t,\lambda), \quad t \geq 0, \ \lambda \in \mathbb{R} \tag{1.1}$$

which was started in [2], [3], [4]. In expression (1.1),

$$J = \begin{pmatrix} I_n & 0 \\ 0 & -I_n \end{pmatrix}, \quad V(t) = \begin{pmatrix} 0 & k(t) \\ k(t)^* & 0 \end{pmatrix},$$

where the function k, called the *potential*, is $\mathbb{C}^{n\times n}$-valued and with entries in $\mathbf{L}_1(0,\infty)$, and the solution $f(t,\lambda)$ is $\mathbb{C}^{2n\times m}$-valued.

Such differential systems appear in a number of situations, of which we mention:

(1) The symmetric two-components wave equation (see [18, Section 2], [7]).
(2) The Schödinger equation; see [18, Section 2.4].
(3) The linear estimation of stationary stochastic processes; see [18, Section 5].

AMS Subject Classification (1991): 34L25, 81U40, 47A56.

Associated to the differential expression (1.1) are a number of functions of λ, which we will call the *characteristic spectral functions* of the canonical system; these are:

(1) The scattering function $S(\lambda)$.
(2) The spectral function $W(\lambda)$.
(3) The Weyl function $N(\lambda)$.
(4) The reflection coefficient function $R(\lambda)$.
(5) The asymptotic equivalence matrix function $V(\lambda)$.

Direct problems consist in computing these functions from the potential while inverse problems go the other way around. These problems have a long history, which originates with the works of I. M. Gelfand and B. M. Levitan [11], M. G. Kreĭn (for the inverse spectral problem $W(\lambda)$; see [15]) and V. A. Marchenko (for the inverse scattering problem; see [16]). We refer to [10] for a review of the literature.

This paper is part of a series where one studies these problems, and in particular wishes to obtain direct proofs and explicit formulas when the function $k(t)$ is of the special form

$$(1.2)\qquad k(t) = -2ce^{ita}\left(I_p + \Omega\left(Y - e^{-2ita^*}Ye^{2ita}\right)\right)^{-1}(b + i\Omega c^*),$$

where $(a,b,c) \in \mathbb{C}^{p\times p} \times \mathbb{C}^{p\times n} \times \mathbb{C}^{n\times p}$ is a minimal triple of matrices, this means

$$(1.3)\qquad \bigcap_{\ell=0}^{\infty} \ker ca^{\ell} = \{0\}, \quad \text{and} \quad \bigcup_{\ell=0}^{\infty} \operatorname{Im} a^{\ell}b = \mathbb{C}^p,$$

and where Ω and Y belong to $\mathbb{C}^{p\times p}$ and are the solutions of the Lyapunov equations

$$(1.4)\qquad i(\Omega a^{\times *} - a^{\times}\Omega) = bb^*,$$

$$(1.5)\qquad -i(Ya - a^*Y) = c^*c.$$

The minimality conditions (1.3) imply that the matrices Ω and Y are strictly positive, and it is shown in [2] that $\left(I_p + \Omega\left(Y - e^{-2ita^*}Ye^{2ita}\right)\right)$ is invertible for all positive t.

This class of potentials was introduced in [2] and called in [12] strictly pseudo-exponential potentials. They correspond, as we will recall in the sequel, to *rational* characteristic spectral functions. As a simple example, set $n = p = 1$ and $a = i\alpha$ (with $\alpha > 0$), $b = 1$ and $c = c^* \neq 0$ and $\alpha > 0$. Then,

$$\Omega = \frac{1}{2\alpha}, \qquad Y = \frac{c^2}{2\alpha}.$$

We then have $e^{2ita} = e^{-2ita^*} = e^{-2\alpha t}$,

$$(I_p + \Omega(Y - e^{-2ita^*} Y e^{2ita})) = 1 + \frac{1}{2\alpha}\left(\frac{c^2}{2\alpha} - e^{-4\alpha t}\frac{c^2}{2\alpha}\right) = 1 + \frac{c^2}{4\alpha^2}(1 - e^{-4\alpha t})$$

and the potential is given by

$$k(t) = -\frac{2ce^{-\alpha t}(1 + \frac{ic}{2\alpha})}{1 + \frac{c^2}{4\alpha^2}(1 - e^{-4\alpha t})}. \tag{1.6}$$

It is the potential corresponding to the scattering function

$$S(\lambda) = \left(1 + \frac{c}{\lambda - i\alpha}\right)\left(1 + \frac{c}{\lambda + i\alpha}\right)^{-1}.$$

See [3], [5]; the definition of the scattering function is recalled in the next section.

The purpose of this paper is twofolds. We first relate the various characteristic spectral functions. Then we study inverse and direct problems associated to the Weyl function and to the reflection coefficient function.

The paper contains four sections besides the introduction. In the second section, which is essentially of a review nature and based on the papers [5], [12], we give the definitions and properties of the characteristic spectral functions. The various relationships between these functions are considered in the third section. Minimal realizations of the characteristic spectral functions are studied in the fourth section, while in the last section we study inverse problems.

It is our pleasure to thank M. A. Kaashoek and A. L. Sakhnovich for their useful remarks.

2. Definitions and properties

2.1. Special solutions of the equation (1.1). Equation (1.1) has a unique $\mathbb{C}^{2n\times 2n}$-valued solution $U(t,\lambda)$ such that

$$\lim_{t\longrightarrow+\infty} e^{-i\lambda tJ}U(t,\lambda) = I_{2n}. \tag{2.1}$$

This is a general fact for summable potentials $k(t)$; see [10, Theorem 9.2 p. 213–214]. A direct proof with an explicit formula for $U(t,\lambda)$ in the case of potentials of the form (1.2) is given in [5]. We now recall the following theorem, a proof of which can be found in [5]:

THEOREM 2.1. *Assume given $k(t)$ in the form (1.2). Let $Q(t,y)$ be given by*

$$Q(t,y) = Fe^{tT}(I_{2p} - e^{tT}Ze^{tT})^{-1}e^{yT}G, \tag{2.2}$$

where (F,G,T) are defined by

$$F = i\begin{pmatrix} -c & 0 \\ 0 & f_1 \end{pmatrix}, \quad T = \begin{pmatrix} ia & 0 \\ 0 & -ia^* \end{pmatrix}, \quad G = \begin{pmatrix} 0 & f_1^* \\ c^* & 0 \end{pmatrix}, \tag{2.3}$$

where $f_1 = (b^* - ic\Omega)(I_p + Y\Omega)^{-1}$ and Y and Ω are the solutions of the Lyapunov equations (1.4) and (1.5) and where Z is the unique solution of the matrix equation

$$TZ + ZT = -GF. \tag{2.4}$$

Then the matrix function

$$U(t,\lambda) = e^{i\lambda Jt} + \int_t^\infty Q(t,u)e^{i\lambda Ju}\,du \tag{2.5}$$

is the unique solution of (1.1) with the potential as in (1.2) subject to the condition (2.1). Furthermore, the $\mathbb{C}^{n\times n}$-valued blocks in the decomposition of the matrix-function $U(0,\lambda) = (U_{ij}(0,\lambda))$ are given by

$$U_{11}(0,\lambda) = I_n + ic\Omega(\lambda I_p - a^*)^{-1}c^* \tag{2.6}$$

$$U_{21}(0,\lambda) = (-b^* + ic\Omega)(\lambda I_p - a^*)^{-1}c^* \tag{2.7}$$

$$U_{12}(0,\lambda) = -c(I_p + \Omega Y)(\lambda I_p - a)^{-1}(I_p + \Omega Y)^{-1}(b + i\Omega c^*) \tag{2.8}$$

$$U_{22}(0,\lambda) = I_n - (ib^*Y + c\Omega Y)(\lambda I_p - a)^{-1}(I_p + \Omega Y)^{-1}(b + i\Omega c^*). \tag{2.9}$$

We remark that the matrix $(I_p + \Omega Y)$ is invertible since the matrices Ω and Y are positive and since

$$\det(I_p + \Omega Y) = \det(I_p + \sqrt{Y}\Omega\sqrt{Y}).$$

We denote by $\Theta(t,\lambda)$ the $\mathbb{C}^{2n\times 2n}$-valued solution of (1.1) subject to $\Theta(0,\lambda) = I_{2n}$. Note that $\Theta(t,\lambda) = U(t,\lambda)U(0,\lambda)^{-1}$, where $U(t,\lambda)$ is the solution of (1.1) subject to (2.1). For any given $t \geq 0$, the function $\lambda \mapsto \Theta(t,\lambda)$ is entire of finite exponential type. An explicit formula for Θ for potentials of the form (1.2) is given in [12] and recalled in [5].

Various properties of $U(t,\lambda)$ for arbitrary potentials k can found in [10, Section 2, pp 147–159].

In each of the following subsections we review the definition and main properties of one of the characteristic spectral functions.

2.2. The asymptotic equivalence matrix function. Let $U(t,\lambda)$ be the $\mathbb{C}^{2n\times 2n}$-valued solution of (1.1) subject to the asymptotic condition (2.1). The function $V(\lambda){:=}U(0,\lambda)$ is called the *asympotic equivalence matrix function*; we refer to [8] and [10, p.156]. The terminology is explained in [10], and we repeat here the explanation given in that paper: let $\lambda\in\mathbb{R}$ and ξ_0 and ξ_1 in $\mathbb{C}^{2n}$. Let $f_0(t,\lambda)=e^{i\lambda tJ}\xi_0$ be the $\mathbb{C}^{2n}$-valued solution to (1.1) corresponding to $k(t)=0$ and $f_0(0,\lambda)=\xi_0$ and let $f_1(t,\lambda)\xi_1$ be corresponding to an arbitrary potential k with $f_1(0,\lambda)=\xi_1$. These two solutions are asymptotic in the sense that

$$\lim_{t\to\infty}\|f_1(t,\lambda)-f_0(t,\lambda)\|=0$$

if and only if

$$\xi_1=V(\lambda)\xi_0.$$

The claim is a direct consequence of the expression $f_1(t,\lambda)=U(t,\lambda)U(0,\lambda)^{-1}\xi_1$ and of the asymptotic property of U.

The function $V(\lambda)$ is defined for real λ and is J-unitary on the real line (recall that $J=\begin{pmatrix} I_n & 0\\ 0 & -I_n\end{pmatrix}$); see [10]. The precise form of $V(\lambda)$ for potential of the form (1.2) will be given in the next section; see formula (3.6).

2.3. The reflection coefficient function. The system (1.1) can be interpreted as the model for a scattering medium (see [7], [18, Section 2]). Then k is called the *local reflexivity coefficient function*, and the medium is characterized by a function of λ called the reflection coefficient function. Its definition follows: the function $V(\lambda)$ is J-unitary on the real line and in particular it holds that

$$V_{11}(\lambda)^*V_{11}(\lambda)-V_{21}(\lambda)^*V_{21}(\lambda)=I_n,\qquad \lambda\in\mathbb{R},$$

where $V=(V_{ij})_{\{i,j=1,2\}}$ is the decomposition of V into $\mathbb{C}^{n\times n}$-valued block. Therefore $\det V_{11}(\lambda)$ does not vanish on the real line. The quotient

$$R(\lambda)=V_{21}(\lambda)V_{11}(\lambda)^{-1} \tag{2.10}$$

is called the *reflection coefficient function.* The inverse problem associated to the reflection coefficient function was considered in [6]. Another solution is given in the present work; see Section .

2.4. The scattering matrix function. The function $U(t,\lambda)$ with the asymptotic property (2.1) is J-unitary on the real line and it follows that $U_{11}(0,\lambda)^*U_{11}(0,\lambda) - U_{21}(0,\lambda)^*U_{21}(0,\lambda) = I_n$, i.e.

$$U_{11}(0,\lambda)^*U_{11}(0,\lambda) = I_n + U_{21}(0,\lambda)^*U_{21}(0,\lambda).$$

In particular, $\det U_{11}(0,\lambda) \neq 0$ and $U_{11}(0,\lambda)^{-1}U_{12}(0,\lambda)$ is a strict contraction, and hence

$$\det(U_{11}(0,\lambda) - U_{21}(0,\lambda)) \neq 0, \qquad \lambda \in \mathbb{R}.$$

THEOREM 2.2. *Set*

$$S(\lambda) = (U_{11}(0,\lambda) - U_{21}(0,\lambda))((U_{22}(0,\lambda) - U_{12}(0,\lambda))^{-1}. \tag{2.11}$$

The function $X(t,\lambda) = U(t,\lambda)\begin{pmatrix} S(\lambda) \\ I_n \end{pmatrix}$ *is the unique* $\mathbb{C}^{2n\times n}$*-valued solution of* (1.1) *with the following properties*:

$$(I_n \quad -I_n)X(0,\lambda) = 0, \tag{2.12}$$

$$(0 \quad I_n)X(t,\lambda) = e^{-i\lambda t}I_n + o(1) \qquad (t \to \infty) \tag{2.13}$$

Proof. Let $X(t,\lambda) = U(t,\lambda)\begin{pmatrix} \alpha(\lambda) \\ \beta(\lambda) \end{pmatrix}$ be a solution of (1.1), where $\alpha(\lambda)$ and $\beta(\lambda)$ are in $\mathbb{C}^{n\times n}$. From

$$\begin{aligned}(0 \quad I_n)U(t,\lambda)\begin{pmatrix} \alpha(\lambda) \\ \beta(\lambda) \end{pmatrix} &= (0 \quad I_n)\left(\begin{pmatrix} e^{i\lambda t}I_n & 0 \\ 0 & e^{-i\lambda t}I_n \end{pmatrix} + o(1)\right)\begin{pmatrix} S(\lambda) \\ I_n \end{pmatrix} \\ &= ((0 \quad e^{-it\lambda}\beta(\lambda)) + 0(1)),\end{aligned}$$

we see that (2.13) holds if and only if $\beta(\lambda) = I_n$. Then,

$$\begin{aligned}(I_n \quad -I_n)X(0,\lambda) &= (I_n \quad -I_n)U(0,\lambda)\begin{pmatrix} \alpha(\lambda) \\ \beta(\lambda) \end{pmatrix} \\ &= (U_{11}(0,\lambda) - U_{21}(0,\lambda) \quad U_{12}(0,\lambda) - U_{22}(0,\lambda))\begin{pmatrix} \alpha(\lambda) \\ I_n \end{pmatrix}\end{aligned}$$

implies that (2.12) holds if and only if $\alpha(\lambda) = S(\lambda)$. □

Let $X(t,\lambda)$ be as in the previous theorem. One has

$$
\begin{aligned}
(I_n \;\; 0)X(t,\lambda) &= (I_n \;\; 0)U(t,\lambda)\begin{pmatrix} S(\lambda) \\ I_n \end{pmatrix} \\
&= (I_n \;\; 0)\left(\begin{pmatrix} e^{i\lambda t}I_n & 0 \\ 0 & e^{-i\lambda t}I_n \end{pmatrix} + o(1)\right)\begin{pmatrix} S(\lambda) \\ I_n \end{pmatrix} \\
&= ((e^{it\lambda}I_n \;\; 0) + o(1))\begin{pmatrix} S(\lambda) \\ I_n \end{pmatrix} \\
&= e^{it\lambda}S(\lambda) + o(1) \qquad (t \to \infty).
\end{aligned} \tag{2.14}
$$

The function S is called the *scattering matrix function.* The scattering function S has the following properties: it belongs to the Wiener algebra and $S(\infty) = I_n$, i.e. there exists a function $\sigma \in \mathbf{L}_1^{n\times n}(\mathbb{R})$ such that

$$S(\lambda) = I_n + \int_{-\infty}^{\infty} e^{i\lambda u}\sigma(u)\,du,$$

and it admits a Wiener–Hopf factorization

$$S = S_- S_+ \tag{2.15}$$

where S_- and its inverse are in the Wiener algebra $\mathcal{W}_-^{n\times n}$:

$$\mathcal{W}_-^{n\times n} = \left\{F(\lambda) = c + \int_{-\infty}^{0} e^{i\lambda u} f(u)\,du \; ; \; c \in \mathbb{C}^{n\times n}, \;\; f \in \mathbf{L}_1^{n\times n}(-\infty, 0)\right\}$$

and S_+ and its inverse are in the Wiener algebra $\mathcal{W}_+^{n\times n}$:

$$\mathcal{W}_+^{n\times n} = \left\{F(\lambda) = c + \int_{0}^{\infty} e^{i\lambda u} f(u)\,du \; ; \; c \in \mathbb{C}^{n\times n}, \;\; f \in \mathbf{L}_1^{n\times n}(0, \infty)\right\}.$$

For a proof of these facts in the setting of strictly pseudo-exponential potentials, see [5].

The next two characteristic spectral functions which we define are expressed via the solution $\Theta(t,\lambda)$ of (1.1) subject to the initial condition $\Theta(0,\lambda) = I_{2n}$.

2.5. The spectral function. The operator

$$Hf(t) = -iJ\frac{\mathrm{d}\, f}{\mathrm{d}\, t}(t) - V(t)f(t)$$

restricted to the space of $\mathbb{C}^{2n}$-valued absolutely continuous functions with entries in $\mathbf{L}_2$ and such that

$$(I_n \;\; -I_n)f(0) = 0$$

is self-adjoint. A positive function $W:\mathbb{R}\to\mathbb{C}^{n\times n}$ is called *spectral function* if there is a unitary map U from $\mathbf{L}_2^n$ onto $\mathbf{L}_2^n(W)$ mapping H onto the operator of multiplication by the variable in $\mathbf{L}_2^n(W)$. The function

$$W(\lambda)=(V_{22}(\lambda)-V_{12}(\lambda))^{-*}(V_{22}(\lambda)-V_{12}(\lambda))^{-1} \tag{2.16}$$

is a spectral function, the map U being given by

$$F(\lambda)=\frac{1}{\sqrt{2\pi}}\int_0^\infty (I_n \ \ I_n)\Theta(t,\lambda)^* f(t)\,dt.$$

See [12] for a proof for potentials of the form (1.2). For a proof in the setting of general potentials, see [10, Section 5, pp 175–186], and in particular Theorem 5.2 p. 182 of that paper.

2.6. The Weyl coefficient function. The *Weyl coefficient function* $N(\lambda)$ is defined in the open upper half plane; it is the unique $\mathbb{C}^{n\times n}$-valued function such that

$$\int_0^\infty (iN(\lambda)^* \ \ I_n)\begin{pmatrix} I_n & I_n\\ I_n & -I_n\end{pmatrix}\Theta(t,\lambda)^*\Theta(t,\lambda)\begin{pmatrix} I_n & I_n\\ I_n & -I_n\end{pmatrix}\begin{pmatrix} -iN(\lambda)\\ I_n\end{pmatrix}<\infty$$

for $i(\lambda-\lambda^*)>0$.

In the setting of differential expressions (1.1), the function N was introduced in [13]. The motivation comes from the theory of the Sturm–Liouville equation. The Weyl function is analytic in the open lower half-plane and has a nonnegative imaginary part there.

PROPOSITION 2.3. *The Weyl coefficient is given by the formula*

$$N(\lambda)=i\cdot(U_{12}(0,\lambda)+U_{22}(0,\lambda))(U_{12}(0,\lambda)-U_{22}(0,\lambda))^{-1}. \tag{2.17}$$

Proof. We first look for a $\mathbb{C}^{n\times 2n}$-valued function $P(\lambda)$ such that $t\mapsto P(\lambda)\Theta(t,\lambda)^*$ has square summable entries for $\lambda\in\mathbb{C}^+$. Let $U(\lambda,t)$ be the solution of the differential system (1.1) subject to the asymptotic condition (2.1). Then, $U(t,\lambda)=\Theta(t,\lambda)\cdot U(0,\lambda)$. We thus require the entries of the function

$$t\mapsto P(\lambda)U(0,\lambda)^{-*}U(t,\lambda) \tag{2.18}$$

to be square summable. By definition of U, it is necessary for $P(\lambda)U(0,\lambda)^{-*}$ to be of the form $(0,\ p(\lambda))$ where $p(\lambda)$ is $\mathbb{C}^{n\times n}$-valued. It follows from the definition of $U(0,\lambda)$ that one can take

$$P(\lambda)=(0,\ \ I_p)U(0,\lambda)^*=(U_{12}(0,\lambda)^*,\ \ U_{22}(0,\lambda)^*)$$

and hence the necessity condition. Conversely, we have to show that the function (2.18) has indeed summable entries. But this is just doing the above argument backwards.

□

For more on the properties of this function, we refer to [17].

These various functions are related to each other, and to know one of them amounts to know all of them. The *rational case* corresponds to the situation where one of them (and hence all the others, as will be explained in Section 4) is rational.

3. The relationships between the characteristic spectral functions

Any of the functions in the spectral domain determines all the others, as follows from the next theorem:

THEOREM 3.1. *Assume given a differential system of the form* (1.1) *with potential* $k(t)$ *of the form* (1.2). *Assume* $W(\lambda)$, $V(\lambda)$, $R(\lambda)$, $S(\lambda)$ *and* $N(\lambda)$ *are the characteristic spectral functions of* (1.1), *and let* $S = S_- S_+$ *be the spectral factorization of the spectral function S. Then, the connections between these functions are*

$$W(\lambda) = S_-(\lambda)^{-1} S_-(\lambda)^{-*} = S_+(\lambda)^{-1} S_+(\lambda)^{-*}, \tag{3.1}$$

$$W(\lambda) = \operatorname{Im} N(\lambda), \tag{3.2}$$

$$S(\lambda) = S_-(\lambda) S_+(\lambda), \tag{3.3}$$

$$R(\lambda) = (iN(\lambda)^* - I_n)(iN(\lambda)^* + I_n)^{-1}, \tag{3.4}$$

$$N(\lambda) = i(I_n + R(\lambda)^*)(I_n - R(\lambda)^*)^{-1}, \tag{3.5}$$

$$V(\lambda) = \frac{1}{2} \begin{pmatrix} (iN(\lambda)^* + I_n) S_-(\lambda)^* & (-iN(\lambda) - I_n) S_+(\lambda)^{-*} \\ (iN(\lambda)^* - I_n) S_-(\lambda)^* & (-iN(\lambda) + I_n) S_+(\lambda)^{-*} \end{pmatrix} \tag{3.6}$$

for $\lambda \in \mathbb{R}$.

Proof. The fact that $W = S_-^{-1} S_-^{-*}$ is a spectral function and the relationship (3.2) between the spectral function and the Weyl coefficient function are proved in [12, Theorem 3.1] for strictly pseudo-exponential potentials. The fact that $S(\lambda)$ given by (3.3) is the spectral function is proved in [5] for pseudo-exponential potentials. The formula (3.4) follows from (3.6), and we now prove (3.6): it is shown in [5] that the $\mathbb{C}^{n\times n}$-valued blocks in the decomposition of the matrix-function $U(0,\lambda) = (U_{ij}(0,\lambda))$ are given by

$$U_{11}(0,\lambda) = I_n + ic\Omega(\lambda I_p - a^*)^{-1}c^* \tag{3.7}$$
$$U_{21}(0,\lambda) = (-b^* + ic\Omega)(\lambda I_p - a^*)^{-1}c^* \tag{3.8}$$
$$U_{12}(0,\lambda) = -c(I_p + \Omega Y)(\lambda I_p - a)^{-1}(I_p + \Omega Y)^{-1}(b + i\Omega c^*) \tag{3.9}$$
$$U_{22}(0,\lambda) = I_n - (ib^*Y + c\Omega Y)(\lambda I_p - a)^{-1}(I_p + \Omega Y)^{-1}(b + i\Omega c^*). \tag{3.10}$$

We thus have to prove that these entries are equal to the entries in (3.6), i.e. that we have

$$U_{11}(0,\lambda) = (iN(\lambda)^* + I_n)S_-(\lambda)^* \tag{3.11}$$
$$U_{21}(0,\lambda) = (iN(\lambda)^* - I_n)S_-(\lambda)^* \tag{3.12}$$
$$U_{12}(0,\lambda) = (-iN(\lambda) - I_n)S_+(\lambda)^{-*} \tag{3.13}$$
$$U_{22}(0,\lambda) = (-iN(\lambda) + I_n)S_+(\lambda)^{-*}. \tag{3.14}$$

Let us define

$$\phi(\lambda) = I_n + 2(-b^* + ic\Omega))(\lambda I_p - a^{\times *})^{-1}c^*. \tag{3.15}$$

We claim that for all λ, μ in the resolvent set of $(\lambda I_p - a^{\times *})$ it holds that

$$\begin{aligned} &\frac{\phi(\lambda) + \phi(\mu)^*}{2} \\ &\quad = S_-(\mu^*)^{-1}S_-(\lambda^*)^{-*} + \frac{\lambda - \mu^*}{2i}c(\mu^* I_p - a^\times)^{-1}\Omega(\lambda I_p - a^{\times *})^{-1}c^*. \end{aligned} \tag{3.16}$$

Indeed, we have

$$\begin{aligned} \frac{\phi(\lambda) + \phi(\mu)^*}{2} &= I_n + (-b^* + ic\Omega)(\lambda I_p - a^{\times *})^{-1}c^* + c(\mu^* I_p - a^\times)^{-1}(-b - i\Omega c^*) \\ &= I_n + i\left(c\Omega(\lambda I_p - a^{\times *})^{-1}c^* - c(\mu^* I_p - a^\times)^{-1}\Omega c^*\right) - \\ &\quad - b^*(\lambda I_p - a^{\times *})^{-1}c^* - c(\mu^* I_p - a^\times)^{-1}b \\ &= I_n + ic(\mu^* I_p - a^\times)^{-1}\left((\mu^* - a^\times)\Omega - \Omega(\lambda I_p - a^{\times *})\right)(\lambda I_p - a^{\times *})^{-1}c^* - \\ &\quad - b^*(\lambda I_p - a^{\times *})^{-1}c^* - c(\mu^* I_p - a^\times)^{-1}b. \end{aligned}$$

In view of the Lyapunov equation (1.4) we have

$$(\mu^* I_p - a^\times)\Omega - \Omega(\lambda I_p - a^{\times *}) = (\mu^* - \lambda)\Omega - ibb^*, \tag{3.17}$$

and therefore, we have

$$\begin{aligned}\frac{\phi(\lambda)+\phi(\mu)^*}{2} &= I_n + ic(\mu^* I_p - a^\times)^{-1}\left((\mu^*-\lambda)\Omega - ibb^*\right)(\lambda I_p - a^{\times *})^{-1}c^* - \\ &\quad - b^*(\lambda I_p - a^{\times *})^{-1}c^* - c(\mu^* I_p - a^\times)^{-1}b \\ &= I_n + i(\mu^*-\lambda)c(\mu^* I_p - a^\times)^{-1}\Omega(\lambda I_p - a^{\times *})^{-1}c^* + \\ &\quad + c(\mu^* I_p - a^\times)^{-1}bb^*(\lambda I_p - a^{\times *})^{-1}c^* - \\ &\quad - b^*(\lambda I_p - a^{\times *})^{-1}c^* - c(\mu^* I_p - a^\times)^{-1}b,\end{aligned}$$

from which follows (3.16). From (3.16) and equations (3.1) and (3.2) we see that

$$N(\lambda) = i\phi(\lambda^*)^* \tag{3.18}$$

is the Weyl coefficient. Furthermore, in [3] and [5] is shown that a minimal realization of S_+^{-*} is given by

$$S_+(\lambda)^{-*} = I_n - (ib^*Y - c)(\lambda I_p - a)^{-1}(I_p + \Omega Y)^{-1}(b + i\Omega c^*). \tag{3.19}$$

With these equations at hand, we can now prove (3.11)–(3.14). In all the computations below we take λ real, i.e. we have $N(\lambda) = i\phi(\lambda)^*$. Let us begin with (3.11):

$$\begin{aligned}\frac{iN(\lambda)^* + I_n}{2}S_-(\lambda)^* &= (I_n + (-b^* + ic\Omega)(\lambda I_p - a^{\times *})^{-1}c^*)(I_n + b^*(\lambda I_p - a^*)^{-1}c^*) \\ &= I_n + (-b^* + ic\Omega)(\lambda I_p - a^{\times *})^{-1}c^* + b^*(\lambda I_p - a^*)^{-1}c^* + \\ &\quad + (-b^* + ic\Omega)(\lambda I_p - a^{\times *})^{-1}c^*b^*(\lambda I_p - a^*)^{-1}c^*.\end{aligned}$$

Writing

$$c^*b^* = a^* - \lambda I_p + \lambda_p - a^{\times *} \tag{3.20}$$

we have

$$\begin{aligned}&\frac{iN(\lambda)^* + I_n}{2}S_-(\lambda)^* \\ &\quad = I_n + (-b^* + ic\Omega)(\lambda I_p - a^{\times *})^{-1}c^* + b^*(\lambda I_p - a^*)^{-1}c^* + \\ &\qquad + (-b^* + ic\Omega)(\lambda I_p - a^{\times *})^{-1}(\lambda I_p - a^{\times *} - \lambda I_p + a^*)(\lambda I_p - a^*)^{-1}c^* \\ &\quad = I_n + (-b^* + ic\Omega)(\lambda I_p - a^{\times *})^{-1}c^* + b^*(\lambda I_p - a^*)^{-1}c^* + \\ &\qquad + (-b^* + ic\Omega)(\lambda I_p - a^*)^{-1}c^* - (-b^* + ic\Omega)(\lambda I_p - a^{\times *})^{-1}c^* \\ &\quad = I_n + ic\Omega(\lambda I_p - a^*)^{-1}c^* = U_{11}(0,\lambda).\end{aligned}$$

Next we turn to (3.12), and use as above (3.20) to get from the second equality sign to the third:

$$\begin{aligned}
&\frac{(iN(\lambda)^* - I_n)}{2} S_-(\lambda)^* \\
&= (-b^* + ic\Omega)(\lambda I_p - a^{\times *})^{-1} c^* (I_n + b^* (\lambda I_p - a^*)^{-1} c^*) \\
&= (-b^* + ic\Omega)(\lambda I_p - a^{\times *})^{-1} c^* + (-b^* + ic\Omega)(\lambda I_p - a^{\times *})^{-1} c^* b^* (\lambda I_p - a^*)^{-1} c^* \\
&= (-b^* + ic\Omega)(\lambda I_p - a^{\times *})^{-1} c^* + \\
&\qquad + (-b^* + ic\Omega)(\lambda I_p - a^{\times *})^{-1} (a^* - \lambda I_p + \lambda I_p - a^{\times *})(\lambda I_p - a^*)^{-1} c^* \\
&= (-b^* + ic\Omega)(\lambda I_p - a^*)^{-1} c^* \\
&= U_{21}(0, \lambda).
\end{aligned}$$

The next two computations are a bit more involved; we recall that in [5] the following identity is proved:

$$a^{\times} + (b + i\Omega c^*)(c - ib^* Y)(I_p + \Omega Y)^{-1} = (I_p + \Omega Y) a (I_p + \Omega Y)^{-1}. \tag{3.21}$$

With (3.19) and this identity in hands, we turn to the proof of (3.13):

$$\begin{aligned}
&\frac{-iN(\lambda) - I_n}{2} S_+(\lambda)^{-*} \\
&= c(\lambda I_p - a^{\times})^{-1}(-b - i\Omega c^*)(I_n - (ib^* Y - c)(\lambda I - a)^{-1}(I_p + \Omega Y)^{-1}(b + i\Omega c^*) \\
&= -c(\lambda I_p - a^{\times})^{-1}(b + i\Omega c^*) + \\
&\qquad + c(\lambda I_p - a^{\times})^{-1}(b + i\Omega c^*)(ib^* Y - c)(\lambda I_p - a)^{-1}(I_p + \Omega Y)^{-1}(b + i\Omega c^*) \\
&= c(\lambda I_p - a^{\times})^{-1} \left(-(I_p + \Omega Y)(\lambda I_p - a) + a^{\times}(I_p + \Omega Y) - (I_p + \Omega Y) a\right) \times \\
&\qquad \times (\lambda I_p - a)^{-1}(I_p + \Omega Y)^{-1}(b + i\Omega c^*) \\
&= c(\lambda I_p - a^{\times})^{-1}(a^{\times} - \lambda I_p)(I + \Omega Y)(\lambda I_p - a)^{-1}(I_p + \Omega Y)^{-1}(b + i\Omega c^*) \\
&= U_{12}(0, \lambda).
\end{aligned}$$

Finally we check (3.14):

$$\begin{aligned}
\frac{-iN(\lambda) + I_n}{2} S_+(\lambda)^{-*} &= (I_n + c(\lambda I_p - a^{\times})^{-1}(-b - i\Omega c^*)) \times \\
&\qquad \times (I_n - (ib^* Y - c)(\lambda I_p - a)^{-1}(I_p + \Omega Y)^{-1}(b + i\Omega c^*)) \\
&= I_n - (ib^* Y - c)(\lambda I_p - a)^{-1}(I_p + \Omega Y)^{-1}(b + i\Omega c^*) + \\
&\qquad + c(\lambda I_p - a^{\times})^{-1}(-b - i\Omega c^*) \times \\
&\qquad \times (I_n - (ib^* Y - c)(\lambda I - a)^{-1}(I_p + \Omega Y)^{-1}(b + i\Omega c^*))
\end{aligned}$$

$$
\begin{aligned}
&= I_n - (ib^*Y - c)(\lambda I - a)^{-1}(I_p + \Omega Y)^{-1}(b + i\Omega c^*) - \\
&\quad - c(I + \Omega Y)(\lambda I - a)^{-1}(I_p + \Omega Y)^{-1}(b + i\Omega c^*) \\
&= I_n - (ib^*Y + c\Omega)(\lambda I - a)^{-1}(I_p + \Omega Y)^{-1}(b + i\Omega c^*) \\
&= U_{22}(0, \lambda).
\end{aligned}
$$

□

As a corollary of (3.16) we have:

COROLLARY 3.2. *Let ϕ be defined by* (3.15). *The spectrum of the matrix $a + bc + 2i\Omega c^*c$ lies in $\mathbb{C}_+$.*

Proof. We first check that the realization (3.15) is minimal. Indeed, deg $\phi \leq p$. Setting in (3.16) $\mu = \lambda^*$, we obtain

$$
\frac{\phi(\lambda) + \phi(\lambda^*)^*}{2} = S_-(\lambda)^{-1}S_-(\lambda^*)^{-*}, \tag{3.22}
$$

and hence, with $\phi^\sharp(\lambda) = \phi(\lambda^*)^*$, we have deg $(\phi + \phi^\sharp) = 2p$. Since deg $\phi =$ deg $\phi^\sharp \leq p$, we conclude that

$$
2p = \text{deg } (\phi + \phi^\sharp) \leq \text{deg } \phi + \text{deg } \phi^\sharp = 2\text{deg } \phi \leq 2p.
$$

Hence deg $\phi = p$ and the realization (3.15) is minimal. Thus, the realization

$$
\phi^{-1}(\lambda) = I_n - 2(-b^* + ic\Omega)(\lambda I_p - (a^* + c^*b^* - 2ic^*c\Omega))^{-1}c^* \tag{3.23}
$$

is a minimal realization of ϕ^{-1}, and the spectrum of $(a^* + c^*b^* - 2ic^*c\Omega)$ consists of the singularities of ϕ. It follows from (3.16) that Re $\phi(\lambda) > 0$ for λ in the closed upper half-plane. In particular ϕ^{-1} is analytic there and hence the claim on the spectrum of $(a^* + c^*b^* - 2ic^*c\Omega)$ follows.

□

4. Minimal realizations of the characteristic spectral functions

4.1. Minimal realizations of the Weyl coefficient function. Formula (3.15) gives a characterization of the minimal realizations of the Weyl coefficient function.

THEOREM 4.1. *Let N be a $\mathbb{C}^{n\times n}$-valued rational function analytic at infinity, and let $N(\lambda)=\mathcal{D}+\mathcal{C}(\lambda I-\mathcal{A})^{-1}\mathcal{B}$ be a minimal realization of N. Then the following are equivalent:*

(1) *N is the Weyl coefficient function of a canonical system (1.1).*

(2) *$\mathcal{D}=iI_n$, $\sigma(\mathcal{A})$ is in the open upper half plane and there exists a solution $\Omega>0$ to the Riccati equation*

$$i(\Omega\mathcal{A}^*-\mathcal{A}\Omega)=\left(\frac{\mathcal{B}}{2}-\Omega\mathcal{C}^*\right)\left(\frac{\mathcal{B}}{2}-\Omega\mathcal{C}^*\right)^* \tag{4.1}$$

such that

$$\sigma\left(\mathcal{A}+i\left(\frac{\mathcal{B}}{2}-\Omega\mathcal{C}^*\right)\mathcal{C}\right)\subset\mathbb{C}_+. \tag{4.2}$$

(3) *N is analytic in the closed upper half-plane, $N(\infty)=iI_n$ and* Im $N(\lambda)>0$ *on the real line.*

Proof. Assume first that N is the Weyl coefficient function of a canonical system (1.1) with strictly pseudo-exponential potential $k(t)$ and let $S_-(\lambda)=I_n+c(\lambda I_p-a)^{-1}b$ be a minimal realization of the left spectral factor of the scattering matrix. A realization of $\phi(\lambda)=-iN(\lambda)$ is then given by (3.15). We have seen in the proof of Corollary 3.2 that this realization is minimal. Thus, a minimal realization of $N(\lambda)=i\phi(\lambda)^*$ is given by

$$N(\lambda)=iI_n+\mathcal{C}(\lambda I_p-\mathcal{A})^{-1}\mathcal{B},$$

where we have set

$$\mathcal{A}=a^\times, \tag{4.3}$$

$$\mathcal{B}=2(-ib+\Omega c^*), \tag{4.4}$$

$$\mathcal{C}=c. \tag{4.5}$$

We thus have

$$c=\mathcal{C}, \tag{4.6}$$

$$b=i\left(\frac{\mathcal{B}}{2}-\Omega\mathcal{C}^*\right) \tag{4.7}$$

and

$$a=\mathcal{A}+bc=\mathcal{A}-i\left(\frac{\mathcal{B}}{2}-\Omega\mathcal{C}^*\right)\mathcal{C}. \tag{4.8}$$

With these expressions, the Lyapunov equation (1.4) becomes the Riccati equation (4.1) and the condition on the spectrum of $a - bc$ translates into the spectrum condition (4.2). This completes the proof that (1) implies (2).

We now prove that (2) implies (3): starting with a realization of N with the properties stated in (2), build matrices a, b, c via (4.8), (4.7) and (4.6), and set $S_-(\lambda) = I_n + c(\lambda I_p - a)^{-1}b$. By construction, S_- is analytic and invertible in the closed lower half-plane and (3.16) holds, i.e. with $N = i\phi$,

$$\frac{-iN(\lambda) + iN(\mu)^*}{2} = S_-(\mu^*)^{-1}S_-(\lambda^*)^{-*} + \frac{\lambda - \mu^*}{2i} c(\mu^* I_p - a^\times)^{-1}\Omega(\lambda I_p - a^{\times *})^{-1}c^*.$$

The function N has therefore a strictly positive imaginary part in the closed lower half-plane, and in particular on the real line.

Finally, we prove that (3) implies (1): by the hypothesis on N, there is a function S_- analytic and invertible in the closed lower half-plane such that $\text{Im } N(\lambda) = S_-(\lambda)^{-1}S_-(\lambda)^{-*}$ and $S(\infty) = I_n$. Let $S_-(\lambda) = I_n + c(\lambda I_p - a)^{-1}b$ be a minimal realization of S. N is the Weyl coefficient associated to the potential (1.2). □

The Riccati equation (4.1) has in general more than one strictly invertible solution; we do not know if the spectral condition (4.2) fixes Ω uniquely. The solution of the inverse problem does not depend on the choice of Ω, as we will see in the next section.

Let us take as example the potential (1.6). Then,

$$\mathcal{A} = i\alpha - c, \quad \mathcal{B} = 2\Big(-i + \frac{c}{2\alpha}\Big), \quad \mathcal{C} = c,$$

and

$$R(\lambda) = i + \frac{2(-i + \frac{c}{2\alpha})c}{\lambda - (i\alpha - c)}.$$

The Riccati equation (4.1) becomes:

$$\Omega = \frac{1}{2\alpha} + \frac{c^2}{2\alpha}\Big(\frac{1}{2\alpha} - \Omega\Big)^2, \tag{4.9}$$

which can be rewritten as

$$0 = \Big(\frac{1}{2\alpha} - \Omega\Big)\Big(1 + \frac{c^2}{2\alpha}\Big(\frac{1}{2\alpha} - \Omega\Big)\Big),$$

and has two solutions: $\Omega_1 = \frac{1}{2\alpha}$ and $\Omega_2 = \frac{1}{2\alpha} + \frac{2\alpha}{c^2}$. Only the first one has the spectral property (4.2).

As another example, take

$$N(\lambda) = i\frac{\lambda - 3i + 1}{\lambda - 3i} = i + \frac{i}{\lambda - 3i}.$$

Then, $\operatorname{Im} N(\lambda) = 1 + \frac{\operatorname{Re}\lambda}{|\lambda - 3i|^2}$ and

$$\mathcal{A} = 3i, \quad \mathcal{B} = 1, \quad \mathcal{C} = i, \quad \mathcal{D} = i.$$

The Riccati equation (4.1) becomes

$$\Omega^2 - 6\Omega + \frac{1}{4} = 0.$$

It has two strictly positive solutions $\Omega_\pm = \frac{6 \pm \sqrt{35}}{2}$. Only one of them (namely, Ω_-) has the property (4.2).

4.2. Minimal realizations of the reflection coefficient function. The equivalence between (2) and (3) in the next theorem is a consequence of [14, Theorem 3.2 p. 231].

THEOREM 4.2. *Let R be a $\mathbb{C}^{n\times n}$-valued rational function, analytic at infinity with $R(\infty) = 0$ and let $R(\lambda) = \mathbf{C}(\lambda I - \mathbf{A})^{-1}\mathbf{B}$ be a minimal realization of R. The following are equivalent:*

(1) *R is the reflection coefficient function of a canonical system of the form* (1.1) *with rational spectral data.*

(2) *Both $\sigma(\mathbf{A})$ and $\sigma(\mathbf{A} + \mathbf{BC}) \subset \mathbb{C}_+$ and there is a strictly positive solution Ω to the equation*

$$i(\Omega\mathbf{A}^* - \mathbf{A}\Omega) = \mathbf{BB}^* + \Omega\mathbf{C}^*\mathbf{C}\Omega \tag{4.10}$$

such that

$$\sigma(\mathbf{A} - i\Omega\mathbf{C}\mathbf{C}^*) \subset \mathbb{C}_+. \tag{4.11}$$

(3) *R is analytic in the closed lower half-plane and takes strictly contractive values on the real line:* $\sup_{\lambda\in\mathbb{R}} \|R(\lambda)\| < 1$.

Proof. Assume first that R is the reflection coefficient function of a canonical system (1.1), and let $S_-(\lambda) = I_n + c(\lambda I - a)^{-1}b$ be a minimal realization of the left spectral factor S_- of the scattering function associated to (1.1). Then the Lyapunov equation (1.4) has a unique solution, which moreover is strictly positive (since $\sigma(a - bc) \subset \mathbb{C}_+$). Define ϕ by (3.15). Then as seen in the proof of Corollary 3.2, the realization is minimal.

For real λ we have

$$
\begin{aligned}
R(\lambda) &= (\phi(\lambda)^* - I_n)(\phi(\lambda)^* + I_n)^{-1} \\
&= I_n - 2(I_n + \phi(\lambda)^*)^{-1} \\
&= I_n - (I_n - c(\lambda I_p - a^\times)^{-1}(b + i\Omega c^*))^{-1} \qquad (4.12)\\
&= I_n - (I_n + c(\lambda I_p - \mathbf{A})^{-1}(b + i\Omega c^*) \\
&= -c(\lambda I_p - \mathbf{A})^{-1}(b + i\Omega c^*),
\end{aligned}
$$

where we have set

$$
\mathbf{A} = a^\times + (b + i\Omega c^*)c = a + i\Omega c^* c. \qquad (4.13)
$$

Set

$$
\mathbf{B} = b + i\Omega c^*, \qquad \mathbf{C} = -c. \qquad (4.14)
$$

The realization $R(\lambda) = \mathbf{C}(\lambda I_p - \mathbf{A})^{-1}\mathbf{B}$ is minimal since by (4.12) deg R = deg $\phi = p$. Furthermore, $\mathbf{A} + \mathbf{BC} = a - bc$ has its spectrum in the open upper half-plane since $S_-(\lambda) = I_n - c(\lambda I_p - a)^{-1}b$ is a minimal realization of a spectral factor. Similarly, the spectrum of $\mathbf{A}$ is inside $\mathbb{C}_+$ since R is analytic in $\mathbb{C}_-$ and since the realization of R is minimal. Finally the Lyapunov equation (1.4) can be rewritten as

$$
i(\Omega(\mathbf{A} + \mathbf{BC})^* - (\mathbf{A} + \mathbf{BC})\Omega) = (\mathbf{B} + i\Omega\mathbf{C}^*)(\mathbf{B} + i\Omega\mathbf{C}^*)^*, \qquad (4.15)
$$

which leads to (4.10). Property (4.11) follows from (4.13).

We now prove the implication (2) $\Rightarrow$ (1): Given a minimal realization of the function R with the given properties, we set

$$
a = \mathbf{A} - i\Omega\mathbf{C}^*\mathbf{C} \qquad (4.16)
$$

$$
b = \mathbf{B} + i\Omega\mathbf{C}^* \qquad (4.17)
$$

$$
c = -\mathbf{C}, \qquad (4.18)
$$

and ϕ by (3.15). Then the function $S_-(\lambda) = I_n + c(\lambda I_p - a)^{-1}b$ is analytic and invertible in the closed lower half-plane and (3.22) holds. Thus R is the reflection coefficient function associated to the weight function Re ϕ.

We now prove the equivalence between (1) and (3). Assume that (1) holds and let $W(\lambda) = \text{Re } \phi(\lambda)$ be the associated weight function. Formula (3.2) implies that R is analytic in $\mathbb{C}_-$. Since there is $\epsilon > 0$ independent of λ such that $W(\lambda) \geq \epsilon$ we deduce that $\sup_{\mathbb{R}\cup\mathbb{C}_-} \|R(\lambda)\| < 1$ and that R is also analytic on the real line. Conversely, if R is as in (3), set $\phi(\lambda) = (R(\lambda)^* - I_n)(R(\lambda)^* + I_n)^{-1}$ and $W(\lambda) = \text{Re } \phi(\lambda)$. R is the reflection coefficient associated to W.

□

As in the previous section, the Riccati equation (4.10) has in general more than one strictly invertible solution; we do not know if the spectral condition (4.11) fixes Ω uniquely. The solution of the inverse problem does not depend on the choice of Ω, as we will see.

We note that the characterization of minimal realizations of matrix-valued rational functions contractive on the real line and strictly contractive at infinity, and the Riccati equation (4.10) appears first in [14]; there, equation (4.10) is called the *state characteristic equation.*

COROLLARY 4.3. *If R is a reflection coefficient function so is $-R$. If $R(\lambda) = \mathbf{C}(\lambda I_p - \mathbf{A})^{-1}\mathbf{B}$ is a minimal realization of R with associated matrix Ω, $-R(\lambda) = -\mathbf{C}(\lambda I_p - \mathbf{A})^{-1}\mathbf{B}$ is a minimal realization with same associated matrix.*

Proof. The first claim directly follows from (3). To check the second claim, we have to verify that the conditions in (2) are still in force with $-\mathbf{B}$ instead of $\mathbf{B}$ and same Ω. The claim on the spectrum of $\mathbf{A} - \mathbf{BC}$ follows from Corollary 3.2.

□

In Section we will see that if $R(\lambda)$ is the reflection coefficient associated to the potential $k(t)$, then $-R(\lambda)$ corresponds to the potential $-k(t)$.

We conclude this section with some examples: first, consider the potential (1.6). Then,

$$\mathbf{A} = i\left(\alpha + \frac{c^2}{2\alpha}\right), \quad \mathbf{B} = 1 + \frac{ic}{2\alpha}, \quad \mathbf{C} = -c.$$

The Riccati equation (4.10) becomes (4.9), and the discussion is as the discussion following (4.9).

Next, take $\mathbf{B} = \mathbf{C} = 1$ and $\mathbf{A} = ia$ with $a > 1$. Then clearly $R(\lambda) = \frac{1}{\lambda - ia}$ takes strictly contractive values in the closed lower half-plane. The Riccati equation

(4.10) becomes

$$\Omega^2 - 2a\Omega + 1 = 0, \tag{4.19}$$

and has two solutions, both strictly positive numbers:

$$\Omega_1 = a + \sqrt{a^2 - 1}, \qquad \Omega_2 = a - \sqrt{a^2 - 1}.$$

Only Ω_2 has the property that $\sigma(\mathbf{A} - i\Omega\mathbf{C}^*\mathbf{C}) \subset \mathbb{C}_+$.

When $a = 1$, the function R is not a strict contraction on the real line ($R(0) = -i$ has modulus 1) and the Riccati equation has no solution for which $\sigma(\mathbf{A} - i\Omega\mathbf{C}^*\mathbf{C}) \subset \mathbb{C}_+$.

4.3. Minimal realization of the asympotically equivalence matrix function. Minimal realizations of rational functions analytic at infinity and taking J-unitary values on the real line were studied in [1], where the following theorem is proved (for functions J-unitary on the imaginary axis; see [1, Theorem 2.8 p. 192]):

THEOREM 4.4. *Let R be a $\mathbb{C}^{2n\times 2n}$-valued rational function analytic at infinity and let $R(\lambda) = D + C(\lambda I_P - A)^{-1}B$ be a minimal realization of R. Then, R takes J-unitary values on the real line if and only if there exists an hermitian invertible matrix $H \in \mathbb{C}^{P\times P}$ solution of the system of equations*

$$i(A^*H - HA) = C^*JC \tag{4.20}$$

$$C = iJB^*H. \tag{4.21}$$

The matrix H is uniquely defined by the minimal realization of R and is called the *associated hermitian matrix* to the minimal realization (see [1]).

THEOREM 4.5. *Let $k(t)$ be given in the form* (1.2) *and let $U(t, \lambda)$ be the $\mathbb{C}^{2n\times 2n}$-valued solution of* (1.1) *subject to* (2.1). *Then a minimal realization of the asymptotic equivalence matrix $V(\lambda) = U(0, \lambda)$ is given by $U(0, \lambda) = I_{2n} + C(\lambda I_{2p} - A)^{-1}B$, where*

$$A = \begin{pmatrix} a^* & 0 \\ 0 & a \end{pmatrix}, \quad B = \begin{pmatrix} c^* & 0 \\ 0 & (I_p + \Omega Y)^{-1}(b + i\Omega c^*) \end{pmatrix} \tag{4.22}$$

and

$$C = \begin{pmatrix} ic\Omega & -c(I_p + \Omega Y) \\ -b^* + ic\Omega & -ib^*Y - c\Omega Y \end{pmatrix}, \tag{4.23}$$

and the associated hermitian matrix is given by

$$H = \begin{pmatrix} \Omega & i(I_p + \Omega Y) \\ -i(I_p + Y\Omega) & (I_p + Y\Omega)Y \end{pmatrix}. \tag{4.24}$$

Proof. We obtain from (3.7)–(3.8) that

$$\begin{pmatrix} U_{11}(0,\lambda) \\ U_{21}(0,\lambda) \end{pmatrix} = \begin{pmatrix} I_n \\ 0 \end{pmatrix} + \begin{pmatrix} ic\Omega \\ -b^* + ic\Omega \end{pmatrix} (\lambda I_p - a^*)^{-1} c^*, \tag{4.25}$$

while (3.9)–(3.10) lead to

$$\begin{pmatrix} U_{12}(0,\lambda) \\ U_{22}(0,\lambda) \end{pmatrix} = \begin{pmatrix} 0 \\ I_n \end{pmatrix} + \begin{pmatrix} -c(I_p + \Omega Y) \\ -ib^*Y - c\Omega Y \end{pmatrix} (\lambda I_p - a)^{-1}(I_p + \Omega Y)^{-1}(b + i\Omega c^*).$$

Combining this equation and (4.25), we obtain the realization (4.22)–(4.23). We now show that the realization is minimal. First we note that deg $U \le 2n$. From [1, Theorem 2.4 p. 185], we know that

$$\deg U = \dim \ \ell.s. \left\{ \frac{J - U(\lambda)JU(\mu)^*}{\lambda - \mu^*}\xi \right\} \tag{4.26}$$

where $\ell.s.$ stands for linear span and where ξ runs in $\mathbb{C}^n$ and μ varies in $\mathbb{C}$. From the equation

$$S(\lambda) = (U_{11}(0,\lambda) - U_{21}(0,\lambda))^{-1}(U_{22}(0,\lambda) - U_{12}(0,\lambda))$$

we have

$$\begin{aligned} \frac{I_n - S(\lambda)S(\mu)^*}{\lambda - \mu^*} &= (U_{11}(0,\lambda) - U_{21}(0,\lambda))^{-1}(I_n \quad -I_n) \times \\ &\quad \times \frac{J - U(\lambda)JU(\mu)^*}{\lambda - \mu^*} \begin{pmatrix} I_n \\ -I_n \end{pmatrix} (U_{11}(0,\mu) - U_{21}(0,\mu))^{-*} \end{aligned}$$

and thus, still by [1, Theorem 2.4 p. 185],

$$\deg S \le \deg U. \tag{4.27}$$

But deg $S = 2n$, and hence $2n = \deg S \le \deg U \le 2n$. Hence deg $U \le 2n$ and the realization is minimal.

We now check that H is the associated hermitian matrix. We first prove (4.21):

$$\begin{aligned} iJB^*H &= \begin{pmatrix} ic & 0 \\ 0 & -i(b^* - ic\Omega)(I_p + Y\Omega)^{-1} \end{pmatrix} \begin{pmatrix} \Omega & i(I_p + \Omega Y) \\ -i(I_p + Y\Omega) & (I_p + Y\Omega)Y \end{pmatrix} \\ &= \begin{pmatrix} ic\Omega & -c(I_p + \Omega Y) \\ -b^* + ic\Omega & -ib^*Y - c\Omega Y \end{pmatrix} \\ &= C. \end{aligned}$$

We now turn to the Lyapunov equation (4.20):

$$\begin{aligned} i(A^*H-HA) &= i\left(\begin{pmatrix} a & 0 \\ 0 & a^* \end{pmatrix}\begin{pmatrix} \Omega & i(I_p+\Omega Y) \\ -i(I_p+Y\Omega) & (I_p+Y\Omega)Y \end{pmatrix} - \right. \\ &\qquad \left. -\begin{pmatrix} \Omega & i(I_p+\Omega Y) \\ -i(I_p+Y\Omega) & (I_p+Y\Omega)Y \end{pmatrix}\begin{pmatrix} a^* & 0 \\ 0 & a \end{pmatrix}\right) \\ &= i\begin{pmatrix} a\Omega-\Omega a^* & \Omega Ya - a\Omega Y \\ a^*Y\Omega - Y\Omega a^* & i(a^*(I_p+Y\Omega)Y-(I_p+Y\Omega)Ya) \end{pmatrix}. \end{aligned}$$

To show that this expression is equal to C^*JC we have to check three equalities corresponding to the block entries of C^*JC, namely

$$i(a\Omega-\Omega a^*) = \Omega c^*c\Omega - (-b^*+ic\Omega)^*(-b^*+ic\Omega) \tag{4.28}$$

$$\Omega Ya - a\Omega Y = i\Omega c^*c(I_p+\Omega Y) + (-b^*+ic\Omega)^*(ib^*Y+c\Omega Y) \tag{4.29}$$

and

$$\begin{aligned} &i(a^*(I_p+Y\Omega)Y-(I_p+Y\Omega)Ya) \\ &\qquad = (I_p+Y\Omega)c^*c(I_p+\Omega Y) - Y(ib-\Omega c)^*(ib-\Omega c)Y. \end{aligned} \tag{4.30}$$

We begin with (4.28): it is proved in [3, p. 13] that

$$(-b^*+ic\Omega)^*(-b^*+ic\Omega) = -i((I_p+\Omega Y)a\Omega - \Omega a^*(I_p+Y\Omega)). \tag{4.31}$$

Plugging this expression in (4.28), we see that to prove (4.28) amounts to prove that

$$\begin{aligned} i(a\Omega-\Omega a^*) &= (\Omega c^*c\Omega + i((I_p+\Omega Y)a\Omega - \Omega a^*(I_p+Y\Omega))) \\ &= \Omega c^*c\Omega + ia\Omega + i\Omega Ya\Omega - i\Omega a^* - i\Omega a^*Y\Omega, \end{aligned}$$

i.e. $0 = \Omega(c^*c + i(Ya - a^*Y))\Omega$, which is just the Lyapunov equation (1.5). We now turn to the proof of (4.29): taking into account (4.31), (4.29) is equivalent to

$$\Omega Ya - a\Omega Y = i\Omega c^*c(I_p+\Omega Y) - i((I_p+\Omega Y)a\Omega - \Omega a^*(I_p+Y\Omega))(-iY).$$

Using (1.5) this last expression can be rewritten as

$$\Omega Ya - a\Omega Y = \Omega(Ya - a^*Y)(I_p+\Omega Y) - ((I_p+\Omega Y)a\Omega - \Omega a^*(I_p+Y\Omega))Y.$$

This in turn is readily verified. Finally we prove (4.30): taking into account the Lyapunov equation (1.5) and equality (4.31), we have to prove that

$$\begin{aligned} &i(a^*Y(I_p+\Omega Y) - Y(I_p+\Omega Y)a) \\ &\quad = (I_p+Y\Omega)(-i)(Ya-a^*Y)(I_p+\Omega Y) + iY((I_p+\Omega Y)a\Omega - \Omega a^*(I_p+Y\Omega))Y, 0 \end{aligned}$$

i.e., after dividing by i both sides,

$$
\begin{aligned}
&(a^*Y(I_p+\Omega Y)-Y(I_p+\Omega Y)a)\\
&\quad=-(I_p+Y\Omega)(Ya-a^*Y)(I_p+\Omega Y)+Y((I_p+\Omega Y)a\Omega-\Omega a^*(I_p+Y\Omega))Y,
\end{aligned}
$$

which is easily checked. □

Using [9, formula (0.3) p. 3] we obtain the following formula for H, which can also be checked easily directly:

$$
H=\begin{pmatrix} I_p & 0\\ -i(\Omega^{-1}+Y) & I_p\end{pmatrix}\begin{pmatrix}\Omega & 0\\ 0 & -(\Omega^{-1}+Y)\end{pmatrix}\begin{pmatrix} I_p & 0\\ -i(\Omega^{-1}+Y) & I_p\end{pmatrix}^*. \tag{4.32}
$$

For the potential (1.6) we have:

$$
A=\begin{pmatrix}-i\alpha & 0\\ 0 & i\alpha\end{pmatrix},\quad B=\begin{pmatrix} c & 0\\ 0 & \frac{1+\frac{ic}{2\alpha}}{1+\frac{c^2}{4\alpha^2}}\end{pmatrix},
$$

and

$$
C=\begin{pmatrix}\frac{ic}{2\alpha} & -c(1+\frac{c^2}{4\alpha^2})\\ -1+i\frac{c}{2\alpha} & -\frac{ic^2}{2\alpha}-\frac{c^3}{4\alpha^2}\end{pmatrix}.
$$

5. Inverse problems

The inverse problem for the scattering function has been solved in [3] and in [5]. If one knows a minimal realization $S_-(\lambda)=I_n+c(\lambda I_p-a)^{-1}b$ of the spectral factor S_- of the scattering function, the potential $k(t)$ is then given by formula (1.2). The principle in this section is very simple: starting from a minimal realization of either the Weyl function or the reflection coefficient function or the asymptotic equivalence function, we use the results of the previous sections and compute in each case a minimal realization of S_- and use formula (1.2).

5.1. Inverse problem associated to the Weyl coefficient function. We begin with the Weyl coefficient:

THEOREM 5.1. *Let N be a $\mathbb{C}^{n\times n}$-valued rational function analytic on the closed upper half-plane with* Im $N(\lambda) > 0$ *there, and let $N(\lambda) = iI_n + \mathcal{C}(\lambda I - \mathcal{A})^{-1}\mathcal{B}$ be a minimal realization of N with associated matrix Ω. Then, N is the Weyl reflection coefficient for the system* (1.1) *with potential*

$$\begin{aligned} k(t) = &- 2\mathcal{B}^* e^{it(\mathcal{A}^* - i(\Omega\mathcal{B} + \frac{\mathcal{C}^*}{2})\mathcal{B}^*)} \times \\ &\times \left(I_p + \Omega\left(Y - e^{-2it(\mathcal{A}^* - i(\Omega\mathcal{B}+\frac{\mathcal{C}^*}{2})\mathcal{B}^*)^*} Y e^{2it(\mathcal{A}^* - i(\Omega\mathcal{B}+\frac{\mathcal{C}^*}{2})\mathcal{B}^*)}\right)\right)^{-1} \mathcal{C}, \end{aligned} \tag{5.1}$$

where Y is the solution of the Lyapunov equation

$$-i(Y(\mathcal{A}^* - i(\Omega\mathcal{B} + \mathcal{C}^*/2)\mathcal{B}^*) - (\mathcal{A}^* - i(\Omega\mathcal{B} + \mathcal{C}^*/2)\mathcal{B}^*)^* Y) = \mathcal{B}\mathcal{B}^*. \tag{5.2}$$

Proof. Given the realization $N(\lambda) = iI_n + \mathcal{C}(\lambda I - \mathcal{A})^{-1}\mathcal{B}$ with an associated matrix Ω, define matrices a, b, c by (4.8)–(4.6). The matrix function $S_-(\lambda) = I_n + c(\lambda I - a)^{-1}b$ is analytic and invertible in the closed lower half-plane and satisfies

$$\text{Im } N(\lambda) = S_-(\lambda)^{-1} S_-(\lambda)^{-*}.$$

The weight function $W(\lambda) = S_-(\lambda)^{-1} S_-(\lambda)^{-*}$ is associated uniquely to the potential (5.1), and this shows that the solution is independent of the choice of Ω with the given spectral constraint. Indeed, for two solutions Ω_1 and Ω_2 with the required properties, denote the corresponding spectral factors $S_{1,-} = S_{2,-}$. Then,

$$\text{Im } N(\lambda) = S_{1,-}(\lambda)^{-1} S_{1,-}(\lambda)^{-*} = S_{2,-}(\lambda)^{-1} S_{2,-}(\lambda)^{-*},$$

and by uniqueness of the spectral factor with value at infinity I_n we have that $S_{1,-} = S_{2,-}$.

□

5.2. Inverse problem associated to the reflection coefficient function. The problem we address in this section is the following: given a rational function R satisfying the hypothesis of Theorem 4.2, find the potential $k(t)$ such that R is the reflection coefficient function of the canonical differential system (1.1). It was treated in a different way in [6].

THEOREM 5.2. *Let R be a $\mathbb{C}^{n\times n}$-valued rational function, analytic in the closed lower half-plane and taking strictly contractive values on the real line, with $R(\infty) = 0$, and let $R(\lambda) = \mathbf{C}(\lambda I - \mathbf{A})^{-1}\mathbf{B}$ be a minimal realization of R, with associated positive matrix Ω. Then, R is the reflection coefficient function for the system* (1.1) *with potential*

$$k(t) = -2\mathbf{C}e^{it(\mathbf{A}-i\Omega\mathbf{C}\mathbf{C}^*)}\times \times\left(I_p+\Omega\left(Y-e^{-2it(\mathbf{A}-i\Omega\mathbf{C}\mathbf{C}^*)^*}Ye^{2it(\mathbf{A}-i\Omega\mathbf{C}\mathbf{C}^*)}\right)\right)^{-1}\mathbf{B}, \tag{5.3}$$

where Y is the solution of the Lyapunov equation

$$i(Y(\mathbf{A}-i\Omega\mathbf{C}^*\mathbf{C})-(\mathbf{A}^*+i\mathbf{C}^*\mathbf{C}\Omega)Y)=\mathbf{C}^*\mathbf{C}. \tag{5.4}$$

Proof. Starting from a minimal realization of R we define a, b and c by (4.16)–(4.18), and $S_-(\lambda) = I_n + c(\lambda I_p - a)^{-1}b$. The proof then goes as in the proof of Theorem 5.1. □

COROLLARY 5.3. *If k is the potential of the form* (1.2) *associated to the reflection coefficient R, then $-k$ corresponds to $-R$.*

Proof. Replacing $\mathbf{B}$ by $-\mathbf{B}$ does not change the Riccati equation (4.10). □

It follows from (3.5) that the reflection coefficient $-R$ corresponds to the Weyl coefficient $-\frac{1}{iN}$ if N is the Weyl coefficient associated to R. Hence, we have:

COROLLARY 5.4. *If k is the potential of the form* (1.2) *associated to the Weyl coefficient N, then $-k$ corresponds to the Weyl coefficient $-\frac{1}{iN}$.*

5.3. Inverse problem for the scattering function. The following result appears in [3] and is proved there using Marchenko's approach to the inverse spectral problem. For a direct proof, see [5].

THEOREM 5.5. *Let S be a $\mathbb{C}^{n\times n}$-valued rational function analytic at infinity with $S(\infty) = I_n$. Assume that S takes unitary values on the real line and admits a Wiener–Hopf factorization. Then S is the scattering function of a differential operator of the form* (1.1)*. Let $S(\lambda) = I_n + C(\lambda I_m - A)^{-1}B$ be a minimal realization of S. Then the potential is equal to*

$$(5.5)\qquad k(t) = -2CP\left(\left(Pe^{-2itA}P - PX_2P^*e^{-2itA^*}P^*X_1P\right)|_{\mathrm{Im}\ P}\right)^{-1}PB.$$

In this expression, P denotes the Riesz projection corresponding to the eigenvalues of A in the open upper half plane $\mathbb{C}_+$ and X_1 and X_2 are such that

$$(5.6)\qquad i((P^*X_1P)(AP) - (AP)^*(P^*X_1P)) = -P^*C^*CP$$

and

$$(5.7)\qquad i((AP)(PX_2P^*) - (PX_2P^*)(AP)^* = -PBB^*P^*.$$

5.4. Inverse problem for the spectral function. The following result appears in [2] and is proved there using Kreĭn's approach to the inverse spectral problem. For a direct proof, see [12].

THEOREM 5.6. *Let W be a $\mathbb{C}^{n\times n}$-valued rational function analytic at infinity and on the real line and assume that $W(\infty) = I_n$ and $W(\lambda) > 0$ for all $\lambda \in \mathbb{R}$. Then W is the spectral function of a differential operator of the form* (1.1)*. Let $W(\lambda) = I_n + C(\lambda I_m - A)^{-1}B$ be a minimal realization of W. The function k is given by the formula*

$$(5.8)\qquad k(t) = 2C(Pe^{-2itA^\times}|_{\mathrm{Im}\ P})^{-1}PB.$$

In this expression, $A^\times = A - BC$ and P is the Riesz projection corresponding to the eigenvalues of A in the open upper half-plane.

References

[1] D. ALPAY and I. GOHBERG, Unitary rational matrix functions, *Operator Theory: Advances and Applications* **33**, Birkhäuser Verlag, Basel, 1988, 175–222.

[2] D. ALPAY and I. GOHBERG, Inverse spectral problem for differential operators with rational scattering matrix functions, *J. Differential Equations*, **118** (1995), 1–19.

[3] D. ALPAY and I. GOHBERG, Inverse scattering problem for differential operators with rational scattering matrix functions, *Operator Theory: Advances and Applications* **90**, Birkhäuser Verlag, Basel, 1996, 1–18.

[4] D. ALPAY and I. GOHBERG, Potentials associated to rational weights, *Operator Theory: Advances and Applications* **98**, Birkhäuser Verlag, Basel, 1997, 23–40.

[5] D. ALPAY, I. GOHBERG, R. KAASHOEK and A. SAKHNOVICH, Direct and inverse scattering problem for canonical systems with a strictly pseudo-exponential potential, *Preprint* (1998).

[6] D. ALPAY, I. GOHBERG and L. SAKHNOVICH, Inverse scattering for continuous transmission lines with rational reflection coefficient function, *Operator Theory: Advances and Applications* **87**, Birkhäuser Verlag, Basel, 1996, 1–16.

[7] A. BRUCKSTEIN, B. LEVY and T. KAILATH, Differential methods in inverse scattering, *SIAM J. Appl. Math.*, **45** (1985), 312–335.

[8] YU. L. DALECKII and M. G. KREĬN, *Stability solutions of differential equations in Banach spaces,* Translations of mathematical monographs, American Mathematical Society, Providence, Rhode Island, 1974.

[9] H. DYM, *J contractive matrix functions, reproducing kernel spaces and interpolation,* CBMS Lecture Notes **71**, Amer. Math. Soc., Providence R.I., 1989.

[10] H. DYM and A. IACOB, Positive definite extensions, canonical equations and inverse problems, *Operator Theory: Advances and Applications* **12**, Birkhäuser Verlag, Basel, 1984, 141–240.

[11] I. M. GELFAND and B. M. LEVITAN, On the determination of a differential equation from its spectral function, *Izv. Akad. Nauk SSSR ser. math.*, **15** (1951), 309–360.

[12] I. GOHBERG, M. A. KAASHOEK and A. L. SAKHNOVICH, Canonical systems with rational spectral densities: explicit formulas and applications, *Math. Nachrichten*, to appear.

[13] I. GOHBERG, M. A. KAASHOEK and A. L. SAKHNOVICH, Pseudo-canonical systems with rational Weyl functions: explicit formulas and applications, *J. Differential Equations*, **146** (1998), 375–398.

[14] I. GOHBERG and S. RUBINSTEIN, Proper contractions and their unitary minimal completions, *Topics in interpolation theory of rational matrix-valued functions,* Operator Theory: Advances and Applications **33**, Birkhäuser Verlag, Basel, 1988, 223–247.

[15] M. G. KREĬN, *Topics in differential and integral equations and operator theory,* Operator Theory: Advances and Applications **7**, Birkhäuser Verlag, 1983.

[16] V. A. MARCHENKO, *Sturm–Liouville operators and applications,* Operator Theory: Advances and Applications **22**, Birkhäuser Verlag, Basel, 1986.

[17] LEV A. SAKHNOVICH, Spectral problems for canonical systems of equations on the axis, *Russian J. Math. Phys.*, **2** (1995), 517–526.

[18] A. YAGLE and B. LEVY, The Schur algorithm and its applications, *Acta Applicandae Mathematicae*, **3** (1985), 255–284.

D. ALPAY, Department of Mathematics, Ben–Gurion University of the Negev, POB 653. 84105 Beer-Sheva, Israel

I. GOHBERG, School of Mathematical Sciences, The Raymond and Beverly Sackler Faculty of Exact Sciences, Tel–Aviv University, Tel–Aviv, Ramat–Aviv 69989, Israel

Operator Theory: Advances and Applications, Vol. 127, 29–39

Construction of Schwarz norms

T. Ando*

To the memory of the late Professor Béla Szőkefalvi-Nagy

We present a general method of construction of Schwarz norms on the space of bounded linear operators on a Hilbert space.

1. Introduction

Let us consider a Hilbert space $\mathcal{H}$ with inner product $\langle \cdot, \cdot \rangle$, and the space $\mathcal{B}(\mathcal{H})$ of bounded linear operators on $\mathcal{H}$ with the induced (operator) norm $||\cdot||$.

An operator $T \in \mathcal{B}(\mathcal{H})$ with $||T|| \leq 1$ is called a *contraction*. The unit ball $\mathcal{B}_1 \equiv \{T; ||T|| \leq 1\}$ has some similarity with the closed unit disc of the complex plane. One aspect of this similarity was discovered by von Neumann [15]. It says that for any contraction T and for any complex valued function $f(z)$, defined and analytic on the closed unit disc, the following norm inequality holds:

$$||f(T)|| \leq ||f||_\infty \equiv \max_{|z|\leq 1} |f(z)|.$$

Here the operator $f(T)$ is defined by the usual functional calculus. This inequality is called the *von Neumann inequality*. The original proof of von Neumann was based on successive approximation of a bounded analytic function by compositions of fractional-linear functions according to the so-called Schur algorithm.

For a unitary operator U one can use its spectral representation

$$f(U) = \int_0^{2\pi} f(e^{i\theta})\, dE(\theta),$$

AMS Subject Classification (1991): 47A30, 47A20, 47A63.

* Research supported in part by Grant-in-Aid for Scientific Research (C) 10640183.

where $E(\theta)$ $(0 \le \theta \le 2\pi)$ is the spectral measure of U. The von Neumann inequality in this case follows from the norm identity

$$||f(U)x||^2 = \int_0^{2\pi} |f(e^{i\theta})|^2 \, d||E(\theta)x||^2.$$

Sz.-Nagy [16] gave a proof of the von Neumann inequality, based on his theory of the unitary dilation of a contraction. It says that for any contraction T there is a unitary operator U on a Hilbert superspace $\mathcal{K} \supset \mathcal{H}$ such that

$$T^n = P_{\mathcal{H}} U^n|_{\mathcal{H}} \quad (n = 0, 1, 2, \ldots),$$

where $P_{\mathcal{H}}$ is the orthoprojection from $\mathcal{K}$ to $\mathcal{H}$. This impies that for any complex valued function $f(z)$, defined and analytic on the closed unit disc,

$$f(T) = P_{\mathcal{H}} f(U)|_{\mathcal{H}}.$$

Since $||P_{\mathcal{H}}|| \le 1$, the von Neumann inequality for T follows from that for the unitary operator U.

For a bounded invertible operator S the new norm $||T||_S \equiv ||STS^{-1}||$ on $\mathcal{B}(\mathcal{H})$ also satisfies the von Neumann inequality. This is because the norm $||\cdot||_S$ is nothing but the (operator) norm for the Hilbert space $(\mathcal{H}, \langle S\cdot, S\cdot\rangle)$. Foiaş [9] showed a converse. More precisely, if the operator norm $|||\cdot|||$ on the space of bounded linear operators on a normed vector space $\mathcal{L}$ satisfies the von Neumann inequality, then the normed vector space $\mathcal{L}$ is isometric-isomorphic to a Hilbert space.

For $T \in \mathcal{B}(\mathcal{H})$, besides the operator norm $||T||$, there is another familiar norm $w(T)$, called *numerical radius*, defined by

$$w(T) \equiv \sup\{|\langle Tx, x\rangle| \ ; \ ||x|| = 1\}.$$

From the standpoint of physics the numerical radius is more natural than the operator norm because for a unit vector x the quantity $\langle Tx, x\rangle$ is considered to describe the expectation value of the physical observable, represented by T, at the state x.

It is immediate to see that

$$w(T) \le ||T|| \le 2w(T).$$

Let us call T a *numerical radius contraction* if $w(T) \le 1$.

It was a surprise to see that Berger [5] (see also Berger and Stampfli [6]) pointed out a strong connection between a numerical radius contraction and the theory of unitary dilation. More precisely, it was shown that T is a numerical radius contraction if and only if there is a unitary operator U on a Hilbert superspace $\mathcal{K} \supset \mathcal{H}$ such that

$$T^n = 2P_{\mathcal{H}}U^n|_{\mathcal{H}} \quad (n = 1, 2, \ldots).$$

Sz.-Nagy and Foiaş [17] immediately generalized this observation by introducing the notion of *unitary ρ-dilation.* More precisely, for $\rho > 0$ an operator T is said to admit a unitary ρ-dilation if there is a unitary operator U on a Hilbert superspace $\mathcal{K} \supset \mathcal{H}$ such that

$$T^n = \rho P_{\mathcal{H}}U^n|_{\mathcal{H}} \quad (n = 1, 2, \ldots).$$

Let $\mathcal{C}_\rho$ denote the class of operators which admit unitary ρ-dilations. Sz.-Nagy and Foiaş [17] gave various characterizations for an operator T to be in $\mathcal{C}_\rho$. One can define the *ρ-radius* $w_\rho(T)$ of an operator T by

$$w_\rho(T) \equiv \inf\left\{\lambda > 0\,;\ \frac{1}{\lambda}T \in \mathcal{C}_\rho\right\}.$$

The classes $\mathcal{C}_\rho$ were introduced by Sz.-Nagy and Foiaş [17] and by Williams [19](with a different point of view, and notations). The corresponding operator radii $w_\rho(\cdot)$ were introduced by Williams [19] and by Holbrook [11], and subsequently the ρ-phenomena were studied in a number of papers (e.g. [2], [3], [11]). Among others it has been shown that in general $w_\rho(T)$ is merely a quasi-norm of T, that is, the triangular inequality is not always satisfied. It becomes a norm only for $0 < \rho \leq 2$, and the following relations hold:

$$\rho w_\rho(T) = (2 - \rho)w_{2-\rho}(T) \quad (0 < \rho < 2).$$

Further $w_\rho(T)$ is a decreasing function of ρ with $w_1(T) = ||T||$ and $w_2(T) = w(T)$. Therefore only $w_\rho(\cdot)$ for $1 \leq \rho \leq 2$ are of interest from the standpoint of norms, and

$$||T|| \geq w_\rho(T) \geq w(T) \qquad (1 \leq \rho \leq 2).$$

Suppose that a complex valued function $f(z)$ is defined and analytic on the closed unit disc, and $f(0) = 0$. If T admits a unitary ρ-dilation U, then through a series expansion of $f(z)^n$ we can see

$$f(T)^n = \rho P_{\mathcal{H}}f(U)^n|_{\mathcal{H}} \qquad n = 1, 2, \ldots.$$

Since $f(U)/||f||_\infty$ is a contraction, there is a unitary operator V on a Hilbert superspace $\mathcal{L} \supset \mathcal{K}$ such that

$$\frac{f(U)^n}{||f||_\infty^n} = P_{\mathcal{K}}V^n|_{\mathcal{K}} \qquad n = 0, 1, 2, \ldots$$

where $P_{\mathcal{K}}$ is the orthoprojection from $\mathcal{L}$ to $\mathcal{K}$. Then it follows that

$$\frac{f(T)^n}{||f||_\infty^n} = \rho \tilde{P}_{\mathcal{H}} V^n|_{\mathcal{H}}$$

where $\tilde{P}_{\mathcal{H}}$ is the orthoprojection from $\mathcal{L}$ to $\mathcal{H}$. This implies by definition

$$w_\rho(f(T)) \leq ||f||_\infty.$$

The monograph [18] contains a nice chapter on this subject.

Williams [19] called a norm $|||\cdot|||$ on $\mathcal{B}(\mathcal{H})$, equivalent to the operator norm, a *Schwarz norm* if $|||T||| \leq 1$ implies $||| f(T)||| \leq ||f||_\infty$ for any complex valued function $f(z)$, defined and analytic on the closed unit disc, with $f(0) = 0$. The foregoing observation shows that ρ-radii $w_\rho(\cdot)$ $(1 \leq \rho \leq 2)$ are Schwarz norms.

Holbrook [12] pointed out, as a consequence of a more general result, that each invertible contraction S gives rise to a Schwarz norm by $|||T||| \equiv ||S^{-1}T||$.

In this short paper we will present a general method of construction of Schwarz norms, which contain ρ-radii for $1 \leq \rho \leq 2$.

2. Construction of Schwarz norms

As usual, $A \geq 0$ means that A is *positive* (*semi-definite*). The order relation $A \geq B$ for selfadjoint A, B is defined as $A - B \geq 0$. Therefore, for a selfadjoint operator C to be a contraction is characterized by $-I \leq C \leq I$, where I is the identity operator.

Ando [1] gave a different type of characterization of a numerical radius contraction. It says that $w(T) \leq 1$ if and only if there is a selfadjoint contraction C, that is $-I \leq C \leq I$, such that

$$\begin{bmatrix} I+C & T \\ T^* & I-C \end{bmatrix} \geq 0.$$

Here a 2×2 matrix of operators is considered on the direct sum $\mathcal{H} \oplus \mathcal{H}$.

With $A \equiv (I-C)/2$, the above inequality is written in the form that $0 \leq A \leq I$ and

$$\begin{bmatrix} 2(I-A) & T \\ T^* & 2A \end{bmatrix} \geq 0,$$

or equivalently, there are a positive contraction A and a contraction W such that

$$T = 2(I-A)^{1/2}\cdot W\cdot A^{1/2}.$$

This observation was extended to the case of general ρ-radii by Durszt [8] and Ando and Okubo [3]. It says that $w_\rho(T)\le 1$ if and only if there is a positive contraction A such that

$$\begin{bmatrix} \rho(I-A)\{I-\rho(2-\rho)A\}^{-1} & T \\ T^* & \rho A \end{bmatrix} \ge 0,$$

or equivalently, there are a positive contraction A and a contraction W such that

$$T = \rho\{(I-A)[I-\rho(2-\rho)A]^{-1}\}^{1/2}\cdot W\cdot A^{1/2}.$$

We have to single out a characteristic property of the functions

$$\varphi_\rho(t) \equiv \frac{\rho(1-t)}{1-\rho(2-\rho)t} \quad (1<\rho<2), \quad \varphi_1(t) = 1, \quad \text{and} \quad \varphi_2(t) = 2(1-t).$$

Recall that a real valued function $\varphi(t)$ on $[0,1]$ is said to be *operator-concave* if for any positive contractions A, B and $\alpha,\beta\ge 0$ with $\alpha+\beta=1$

$$\alpha\varphi(A)+\beta\varphi(B) \le \varphi(\alpha A+\beta B).$$

Here, of course, $\varphi(A),\varphi(B)$ are defined by the functional calculus.

Related notions are operator-convexity and operator-monotonicity of a function. The function $\varphi(t)$ is *operator-convex* if $-\varphi(t)$ is operator-concave. It is *operator-monotone* if $0\le A\le B\le I$ implies $\varphi(A)\le\varphi(B)$.

There is a deep theory of operator-monotone fucntions, initiated by Löwner [14]. Modern developments are found in Bender and Sherman [4], Donoghue [7], and Hansen and Pedersen [10].

Typical examples of non-negative operator-concave functions on $[0,1]$ are positive constants, t, $(1-t)/(1-\lambda t)$ for $0\le\lambda<1$, and their weighted (continuous) sums. Thus we can conclude that the functions $\varphi_\rho(t)$ are non-negative operator-concave.

One more point to mention is that the function

$$\psi_\rho(t)\equiv\rho t \qquad (0\le t\le 1)$$

is obviously operator-concave, and satisfies:

$$\max_{0<t<1}\varphi_\rho(t)\psi_\rho(t) = 1.$$

Conversely, given a pair $(\varphi(t), \psi(t))$ of non-negative operator-concave functions $\varphi(t)$ and $\psi(t)$ on $[0,1]$, let us consider the class $\mathcal{B}_{(\varphi,\psi)}$ of operators T for which there is a positive contraction A such that

$$\begin{bmatrix} \varphi(A) & T \\ T^* & \psi(A) \end{bmatrix} \geq 0,$$

or equivalently, there are a positive contraction A and a contraction W such that

$$T = \varphi(A)^{1/2} \cdot W \cdot \psi(A)^{1/2}.$$

THEOREM 2.1. *If non-negative operator-concave functions $\varphi(t), \psi(t)$ on $[0,1]$ satisfy the condition*

$$0 < \max_{0\leq t\leq 1} \varphi(t)\psi(t) \leq 1,$$

then the functional

$$\|T\|_{(\varphi,\psi)} \equiv \inf\left\{\lambda > 0 : \frac{1}{\lambda}T \in \mathcal{B}_{(\varphi,\psi)}\right\}$$

is a Schwarz norm.

Proof. Absolute homogeneity of the functional is obvious. For the triangular inequality, we have to show that $\mathcal{B}_{(\varphi,\psi)}$ is a convex set. Take $S, T \in \mathcal{B}_{(\varphi,\psi)}$. Since by definition there are positive contractions A, B such that

$$\begin{bmatrix} \varphi(A) & S \\ S^* & \psi(A) \end{bmatrix} \geq 0, \quad \begin{bmatrix} \varphi(B) & T \\ T^* & \psi(B) \end{bmatrix} \geq 0,$$

for any $\alpha, \beta \geq 0$ with $\alpha + \beta = 1$, by operator-concavity of $\varphi(t)$ and $\psi(t)$ we can see

$$\begin{bmatrix} \varphi(\alpha A + \beta B) & \alpha S + \beta T \\ (\alpha S + \beta T)^* & \psi(\alpha A + \beta B) \end{bmatrix} \geq \alpha \begin{bmatrix} \varphi(A) & S \\ S^* & \psi(A) \end{bmatrix} + \beta \begin{bmatrix} \varphi(B) & T \\ T^* & \psi(B) \end{bmatrix} \geq 0,$$

showing that $\alpha S + \beta T \in \mathcal{B}_{(\varphi,\psi)}$. Therefore $\mathcal{B}_{(\varphi,\psi)}$ is a convex set.

There is $t_0 \in [0,1]$ with $\varphi(t_0), \psi(t_0) > 0$. Then with

$$\gamma_1 \equiv \min\{\varphi(t_0), \psi(t_0)\}$$

and $A \equiv t_0 I$ we can see

$$\begin{bmatrix} \varphi(A) & \frac{\gamma_1}{||T||}T \\ \frac{\gamma_1}{||T||}T^* & \psi(A) \end{bmatrix} \geq \gamma_1 \begin{bmatrix} I & \frac{1}{||T||}T \\ \frac{1}{||T||}T^* & I \end{bmatrix} \geq 0,$$

which shows

$$||T||_{(\varphi,\psi)} \leq \frac{||T||}{\gamma_1} < \infty.$$

Let

$$\gamma_2 \equiv \max\{\sup_{0\leq t\leq 1} \varphi(t), \sup_{0\leq t\leq 1} \psi(t)\}.$$

Then since $0 \leq \varphi(A), \psi(A) \leq \gamma_2 I$, we see

$$\begin{bmatrix} \gamma_2 I & T \\ T^* & \gamma_2 I \end{bmatrix} \geq \begin{bmatrix} \varphi(A) & T \\ T^* & \psi(A) \end{bmatrix}.$$

This inequality shows that

$$\gamma_2 ||T||_{(\varphi,\psi)} \geq ||T||.$$

Therefore the functional $||\cdot||_{(\varphi,\psi)}$ becomes a norm on $\mathcal{B}(\mathcal{H})$, equivalent to the operator norm.

It remains to prove the Schwarz property. If $||T||_{(\varphi,\psi)} < 1$, by definition there are a positive contraction A and a contraction W such that

$$T = \varphi(A)^{1/2} \cdot W \cdot \psi(A)^{1/2}.$$

Let

$$X \equiv W \cdot \psi(A)^{1/2}\varphi(A)^{1/2} = W \cdot (\varphi \cdot \psi)(A)^{1/2}.$$

Since $(\varphi\cdot\psi)(t) \leq 1$ by assumption, X is a contraction. Take a complex valued function $f(z)$, defined and analytic on the closed unit disc, with $f(0) = 0$. According to the classical Schwarz lemma the function $g(z)$, defined by

$$g(z) \equiv \frac{f(z)}{z} \quad (z \neq 0) \quad \text{and} \quad g(0) \equiv f'(0),$$

is analytic on the closed unit disc, and $||f||_\infty = ||g||_\infty$. Now since

$$T^{n+1} = \varphi(A)^{1/2} \cdot X^n W \cdot \psi(A)^{1/2} \quad (n = 0, 1, \ldots),$$

it is seen through a series expansion of $g(z)$ that

$$f(T) = \varphi(A)^{1/2} \cdot g(X)W \cdot \psi(A)^{1/2}.$$

Since by the von Neumann inequality

$$||g(X)W|| \leq ||g(X)|| \leq ||g||_\infty = ||f||_\infty,$$

we can conclude

$$||f(T)||_{(\varphi,\psi)} \leq ||f||_\infty.$$

This completes the proof. □

For different pairs (φ_1, ψ_1) and (φ_2, ψ_2) of positive operator-concave functions the classes of operators $\mathcal{B}_{(\varphi_1,\psi_1)}$ and $\mathcal{B}_{(\varphi_2,\psi_2)}$ may coincide with each other.

3. Properties of $||\cdot||_{(\varphi,\psi)}$

The norm $||\cdot||_{(\varphi,\psi)}$ is invariant under *unitary similarity*:

$$||U^*TU||_{(\varphi,\psi)} = ||T||_{(\varphi,\psi)} \qquad \text{(unitary } U).$$

But in general it is not invariant under formation of adjoints. A general relation here is

$$||T^*||_{(\varphi,\psi)} = ||T||_{(\psi,\varphi)}.$$

As a consequence, if both $\varphi(t)$ and $\psi(t)$ are bijection from the interval $[0,1]$ to an interval $[0,\kappa]$ for some $\kappa > 0$, and

$$\varphi^{(-1)} \circ \psi = \psi^{(-1)} \circ \varphi$$

then the norm $||\cdot||_{(\varphi,\psi)}$ is invariant under formation of adjoints. In fact, if for a positive contraction A and a contraction W

$$T = \varphi(A)^{1/2} \cdot W \cdot \psi(A)^{1/2},$$

then by assumption the operator

$$B \equiv \varphi^{(-1)}(\psi(A)) = \psi^{(-1)}(\varphi(A))$$

is a positive contraction and

$$T = \psi(B)^{1/2} \cdot W \cdot \varphi(B)^{1/2},$$

which implies $T^* \in \mathcal{B}_{(\varphi,\psi)}$. Therefore $\mathcal{B}_{(\varphi,\psi)}$ is invariant under formation of adjoints.

It is immediate to see that the above relation occurs for the case of $(\varphi_\rho, \psi_\rho)$ with $\kappa = \rho$.

There is an extensive study by Holbrook and Fong [13] of general norms on $\mathcal{B}(\mathcal{H})$ which are invariant under unitary similarity. Though the following result can be derived from their general theory, in the present case we can give a simple proof.

THEOREM 3.1. *Let $\varphi(t), \psi(t)$ be non-negative operator-concave functions on $[0,1]$ such that*

$$\max_{0\le t\le 1} \varphi(t)\psi(t) \;=\; 1.$$

Then for any operator $T \in \mathcal{B}(\mathcal{H})$

$$||T|| \;\ge\; ||T||_{(\varphi,\psi)} \;\ge\; w(T).$$

Proof. The first inequality is almost trivial. In fact, if $||T|| = 1$, take $t_0 \in [0,1]$ such that

$$\varphi(t_0)\psi(t_0) \;=\; 1.$$

The existence of such t_0 is guaranteed by assumption. With $A \equiv t_0 I$ and $W \equiv T$, we have

$$T \;=\; \varphi(A)^{1/2} \cdot W \cdot \psi(A)^{1/2}.$$

Therefore $||T||_{(\varphi,\psi)} \le 1$.

To prove the second inequality, we may assume that $||T||_{(\varphi,\psi)} = 1$, and there are a positive contraction A and a contraction W such that

$$T \;=\; \varphi(A)^{1/2} \cdot W \cdot \psi(A)^{1/2}.$$

Then for any unit vector x

$$|\langle Tx, x\rangle| \;=\; |\langle W\psi(A)^{1/2}x, \varphi(A)^{1/2}x\rangle| \;\le\; \sqrt{\langle\varphi(A)x, x\rangle \cdot \langle\psi(A)x, x\rangle}.$$

Let

$$A \;=\; \int_0^1 t\, dE(t)$$

be the spectral representation of the positive contraction A. Since

$$\langle\varphi(A)x, x\rangle \;=\; \int_0^1 \varphi(t)\, d\langle E(t)x, x\rangle$$

and

$$\int_0^1 d\langle E(t)x, x\rangle \;=\; \langle x, x\rangle \;=\; 1,$$

by (operator) concavity of $\varphi(t)$ we have

$$\langle\varphi(A)x, x\rangle \;\le\; \varphi\Big(\int_0^1 t\, d\langle E(t)x, x\rangle\Big) \;=\varphi(\langle Ax, x\rangle),$$

and similarly

$$\langle \psi(A)x, x\rangle \leq \psi(\langle Ax, x\rangle).$$

Therefore we have

$$|\langle Tx, x\rangle| \leq \sqrt{\varphi(\langle Ax, x\rangle) \cdot \psi(\langle Ax, x\rangle)}.$$

Now since

$$0 \leq \langle Ax, x\rangle \leq \langle x, x\rangle = 1,$$

by our assumption on the product $\varphi(t) \cdot \psi(t)$ we can conclude

$$|\langle Tx, x\rangle| \leq \max_{0\leq t\leq 1} \varphi(t)\psi(t) = 1.$$

This completes the proof.

□

References

[1] T. ANDO, Structure of operators with numerical radius one, *Acta Sci. Math. (Szeged)*, **34** (1973), 11–15.

[2] T. ANDO and K. NISHIO, Convexity properties of operator radii associated with unitary ρ-dilations, *Michigan Math. J.*, **20** (1973), 303–307.

[3] T. ANDO and K. OKUBO, Constants related to operators of class C_ρ, *Manuscripta Math.*, **16** (1975), 385–394.

[4] J. BENDER and S. SHERMAN, Monotone and convex operator functions, *Trans. Amer. Math. Soc.*, **79** (1955), 58–7_.

[5] C. A. BERGER, A strange dilation theorem (abstract), *Amer. Math. Soc. Notice*, **12** (1965), 590.

[6] C. A. BERGER and J. G. STAMPFLI, Norm relations and skew dilations, *Acta Sci. Math. (Szeged)*, **28** (1967), 191–195.

[7] W. DONOGHUE, *Monotone Matrix Functions and Analytic Continuation*, Springer–Verlag, New York, 1974.

[8] E. DURSZT, Factorization of operators in C_ρ classes, *Acta Sci. Math. (Szeged)*, **37** (1975), 195–200.

[9] C. FOIAŞ, Sur certains théorèms de J. von Neumann concernant les ensembles spectraux, *Acta Sci. Math. (Szeged)*, **18** (1957), 15–20.

[10] F. HANSEN and G.K. PEDERSEN, Jensen's inequality for operators and Löwner's theorem, *Math. Ann.*, **258** (1982), 229–241.

[11] J. A. R. HOLBROOK, On the power bounded operators of Sz.-Nagy and Foiaş, *Acta Sci. Math. (Szeged)*, **29** (1968), 299–310.

[12] J. A. R. HOLBROOK, Inequalities of von Neumann type for small matrices, *Function Spaces* (ed. K. Jarosz), Marcel Dekker, 1992, 273–280.

[13] J. A. R. HOLBROOK and C.-K. FONG, Unitary invariant operator norms, *Canad. J. Math.*, **35** (1983), 274–299.

[14] K. LÖWNER, Über monotone Matrixfunktionen, *Math. Z.*, **38** (1934), 177–216.

[15] J. VON NEUMANN, Eine Spektraltheorie für allgemeine Operatoren eines unitären Raumes, *Math. Nach.*, **4** (1951), 258–281.

[16] B. SZ.-NAGY, *Extensions of linear transformations in Hilbert space which extend beyond this space*, Budapest, 1955.

[17] B. SZ.-NAGY and C. FOIAŞ, On certain classes of power bounded operators in Hilbert space, *Acta Sci. Math. (Szeged)*, **27** (1966), 17–25.

[18] B. SZ.-NAGY and C. FOIAŞ, *Harmonic Analysis of Operators in Hilbert Spaces*, North-Holland, 1970.

[19] J. WILLIAMS, Schwarz norms for operators, *Pacific J. Math.*, **21** (1968), 181–188.

T. ANDO, Hokusei Gakuen University, Sapporo 004-8631, Japan;
e-mail: ando@hokusei.ac.jp

Operator Theory: Advances and Applications, Vol. 127, 41–81

On the class of extremal extensions of a nonnegative operator

Yu. M. Arlinskiĭ, S. Hassi, Z. Sebestyén
and H. S. V. de Snoo

Dedicated to the memory of Béla Szőkefalvi-Nagy

A nonnegative selfadjoint extension $\widetilde{A}$ of a nonnegative operator A is called extremal if $\inf\{(\widetilde{A}(\varphi - f), \varphi - f) : f \in \operatorname{dom} A\} = 0$ for all $\varphi \in \operatorname{dom} \widetilde{A}$. A new construction of all extremal extensions of a nonnegative densely defined operator will be presented. It employs a fixed auxiliary Hilbert space to factorize each extremal extension. Various functional-analytic interpretations of extremal extensions are studied and some new types of characterizations are obtained. In particular, a purely analytic description of extremal extensions is established, based on a class of functions introduced by M. G. Kreĭn and I. E. Ovčarenko.

1. Introduction

Let A be a densely defined nonnegative operator in a separable Hilbert space $\mathfrak{H}$ with inner product $(\cdot, \cdot)$; its defect numbers $n_\pm(A)$ are necessarily equal and in general infinite. There exist two nonnegative selfadjoint extensions, the Friedrichs

---

AMS Subject Classification (1991): 47A63, 47B25, 47B65.
The present research was partly supported by the Dutch Organization for Scientific Research NWO (SWON). The second author was supported by the research funds of the University of Helsinki and by the Academy of Finland. The fourth author was supported by the Faculty of Mathematics and Natural Sciences of the University of Groningen.

extension A_F and the Kreĭn–von Neumann extension A_N, with the property that every nonnegative selfadjoint extension $\widetilde{A}$ of A satisfies the inequalities

$$(1.1)\quad ((A_F - x)^{-1}h, h) \le ((\widetilde{A} - x)^{-1}h, h) \le ((A_N - x)^{-1}h, h), \quad h \in \mathfrak{H}, \quad x \le 0.$$

The Friedrichs extension A_F is characterized by

$$(1.2)\qquad A_F = \{\, \{f, g\} \in A^* : f \in \mathfrak{D}[A] \,\},$$

where the energy space $\mathfrak{D}[A]$ ($= \operatorname{dom} A_F^{1/2}$) denotes the completion of $\operatorname{dom} A$ under the inner product $(f, g) + (Af, g)$, $f, g \in \operatorname{dom} A$. In other words, $\mathfrak{D}[A]$ is the set of all $f \in \mathfrak{H}$ for which there exists a sequence $f_n \in \operatorname{dom} A$ such that

$$f_n \to f, \quad (A(f_n - f_m), f_n - f_m) \to 0, \quad n, m \to \infty.$$

Moreover, A_F is the unique selfadjoint extension H of A for which $\operatorname{dom} H \subset \mathfrak{D}[A]$. The Kreĭn–von Neumann extension A_N and the Friedrichs extension A_F are formally related by $A_N = ((A^{-1})_F)^{-1}$, cf. [4], [11], [17], which leads to

$$(1.3)\qquad A_N = \{\, \{f, g\} \in A^* : g \in \mathfrak{R}[A] \,\},$$

where $\mathfrak{R}[A]$ denotes the completion of $\operatorname{ran} A$ under the inner product $(f, g) + (Af, g)$, $f, g \in \operatorname{dom} A$, so that $\mathfrak{R}[A]$ ($= \operatorname{ran} A_N^{1/2}$) is the set of all $g \in \mathfrak{H}$ for which there exists a sequence $f_n \in \operatorname{dom} A$ such that

$$Af_n \to g, \quad (A(f_n - f_m), f_n - f_m) \to 0, \quad n, m \to \infty.$$

Likewise, A_N is the unique selfadjoint extension H of A for which $\operatorname{ran} H \subset \mathfrak{R}[A]$. Each nonnegative selfadjoint extension $\widetilde{A}$ of A satisfies

$$(1.4)\quad \operatorname{dom} A_F^{1/2} \subset \operatorname{dom} \widetilde{A}^{1/2} \subset \operatorname{dom} A_N^{1/2}, \quad \operatorname{ran} A_N^{1/2} \subset \operatorname{ran} \widetilde{A}^{1/2} \subset \operatorname{ran} A_F^{1/2}.$$

If $\operatorname{dom} A_F^{1/2} = \widetilde{A}^{1/2}$, then $\widetilde{A} = A_F$, and if $\operatorname{ran} \widetilde{A}^{1/2} = \operatorname{ran} A_N^{1/2}$, then $\widetilde{A} = A_N$.

The Friedrichs extension A_F and the Kreĭn–von Neumann extension A_N are examples of extremal extensions of A. By definition, a selfadjoint extension $\widetilde{A}$ of $A \ge 0$ belongs to the class $\mathbf{E}(A)$ of extremal extensions of A if it is nonnegative and satisfies

$$(1.5)\quad \inf\{\, (\widetilde{A}(\varphi - f), \varphi - f) : f \in \operatorname{dom} A \,\} = 0 \quad \text{for all } \varphi \in \operatorname{dom} \widetilde{A}.$$

The notion of an extremal extension for the case of proper m-accretive extensions of a symmetric operator was introduced by Yu. M. Arlinskiĭ and E. R. Tsekanovskiĭ in [10] via Cayley transforms. The definition (1.5) above of the class $\mathbf{E}(A)$ was suggested in [5], and extended in [6]–[9] to the case of m-accretive extensions of a sectorial operator or, more generally, of a sectorial linear relation. In these papers various characterizations of extremal extensions were obtained. Extremal selfadjoint extensions for nonnegative operators of the form B^2, where B is simple and maximal symmetric, were studied by S. Kuzhel in [25], [26] in the context of an abstract wave equation; instead of extremal extensions the terminology of unperturbed operators was used.

The purpose of the present paper is to offer some new characterizations of the extremal extensions of a nonnegative densely defined operator A, both in a functional-analytic and in a function-theoretic manner. In particular, it will be shown that all extremal extensions can be characterized by means of a factorization over a fixed auxiliary Hilbert space. A construction of the Kreĭn–von Neumann and the Friedrichs extensions via such factorizations has been earlier given by V. Prokaj, Z. Sebestyén and J. Stochel [30], [32]. By a systematic use of semibounded sesquilinear forms their arguments are extended to the present situation. The description of the class $\mathbf{E}(A)$ of extremal extensions is simplified in the case when A_F and A_N are transversal extensions of A, i.e., $A_F \dot{+} A_N = A^*$. It turns out that in this case each extremal extension $\widetilde{A}$ of A can be interpreted as the Friedrichs extension S_F of an intermediate nonnegative extension $S \subset A_N$, or as the Kreĭn–von Neumann extension K_N of an intermediate nonnegative extension $K \subset A_F$. The translation of these facts by means of the theory of abstract boundary value spaces (boundary pairs and boundary triplets), see [5], [13], [14], [15], [20], [27], gives a description of the class $\mathbf{E}(A)$ by means of abstract boundary conditions involving orthogonal projections in the parameter space, cf. [5]. The analytic descriptions concerning extremal extensions are investigated by means of the associated Weyl functions. With the methods recently developed in [12], the Weyl function of $\widetilde{A} \in \mathbf{E}(A)$ can be decomposed by means of the Weyl functions induced by the intermediate extensions S and K. These ideas combined with a class of functions introduced by M. G. Kreĭn and I. E. Ovčarenko in [22]–[24] lead finally to some purely analytic characterizations of extremal extensions. The results are applied to some second order differential operators. In particular, for point interactions on the real line, cf. [2], [3], the extremal extensions and their unitary equivalence, cf. [25], [26], are treated in detail.

2. Semibounded forms

Some of the results in this paper will be expressed in terms of semibounded (sesquilinear) forms $\mathfrak{t} = \mathfrak{t}[\cdot,\cdot]$. For the convenience of the reader some useful notions from [19] are recalled. The notation $\mathfrak{t}[u]$ will be used to denote $\mathfrak{t}[u,u]$, $u \in \operatorname{dom}\mathfrak{t}$. The inclusion $\mathfrak{t}_1 \subset \mathfrak{t}_2$ for semibounded forms $\mathfrak{t}_1$ and $\mathfrak{t}_2$ is defined by

$$\operatorname{dom}\mathfrak{t}_1 \subset \operatorname{dom}\mathfrak{t}_2, \quad \mathfrak{t}_1[u] = \mathfrak{t}_2[u], \quad u \in \operatorname{dom}\mathfrak{t}_1.$$

The semibounded form $\mathfrak{t}$ is closed if

$$u_n \to u, \quad \mathfrak{t}[u_n - u_m] \to 0, \quad u_n \in \operatorname{dom}\mathfrak{t}, \quad u \in \mathfrak{H}, \quad m,n \to \infty,$$

imply that $u \in \operatorname{dom}\mathfrak{t}$ and $\mathfrak{t}[u_n - u] \to 0$. The semibounded form $\mathfrak{t}$ is closable if it has a closed extension; in this case the closure of $\mathfrak{t}$ is the smallest closed extension of $\mathfrak{t}$. The inequality $\mathfrak{t}_1 \geq \mathfrak{t}_2$ for semibounded forms $\mathfrak{t}_1$ and $\mathfrak{t}_2$ is defined by

$$\operatorname{dom}\mathfrak{t}_1 \subset \operatorname{dom}\mathfrak{t}_2, \quad \mathfrak{t}_1[u] \geq \mathfrak{t}_2[u], \quad u \in \operatorname{dom}\mathfrak{t}_1.$$

In particular, $\mathfrak{t}_1 \subset \mathfrak{t}_2$ implies $\mathfrak{t}_1 \geq \mathfrak{t}_2$. If the forms $\mathfrak{t}_1$ and $\mathfrak{t}_2$ are closable, the inequality $\mathfrak{t}_1 \geq \mathfrak{t}_2$ is preserved by their closures, cf. [19, p. 330].

The first representation theorem states that for each densely defined, closed, semibounded form $\mathfrak{t}$ there exists a unique semibounded selfadjoint operator H, with the same lower bound as $\mathfrak{t}$, such that $\operatorname{dom}H \subset \operatorname{dom}\mathfrak{t}$ and $\mathfrak{t}[u,v] = (Hu,v)$ for all $u \in \operatorname{dom}H$ and $v \in \operatorname{dom}\mathfrak{t}$. Conversely, each semibounded selfadjoint operator H gives rise to such a densely defined, closed, semibounded form $\mathfrak{t}$, cf. [19, pp. 322, 323].

For semibounded selfadjoint operators H_1 and H_2, the inequality $H_1 \geq H_2$ is defined in terms of the corresponding forms $\mathfrak{t}_1$ and $\mathfrak{t}_2$ by $\mathfrak{t}_1 \geq \mathfrak{t}_2$. If γ_1 and γ_2 are the lower bounds of H_1 and H_2, then the inequality $H_1 \geq H_2$ implies that $\gamma_1 \geq \gamma_2$ and for all $x < \gamma_2$:

$$(H_1 - x)^{-1} \leq (H_2 - x)^{-1}. \tag{2.1}$$

Conversely, if the last inequality is satisfied for some $x < \min(\gamma_1,\gamma_2)$, then $H_1 \geq H_2$, cf. [19, p. 330].

If the densely defined closed semibounded form $\mathfrak{t}$ is nonnegative, the first representation theorem is amplified by the second representation theorem which asserts that then $\operatorname{dom}\mathfrak{t} = \operatorname{dom}H^{1/2}$ and $\mathfrak{t}[u,v] = (H^{1/2}u, H^{1/2}v)$ for all $u,v \in \operatorname{dom}\mathfrak{t}$. Hence, if H_1 and H_2 are nonnegative selfadjoint operators, then $H_1 \geq H_2$ is equivalent to

$$\operatorname{dom}H_1^{1/2} \subset \operatorname{dom}H_2^{1/2}, \quad \|H_1^{1/2}u\| \geq \|H_2^{1/2}u\|, \quad u \in \operatorname{dom}H_1^{1/2}, \tag{2.2}$$

cf. [19, pp. 331, 332]. The following lemma gives a useful equivalent statement.

LEMMA 2.1. *Let H_1 and H_2 be nonnegative selfadjoint operators. Then $H_1 \geq H_2$ is equivalent to*

$$\text{(2.3)} \qquad \operatorname{dom} H_1 \subset \operatorname{dom} H_2^{1/2}, \quad (H_1 u, u) \geq \|H_2^{1/2} u\|^2, \quad u \in \operatorname{dom} H_1.$$

Proof. Clearly, (2.2) implies (2.3). To prove the converse implication observe that (2.3) implies $\|H_1^{1/2}u\|^2 \geq \|H_2^{1/2}u\|^2$ for all $u \in \operatorname{dom} H_1$. For each $u \in \operatorname{dom} H_1^{1/2}$ there is a sequence $(u_n) \subset \operatorname{dom} H_1$ such that

$$u_n \to u, \quad H_1^{1/2} u_n \to H_1^{1/2} u, \quad n \to \infty.$$

Hence, $(H_2^{1/2} u_n)$ is a Cauchy sequence. Since $H_2^{1/2}$ is closed, it follows that $u \in \operatorname{dom} H_2^{1/2}$ and $H_2^{1/2} u_n \to H_2^{1/2} u$, $n \to \infty$. Therefore, $\|H_1^{1/2}u\| \geq \|H_2^{1/2}u\|$. □

Let S be a densely defined nonnegative operator in a Hilbert space $\mathfrak{H}$. Then the form

$$\mathfrak{t}[u, v] = (Su, v), \quad u, v \in \operatorname{dom} S,$$

is densely defined and semibounded; moreover, it is closable and the selfadjoint operator associated to the closure of $\mathfrak{t}$ is the Friedrichs extension of S. The Friedrichs extension of the closure of S coincides with the Friedrichs extension of S itself, cf. [19, pp. 323, 325].

Now let S be a densely defined operator from the Hilbert space $\mathfrak{H}$ into another Hilbert space $\mathfrak{H}'$. Then

$$\text{(2.4)} \qquad \mathfrak{t}[u, v] = (Su, Sv), \quad u, v \in \operatorname{dom} \mathfrak{t} = \operatorname{dom} S,$$

defines a nonnegative symmetric form which is densely defined. The form $\mathfrak{t}$ is closed if and only if S is closed, and then the selfadjoint operator determined by the form $\mathfrak{t}$ is given by S^*S. The form $\mathfrak{t}$ is closable if and only if the operator S is closable, in which case the closures $\tilde{\mathfrak{t}}$ and $\tilde{S}$ are connected by

$$\tilde{\mathfrak{t}}[u, v] = (\tilde{S}u, \tilde{S}v), \quad u, v \in \operatorname{dom} \tilde{\mathfrak{t}} = \operatorname{dom} \tilde{S},$$

cf. [19, pp. 315, 317, 326]. Moreover, the selfadjoint operator $\tilde{S}^*\tilde{S} = S^*S^{**}$ corresponding to $\tilde{\mathfrak{t}}$ is the closure of the densely defined nonnegative operator S^*S.

If the semibounded form $\mathfrak{t}$ and the selfadjoint operator H are connected via the first representation theorem, then the form $\mathfrak{t}$ will also be denoted by $H[\cdot, \cdot]$ and the form domain by $\operatorname{dom}[H]$. Hence, if H is nonnegative, then $\operatorname{dom}[H] = \operatorname{dom} H^{1/2}$.

Some of the above results have natural interpretations in terms of relations when the condition of a dense domain is removed. It suffices for the present purposes to mention the following. If $\mathfrak{t}$ is a closed semibounded form in a Hilbert space $\mathfrak{H}$, then the form $\mathfrak{t}$ is densely defined in the Hilbert space $\mathfrak{H}_s = \overline{\operatorname{dom} \mathfrak{t}}$ and it is still closed and semibounded. According to the first representation theorem there exists a selfadjoint operator H_s in $\mathfrak{H}_s$ with the same lower bound as $\mathfrak{t}$, so that $\mathfrak{t}[f,g] = (H_s f, g)$, $f \in \operatorname{dom} H_s$, $g \in \mathfrak{H}_s$. Then $H = (\{0\} \oplus (\mathfrak{H} \ominus \mathfrak{H}_s)) \oplus H_s$ is a semibounded selfadjoint relation in $\mathfrak{H}$ whose multivalued part $\operatorname{mul} H$ is given by $\mathfrak{H} \ominus \mathfrak{H}_s$. Since every (semibounded) selfadjoint relation H in $\mathfrak{H}$ admits a unique decomposition of the form $H = (\{0\} \oplus \operatorname{mul} H) \oplus H_s$ with H_s a (semibounded) selfadjoint operator in $\mathfrak{H} \ominus \operatorname{mul} H$, it follows from the first representation theorem that there is a one-to-one correspondence between all closed semibounded forms in $\mathfrak{H}$ and all semibounded selfadjoint relations in $\mathfrak{H}$. There is also an interpretation of the second representation theorem. In fact, each nonnegative selfadjoint relation H has a unique nonnegative selfadjoint square root relation given by $H^{1/2} = (\{0\} \oplus \operatorname{mul} H) \oplus H_s^{1/2}$. Observe, that the inequality $H_1 \geq H_2$ for semibounded selfadjoint relations can be defined as in (2.1), or equivalently via

$$\operatorname{dom} H_1^{1/2} \subset \operatorname{dom} H_2^{1/2}, \quad \|(H_1)_s^{1/2} u\| \geq \|(H_2)_s^{1/2} u\|, \quad u \in \operatorname{dom} H_1^{1/2},$$

cf. [11], but now it also makes sense in terms of the corresponding forms.

3. A factorization of Friedrichs and Kreĭn–von Neumann extensions

Let A be a closed, nonnegative, densely defined operator in a Hilbert space $\mathfrak{H}$. The Hilbert space $\mathfrak{H}_A$ is the completion of $\operatorname{ran} A$ under the inner product

$$\langle Af, Ag \rangle = (Af, g), \quad f, g \in \operatorname{dom} A. \tag{3.1}$$

The mappings $J: \mathfrak{H}_A \to \mathfrak{H}$, $\operatorname{dom} J = \operatorname{ran} A$, and $Q: \mathfrak{H} \to \mathfrak{H}_A$, $\operatorname{dom} Q = \operatorname{dom} A$, defined by

$$J(Af) = Af, \quad Qf = Af, \quad f \in \operatorname{dom} A, \tag{3.2}$$

are densely defined and closable. Therefore, their adjoints are densely defined operators. The equalities

$$(J(Ag), f) = (Ag, f) = (g, Af) = (g, Qf), \quad f, g \in \operatorname{dom} A,$$

show that $\operatorname{dom} A \subset \operatorname{dom} J^*$, $\operatorname{ran} A \subset \operatorname{dom} Q^*$, and

$$J = Q^* \upharpoonright \operatorname{ran} A, \quad Q = J^* \upharpoonright \operatorname{dom} A. \tag{3.3}$$

The factorizations in the next proposition can be found in [30], [32]; the present proof is based directly on the properties (1.2), (1.3) of the Friedrichs extension and the Kreĭn–von Neumann extension of A.

PROPOSITION 3.1. *Let A be a closed, nonnegative, densely defined operator in $\mathfrak{H}$, and let J and Q be defined by* (3.2). *Then:*

(i) *The Kreĭn–von Neumann extension A_N of A and the corresponding closed form $A_N[\cdot,\cdot]$ are given by $A_N = J^{**}J^*$ and*

$$A_N[f,g] = (J^*f, J^*g), \quad f,g \in \operatorname{dom} J^* = \operatorname{dom} A_N^{1/2};$$

(ii) *The Friedrichs extension A_F of A and the corresponding closed form $A_F[\cdot,\cdot]$ are given by $A_F = Q^*Q^{**}$ and*

$$A_F[f,g] = (Q^{**}f, Q^{**}g), \quad f,g \in \operatorname{dom} Q^{**} = \operatorname{dom} A_F^{1/2}.$$

Proof. (i) It follows from (3.2) and (3.3), that

$$J^{**}J^*f = J^{**}(Af) = J(Af) = Af, \quad f \in \operatorname{dom} A.$$

Hence, the operator $J^{**}J^*$ is a nonnegative selfadjoint extension of A. Since $\operatorname{ran}(J^{**}J^*)^{1/2} = \operatorname{ran} J^{**}$, for every $g \in \operatorname{ran}(J^{**}J^*)^{1/2}$ there is a sequence $(f_n) \subset \operatorname{dom} J$ such that $\|g - J(Af_n)\| = \|g - Af_n\| \to 0$ and

$$\|Af_n - Af_m\|^2_{\mathfrak{H}_A} = (A(f_n - f_m), f_n - f_m) \to 0, \quad n,m \to \infty.$$

This means that $\operatorname{ran}(J^{**}J^*)^{1/2} = \operatorname{ran} A_N^{1/2}$, which implies $J^{**}J^* = A_N$. The closure of $(A_N\cdot,\cdot)$ is given by $A_N[f,g] = (J^*f, J^*g)$, $f,g \in \operatorname{dom} J^* = \operatorname{dom} A_N^{1/2}$, cf. the statements concerning (2.4).

(ii) The operator Q^*Q^{**} is a nonnegative selfadjoint extension of A, since

$$Q^*Q^{**}f = Q^*J^*f = J^{**}(Af) = J(Af) = Af, \quad f \in \operatorname{dom} A.$$

Moreover, $f \in \operatorname{dom}(Q^*Q^{**})^{1/2} = \operatorname{dom} Q^{**}$ means that there is a sequence $(f_n) \subset \operatorname{dom} Q = \operatorname{dom} A$, such that $\|f - f_n\| \to 0$ and

$$\|Qf_n - Qf_m\|^2_{\mathfrak{H}_A} = (A(f_n - f_m), f_n - f_m) \to 0, \quad n,m \to \infty.$$

Therefore, $\operatorname{dom}(Q^*Q^{**})^{1/2} = \operatorname{dom} A_F^{1/2}$, and consequently $Q^*Q^{**} = A_F$. In addition, $A_F[f,g] = \langle Q^{**}f, Q^{**}g\rangle$, $f,g \in \operatorname{dom} Q^{**}$, cf. the statements concerning (2.4).

□

The significance of the Friedrichs and Kreĭn–von Neumann extensions was observed by M. G. Kreĭn in [21]: a selfadjoint extension $\widetilde{A}$ of a densely defined operator $A \geq 0$ is nonnegative if and only if $A_N \leq \widetilde{A} \leq A_F$, or equivalently, if and only if (1.1) is satisfied. The same characterization, when the ordering is defined via (1.1), is still valid for selfadjoint extensions of nondensely defined operators or relations, cf. [11], [14]. As the smallest nonnegative extension of A the Kreĭn–von Neumann extension A_N has the largest square root domain $\operatorname{dom} A_N^{1/2} = \operatorname{dom} A_N[\cdot,\cdot]$, cf. (1.4). This domain will be used in the description of other selfadjoint extensions. Therefore some useful relations are recalled. The domain of $A_N[\cdot,\cdot]$ can be characterized as follows (see [4]):

$$\operatorname{dom} A_N^{1/2} = \left\{ \varphi \in \mathfrak{H} : \sup_{f \in \operatorname{dom} A} \frac{|(\varphi, Af)|^2}{(Af, f)} < \infty \right\}. \tag{3.4}$$

In fact, the supremum in this formula is equal to $A_N[\varphi] = \|A_N^{1/2}\varphi\|^2$. These relations also yield the equality

$$\inf\{ A_N[\varphi - f] : f \in \operatorname{dom} A \} = 0 \quad \text{for all} \quad \varphi \in \operatorname{dom} A_N^{1/2}. \tag{3.5}$$

Moreover, $\operatorname{dom} A_N^{1/2}$ has the following direct sum decomposition:

$$\operatorname{dom} A_N^{1/2} = \operatorname{dom} A_F^{1/2} \dotplus \left(\ker(A^* + I) \cap \operatorname{dom} A_N^{1/2}\right). \tag{3.6}$$

If A is positive definite, then there is a direct sum decomposition of the form

$$\operatorname{dom} A_N^{1/2} = \operatorname{dom} A_F^{1/2} \dotplus \ker A^*, \tag{3.7}$$

cf. e.g. [21], which leads to

$$A_N[f] = A_F[\mathcal{P}_F f], \quad f \in \operatorname{dom} A_N^{1/2}, \tag{3.8}$$

where $\mathcal{P}_F$ is the corresponding projection from $\operatorname{dom} A_N^{1/2}$ onto $\operatorname{dom} A_F^{1/2}$.

4. A construction of extremal extensions

The construction in the previous section (see Proposition 3.1) can be extended by associating a nonnegative selfadjoint extension to each subspace $\mathfrak{L}$ with $\operatorname{dom} A \subset \mathfrak{L} \subset \operatorname{dom} J^* = \operatorname{dom} A_N^{1/2}$. Define the restriction operator $R_{\mathfrak{L}}$ by

$$R_{\mathfrak{L}} := J^* \restriction \mathfrak{L}, \quad \operatorname{dom} A \subset \mathfrak{L} \subset \operatorname{dom} A_N^{1/2}. \tag{4.1}$$

Then $R_{\mathfrak{L}}$ is a densely defined closable operator from $\mathfrak{H}$ into $\mathfrak{H}_A$.

Proposition 4.1. *Let $\mathfrak{L}$ be a subspace of $\mathfrak{H}$ such that* $\operatorname{dom} A \subset \mathfrak{L} \subset \operatorname{dom} A_N^{1/2}$, *and let $R_{\mathfrak{L}}$ be defined by* (4.1).

(i) *The operator $R_{\mathfrak{L}}$ is closed if and only if the restriction to $\mathfrak{L}$ of the form $A_N[\cdot,\cdot]$ is closed.*

(ii) *The operator $\widetilde{A}_{\mathfrak{L}} = R_{\mathfrak{L}}^* R_{\mathfrak{L}}^{**}$ is a nonnegative selfadjoint extension of A and the corresponding closed form $\widetilde{A}_{\mathfrak{L}}[\cdot,\cdot]$ is given by*

$$\widetilde{A}_{\mathfrak{L}}[f,g] = \langle R_{\mathfrak{L}}^{**} f, R_{\mathfrak{L}}^{**} g\rangle, \quad f,g \in \operatorname{dom} R_{\mathfrak{L}}^{**} = \operatorname{dom} \widetilde{A}_{\mathfrak{L}}^{1/2}. \tag{4.2}$$

(iii) *If* $\operatorname{dom} A \subset \mathfrak{L} \subset \mathfrak{M} \subset \operatorname{dom} A_N^{1/2}$, *then*

$$\widetilde{A}_{\mathfrak{L}} = R_{\mathfrak{L}}^* R_{\mathfrak{L}}^{**} \geq R_{\mathfrak{M}}^* R_{\mathfrak{M}}^{**} = \widetilde{A}_{\mathfrak{M}}.$$

Proof. (i) Observe that the definition of $R_{\mathfrak{L}}$ implies that

$$\langle R_{\mathfrak{L}} f, R_{\mathfrak{L}} g\rangle = \langle J^* f, J^* g\rangle = \langle A_N^{1/2} f, A_N^{1/2} g\rangle, \quad f,g \in \mathfrak{L}.$$

Hence, it follows from the statements concerning (2.4) that $R_{\mathfrak{L}}$ is closed if and only if the restriction to $\mathfrak{L}$ of the form $A_N[\cdot,\cdot]$ is closed.

(ii) The following identity is easily verified:

$$R_{\mathfrak{L}}^* R_{\mathfrak{L}}^{**} f = R_{\mathfrak{L}}^* J^* f = J^{**} A f = J A f = A f, \quad f \in \operatorname{dom} A.$$

Hence, the operator $\widetilde{A}_{\mathfrak{L}} = R_{\mathfrak{L}}^* R_{\mathfrak{L}}^{**}$ is a nonnegative selfadjoint extension of A. Clearly, the closed form associated to $R_{\mathfrak{L}}^* R_{\mathfrak{L}}^{**}$ is given by (4.2), cf. the statements concerning (2.4).

(iii) Observe that the inclusion $\mathfrak{L} \subset \mathfrak{M}$ implies the inclusion of the corresponding forms $\mathfrak{t}_{\mathfrak{L}} \subset \mathfrak{t}_{\mathfrak{M}}$, and hence $\mathfrak{t}_{\mathfrak{L}} \geq \mathfrak{t}_{\mathfrak{M}}$. This inequality is preserved for the corresponding operators.

□

The subspace $\mathfrak{L}$ determines the selfadjoint extension $\widetilde{A}_{\mathfrak{L}} = R_{\mathfrak{L}}^* R_{\mathfrak{L}}^{**}$ of A. In fact, $\mathfrak{L}$ is a core for the corresponding form $\widetilde{A}_{\mathfrak{L}}[\cdot,\cdot]$. Hence, Proposition 3.1 can be strengthened in the following manner.

Corollary 4.2. *Let the assumptions be as in Proposition* 4.1. *Then:*
(i) *for every* $\mathfrak{L}$ *with* $\operatorname{dom} A_N \subset \mathfrak{L} \subset \operatorname{dom} A_N^{1/2}$, *one has* $R_{\mathfrak{L}}^* R_{\mathfrak{L}}^{**} = A_N$;
(ii) *for every* $\mathfrak{L}$ *with* $\operatorname{dom} A \subset \mathfrak{L} \subset \operatorname{dom} A_F^{1/2}$, *one has* $R_{\mathfrak{L}}^* R_{\mathfrak{L}}^{**} = A_F$.

The decomposition (3.6) implies that for every subspace $\mathfrak{L} \subset \mathfrak{H}$ such that $\operatorname{dom} A \subset \mathfrak{L} \subset \operatorname{dom} A_N^{1/2}$, there exists a direct sum decomposition

$$\mathfrak{L} = \mathfrak{M} \dotplus \mathfrak{N}, \quad \mathfrak{M} \subset \operatorname{dom} A_F^{1/2}. \tag{4.3}$$

Hence, the summand $\mathfrak{M}$ can be deleted in the construction of $\widetilde{A}_{\mathfrak{L}}$.

Corollary 4.3. *Let the subspace* $\mathfrak{L}$ *with* $\operatorname{dom} A \subset \mathfrak{L} \subset \operatorname{dom} A_N^{1/2}$ *have a direct sum decomposition of the form* (4.3). *Then*

$$R_{\mathfrak{L}}^* R_{\mathfrak{L}}^{**} = R_{\mathfrak{N}}^* R_{\mathfrak{N}}^{**}. \tag{4.4}$$

The next theorem gives a characterization of the nonnegative selfadjoint extensions of A which allow a factorization of the form (4.2).

Theorem 4.4. *Let* $A \geq 0$ *be a closed densely defined operator in* $\mathfrak{H}$. *Then for each nonnegative selfadjoint extension* $\widetilde{A}$ *of* A *the following statements are equivalent:*
(i) $\widetilde{A} = R_{\mathfrak{L}}^* R_{\mathfrak{L}}^{**}$ *for some* $\mathfrak{L}$ *with* $\operatorname{dom} A \subset \mathfrak{L} \subset \operatorname{dom} A_N^{1/2}$;
(ii) $\widetilde{A}$ *is an extremal extension of* A (*i.e.* $\widetilde{A} \in \mathbf{E}(A)$);
(iii) *the form associated to* $\widetilde{A}$ *satisfies* $\widetilde{A}[f,g] = A_N[f,g]$, $f,g \in \operatorname{dom} \widetilde{A}^{1/2}$.
In particular, there is a one-to-one correspondence between the closed restrictions $\mathfrak{t}$ *of* $A_N[\cdot,\cdot]$ *and the extremal extensions* $\widetilde{A} \in \mathbf{E}(A)$ *given by*

$$\mathfrak{t}[f,g] = \langle R_{\mathfrak{L}} f, R_{\mathfrak{L}} g\rangle, \quad f,g \in \operatorname{dom} \mathfrak{t} = \mathfrak{L} = \operatorname{dom} \widetilde{A}^{1/2}, \quad \operatorname{dom} A_F^{1/2} \subset \mathfrak{L} \subset \operatorname{dom} A_N^{1/2}.$$

Proof. First it is shown that (i) implies (ii). Let $\widetilde{A} = R_{\mathfrak{L}}^* R_{\mathfrak{L}}^{**}$ with $R_{\mathfrak{L}}$ be defined by (4.1). Then with $\varphi \in \operatorname{dom} \widetilde{A}$ and $f \in \operatorname{dom} A$

$$\begin{aligned}(\widetilde{A}(\varphi - f), \varphi - f) &= \langle R_{\mathfrak{L}}^{**}(\varphi - f), R_{\mathfrak{L}}^{**}(\varphi - f)\rangle \\ &= \langle J^*(\varphi - f), J^*(\varphi - f)\rangle = \|J^*\varphi - Af\|_{\mathfrak{H}_A}^2.\end{aligned}$$

By definition $\mathfrak{H}_A$ is the completion of $\operatorname{ran} A$ with respect to the norm $\|\cdot\|_{\mathfrak{H}_A}^2$. Therefore, for all $\varphi \in \operatorname{dom} \widetilde{A}$,

$$\inf\{\, \|J^*\varphi - Af\|_{\mathfrak{H}_A} : f \in \operatorname{dom} A\,\} = 0,$$

so that $\widetilde{A}$ is extremal.

To prove that (ii) implies (iii), assume that $\widetilde{A}$ is extremal. This together with the inequality $\widetilde{A} \geq A_N$ shows that for every $\varphi \in \operatorname{dom} \widetilde{A}$ and every $\varepsilon > 0$ there is an element $f_\varepsilon \in \operatorname{dom} A$, such that

$$\varepsilon > (\widetilde{A}(\varphi - f_\varepsilon), \varphi - f_\varepsilon) \geq \|A_N^{1/2}(\varphi - f_\varepsilon)\|^2 = \|J^*\varphi - Af_\varepsilon\|^2_{\mathfrak{H}_A} \geq 0.$$

This leads to

$$\varepsilon > (\widetilde{A}\varphi, \varphi) - \langle J^*\varphi, J^*\varphi\rangle \geq 0,$$

and consequently

$$(\widetilde{A}\varphi, \varphi) = \langle J^*\varphi, J^*\varphi\rangle = (A_N^{1/2}\varphi, A_N^{1/2}\varphi),$$

for all $\varphi \in \operatorname{dom} \widetilde{A}$. By the polarization identity the form $(\widetilde{A}u, v)$, $u, v \in \operatorname{dom} \widetilde{A}$, and hence also its closure $\widetilde{A}[\cdot,\cdot]$ is a restriction of $A_N[\cdot,\cdot]$.

Next it is shown that (iii) implies (i). Let $\widetilde{A}$ be a nonnegative selfadjoint extension of A such that $\widetilde{A}[u,v] \subset A_N[u,v]$, $u, v \in \operatorname{dom}[\widetilde{A}] =: \mathfrak{L}$. Since $\widetilde{A}[u,v]$ is closed, one obtains

$$\widetilde{A}[u,v] = \langle J^* \restriction \mathfrak{L}u, J^* \restriction \mathfrak{L}v\rangle = \langle R_{\mathfrak{L}}u, R_{\mathfrak{L}}v\rangle = \langle R_{\mathfrak{L}}^{**}u, R_{\mathfrak{L}}^{**}v\rangle.$$

Hence, the operator $\widetilde{A}$ associated to $\widetilde{A}[u,v]$ coincides with $R_{\mathfrak{L}}^* R_{\mathfrak{L}}^{**}$.

It remains to prove the one-to-one correspondence. Let $\mathfrak{t}$ be a closed restriction of the form $A_N[\cdot,\cdot]$ on the domain $\mathfrak{L} = \operatorname{dom}\mathfrak{t}$ and $\operatorname{dom} A_F^{1/2} \subset \mathfrak{L} \subset \operatorname{dom} A_N^{1/2}$. Then

$$\mathfrak{t}[f,g] = A_N[f,g] = \langle J^*f, J^*g\rangle = \langle R_{\mathfrak{L}}f, R_{\mathfrak{L}}g\rangle, \quad f, g \in \mathfrak{L}.$$

Since the form $\mathfrak{t}$ is closed, the operator $R_{\mathfrak{L}}\colon \mathfrak{H} \to \mathfrak{H}_A$ is closed. Therefore, $R_{\mathfrak{L}} = R_{\mathfrak{L}}^{**}$ and $\widetilde{A} = R_{\mathfrak{L}}^* R_{\mathfrak{L}} \in \mathbf{E}(A)$. Conversely, let $\widetilde{A}$ be an extremal extension of A. The definition in (1.5) shows that

$$\inf\{\, \widetilde{A}[\varphi - f] : f \in \operatorname{dom} A \,\} = 0 \quad \text{for all } \varphi \in \operatorname{dom} \widetilde{A}^{1/2}.$$

Therefore, for each $\varphi \in \operatorname{dom} \widetilde{A}^{1/2}$ there exists a sequence $(f_n) \subset \operatorname{dom} A$, such that

$$0 = \lim_{n\to\infty} \widetilde{A}[\varphi - f_n] \geq \lim_{n\to\infty} A_N[\varphi - f_n]\,(\geq 0).$$

Hence,

$$\widetilde{A}[\varphi] = \lim_{n\to\infty} (\widetilde{A}f_n, f_n) = \lim_{n\to\infty} (Af_n, f_n) = A_N[\varphi].$$

Consequently, the form $\widetilde{A}[\cdot,\cdot]$ is a closed restriction of $A_N[\cdot,\cdot]$ to the domain $\mathfrak{L} = \operatorname{dom} \widetilde{A}^{1/2}$ and $\operatorname{dom} A_F^{1/2} \subset \mathfrak{L} \subset \operatorname{dom} A_N^{1/2}$.

□

It is also possible to characterize the extremality of $\widetilde{A}$ by means of the operator $R_{\mathfrak{L}}^* R_{\mathfrak{L}}^{**}$ constructed with $\mathfrak{L} = \operatorname{dom} \widetilde{A}$.

PROPOSITION 4.5. *Let $\widetilde{A} \geq 0$ be a selfadjoint extension of A and let $\mathfrak{L} = \operatorname{dom} \widetilde{A}$. Then:*

(i) *The inequality $R_{\mathfrak{L}}^{*} R_{\mathfrak{L}}^{**} \leq \widetilde{A}$ holds.*

(ii) *If H is an extremal extension of A such that*

$$R_{\mathfrak{L}}^{*} R_{\mathfrak{L}}^{**} \leq H \leq \widetilde{A},$$

then $H = R_{\mathfrak{L}}^{*} R_{\mathfrak{L}}^{**}$.

(iii) *The equality $R_{\mathfrak{L}}^{*} R_{\mathfrak{L}}^{**} = \widetilde{A}$ holds if and only if $\widetilde{A}$ is extremal.*

Proof. (i) Denote $\widetilde{A}_{\mathfrak{L}} = R_{\mathfrak{L}}^{*} R_{\mathfrak{L}}^{**}$. Clearly, $\operatorname{dom} \widetilde{A} = \mathfrak{L} \subset \operatorname{dom} \widetilde{A}_{\mathfrak{L}}^{1/2}$. Moreover, the second representation theorem and the inequality $A_N \leq \widetilde{A}$ imply

$$\|\widetilde{A}_{\mathfrak{L}}^{1/2} f\|^2 = \|R_{\mathfrak{L}}^{**} f\|_{\mathfrak{H}_A}^2 = \|A_N^{1/2} f\|^2 \leq \|\widetilde{A}^{1/2} f\|^2 = (\widetilde{A} f, f), \quad f \in \operatorname{dom} \widetilde{A}.$$

Therefore, $\widetilde{A}_{\mathfrak{L}} \leq \widetilde{A}$ by Lemma 2.1.

(ii) Let H be an extremal extension of A such that $R_{\mathfrak{L}}^{*} R_{\mathfrak{L}}^{**} \leq H \leq \widetilde{A}$. By Theorem 4.4 it follows that

$$\langle R_{\mathfrak{L}}^{**} f, R_{\mathfrak{L}}^{**} g \rangle = (H^{1/2} f, H^{1/2} g), \quad f, g \in \operatorname{dom} \widetilde{A}.$$

Since $\operatorname{dom} \widetilde{A}$ is a core for both forms the equality $R_{\mathfrak{L}}^{*} R_{\mathfrak{L}}^{**} = H$ follows.

(iii) If $R_{\mathfrak{L}}^{*} R_{\mathfrak{L}}^{**} = \widetilde{A}$, then $\widetilde{A}$ is extremal by Theorem 4.4. Conversely, if $\widetilde{A}$ is extremal, then (i) and (ii) with $H = \widetilde{A}$ imply that $R_{\mathfrak{L}}^{*} R_{\mathfrak{L}}^{**} = \widetilde{A}$. □

Clearly, A_F and A_N are extremal extensions of A. If the defect numbers of A are $(1,1)$, then the direct sum decomposition (3.6) shows that A_F and A_N are the only extremal extensions of A. When the defect numbers of A are greater than one, the class $\mathbf{E}(A)$ contains infinitely many selfadjoint nonnegative extremal extensions of A. Observe that if $\widetilde{A}$ is an extremal extension of A and $\operatorname{dom} \widetilde{A}^{1/2} = \operatorname{dom} A_N^{1/2}$ then $\widetilde{A} = A_N$.

5. Boundary triplets and boundary pairs

Let A be a densely defined, closed, symmetric operator in $\mathfrak{H}$. The selfadjoint extensions H_1 and H_2 are said to be disjoint if $\operatorname{dom} H_1 \cap \operatorname{dom} H_2 = \operatorname{dom} A$, and transversal if $\operatorname{dom} H_1 + \operatorname{dom} H_2 = \operatorname{dom} A^*$. Observe, that the transversality of H_1 and H_2 implies their disjointness.

In the rest of this paper notions from the theory of abstract boundary value spaces will be needed, cf. [5], [13], [14], [15], [20], [27]. A triplet $\Pi = \{\mathcal{H}, \Gamma_0, \Gamma_1\}$ of a Hilbert space $\mathcal{H}$ with $\dim \mathcal{H} = n_\pm(A)$ and two linear mappings Γ_j, $j = 0, 1$, from $\operatorname{dom} A^*$ into $\mathcal{H}$ is said to be a boundary triplet for A^*, if the mapping $\Gamma = (\Gamma_0, \Gamma_1)\colon f \to \{\Gamma_0 f, \Gamma_1 f\}$ from $\operatorname{dom} A^*$ onto $\mathcal{H} \oplus \mathcal{H}$ is surjective, and the abstract Green's identity

$$(5.1) \qquad (A^* f, g) - (f, A^* g) = (\Gamma_1 f, \Gamma_0 g)_{\mathcal{H}} - (\Gamma_0 f, \Gamma_1 g)_{\mathcal{H}}$$

holds for all $f, g \in \operatorname{dom} A^*$. To each boundary triplet Π one may naturally associate two transversal selfadjoint extensions A_i of A by $A_i = A^* \upharpoonright \ker \Gamma_i$, $i = 0, 1$. The mapping $\Gamma\colon f \to \{\Gamma_0 f, \Gamma_1 f\}$ from $\operatorname{dom} A^*$ onto $\mathcal{H} \oplus \mathcal{H}$ establishes a one-to-one correspondence between the set of all closed extensions $\widetilde{A}_\Theta \subset A^*$ of A and the set $\widetilde{\mathcal{C}}(\mathcal{H})$ via

$$(5.2) \qquad \operatorname{dom} \widetilde{A}_\Theta = \Gamma^{-1}\Theta = \{ f \in \operatorname{dom} A^* : \Gamma f \in \Theta \}, \quad \widetilde{A}_\Theta := A^* \upharpoonright \operatorname{dom} \widetilde{A}_\Theta.$$

Here the sets of all closed linear relations, closed linear operators, and bounded linear operators in $\mathcal{H}$ are denoted by $\widetilde{\mathcal{C}}(\mathcal{H})$, $\mathcal{C}(\mathcal{H})$, and $[\mathcal{H}]$, respectively. Recall, that $\Theta \in \mathcal{C}(\mathcal{H})$ ($\Theta \in [\mathcal{H}]$) if and only if the extensions $\widetilde{A}_\Theta$ and A_0 are disjoint (transversal).

The γ-field $\gamma(\lambda)$ and the Weyl function $M(\lambda)$ associated to the boundary triplet $\Pi = \{\mathcal{H}, \Gamma_0, \Gamma_1\}$ are defined by

$$(5.3) \qquad \gamma(\lambda) = (\Gamma_0 \upharpoonright \mathfrak{N}_\lambda)^{-1}, \quad M(\lambda) = \Gamma_1 \gamma(\lambda), \quad \lambda \in \rho(A_0).$$

Here $\mathfrak{N}_\lambda = \ker(A^* - \lambda)$, $\lambda \in \mathbb{C}$, denotes the defect subspace of A.

Let $\widetilde{A}_\Theta$ be a selfadjoint extension of A corresponding to $\Theta = \Theta^* \in \widetilde{\mathcal{C}}(\mathcal{H})$, see (5.2). Then the resolvents of $\widetilde{A}_\Theta$ and $A_0 = \ker \Gamma_0$ are connected via Kreĭn's formula:

$$(5.4) \qquad (\widetilde{A}_\Theta - \lambda)^{-1} = (A_0 - \lambda)^{-1} + \gamma(\lambda)(\Theta - M(\lambda))^{-1}\gamma(\bar{\lambda})^*.$$

Moreover, $\lambda \in \rho(\widetilde{A}_\Theta)$ if and only if $0 \in \rho(\Theta - M(\lambda))$, cf. [13].

Let A be a closed nonnegative operator in $\mathfrak{H}$. In what follows, the Friedrichs extension A_F and the Kreĭn–von Neumann extension A_N of A are assumed to be transversal, i.e., $\operatorname{dom} A_F + \operatorname{dom} A_N = \operatorname{dom} A^*$. A simple necessary and sufficient condition for the transversality of A_F and A_N is the inclusion $\operatorname{dom} A^* \subset \operatorname{dom} A_N^{1/2}$, cf. [28]. In this case one can define a boundary triplet $\Pi = \{\mathcal{H}, \Gamma_0, \Gamma_1\}$ for A^* such that $A_F = A^* \upharpoonright \ker \Gamma_0$ and $A_N = A^* \upharpoonright \ker \Gamma_1$; such a triplet Π is said to be a basic boundary triplet for A. The notion of a basic boundary triplet was defined in [5] as a generalization of a positive boundary value space introduced by A. N. Kochubeĭ in [20] for a positive definite symmetric operator. If the boundary triplet Π is basic, the extension $\widetilde{A}_\Theta$ in (5.2) associated to $\Theta \in \widetilde{\mathcal{C}}(\mathcal{H})$ is nonnegative if and only if Θ is a nonnegative linear relation in $\mathcal{H}$ (see [5]).

For a basic boundary triplet $\Pi = \{\mathcal{H}, \Gamma_0, \Gamma_1\}$ of A^* the following equality holds (see [5], [6]):

$$(A^* f, g) - (\Gamma_1 f, \Gamma_0 g)_\mathcal{H} = A_N[f, g], \quad f, g \in \operatorname{dom} A^*. \tag{5.5}$$

To see this, let $g = f$ and decompose $f = f_1 + f_2$ with $f_1 \in \operatorname{dom} A_N$ and $f_2 \in \operatorname{dom} A_F$. It follows from (5.1) that

$$\omega(f, f) := (A^* f, f) - (\Gamma_1 f, \Gamma_0 f)_\mathcal{H} = (A_N f_1, f_1) + (A_F f_2, f_2) + 2\mathrm{Re}\,(A_N f_1, f_2).$$

Since $A_N[h] = A_F[h]$ for all $h \in \operatorname{dom} A_F$, one obtains

$$\omega(f, f) = A_N[f_1] + A_N[f_2] + 2\mathrm{Re}\, A_N[f_1, f_2] = A_N[f_1 + f_2] = A_N[f].$$

In particular, (5.2) and (5.5) show that

$$(\widetilde{A}_\Theta f, f) = (A^* f, f) = A_N[f] + (\Gamma_1 f, \Gamma_0 f)_\mathcal{H}, \quad \Gamma f \in \Theta,$$

which in view of (3.5) implies

$$\begin{aligned} 0 &\le \inf\{\, (\widetilde{A}_\Theta(f - \varphi), f - \varphi) : \varphi \in \operatorname{dom} A \,\} \\ &= \inf\{\, A_N[f - \varphi] : \varphi \in \operatorname{dom} A \,\} + (\Gamma_1 f, \Gamma_0 f)_\mathcal{H} \\ &= (\Gamma_1 f, \Gamma_0 f)_\mathcal{H}. \end{aligned}$$

It follows from (5.5) that the mapping $\Gamma_0 \colon \operatorname{dom} A_N \to \mathcal{H}$ is continuous with respect to the topology of $A_N[\cdot,\cdot]$. Hence, it admits a continuation $\overline{\Gamma}_0 \colon \operatorname{dom} A_N^{1/2} \to \mathcal{H}$. Clearly, $\ker \overline{\Gamma}_0 = \operatorname{dom} A_F^{1/2}$. The pair $\{\mathcal{H}, \overline{\Gamma}_0\}$ is called a boundary pair for A, cf. [6].

Theorem 5.1. ([6]). *Let $\{\mathcal{H}, \overline{\Gamma}_0\}$ be a boundary pair for $A \ge 0$. There is a one-to-one correspondence between all closed forms $\widetilde{A}[\cdot,\cdot]$ associated with nonnegative selfadjoint extensions $\widetilde{A}$ of A and all closed nonnegative forms $\widetilde{\omega}$ in $\mathcal{H}$ given by*

$$\begin{cases} \operatorname{dom} \widetilde{\omega} = \overline{\Gamma}_0(\operatorname{dom} \widetilde{A}^{1/2}), \\ \widetilde{A}[f, g] = A_N[f, g] + \widetilde{\omega}[\overline{\Gamma}_0 f, \overline{\Gamma}_0 g], \quad f, g \in \operatorname{dom} \widetilde{A}^{1/2}. \end{cases} \tag{5.6}$$

Proof. Let $\widetilde{A}$ be a nonnegative selfadjoint extension of A. Then $\operatorname{dom}\widetilde{A}^{1/2} \subset \operatorname{dom} A_N^{1/2}$ and $\widetilde{A}[f] \geq A_N[f]$ for all $f \in \operatorname{dom}\widetilde{A}^{1/2}$. Define a sesquilinear form $\widetilde{a}$ on $\operatorname{dom}\widetilde{A}^{1/2}$ via

$$\widetilde{a}[f,g] = \widetilde{A}[f,g] - A_N[f,g].$$

Clearly, this form is nonnegative and $\widetilde{a}[f] = 0$ for all $f \in \operatorname{dom}\widetilde{A}_F^{1/2}$. The form $\widetilde{a}$ is closed in the Hilbert space $\operatorname{dom} A_N^{1/2}$ equipped with the graph norm. In fact, if $(f_n) \subset \operatorname{dom}\widetilde{A}^{1/2}$ is a sequence, such that

$$\lim_{n\to\infty} f_n = f, \quad \lim_{n\to\infty} A_N^{1/2} f_n = A_N^{1/2} f \quad \text{and} \quad \lim_{n,m\to\infty} \widetilde{a}[f_n - f_m] = 0,$$

then, by definition

$$\lim_{n,m\to\infty} \widetilde{A}[f_n - f_m] = 0.$$

Consequently, $f \in \operatorname{dom}\widetilde{A}^{1/2}$ and $\widetilde{a}[f - f_n] \to 0$ as $n \to \infty$. Define

$$\widetilde{\omega}[\overline{\Gamma}_0 f, \overline{\Gamma}_0 g] := \widetilde{a}[f,g].$$

Then the form $\widetilde{\omega}$ is closed and nonnegative in $\mathcal{H}$, and it satisfies

$$\widetilde{A}[f,g] = A_N[f,g] + \widetilde{\omega}[\overline{\Gamma}_0 f, \overline{\Gamma}_0 g], \quad \operatorname{dom}\widetilde{A}^{1/2} = \overline{\Gamma}_0^{-1}(\operatorname{dom}\widetilde{\omega}).$$

To see the converse, let $\widetilde{\omega}$ be a closed nonnegative form in $\mathcal{H}$. Define the form

$$\widetilde{\tau}[f,g] := A_N[f,g] + \widetilde{\omega}[\overline{\Gamma}_0 f, \overline{\Gamma}_0 g], \quad \operatorname{dom}\widetilde{\tau} = \overline{\Gamma}_0^{-1}(\operatorname{dom}\widetilde{\omega}).$$

It is easy to check that $\widetilde{\tau}$ is a closed, densely defined nonnegative form in the Hilbert space $\mathfrak{H}$. Let $\widetilde{A}$ be the nonnegative selfadjoint operator associated to $\widetilde{\tau}$ according to the first representation theorem. Then

$$\widetilde{\tau}[h,g] = (Ah, g) \quad \text{for all } h \in \operatorname{dom} A,\ g \in \operatorname{dom}\widetilde{\tau}.$$

It follows that $h \in \operatorname{dom}\widetilde{A}$ and $\widetilde{A}h = Ah$. Thus, $\widetilde{A}$ is a nonnegative selfadjoint extension of A, and according to the second representation theorem $\operatorname{dom}\widetilde{A}^{1/2} = \operatorname{dom}\widetilde{\tau}$.

□

Let the nonnegative selfadjoint extension $\widetilde{A}$ of A and the nonnegative selfadjoint relation $\widetilde{\Theta} \in \widetilde{\mathcal{C}}(\mathcal{H})$ be connected by (5.2). Then $\widetilde{\Theta}$ is the nonnegative selfadjoint relation associated with the nonnegative form $\widetilde{\omega}$ in (5.6), which can be rewritten more explicitly as follows

$$\widetilde{\omega}[\overline{\Gamma}_0 f, \overline{\Gamma}_0 g] = \widetilde{\Theta}_s[\overline{\Gamma}_0 f, \overline{\Gamma}_0 g], \quad \overline{\Gamma}_0 f, \overline{\Gamma}_0 g \in \operatorname{dom}\widetilde{\omega} = \operatorname{dom}\widetilde{\Theta}_s^{1/2},$$

where $\widetilde{\Theta}_s$ is the operator part of $\widetilde{\Theta}$. The description of the closed forms associated to selfadjoint extensions $\widetilde{A}$ of A in Theorem 5.1 together with Theorem 4.4 shows that $\widetilde{A}$ belongs to $\mathbf{E}(A)$ if and only if the corresponding form $\widetilde{\omega}$ is defined on a subspace of $\mathcal{H}$ and vanishes on it.

6. Extremal extensions and symmetric restrictions

The main result in this section is a characterization of extremal extensions by means of the Friedrichs and Kreĭn–von Neumann extensions of certain intermediate symmetric extensions of the original nonnegative operator A.

Proposition 6.1. *Let A be a nonnegative, closed, symmetric operator in $\mathfrak{H}$ with a basic boundary triplet $\{\mathcal{H},\Gamma_0,\Gamma_1\}$ and let $S=\widetilde{A}_\Theta$, $\Theta\in\widetilde{\mathcal{C}}(\mathcal{H})$, be a closed nonnegative symmetric extension of A. Then there is a one-to-one correspondence between the class $\mathbf{E}(S)$ and the class $\mathbf{E}(\Theta)$ given by*

$$\widetilde{S}=\widetilde{A}_{\widetilde{\Theta}},\quad \widetilde{\Theta}=\mathbf{\Gamma}(\operatorname{dom}\widetilde{S})\in\mathbf{E}(\Theta).$$

In particular, the Friedrichs and the Kreĭn–von Neumann extensions of S and Θ are related by

$$S_F=\widetilde{A}_{\Theta_F},\quad S_N=\widetilde{A}_{\Theta_N}.$$

Proof. By (5.2), the identity $\widetilde{S}=\widetilde{A}_{\widetilde{\Theta}}$ gives a one-to-one correspondence between the selfadjoint extensions $\widetilde{S}$ of S and the selfadjoint extensions $\Gamma(\operatorname{dom}\widetilde{S})=\widetilde{\Theta}$ of Θ. Moreover, $\widetilde{S}$ is a nonnegative extension of S if and only if $\widetilde{\Theta}$ is a nonnegative extension of Θ. Let $\widetilde{\Theta}$ be a nonnegative selfadjoint extension of Θ. Then $\Theta_N\le\widetilde{\Theta}\le\Theta_F$ and Theorem 5.1 shows that

$$\widetilde{A}_{\Theta_N}\le\widetilde{A}_{\widetilde{\Theta}}\le\widetilde{A}_{\Theta_F}.$$

This implies $S_F=\widetilde{A}_{\Theta_F}$ and $S_N=\widetilde{A}_{\Theta_N}$. If $\widetilde{\Theta}$ is an extremal extension of Θ, then according to Theorem 4.4, the form $\widetilde{\Theta}[\cdot,\cdot]$ is a closed restriction of the form $\Theta_N[\cdot,\cdot]$. Consequently, the form $\widetilde{A}_{\widetilde{\Theta}}[\cdot,\cdot]$ is a closed restriction of the form $\widetilde{A}_{\Theta_N}[\cdot,\cdot]=S_N[\cdot,\cdot]$. Again by Theorem 4.4 one concludes that $\widetilde{S}=\widetilde{A}_{\widetilde{\Theta}}$ is an extremal extension of S. The converse part is proved similarly.

□

The assertion of Proposition 6.1 about the Friedrichs and the Kreĭn–von Neumann extensions was proved also in [12, Proposition 5.2] by means of (5.4).

COROLLARY 6.2. *Let A be a nonnegative, closed, densely defined, symmetric operator in $\mathfrak{H}$ with a basic boundary triplet $\{\mathcal{H},\Gamma_0,\Gamma_1\}$.*

(i) *For a closed symmetric restriction S of A_N with $A\subset S$ one has*

$$S_N = A_N, \quad S_F = \widetilde{A} = A^*\upharpoonright \Gamma^{-1}(\mathcal{H}_1 \oplus \mathcal{H}_1^{\perp}), \tag{6.1}$$

where $\mathcal{H}_1 = \Gamma_0(\operatorname{dom} S)$.

(ii) *For a closed symmetric restriction K of A_F with $A\subset K$ one has*

$$K_F = A_F, \quad K_N = \widetilde{A} = A^*\upharpoonright \Gamma^{-1}(\mathcal{H}_2^{\perp} \oplus \mathcal{H}_2), \tag{6.2}$$

where $\mathcal{H}_2 = \Gamma_1(\operatorname{dom} K)$.

Proof. For $S\subset A_N$ let $\mathcal{H}_1$ be the closed subspace defined by $\mathcal{H}_1 = \Gamma_0(\operatorname{dom} S)$. Then $\Gamma(\operatorname{dom} S) = \mathcal{H}_1 \oplus \{0\} =: \Theta$. It is an immediate consequence of the definitions of the Friedrichs and the Kreĭn–von Neumann extensions of a symmetric relation that

$$\Theta_F = \mathcal{H}_1 \oplus \mathcal{H}_1^{\perp}, \quad \Theta_N = \mathcal{H} \oplus \{0\}.$$

Now by Proposition 6.1, $\operatorname{dom} S_N = \Gamma^{-1}\Theta_N$ and $\operatorname{dom} S_F = \Gamma^{-1}\Theta_F$ and this is equivalent to (6.1). Hence, (i) has been proved.

Similarly one obtains (ii) by using the fact that with $K\subset A_F$, the Friedrichs and the Kreĭn–von Neumann extensions of $\Theta' := \Gamma(\operatorname{dom} K) = \{0\}\oplus\mathcal{H}_2$ are given by

$$\Theta'_F = \{0\}\oplus\mathcal{H}, \quad \Theta'_N = \mathcal{H}_2^{\perp}\oplus\mathcal{H}_2.$$

□

The following statement is an easy consequence of the definition of the Friedrichs extension.

LEMMA 6.3. *Let $H \geq 0$ and let S be a densely defined restriction of H. Then the operator associated to the closure in $H[\cdot,\cdot]$ of the form $(S\cdot,\cdot)$ is the Friedrichs extension S_F of S.*

Proof. The closed form associated to $(S\cdot,\cdot)$ gives rise to the Friedrichs extension S_F. Since S is a restriction of H, one has $(S\cdot,\cdot)\subset H[\cdot,\cdot]$ and the corresponding topologies are equal on $\operatorname{dom} S$, so that the closure of $(S\cdot,\cdot)$ in $H[\cdot,\cdot]$ coincides with $S_F[\cdot,\cdot]$.

□

The inclusion $S\subset H$ is essential for the result in Lemma 6.3 as becomes clear in the sequel. The next theorem characterizes the class $\mathbf{E}(A)$ of extremal extensions as Friedrichs and Kreĭn–von Neumann extensions of certain intermediate symmetric extensions of A.

Theorem 6.4. *Let $A \geq 0$ be a closed densely defined operator in $\mathfrak{H}$ and assume that the Friedrichs extension A_F and the Kreĭn–von Neumann extension A_N are transversal. Then the following statements are equivalent:*

(i) $\widetilde{A} \in \mathbf{E}(A)$;

(ii) $\widetilde{A} = S_F$ *for some closed symmetric operator* S *with* $A \subset S \subset A_N$;

(iii) $\widetilde{A} = K_N$ *for some closed symmetric operator* K *with* $A \subset K \subset A_F$.

Proof. To prove that (i) implies (ii) assume that $\widetilde{A} \in \mathbf{E}(A)$. By Proposition 4.5 the identity $\widetilde{A} = R_{\mathfrak{L}}^* R_{\mathfrak{L}}^{**}$ holds with $\mathfrak{L} = \operatorname{dom} \widetilde{A}$. Corollary 4.3 shows that $\mathfrak{L}$ can be replaced by $\mathfrak{L} + \operatorname{dom} A_F$, and consequently also by $\mathfrak{L}_1$, the projection of $\operatorname{dom} \widetilde{A}$ into $\operatorname{dom} A_N$ along $\operatorname{dom} A_F$, so that $\widetilde{A} = R_{\mathfrak{L}_1}^* R_{\mathfrak{L}_1}^{**}$. In particular, $\mathfrak{L}_1 \subset \operatorname{dom} \widetilde{A} \cap \operatorname{dom} A_N$. Let $S = A_N \upharpoonright \mathfrak{L}_1$. According to Lemma 6.3, the operator associated to the closure of $(S\cdot,\cdot)$ in $A_N[\cdot,\cdot]$ is the Friedrichs extension S_F of S. This means that $\widetilde{A} = S_F$.

The equivalence of (ii) and (iii) is seen as follows. Let $\Pi = \{\mathcal{H}, \Gamma_0, \Gamma_1\}$ be a basic boundary triplet for A^*. By Corollary 6.2, $\operatorname{dom} \widetilde{A} = \operatorname{dom} S_F = \Gamma^{-1}(\mathcal{H}_1 \oplus \mathcal{H}_1^{\perp})$, where $\mathcal{H}_1 = \Gamma_0(\operatorname{dom} S)$ and, moreover, $\widetilde{A} = K_N$ for the symmetric restriction K, $\operatorname{dom} K = \Gamma_0^{-1}(\mathcal{H}_1^{\perp})$ of A_F.

Finally, it is shown that (ii) implies (i). Assume that $\widetilde{A} = S_F$ for some S with $A \subset S \subset A_N$. Then, as the Friedrichs extension of S, $\widetilde{A}$ is an extremal extension of S. By Theorem 4.4 the form $\widetilde{A}[\cdot,\cdot]$ corresponding to $\widetilde{A}$ is a restriction of the form $S_N[\cdot,\cdot]$ associated to the Kreĭn–von Neumann extension S_N of S. By Corollary 6.2, $S_N = A_N$ and consequently, $S_N[\cdot,\cdot] = A_N[\cdot,\cdot]$, so that $\widetilde{A}[\cdot,\cdot] \subset A_N[\cdot,\cdot]$ and $\widetilde{A} \in \mathbf{E}(A)$ by Theorem 4.4.

□

The proof of Theorem 6.4 shows that, in fact, for every $S \subset A_N$ the corresponding Friedrichs extension S_F is an extremal extension of A, without the assumption on the transversality of A_F and A_N. This follows also from the characterization of $\mathbf{E}(A)$ in Proposition 4.5: if $\widetilde{A} = S_F$ for some $S \subset A_N$, then the operator associated to the closure of $A_N[x,y]$, $x, y \in \operatorname{dom} S$, is of the form $R_{\mathfrak{L}}^* R_{\mathfrak{L}}^{**}$ with $\mathfrak{L} = \operatorname{dom} S$ and hence belongs to $\mathbf{E}(A)$ by Theorem 6.4. Moreover, by Lemma 6.3 the same closure has also the representing operator S_F, so that $S_F = R_{\mathfrak{L}}^* R_{\mathfrak{L}}^{**} \in \mathbf{E}(A)$. It can be shown that without the assumption of transversality the equivalence of the statements in Theorem 6.4 does not hold in general.

The following fact concerning the lower bounds of extremal extensions of a positive definite operator A is a direct consequence of Theorem 6.4.

Corollary 6.5. *Let $A > 0$ be a densely defined positive definite operator in $\mathfrak{H}$. Then the lower bound $\mu(\widetilde{A})$ of an extremal extension $\widetilde{A}$ of A satisfies*

$$\mu(\widetilde{A}) = 0 \quad \text{if and only if} \quad \widetilde{A} \in \mathbf{E}(A) \setminus \{A_F\}. \tag{6.3}$$

Proof. Since A is positive definite,

$$\operatorname{dom} A_N = \operatorname{dom} A + \ker A_N, \quad \ker A_N = \ker A^*.$$

If $\widetilde{A} \in \mathbf{E}(A)\backslash\{A_F\}$, then $\widetilde{A} = S_F$ for some symmetric extensions $S \subset A_N$, $S \neq A$, of A. Therefore $\ker \widetilde{A} \neq \{0\}$ and since $\mu(A_F) = \mu(A) > 0$, the statement (6.3) follows.

□

The next result has some independent interest.

PROPOSITION 6.6. *Let $A \geq 0$ be a densely defined operator in $\mathfrak{H}$ and let H_1 and H_2 be nonnegative selfadjoint extensions of A. If the graphs $H_j \dot{+} A_F$ and $H_j \dot{+} A_N$, $j = 1, 2$, are closed, then:*
(i) *$H_1 \dot{+} A_F \subset H_2 \dot{+} A_F$ if and only if $\operatorname{dom} H_1^{1/2} \subset \operatorname{dom} H_2^{1/2}$;*
(ii) *$H_1 \dot{+} A_F = H_2 \dot{+} A_F$ if and only if $\operatorname{dom} H_1^{1/2} = \operatorname{dom} H_2^{1/2}$;*
(iii) *$H_1 \dot{+} A_N \subset H_2 \dot{+} A_N$ if and only if $\operatorname{ran} H_1^{1/2} \subset \operatorname{ran} H_2^{1/2}$;*
(iv) *$H_1 \dot{+} A_N = H_2 \dot{+} A_N$ if and only if $\operatorname{ran} H_1^{1/2} = \operatorname{ran} H_2^{1/2}$.*

Proof. Let $K_j = H_j \cap A_F$ and $S_j = H_j \cap A_N$, $j = 1, 2$. Since $H_j \dot{+} A_F$ and $H_j \dot{+} A_N$ are assumed to be closed, one has $K_j^* = H_j \dot{+} A_F$ and $S_j^* = H_j \dot{+} A_N$, and the corresponding images $\Theta_j = \Gamma(\operatorname{dom} K_j^*)$, $\Upsilon_j = \Gamma(\operatorname{dom} S_j^*)$ belong to $\widetilde{\mathcal{C}}(\mathcal{H})$. Observe, that $\Gamma(\operatorname{dom} A_F) = \{0\} \oplus \mathcal{H}$ and $\Gamma(\operatorname{dom} A_N) = \mathcal{H} \oplus \{0\}$, so that

$$\Theta_j = \Gamma(\operatorname{dom} H_j) \dot{+} (\{0\} \oplus \mathcal{H}), \quad \Upsilon_j = \Gamma(\operatorname{dom} H_j) \dot{+} (\mathcal{H} \oplus \{0\}).$$

It is easy to check that the closedness of Θ_j is equivalent to the closedness of its domain $\operatorname{dom} \Theta_j$, and similarly that the closedness of Υ_j is equivalent to the closedness of its range $\operatorname{ran} \Upsilon_j$. This means that $\operatorname{dom} \Theta_j^{1/2} = \operatorname{dom} \Theta_j$ and $\operatorname{ran} \Upsilon_j^{1/2} = \operatorname{ran} \Upsilon_j$. Therefore, the inclusions $K_1^* \subset K_2^*$ and $S_1^* \subset S_2^*$ are equivalent to the inclusions

$$\operatorname{dom} \Theta_1^{1/2} = \operatorname{dom} \Theta_1 = \Gamma_0(\operatorname{dom} H_1) \subset \Gamma_0(\operatorname{dom} H_2) = \operatorname{dom} \Theta_2 = \operatorname{dom} \Theta_2^{1/2}$$

and

$$\operatorname{ran} \Upsilon_1^{1/2} = \operatorname{ran} \Upsilon_1 = \Gamma_1(\operatorname{dom} H_1) \subset \Gamma_1(\operatorname{dom} H_2) = \operatorname{ran} \Upsilon_2 = \operatorname{ran} \Upsilon_2^{1/2}.$$

The statements (i) and (ii) follow now from the description of the forms $H_j[\cdot, \cdot]$ and their domains $\operatorname{dom} H_j^{1/2}$, $j = 1, 2$, in Theorem 5.1. The statements (iii) and (iv) are obtained from the statements (i) and (ii) by inversion of the graphs.

□

If the defect numbers of A are finite then the statement (i) in Proposition 6.6 can be formulated as follows:

$$H_1 \cap A_F \supset H_2 \cap A_F \quad \text{if and only if} \quad \operatorname{dom} H_1^{1/2} \subset \operatorname{dom} H_2^{1/2}.$$

Analogous observations hold for (ii)–(iv). In [16] and [17] similar statements were given for the selfadjoint extensions of a not necessarily semibounded operator A with defect numbers $(1,1)$. The invariance properties concerning the domains and ranges in (ii) and (iv) are reflected by the fact that the above construction results in selfadjoint extensions of the form S_F and K_N as given in Theorem 6.4.

7. Extremal extensions and basic boundary triplets

In this section extremal extensions of A are characterized in terms of boundary triplets. The first result describes the class $\mathbf{E}(A)$ of extremal extensions of $A \geq 0$ by means of the parameters $\Theta = \Theta^* \in \widetilde{\mathcal{C}}(\mathcal{H})$ in (5.2) corresponding to a basic boundary triplet for A^*; it was proved earlier in another form in [5].

Proposition 7.1. *Let $A \geq 0$ be a closed, densely defined operator in $\mathfrak{H}$ with a basic boundary triplet $\{\mathcal{H}, \Gamma_0, \Gamma_1\}$. Then the identity* (5.2) *defines a one-to-one correspondence between all extremal extensions $\widetilde{A} \in \mathbf{E}(A)$ and all selfadjoint relations $\Theta \in \widetilde{\mathcal{C}}(\mathcal{H})$ of the form*

$$\Theta = \{\, \{Ph, (I-P)h\} : h \in \mathcal{H}\,\}, \quad P = P^* = P^2 \in [\mathcal{H}]. \tag{7.1}$$

Proof. By Theorem 6.4 the extremal extensions $\widetilde{A} \in \mathbf{E}(A)$ are in one-to-one correspondence with the closed symmetric restrictions S of A_N, $A \subset S$, via $\widetilde{A} = S_F$. In view of Proposition 6.1 this leads to the following representation for $\Gamma(\operatorname{dom} \widetilde{A})$:

$$\Gamma(\operatorname{dom} \widetilde{A}) = \mathcal{H}_1 \oplus \mathcal{H}_1^{\perp},$$

where $\mathcal{H}_1$ is a closed subspace of $\mathcal{H}$; cf. the proof of Theorem 6.4. Now with P the orthogonal projection onto $\mathcal{H}_1$ in $\mathcal{H}$, the representation (7.1) follows. □

By means of P the extension $\widetilde{A} \in \mathbf{E}(A)$, $\operatorname{dom} \widetilde{A} = \Gamma^{-1}\Theta$, can be expressed via

$$\operatorname{dom} \widetilde{A} = \{\, f \in \operatorname{dom} A^* : (I-P)\Gamma_0 f = 0, \quad P\Gamma_1 f = 0\,\}. \tag{7.2}$$

Here the case $P=0$ corresponds to A_F, while the case $P = I$ corresponds to A_N.

To a selfadjoint extensions $\widetilde{A}$ of A associate the following closed symmetric restrictions K and S of $\widetilde{A}$:

$$K = \widetilde{A} \cap A_F \,(\subset A_F) \qquad S = \widetilde{A} \cap A_N \,(\subset A_N). \tag{7.3}$$

The following result is a consequence of Theorem 6.4.

PROPOSITION 7.2. *Let $A \geq 0$ be a closed, densely defined operator in $\mathfrak{H}$ with a basic boundary triplet $\Pi = \{\mathcal{H}, \Gamma_0, \Gamma_1\}$. Let $\widetilde{A}$ be an extremal extension of A given by (7.2) with P an orthogonal projection in $\mathcal{H}$, and let K and S be defined by (7.3).*

(i) *The symmetric restriction K and its adjoint K^* are given by*

$$\begin{aligned} \operatorname{dom} K &= \{ f \in \operatorname{dom} A^* : \Gamma_0 f = 0, \quad P\Gamma_1 f = 0 \}, \\ \operatorname{dom} K^* &= \{ f \in \operatorname{dom} A^* : (I-P)\Gamma_0 f = 0 \}. \end{aligned} \tag{7.4}$$

The triplet $\Pi_K = \{P\mathfrak{H}, \Gamma_0', \Gamma_1'\}$ with $\Gamma_j' = P\Gamma_j \upharpoonright \operatorname{dom} K^*$, $j = 0, 1$, defines a basic boundary triplet for K^* with

$$\ker \Gamma_0' = \operatorname{dom} A_F \, (= \operatorname{dom} K_F), \quad \ker \Gamma_1' = \operatorname{dom} \widetilde{A} \, (= \operatorname{dom} K_N).$$

(ii) *The symmetric restriction S and its adjoint S^* are given by*

$$\begin{aligned} \operatorname{dom} S &= \{ f \in \operatorname{dom} A^* : (I-P)\Gamma_0 f = 0, \quad \Gamma_1 f = 0 \}, \\ \operatorname{dom} S^* &= \{ f \in \operatorname{dom} A^* : P\Gamma_1 f = 0 \}. \end{aligned} \tag{7.5}$$

The triplet $\Pi_S = \{(I-P)\mathfrak{H}, \Gamma_0'', \Gamma_1''\}$ with $\Gamma_j'' = (I-P)\Gamma_j \upharpoonright \operatorname{dom} S^*$, $j = 0, 1$, defines a basic boundary triplet for S^* with

$$\ker \Gamma_0'' = \operatorname{dom} \widetilde{A} \, (= \operatorname{dom} S_F), \quad \ker \Gamma_1'' = \operatorname{dom} A_N \, (= \operatorname{dom} S_N).$$

Proof. Assume that $\widetilde{A} \in \mathbf{E}(A)$. Theorem 6.4 implies that

$$\Gamma(\operatorname{dom} K) = \{0\} \oplus \mathcal{H}_2, \quad \Gamma(\operatorname{dom} S) = \mathcal{H}_1 \oplus \{0\},$$

where the subspaces $\mathcal{H}_1$ and $\mathcal{H}_2$ are connected by $\mathcal{H}_2 = \mathcal{H}_1^{\perp}$. Therefore,

$$\begin{aligned} \operatorname{dom} K^* &= \Gamma^{-1}\left((\{0\} \oplus \mathcal{H}_2)^*\right) = \Gamma^{-1}(\mathcal{H}_1 \oplus \mathcal{H}), \\ \operatorname{dom} S^* &= \Gamma^{-1}\left((\mathcal{H}_1 \oplus \{0\})^*\right) = \Gamma^{-1}(\mathcal{H} \oplus \mathcal{H}_2). \end{aligned} \tag{7.6}$$

This leads to (7.4), (7.5) with P the orthogonal projection onto $\mathcal{H}_1$.

The description of K^* in (7.4) shows that, for every $f \in \operatorname{dom} K^*$, the equality $P\Gamma_0 f = \Gamma_0 f$ holds. Hence for all $f, g \in \operatorname{dom} K^*$,

$$(\Gamma_1' f, \Gamma_0' g)_{\mathcal{H}_1} - (\Gamma_0' f, \Gamma_1' g)_{\mathcal{H}_1} = (\Gamma_1 f, \Gamma_0 g)_{\mathcal{H}} - (\Gamma_0 f, \Gamma_1 g)_{\mathcal{H}}.$$

Thus, Green's formula (5.1) for Π_K is obtained from the corresponding formula for Π, while the surjectivity of the mapping $\Gamma' = (\Gamma_0', \Gamma_1')$: $\operatorname{dom} K^* \to \mathcal{H}_1 \oplus \mathcal{H}_1$ follows from the surjectivity of Γ: $\operatorname{dom} A^* \to \mathcal{H} \oplus \mathcal{H}$. Hence, Π_K is a boundary triplet for

K^*. Similarly one proves that Π_S is a boundary triplet for S^*. The description of K^* in (7.4) shows that $\ker \Gamma_0' = \operatorname{dom} A_F$, $\ker \Gamma_1' = \operatorname{dom} \widetilde{A}$, and the description of S^* in (7.5) shows that $\ker \Gamma_0'' = \operatorname{dom} \widetilde{A}$, $\ker \Gamma_1' = \operatorname{dom} A_N$. It is a consequence of the equalities (6.1), (6.2) in Corollary 6.2 that Π_K and Π_S are basic boundary triplets for K and S.

□

Proposition 7.2 is related to the construction of boundary triplets and Weyl functions for arbitrary symmetric intermediate (not necessarily nonnegative) extensions H of A in [12]. Proposition 7.2 shows how in the situation of extremal extensions of A the boundary triplets Π_K, Π_S are connected to each other when the intermediate extensions K, S are defined by (7.3). It also shows that transversality of the Friedrichs and the Kreĭn–von Neumann extensions is preserved for the intermediate extensions $K \subset A_F$ and $S \subset A_N$.

The description of extremal extensions of A in Proposition 7.1 is via the set of extremal extensions belonging to $\mathbf{E}(\Theta)$, with $\Theta = \{\{0,0\}\} \in \widetilde{\mathcal{C}}(\mathcal{H})$; cf. Proposition 6.1. When $\Theta = \{\{0,0\}\}$, the class $\mathbf{E}(\Theta)$ can be identified with the extreme points on the operator interval $[0, I]$. There is an analogous interpretation via 'Cayley' transforms in the original Hilbert space $\mathfrak{H}$.

Let A be a nonnegative linear relation and let S be defined by the transform $S = (I-A)(I+A)^{-1}$. The operator S is a nondensely defined symmetric contraction and there is a one-to-one correspondence between all nonnegative selfadjoint extensions of A and all selfadjoint contractive extensions of S [21]–[24]. The last set forms an operator interval $[S_\mu, S_M]$, where

$$S_\mu = (I - A_F)(I + A_F)^{-1}, \quad S_M = (I - A_N)(I + A_N)^{-1}.$$

This result, due to M. G. Kreĭn in [21], can be expressed more explicitly by the formula

$$\widetilde{S} = S_\mu + (S_M - S_\mu)^{1/2} \widetilde{X} (S_M - S_\mu)^{1/2},$$

which gives a one-to-one correspondence between all selfadjoint contractive extensions $\widetilde{S}$ of S and all nonnegative selfadjoint contractions $\widetilde{X}$ in the subspace $\mathfrak{M} := \overline{\operatorname{ran}}(S_M - S_\mu)$. The selfadjoint contractive extensions S_μ and S_M satisfy the following identities

$$(I + S_\mu)_{\mathfrak{M}} = (I - S_M)_{\mathfrak{M}} = 0.$$

Here $C_{\mathfrak{M}}$ denotes the shorted operator associated to a bounded nonnegative selfadjoint operator C and the subspace $\mathfrak{M}$ [21]. The equivalence of the following assertions was proved in [10]:

(i) $\widetilde{A} = (I - \widetilde{S})(I + \widetilde{S})^{-1}$ is an extremal extension of A;

(ii) $\widetilde{S} = S_\mu + (S_M - S_\mu)^{1/2}\widetilde{P}(S_M - S_\mu)^{1/2}$, $\widetilde{P}$ an orthogonal projection in $\mathfrak{M}$;
(iii) $\widetilde{S}$ is an extreme point of the operator interval $[S_\mu, S_M]$;
(iv) $(I - \widetilde{S}^2)_{\mathfrak{M}} = 0$.

In [10] also extreme points of the set of all quasi-selfadjoint contractive extensions of a symmetric contraction S were characterized via the identities $(I - \widetilde{S}^* S)_{\mathfrak{M}} = 0$ and $(I - \widetilde{S}S^*)_{\mathfrak{M}} = 0$. Various characterizations of extremal m-sectorial extensions of a sectorial linear relation and its Cayley transforms were obtained in [6]–[9].

8. Extremal extensions and Weyl functions

In this section the Weyl functions (or equivalently Q-functions) associated to an extremal extension will be characterized. For this purpose, recall some notions due to M. G. Kreĭn and I. E. Ovčarenko, cf. [22], [23], [24]. The class of Q_M-functions consists of Nevanlinna functions $Q(z)$ which have a holomorphic continuation to the negative real axis, and have the following properties:
(i) $Q^{-1}(z) \in [\mathcal{H}]$, $z \in \operatorname{Ext}[0,\infty)$;
(ii) $\lim_{x\uparrow 0}(Q(x)h,h) = +\infty$ for all $h \in \mathcal{H}\setminus\{0\}$;
(iii) $\text{s-lim}_{x\downarrow -\infty} Q(x) = 0$;
(iv) $\lim_{x\downarrow -\infty} x(Q(x)h,h) = -\infty$ for all $h \in \mathcal{H}\setminus\{0\}$.

Similarly, the class of Q_μ-functions consists of Nevanlinna functions $Q(z)$ which have a holomorphic continuation to the negative real axis, and have the following properties:
(i) $Q^{-1}(z) \in [\mathcal{H}]$, $z \in \operatorname{Ext}[0,\infty)$;
(ii) $\text{s-lim}_{x\uparrow 0} Q(x) = 0$;
(iii) $\lim_{x\downarrow -\infty}(Q(x)h,h) = -\infty$ for all $h \in \mathcal{H}\setminus\{0\}$;
(iv) $\text{s-lim}_{x\downarrow -\infty} Q(x)/x = 0$.

Let $\widetilde{A}$ be an extremal extension of A given by (7.2). The orthogonal projection P defining the extension $\widetilde{A}$ in (7.2) induces a natural decomposition for $\mathcal{H}$ and the Weyl function $M(\lambda)$:

$$(8.1) \qquad \mathcal{H} = \mathcal{H}_1 \oplus \mathcal{H}_2, \quad M(\lambda) = (M_{ij}(\lambda))_{i,j=1}^2, \quad \mathcal{H}_1 = \operatorname{ran} P, \quad \mathcal{H}_2 = \ker P,$$

where $M_{ij}(\lambda) = P_{\mathcal{H}_i} M(\lambda){\upharpoonright}\mathcal{H}_j$, $i,j = 1,2$. The next result expresses the specific form of the Weyl functions associated to extremal extensions of A. Moreover, it identifies the Weyl functions $M_K(\lambda)$ and $M_S(\lambda)$ for K and S in Proposition 7.2. In fact, $M_K(\lambda)$ is a compression and $M_S(\lambda)$ is a Schur complement of $M(\lambda)$ of the original Weyl function $M(\lambda)$. In [12] it was shown that in general such functions are related to certain intermediate symmetric extensions of the original symmetric (not necessarily semibounded) operator A and their boundary triplets were constructed.

For the convenience of the reader a complete proof of the next proposition will be given.

PROPOSITION 8.1. *Let Π be a basic boundary triplet for A^* with the Weyl function $M(\lambda)$ decomposed by* (8.1). *Let $\widetilde{A}$ be an extremal extension of A given by* (7.2), *and let K, S be given by* (7.3) *with boundary triplets Π_K, Π_S as in Proposition* 7.2. *Then:*

(i) *the Weyl function $\widetilde{M}(\lambda)$ corresponding to $\widetilde{A}$ has the form*

$$\widetilde{M}(\lambda) = \begin{pmatrix} -M_{11}(\lambda)^{-1} & -M_{11}(\lambda)^{-1}M_{12}(\lambda) \\ -M_{21}(\lambda)M_{11}(\lambda)^{-1} & M_{22}(\lambda) - M_{21}(\lambda)M_{11}(\lambda)^{-1}M_{12}(\lambda) \end{pmatrix}; \tag{8.2}$$

(ii) *the Weyl function $M_K(\lambda)$ associated to Π_K is a Q_μ-function in $\mathcal{H}_1$ given by*

$$M_K(\lambda) = M_{11}(\lambda); \tag{8.3}$$

(iii) *the Weyl function $M_S(\lambda)$ associated to Π_S is a Q_μ-function in $\mathcal{H}_2$ given by*

$$M_S(\lambda) = M_{22}(\lambda) - M_{21}(\lambda)M_{11}(\lambda)^{-1}M_{12}(\lambda). \tag{8.4}$$

Proof. (i) Introduce the fundamental symmetry J and a J-unitary operator $W = (W_{ij})_{i,j=0}^1$ in $\mathcal{H} \oplus \mathcal{H}$ via

$$J = i\begin{pmatrix} 0 & -I \\ I & 0 \end{pmatrix}, \quad W = \begin{pmatrix} I-P & -P \\ P & I-P \end{pmatrix}. \tag{8.5}$$

Then the transform $\widetilde{\Gamma} = W\Gamma$ of the boundary mappings Γ_0, Γ_1 gives rise to a new boundary triplet $\widetilde{\Pi} = \{\mathcal{H}, \widetilde{\Gamma}_0, \widetilde{\Gamma}_1\}$ for A^* with

$$\begin{aligned} \widetilde{\Gamma}_0 &= W_{00}\Gamma_0 + W_{01}\Gamma_1 = (I-P)\Gamma_0 - P\Gamma_1, \\ \widetilde{\Gamma}_1 &= W_{10}\Gamma_0 + W_{11}\Gamma_1 = P\Gamma_0 + (I-P)\Gamma_1. \end{aligned} \tag{8.6}$$

This implies that the basic selfadjoint extension corresponding to $\ker \widetilde{\Gamma}_0$ coincides with $\widetilde{A}$ defined by (7.2), so that the Weyl function of the boundary triplet $\widetilde{\Pi}$ is associated to $\widetilde{A}$. It is given by (see [14])

$$\begin{aligned} \widetilde{M}(\lambda) &= (W_{10} + W_{11}M(\lambda))(W_{00} + W_{01}M(\lambda))^{-1} \\ &= (P + (I-P)M(\lambda))((I-P) - PM(\lambda))^{-1} \\ &= \begin{pmatrix} I & 0 \\ M_{21}(\lambda) & M_{22}(\lambda) \end{pmatrix} \begin{pmatrix} -M_{11}(\lambda) & -M_{12}(\lambda) \\ 0 & I \end{pmatrix}^{-1}. \end{aligned}$$

This gives the expression (8.2) for $\widetilde{M}(\lambda)$.

(ii) & (iii) By definition (5.3) the Weyl function associated to the boundary triplet Π_K is given by

$$\begin{aligned} M_K(\lambda) &= \Gamma_1'(\Gamma_0' \restriction \mathfrak{N}_\lambda(K^*))^{-1} = P\Gamma_1(P\Gamma_0 \restriction \mathfrak{N}_\lambda(K^*))^{-1} \\ &= P\Gamma_1(\Gamma_0 \restriction \mathfrak{N}_\lambda(A^*))^{-1} \restriction \mathcal{H}_1 = PM(\lambda) \restriction \mathcal{H}_1 = M_{11}(\lambda), \end{aligned}$$

where the identity $P\Gamma_0 f = \Gamma_0 f$, $f \in \operatorname{dom} K^*$, has been used. This proves (8.3).

To see (8.4) apply the previous argument to the boundary triplet $\widetilde{\Pi}$ with $(I-P)$ instead of P. In view of (8.6) the symmetric operator in (7.4) is determined by

$$\begin{aligned} \operatorname{dom} \widetilde{K} &= \left\{ f \in \operatorname{dom} A^* : \widetilde{\Gamma}_0 f = 0, \quad (I-P)\widetilde{\Gamma}_1 f = 0 \right\} \\ &= \{ f \in \operatorname{dom} A^* : (I-P)\Gamma_0 f = 0, \quad \Gamma_1 f = 0 \}, \end{aligned}$$

and hence it coincides with S in (7.5). The corresponding Weyl function $M_S(\lambda)$ is now the compression to $\mathcal{H}_2$ of $\widetilde{M}(\lambda)$ in (8.2). Thus, (8.4) has been shown. It is known that the Weyl function $M(\lambda)$ associated to a basic boundary triplet Π is a Q_μ-function, see [22], cf. also [5]. By Proposition 7.2 the boundary triplets Π_K and Π_S are basic. Therefore also the functions $M_K(\lambda)$ and $M_S(\lambda)$ are Q_μ-functions. □

The description of the Weyl functions of the extremal extensions in (8.2) yields the following result concerning the unitary equivalence of extremal extensions of A; it is connected with some results of S. Kuzhel in [26] (cf. also the last section of the present paper).

Corollary 8.2. *Let $\widetilde{A}_1$ and $\widetilde{A}_2$ be extremal extensions of $A \geq 0$ determined by the projectors P_1 and P_2 via (7.2), and assume that $M(\lambda)$ has the form $M(\lambda) = m(\lambda)I$, where $m(z)$ is a scalar Nevanlinna function. Then the extensions $\widetilde{A}_1$ and $\widetilde{A}_2$ are unitarily equivalent, when P_1 and P_2 are unitarily equivalent.*

Proof. If P_1 and P_2 are unitarily equivalent, then in view of (8.2) the Weyl functions $M_1(\lambda)$ and $M_2(\lambda)$ corresponding to $\widetilde{A}_1$ and $\widetilde{A}_2$ are also unitarily equivalent. Therefore, (as Q-functions) they give rise to unitarily equivalent minimal model spaces for the pairs $\{A, \widetilde{A}_1\}$ and $\{A, \widetilde{A}_2\}$. In particular, the corresponding selfadjoint extensions $\widetilde{A}_1$ and $\widetilde{A}_2$ are unitarily equivalent. □

The following theorem gives a purely analytic characterization for the class of extremal extensions, in terms of the Weyl functions of A.

THEOREM 8.3. *Let A be a simple, densely defined, nonnegative operator in $\mathfrak{H}$ with the Friedrichs extension A_F and the Kreĭn–von Neumann extension A_N. If A_F and A_N are transversal extensions of A, then every extremal extension $\widetilde{A} \in \mathbf{E}(A)$ has a Weyl function $\widetilde{M}(\lambda)$ with the following properties:*

(i) *$\widetilde{M}(\lambda)$ is holomorphic on* $\operatorname{Ext}[0,\infty)$;

(ii) *$\widetilde{M}(\lambda)^{-1} \in [\mathcal{H}]$, $\lambda \in \operatorname{Ext}[0,\infty)$;*

(iii) *there exists an orthogonal decomposition $\mathcal{H} = \mathcal{H}_1 \oplus \mathcal{H}_2$, such that $\widetilde{M}_{11}(\lambda) = P_1\widetilde{M}(\lambda)\upharpoonright \mathcal{H}_1$ is a Q_M-function and $\widetilde{M}_{22}(\lambda) = P_2\widetilde{M}(\lambda)\upharpoonright \mathcal{H}_2$ is a Q_μ-function.*

Conversely, if $\widetilde{A}$ is a selfadjoint extension of A whose Weyl function $\widetilde{M}(\lambda)$ has the properties (i)–(iii), *then A_F and A_N are transversal extensions of A and $\widetilde{A}$ belongs to the class $\mathbf{E}(A)$.*

Proof. To prove the first part of the theorem, assume that $A_F \dot{+} A_N = A^*$ and let Π be a basic boundary triplet for A^*. Then by Proposition 8.1 the projection P associated to $\widetilde{A} \in \mathbf{E}(A)$ in (7.2) induces an orthogonal decomposition $\mathcal{H} = \operatorname{ran} P \oplus \ker P$, such that the corresponding Weyl function $\widetilde{M}(\lambda) = (\widetilde{M}_{ij}(\lambda))_{i,j=0}^1$ has the properties in (8.2)–(8.4). Now (i) follows from the fact that $\widetilde{A} \geq 0$. The property (ii) is a consequence of the Frobenius formula for the inverse of $\widetilde{M}(\lambda)$ in (8.2), since $M_{11}(\lambda)$ and $M_{22}(\lambda)$ as Q_μ-functions have bounded inverses, in which case

$$(8.7)\qquad -\widetilde{M}(\lambda)^{-1} = \begin{pmatrix} M_{11}(\lambda) - M_{12}(\lambda)M_{22}(\lambda)^{-1}M_{21}(\lambda) & M_{12}(\lambda)M_{22}(\lambda)^{-1} \\ M_{22}(\lambda)^{-1}M_{21}(\lambda) & -M_{22}(\lambda)^{-1} \end{pmatrix}.$$

The property (iii) follows from (8.2), since $M_{11}(\lambda)$ is a Q_μ-function if and only if $-M_{11}(\lambda)^{-1}$ is a Q_M-function.

Now, assume that $\widetilde{A}$ is a selfadjoint extension of A with a Weyl function $\widetilde{M}(\lambda)$ satisfying the properties (i)–(iii). Let $\widetilde{\Pi} = \{\mathcal{H}, \widetilde{\Gamma}_0, \widetilde{\Gamma}_1\}$, $\ker\widetilde{\Gamma}_0 = \widetilde{A}$, be the corresponding boundary triplet. Since A is simple, the assumption (i) implies that $\widetilde{A} \geq 0$. Let P be the orthogonal projection corresponding to the decomposition of $\mathcal{H}$ in (iii). Let J and W be given by (8.5) and define a new boundary triplet $\Pi = \{\mathcal{H}, \Gamma_0, \Gamma_1\}$ for A^* via

$$(8.8)\qquad \Gamma_0 = (I-P)\widetilde{\Gamma}_0 - P\widetilde{\Gamma}_1, \quad \Gamma_1 = P\widetilde{\Gamma}_0 + (I-P)\widetilde{\Gamma}_1.$$

Then the Weyl function $M(\lambda)$ associated to Π has the form on the right side of (8.2) (with $M_{ij}(\lambda)$ replaced by $\widetilde{M}_{ij}(\lambda)$). By assumption (iii) $\widetilde{M}_{11}(\lambda)$ is a Q_M-function, so that $-\widetilde{M}_{11}(\lambda)^{-1} \in [\mathcal{H}_1]$ is a Q_μ-function. In particular,

$$\lim_{x\downarrow-\infty} (M_{11}(x)h_1, h_1) = \lim_{x\downarrow-\infty} -\left(\widetilde{M}_{11}(x)^{-1}h_1, h_1\right) = -\infty$$

holds for every $h_1 \in \mathcal{H}_1\backslash\{0\}$. Moreover, $-\widetilde{M}_{11}(x)^{-1} \leq 0$ for all $x < 0$. Since $\widetilde{M}(x) = \widetilde{M}(x)^*$, $x < 0$, and $\widetilde{M}_{22}(\lambda)$ is a Q_μ-function one obtains

$$\begin{aligned}\lim_{x\downarrow-\infty}(M_{22}(x)h_2,h_2) &= \lim_{x\downarrow-\infty}\left((\widetilde{M}_{22}(x)-\widetilde{M}_{21}(x)\widetilde{M}_{11}(x)^{-1}\widetilde{M}_{12}(x))h_2,h_2\right)\\ &\leq \lim_{x\downarrow-\infty}(\widetilde{M}_{22}(x)h_2,h_2) = -\infty\end{aligned}$$

for every $h_2 \in \mathcal{H}_2\backslash\{0\}$. Consequently, the new Weyl function $M(\lambda)$ is holomorphic on $\operatorname{Ext}[0,\infty)$ and satisfies

$$\lim_{x\downarrow-\infty}(M(x)h,h) = -\infty$$

for all $h \in \mathcal{H}\backslash\{0\}$. According to [13, Proposition 4] this means that $\ker\Gamma_0 = \operatorname{dom} A_F$.

Consider the boundary triplet $\{\mathcal{H},\Gamma_1,-\Gamma_0\}$. The corresponding Weyl function is $-M(\lambda)^{-1}$. The assumptions (ii) and (iii) imply that $-M(\lambda)^{-1}$ is bounded and holomorphic on $\operatorname{Ext}[0,\infty)$. Moreover, it has the representation on the right side of (8.7) (with $M_{ij}(\lambda)$ replaced by $\widetilde{M}_{ij}(\lambda)$). By assumption (iii) $\widetilde{M}_{22}(\lambda)$ is a Q_μ-function, so that $-\widetilde{M}_{22}(\lambda)^{-1} \in [\mathcal{H}_2]$ is a Q_M-function. In particular,

$$\lim_{x\uparrow0} -\left(\widetilde{M}_{22}(x)^{-1}h_2,h_2\right) = +\infty$$

holds for every $h_2 \in \mathcal{H}_2\backslash\{0\}$. Moreover, $-\widetilde{M}_{22}(x)^{-1} \geq 0$ for all $x < 0$. Since $\widetilde{M}(x) = \widetilde{M}(x)^*$, $x < 0$, and $\widetilde{M}_{11}(\lambda)$ is a Q_M-function one obtains

$$\lim_{x\uparrow0}\left((\widetilde{M}_{11}(x)-\widetilde{M}_{12}(x)\widetilde{M}_{22}(x)^{-1}\widetilde{M}_{21}(x))h_1,h_1\right) \geq \lim_{x\uparrow0}(\widetilde{M}_{11}(x)h_1,h_1) = +\infty$$

for every $h_1 \in \mathcal{H}_1\backslash\{0\}$. Consequently, the function $-M(\lambda)^{-1}$ satisfies

$$\lim_{x\uparrow0} -(M(x)^{-1}h,h) = +\infty$$

for all $h \in \mathcal{H}\backslash\{0\}$. According to [13, Proposition 4] this means that $\ker\Gamma_1 = \operatorname{dom} A_N$. Consequently, Π defines a basic boundary triplet for A^*, and therefore A_F and A_N are transversal extensions of A. Finally, observe that the identities in (8.8) yield

$$\operatorname{dom}\widetilde{A} = \ker\widetilde{\Gamma}_0 = \{\, f \in \operatorname{dom}A^* : (I-P)\Gamma_0 f = 0, \quad P\Gamma_1 f = 0\,\}.$$

Hence, by Proposition 7.1 (see also (7.2)) $\widetilde{A}$ is an extremal extension of A. This completes the proof.

□

Theorem 8.3 combined with the integral representations of Q_M- and Q_μ-functions given in [24] yields the following characterization concerning the transversality of A_F and A_N, and the class of extremal extensions of A.

Corollary 8.4. *Let the assumptions for A be as in Theorem 8.3. Then A_F and A_N are transversal if and only if there exists a Weyl function $\widetilde{M}(\lambda)$ for A with the following properties:*
(i) $\widetilde{M}(\lambda)^{-1} \in [\mathcal{H}]$, $\lambda \in \operatorname{Ext}[0,\infty)$;
(ii) *there exists an orthogonal decomposition $\mathcal{H} = \mathcal{H}_1 \oplus \mathcal{H}_2$, such that*

$$\widetilde{M}_{11}(\lambda) = \int_0^\infty \frac{d\Sigma_{11}(t)}{t-\lambda}, \quad \widetilde{M}_{22}(\lambda) = \lambda \int_0^\infty \frac{d\Sigma_{22}(t)}{t-\lambda},$$

where $\Sigma_{11}(t) \in [\mathcal{H}_1]$ and $\Sigma_{22}(t) \in [\mathcal{H}_2]$ are nondecreasing functions on $[0,\infty)$ satisfying for all $g \in \mathcal{H}\backslash\{0\}$

$$\int_0^1 \frac{d(\Sigma_{11}(t)g,g)}{t} = \infty, \quad \Sigma_{22}(+0) = \Sigma_{22}(0),$$
$$\int_1^\infty \frac{d(\Sigma_{jj}(t)g,g)}{t} < \infty, \quad \int_0^\infty d(\Sigma_{jj}(t)g,g) = \infty, \quad j=1,2.$$

In this case $\widetilde{M}(\lambda)$ is a Weyl function corresponding to an extremal extension $\widetilde{A}$ of A and every $\widetilde{A} \in \mathbf{E}(A)$ has a Weyl function satisfying (i) *and* (ii).

9. Factorization of the underlying symmetric operator

The construction of extremal extensions of A in Sections 3, 4 was based on the auxiliary Hilbert space $\mathfrak{H}_A$ and it resulted in a factorization of all extremal extensions $\widetilde{A}$ of A. The definition of the mappings J and Q in (3.2) shows that A itself admits a factorization $A = JQ$. When the mappings J and Q are replaced by their closures, one obtains a factorization $H = J^{**}Q^{**}$ for $H = A_F \cap A_N$ ($\supset A$). Observe that $Q^{**} \subset J^*: \mathfrak{H} \to \mathfrak{H}_A$. In this section the procedure used above is inverted by introducing a Hilbert space $\mathfrak{H}'$ and two closed mappings $L_j: \mathfrak{H} \to \mathfrak{H}'$ with $L_1 \subset L_2$ and by letting $A = L_2^* L_1$. The considerations below are strongly motivated by [9], [25], [26].

Let L_1 and L_2 be densely defined, closed operators from a Hilbert space $\mathfrak{H}$ into another Hilbert space $\mathfrak{H}'$, and assume that $L_1 \subset L_2$. Define the forms $\mathfrak{t}_1$ and $\mathfrak{t}_2$ by

$$\mathfrak{t}_j[f,g] = (L_j f, L_j g), \quad f,g \in \operatorname{dom} L_j, \quad j=1,2.$$

The forms $\mathfrak{t}_1$ and $\mathfrak{t}_2$ are densely defined, closed and nonnegative in $\mathfrak{H}$ with $\operatorname{dom}\mathfrak{t}_j = \operatorname{dom} L_j$. They satisfy $\mathfrak{t}_1 \subset \mathfrak{t}_2$, so that $\mathfrak{t}_1 \geq \mathfrak{t}_2$ and hence $H_1 \geq H_2$, where $H_j = L_j^* L_j$, $j=1,2$, are the corresponding nonnegative selfadjoint operators. The operator $A = H_1 \cap H_2$ is closed and nonnegative in $\mathfrak{H}$. However, A need not be densely defined. For instance, if L is a closed, densely defined symmetric operator in $\mathfrak{H}$ and $L_1 = L$, $L_2 = L^*$, then

$$A = L^*L \cap LL^* = L^2.$$

In this case it may happen that $\operatorname{dom} A = \operatorname{dom} L^2 = \{0\}$, cf. [29]. If $A = H_1 \cap H_2$ is densely defined, its Friedrichs extension A_F is an operator which coincides with $L_1^* L_1$ precisely when $\operatorname{dom} A$ is a core for $\mathfrak{t}_1$. The next result is closely connected with [9, Proposition 1.3].

THEOREM 9.1. *Let L_1 and L_2 be densely defined, closed operators from a Hilbert space $\mathfrak{H}$ into another Hilbert space $\mathfrak{H}'$, and assume that $L_1 \subset L_2$. Let P denote the orthogonal projection from $\mathfrak{H}'$ onto $\overline{\operatorname{ran}}\, L_1$. Then the closed nonnegative operator $A = L_1^* L_1 \cap L_2^* L_2$ is equal to*

$$A = L_2^* L_1.$$

Assume that $\operatorname{dom} A$ *is a core for* $\mathfrak{t}_1$. *Then A is densely defined and, moreover:*

(i) *The Friedrichs extension of A is given by $A_F = L_1^* L_1$ and*

$$A_F[f] = \langle Q^{**}f, Q^{**}f\rangle = \|L_1 f\|^2, \quad f \in \operatorname{dom} A_F^{1/2} = \operatorname{dom} Q^{**} = \operatorname{dom} L_1. \tag{9.1}$$

(ii) *The operator $\widetilde{A} = L_2^* \overline{PL_2}$ is an extremal extension of A, such that*

$$\widetilde{A} \cap A_F = A, \tag{9.2}$$

and

$$\widetilde{A}[f] = \langle J^*f, J^*f\rangle = \|\overline{PL_2} f\|^2, \quad f \in \operatorname{dom} \widetilde{A}^{1/2} = \operatorname{dom} \overline{PL_2} \subset \operatorname{dom} J^*, \tag{9.3}$$

where $\overline{PL_2}$ denotes the closure of PL_2.

(iii) *The following statements are equivalent:*

(a) $\operatorname{dom} A^* \subset \operatorname{dom} \overline{PL_2}$;

(b) *$\widetilde{A}$ and A_F are transversal;*

(c) *A_N and A_F are transversal.*

If one of the conditions in (iii) *is satisfied, then $\widetilde{A}$ coincides with the Kreĭn–von Neumann extension A_N of A.*

Proof. It is easy to check that $L_1^*L_1 \cap L_2^*L_2 = L_2^*L_1$, which gives the first statement. Now assume that $\operatorname{dom} A$ is a core for $\mathfrak{t}_1$. Then clearly A is densely defined. The operator $L_1^*L_1$ associated to $\mathfrak{t}_1$ gives the Friedrichs extension A_F of A and (9.1) follows from Proposition 3.1 and the definition of Q in (3.2). This proves (i).

To prove (ii) observe that for all $\varphi \in \operatorname{dom} A$,

$$\langle J^*f, A\varphi\rangle = (f, J(A\varphi)) = (f, A\varphi) = (f, L_2^*L_1\varphi), \quad f \in \operatorname{dom} J^*,$$

and $\langle A\varphi, A\varphi\rangle = (A\varphi, \varphi) = \|L_1\varphi\|^2$. Since $\operatorname{dom} A$ is a core for $\mathfrak{t}_1$, the graph of the restriction of L_1 to $\operatorname{dom} A$ is dense in the graph of L_1. It follows that for every $f \in \operatorname{dom} L_2$

$$\|PL_2f\|^2 = \sup_{\varphi\in\operatorname{dom} L_1} \frac{|(PL_2f, L_1\varphi)|^2}{\|L_1\varphi\|^2} = \sup_{\varphi\in\operatorname{dom} A} \frac{|(f, L_2^*L_1\varphi)|^2}{\|L_1\varphi\|^2} < \infty, \tag{9.4}$$

so that $f \in \operatorname{dom} J^*$ and $\|PL_2f\|^2 = \|J^*f\|_A^2$. Hence, $\operatorname{dom} L_2 \subset \operatorname{dom} J^*$ and Proposition 3.1 shows that

$$(PL_2f, PL_2g) = \langle J^*f, J^*g\rangle = A_N[f,g], \quad f, g \in \operatorname{dom} L_2.$$

The form (PL_2f, PL_2g) is closable. Its closure is given by (9.3) and it belongs to $A_N[\cdot,\cdot]$. According to Theorem 4.4 $\widetilde{A}$ is an extremal extension of A. Observe that if $f \in \operatorname{dom} A_F \cap \operatorname{dom} \widetilde{A}$ then $\overline{PL_2}f = L_1f$ and $\widetilde{A}f = Af$. Therefore, $\operatorname{dom} A_F \cap \operatorname{dom} \widetilde{A} \subset \operatorname{dom} A$ and (9.2) follows.

To prove (iii) it is first shown that (b) and (c) are equivalent. That (b) implies (c) follows from the fact that the existence of transversal nonnegative extensions implies the transversality of the corresponding Friedrichs and Krein–von Neumann extension, cf. e.g. [5, Proposition 2]. Conversely, if A_F and A_N are transversal then (9.2) together with Theorem 6.4 forces that $\widetilde{A} = A_N$, so that (b) is automatically satisfied. Now assume that (a) holds. Then also the inclusion $\operatorname{dom} A^* \subset \operatorname{dom} A_N^{1/2}$ holds and this is equivalent to the transversality of A_F and A_N. Hence (a) implies (c). Moreover, if (c) holds then $A_N = \widetilde{A}$ and

$$\operatorname{dom} A^* \subset \operatorname{dom} A_N^{1/2} = \operatorname{dom} \widetilde{A}^{1/2} = \operatorname{dom} \overline{PL_2},$$

so that (a) follows. This completes the proof. □

It is clear from (9.3) that the equality $\widetilde{A} = A_N$ holds if and only if $\operatorname{dom} A_N \subset \operatorname{dom} \overline{PL_2}$. In this case, A_F and $A_N = \widetilde{A}$ need not be transversal. If $\widetilde{A} = A_N$, Theorems 4.4 and 9.1 imply that $\mathbf{E}(A)$ can be described as the set of all nonnegative selfadjoint operators of the form $\widetilde{A} = \widetilde{L}^*\widetilde{L}$, where $\widetilde{L}$ is a closed operator from $\mathfrak{H}$ into $\mathfrak{H}'$, such that $L_1 \subset \widetilde{L} \subset \overline{PL_2}$.

Observe, that the following restriction of L_2 is closed:

$$L_2' = \{ (f, L_2 f) : f \in \operatorname{dom} L_2, \quad L_2 f \in \operatorname{ran} P = \overline{\operatorname{ran}} \, L_1 \}. \tag{9.5}$$

Thus if, for instance, L_2/L_2' is finite-dimensional, then PL_2 is closed and the expressions in Theorem 9.1 simplify accordingly.

Consider the Hilbert space $\mathfrak{H}' = \mathfrak{H}_A$ as constructed in Section 3. Let the operators L_1 and L_2 defined by

$$L_1 = Q^{**}, \quad L_2 = J^*.$$

Clearly, $L_1 \subset L_2$. Since $\operatorname{ran} Q \,(= \operatorname{ran} A)$ is dense in $\mathfrak{H}_A$, one has $P = I$ and $PL_2 = L_2$ is closed, so that $L_1^* L_1 = Q^* Q^{**}$ and $L_2^* P L_2^* = J^{**} J^*$ are the Friedrichs and the Kreĭn–von Neumann extensions of

$$H := L_2^* L_1 = J^{**} Q^{**} = Q^* Q^{**} \cap J^{**} J^* = A_F \cap A_N \ (\supset A).$$

It is clear from (9.3) that $\mathbf{E}(H) = \mathbf{E}(A)$.

The situation appearing in the next proposition has been considered by S. Kuzhel in [25], [26], where the underlying extremal extensions were called unperturbed extensions of A.

PROPOSITION 9.2. *Let L be a simple, maximal, symmetric operator in $\mathfrak{H}$ with defect numbers $(n, 0)$ or $(0, n)$, $n \in \mathbb{N} \cup \{\infty\}$. Then $A = L^2$ is a closed, nonnegative, densely defined, symmetric operator with defect numbers (n, n). The Friedrichs extension of A is given by $A_F = L^* L$ and*

$$A_F[f] = \langle Q^{**} f, Q^{**} f \rangle = \|Lf\|^2, \quad f \in \operatorname{dom} A_F^{1/2} = \operatorname{dom} Q^{**} = \operatorname{dom} L.$$

The Kreĭn–von Neumann extension of A is given by $A_N = LL^*$ and

$$A_N[f] = \langle J^* f, J^* f \rangle = \|L^* f\|^2, \quad f \in \operatorname{dom} A_N^{1/2} = \operatorname{dom} J^* = \operatorname{dom} L^*.$$

Proof. Clearly $\ker L = \{0\}$ and $\overline{\operatorname{ran}} L = \mathfrak{H}$. Moreover, $\operatorname{dom} L^2$ is a core for $\mathfrak{t}_1[\cdot,\cdot] = (L\cdot, L\cdot)$. To see this assume that $(Lf, Lh) + (f,h) = 0$ for some $f \in \operatorname{dom} L$ and for every $h \in \operatorname{dom} L^2$. Then $(f, (L^2 + I)h) = 0$ and since

$$L^2 + I = (L - iI)(L + iI) = (L + iI)(L - iI),$$

the assumption about the defect numbers implies that $f \in (\operatorname{ran}(L \mp iI))^\perp = \ker(L^* \pm iI)$, so that $Lf = \pm if$; a contradiction with the symmetry of L. Since $L \subset L^*$, the first statement follows from Theorem 9.1 by taking $L_1 = L$ and $L_2 = L^*$.

According to [26, Lemma 1.1], the adjoint A^* of A is given by $(L^2)^* = (L^*)^2$. Therefore, $\operatorname{dom} A^* \subset \operatorname{dom} L^* = \operatorname{dom} L_2$. Since $\overline{\operatorname{ran}} L = \mathfrak{H}$, $PL_2 = L_2 = L^*$ is closed and the statement concerning A_N follows now from Theorem 9.1.

□

Theorem 4.4 implies that there is a one-to-one correspondence between all extremal extensions $\widetilde{A} \in \mathbf{E}(A)$ of A and all closed intermediate extensions $\widetilde{L}$, $L \subset \widetilde{L} \subset L^*$, given by $\widetilde{A} = \widetilde{L}^* \widetilde{L}$. The following result from [9] has a similar nature.

PROPOSITION 9.3. *Let L_1, L_2 be closed, densely defined operators acting from a Hilbert space $\mathfrak{H}$ into a Hilbert space $\mathfrak{H}'$. Assume that*

$$L_1 \subset L_2 \quad \text{and} \quad n = \dim\Big(\operatorname{dom} L_2/\operatorname{dom} L_1\Big) < \infty. \tag{9.6}$$

Let P be the orthogonal projection from $\mathfrak{H}'$ onto $\overline{\operatorname{ran}} L_1$. Then $A = L_2^* L_1$ is a closed, densely defined nonnegative operator with defect numbers (n,n). The Friedrichs extension of A is given by $A_F = L_1^* L_1$ and

$$A_F[f] = \langle Q^{**}f, Q^{**}f\rangle = \|L_1 f\|^2, \quad f \in \operatorname{dom} A_F^{1/2} = \operatorname{dom} Q^{**} = \operatorname{dom} L_1.$$

The Kreĭn–von Neumann extension of A is given by $A_N = L_2^* P L_2$ and

$$A_N[f] = \langle J^* f, J^* f\rangle = \|PL_2 f\|^2, \quad f \in \operatorname{dom} A_N^{1/2} = \operatorname{dom} J^* = \operatorname{dom} L_2.$$

Proof. First it is shown that A has finite defect numbers (n,n) and that $\operatorname{dom} A$ is a core for $\mathfrak{t}_1$. Since $A \geq 0$ the defect numbers of A are equal, say (k,k), and $H_j = L_j^* L_j$, $j = 1,2$, is a nonnegative selfadjoint extension of A. Moreover, $A = H_1 \cap H_2 = L_2^* L_1$ is at most an n-dimensional restriction of $L_2^* L_2$, so that $k \leq n < \infty$. Hence, H_1 and H_2 are transversal extensions of A and since $\operatorname{dom} H_1^{1/2} = \operatorname{dom} L_1$ and $\operatorname{dom} H_2^{1/2} = \operatorname{dom} L_2$, the assumption (9.6) implies that $k \geq n$. Therefore,

$k = n$ and a dimension argument shows that $\operatorname{dom} A$ is dense in $\operatorname{dom} L_1$ with respect to the graph norm of L_1. In view of (9.5) the operator PL_2 is closed. It follows from (9.2) and the finiteness of the defect numbers that $\widetilde{A}$ and A_F are transversal. Hence, the remaining statements are obtained from Theorem 9.1.

□

In view of Theorem 4.4, all nonnegative selfadjoint extremal extensions of $A = L_2^* L_1$ have the form $\widetilde{A} = \widetilde{L}^* \widetilde{L}$, where $\widetilde{L}$ is a closed operator from $\mathfrak{H}$ into $\mathfrak{H}'$, such that $L_1 \subset \widetilde{L} \subset PL_2$.

10. Some applications

The results in the previous sections will be used to consider some concrete examples of second order differential operators: in particular, their extremal extensions are characterized by means of boundary conditions and the associated Titchmarsh–Weyl coefficients.

10.1. The differential operator $-D^2$ on $[0,1]$. Let $\mathfrak{H} = L_2[0,1]$ and denote the usual inner product by $(\cdot,\cdot)$. Let

$$W_2^1[0,1], \quad W_2^2[0,1], \quad \overset{0}{W}{}_2^1[0,1], \quad \overset{0}{W}{}_2^2[0,1]$$

be the standard Sobolev spaces. The second order differential operator

$$A = -\frac{d^2}{dx^2}, \quad \operatorname{dom} A = \overset{0}{W}{}_2^2[0,1],$$

is closed and positive definite in $\mathfrak{H}$; it has defect numbers $(2,2)$ and the lower bound $\mu(A) = \pi^2$. The adjoint of A is given by

$$A^* = -\frac{d^2}{dx^2}, \quad \operatorname{dom} A^* = W_2^2[0,1],$$

so that

$$\ker A^* = \operatorname{span}\{\,1, x\,\}.$$

The operators

$$L_1 = i\frac{d}{dx}, \quad \operatorname{dom} L_1 = \overset{0}{W}{}_2^1[0,1] \quad \text{and} \quad L_2 = i\frac{d}{dx}, \quad \operatorname{dom} L_1 = W_2^1[0,1],$$

satisfy the assumptions of Proposition 9.3 with $n = 2$. Clearly $L_2 = L_1^*$, so that $A = L_1^2$. The Friedrichs extension of A is given by

$$A_F = L_1^* L_1 = -\frac{d^2}{dx^2}, \quad \operatorname{dom} A_F = \overset{0}{W}{}_2^1[0,1] \cap W_2^2[0,1],$$

and

$$A_F[u,v] = \int_0^1 u'(x)\overline{v'(x)}\,dx = (L_1 u, L_1 v), \quad u,v \in \operatorname{dom} A_F^{1/2} = \overset{0}{W}{}_2^1[0,1].$$

Let P be the orthogonal projection from $\mathfrak{H}$ onto $\overline{\operatorname{ran}}\, L_1$, $\ker P = \ker L_2 = \operatorname{span}\{\mathbf{1}\}$. Clearly, $(I-P)u' = (u(1)-u(0))\mathbf{1}$ so that $A_N = L_2^* P L_2$ is the restriction of A^* to

$$\operatorname{dom} A_N = \{\, u \in W_2^2[0,1] : u'(0) = u'(1) = u(1) - u(0) \,\}.$$

According to (3.7)

$$\operatorname{dom} A_N^{1/2} = \operatorname{dom} A_F^{1/2} + \ker A^* = W_2^1[0,1]$$

and $\mathcal{P}_F u(x) = u(x) - \big(u(1)-u(0)\big)x - u(0)$ for all $u(x) \in W_2^1[0,1]$. Hence,

$$A_N[u,v] = \int_0^1 u'(x)\overline{v'(x)}\,dx - \big(u(1)-u(0)\big)\big(\overline{v(1)} - \overline{v(0)}\big), \quad u,v \in W_2^1[0,1]$$

by (3.8). Let

$$\Gamma_0 u = \begin{pmatrix} u(0) \\ u(1) \end{pmatrix}, \ \Gamma_1 u = \begin{pmatrix} u'(0) - u(1) + u(0) \\ -u'(1) + u(1) - u(0) \end{pmatrix}, \quad u,v \in W_2^2[0,1].$$

It is straightforward to check that the triplet $\{\mathbb{C}^2, \Gamma_0, \Gamma_1\}$ forms a basic boundary triplet for A^*. By Proposition 7.1 all extremal extensions $\widetilde{A} \in \mathbf{E}(A)$ (apart from A_F and A_N) can be parametrized by the one-dimensional subspaces spanned by the vectors $c = (c_1, c_2)^\top \in \mathbb{C}^2$, $|c| = 1$, $c_1 \in \mathbb{R}$, as restrictions of A^* via the boundary conditions

$$\begin{aligned} &c_2 u(0) = c_1 u(1), \\ &c_1\big(u'(0) - u(1) + u(0)\big) = \bar{c}_2\big(u'(1) - u(1) + u(0)\big). \end{aligned} \tag{10.1}$$

The forms determined by the operators Q^{**} and J^* in Section 3 can be expressed as follows

$$\langle Q^{**}u, Q^{**}u\rangle = \|L_1 u\|^2 = \int_0^1 |u'(x)|^2\,dx, \quad u \in \overset{0}{W}{}_2^1[0,1],$$

$$\langle J^* u, J^* u\rangle = \|L_1 \mathcal{P}_F u\|^2 = \int_0^1 |u'(x)|^2\,dx - |u(1)-u(0)|^2, \quad u \in W_2^1[0,1].$$

The extremal extensions $\widetilde{A}$ in (10.1) are given by

$$\operatorname{dom} \widetilde{A}^{1/2} = \overset{0}{W}{}_2^1[0,1] + \left\{ b((c_2 - c_1)x + c_1) : b \in \mathbb{C} \right\}.$$

By Corollary 6.5 $\mu(\widetilde{A}) = 0$ for all $\widetilde{A} \in \mathbf{E}(A)\backslash\{A_F\}$, which is also easy to check directly from the formulas above.

10.2. A one-dimensional Schrödinger operator on the halfline. Consider the second order differential operator A in $L_2(0,\infty;\mathcal{H})$, generated by the differential expression $l(y) = -y'' + q(x)y$ and the boundary conditions $y'(0) = y(0) = 0$, with an $n \times n$-matrix potential $q(x)$ such that the minimal operator A is nonnegative and in the limit point case at ∞. The aim is to find the Weyl functions $\widetilde{M}(\lambda)$, $M_K(\lambda)$, and $M_S(\lambda)$ induced by the extremal extensions $\widetilde{A} \in \mathbf{E}(A)$. Let $\Pi = \{\mathbb{C}^n, \Gamma_0, \Gamma_1\}$ be a boundary triplet for A^* determined by $\Gamma_0 y = y(0)$ and $\Gamma_1 y = y'(0)$. Then $\ker \Gamma_0$ corresponds to the Friedrichs extension A_F of A. By using the Weyl function $M(\lambda)$ of Π the transversality of A_N and A_F can be checked by using

$$\ker \Gamma_1 = \operatorname{dom} A_N \quad \text{if and only if} \quad M(0) = 0. \tag{10.2}$$

To describe the extremal extensions of A let P be an orthogonal projection in $\mathbb{C}^n$ associated to $\widetilde{A} \in \mathbf{E}(A)$, cf. (7.2). Decompose $\mathbb{C}^n = \mathfrak{H}_1 \oplus \mathfrak{H}_2$ and $M(\lambda) = (M_{ij}(\lambda))_{i,j=1}^2$ according to $\mathcal{H}_1 = \operatorname{ran} P$, $\mathcal{H}_2 = \ker P$.

Let (10.2) be satisfied. Then Proposition 7.1 yields the following description for the class $\mathbf{E}(A)$ of the extremal extensions of A:

$$\operatorname{dom} \widetilde{A} = \{\, f \in W_2^2(0,\infty) : \, f_2(0) = 0, \quad f_1'(0) = 0 \,\},$$

where $f_1 = Pf$ and $f_2 = (I - P)f$. By Proposition 7.2 the symmetric extension $K \geq 0$ in (7.3) is determined by

$$\operatorname{dom} K = \{\, f \in W_2^2(0,\infty) : \, f(0) = 0, \quad f_1'(0) = 0 \,\}$$

and the symmetric extension $S \geq 0$ in (7.3) is determined by

$$\operatorname{dom} S = \{\, f \in W_2^2(0,\infty) : \, f_2(0) = 0, \quad f'(0) = 0\}.$$

The Weyl function $M_K(\lambda)$ for K is the compression of $M(\lambda)$ to $\mathcal{H}_1$ in (8.3) and the Weyl function $M_S(\lambda)$ for S is the Schur complement of $M(\lambda)$ in (8.4). In the particular case of the zero potential $q(x) \equiv 0$, the Weyl function associated to Π is of the form $M(\lambda) = i\sqrt{\lambda} I_{\mathcal{H}}$. The Weyl function for $\widetilde{A}$ in (7.2) takes the form

$$\widetilde{M}(\lambda) = \frac{i}{\sqrt{\lambda}} I_{\mathcal{H}_1} \oplus i\sqrt{\lambda} I_{\mathcal{H}_2},$$

and the Weyl functions associated to K and S are

$$M_K(\lambda) = i\sqrt{\lambda} I_{\mathcal{H}_1}, \quad M_S(\lambda) = i\sqrt{\lambda} I_{\mathcal{H}_2}.$$

Clearly, $M(\lambda)$, $M_K(\lambda)$ and $M_S(\lambda)$ all are Q_μ-functions, while $\widetilde{M}(\lambda)$ is an orthogonal sum of a Q_M-function in $\mathcal{H}_1$ and a Q_μ-function in $\mathcal{H}_2$.

10.3. A one-dimensional Schrödinger operator with point-interactions. In the present example one point-interactions on the line will be treated, cf. [2], [3]. In [9] all m-sectorial Schrödinger operators with vertex at the origin corresponding to δ and δ' one point-interactions on the line were parametrized. Here a description of all nonnegative extremal quantum Hamiltonians corresponding to δ-, δ'-interactions is given and their unitary equivalence is treated.

Let $\mathfrak{H} = L_2(\mathbb{R})$ and let $y \in \mathbb{R}$ be fixed. In addition to $W_2^1(\mathbb{R})$, $W_2^2(\mathbb{R})$ the following Sobolev spaces are needed:

$$\begin{aligned} &W_2^k(-\infty, y), \quad W_2^k(y, \infty), \quad \overset{0}{W}{}_2^k(-\infty, y), \quad \overset{0}{W}{}_2^k(y, \infty), \\ &W_2^k(\mathbb{R} \setminus \{y\}) = W_2^k(-\infty, y) \oplus W_2^k(y, \infty), \\ &\overset{0}{W}{}_2^k(\mathbb{R} \setminus \{y\}) = \overset{0}{W}{}_2^k(-\infty, y) \oplus \overset{0}{W}{}_2^k(y, \infty), \quad k = 1, 2. \end{aligned}$$

Consider in $\mathfrak{H}$ the following restrictions L_n, $n = 0, 1, 2$, of the operator $i\frac{d}{dx}$:

$$\operatorname{dom} L_0 = \overset{0}{W}{}_2^1(\mathbb{R} \setminus \{y\}), \quad \operatorname{dom} L_1 = W_2^1(\mathbb{R}), \quad \operatorname{dom} L_2 = W_2^1(\mathbb{R} \setminus \{y\}).$$

Then L_0 is a symmetric operator with defect numbers $(1, 1)$ and it coincides with orthogonal sum of two simple, symmetric operators with defect numbers $(0, 1)$ and $(1, 0)$, L_1 is a selfadjoint extension of L_0, and $L_2 = L_0^*$. In addition $\overline{\operatorname{ran}}\, L_0 = \mathfrak{H}$. Let

$$A = L_0^2 = -\frac{d^2}{dx^2}, \quad \operatorname{dom} A = \overset{0}{W}{}_2^2(\mathbb{R} \setminus \{y\}).$$

Then A is simple, and by applying Proposition 9.3 to the pair $L_0 \subset L_0^*$ it is seen that A is a nonnegative symmetric operator with defect numbers $(2, 2)$. Moreover, the Friedrichs extension of A is an operator given by

$$A_F = L_0^* L_0 = -\frac{d^2}{dx^2}, \quad \operatorname{dom} A_F = \left\{ u(x) \in \overset{0}{W}{}_2^1(\mathbb{R} \setminus \{y\}) : u'(x) \in W_2^1(\mathbb{R} \setminus \{y\}) \right\},$$

and the Kreĭn–von Neumann extension of A is given by

$$A_N = L_0 L_0^* = -\frac{d^2}{dx^2}, \quad \operatorname{dom} A_N = \left\{ u(x) \in W_2^1(\mathbb{R} \setminus \{y\}) : u'(x) \in \overset{0}{W}{}_2^1(\mathbb{R} \setminus \{y\}) \right\}.$$

All nonnegative selfadjoint extremal extensions of A (apart from A_F and A_N) take the form $\widetilde{A} = \widetilde{L}^*\widetilde{L}$, where $\widetilde{L}$ is a closed intermediate extension of L_0 with $L_0 \subset \widetilde{L} \subset L_0^*$. Clearly, every intermediate extension $\widetilde{L}$ of L_0 can be described as follows:

$$\operatorname{dom} \widetilde{L} = \left\{ u(x) \in W_2^1(\mathbb{R} \setminus \{y\}) : zu(y+) = u(y-) \right\}, \quad \widetilde{L} = i\frac{d}{dx},$$

where $z \in \mathbb{C} \cup \{\infty\}$. The adjoint operator has the form

$$\operatorname{dom} \widetilde{L}^* = \left\{ u(x) \in W_2^1(\mathbb{R} \setminus \{y\}) : u(y+) = \bar{z}u(y-) \right\}, \quad \widetilde{L}^* = i\frac{d}{dx}.$$

The extension $\widetilde{L}$ is selfadjoint if and only if $|z| = 1$. A description of the operators $\widetilde{A}_z \in \mathbf{E}(A)\backslash\{A_F, A_N\}$, $z \in \mathbb{C} \cup \{\infty\}$, is obtained via the restrictions of A^* to the domains

$$\operatorname{dom} \widetilde{A}_z = \left\{ u(x) \in W_2^2(\mathbb{R} \setminus \{y\}) : zu(y+) = u(y-),\ u'(y+) = \bar{z}u'(y-) \right\}. \tag{10.3}$$

For the associated closed forms one obtains

$$\operatorname{dom} [\widetilde{A}_z] = \left\{ u(x) \in W_2^1(\mathbb{R} \setminus \{y\}) : zu(y+) = u(y-) \right\},$$

$$\widetilde{A}_z[u, v] = \int_{-\infty}^{+\infty} u'(x)\overline{v'(x)}\, dx.$$

In particular the value $z = 1$ corresponds to the extremal extension

$$\widetilde{A}_1 = L_1^2 = -\frac{d^2}{dx^2}, \quad \operatorname{dom} \widetilde{A}_1 = W_2^2(\mathbb{R}).$$

The above parametrization can be obtained also from Proposition 7.1 by introducing a boundary triplet for A^*. Define the mappings Γ_0, Γ_1 from $\operatorname{dom} A^*$ into $\mathbb{C}^2$ via

$$\Gamma_0 u = \begin{pmatrix} u(y+) \\ u(y-) \end{pmatrix}, \quad \Gamma_1 u = \begin{pmatrix} u'(y+) \\ -u'(y-) \end{pmatrix}, \quad u, v \in W_2^2(\mathbb{R} \setminus \{y\}).$$

Then $\{\mathbb{C}^2, \Gamma_0, \Gamma_1\}$ forms a basic boundary triplet for A^*. It is straightforward to check that the corresponding Weyl function has the form

$$M(\lambda) = \begin{pmatrix} i\sqrt{\lambda} & 0 \\ 0 & i\sqrt{\lambda} \end{pmatrix}.$$

By Corollary 8.2 all extremal extensions in the class $\mathbf{E}(A)\backslash\{A_F, A_N\}$ are mutually unitarily equivalent.

Let us show that also the operators A_F and A_N are unitarily equivalent to $\widetilde{A}_1$. The operator $\widetilde{A}_1$ is the Friedrichs extension of the symmetric operator S with defect numbers $(1,1)$, which is defined by

$$S := L_0 L_1 = -\frac{d^2}{dx^2}, \quad \operatorname{dom} S = \left\{ u(x) \in W_2^1(\mathbb{R}) : u'(x) \in \overset{0}{W}{}_2^1(\mathbb{R} \setminus \{y\}) \right\}.$$

Clearly, S is a simple nonnegative symmetric extension of A with adjoint

$$S^* = L_1 L_0^* = -\frac{d^2}{dx^2}, \quad \operatorname{dom} S^* = \left\{ u(x) \in W_2^2(\mathbb{R} \setminus \{y\}) : u'(y+) = u'(y-) \right\}.$$

Moreover, $S_N = L_0 L_0^* = A_N$. Define

$$\Gamma_0' u = \frac{u(y+) - u(y-)}{\sqrt{2}}, \quad \Gamma_1' u = \frac{u'(y+) + u'(y-)}{\sqrt{2}}, \quad u(x) \in \operatorname{dom} S^*.$$

By Proposition 7.2 the triplet $\{\mathbb{C}, \Gamma_0', \Gamma_1'\}$ is a basic boundary triplet for S^*. Proposition 8.1 shows that the corresponding Weyl function is

$$M_S(\lambda) = i\sqrt{\lambda}.$$

Next consider the simple nonnegative operator $K := L_1 L_0$. The Friedrichs and the Kreĭn–von Neumann extensions are given by

$$K_F = L_0^* L_0 = A_F, \quad K_N = L_1^2 = \widetilde{A}_1.$$

Again applying Propositions 7.2 and 8.1 one obtains the Weyl function $M_K(\lambda)$:

$$M_K(\lambda) = i\sqrt{\lambda}.$$

It follows that the pairs of operators $\{S, S_F\}$ and $\{K, K_F\}$ are unitarily equivalent. Therefore also $\widetilde{A}_1 = S_F$ and $A_F = K_F$ are unitarily equivalent. Similarly, $S_N = A_N$ and $K_N = \widetilde{A}_1$ are unitarily equivalent. Thus, all nonnegative extremal extensions of the operator A are mutually unitarily equivalent.

The relations (10.3) in the case $y = 0$ were obtained by S. Kuzhel in [25], [26]. He also established the fact that all nonnegative extremal (unperturbed) extensions of the square of a simple maximal symmetric operator are mutually unitarily equivalent. The proof of this result in [25], [26] is based on the existence of wave operators and C. Putnam's result in [31] about the unitary equivalence in $L_2(\mathbb{R}_+)$ of the following operators (see the previous subsection):

$$A_F = -\frac{d^2}{dx^2}, \quad \operatorname{dom} A_F = \left\{ u \in W_2^2(\mathbb{R}_+) : u(0) = 0 \right\},$$
$$A_N = -\frac{d^2}{dx^2}, \quad \operatorname{dom} A_N = \left\{ u \in W_2^2(\mathbb{R}_+) : u'(0) = 0 \right\},$$

which are the Friedrichs and Kreĭn–von Neumann extensions of

$$A = -\frac{d^2}{dx^2}, \quad \operatorname{dom} A = \overset{0}{W}{}_2^2(\mathbb{R}_+).$$

References

[1] N. I. Achieser and I. M. Glasmann, *Theorie der linearen Operatoren im Hilbertraum*, Akademie Verlag, Berlin, 1981.

[2] S. Albeverio, F. Gesztesy, R. Høegh-Krohn and H. Holden, *Solvable models in quantum mechanics*, Springer-Verlag, Berlin, 1988.

[3] S. Albeverio and P. Kurasov, *Singular perturbations of differential operators*, London Math. Soc. Lecture Note Series **271**, Cambridge University Press, Cambridge, 2000.

[4] T. Ando and K. Nishio, Positive selfadjoint extensions of positive symmetric operators, *Tohóku Math. J.*, **22** (1970), 65–75.

[5] Yu. M. Arlinskiĭ, Positive spaces of of boundary values and sectorial extensions of nonnegative symmetric operators, *Ukrainian Math. J.*, **40** (1988), 8–15.

[6] Yu. M. Arlinskiĭ, Maximal sectorial extensions and closed forms associated with them, *Ukrainian Math. J.*, **48** (1996), 723–739.

[7] Yu. M. Arlinskiĭ, Extremal extensions of sectorial linear relations, *Math. Studii*, **7** (1997), 81–96.

[8] Yu. M. Arlinskiĭ, On a class of nondensely defined contractions and their extensions, *J. Math. Sciences*, **79** (1999), 4391–4419.

[9] Yu. M. Arlinskiĭ, Abstract boundary conditions for maximal sectorial extensions of sectorial operators, *Math. Nachr.*, **209** (2000), 5–35.

[10] Yu. M. Arlinskiĭ and E. R. Tsekanovskiĭ, Quasi selfadjoint contractive extensions of Hermitian contractions, *Teor. Funkts., Funkts. Anal. Prilozhen*, **50** (1988), 9–16.

[11] E. A. Coddington and H. S. V. de Snoo, Positive selfadjoint extensions of positive symmetric subspaces, *Math. Z.*, **159** (1978), 203–214.

[12] V. A. Derkach, S. Hassi, M. M. Malamud and H. S. V. de Snoo, Generalized resolvents of symmetric operators and admissibility, *Methods of Functional Analysis and Topology*, **6** (2000), 24–55.

[13] V. A. Derkach and M. M. Malamud, Generalized resolvents and the boundary value problems for hermitian operators with gaps, *J. Funct. Anal.*, **95** (1991), 1–95.

[14] V. A. Derkach and M. M. Malamud, The extension theory of hermitian operators and the moment problem, *J. Math. Sciences*, **73** (1995), 141–242.

[15] V. I. Gorbachuk, M. L. Gorbachuk and A. N. Kochubeĭ, Extension theory of symmetric operators and boundary value problems, *Ukr. Mat. Z.*, **41** (1989), 1298–1313.

[16] S. Hassi, M. Kaltenbäck and H. S. V. de Snoo, Triplets of Hilbert spaces and Friedrichs extensions associated with the subclass $\mathbf{N_1}$ of Nevanlinna functions, *J. Operator Theory*, **37** (1997), 155–181.

[17] S. Hassi, M. Kaltenbäck and H. S. V. de Snoo, Generalized Kreĭn–von Neumann extensions and associated operator models, *Acta Sci. Math. (Szeged)*, **64** (1998), 627–655.

[18] I. S. Kac and M. G. Kreĭn, *R*-functions — analytic functions mapping the upper halfplane into itself, *Supplement* to the Russian edition of F. V. Atkinson, *Discrete and continuous boundary problems*, Mir, Moscow 1968 (Russian); English translation: *Amer. Math. Soc. Transl. Ser.* 2, **103** (1974), 1–18.

[19] T. Kato, *Perturbation theory for linear operators*, Springer-Verlag, Berlin, 1980.

[20] A. N. Kochubeĭ, On extensions of positive definite symmetric operators, *Dokl. Akad. Science Ukraine Ser A*, **3** (1979), 168–171.

[21] M. G. Kreĭn, Theory of selfadjoint extensions of semibounded operators and its applications I, II, *Mat. Sb.*, **20, 21** (1947), 431–495, 365–404.

[22] M. G. Kreĭn and I. E. Ovčarenko, On generalized resolvents and resolvent matrices of positive Hermitian operators, *Soviet Math. Dokl.*, **17** (1976), 1705–1709.

[23] M. G. Kreĭn and I. E. Ovčarenko, On *Q*-functions and SC-resolvents of nondensely defined Hermitian contractions, *Siberian Math. Zh.*, **18** (1977), 728–746.

[24] M. G. Kreĭn and I. E. Ovčarenko, Inverse problems for *Q*-functions and resolvent matrices of positive Hermitian operators, *Soviet Math. Dokl.*, **19** (1978), 1131–1134.

[25] S. Kuzhel, Abstract wave equation; definition and properties of solutions, *Preprint* 96.14, Institute of Mathematics of National Academy of Sciences of Ukraine, Kiev, 1996, 44 pp.

[26] S. Kuzhel, On some properties of unperturbed operators, *Methods of Functional Analysis and Topology*, **3** (1997), 82–87.

[27] V. E. Lyantse and O. G. Storozh, *Methods of the theory of unbounded operators*, Naukova Dumka, Kiev, 1983.

[28] M. M. Malamud, On some classes of Hermitian operators with gaps, *Ukr. Mat. Z.*, **44** (1992), 215–234.

[29] M. A. Naimark, On the square of a closed symmetric operator, *Doklady Akad. Nauk SSSR (N.S.)*, **26** (1940) 806–870; ibid., **28** (1940), 207–208.

[30] V. Prokaj and Z. Sebestyén, On Friedrichs extensions of operators, *Acta Sci. Math. (Szeged)*, **62** (1996), 243–246.

[31] C. Putnam, Continuous spectra and unitary equivalence, *Pacific J. Math.*, **7** (1957), 993–995.

[32] Z. Sebestyén and J. Stochel, Restrictions of positive self-adjoint operators, *Acta Sci. Math. (Szeged)*, **55** (1991), 149–154.

Yu. M. Arlinskiĭ, Department of Higher and Applied Mathematics, East Ukrainian University, Kvartal Molodyozhny, 20-A, 348034 Lugansk, Ukraine;
e-mail: yma@vugu.lugansk.ua

S. Hassi, Department of Mathematics and Statistics, University of Vaasa, P.O. Box 700, FIN-65101 Vaasa, Finland; *e-mail*: hassi@cc.helsinki.fi

Z. Sebestyén, Department of Applied Analysis, Eötvös Loránd University, Múzeum krt. 6–8, H-1088 Budapest, Hungary; *e-mail*: sebestyenzol@ludens.elte.hu

H. S. V. de Snoo, Department of Mathematics, University of Groningen, Postbus 800, 9700 AV Groningen, Nederland; *e-mail*: desnoo@math.rug.nl

Operator Theory: Advances and Applications, Vol. 127, 83–96

On Livšic–Brodskii nodes with strongly regular J-inner characteristic matrix functions in the Hardy class

ZOYA D. AROVA

The class of strongly regular J-inner matrix functions (m.f.'s) $U_{sR}(J)$ was introduced and investigated in the paper [ArD]. In this paper we investigate the Livšic–Brodskii nodes with strongly regular J-inner characteristic m.f.'s of Hardy class. In particular, we prove that the basic operator of such a node is similar to a dissipative operator.

1. Introduction

The notions of a Livšic–Brodskii node (LB node) and its characteristic function arose in the development of the spectral theory of nonselfadjoint operators in Hilbert space and they are significant in this theory [Br], [L]. An LB node corresponds to a conservative linear continuous time-invariant transmission (or scattering) system and the characteristic function of such a node is the transfer function of this system [L], [A].

We will have interest in the nodes and corresponding systems in which the exterior space is finite dimensional. In this case the characteristic function is defined by a m.f., which belongs to the Potapov class $P(J)$ of meromorphic J-contractive m.f.'s in the upper half plane $\mathbb{C}_+ := \{z : \operatorname{Im} z > 0\}$, where J is a signature matrix.

We consider LB nodes Σ with characteristic m.f.'s Θ_Σ from the class $U(J)(\subset P(J))$ of J-inner m.f.'s, i.e. m.f.'s for which the corresponding transmission conservative systems are lossless. Moreover, we will investigate the LB nodes with characteristic m.f.'s from the class $U_{sR}(J)$ of strongly regular J-inner m.f.'s. This

---

AMS Subject Classification (1991): 47A48, 93C45, 93B05, 93B07.

class was introduced and investigated in the paper [ArD]. We will consider the case when $\Theta = \Theta_\Sigma$ satisfies the condition

$$\frac{\Theta(z) - I}{z} \in H_2^{m\times m}(\mathbb{C}_+). \tag{1.1}$$

One can check that the condition (1.1) is equivalent to the following

$$\Theta \in H_2^{m\times m}\Big(\frac{d\mu}{1+\mu^2}\Big).$$

The last condition means that $\Theta(z)$ is a holomorphic $m \times m$ m.f. in $\mathbb{C}_+$ and

$$\sup_{\nu>0} \int_{-\infty}^{+\infty} \operatorname{tr}(\Theta^*(\mu + i\nu)\Theta(\mu + i\nu))\frac{d\mu}{1+\mu^2} < \infty.$$

This is equivalent to the following

$$(z+i)^{-1}\Theta(z) \in H_2^{m\times m}.$$

The investigation of an LB node Σ with characteristic function $\Theta = \Theta_\Sigma \in U_{sR}(J)$, for which the condition (1.1) is not satisfied will be presented in another paper. In Section 2 we recall some necessary definitions and facts about LB nodes and strongly regular J-inner m.f.. In Section 3 we give a full characterization of an LB node Σ for which the characteristic function $\Theta_\Sigma \in U_{sR}(J)$ satisfies the condition (1.1). In particular, we prove that the basic operator A of such a node is similar to a dissipative operator and that the evolution semigroup $T(t) = e^{iAt}$ is bistable. In Section 4 we obtain a series of equivalent conditions for this class of nodes in terms of linear time-invariant systems theory.

2. Preliminaries

2.1. Strongly regular J-inner matrix functions. Let J be an $m \times m$ signature matrix with complex entries, i.e.,

$$J^* = J^{-1} = J \qquad \text{and} \qquad J \neq \pm I.$$

An $m \times m$ m. f. $\Theta(z)$ that is meromorphic inside the upper half plane $\mathbb{C}_+$ is called J-contractive if

$$\Theta^*(z)J\Theta(z) \leq J$$

at points in $\mathbb{C}_+$ at which Θ is holomorphic. At such points we also have

$$\Theta(z)J\Theta^*(z) \leq J.$$

We denote the class of J-contractive m.f.'s by $\mathcal{P}(J)$. It is known (see [Dy]) that a m.f. from $\mathcal{P}(J)$ has boundary values a.e., i.e.,

$$\Theta(\mu) = \lim_{\nu \downarrow 0} \Theta(\mu + i\nu), \qquad \text{a.e.} \quad \mu \in \mathbb{R}.$$

If these boundary values are J-unitary a.e., i.e., if

$$\Theta^*(\mu)J\Theta(\mu) = J, \qquad \text{a.e.} \quad \mu \in \mathbb{R},$$

then $\Theta(z)$ is called J-inner.

We denote the class of J-inner m.f.'s by $U(J)$.

For $\Theta \in U(J)$ we denote the Reproducing Kernel Hilbert Space (RKHS) with the Reproducing Kernel (RK)

$$K_\Theta(z, \zeta) := \frac{J - \Theta(z)J\Theta^*(\zeta)}{-2\pi i(z - \overline{\zeta})} \qquad (z, \zeta \in \mathbb{C}_+)$$

by $\mathcal{H}(\Theta)$. This is a Hilbert space of $\mathbb{C}^m$-valued functions $h(z)$ meromorphic in $\mathbb{C}_+$ such that the linear manifold $\mathcal{L}$ consisting of functions

$$h(z) = \sum_{k=1}^{N} \frac{J - \Theta(z)J\Theta^*(\zeta_k)}{-2\pi i(z - \overline{\zeta_k})} \xi_k, \tag{2.1}$$

where $N \geq 1$, $\xi_k \in \mathbb{C}^n$ and ζ_k are points in $\mathbb{C}_+$ at which $\Theta(z)$ is holomorphic, is dense in the space $\mathcal{H}(\Theta)$, and for $h \in \mathcal{L}$

$$\|h\|^2_{\mathcal{H}(\Theta)} = \sum_{i=1}^{N} \sum_{k=1}^{N} \xi_i^* K_\Theta(z_i, z_k)\xi_k \tag{2.2}$$

holds.

A J-inner matrix function $\Theta(z)$ is a strongly regular matrix function ($\Theta \in U_{sR}(J)$) iff for all vectors $x \in \mathcal{H}(\Theta)$ the boundary values $x(\mu) \in L_2^m$, i.e. iff $\mathcal{H}(\Theta) \subset L_2^m$, see [ArD], Theorem 6.7. From this we have that for $\Theta \in U(J)$

$$\Theta \in U_{sr}(J) \bigcap H_2^{m\times m}\Big(\frac{d\mu}{1+\mu^2}\Big) \Longleftrightarrow \mathcal{H}(\Theta) \subset H_2^m. \tag{2.3}$$

We will investigate LB nodes with $\Theta_\Sigma = \Theta \in U(J)$ for which

$$\mathcal{H}(\Theta) \subset H_2^m. \tag{2.4}$$

2.2. Livšic–Brodskii nodes. We denote the space of linear bounded operators acting from a Hilbert space X_1 into a Hilbert space X_2 by $[X_1, X_2]$, and let $[X] := [X, X]$.

The colligation $\Sigma = [A, F; X, Y; J]$, where X and Y are separable Hilbert spaces, $A \in [X]$, $F \in [X, Y]$, $J = J^* = J^{-1} \in [Y]$ and

$$A - A^* = 2iF^*JF,$$

is called a Livsic–Brodskii node. The operator function

$$\Theta_\Sigma = I + 2izF(I - zA)^{-1}F^*J$$

is called the characteristic function of this node Σ. If $J = I$, the node Σ is called dissipative. The node Σ is called simple if $\bigcap_{n\geq 0} \operatorname{Ker}(FA^n) = \{0\}$.

Let $\Sigma = [A, F; X, Y; J]$ be a given LB node. Then such a node Σ corresponds to a continuous time-invariant transmission system

$$\begin{cases} \dfrac{dx}{dt} = iAx(t) - \sqrt{2}F^*Ju(t) \\ y(t) = \sqrt{2}Fx(t) + u(t) \end{cases} \tag{2.5}$$

with conservativeness condition

$$(Ju, u) - (Jy, y) = \frac{d}{dt}\|x\|^2.$$

The function $\Theta_\Sigma(-\frac{1}{z})$ is the transfer function of the system (2.5) which corresponds to the LB node Σ. A conservative transmission system Σ is lossless if its transfer function belongs to $U(J)$. A transmission system which corresponds to a simple LB node is minimal, i.e., it is controllable and observable:

$$X = X_\Sigma^c \qquad \text{and} \qquad X = X_\Sigma^0,$$

where

$$X_\Sigma^c = \bigvee_{k\geq 0} A^k F^* Y, \qquad X_\Sigma^0 = \bigvee_{k\geq 0} (A^*)^k F^* Y$$

($\bigvee_{k>0} \mathcal{L}_k$ is the closed linear span of the sets $\mathcal{L}_k \subset X, k \geq 0$).

Two LB nodes $\Sigma_i = [A_i, F_i; X_i, Y; J]$, $i = 1, 2$, are called unitarily equivalent if

$$A_2 = R^{-1}A_1R, \qquad F_2 = F_1R,$$

where $R \in [X_2, X_1]$ is a unitary map from X_2 onto X_1.

Let $P^0(J)$ (resp. $U^0(J)$) be subclass of m.f.'s $\Theta \in P(J)$ (resp. $U(J)$) such that:

1) $z = 0$ is a holomorphic point of $\Theta(z)$ and $\Theta(0) = I$,

2) $\Theta^*(\mu)J\Theta(\mu) = J$ for all $\mu \in (-\delta_\Theta, \delta_\Theta)$, where $\delta_\Theta > 0$.

Usually, in the theory of LB nodes, $\Theta_\Sigma(\frac{1}{z})$ is called the characteristic function of the node Σ. Keeping this difference in mind, we can formulate the known results about characteristic functions of LB nodes (see [Br], Ch. 1,Theorem 9.4) as follows:

THEOREM. 1) *Let $\Sigma = [A, F; X, \mathbb{C}^m; J]$ be an LB node, and Θ_Σ be the characteristic m.f. of this node. Then $\Theta_\Sigma \in P^0(J)$.*

2) *Let m.f. $\Theta \in P^0(J)$. Then there exists a simple LB node such that $\Theta_\Sigma(z) = \Theta(z)$, and this node is determined by $\Theta(z)$ up to unitary equivalence.*

3. The main criterion

Let $\Theta(z) \in U^0(J)$. Then $\Theta(z) \in H_2^{m\times m}(\frac{d\mu}{1+\mu^2})$ if and only if (1.1) holds. Now we give a full characterization of simple LB nodes Σ for which the characteristic function $\Theta(z) = \Theta_\Sigma(z)$ satisfies the condition:

$$\Theta(z) \in U_{sR}(J) \bigcap H_2^{m\times m}\Big(\frac{d\mu}{1+\mu^2}\Big). \tag{3.1}$$

In particular, we prove that the basic operator A of such a node Σ is similar to a dissipative operator A_0. The following statement is essential to the proof.

LEMMA 1. *Let X_0 be a subspace in the space H_2^m such that every $x(z) \in X_0$ is holomorphic at the point $z = 0$ and $z^{-1}(x(z) - x(0)) \in X_0$. Then*

1) *the operator A_0 in X_0 defined by formula*

$$(A_0x)(z) = z^{-1}(x(z) - x(0)), \quad x \in X_0, \tag{3.2}$$

is a bounded operator, i.e., $A_0 \in [X_0]$;

2) *the operator $F_0 : X_0 \longrightarrow \mathbb{C}^m$ defined by formula*

$$F_0x = x(0), \quad x \in X_0, \tag{3.3}$$

is a bounded operator, i.e., $F_0 \in [X_0, \mathbb{C}^m]$;

3)

$$A_0 - A_0^* = 2iF_0^* F_0, \tag{3.4}$$

i.e., the colligation $\Sigma_0 = [A_0, F_0; X_0, \mathbb{C}^m; I]$ *is a dissipative LB node;*

4) iA_0 *is the generator of the continuous semigroup* $T_0(t) = \exp(iA_0 t)$ *such that* $\text{s-lim}_{t\to+\infty} T_0(t) = 0$, $\text{s-lim}_{t\to+\infty} T_0^*(t) = 0$;

5) $X_0 = H_2(\mathbb{C}^m) \ominus B(z)H_2(\mathbb{C}^m)$, *where* $B(z)$ *is an* $m \times m$ *inner m.f., the characteristic function of the dissipative node* Σ_0, *i.e.,*

$$B(z) = I + 2izF_0(I - zA_0)^{-1}F_0^*.$$

The proof of this lemma is obtained by consideration of the semigroup

$$T(t) := P_{H_2^m} M_{e_t}^* | H_2^m, \quad t \geq 0,$$

where M_{e_t} is the semigroup of isometric operators of multiplication by $e_t(z) = e^{itz}$ in H_2^m. The generator B of the semigroup $T(t)$ is a closed antiaccretive operator ($\text{Re}(Bx, x) \leq 0$). Its domain of definition is dense in H_2^m. The operator B has the inverse operator B^{-1} with domain of definition

$$\mathcal{D}_{B^{-1}} = \left\{ \hat{x}(z) \in H_2^m : \ \frac{\hat{x}(z) - c_x}{-iz} \in H_2^m \text{ for some } \ c_x \in \mathbb{C}^m \right\},$$

and

$$B^{-1}\hat{x} = \frac{\hat{x}(z) - c_x}{-iz}, \ \hat{x} \in \mathcal{D}_{B^{-1}}.$$

If X_0 satisfies the conditions of the lemma, then

$$X_0 \subset \mathcal{D}_{B^{-1}} \quad \text{and} \quad A_0 = -iB^{-1}|X_0.$$

It is known that $T(t) \to 0$ $(t \to +\infty)$ in the strong sense. This implies that the semigroup with the generator B^{-1} has this property too. Consequently, the semigroup $T_0(t) = \exp(iA_0 t)$ has this property, as well. Because B^{-1} is a closed operator, its restriction A_0 onto X_0 is closed operator. Since X_0 is a closed subspace in H_2^m the Banach Theorem guarantees that A_0 is a bounded operator. One can check that the relation (3.4) holds for the operators F_0 and A_0 defined by formulas (3.2) and (3.3). This implies that $F_0 \in [X_0, \mathbb{C}^m]$. As $T_0(t) \to 0$ $(t \to \infty)$, the characteristic m.f. $B(z) = \Theta_{\Sigma_0}$ of the dissipative node Σ_0 is inner, i.e. its boundary values $B(\mu)$ are isometric a.e. on $\mathbb{R}$. Moreover, since $B(z)$ is an $m \times m$ m.f., this m.f. is biinner ($B(z) \in U(I)$). This implies that for the semigroup $T_0(t)$ we have the property $T_0^*(t) \to 0$ $(t \to +\infty)$ (see [SzNF]). The node Σ_0 is the functional model of a simple dissipative node with a biinner characteristic m.f. $B(z)$. Therefore, $X_0 = H(B) = H_2^m \ominus BH_2^m$.

LEMMA 2. *The following is true*:

$$H_1(z) := \frac{1}{z}H\Big(-\frac{1}{z}\Big) \in H_2^m \Longleftrightarrow H(z) \in H_2^m$$

and

$$\forall H(z) \in H_2^m \; : \quad \|H_1(\cdot)\|_{H_2^m} = \|H(\cdot)\|_{H_2^m}.$$

Proof. Let $H(z) \in H_2^m$. Then we have:

$$1)\; H(\mu) \in L_2^m(\mathbb{R}), \qquad \text{and} \qquad 2)\; H(z) = \frac{1}{2\pi i}\int_{-\infty}^{+\infty} \frac{H(\mu)}{\mu - z}\, d\mu, \qquad \text{Im} z > 0.$$

From 1) we get

$$\|H(\cdot)\|_{H_2^m} = \int_{-\infty}^{\infty} \operatorname{tr}(H^*(\mu)H(\mu))\, d\mu < \infty.$$

Let $\nu = -\frac{1}{\mu}$. Then we have

$$\|H(\cdot)\|_{H_2^m} = \int_{-\infty}^{+\infty} \operatorname{tr}\Big(\frac{1}{\nu}H^*\Big(-\frac{1}{\nu}\Big)\Big)\Big(\frac{1}{\nu}H\Big(-\frac{1}{\nu}\Big)\Big)\, d\nu < \infty,$$

and consequently

$$H_1(\nu) = \frac{1}{\nu}H\Big(-\frac{1}{\nu}\Big) \in L_2^m(\mathbb{R}) \;\; \text{and} \;\; \|H_1(\cdot)\|_{L_2^m} = \|H(\cdot)\|_{H_2^m}. \tag{3.5}$$

But

$$H\Big(-\frac{1}{z}\Big) = \frac{1}{2\pi i}\int_{-\infty}^{+\infty} \frac{H(\mu)}{\mu + 1/z}\, d\mu = \frac{z}{2\pi i}\int_{-\infty}^{+\infty} \frac{H(-1/\nu)}{\nu^2(z(-1/\nu)+1)}\, d\nu$$
$$= \frac{z}{2\pi i}\int_{-\infty}^{+\infty} \frac{H(-1/\nu)}{\nu}\frac{1}{(\nu - z)}\, d\nu.$$

Finally we have

$$H_1(z) = \frac{1}{2\pi i}\int_{-\infty}^{+\infty} \frac{H_1(\nu)}{(\nu - z)}\, d\nu. \tag{3.6}$$

From (3.5) and (3.6) we obtain that $H_1(z) \in H_2^m$. Conversely, if $H_1(z) \in H_2^m$ then $H(z) = \frac{1}{z}H_1(-1/z)$, and consequently, $H(z) \in H_2^m$.

□

LEMMA 3. *Let $\Sigma = [A_0, F_0; X_0, \mathbb{C}^m; I]$ be a dissipative node. Then*

$$H(z) = F_0(I - zA_0)^{-1}x_0 \in H_2^m.$$

Proof. Let

$$H_1(z) = \frac{1}{z}H\Big(-\frac{1}{z}\Big) = F_0(zI + A_0)^{-1}x_0.$$

We have that iA_0 is the generator of the semigroup $T_0(t) = e^{iA_0t}$. Then we have

$$F_0(zI + A_0)^{-1}x_0 = i\int_0^\infty e^{izt}F_0e^{iA_0t}\,dt, \qquad z \in \mathbb{C}_+.$$

We will prove that $F_0e^{iA_0t}x_0 \in L_2^m(\mathbb{R}_+)$ i.e.

$$\int_0^\infty (e^{-iA_0^*t}F_0^*F_0e^{iA_0t}x_0, x_0)\,dt < \infty. \tag{3.7}$$

Indeed,

$$\begin{aligned}
2\int_0^T &(e^{-iA_0^*t}F_0^*F_0e^{iA_0t}x_0, x_0)\,dt \\
&= -i\int_0^T (e^{-iA_0^*t}((A_0 - A_0^*))e^{iA_0t}x_0, x_0)\,dt \\
&= -i\int_0^T [(A_0e^{iA_0t}x_0, e^{iA_0t}x_0) - (e^{iA_0t}x_0, A_0e^{iA_0t}x_0)]\,dt \\
&= -\int_0^T \frac{d}{dt}\|e^{iA_0t}x_0\|^2dt = \|x_0\|^2 - \|e^{iA_0T}x_0\|^2.
\end{aligned}$$

From this we obtain

$$\int_0^T (e^{-iA_0^*t}F_0^*F_0e^{iA_0t}x_0, x_0)\,dt \le \|x_0\|^2,$$

and (3.7) holds. Consequently, $F_0e^{iA_0t}x_0 \in L_2^m(\mathbb{R}_+)$ and $H_1(z) \in H_2^m$. By Lemma 2 $H(z) \in H_2^m$, too.

□

THEOREM 1. *Let $\Sigma = [A, F; X, Y; J]$ be an LB node with $\Theta_\Sigma \in U(J)$.*

1) *Let Σ be a simple node and $\Theta_\Sigma \in U_{sR}(J) \cap H_2^{m\times m}(\frac{d\mu}{1+\mu^2})$. Then there exists a simple dissipative Livsic–Brodskii node $\Sigma_0 = [A_0, F_0; X_0, \mathbb{C}^m, I]$ such that*

$$A = R^{-1}A_0R, \qquad F = F_0R, \tag{3.8}$$

where $R \in [X, X_0]$, $R^{-1} \in [X_0, X]$.

2) *Let the relations (3.8) hold where $\Sigma_0 = [A_0, F_0; X_0, \mathbb{C}^m; I]$ is a dissipative node, and $R \in [X, X_0]$, $R^{-1} \in [X_0, X]$. Then $\Theta_\Sigma \in U_{sR}(J) \cap H_2^{m\times m}(\frac{d\mu}{1+\mu^2})$.*

Proof. 1) Let us have $\Theta = \Theta_\Sigma \in U_{sR}(J) \cap H_2^{m\times m}\ (\frac{d\mu}{1+\mu^2})$ for a simple LB node Σ. Then $\Theta \in U^0(J)$. We can consider the functional model of the simple LB node $\dot\Sigma = [\dot A, \dot F; \mathcal{H}(\Theta), \mathbb{C}^m; J]$ with the characteristic m.f. $\Theta_{\dot\Sigma}(z) = \Theta(z)$, where $\mathcal{H}(\Theta)$ is the RKHS with RK $K_\Theta(z,\zeta)$ and the operators $\dot A \in [\mathcal{H}(\Theta)]$ and $\dot F \in [\mathcal{H}(\Theta), \mathbb{C}^m]$ are defined by the formulas:

$$\dot A x = \frac{x(z) - x(0)}{z} \quad \text{and} \quad \dot F x = \sqrt{\pi}\, x(0)$$

(see [ArD]), and from the condition (2.3) we have that $\mathcal{H}(\Theta) \subset H_2^m$. Moreover, in this case $\mathcal{H}(\Theta)$ is a closed linear manifold in H_2^m, see [ArD] Theorem 6.7. Denote this subspace in H_2^m by X_0. Let A_0 and F_0 be defined by the formulas:

$$(A_0x)(z) = z^{-1}(x(z) - x(0)), \qquad F_0x = \sqrt{\pi}\, x(0) \qquad \text{for } x \in X_0.$$

Then from Lemma 1 we have $F_0 \in [X_0, \mathbb{C}^m]$, $A_0 \in [X_0]$, and $\Sigma_0 = [A_0, F_0; X_0, \mathbb{C}^m, I]$ is a simple dissipative LB node. Now let the operator R be such that $(Rx)(z) = x(z)$ where in the left-hand side $x \in \mathcal{H}(\Theta)$ and on the right-hand side $x(z) \in X_0$. By Theorem 6.7 [ArD],

$$\gamma_1 \|x\|_{H_2^m} \le \|x\|_{\mathcal{H}(\Theta)} \le \gamma_2 \|x\|_{H_2^m}, \quad \forall x \in \mathcal{H}(\Theta), \tag{3.9}$$

for some constants $\gamma_2 > \gamma_1 > 0$. Then $R \in [\mathcal{H}(\Theta), X_0]$ and $R^{-1} \in [X_0, \mathcal{H}(\Theta)]$,

$$\dot A = R^{-1} A_0 R, \qquad \dot F = F_0 R.$$

But by the theorem formulated in the Section 2 the nodes Σ and $\dot\Sigma$ are unitarily equivalent.

2) Let $\Sigma = [A, F; X, \mathbb{C}^m; J]$ be an LB node with $\Theta_\Sigma(z) = \Theta(z) \in U(J)$. Let a dissipative LB node $\Sigma_0 = [A_0, F_0; X_0, \mathbb{C}^m; I]$ exist such that the relations (3.8) hold. Then from Lemma 3 we have that for all $x \in X$:

$$F(I - zA)^{-1}x = F_0(I - zA_0)^{-1}Rx \in H_2^m \tag{3.10}$$

and hence for the vector $h(z) \in \mathcal{H}(\Theta)$ such that $h(z) = \sum_{j=1}^n K_\Theta(z, w_j)\xi_j$ we obtain:

$$h(z) = F(I - zA)^{-1} \sum_{j=1}^n (I - \overline{w}_j A^*)^{-1} F^* \xi_j \in H_2^m. \tag{3.11}$$

Here we used the identity for a LB node Σ with $\Theta_\Sigma = \Theta$:

$$K_\Theta(z, w) = F(I - zA)^{-1}(I - \overline{w}A^*)^{-1}F^*.$$

But the linear manifold $\mathcal{L}$ of the considered vectors $h(z)$ is dense in $\mathcal{H}(\Theta)$. For all such $h(z)$ we have

$$\|h\|^2_{\mathcal{H}(\Theta)} = \|x\|^2,$$

where

$$x = \sum_{j=1}^{n}(I - \overline{w}_j A^*)^{-1}F^*\xi_j, \qquad h(z) = F(I - zA)^{-1}x. \tag{3.12}$$

This gives that formula (3.12) defines an isometric map of the subspace $X_\Sigma^0 := \bigvee_{\operatorname{Im} w>0}(I - \overline{w}A^*)^{-1}F^*\mathbb{C}^m$ onto $\mathcal{H}(\Theta)$. Consequently, for each $h(z) \in \mathcal{H}(\Theta)$ we have (3.12) with some $x \in X$. Taking (3.10) and (2.3) into account we obtain the condition (3.1).

4. Some properties of the considered LB nodes

4.1. Bistableness.

THEOREM 2. *Let $\Sigma = [A, F; X, \mathbb{C}^m; J]$ be a simple LB node with $\Theta_\Sigma = \Theta$. Let* (3.1) *hold. Then the semigroup $T(t) = e^{iAt}$ is bistable, i.e.,*

$$T(t) \longrightarrow 0, \qquad T^*(t) \longrightarrow 0 \quad (t \to +\infty). \tag{4.1}$$

Proof. This theorem follows from the proofs of Theorem 1 and Lemma 1. □

4.2. Some properties in terms of systems theory. In this subsection we will use some notations and results of systems theory adapted from the paper [Fu].

The system (2.5) with $Y = \mathbb{C}^m$ is said to be continuously observable if the operator

$$\Omega: X \to L_2^m(\mathbb{R}_+)$$

defined by the formula

$$\Omega x = \sqrt{2}Fe^{iAt}x, \qquad 0 \le t < +\infty, \tag{4.2}$$

is linear and bounded, i.e., if $\Omega \in [X, L_2^m(\mathbb{R}_+)]$. That is equivalent to the boundedness of the observability Grammian

$$Q := \Omega^*\Omega = 2\int_0^\infty e^{-iA^*t}F^*Fe^{iAt}\,dt \in [X].$$

The system (2.5) is called continuously and exactly observable if:

1) the observability operator Ω is well defined by formula (4.2) as an operator from $[X, L_2^m(\mathbb{R}_+)]$;
2) the adjoint operator $\Omega^* \in [L_2^m(\mathbb{R}_+), X]$ is surjective.

The operator Ω^* acts as follows:

$$\Omega^* h = \sqrt{2}\int_0^\infty e^{-iA^*t}F^*h(t)\,dt, \qquad h \in L_2^m(\mathbb{R}_+).$$

If we keep in mind the relation (4.2), we see that a system (2.5) is continuously and exactly observable if and only if its observability Grammian Q is a well defined invertible (strongly positive) operator on X, i.e., $Q \in [X]$ and $Q^{-1} \in [X]$.

THEOREM 3. *Let $\Sigma = [A, F; X, \mathbb{C}^m; J]$ be a simple LB node with J-inner characteristic m.f. $\Theta = \Theta_\Sigma$. Then the following conditions are equivalent:*

(i) $\Theta_\Sigma \in U_{sR}(J) \cap H_2^{m\times m}(\frac{d\mu}{1+\mu^2})$;

(ii) *the system Σ is continuously observable;*

(iii) *the Lyapunov equation*

$$HA - A^*H = 2iF^*F \tag{4.3}$$

has a bounded nonnegative solution $H(\in [X])$.

Proof.

Step 1. (i)$\Longrightarrow$(ii). By Theorem 1 we have (3.10). Then, by Lemma 3, $F_0(zI + A_0)^{-1}x \in H_2^m$ holds, and this gives $F(zI + A)^{-1}x = i\int_0^\infty e^{izt}Fe^{iAt}xdt \in H_2^m$. From this we have $Fe^{iAt} \in L_2^m$, i.e., $\int_0^\infty (e^{-iA^*t}F^*Fe^{iAt}x, x)dt < \infty$. Therefore,

$$Q = 2\int_0^\infty e^{-iA^*t}F^*Fe^{iAt}dt < \infty,$$

i.e., Q is a well defined bounded operator on X, and consequently, the system $\Sigma = [A, F; X, \mathbb{C}^m; J]$ is continuously observable.

Step 2. (ii)$\Longrightarrow$(i). Let the system Σ be continuously observable. Then we have $\int_0^\infty \|Fe^{iAt}x\|^2\,dt < \infty$, i.e., $Fe^{iAt}x \in L_2^m(0,\infty)$. Consequently, $\int_0^\infty e^{izt}Fe^{iAt}x\,dt = iF(zI + A)^{-1}x \in H_2^m$ $\forall x \in X$.

From Lemma 2 we obtain $F(I - zA)^{-1}x \in H_2^m$. But the operator: $x \longrightarrow F(I - zA)^{-1}x$ maps X onto $\mathcal{H}(\Theta)$. Thus, $\mathcal{H}(\Theta) \subset H_2^m$. Consequently, the property (i) holds.

Step 3. (ii)$\Longrightarrow$(iii). Let the system Σ be continuously observable, i.e.,

$$Q = 2\int_0^\infty e^{-iA^*t}F^*Fe^{iAt}\,dt \quad \in [X].$$

Then

$$\begin{aligned}(QAx,x) &= 2\int_0^\infty (e^{-iA^*t}F^*Fe^{iAt}Ax,x)\,dt\\ &= 2(-iFe^{iAt}x,Fe^{iAt}x)|_0^{+\infty} + 2\int_0^\infty (Fe^{iAt}x,FAe^{iAt}x)\,dt\\ &= 2(iFx,Fx) + 2\int_0^\infty (A^*e^{-iA^*t}F^*Fe^{iAt}x,x)\,dt - 2i\lim_{T\to\infty}\|Fe^{iAT}x\|^2\\ &= (A^*Qx,x) + 2(iFx,Fx) - 2i\lim_{T\to\infty}\|Fe^{iAT}x\|^2.\end{aligned}$$

But we have (ii)$\Rightarrow$(i), hence Theorem 2 takes place, and

$$\lim_{T\to+\infty}\|Fe^{iAT}x\|^2 = 0.$$

This gives

$$(QAx,x) - (A^*Qx,x) = 2(iF^*Fx,x) \quad \text{and} \quad QA - A^*Q = 2iF^*F,$$

i.e., $H = Q \geq 0$ is a solution of equation (4.3).

Step 4. (iii)$\Longrightarrow$(ii). Let the Lyapunov equation have a solution $H \geq 0, H \in [X]$. From the equality

$$\frac{d}{dt}(He^{iAt}x,e^{iAt}x) = (iHAe^{iAt}x,e^{iAt}x) + (He^{iAt}x,iAe^{iAt}x)$$

we will have

$$\begin{aligned}(He^{iAT}x,e^{iAT}x) - (Hx,x) &= i\int_0^T ((HA-A^*H)e^{iAt}x,e^{iAt}x)dt\\ &= -2\int_0^T \|Fe^{iAt}x\|^2\,dt,\\ (Hx,x) &= 2\int_0^T \|Fe^{iAt}x\|^2\,dt + (He^{iAT}x,e^{iAT}x),\\ (Hx,x) &\geq 2\int_0^\infty \|Fe^{iAt}x\|^2\,dt,\end{aligned} \tag{4.4}$$

and therefore the system Σ is continuously observable, i.e. the propery (ii) holds. Let us remark that if $\lim_{T\to\infty}(He^{iAT}x,e^{iAT}x) = 0$ then the Lyapunov equation has the unique solution $H = \int_0^\infty e^{-iAt}F^*Fe^{iAt}\,dt = Q$.

□

THEOREM 4. *Let $\Sigma = [A, F; X, \mathbb{C}^m; J]$ be a simple LB node with $\Theta_\Sigma \in U(J)$. If one of the equivalent conditions* (i)–(iii) *of Theorem* 3 *holds, then the corresponding system* (2.5) *has the following properties*:

1) *Σ is continuously and exactly observable, i.e., the observability Grammian Q is not only bounded, but also has a bounded inverse operator.*

2) *The Lyapunov equation has a unique solution and this solution is a positive bounded and boundedly invertible operator, moreover it is the observability Grammian.*

3) *The evolution semigroup $T(t) = e^{iAt}$ satisfies the condition*:

$$\underset{t\to+\infty}{\text{s-lim}}\, T(t) = 0, \qquad \underset{t\to+\infty}{\text{s-lim}}\, T^*(t) = 0.$$

Proof.

Step 1. We have property 3) by Theorem 2.

Step 2. Let $H \in [X]$ be a solution of the Lyapunov equation (4.3). Then Theorem 2 gives $\lim_{T\to\infty} (He^{iAt}x, e^{iAt}x) = 0$. Applying this to equality (4.4) we obtain $(Hx, x) = 2\int_0^\infty \|Fe^{iAt}x\|^2\, dt$, i.e. $H = 2\int_0^\infty e^{-iAt}F^*Fe^{iAt}\, dt = Q$, and consequently the solution H is unique and it is the observability Grammian.

Step 3. Let R be the operator considered in Theorem 1. Let $H_1 = R^*R$. Then via (2.5) we obtain

$$\begin{aligned} H_1A - A^*H_1 &= R^*RA - A^*R^*R = R^*RR^{-1}A_0R - R^*A_0^*(R^*)^{-1}R^*R \\ &= R^*A_0R - R^*A_0^*R = 2iR^*F_0^*F_0R = 2iF^*F. \end{aligned}$$

Thus, $H_1 = R^*R$ is a solution of the Lyapunov equation(4.3). This solution has a bounded inverse $H_1^{-1} \in [X]$. However, it was proved above that the Lyapunov equation under consideration has a unique solution, and this solution is the Grammian Q. Therefore, we see that $Q = H$ and $Q^{-1} \in [X]$, i.e. Q is strictly positive.

□

ACKNOWLEDGEMENT. The author wishes to express her gratitude to the referees for their useful remarks and for correcting the English in the paper.

References

[A] D. Z. AROV, Passive linear stationary dynamic systems, *Sib. Math. Zh.*, **20** (1977), 211–228.

[ArD] D. Z. AROV and H. DYM, J-inner matrix functions, interpolation and inverse problems. I: Foundation, *Integral Equations Operator Theory,* **29** (1997), 373–454.

[Br] M. S. BRODSKII, *Triangular and Jordan representations of linear operators*, Nauka, 1969.

[Dy] HARRY DYM, *J-contractive matrix functions,reproducing kernel Hilbert spaces and interpolation,* Conference Board of the Mathematical Sciences Regional Conference series in Mathematics, Providence, Rhode Island, 1989.

[Fu] P. A. FUHRMANN, Exact controllability and observability and realization theory, *J. Math. Anal. Appl.*, **53** (1976), 377–392.

[SzNF] B. SZ.-NAGY and C. FOIAS, *Harmonic analysis of operators on Hilbert space,* North-Holland, Amsterdam, 1970.

[L] M. S. LIVŠIC, *Operators, oscillations, waves. Open systems*, Nauka, 1966.

Z. D. AROVA, Department of Mathematics, Odessa State Academy of Food Technology, Kanatnaya 112, Odessa, Ukraine

Operator Theory: Advances and Applications, Vol. 127, 97–112

Scalar perturbations of the Sz.-Nagy–Foias characteristic function

Bhaskar Bagchi and Gadadhar Misra

When T is a completely non-unitary (cnu) contraction with Sz.-Nagy–Foias characteristic function θ and μ is a scalar $(0 < \mu < 1)$ then general theory implies the existence of a cnu contraction $T[\mu]$ with characteristic function $\mu\theta$. How to describe $T[\mu]$ in terms of T? In this paper, we find a simple answer in case T is in the class $C_{\cdot 0}$. An operator is called homogeneous if its spectrum is contained in the closed unit disc and all the bi-holomorphic automorphisms of the unit disc lift to automorphisms of the operator T modulo unitary equivalence. When T is homogeneous, so is $T[\mu]$. We find explicit formulae for the characteristic functions of the (homogeneous) twisted Bergman shifts – these have product formulae involving the discrete series projective representations of the Möbius group. These formulae lead to an explicit description of the Sz.-Nagy–Foias models of these weighted shifts. Combining our main results with an analytic continuation argument, we find a three-parameter family of homogeneous operators.

1. Introduction

1.1. All Hilbert spaces in this paper are separable Hilbert spaces over the complex numbers. For positive integers k, $\mathbb{C}^k$ will denote the k-dimensional Hilbert space with the standard inner product. All operators are bounded linear operators between Hilbert spaces. For Hilbert spaces $\mathcal{H}$, $\mathcal{K}$, $\mathcal{B}(\mathcal{H},\mathcal{K})$ (respectively $\mathcal{B}(\mathcal{H})$) will denote the Banach space of all operators from $\mathcal{H}$ to $\mathcal{K}$ (respectively from $\mathcal{H}$ to $\mathcal{H}$) with the operator norm. $\mathbb{C}^{k\times k}$ will denote $\mathcal{B}(\mathbb{C}^k)$ viewed as the space of $k \times k$ matrices (with respect to the standard basis). $\mathcal{U}(\mathcal{H})$ will denote the Borel group of all unitary operators on the Hilbert space $\mathcal{H}$ with the Borel structure induced by the strong operator topology. An operator T is called a contraction if $\|T\| \leq 1$.

AMS Subject Classification (1991): 47A20, 47A45.

It is called a pure contraction if $\|Tx\| < \|x\|$ for all nonzero vectors x. T is called a completely non unitary (cnu) operator if it has no unitary part, i.e., if T has no non-trivial reducing subspace $\mathcal{M}$ such that the restriction $T|_{\mathcal{M}}$ is unitary.

1.2. Let $T \in \mathcal{B}(\mathcal{H})$ be a cnu contraction. Let $\mathcal{K}$ and $\mathcal{L}$ denote the closures of the ranges of $D_T := (I - T^*T)^{1/2}$ and of $D_{T^*} := (I - TT^*)^{1/2}$, respectively. $\mathcal{K}$, $\mathcal{L}$ are called the defect spaces of T, and their dimensions are called its defect indices. D_T and D_{T^*} are called the defect operators associated with T.

1.3. In [7, Chapter VI], Sz.-Nagy and Foias associate with each cnu contraction T a pure contraction valued analytic function θ_T, called the characteristic function of T. Namely, letting $\mathcal{K}$ and $\mathcal{L}$ be the defect spaces of T, one defines $\theta_T: \mathbb{D} \to \mathcal{B}(\mathcal{K}, \mathcal{L})$ as follows. For z in the unit disc $\mathbb{D}$,

$$\theta_T(z) = (-T + zD_{T^*}(I - zT^*)^{-1}D_T)\,|_{\mathcal{K}}\,. \tag{1.1}$$

It is easy to verify that the right hand side of (1.1) maps $\mathcal{K}$ into $\mathcal{L}$; hence it defines a contraction $\theta_T(z) \in \mathcal{B}(\mathcal{K}, \mathcal{L})$. In particular, $-T$ maps $\mathcal{K}$ into $\mathcal{L}$ and hence defines an element $\theta_T(0)$ of $\mathcal{B}(\mathcal{K}, \mathcal{L})$. It is easy to see that $\theta_T(0)$, thus defined, is a pure contraction. However, if a contraction valued analytic function θ is pure at some point in its domain, then, applying the strong maximum modulus principle to the Hilbert space valued analytic function $z \to \theta(z)x$ for a fixed but arbitrary nonzero vector x, one sees that θ must be pure contraction valued throughout its domain. Thus the characteristic function θ_T of T is indeed a pure contraction valued analytic function from $\mathbb{D}$ into $\mathcal{B}(\mathcal{K}, \mathcal{L})$. Note that if T is a cnu contraction then so is its adjoint T^*. If θ is the characteristic function of T, then the characteristic function of T^* is θ^*. Here $*$ is the usual involution on analytic functions defined by $\theta^*(z) = \theta(\bar{z})^*$, $z \in \mathbb{D}$.

1.4. In the terminology of Sz.-Nagy and Foias, two pure contractions C_1 and C_2 *coincide* if there exist unitary operators U, V (between appropriate Hilbert spaces) such that $C_2 = UC_1V$. Two pure contraction valued analytic functions $\theta_1: \mathbb{D} \to \mathcal{B}(\mathcal{K}_1, \mathcal{L}_1)$ and $\theta_2: \mathbb{D} \to \mathcal{B}(\mathcal{K}_2, \mathcal{L}_2)$ *coincide* if there exist unitaries $U: \mathcal{L}_1 \to \mathcal{L}_2$, $V: \mathcal{K}_2 \to \mathcal{K}_1$ such that $\theta_2(z) = U\theta_1(z)V$ for all $z \in \mathbb{D}$. (Of course, this is stronger than merely requiring that $\theta_1(z)$ and $\theta_2(z)$ coincide for each $z \in \mathbb{D}$.) It is easy to see that if the cnu contractions T_1 and T_2 are unitarily equivalent then their characteristic functions coincide. The power of the Sz.-Nagy–Foias theory lies in the fact that the converse statement is also correct. Thus the characteristic function, modulo coincidence, is a complete unitary invariant for cnu contractions.

1.5. This indicates that it should be possible to recover a cnu contraction, up to unitary equivalence, from its characteristic function. Sz.-Nagy–Foias showed that this is indeed the case. Indeed, to any pure contraction valued analytic function $\theta: \mathbb{D} \to \mathcal{B}(\mathcal{K}, \mathcal{L})$, they associate a model operator T_θ in $\mathcal{B}(\mathcal{H})$ as follows. Let $H^2 = H^2(\mathbb{D})$ be the usual Hilbert space of analytic functions on the unit disc $\mathbb{D}$ with square integrable boundary values and let $L^2 = L^2(\mathbb{T})$ be the L^2-space of all square integrable (with respect to the normalised arc length measure) measurable functions on the unit circle $\mathbb{T}$. Let $H^2_{\mathcal{L}} = H^2 \otimes \mathcal{L}$ and $L^2_{\mathcal{K}} = L^2 \otimes \mathcal{K}$ be the Hilbert spaces of $\mathcal{L}$ valued analytic functions on $\mathbb{D}$ with square integrable boundary values on $\mathbb{T}$ and $\mathcal{K}$ valued square integrable measurable functions on $\mathbb{T}$, respectively. Define $H^2_{\mathcal{K}}$ similarly. Define $\Theta: H^2_{\mathcal{K}} \to H^2_{\mathcal{L}}$ and $\Delta: L^2_{\mathcal{K}} \to L^2_{\mathcal{K}}$ by

$$(1.2) \qquad (\Theta f)(z) = \theta(z) f(z), \quad (\Delta g)(w) = (I - \theta(w)^*\theta(w))^{1/2} g(w),$$

for $z \in \mathbb{D}$, $w \in \mathbb{T}$, $f \in H^2_{\mathcal{K}}$, $g \in L^2_{\mathcal{K}}$. (From the classical theory of scalar valued bounded analytic functions, it is easy to deduce that the contraction valued analytic function θ has radial limits at almost all points of the unit circle $\mathbb{T}$. In the definition of Δ in (1.2), we have used the same letter θ to denote the boundary value as well.)

Let $\operatorname{ran}(\Delta)$ denote the closure in $L^2_{\mathcal{K}}$ of the range of Δ. Let $\mathcal{V}$ be the closure in $H^2_{\mathcal{L}} \oplus \operatorname{ran}(\Delta)$ of $\{\Theta f \oplus \Delta f : f \in H^2_{\mathcal{K}}\}$ and let $\mathcal{H}$ be the orthocomplement of $\mathcal{V}$ in $H^2_{\mathcal{L}} \oplus \operatorname{ran}(\Delta)$. This defines a subspace $\mathcal{H}$ of $H^2_{\mathcal{L}} \oplus L^2_{\mathcal{K}}$. Let M be the multiplication operator defined on $H^2_{\mathcal{L}} \oplus L^2_{\mathcal{K}}$ by $M(f \oplus g) = f_1 \oplus g_1$, where $f_1(z) = zf(z), z \in \mathbb{D}$ and $g_1(w) = wg(w), w \in \mathbb{T}$. Finally, let T_θ be the compression of M to $\mathcal{H}$. Then it is shown in Theorem VI.3.1 of [7] that T_θ is a cnu contraction and the characteristic function of T_θ coincides with θ. Thus every pure contraction valued analytic function on $\mathbb{D}$ is indeed the characteristic function of a cnu contraction. In view of this bijection between unitary equivalence classes of cnu contractions and coincidence classes of pure contraction valued analytic functions on $\mathbb{D}$, we shall identify two such functions if they coincide.

1.6. For $w \in \mathbb{D}$, let $\varphi_w: \mathbb{D} \to \mathbb{D}$ denote the Möbius map

$$(1.3) \qquad \varphi_w(z) = \frac{w - z}{1 - \bar{w}z}, \quad z \in \mathbb{D}.$$

Each φ_w is a biholomorphic map from $\mathbb{D}$ onto $\mathbb{D}$ which is its own inverse. Let Möb denote the collection of all the maps $\alpha\varphi_w$, $\alpha \in \mathbb{T}, w \in D$. It is well known that Möb is a group under composition; indeed, it is the group of all biholomorphic automorphisms of the unit disc. The Möbius group is intimately related to the Sz.-Nagy–Foias theory via the following easy fact (which is essentially the result (1.7) in [7, p. 240]): if T is a cnu contraction with characteristic function θ then for any $\varphi \in$ Möb, $\varphi(T)$ is a cnu contraction whose characteristic function coincides with $\theta \circ \varphi^{-1}$. In symbols,

$$(1.4) \qquad \theta_{\varphi(T)} = \theta_T \circ \varphi^{-1}.$$

1.7. Recall from [1] that T is said to be a *homogeneous operator* if $\varphi(T)$ is unitarily equivalent to T for every φ in Möb for which $\varphi(T)$ is well defined (i.e., for every φ which is analytic in a neighborhood of the spectrum of T). It was shown in [1, Lemma 2.2] that if T is homogeneous then the spectrum of T is either the unit circle or the unit disc. Thus, in either case, $\varphi(T)$ is defined (and unitarily equivalent to T) for all φ in Möb.

The notion of homogeneous operators was introduced informally in [5], and since then it has been studied by many authors. (Complete references may be found in [2].) Our long term goal is to classify all homogeneous operators. Note that any contraction may be written uniquely as the direct sum of its cnu part and the unitary part. Further, it is clear that a contraction is homogeneous if and only if both its parts are homogeneous. Finally, using the spectral theorem, it is easy to determine all homogeneous unitary operators. Namely, if T is a homogeneous unitary, then the scalar spectral measure of T must be quasi invariant under Möb and its multiplicity function must be stable under Möb. This implies that the only homogeneous unitaries are direct integrals of copies of the (unweighted) bilateral shift. This reduces the problem of classifying homogeneous contractions to that of classifying cnu homogeneous contractions.

In view of (1.4), it is immediate that a cnu contraction T is homogeneous if and only if its characteristic function θ coincides with $\theta \circ \varphi^{-1}$ for every φ in Möb. This raises the serious possibility that it may be possible to classify all homogeneous cnu contractions using the theory outlined above. However, this is not the objective of this paper.

1.8. Instead, we exhibit in this paper how the Sz.-Nagy–Foias theory may be applied to create various continuums of new homogeneous operators starting from known ones. One basic construction is the following. Let T be a cnu contraction with characteristic function θ. Then, for any scalar μ in the range $0 < \mu \leq 1$, $\mu\theta$ is clearly a pure contraction valued analytic function, and hence it is the characteristic function of a cnu contraction $T[\mu]$. Clearly, the operators $T[\mu], 0 < \mu \leq 1$ are mutually unitarily inequivalent except in the case when θ is the constant function identically equal to zero. From the characterization of homogeneity in terms of characteristic functions presented in Section 1.7 above, it is immediate that if T is a homogeneous cnu contraction then $T[\mu]$ is homogeneous for every μ in $(0, 1)$.

1.9. For the above to be a useful construction, we should have a usable description of $T[\mu]$ in terms of T. Of course, the Sz.-Nagy–Foias model (presented in Section 1.5 above) provides an answer. This is, however, not so useful since the model is complicated. However, there is one case in which the Sz.-Nagy–Foias model is remarkably simple, namely, for the cnu operators in class $C_{\cdot 0}$. Recall that the cnu contraction T is said to be in the class $C_{\cdot 0}$ if $T^{*n} \to 0$ strongly as $n \to \infty$. In Proposition VI.3.5 of [7] it is shown that a cnu contraction T is in the class $C_{\cdot 0}$ if and only if its characteristic function θ is *inner* in the sense that θ (more precisely, the boundary value of θ) is isometry valued at almost all points of the unit circle. (Equivalently, the operator Θ of (1.2) is an isometry.) This implies that the operator Δ of (1.2) is the zero operator, and hence the Sz.-Nagy–Foias model for T simplifies as follows. If $T \in C_{\cdot 0}$ with $\theta_T = \theta$, M is the multiplication operator (by the coordinate function z) on $H^2_{\mathcal{L}}$ and $\mathcal{M}$ is the M-invariant subspace corresponding to the inner function θ (i.e., $\mathcal{M}$ is the range of Θ), then T is unitarily equivalent to the compression of M to $\mathcal{M}^{\perp}$. Thus the operators in the class $C_{\cdot 0}$ are precisely the compressions of unweighted unilateral shifts (with multiplicity) to coinvariant subspaces. Further, for such operators T, the characteristic function θ_T is the inner function involved in the description of the complementary invariant subspace.

In Section 2 of this paper, we show that for $T \in C_{\cdot 0}$ and $0 < \mu < 1$, $T[\mu]$ has a model which is almost as simple as the Sz.-Nagy–Foias model for T. Namely, $T[\mu]$ is a "slight perturbation" of the operator $M \oplus N^*$, where M, N are multiplication operators on $H^2_{\mathcal{L}}$ and $H^2_{\mathcal{K}}$, respectively.

1.10. It is clear that, in order to apply the above to create new examples of homogeneous operators, one must begin with a supply of homogeneous operators in the class $C_{\cdot 0}$ for which the characteristic functions are "known" in a sufficiently nice and usable form. Unfortunately, there is hardly any example in the literature of cnu operators with "known" characteristic functions. (Of course, one could write down the formula (1.1) in any particular case, but this is almost never very useful!) One exception is the unweighted unilateral shift. This is a homogeneous contraction in the class $C_{\cdot 0}$, and its characteristic function is as simple as can be – the identically zero function. But this is too trivial. A slightly less trivial example that we can think of is the operator N on $L^2(\mathbb{D})$ given by $Nf(z) = zf(z)$. By [1, Proposition 2.3], N is homogeneous. Also, clearly, N is in the class $C_{\cdot 0}$. Since N is normal, it is easy to simplify the formula (1.1) in this case to obtain the following simple but interesting formula for the characteristic function $\theta_N : \mathbb{D} \to \mathcal{B}(L^2(\mathbb{D}))$:

$$(1.5) \qquad (\theta_N(z)f)(w) = -\varphi_w(z)f(w), \quad z, w \in \mathbb{D}, f \in L^2(\mathbb{D}).$$

Here φ_w is as in (1.3). The formula (1.5) indicates yet another mysterious link between characteristic function and the Möbius group. Unfortunately, we are unable to use formula (1.5) to obtain explicit descriptions of the operators $N[\mu]$, $0 < \mu < 1$.

1.11. Notation. For any real number x and integer $n \geq 0$, $\binom{x}{n}$ will denote the coefficient of t^n in the Taylor expansion of $(1+t)^x$ around $t = 0$. For positive integers x, this agrees with the usual notation for the binomial coefficients.

The earliest known examples (cf. [5]) of homogeneous operators are the twisted Bergman shifts $M^{(\lambda)}$, $\lambda > 0$. $M^{(\lambda)}$ may most simply be described as the unilateral weighted shifts with weight sequence $\sqrt{\frac{n+1}{n+\lambda}}$, $n \geq 0$. Thus in particular, $M^{(1)}$ is the unweighted unilateral shift, while $M^{(2)}$ is the Bergman shift. Clearly, $M^{(\lambda)}$ is a contraction if and only if $\lambda \geq 1$. Let $\mathcal{H}^{(\lambda)}$ denote the functional Hilbert space of analytic functions on $\mathbb{D}$ with reproducing kernel $(z,w) \to (1 - z\bar{w})^{-\lambda}$. Then $M^{(\lambda)}$ may alternatively be described as the operator of multiplication by the coordinate function z on $\mathcal{H}^{(\lambda)}$. (Indeed, let $e_n^{(\lambda)}$, $n \geq 0$, be the elements of $\mathcal{H}^{(\lambda)}$ defined by $e_n^{(\lambda)}(z) = \sqrt{\binom{n+\lambda-1}{n}} z^n$, $z \in \mathbb{D}$. Then $\{e_n^{(\lambda)} : n \geq 0\}$ is an orthonormal basis of $\mathcal{H}^{(\lambda)}$, and multiplication by z acts as the weighted shift with the indicated weight sequence with respect to this basis.) This description shows that, for $\lambda \geq 1$, $M^{(\lambda)}$ is actually in the class $C_{\cdot 0}$.

In Section 3 of this paper, we explicitly determine the characteristic function of $M^{(\lambda)}$, $\lambda > 1$. To describe these, recall that a projective representation π of Möb on a Hilbert space $\mathcal{H}$ is a Borel map from Möb into the group $\mathcal{U}(\mathcal{H})$ of unitary operators on $\mathcal{H}$ such that $\pi(\varphi_1\varphi_2) = m(\varphi_1,\varphi_2)\pi(\varphi_1)\pi(\varphi_2)$ for all φ_1, φ_2 in Möb. Here $m(\varphi_1,\varphi_2)$ is a unimodular constant, depending on φ_1, φ_2. The function $m\colon \text{Möb} \times \text{Möb} \to \mathbb{T}$ is called the multiplier associated with π. For $\lambda > 0$, D_λ will denote the holomorphic discrete series (projective) representation of Möb on $\mathcal{H}^{(\lambda)}$ given by

$$(1.6) \qquad (D_\lambda(\varphi^{-1})f)(z) = \varphi'(z)^{(\lambda/2)} f(\varphi(z))$$

for $z \in \mathbb{D}$, $f \in \mathcal{H}^{(\lambda)}$, $\varphi \in$ Möb. (Warning: D_λ is an "ordinary" representation, i.e., the associated multiplier is the constant function 1, only in the cases where λ is an even integer. The adjective "discrete series" is usually reserved for these cases. We are using this word in a generalised sense.)

It is well known, and easy to see, that the representations D_λ are intimately related to the operator $M^{(\lambda)}$. Indeed, D_λ is *associated* with $M^{(\lambda)}$ in the sense that

$$(1.7) \qquad \varphi(M^{(\lambda)}) = D_\lambda(\varphi)^* M^{(\lambda)} D_\lambda(\varphi), \quad \varphi \in \text{Möb},$$

for $\lambda > 0$. This is easy to verify and contains a direct verification of the homogeneity of $M^{(\lambda)}$, $\lambda > 0$. One surprising finding of this paper is that, for $\lambda > 1$, the representations $D_{\lambda-1}$ and $D_{\lambda+1}$ are also intimately related to $M^{(\lambda)}$. Indeed, the characteristic function of $M^{(\lambda)}$ coincides with the function $\theta_\lambda\colon \mathbb{D} \to \mathcal{B}(\mathcal{H}^{(\lambda+1)}, \mathcal{H}^{(\lambda-1)})$ given by

$$(1.8) \qquad \theta_\lambda(z) = \frac{1}{\sqrt{\lambda(\lambda-1)}} D_{\lambda-1}(\varphi_z)^* \partial^* D_{\lambda+1}(\varphi_z).$$

Here ∂^* is the adjoint of the differentiation operator $\partial: \mathcal{H}^{(\lambda-1)} \to \mathcal{H}^{(\lambda+1)}$ and φ_z is the involutive element of Möb defined in Section 1.6.

Using (1.8) it is easy to calculate the Sz.-Nagy–Foias model for $M^{(\lambda)}$. This is done in Section 3 and it turns out that the model is as follows. Let us identify, as usual, the tensor product $\mathcal{H}^{(1)} \otimes \mathcal{H}^{(\lambda-1)}$ with a space of functions on the bidisc $\mathbb{D} \times \mathbb{D}$, and let $\mathcal{M}$ be the subspace of functions vanishing on the diagonal $\{(z,z) : z \in \mathbb{D}\}$. Let M_1 be the multiplication by the first coordinate on $\mathcal{H}^{(1)} \otimes \mathcal{H}^{(\lambda-1)}$. That is, $M_1 = M^{(1)} \otimes I$. Then the Sz.-Nagy–Foias model of $M^{(\lambda)}$ is the compression of M_1 to $\mathcal{M}^\perp$. In particular, this shows that $M^{(\lambda)}$ is unitarily equivalent to this compression; but that is a special case of a well known description (arising from the theory of module tensor products: see Theorem 1.1 in [4]) of $M^{(\lambda_1+\lambda_2)}$ as the compression of $M^{(\lambda_1)} \otimes I$ on $\mathcal{H}^{(\lambda_1)} \otimes \mathcal{H}^{(\lambda_2)}$ to the orthocomplement of functions vanishing on the diagonal. The interesting point is that at least a special case of this construction is implicit in the Sz.-Nagy–Foias theory: the minimal isometric dilation of $M^{(\lambda)}$ ($\lambda > 1$) is $M^{(1)} \otimes I$ on $\mathcal{H}^{(1)} \otimes \mathcal{H}^{(\lambda-1)}$ in a natural way.

1.12. With an explicit description of the characteristic functions of $M^{(\lambda)}$, $\lambda > 1$, available, the path is now open for the application of the results of Section 2 to explicitly determine the homogeneous operators $M^{(\lambda)}[\mu]$, $\lambda > 1$, $0 < \mu < 1$. This is done in Section 4. Indeed, in conjunction with Proposition 2.4 of [1] and an analytic continuation argument, this yields a new three parameter family of homogeneous operators. This is a record, the previous record being held by the two parameter family of homogeneous operators found by Wilkins in [8].

2. The models for scalar multiples of inner functions

2.1. NOTATION. Throughout, T is a cnu contraction in the class $C_{\cdot 0}$, and θ is its characteristic function. Thus, for a suitable Hilbert space $\mathcal{L}$, T may be identified with the compression of M to $\mathcal{M}^\perp$, where $M: H^2_{\mathcal{L}} \to H^2_{\mathcal{L}}$ is multiplication by the coordinate function, and $\mathcal{M}$ is an invariant subspace for M (corresponding to the inner function θ). Let $M = \begin{pmatrix} M_{11} & 0 \\ M_{21} & M_{22} \end{pmatrix}$ be the block matrix representation of M corresponding to the decomposition $H^2_{\mathcal{L}} = \mathcal{M}^\perp \oplus \mathcal{M}$. (Thus, in particular, $M_{11} = T$ and M_{22} is the restriction of M to $\mathcal{M}$.) Finally, let $\mathcal{K}$ denote the cokernel (i.e., the orthocomplement of the range) of M_{22}, and let $N: H^2_{\mathcal{K}} \to H^2_{\mathcal{K}}$ be multiplication by the coordinate function, and let $E: H^2_{\mathcal{K}} \to \mathcal{M}$ be defined by $Ef = f(0) \in \mathcal{K}$. (Remember $\mathcal{K} \subseteq \mathcal{M}$.)

THEOREM 2.1. *Let T be a cnu contraction in $C_{\cdot 0}$ with characteristic function θ. Let μ be a scalar in the range $0 < \mu < 1$, and put $\delta = +\sqrt{1-\mu^2}$. In terms of the above notation, let $T[\mu]: H^2_{\mathcal{L}} \oplus H^2_{\mathcal{K}} \to H^2_{\mathcal{L}} \oplus H^2_{\mathcal{K}}$ be the operator whose block matrix representation with respect to the decomposition $\mathcal{M}^\perp \oplus \mathcal{M} \oplus H^2_{\mathcal{K}}$ is given by*

$$T[\mu] = \begin{pmatrix} M_{11} & 0 & 0 \\ \delta M_{21} & M_{22} & \mu E \\ 0 & 0 & N^* \end{pmatrix}.$$

Then the characteristic function of $T[\mu]$ coincides with $\mu\theta$. (Of course, this includes the fact that $T[\mu]$, thus defined, is a cnu contraction.)

Proof. We begin with the Sz.-Nagy–Foias model $T_{\mu\theta}$ and reduce it to the above description. Let $\tilde{M}$ be the multiplication operator on $L^2_{\mathcal{K}}$, where, for now, $\mathcal{K}$ is the first defect space of $T_{\mu\theta}$ (we shall later identify $\mathcal{K}$ with $\operatorname{coker}(M_{22})$). Recall that $T_{\mu\theta}$ is the compression of $M \oplus \tilde{M}$ to the subspace $\mathcal{H}$ of $H^2_{\mathcal{L}} \oplus L^2_{\mathcal{K}}$, where in our case (since θ is inner), $\mathcal{H}^\perp = \{(\mu\Theta f, \delta f) : f \in H^2_{\mathcal{K}}\}$. Let $\sim$ denote the unitary operator from $H^2_{\mathcal{K}}$ onto $L^2_{\mathcal{K}} \ominus H^2_{\mathcal{K}}$ given by $\tilde{g}(z) = z^{-1}g(z^{-1})$, $z \in \mathbb{T}, g \in H^2_{\mathcal{K}}$. (Here we have used the usual identification of the elements of $H^2_{\mathcal{K}}$ with their boundary values. Thus, for now, we view all Hilbert spaces in sight as Hilbert spaces of functions on the unit circle $\mathbb{T}$.) Then it is trivial to verify that for $g \in H^2_{\mathcal{L}}$ and $h \in H^2_{\mathcal{K}}$, $(g, -\delta^{-1}\mu\Theta^* g + \tilde{h})$ is orthogonal to all elements of $\mathcal{H}^\perp$ and hence is in $\mathcal{H}$. Conversely, if $(g,k) \in H^2_{\mathcal{L}} \oplus L^2_{\mathcal{K}}$ is orthogonal to $(\mu\Theta f, \delta f)$ for every $f \in H^2_{\mathcal{K}}$ then $k + \delta^{-1}\mu\Theta^* g$ is orthogonal to all the elements of $H^2_{\mathcal{K}}$ and hence is of the form $\tilde{h}$ for a uniquely determined $h \in H^2_{\mathcal{K}}$. Thus $k = -\delta^{-1}\mu\Theta^* g + \tilde{h}$ in this case. This shows

$$\mathcal{H} = \{(g, -\delta^{-1}\mu\Theta^* g - \tilde{h}) : g \in H^2_{\mathcal{L}}, h \in H^2_{\mathcal{K}}\}.$$

It follows that we have the orthogonal decomposition

$$\mathcal{H} = \mathcal{H}_1 \oplus \mathcal{H}_2 \oplus \mathcal{H}_3,$$

where $\mathcal{H}_1 = \{(g, -\delta^{-1}\mu\Theta^* g) : g \in \mathcal{M}\}$, $\mathcal{H}_2 = \{(g, 0) : g \in \ker\Theta^*\}$ and $\mathcal{H}_3 = \{(0, \tilde{h}) : h \in H^2_{\mathcal{K}}\}$. (Here, as in Notation 2.1, $\mathcal{M}$ is the range closure of Θ. So $\ker\Theta^* = \mathcal{M}^\perp$.) To compute $T_{\mu\theta}$, it suffices to compute its action on these three subspaces.

Notice that for $g \in \mathcal{M}, (\Theta^* g)(z) = \Theta(z)^* g(z)$, $z \in \mathbb{T}$. (This follows since θ is isometry valued on $\mathbb{T}$.) It follows that $\tilde{M}\Theta^* g = \Theta^* M g$ for $g \in \mathcal{M}$. Therefore, $\mathcal{H}_1$ is invariant under $M \oplus \tilde{M}$. Hence

$$T_{\mu\theta}(g, -\delta^{-1}\mu\Theta^* g) = (Mg, -\delta^{-1}\mu\Theta^* Mg), \quad g \in \mathcal{M}, \tag{2.1}$$

gives the action of $T_{\mu\theta}$ on $\mathcal{H}_1$.

Next we show that the action of $T_{\mu\theta}$ on $\mathcal{H}_2$ is given by

$$\text{(2.2)}\quad T_{\mu\theta}(g,0) = (\delta^2\Theta E\Theta^* Mg, -\delta\mu E\Theta^* Mg) + (Mg - \Theta E\Theta^* Mg, 0), \quad g \in \mathcal{M}^\perp.$$

Indeed, since Θ is an isometry, the first term in the right hand side of (2.2) is in $\mathcal{H}_1$. Therefore, to show that the right hand side is in $\mathcal{H}$, it is enough to establish that its second term is in $\mathcal{H}_2$, i.e., that $\mathcal{M}^\perp$ is invariant under $(I-\Theta E\Theta^*)M$. Since Θ is an isometry, this amounts to showing that $\mathcal{M}^\perp$ is contained in the kernel of $(I-E)\Theta^* M$. Since the kernel of $I-E$ is $\mathcal{K}$ (identified with the subspace of constant functions in $H^2_{\mathcal{K}}$), we need to show that $\Theta^* M$ maps $\mathcal{M}^\perp$ into $\mathcal{K}$. But this is obvious from the fact that, for $g \in H^2_{\mathcal{L}}$, $\Theta^* g$ is the projection to $H^2_{\mathcal{K}}$ of the element $z \mapsto \theta(z)^* g(z)$ of $L^2_{\mathcal{K}}$. Thus the right hand side of (2.2) is indeed in $\mathcal{H}$. So, to complete the proof of (2.2), it is enough to verify that the difference between this right hand side and $(Mg, 0)$ is in $\mathcal{H}^\perp$. But this difference is $(\mu^2\Theta E\Theta^* Mg, \delta\mu E\Theta^* Mg) = (\mu\Theta f, \delta f)$, where $f = \mu E\Theta^* Mg$. But this is indeed in $\mathcal{H}^\perp$ as is clear from the initial description of $\mathcal{H}^\perp$.

Finally, we claim that the action of $T_{\mu\theta}$ on $\mathcal{H}_3$ is given by

$$\text{(2.3)}\qquad T_{\mu\theta}(0,\tilde{h}) = (-\delta\mu\Theta Eh, \mu^2 Eh + \widetilde{N^* h}), \qquad h \in H^2_{\mathcal{K}}.$$

Indeed, since $(-\delta\mu\Theta Eh, \mu^2 Eh)$ is in $\mathcal{H}_1$ and $(0, \widetilde{N^* h})$ is in $\mathcal{H}_3$, the right hand side of (2.3) is in $\mathcal{H}$. So, to prove (2.3), it is enough to show that the difference of the right hand side of (2.3) and $(0, \tilde{M}\tilde{h})$ is in $\mathcal{H}^\perp$. But this difference equals $(\delta\mu\Theta Eh, \tilde{M}\tilde{h} - \widetilde{N^* h} - \mu^2 Eh)$. Using $\mu^2 + \delta^2 = 1$ and the trivial identity $\tilde{M}\tilde{h} - \widetilde{N^* h} = Eh$ for $h \in H^2_{\mathcal{K}}$, this vector equals $(\delta\mu\Theta Eh, \delta^2 Eh) = (\mu\Theta f, \delta f)$, where $f = \delta Eh$. This is in $\mathcal{H}^\perp$ by definition.

This completes our description of $T_{\mu\theta}$. Now define $U\colon H^2_{\mathcal{L}} \oplus H^2_{\mathcal{K}} \to \mathcal{H}$ by

$$Ug = \begin{cases} (\delta g, -\mu\Theta^* g) & \text{if } g \in \mathcal{M} \\ (g, 0) & \text{if } g \in \mathcal{M}^\perp \\ (0, \tilde{g}) & \text{if } g \in H^2_{\mathcal{K}}. \end{cases}$$

Since Θ is an isometry and $\delta^2 + \mu^2 = 1$, U is clearly a unitary operator. Put $T[\mu] = U^* T_{\mu\theta} U$. Since $T[\mu]$ is unitarily equivalent to $T_{\mu\theta}$, by the Sz.-Nagy–Foias theory it is a cnu contraction whose characteristic function coincides with $\mu\theta$. The formule (2.1), (2.2), (2.3) transform to the following description of $T[\mu]$:

$$T[\mu]g = \begin{cases} Mg & \text{for } g \in \mathcal{M} \\ Mg - (1-\delta)\Theta E\Theta^* Mg & \text{for } g \in \mathcal{M}^\perp \\ (-\mu\Theta Eg) \oplus (N^* g) & \text{for } g \in H^2_{\mathcal{K}}. \end{cases}$$

Since $-\Theta$ is an isometry, it defines a unitary operator from $H^2_{\mathcal{K}}$ onto $\mathcal{M}$ which carries the range of E (viz. the subspace of $H^2_{\mathcal{K}}$ consisting of constant functions) onto coker(M_{22}). If we identify $\mathcal{K}$ with coker(M_{22}) via this map, then the above becomes the description of $T[\mu]$ given in the statement of the theorem.

□

2.3. REMARK. From the description of $T[\mu]$ given in the above theorem it is clear that both $\mathcal{M}$ and $\mathcal{M} \oplus H^2_{\mathcal{K}}$ are invariant for $T[\mu]$. The compression of $T[\mu]$ to the coinvariant subspace $\mathcal{M}^\perp$ is $M_{11} = T$. Also, it is easy to verify that the restriction of $T[\mu]$ to the invariant subspace $\mathcal{M} \oplus H^2_{\mathcal{K}}$ is the direct sum of dim $\mathcal{K}$ copies of the bilateral weighted shift T_μ with weight sequence $\{\ldots, 1, 1, \mu, 1, 1, \ldots\}$. By [1], the characteristic function of $T[\mu]|_{\mathcal{M}\oplus H^2_{\mathcal{K}}}$ is the constant function $\mu I_{\mathcal{K}}$. Since $\mu\theta = (\theta)(\mu I_{\mathcal{K}})$ is an instance of inner-outer factorization, the above theorem may be viewed as an elaboration of Theorem VI.1.1 in [7].

3. The characteristic function of the twisted Bergman shifts

Recall that for $\lambda > 0$, the twisted Bergman shift $M^{(\lambda)}$ is the multiplication operator on $\mathcal{H}^{(\lambda)}$ and D_λ denotes the discrete series representation of Möb on $\mathcal{H}^{(\lambda)}$. We now prove:

THEOREM 3.1. *For $\lambda > 1$, the characteristic function of $M^{(\lambda)}$ coincides with the function $\theta_\lambda \colon \mathbb{D} \to \mathcal{B}(\mathcal{H}^{(\lambda+1)}, \mathcal{H}^{(\lambda-1)})$ defined by*

$$\theta_\lambda(z) = \frac{1}{\sqrt{\lambda(\lambda-1)}} D_{\lambda-1}(\varphi_z)^* \partial^* D_{\lambda+1}(\varphi_z).$$

(Here ∂^* is the adjoint of the differentiation operator $\partial \colon \mathcal{H}^{(\lambda-1)} \to \mathcal{H}^{(\lambda+1)}$.)

Proof. It is slightly easier to prove that the characteristic function θ^*_λ of the adjoint $M^{(\lambda)^*}$ is given by the formula

$$\theta^*_\lambda(z) = \frac{1}{\sqrt{\lambda(\lambda-1)}} D_{\lambda+1}(\varphi^*_z)^* \partial D_{\lambda-1}(\varphi^*_z), \tag{3.1}$$

where $\varphi^*_z(w) = \overline{\varphi_z(\bar{w})}$. Notice that $M^{(\lambda)}$ is a pure contraction, so that its defect spaces are both equal to the whole of $\mathcal{H}^{(\lambda)}$. Define the unitary operators $\Gamma \colon \mathcal{H}^{(\lambda)} \to \mathcal{H}^{(\lambda+1)}$ and $\Lambda \colon \mathcal{H}^{(\lambda-1)} \to \mathcal{H}^{(\lambda)}$ by $\Gamma(e^{(\lambda)}_n) = e^{(\lambda+1)}_n$, $\Lambda(e^{(\lambda-1)}_n) = e^{(\lambda)}_n$ for $n \geq 0$. (Here $\{e^{(\lambda)}_n\}$, for instance, is the standard orthonormal basis of $\mathcal{H}^{(\lambda)}$ defined in

Section 1.11.) Define $\theta: \mathbb{D} \to \mathcal{B}(\mathcal{H}^{(\lambda-1)}, \mathcal{H}^{(\lambda+1)})$ by $\theta(z) = -\Gamma\tilde{\theta}(z)\Lambda$, where $\tilde{\theta}: \mathbb{D} \to \mathcal{B}(\mathcal{H}^{(\lambda)})$ is the characteristic function of $M^{(\lambda)^*}$. Clearly, θ coincides with $\tilde{\theta}$. Now, one verifies that θ_λ^* defined by (3.1) is also given by

$$(\theta_\lambda^*(z)f)(w) = \frac{1}{\sqrt{\lambda(\lambda-1)}} f'(w) - \sqrt{\frac{\lambda-1}{\lambda}} \frac{z}{1-zw} f(w) \tag{3.2}$$

for $z, w \in \mathbb{D}$, $f \in \mathcal{H}^{(\lambda-1)}$. To verify this one uses the two trivial identities obtained by differentiating $\varphi_z^{*-1} \circ \varphi_z^* = id$ once and twice and the identity $\varphi_z^{*\prime\prime}(w)/\varphi_z^{*\prime}(w) = 2z/(1-zw)$. To verify that $\tilde{\theta}$ is also given by the right hand side of (3.2), note that (3.2) holds for $f = e_n^{(\lambda-1)}$, $n \geq 0$, and that both sides of (3.2) are bounded operators in f. So, $\theta_\lambda^* = \tilde{\theta}$.

□

THEOREM 3.2. *The Sz.-Nagy–Foias model of $M^{(\lambda)}$ ($\lambda > 1$) is precisely the compression of $M^{(1)} \otimes I$ on $\mathcal{H}^{(1)} \otimes \mathcal{H}^{(\lambda-1)}$ to the orthocomplement of the subspace of functions vanishing on the diagonal. In consequence, the minimal isometric dilation of $M^{(\lambda)}$ is $M^{(1)} \otimes I$.*

Proof. Since $M^{(\lambda)}$, $\lambda \geq 1$, is in the class $C_{\cdot 0}$, its characteristic function θ_λ (given by (1.8)) is an inner function. Hence to describe the model operator T, it suffices to determine the invariant subspace $\mathcal{M}$ of $\mathcal{H}^{(1)} \otimes \mathcal{H}^{(\lambda-1)}$ corresponding to the inner function θ_λ. Recall that $\mathcal{M}$ is the closure of the range of the operator $\Theta_\lambda: \mathcal{H}^{(1)} \otimes \mathcal{H}^{(\lambda+1)} \to \mathcal{H}^{(1)} \otimes \mathcal{H}^{(\lambda-1)}$ given by $(\Theta_\lambda f)(z) = \theta_\lambda(z)f(z)$, $z \in \mathbb{D}$, $f \in \mathcal{H}^{(1)} \otimes \mathcal{H}^{(\lambda+1)}$. (Here we are viewing elements of $\mathcal{H}^{(1)} \otimes \mathcal{H}^{(\lambda+1)}$ as $\mathcal{H}^{(\lambda+1)}$-valued analytic functions on $\mathbb{D}$, but later in this proof we view the same elements as scalar valued analytic functions on the bidisc $\mathbb{D} \times \mathbb{D}$.) Of course, this amounts to calculating $(\ker \Theta_\lambda^*)^\perp$. Note that $\Theta_\lambda^* f$ is the orthogonal projection into $H^{(1)} \otimes H^{(\lambda+1)}$ of the element $(z, \cdot) \mapsto \theta_\lambda^*(\bar{z}) f(z, \cdot)$ of $L^2(\mathbb{T}) \otimes H^{(\lambda+1)}$. Consequently, equation (3.2) implies that

$$(\Theta_\lambda^* f)(z,w) = \frac{1}{\sqrt{\lambda(\lambda-1)}} \frac{\partial}{\partial w} f(z,w) - \sqrt{\frac{\lambda-1}{\lambda}} P\Big((z,w) \mapsto \frac{\bar{z}}{1-\bar{z}w} f(z,w)\Big),$$

where P is the orthogonal projection onto the space $H^{(1)} \otimes H^{(\lambda+1)}$. An easy calculation in terms of the monomial basis leads to the formula

$$(\Theta_\lambda^* f)(z,w) = \frac{1}{\sqrt{\lambda(\lambda-1)}} \frac{\partial}{\partial w} f(z,w) - \sqrt{\frac{\lambda-1}{\lambda}} \frac{f(z,w)-f(w,w)}{z-w}.$$

Therefore the kernel of Θ_λ^* consists of all $f \in H^{(1)} \otimes H^{(\lambda-1)}$ satisfying the differential equation

$$\frac{\partial}{\partial w} f(z,w) = (\lambda-1) \frac{f(z,w)-f(w,w)}{z-w}.$$

Another calculation with the monomial basis shows that the solutions to this equation form the closed linear span of $\{f_k : k = 0, 1, \ldots\}$, where the elements f_k of $H^{(1)} \otimes H^{(\lambda-1)}$ are as in Equation (4.2) below. Since we show in Section 4.1 that this is the orthocomplement of the subspace of functions vanishing on the diagonal, we are done.

$\square$

4. A continuum of new homogeneous operators from the twisted Bergman shifts

4.1. In this section we analyse the twisted Bergman shift $M^{(\lambda)}$ in the light of results of the previous two sections. Fix $\lambda > 1$, and let $\mathcal{L} = \mathcal{H}^{(\lambda-1)}$. So, $H^2_{\mathcal{L}} \cong \mathcal{H}^{(1)} \otimes \mathcal{H}^{(\lambda-1)}$ is naturally identified with a Hilbert space of analytic functions on the bidisc. Let $\mathcal{M}$ be the subspace of all functions in $\mathcal{H}^{(1)} \otimes \mathcal{H}^{(\lambda-1)}$ which vanish on the diagonal $\{(z,w) \in \mathbb{D} \times \mathbb{D} : z = w\}$. Let $M = M^{(1)} \otimes I$ be the operation of multiplication by the first coordinate on $\mathcal{H}^{(1)} \otimes \mathcal{H}^{(\lambda-1)}$. Thus, by Theorem 3.2, $M^{(\lambda)}$ is the compression of M to $\mathcal{M}^{\perp}$. As in Notation 2.1, we write $M = \begin{pmatrix} M_{11} & 0 \\ M_{21} & M_{22} \end{pmatrix}$ with respect to the decomposition $\mathcal{M}^{\perp} \oplus \mathcal{M}$ of $\mathcal{H}^{(1)} \otimes \mathcal{H}^{(\lambda-1)}$. Thus $M_{11} = M^{(\lambda)}$. For integers $k \geq \ell \geq 1$, let $e_{k,\ell} \in \mathcal{H}^{(1)} \otimes \mathcal{H}^{(\lambda-1)}$ be defined by

$$e_{k,\ell}(z,w) = (\lambda-1)^{1/2}\ell^{-1/2}\binom{\ell+\lambda-1}{\ell}^{-1/2} \times \tag{4.1}$$
$$\times \left(\sum_{j=0}^{\ell-1}\binom{j+\lambda-2}{j} z^{k-j}w^{j} - \binom{\ell+\lambda-2}{\ell-1} z^{k-\ell}w^{\ell}\right),$$

$z, w \in \mathbb{D}$. For integers $k \geq 0$, define $f_k \in \mathcal{H}^{(1)} \otimes \mathcal{H}^{(\lambda-1)}$ by

$$f_k(z,w) = \binom{k+\lambda-1}{k}^{-1/2} \sum_{j=0}^{k}\binom{j+\lambda-2}{j} z^{k-j}w^{j}, \quad z, w \in \mathbb{D}. \tag{4.2}$$

We have the orthogonal decomposition $\mathcal{H}^{(1)} \otimes \mathcal{H}^{(\lambda-1)} = \oplus_{k \geq 0}\mathrm{Hom}(k)$, where $\mathrm{Hom}(k)$ is the space of all homogeneous polynomials of degree k (in two variables z, w). In view of the well known identity

$$\sum_{j=0}^{\ell-1}\binom{j+\lambda-2}{j} = \binom{\ell+\lambda-2}{\ell-1}, \tag{4.3}$$

it is easy to verify that $\{e_{k,\ell} : 1 \leq \ell \leq k\}$ is an orthonormal set of vectors in $\mathrm{Hom}(k) \cap \mathcal{M}$ and f_k is a unit vector orthogonal to all $e_{k,\ell}$, $1 \leq \ell \leq k$. Since $\mathrm{Hom}(k) \cap \mathcal{M}$ is clearly a codimension 1 subspace of the $(k+1)$-dimensional space $\mathrm{Hom}(k)$, it follows that $\{f_k : k = 0, 1, 2, \ldots\}$ is an orthonormal basis of $\mathcal{M}^\perp$ and $\{e_{k,\ell} : k = 1, 2, \ldots, \ell = 1, \ldots, k\}$ is an orthonormal basis of $\mathcal{M}$. (The special case $\lambda = 2$ of this computation occurs in [4].)

(To verify (4.3), note that both sides are polynomials in λ, so it is enough to verify it for positive integral values λ. In this case, $\binom{\ell+\lambda-2}{\ell-1}$ is the number of subsets of $\{1, 2, \ldots, \ell + \lambda - 2\}$ of size $\lambda - 1$, while $\binom{j+\lambda-2}{j}$ is the number of such subsets with maximum element $= \lambda + j - 1$.)

It is easy to verify that the action of $M = M^{(1)} \otimes I$ on these basis elements is given by

$$M(e_{k,\ell}) = e_{k+1,\ell}, \quad M(f_k) = \sqrt{\frac{k+1}{k+\lambda}} f_{k+1} + \sqrt{\frac{\lambda-1}{k+\lambda}} e_{k+1,k+1}.$$

This yields the following description of the operators M_{ij}:

$$\begin{aligned} M_{11} :\quad & f_k \to \sqrt{\frac{k+1}{k+\lambda}} f_{k+1}, \quad k \geq 0, \\ M_{21} :\quad & f_k \to \sqrt{\frac{\lambda-1}{k+\lambda}} e_{k+1,k+1}, \quad k \geq 0, \\ M_{22} :\quad & e_{k,\ell} \to e_{k+1,\ell}, \quad \ell \geq 1, k \geq \ell. \end{aligned}$$

(This again identifies M_{11} with $M^{(\lambda)}$.)

In particular, $\mathcal{K} := \mathrm{coker}(M_{22})$ is spanned by $\{e_{\ell,\ell} : \ell \geq 1\}$. Let $\{g_{\ell,m} : \ell \geq 1, m \geq 0\}$ be the orthonormal basis of $H^2_{\mathcal{K}}$ defined by $g_{\ell,m}(w) = w^m e_{\ell,\ell}$, $w \in \mathbb{D}$. With respect to this basis the actions of the operator $E: H^2_{\mathcal{K}} \to \mathcal{L} = \mathcal{H}^{(1)} \otimes \mathcal{H}^{(\lambda-1)}$ given by $Ef = f(0)$ and of the adjoint N^* of the multiplication (by coordinate function) operator N on $H^2_{\mathcal{K}}$ are as follows

$$E : \quad g_{\ell,m} \to \begin{cases} e_{\ell,\ell} & \text{if } m = 0, \ell \geq 1 \\ 0 & \text{if } m \geq 1, \ell \geq 1 \end{cases}$$

$$N^* : \quad g_{\ell,m} \to \begin{cases} 0 & \text{if } m = 0, \ell \geq 1 \\ g_{\ell,m-1} & \text{if } m \geq 1, \ell \geq 1. \end{cases}$$

Thus in view of Theorem 2.1 we have all the ingredients for the description of the operators $M^{(\lambda)}[\mu]$, $\lambda > 1, 0 < \mu < 1$. However, for the sake of clarity, and to be consistent with Remark 2.3, it is better to rename the basis elements as follows. Put

$$h_{k,\ell} = \begin{cases} g_{\ell,-k} & \text{for } k \leq 0, \ell \geq 1 \\ e_{k+\ell-1,\ell} & \text{for } k \geq 1, \ell \geq 1. \end{cases}$$

Thus we get the following description of the operator $M^{(\lambda)}[\mu]$ (with $\delta = \sqrt{1-\mu^2}$): This operator acts on a Hilbert space with orthonormal basis

$$\{f_k : k = 0, 1, 2, \dots\} \cup \{h_{k,\ell} : k = 0, \pm 1, \pm 2, \dots, \ell = 1, 2, \dots\} \tag{4.4}$$

by the rule

$$M^{(\lambda)}[\mu] : \begin{cases} f_k \to \sqrt{\frac{k+1}{k+\lambda}} f_{k+1} + \delta\sqrt{\frac{\lambda-1}{k+\lambda}} h_{1,k+1} \\ h_{k,\ell} \to \begin{cases} \mu h_{k+1,\ell} & \text{if } k = 0, \ell \geq 1 \\ h_{k+1,\ell} & \text{if } k = \pm 1, \pm 2, \dots, \ell \geq 1. \end{cases} \end{cases} \tag{4.5}$$

4.2. Recall that a cnu contraction T is said to be in the class $C_{\cdot 0}$ if $T^{*n}x \to 0$ for all x, and it is said to be in the class $C_{\cdot 1}$ if $T^{*n}x \not\to 0$ for all $x \neq 0$. T is in class $C_{\cdot 0}$ if and only if its characteristic function θ is inner (i.e., θ is almost surely isometry valued on the boundary of $\mathbb{D}$) and T is in class $C_{\cdot 1}$ if and only if its characteristic function θ is outer (i.e., the operator Θ given by (1.1) has dense range). For an arbitrary cnu contraction T on a Hilbert space $\mathcal{H}$, let $\mathcal{M}^\perp$ be the subspace of $\mathcal{H}$ consisting of all x for which $T^{*n}x \to 0$. Clearly, $\mathcal{M}$ is invariant for T. Let T_1 be the compression of T to $\mathcal{M}^\perp$ and let T_2 be the restriction of T to $\mathcal{M}$. Thus, with respect to the decomposition $\mathcal{M}^\perp \oplus \mathcal{M}$ of $\mathcal{H}$, T has a canonical triangularization $\begin{pmatrix} T_1 & 0 \\ X & T_2 \end{pmatrix}$, where T_1 is in $C_{\cdot 0}$ and T_2 is in $C_{\cdot 1}$. T_1 and T_2 are called the $C_{\cdot 0}$ part of T and the $C_{\cdot 1}$ part of T, respectively. Clearly, the bilateral shifts $T_\mu, 0 < \mu < 1$, mentioned in Remark 2.3 are in the class $C_{\cdot 1}$. In view of this remark, it follows that if T is in class $C_{\cdot 0}$ then the $C_{\cdot 0}$ part of the operator $T[\mu]$ of Theorem 2.1 is T while the $C_{\cdot 1}$ part is $\begin{pmatrix} M_{22} & \mu E \\ 0 & N^* \end{pmatrix}$. The latter is the direct sum of $\dim \mathcal{K}$ copies of T_μ.

LEMMA 4.1. *If T is a cnu contraction in the class $C_{\cdot 0}$, then $\varphi(T)$ is a cnu contraction in the same class for every φ in Möb.*

Proof. Let θ be the characteristic function of T. Thus θ is inner. Since φ maps $\mathbb{T}$ onto itself and sends null sets (with respect to arc length measure) to null sets, it follows that $\theta \circ \varphi^{-1}$ is also inner. Hence by (1.3) $\varphi(T)$ is in the class $C_{\cdot 0}$. □

4.3. Let T be an operator on a Hilbert space $\mathcal{H}$ and let π be the representation of Möb on $\mathcal{H}$. Recall from [1] that π is said to be associated to T if the spectrum of T is contained in the closed unit disc and $\pi(\varphi)^* T \pi(\varphi) = \varphi(T)$ for all φ in Möb. Clearly, if T has an associated (projective unitary) representation then T is homogeneous. In [6], the converse statement was shown to hold for irreducible operators.

PROPOSITION 4.2. *Let T be a (homogeneous) cnu contraction with associated projective representation π. Let T_1 and T_2 be the $C_{\cdot 0}$ and $C_{\cdot 1}$ part of T, respectively. Then there are projective representations π_1 and π_2 of Möb such that $\pi = \pi_1 \oplus \pi_2$ and π_j is associated to T_j, for $j = 1, 2$. In consequence, T_1 and T_2 are homogeneous operators. More generally, letting $T = \begin{pmatrix} T_1 & 0 \\ X & T_2 \end{pmatrix}$ be the canonical triangularization of T (cf. Section 4.2 above), one gets that for any scalar $t \geq 0$, the operator $\begin{pmatrix} T_1 & 0 \\ tX & T_2 \end{pmatrix}$ is homogeneous with associated representation π.*

Proof. Let $\mathcal{M}$ be the domain of T_2. It is sufficient to show that $\mathcal{M}$ (or equivalently $\mathcal{M}^\perp$) is reducing subspace for π, so that π breaks up as $\pi_1 \oplus \pi_2$ along $\mathcal{H} = \mathcal{M} \oplus \mathcal{M}^\perp$. The rest of the statement will then be immediate consequence of [1, Proposition 2.4].

Fix φ in Möb. Since $T_1 \in C_{\cdot 0}$, by Lemma 4.1 we get $\varphi(T_1) \in C_{\cdot 0}$. That is, $\varphi(T_1)^{*^n} x = (\varphi^*(T_1))^{*^n} x \to 0$ for all $x \in \mathcal{M}^\perp$. But $\varphi(T_1)^*$ is the restriction of $\varphi(T)^*$ to $\mathcal{M}^\perp$. So we get $\varphi(T)^{*^n} x \to 0$ for each $x \in \mathcal{M}^\perp$. That is, $\pi(\varphi)^* T^{*^n} \pi(\varphi) x \to 0$ as $n \to \infty$. Since $\pi(\varphi)$ is unitary this implies $T^{*^n} \pi(\varphi) x \to 0$. Thus $\pi(\varphi) x \in \mathcal{M}^\perp$. Since this holds for all $x \in \mathcal{M}^\perp$ and all φ in Möb, this means that $\mathcal{M}^\perp$ is invariant (hence reducing) under π.

□

THEOREM 4.3. *For any three real numbers $\lambda > 0, \mu > 0, \delta > 0$, let $M^{(\lambda)}[\mu, \delta]$ be the operator defined by the right hand side of* (4.5). *(So we have removed the restrictions $\lambda > 1, \mu < 1$ and $\delta = \sqrt{1-\mu^2}$).) Then $M^{(\lambda)}[\mu, \delta]$, $\lambda > 0, \mu > 0, \delta > 0$, are mutually unitarily inequivalent homogeneous operators.*

Proof. First take $0 < \mu < 1$, $\delta = \sqrt{1-\mu^2}$. Then the $C_{\cdot 0}$ part of the operator $M^{(\lambda)}[\mu, \delta] = M^{(\lambda)}[\mu]$ is $M^{(\lambda)}$, while the $C_{\cdot 1}$ part is the direct sum of infinitely many copies of T_μ. But $M^{(\lambda)}$ has associated representation D_λ, while T_μ has associated representation $D_1 \oplus D_2$ by a result in [1]. It follows that the $C_{\cdot 1}$ part has associated representation E, where E is the direct sum of infinitely many copies of $D_1 \oplus D_2$. So, by Proposition 4.2, $M^{(\lambda)}[\mu, \delta]$ has associated representation $\pi = D_\lambda \oplus E$. The point is that there is a projective representation π, independent of all parameters, such that π is associated with $M^{(\lambda)}[\mu, \sqrt{1-\mu^2}]$ for each fix μ in the range $0 < \mu < 1$. But replacing the X part in the canonical triangularization of $M^{(\lambda)}[\mu, \sqrt{1-\mu^2}]$ by an arbitrary positive scalar multiple has the effect of replacing $M^{(\lambda)}[\mu, \sqrt{1-\mu^2}]$ by $M^{(\lambda)}[\mu, \delta]$ for an arbitrary $\delta > 0$. So, by Proposition 4.2, $M^{(\lambda)}[\mu, \delta]$ is homogeneous with the same associated projective representation π for all such μ, δ. That is, $\pi(\varphi)^* M^{(\lambda)}[\mu, \delta] \pi(\varphi) = \varphi(M^{(\lambda)}[\mu, \delta])$ for all μ, δ with

$0 < \mu < 1$, $\delta > 0$. But both sides of this equation are real analytic functions of μ for each fixed $\delta > 0$. So, by analytic continuation, the equation holds with the same associated representation π for $\mu > 0, \delta > 0$. (One might object that this argument is incorrect: $\varphi(M^{(\lambda)}[\mu,\delta])$ may not be defined for arbitrary μ, δ. This objection is overruled by noting that, according to [1, Lemma 2.2], to establish homogeneity of an operator T, it is enough to show that $\varphi(T)$ is unitarily equivalent to T for all φ in an arbitrarily small neighborhood of identity in Möb.) We omit the messy proof of the irreducibility and unitary inequivalence of these operators.

□

References

[1] B. BAGCHI and G. MISRA, Constant characteristic functions and homogeneous operators, *J. Operator Theory,* **37** (1997), 51–65.

[2] B. BAGCHI and G. MISRA, Homogeneous operators and projective representations of the Möbius group, a survey, *Proc. Ind. Acad. Sci.* (2001), to appear.

[3] D. N. CLARK and G. MISRA, On some homogeneous contractions and unitary representations of $SU(1,1)$, *J. Operator Theory,* **30** (1993), 109–122.

[4] R. G. DOUGLAS and G. MISRA, Geometric invariants for resolutions of Hilbert modules, *Operator Theory: Advances and Applications,* Birkhäuser Verlag, **104** (1998), 83–112.

[5] G. MISRA, Curvature and discrete series representation of $SL_2(\mathbb{R})$, *J. Integral Equations Operator Theory,* **9** (1986), 452–459.

[6] G. MISRA and N. S. N. SASTRY, Homogeneous tuples of operators and holomorphic discrete series representation of some classical groups, *J. Operator Theory,* **24** (1990), 23–32.

[7] B. SZ.-NAGY and C. FOIAS, *Harmonic analysis of operators on Hilbert space,* North Holland.

[8] D. R. WILKINS, Homogeneous vector bundles and Cowen–Douglas operators, *International J. Math.,* **4** (1993), 503–520.

B. BAGCHI, Indian Statistical Institute, R. V. College Post, Bangalore 560 059, India
G. MISRA, Indian Statistical Institute, R. V. College Post, Bangalore 560 059, India

Operator Theory: Advances and Applications, Vol. 127, 113–126

Inequalities for eigenvalues of sums in a von Neumann algebra

HARI BERCOVICI and WING SUET LI

This paper is dedicated to the memory of Professor Béla Szőkefalvi-Nagy, a founder for operator theory, and a mentor for us

We prove an analogue of the Freede–Thompson inequalities for the sum of two selfadjoint elements of a von Neumann algebra. The result also applies to singular values, and implies earlier results of Grothendieck and Fack.

1. Introduction

Given an $n \times n$ selfadjoint matrix A, denote by $\Lambda(A) = \{\lambda_1(A) \geq \lambda_2(A) \geq \cdots \geq \lambda_n(A)\}$ the sequence of its eigenvalues in decreasing order and repeated according to their multiplicities. If B is another selfadjoint matrix and $C = A + B$, the relationship among the sequences $\Lambda(A)$, $\Lambda(B)$ and $\Lambda(C)$ has been investigated since Weyl's work [7]. Recently, due to a breakthrough by Klyachko [5], complete sets of inequalities satisfied by these sequences were found. One such set of inequalities was conjectured by Horn [4], and that conjecture has now been verified. See Fulton [2] for an account of these results, along with their connections with geometry and invariant theory.

If $\mathfrak{A}$ is a factor of type II_1 (or just a finite von Neumann algebra without minimal projections), and $a \in \mathfrak{A}$ is a selfadjoint element, we can also talk about the 'eigenvalues' of a, and these are embodied in a decreasing function $\lambda_a\colon [0,1] \to \mathbb{R}$. Given selfadjoint elements $a, b, c \in \mathfrak{A}$ with $c = a + b$ one can inquire about the

---

AMS Subject Classification (1991): 46L10, 15A42.

The authors were partially supported by grants from the National Science Foundation.

relationships among the functions λ_a, λ_b and λ_c. In this paper we provide a class of inequalities of the form

$$\int_A \lambda_a(t)\,dt + \int_B \lambda_b(t)\,dt \geq \int_C \lambda_c(t)\,dt,$$

where A, B, and C are Borel sets in $[0,1]$ with equal Lebesgue measure. Our inequalities generalize those given by Freede and Thompson [6] which in turn are closely related to the Horn inequalities. In particular we have

$$\int_0^s \lambda_a(t)\,dt + \int_0^s \lambda_b(t)\,dt \geq \int_0^s \lambda_c(t)\,dt, \quad s \in [0,1].$$

The analogue of this inequality for singular values appears in [3] (Corollary 5) and [1]; we will see that inequalities for singular values follow from the corresponding ones for eigenvalues. Another consequence of our results is the following analogue of Weyl's inequality

$$\lambda_a(s) + \lambda_b(t) \geq \lambda_c(s+t)$$

provided that $s+t \leq 1$ and the functions λ are taken to be right continuous. Again, this result was proved in [1] for singular values.

As in the finite-dimensional case, our inequalities are derived from the existence of projections in $\mathfrak{A}$ which satisfy certain properties. These projections belong to intersections of analogues of the Schubert varieties in the Grassmannian. It may turn out that a complete collection of inequalities can be obtained this way.

2. Eigenvalue inequalities

For the remainder of this paper we fix a factor $\mathfrak{A}$ of type II_1 with normalized faithful normal trace τ. Given a selfadjoint element $a \in \mathfrak{A}$, there is an $\mathfrak{A}$-valued spectral measure E with bounded support in $\mathbb{R}$ such that $a = \int_{-\infty}^{\infty} xE(dx)$. The function $x \mapsto u(x) = \tau(E(x,\infty))$ is nonincreasing, and the closed support $\operatorname{supp}(E)$ is the complement of the largest open set G such that $u'(x) = 0$ for $x \in G$. Clearly u is invertible on $\operatorname{supp}(E)$; call λ_a the unique nonincreasing right-continuous function from $[0,1]$ to $\mathbb{R}$ such that $\lambda_a(u(x)) = x$ for most x in $\operatorname{supp}(E)$. In other words,

$$\lambda_a(t) = \inf\{x : u(x) \leq t\}, \quad t \in [0,1].$$

The spectrum of a coincides with $\operatorname{supp}(E)$ and is therefore the closure of the range of λ_a. It is natural to view λ_a as representing the 'eigenvalues' of a. Of course, the actual eigenvalues of a are values μ for which $\{t : \lambda_a(t) = \mu\}$ is a nondegenerate interval.

Assume now that $t \in [0,1]$ is such that $u(\lambda_a(t)) = t$; for such t we set $e_t = E(\lambda_a(t), \infty)$. It is now easy to see that e_t can in fact be defined for all $t \in [0,1]$ such that $e_t \leq e_{t'}$ for $t < t'$, and $\tau(e_t) = t$ for all t. We can then represent a in terms of its eigenvalue function as

$$a = \int_0^1 \lambda_a(t)\, de_t.$$

The basic tool for deriving eigenvalue inequalities is the following variational inequality whose linear algebraic counterpart is well known (see for instance [8]). We will use $\wedge$ and $\vee$ to denote the usual lattice operations among orthogonal projections. Thus, $p \wedge q$ is the orthogonal projection onto the intersection of the ranges of p and q.

PROPOSITION 2.1. *Fix a selfadjoint element $a \in \mathfrak{A}$ with eigenvalue decomposition $a = \int_0^1 \lambda_a(t)\, de_t$. For every projection p in $\mathfrak{A}$ we have*

$$\tau(ap) \geq \int_0^1 \lambda_a(t)\, d\tau(p \wedge e_t).$$

Proof. Fix a natural number n and points $0 = t_0 < t_1 < \cdots < t_n = 1$. If we set $p_j = p \wedge e_{t_j}$ we have $p_j \leq e_{t_j}$ and $\sum_{j=1}^n (p_j - p_{j-1}) = p$. Now, λ_a is decreasing so that

$$e_{t_j}(a - \lambda_a(t_j))e_{t_j} = \int_0^{t_j} (\lambda_a(t) - \lambda_a(t_j))\, de_t \geq 0$$

and therefore

$$(p_j - p_{j-1})(a - \lambda_a(t_j))(p_j - p_{j-1}) \geq 0.$$

Thus

$$(p_j - p_{j-1})a(p_j - p_{j-1}) \geq \lambda_a(t_j)(p_j - p_{j-1}),$$

and therefore

$$\begin{aligned}\tau(ap) &= \sum_{j=1}^n \tau(a(p_j - p_{j-1})) = \sum_{j=1}^n \tau((p_j - p_{j-1})a(p_j - p_{j-1})) \\ &\geq \sum_{j=1}^n \lambda(t_j)\tau(p_j - p_{j-1}).\end{aligned}$$

The last sum is a Riemann sum corresponding with the integral in the statement of the theorem. The proposition follows by letting the norm of the partition $\{t_0, t_1, \dots, t_n\}$ tend to zero.

□

In order to formulate a min-max characterization for eigenvalues we need the following easy result (cf. also Proposition 4 of [3]).

LEMMA 2.2. *Consider nondecreasing continuous functions* $\varphi_1, \varphi_2\colon [0,1] \to \mathbb{R}$ *such that* $\varphi_1(0) = \varphi_2(0) = 0$, $\varphi_1(1) = \varphi_2(1)$, *and* $\varphi_1(t) \geq \varphi_2(t)$ *for all* $t \in [0,1]$. *We have*

$$\int_0^1 \lambda(t)\, d\varphi_1(t) \geq \int_0^1 \lambda(t)\, d\varphi_2(t)$$

for every nonincreasing function $\lambda\colon [0,1] \to \mathbb{R}$.

Proof. Integration by parts yields

$$\int_0^1 \lambda(t)\, d\varphi(t) = \varphi(1)\lambda(1) + \int_0^1 \varphi(t)\, d(-\lambda(t))$$

when $\varphi(0) = 0$. The lemma follows immediately from this identity.

□

The functions to which we will apply the lemma are of the form $\varphi(t) = \tau(p \wedge e_t)$, where p is a projection. In general, we will reserve the letter p for a generic selfadjoint projection in $\mathfrak{A}$. We will use $|E|$ to denote the Lebesgue measure of a set E.

PROPOSITION 2.3. *Fix a selfadjoint element* $a \in \mathfrak{A}$ *with eigenvalue decomposition* $a = \int_0^1 \lambda_a(t)\, de_t$, *and let* $E \subset [0,1]$ *be a Borel set. We have*

$$\int_E \lambda_a(t)\, dt = \min\{\tau(ap) : \tau(p) = |E|, \tau(p \wedge e_t) \geq |E \cap [0,t]| \text{ for all } t \in [0,1]\}.$$

Proof. Assume first that p is such that $\tau(p) = |E|$ and $\tau(p \wedge e_t) \geq |E \cap [0,t]|$ for $t \in [0,1]$. Then Proposition 2.1 and Lemma 2.2 (with $\varphi_1(t) = \tau(p \wedge e_t)$ and $\varphi_2(t) = |E \cap [0,t]|$, yield

$$\tau(ap) \geq \int_0^1 \lambda_a(t)\, d\varphi_1(t) \geq \int_0^1 \lambda_a(t)\, d\varphi_2(t) = \int_E \lambda_a(t)\, dt.$$

Equality in these relations is achieved for $p = \int_E de_t$.

□

3. Intersections of Schubert varieties

The finite-dimensional analogue of Proposition 2.3 relates eigenvalue inequalities with Schubert varieties. One can formulate an analogue of these ideas in the context of our factor $\mathfrak{A}$. Namely, given a number $\delta \in [0,1]$, consider the collection $\mathfrak{G}(\mathfrak{A},\delta)$ of all projections $p \in \mathfrak{A}$ with $\tau(p) = \delta$. This is an analogue of the Grassmann manifold; it is a space on which the unitary group of $\mathfrak{A}$ acts transitively via the action $(u,p) \mapsto upu^*$. Fix now an increasing family $e = (e_t)_{t\in[0,1]}$ of projections in $\mathfrak{A}$ with $\tau(e_t) = t$, $t \in [0,1]$. Such a collection of projections will be called a flag in $\mathfrak{A}$. Fix also a function $\varphi\colon [0,1] \to [0,1]$ such that $\varphi(0) = 0$ and $0 \le \varphi(t) - \varphi(s) \le t - s$ whenever $0 \le s \le t \le 1$; set $\delta = \varphi(1)$. The set

$$\mathfrak{S}(\varphi,e) = \{p \in \mathfrak{G}(\mathfrak{A},\delta) : \tau(p \wedge e_t) \ge \varphi(t), t \in [0,1]\}$$

is an analogue of the finite-dimensional Schubert varieties. The corresponding Schubert cell would be

$$\sigma(\varphi,e) = \{p \in \mathfrak{G}(\mathfrak{A},\delta) : \tau(p \wedge e_t) = \varphi(t), t \in [0,1]\}.$$

The varieties entering Proposition 2.3 correspond to $\varphi(t) = |A \cap [0,t]|$ for some Borel set $A \subset [0,1]$ with $|A| = \delta$. We will use the notation $\mathfrak{S}(\mathfrak{A},e) = \mathfrak{S}(\varphi,e)$ when φ is of this form. Inequalities for eigenvalues of sums of selfadjoint elements can be obtained when one can show that various Schubert varieties have nonempty intersection. We will produce such nonempty intersections which are related with an operation defined on closed subsets of $[0,\infty)$. Given such a closed set A let us define

$$\psi_A(x) = \inf\{t : |A \cap [0,t]| \ge x\}, \quad x \in [0,\infty).$$

Then $\psi_A(x) = \infty$ for $x > |A|$, and A coincides almost everywhere with the range of ψ_A. The function ψ_A can be described rather explicitly. For each connected component I of the complement $A^c = [0,\infty)\setminus A$ introduce the point $x_I = |A\cap[0,t]|$, where t is an arbitrary point in I. Then we have

$$\psi_A(x) = x + \sum_I |I|\chi_{[x_I,\infty)}(x),$$

where we denoted by χ_B the characteristic function of the set B. Conversely, given a function $\psi\colon [0,\infty) \to [0,\infty]$ such that $\psi(t) - t$ is an increasing, right continuous, pure jump function, then $\psi = \psi_A$ for some closed set A, uniquely determined up to sets of Lebesgue measure zero. Consider now two closed sets A and B with $|A| = |B| > 0$. The function

$$\psi(x) = \psi_A(x) + \psi_B(x) - x, \quad x \in [0,\infty),$$

is increasing, right-continuous, and pure jump, hence $\psi = \psi_C$ for some compact subset C of $[0,\infty)$. Observe that $|C| = |A| = |B|$. We will use the notation $A\widehat{+}B$ for the set C. To get a better understanding of this operation, let us consider a few examples. If $A = [\alpha-\delta,\alpha]$ and $B = [\beta-\delta,\beta]$ then $A\widehat{+}B = [\gamma-\delta,\gamma]$, where $\gamma = \alpha+\beta-\delta$. In general, if α is the essential last point in A (i.e., the least α such that $|A\cap(\alpha,\infty)| = 0$) and β is the essential last point of B, then the essential last point of $A\widehat{+}B$ is $\alpha+\beta-|A|$. If $A' = A\cup A_1$ with A_1 to the right of A (i.e. $x_1 \geq x$ for all $x_1 \in A_1$ and $x\in A$; in symbols, $A_1 \geq A$) and $B' = B\cup B_1$ with $B_1 \geq B$, then $A'\widehat{+}B' = (A\widehat{+}B)\cup C_1$, where $C_1 \geq A\widehat{+}B$ and $|C_1| = |A_1| = |B_1|$. From this it is relatively easy to calculate $A\widehat{+}B$ if $|A| = |B|$, and both sets are finite unions of compact intervals. Indeed, in this case we can find a natural number n and numbers $\alpha_j,\beta_j,\varepsilon_j$, $j = 1,2\dots,n$ such that $\alpha_j-\varepsilon_j \geq \alpha_{j-1}$, $\beta_j-\varepsilon_j \geq \beta_{j-1}$ for $j = 2,3,\dots,n$, and

$$A = \bigcup_{j=1}^{n}[\alpha_j-\varepsilon_j,\alpha_j],\quad B = \bigcup_{j=1}^{n}[\beta_j-\varepsilon_j,\beta_j].$$

The set $C = A\widehat{+}B$ can similarly be written as

$$C = \bigcup_{j=1}^{n}[\gamma_j-\varepsilon_j,\gamma_j],$$

where

$$\gamma_j = \alpha_j+\beta_j-\sum_{i=1}^{j}\varepsilon_i,\quad j = 1,2,\dots,n.$$

It may be interesting to note the spacing between consecutive intervals in the decompositions above:

$$(\gamma_j-\varepsilon_j)-\gamma_{j-1} = [(\alpha_j-\varepsilon_j)-\alpha_{j-1}]+[(\beta_j-\varepsilon_j)-\beta_{j-1}],\quad j = 2,3,\dots n.$$

Consider now the metric $(A_1,A_2)\mapsto |A_1\triangle A_2|$, where $\triangle$ indicates symmetric difference. Compact sets can be approximated arbitrarily well in this metric by finite unions of intervals. The following result will allow us to reduce certain calculations to the case in which the sets are finite unions of intervals.

Proposition 3.1. *The map $(A,B)\mapsto A\widehat{+}B$ is continuous when restricted to closed sets with finite measure. That is, if A and B are given closed sets with $|A| = |B| < \infty$, and $\varepsilon > 0$, there exists $\delta > 0$ such that for any closed sets A', B' satisfying $|A'| = |B'|$, $|A'\triangle A| < \delta$, and $|B'\triangle B| < \delta$, we have $|(A'\widehat{+}B')\triangle(A\widehat{+}B)| < \varepsilon$.*

Proof. First note that, given a closed set A with finite measure, one can find $a > 0$ such that $|A \cap [0,a]|$ is arbitrarily close to $|A|$. Applying this observation to all the sets A, B, A', B' we have to handle, it is easy to see from the preceding observations that it is sufficient to prove the proposition for compact sets A and B. Assume therefore that compact sets A and B with equal measure are given, along with a positive number ε. We can find pairwise disjoint compact intervals $J_1, J_2, \dots, J_N$ such that the set $K = \bigcup_{i=1}^{N} J_i$ contains $A \hat{+} B$ and $|K \setminus (A \hat{+} B)| < \varepsilon$. Let us set $\delta = \varepsilon/N$, and assume that A' and B' are compact sets with equal measure such that $|A' \triangle A| < \delta$ and $|B' \triangle B| < \delta$. It follows easily that for every $x \geq \delta$ we have $\psi_A(x) \geq \psi_{A'}(x-\delta)$, $\psi_{A'}(x) \geq \psi_A(x-\delta)$, $\psi_B(x) \geq \psi_{B'}(x-\delta)$, and $\psi_{B'}(x) \geq \psi_B(x-\delta)$. From the definition of $\hat{+}$ we deduce that for such x we also have $\psi_{A \hat{+} B}(x) \geq \psi_{A' \hat{+} B'}(x-\delta)$ and $\psi_{A' \hat{+} B'}(x) \geq \psi_{A \hat{+} B}(x-\delta)$. This relation can now be converted into

$$||(A \hat{+} B) \cap [0,t]| - |(A' \hat{+} B') \cap [0,t]|| \leq \delta$$

for all $t \geq 0$, in particular

$$||(A \hat{+} B) \cap J_i| - |(A' \hat{+} B') \cap J_i|| \leq 2\delta$$

for $i = 1, 2 \dots, N$. Since

$$\sum_{i=1}^{N} (|J_i| - |(A \hat{+} B) \cap J_i|) < \varepsilon,$$

we conclude that

$$\sum_{i=1}^{N} (|J_i| - |(A' \hat{+} B') \cap J_i|) < \varepsilon + 2N\delta = 3\varepsilon.$$

Thus

$$\begin{aligned}
&|(A' \hat{+} B') \triangle (A \hat{+} B)| \\
&\quad \leq |(A' \hat{+} B') \triangle K| + \varepsilon \\
&\quad \leq |A' \hat{+} B'| - |(A' \hat{+} B') \cap K| + 3\varepsilon + \varepsilon = |A' \hat{+} B'| - |K| + |K \setminus (A' \hat{+} B')| + 4\varepsilon \\
&\quad \leq |A' \hat{+} B'| - |A \hat{+} B| + 7\varepsilon = |A'| - |A| + 7\varepsilon < \delta + 7\varepsilon < 8\varepsilon.
\end{aligned}$$

The lemma follows.

□

It may be amusing to note that the continuity in Proposition 3.1 is not uniform, and it really does not hold for sets of infinite measure. To see this latter fact consider $A = B = \bigcup_{k=0}^{\infty} [2k, 2k+1]$ in which case $A \hat{+} B = \bigcup_{k=0}^{\infty} [3k, 3k+1]$. If we take $A' = B' = A \setminus [0, \delta]$, $\delta > 0$, then $(A \hat{+} B) \triangle (A' \hat{+} B')$ has infinite measure. The lack of uniform continuity can be proved by taking finite unions in this example.

If $A \subset [0,1]$ we will write $1 - A = \{1 - x : x \in A\}$. We can now prove our main result concerning intersections of varieties.

Theorem 3.2. *Let e, f, and g be three flags in $\mathfrak{A}$. Assume that A and B are two compact subsets of $[0,1]$, both of which are finite unions of intervals, such that $|A| = |B|$ and $A\widehat{+}B$ is also contained in $[0,1]$. Then*

$$\mathfrak{S}(1-A, e) \cap \mathfrak{S}(1-B, f) \cap \mathfrak{S}(A\widehat{+}B, g) \neq \emptyset.$$

Proof. As observed earlier, the sets A, B, and $A\widehat{+}B$ can be written as

$$A = \bigcup_{i=1}^{n} [\alpha_i - \varepsilon_i, \alpha_i], \quad B = \bigcup_{i=1}^{n} [\beta_i - \varepsilon_i, \beta_i], \quad A\widehat{+}B = \bigcup_{i=1}^{n} [\gamma_i - \varepsilon_i, \gamma_i],$$

with $\alpha_1 - \varepsilon_1 \geq 0$, $\beta_1 - \varepsilon_1 \geq 0$, $\alpha_i - \varepsilon_i \geq \alpha_{i-1}$, $\beta_i - \varepsilon_i \geq \beta_{i-1}$, and $\gamma_i = \alpha_i + \beta_i - \sum_{j=1}^{i} \varepsilon_j$. Let us observe that a projection p in $\mathfrak{S}(A\widehat{+}B, g)$ satisfies

$$\tau(p) = |A\widehat{+}B| = \sum_{i=1}^{n} \varepsilon_i,$$

and

$$\tau(p \wedge g_{\gamma_i}) \geq \varepsilon_1 + \varepsilon_2 + \cdots + \varepsilon_i \quad \text{for } i = 1, 2, \ldots, n.$$

(Note also that the second condition for $i = n$ simply says that $p \leq g_{\gamma_n}$.) In fact these two conditions are also sufficient for p to belong to $\mathfrak{S}(A\widehat{+}B, g)$. Indeed, for $t \in (\gamma_i, \gamma_{i+1} - \varepsilon_{i+1})$ we have

$$\tau(p \wedge g_t) \geq \tau(p \wedge g_{\gamma_i}) \geq \varepsilon_1 + \varepsilon_2 + \cdots + \varepsilon_i = |A \cap [0,t]|,$$

while for $t \in [\gamma_i - \varepsilon_i, \gamma_i]$

$$p \wedge g_{\gamma_i} \leq p \wedge g_t + g_{\gamma_i} - g_t,$$

whence

$$\varepsilon_1 + \varepsilon_2 + \cdots + \varepsilon_i \leq \tau(p \wedge g_{\gamma_i}) \leq \tau(p \wedge g_t) + \gamma_i - t.$$

Thus $\tau(p \wedge g_t) \geq \varepsilon_1 + \varepsilon_2 + \cdots \varepsilon_i - (\gamma_i - t) = |A \cap [0,t]|$. Analogously, the condition $p \in \mathfrak{S}(1-A, e)$ reduces to $\tau(p) = |A|$ and

$$\tau(p \wedge e_{1-\alpha_i+\varepsilon_i}) \geq \varepsilon_i + \varepsilon_{i+1} + \cdots + \varepsilon_n \quad \text{for } i = 1, 2, \ldots, n,$$

while $p \in \mathfrak{S}(1-B, f)$ reduces to $\tau(p) = |A|$ and

$$\tau(p \wedge f_{1-\beta_i+\varepsilon_i}) \geq \varepsilon_i + \varepsilon_{i+1} + \cdots + \varepsilon_n \quad \text{for } i = 1, 2, \ldots, n.$$

Denote $A_0 = \bigcup_{i=1}^{n-1}[\alpha_i - \varepsilon_i, \alpha_i]$, $B_0 = \bigcup_{i=1}^{n-1}[\beta_i - \varepsilon_i, \beta_i]$, and observe that $A_0 \widehat{+} B_0 = \bigcup_{i=1}^{n-1}[\gamma_i - \varepsilon_i, \gamma_i]$. The theorem is trivial when $n = 0$, and it will be proved by induction on n. The inductive step consists in showing that the conclusion of the theorem is true provided that

$$\mathfrak{S}(1 - A_0, e) \cap \mathfrak{S}(1 - B_0, f) \cap \mathfrak{S}(A_0 \widehat{+} B_0, g) \neq \emptyset.$$

Fix therefore a projection p_0 in this intersection. As noted above, this projection satisfies

$$\tau(p_0) = |A_0| = \sum_{i=1}^{n-1} \varepsilon_i,$$

$$\tau(p_0 \wedge e_{1-\alpha_i+\varepsilon_i}) \geq \varepsilon_i + \varepsilon_{i+1} + \cdots + \varepsilon_{n-1} \quad \text{for } i = 1, 2, \ldots, n-1,$$

$$\tau(p_0 \wedge f_{1-\beta_i+\varepsilon_i}) \geq \varepsilon_i + \varepsilon_{i+1} + \cdots + \varepsilon_{n-1} \quad \text{for } i = 1, 2, \ldots, n-1,$$

and

$$\tau(p_0 \wedge g_{\gamma_i}) \geq \varepsilon_1 + \varepsilon_2 + \cdots + \varepsilon_i \quad \text{for } i = 1, 2, \ldots, n-1.$$

Denote by $\mathcal{P}$ the collection of all projections $p \in \mathfrak{A}$ satisfying the following conditions:

$$p \geq p_0 \text{ and } \eta = \tau(p) - \tau(p_0) \leq \varepsilon_n, \tag{3.3}$$

$$\tau(p \wedge e_{1-\alpha_i+\varepsilon_i}) \geq \varepsilon_i + \varepsilon_{i+1} + \cdots + \varepsilon_{n-1} + \eta \quad \text{for } i = 1, 2, \ldots, n, \tag{3.4}$$

$$\tau(p \wedge f_{1-\beta_i+\varepsilon_i}) \geq \varepsilon_i + \varepsilon_{i+1} + \cdots + \varepsilon_{n-1} + \eta \quad \text{for } i = 1, 2, \ldots, n, \tag{3.5}$$

and

$$p \leq g_{\gamma_n}. \tag{3.6}$$

The set $\mathcal{P}$ is not empty since it contains p_0. The weak* continuity of τ, and the fact that increasing sequences of projections converge in the weak* topology, imply that $\mathcal{P}$ is inductive and therefore it contains maximal elements. Let p be such a maximal element. The proof will be concluded by showing that the number η equals ε_n, and therefore p belongs to $\mathfrak{S}(1 - A, e) \cap \mathfrak{S}(1 - B, f) \cap \mathfrak{S}(A \widehat{+} B, g)$. Assume to the contrary that $\eta < \varepsilon_n$. In this case there must be i_0 and j_0 such that

$$\tau(p \wedge e_{1-\alpha_{i_0}+\varepsilon_{i_0}}) < \varepsilon_{i_0} + \varepsilon_{i_0+1} + \cdots + \varepsilon_n$$

and

$$\tau(p \wedge f_{1-\beta_{j_0}+\varepsilon_{j_0}}) < \varepsilon_{j_0} + \varepsilon_{j_0+1} + \cdots + \varepsilon_n.$$

Such integers certainly exist (e.g., $i_0 = j_0 = 1$); we choose them as large as possible. Given an arbitrary projection q let us set $\pi(q) = (q \vee p) - p$. In order to facilitate further calculations let us note that

$$\begin{aligned}\tau(\pi(e_{1-\alpha_i+\varepsilon_i})) &= \tau(p \vee e_{1-\alpha_i+\varepsilon_i}) - \tau(p)\\ &= \tau(e_{1-\alpha_i+\varepsilon_i}) - \tau(p \wedge e_{1-\alpha_i+\varepsilon_i})\\ &= 1 - \alpha_i + \varepsilon_i - \tau(p \wedge e_{1-\alpha_i+\varepsilon_i}).\end{aligned}$$

Let us also write $\mu_i = \varepsilon_i + \varepsilon_{i+1} + \cdots + \varepsilon_n - \tau(p \wedge e_{1-\alpha_i+\varepsilon_i})$. Note that μ_i may be negative; it certainly is positive for $i = i_0$ and $\mu_1 = \varepsilon_n - \eta$. The preceding formula gives then

$$\tau(\pi(e_{1-\alpha_i+\varepsilon_i})) = \mu_i + 1 - \alpha_i - \varepsilon_{i+1} - \cdots - \varepsilon_n.$$

Of course, $\alpha_i + \varepsilon_{i+1} \leq \alpha_{i+1}$ so that

$$\alpha_i + \varepsilon_{i+1} + \cdots + \varepsilon_n \leq \alpha_n,$$

and we obtain the inequalities

$$\tau(\pi(e_{1-\alpha_i+\varepsilon_i})) \geq \mu_i + 1 - \alpha_n.$$

Analogous relations hold for the flag f and the quantities $\nu_j = \varepsilon_j + \varepsilon_{j+1} + \cdots + \varepsilon_n - \tau(p \wedge f_{1-\beta_j+\varepsilon_j})$. Using these inequalities along with the formula $\gamma_n = \alpha_n + \beta_n - |A|$ we see that

$$\begin{aligned}\tau(\pi(e_{1-\alpha_i+\varepsilon_i})) &+ \tau(\pi(f_{1-\beta_j+\varepsilon_j})) + \tau(\pi(g_{\gamma_n}))\\ &\geq \mu_i + 1 - \alpha_n + \nu_j + 1 - \beta_n + \gamma_n - \tau(p)\\ &= 2 + \mu_i + \nu_j - \tau(p) - \varepsilon_1 - \varepsilon_2 - \cdots - \varepsilon_n.\end{aligned}$$

If, for instance, $j = 1$ and $\mu_i > 0$, the last quantity is

$$\mu_i + 2 - \tau(p) - (|A| - \nu_1) = \mu_i + 2 - \tau(p) - (\varepsilon_1 + \varepsilon_2 + \cdots + \varepsilon_{n-1} + \eta) = \mu_i + 2(1 - \tau(p))$$

which is greater that $2(1 - \tau(p))$. Since the three projections whose traces are added are all less than $1 - p$, we conclude that

$$\pi(e_{1-\alpha_i+\varepsilon_i}) \wedge \pi(f_{1-\beta_1+\varepsilon_1}) \wedge \pi(g_{\gamma_n}) \neq 0$$

provided that $\mu_i > 0$, in particular for $i = i_0$. It follows that we can choose a largest index j_1 such that $j_1 \leq j_0$ and

$$\pi(e_{1-\alpha_{i_0}+\varepsilon_{i_0}}) \wedge \pi(f_{1-\beta_{j_1}+\varepsilon_{j_1}}) \wedge \pi(g_{\gamma_n}) \neq 0.$$

If $j_1 \neq j_0$ and $j_1 < j \leq j_0$, we have

$$\pi(e_{1-\alpha_{i_0}+\varepsilon_{i_0}}) \wedge \pi(f_{1-\beta_j+\varepsilon_j}) \wedge \pi(g_{\gamma_n}) = 0,$$

and the previous inequalities then yield

$$2 + \mu_{i_0} + \nu_j - \tau(p) - \varepsilon_1 - \varepsilon_2 - \cdots - \varepsilon_n \leq 2 - 2\tau(p)$$

or, equivalently,

$$\mu_{i_0} + \nu_j \leq \varepsilon_n - \eta$$

for such values of j. Choose now a projection r such that

$$r \leq \pi(e_{1-\alpha_{i_0}+\varepsilon_{i_0}}) \wedge \pi(f_{1-\beta_{j_1}+\varepsilon_{j_1}}) \wedge \pi(g_{\gamma_n})$$

and $0 < \tau(r) < \mu_{i_0}$. We set $p' = p + r$ and contradict the maximality of p by showing that $p' \in \mathcal{P}$. To do this we need to verify conditions (3.4–6) with p and η replaced by p' and $\eta' = \eta + \tau(r)$, respectively. Condition (3.6) is immediate since $r \leq \pi(g_{\gamma_n}) = g_{\gamma_n} - p$. To verify (3.4) note first that for $i > i_0$ we have

$$\begin{aligned}\tau(p' \wedge e_{1-\alpha_i+\varepsilon_i}) &\geq \tau(p \wedge e_{1-\alpha_i+\varepsilon_i}) \\ &\geq \varepsilon_i + \varepsilon_{i+1} + \cdots + \varepsilon_n \\ &\geq \varepsilon_i + \varepsilon_{i+1} + \cdots + \varepsilon_{n-1} + \eta'\end{aligned}$$

by the choice of i_0. On the other hand, for $i \leq i_0$ we have

$$r \leq p \vee e_{1-\alpha_{i_0}+\varepsilon_{i_0}} \leq p \vee e_{1-\alpha_i+\varepsilon_i}$$

and therefore

$$\begin{aligned}\tau(e_{1-\alpha_i+\varepsilon_i}) - \tau(p' \wedge e_{1-\alpha_i+\varepsilon_i}) &= \tau(p' \vee e_{1-\alpha_i+\varepsilon_i}) - \tau(p') \\ &= \tau(p \vee e_{1-\alpha_i+\varepsilon_i}) - \tau(p') \\ &= \tau(p \vee e_{1-\alpha_i+\varepsilon_i}) - \tau(p) - \tau(r) \\ &= \tau(e_{1-\alpha_i+\varepsilon_i}) - \tau(p \wedge e_{1-\alpha_i+\varepsilon_i}) - \tau(r),\end{aligned}$$

from which we deduce that $\tau(p' \wedge e_{1-\alpha_i+\varepsilon_i}) = \tau(p \wedge e_{1-\alpha_i+\varepsilon_i}) + \tau(r)$. Thus, for such values of i,

$$\begin{aligned}\tau(p' \wedge e_{1-\alpha_i+\varepsilon_i}) &\geq \varepsilon_i + \varepsilon_{i+1} + \cdots + \varepsilon_{n-1} + \eta + \tau(r) \\ &= \varepsilon_i + \varepsilon_{i+1} + \cdots + \varepsilon_{n-1} + \eta',\end{aligned}$$

where we used the inequality (3.4) for the original p. The inequality

$$\tau(p' \wedge f_{1-\beta_j+\varepsilon_j}) \geq \varepsilon_j - \varepsilon_{j+1} + \cdots + \varepsilon_{n-1} + \eta'$$

is verified in the same way provided that either $j > j_0$ or $j \leq j_1$. It remains to verify this inequality when $j_1 < j \leq j_0$. For such values of j we showed earlier that $\mu_{i_0} + \nu_j \leq \varepsilon_n - \eta$. Thus

$$\begin{aligned}\tau(p' \wedge f_{1-\beta_j+\varepsilon_j}) &\geq \tau(p \wedge f_{1-\beta_j+\varepsilon_j}) = \varepsilon_j + \varepsilon_{j+1} + \cdots + \varepsilon_n - \nu_j \\ &\geq \varepsilon_j + \varepsilon_{j+1} + \cdots + \varepsilon_{n-1} + \eta + \mu_{i_0} \\ &\geq \varepsilon_j + \varepsilon_{j+1} + \cdots + \varepsilon_{n-1} + \eta'\end{aligned}$$

by the choice $\tau(r) \leq \mu_{i_0}$. The proof is now complete. □

We can now prove our analogue of the inequalities of Freede and Thompson.

THEOREM 3.7. *Let $A, B \subset [0,1]$ be compact sets such that $A \hat{+} B \subset [0,1]$. Given selfadjoint elements $a, b \in \mathfrak{A}$ we have*

$$\int_{A\hat{+}B} \lambda_{a+b}(t)\, dt \leq \int_A \lambda_a(t)\, dt + \int_B \lambda_b(t)\, dt.$$

Proof. From Proposition 3.1 it follows that it suffices to prove the theorem under the additional assumption that A and B are finite unions of intervals. Let us set $x = -a$, $y = -b$, and $z = a + b$. Then $x + y + z = 0$ and x, y, z have eigenvalue decompositions $x = \int_0^1 \lambda_x(t)\, de_t$, $y = \int_0^1 \lambda_y(t)\, df_t$, $z = \int_0^1 \lambda_z(t)\, dg_t$, with $\lambda_x(t) = -\lambda_a(1-t)$, $\lambda_y(t) = -\lambda_b(1-t)$, $\lambda_z = \lambda_{a+b}$. The inequality to be proved becomes

$$\int_{1-A} \lambda_x(t)\, dt + \int_{1-B} \lambda_y(t)\, dt + \int_{A\hat{+}B} \lambda_z(t)\, dt \leq 0.$$

To prove this choose a projection $p \in \mathfrak{S}(1-A, e) \cap \mathfrak{S}(1-B, f) \cap \mathfrak{S}(A\hat{+}B, g)$. By Proposition 2.1 we have $\int_{1-A} \lambda_x(t)\, dt \leq \tau(px)$, $\int_{1-B} \lambda_y(t)\, dt \leq \tau(py)$, and $\int_{A\hat{+}B} \lambda_z(t)\, dt \leq \tau(pz)$. The desired result follows by adding these three inequalities and noting that $\tau(px) + \tau(py) + \tau(pz) = \tau(p(x+y+z)) = \tau(0) = 0$. □

When $A = B = [0, \alpha]$ we also have $A \hat{+} B = [0, \alpha]$, so we obtain the inequality

$$\int_0^\alpha \lambda_{a+b}(t)\, dt \leq \int_0^\alpha \lambda_a(t)\, dt + \int_0^\alpha \lambda_b(t)\, dt, \quad \alpha \in [0,1],$$

which is analogous to the Ky Fan inequality for sums of selfadjoint matrices. If $x, y > 0$ and $x + y < 1$ we can take $A = [x, x + \varepsilon]$, $B = [y, y + \varepsilon]$, in which case $A \hat{+} B = [x + y, x + y + \varepsilon]$. The inequality provided by Theorem 3.7 can then be divided by ε, and letting ε tend to zero we get the following analogues of Weyl's inequalities:

$$\lambda_{a+b}(x + y) \leq \lambda_a(x) + \lambda_b(y)$$

provided that the functions λ were all chosen right continuous. If we choose $A = [x - \varepsilon, x]$, $B = [y - \varepsilon, y]$ then $A \hat{+} B = [x + y - 2\varepsilon, x + y - \varepsilon]$. The same argument yields inequalities for the left hand limits:

$$\lambda_{a+b}(x + y - 0) \leq \lambda_a(x - 0) + \lambda_b(y - 0).$$

4. Inequalities for singular values

The singular values of an element $x \in \mathfrak{A}$ are simply the eigenvalues of $|x| = \sqrt{x^* x}$. These singular values are denoted $s_x(t)$, $t \in [0, 1]$, and inequalities for singular values were studied earlier by Grothendieck [3] and Fack [1]. An observation due to Wielandt [8] in the matrix case shows that information about singular values can be obtained from the eigenvalues of another associated selfadjoint element. Namely, for every $x \in \mathfrak{A}$ consider the element

$$x_* = \begin{bmatrix} 0 & x \\ x^* & 0 \end{bmatrix}$$

of the algebra $\mathfrak{M}_2(\mathfrak{A})$ of 2×2 matrices over $\mathfrak{A}$. It is easily seen that x_* is unitarily equivalent to $|x| \oplus (-|x|)$ so that

$$\lambda_{x_*}(t) = s_x(2t), \quad t \in \left[0, \frac{1}{2}\right].$$

Since we also have $(x + y)_* = x_* + y_*$ and

$$\left(\frac{1}{2}A\right) \hat{+} \left(\frac{1}{2}B\right) = \frac{1}{2}(A \hat{+} B),$$

we obtain the analogue of Theorem 3.7 for singular values.

THEOREM 3.7. *Let $A, B \subset [0,1]$ be compact sets such that $|A| = |B|$ and $A \hat{+} B \subset [0,1]$. Given elements $x, y \in \mathfrak{A}$ we have*

$$\int_{A\hat{+}B} s_{x+y}(t)\,dt \leq \int_A s_x(t)\,dt + \int_B s_y(t)\,dt.$$

This result is in [3] and [2] for $A = B = [0, \alpha]$

References

[1] T. FACK, Sur la notion de valeur caractéristique, *J. Operator Theory,* **7** (1982), 307–333.

[2] W. FULTON, Eigenvalues, invariant factors, highest weights, and Schubert calculus, *Bull Amer. Math. Soc. (N.S.)*, **37** (2000), 209–249.

[3] A. GROTHENDIECK , Réarrangements de fonctions et inégalités de convexité dans les algèbres de von Neumann munies d'une trace, *Séminaire Bourbaki,* **3** Exp. No. **113** (1959), 127–139.

[4] A. HORN, Eigenvalues of sums of Hermitian matrices, *Pacific J. Math.*, **12** (1962), 225–241.

[5] A. KLYACHKO, Stable bundles, representation theory and Hermitian operators, *Selecta Math.* (N.S.), **4** (1998), 419–445.

[6] R. THOMPSON and L. FREEDE, On the eigenvalues of sums of Hermitian matrices, *Linear Algebra Appl.*, **4** (1971), 369–376.

[7] H. WEYL, Das asymtotische Verteilungsgesetz der Eigenwerte lineare partieller Differentialgleichungen, *Math. Ann.*, **71** (1912), 441–479.

[8] H. WIELANDT, An extremum property of sums of eigenvalues, *Proc. Amer. Math. Soc.*, **6** (1955), 106–110.

H. BERCOVICI, Mathematics Department Indiana University, Bloomington, IN 47405; *e-mail*: bercovic@indiana.edu

W. S. LI, School of Mathematics, Georgia Institute of Technology, Atlanta, GA 30332; *e-mail*: li@math.gatech.edu

Operator Theory: Advances and Applications, Vol. 127, 127–144

Weighted variants of the Three Chains Completion Theorem

A. Biswas, C. Foias and A. E. Frazho

This paper presents and solves a weighted interpolation problem which is the nonstationary equivalent of the weighted commutant lifting theorem introduced in [1]. This leads to a weighted version of the three chains interpolation theorem in [3] as well as to a weighted variant of the Carswell–Schubert intertwining extension problem (see [2]).

1. Introduction

In this note, we solve a weighted, nonstationary interpolation problem which contains a weighted version of the three chains interpolation theory developed in [3] as well as the weighted commutant lifting theorem established in [1]. We will first describe the setup. The orthogonal projection onto a subspace $\mathcal{H}$ of a Hilbert space is denoted by $P_{\mathcal{H}}$. Let $\{\mathcal{H}_k\}_{k\in\mathbb{Z}}$, $\{\mathcal{F}_k\}_{k\in\mathbb{Z}}$ and $\{\mathcal{K}_k\}_{k\in\mathbb{Z}}$ be three sequences of Hilbert spaces satisfying the inclusion relations $\mathcal{F}_k \subset \mathcal{K}_k$ for all $k \in \mathbb{Z}$, where $\mathbb{Z}$ is the set of all integers. Throughout, we will assume that $\mathbb{Z}$ is the underlying indexing set for a sequence, and to simplify the notation, the subscript $k \in \mathbb{Z}$ will be suppressed. In the geometric interpolation problem considered here, the given data is a sequence of operator triplets $\{A_k, T_k, \Gamma_k\}$ of the form

$$A_k: \mathcal{H}_k \to \mathcal{F}_k, \quad T_k: \mathcal{H}_k \to \mathcal{H}_{k+1} \text{ and } \Gamma_k: \mathcal{K}_k \to \mathcal{K}_{k+1} \qquad (k \in \mathbb{Z}). \tag{1.1}$$

Further, it is assumed that the sequence $\{\Gamma_k\}$ satisfies the following *lifting condition*

$$P_{\mathcal{F}_{k+1}}\Gamma_k = P_{\mathcal{F}_{k+1}}\Gamma_k P_{\mathcal{F}_k} \qquad (k \in \mathbb{Z}). \tag{1.2}$$

We say that a sequence of operators $\{B_k\}$ is an *interpolating sequence* for the data $\{A_k, T_k, \Gamma_k\}$ if each B_k maps $\mathcal{H}_k$ into $\mathcal{K}_k$ and satisfies the following two conditions

$$P_{\mathcal{F}_k} B_k = A_k \text{ and } B_k T_{k-1} = \Gamma_{k-1} B_{k-1} \qquad (k \in \mathbb{Z}). \tag{1.3}$$

AMS Subject Classification (1991): 47A20, 47A57.

Here we are interested in finding an interpolating sequence satisfying a certain supplementary constraint. To be precise, we say that a sequence of operator quadruplets $\{A_k, T_k, \Gamma_k, Q_k\}$ is an *admissible nonstationary set of data* if $\{A_k, T_k, \Gamma_k\}$ is a sequence satisfying (1.1), (1.2) and each Q_k is a positive operator on $\mathcal{H}_k$ satisfying

$$T_{k-1}^* Q_k T_{k-1} \geq c_{k-1}^2 Q_{k-1} \qquad (k \in \mathbb{Z}) \tag{1.4}$$

where $c_k = \max\{1, \|\Gamma_k\|\}$. In this case, a sequence $\{B_k\}$ is said to be a *weighted interpolating sequence* for $\{A_k, T_k, \Gamma_k, Q_k\}$ if $\{B_k\}$ is an interpolating sequence for $\{A_k, T_k, \Gamma_k\}$ satisfying the *subordination condition* $B_k^* B_k \leq Q_k$ for all $k \in \mathbb{Z}$. This naturally leads to the following nonstationary interpolation problem: Find necessary and sufficient condition for the existence of a weighted interpolating sequence $\{B_k\}$ for $\{A_k, T_k, \Gamma_k, Q_k\}$. By using the weighted commutant lifting theorem developed in [1], we will present in Section 2 a solution to this nonstationary interpolation problem. Note that if the weights $\{Q_k\}$ are uniformly bounded, that is, $\sup_{k\in\mathbb{Z}} \|Q_k\|$ is finite, then any weighted interpolating sequence $\{B_k\}$ is also uniformly bounded. Moreover, if T_k and Γ_k are inclusions for all k, then the sequence $\{B_k\}$ is strongly convergent. These remarks are used in Section 3 to obtain a weighted version of the three chains completion theorem in [3]. This version is used in Section 4 to generalize the Carswell–Schubert intertwining extension theorem. The proof is modeled after that given in [3] for the Carswell–Schubert theorem. As an application of that generalization, in Section 5, we characterize the solvability of the following factorization problem: given operator valued bounded analytic functions Θ and M on the unit disc $\mathbb{D}$, when does there exist a bounded analytic function G of the same kind such that $G\Theta = F$ on $\mathbb{D}$ and $GG^* \leq M^*M$ on $\partial\mathbb{D}$.

We conclude this introduction with a presentation of the weighted commutant lifting theorem in [1]. Recall that $\mathcal{G}$ is a co-invariant subspace for an operator R on $\mathcal{R}$ if $\mathcal{G}$ is an invariant subspace for R^*. Notice that $\mathcal{G}$ is a co-invariant subspace for R if and only if $P_{\mathcal{G}}R = P_{\mathcal{G}}RP_{\mathcal{G}}$. Now let T be an operator on $\mathcal{H}$. Let U' be a contraction on $\mathcal{K}'$, that is, $\|U'\| \leq 1$. Assume that $\mathcal{H}'$ is a co-invariant subspace for U', or equivalently, $P_{\mathcal{H}'}U' = P_{\mathcal{H}'}U'P_{\mathcal{H}'}$. Let T' be the compression of U' to $\mathcal{H}'$, that is, $T' = P_{\mathcal{H}'}U'|\mathcal{H}'$. Obviously, T' is a contraction. Finally, assume that A is an operator from $\mathcal{H}$ into $\mathcal{H}'$ satisfying $AT = T'A$. We say that B is an *intertwining lifting* of A if B is an operator from $\mathcal{H}$ into $\mathcal{K}'$ satisfying $U'B = BT$ and $P_{\mathcal{H}'}B = A$. A fundamental interpolation problem is to find an intertwining lifting B of A satisfying a certain constraint. One solution to this problem is given by the following weighted commutant lifting theorem, which is a generalization of the Treil–Volberg result [5].

THEOREM 1.1. ([1] Corollary 3.3). *Let $\mathcal{H}'$ be a co-invariant subspace for a contraction U' on $\mathcal{K}'$ and let T' be the compression of U' to $\mathcal{H}'$. Let T be an operator on $\mathcal{H}$ and A an operator intertwining T with T'. If there is a positive operator Q on $\mathcal{H}$ satisfying*

$$A^*A \leq Q \leq T^*QT, \tag{1.5}$$

then there exists an intertwining lifting B of A satisfying $B^*B \leq Q$. Moreover, if U' is an isometry, then there exists a positive operator Q on $\mathcal{H}$ satisfying (1.5).

Finally we want to dedicate this note to the memory of Béla Sz.-Nagy who had and still has a very strong influence on our work.

2. Main result

Assume that $\{A_k, T_k, \Gamma_k, Q_k\}$ is an admissible set of nonstationary data. The main result of this note provides a necessary and sufficient condition for the existence of a weighted interpolating sequence $\{B_k\}$ for $\{A_k, T_k, \Gamma_k, Q_k\}$. In the following lemma, a necessary condition for the existence of a weighted interpolating sequence is established.

LEMMA 2.1. *If $\{B_k\}$ is a weighted interpolating sequence for an admissible nonstationary set of data $\{A_k, T_k, \Gamma_k, Q_k\}$, then*

$$A_k^*A_k \leq Q_k \text{ and } A_kT_{k-1} = P_{\mathcal{F}_k}\Gamma_{k-1}A_{k-1} \qquad (k \in \mathbb{Z}). \tag{2.1}$$

Proof. By employing (1.4), with $h_k \in \mathcal{H}_k$, we have

$$\|A_kh_k\|^2 = \|P_{\mathcal{F}_k}B_kh_k\|^2 \leq \|B_kh_k\|^2 \leq \|Q_k^{1/2}h_k\|^2. \tag{2.2}$$

This implies that the first condition in (2.1) holds. Furthermore, the lifting condition (1.2) along with the second condition in (1.3) imply that

$$\begin{aligned} P_{\mathcal{F}_k}\Gamma_{k-1}A_{k-1} &= P_{\mathcal{F}_k}\Gamma_{k-1}P_{\mathcal{F}_{k-1}}B_{k-1} \\ &= P_{\mathcal{F}_k}\Gamma_{k-1}B_{k-1} = P_{\mathcal{F}_k}B_kT_{k-1} = A_kT_{k-1}. \end{aligned}$$

Therefore, the second condition in (2.1) holds. This completes the proof. □

First, a weak converse of Lemma 2.1 will be established.

LEMMA 2.2. *Let $\{A_k, T_k, \Gamma_k, Q_k\}$ be an admissible nonstationary set of data. Assume that the operators $\{Q_k\}$ and $\{T_k\}$ are uniformly bounded in norm, that is,*

$$\sup\{\|Q_k\|,\ \|T_k\| : k \in \mathbb{Z}\} < \infty. \tag{2.3}$$

If the conditions in (2.1) *hold, then there exists a weighted interpolating sequence $\{B_k\}$ for $\{A_k, T_k, \Gamma_k, Q_k\}$.*

Proof. Recall that $c_k = \max\{1, \|\Gamma_k\|\}$. By replacing T_k and Γ_k by T_k/c_k and Γ_k/c_k respectively, it is easy to see that we may, without loss of generality, assume that Γ_k is a contraction and the admissibility conditions $T_{k-1}^* Q_k T_{k-1} \geq Q_{k-1}$ hold for all k in $\mathbb{Z}$. For the rest of the proof, we will work under these assumptions. Let us introduce the following Hilbert spaces:

$$\begin{aligned} \mathcal{H} &= \{\vec{u} = (u_k)_{-\infty}^{\infty} : u_k \in \mathcal{H}, \text{ and } \|\vec{u}\|^2 = \sum_{-\infty}^{\infty} \|u_k\|^2 < \infty\}, \\ \mathcal{H}' &= \{\vec{f} = (f_k)_{-\infty}^{\infty} : f_k \in \mathcal{F}_k \text{ and } \|\vec{f}\|^2 = \sum_{-\infty}^{\infty} \|f_k\|^2 < \infty\}, \\ \mathcal{K}' &= \{\vec{y} = (y_k)_{-\infty}^{\infty} : y_k \in \mathcal{K}_k \text{ and } \|\vec{y}\|^2 = \sum_{-\infty}^{\infty} \|y_k\|^2 < \infty\}. \end{aligned} \tag{2.4}$$

Let A be the operator from $\mathcal{H}$ into $\mathcal{H}'$ and Q on $\mathcal{H}$ be the diagonal operator defined by

$$\begin{aligned} A((h_k)_{k=-\infty}^{\infty}) &= (A_k h_k)_{k=-\infty}^{\infty} \\ Q((h_k)_{k=-\infty}^{\infty}) &= (Q_k h_k)_{k=-\infty}^{\infty} \qquad (\vec{h} = (h_k)_{-\infty}^{\infty} \in \mathcal{H}). \end{aligned} \tag{2.5}$$

Using the first condition in (2.1) and (2.3), it follows that both A and Q are bounded operators. Moreover, Q is positive. Define the operators T on $\mathcal{H}$ and U' on $\mathcal{K}'$ by

$$T((h_k)_{k=-\infty}^{\infty}) = (T_{k-1} h_{k-1})_{k=-\infty}^{\infty} \text{ and } U'((y_k)_{k=-\infty}^{\infty}) = (\Gamma_{k-1} y_{k-1})_{k=-\infty}^{\infty} \tag{2.6}$$

where $\vec{h} = (h_k)_{-\infty}^{\infty}$ is in $\mathcal{H}$ and $\vec{y} = (y_k)_{-\infty}^{\infty}$ is in $\mathcal{K}'$. From (2.3), it follows that the operator T is bounded and U' is a contraction. Now notice that for any $\vec{y} = (y_k)_{-\infty}^{\infty}$ in $\mathcal{K}'$, we have $P_{\mathcal{H}'}\vec{y} = (P_{\mathcal{F}_k} y_k)_{-\infty}^{\infty}$. This along with definition of U' in (2.6) and (1.2) implies that

$$\begin{aligned} P_{\mathcal{H}'} U' P_{\mathcal{H}'} \vec{y} &= (P_{\mathcal{F}_k} \Gamma_{k-1} P_{\mathcal{F}_{k-1}} y_{k-1})_{k=-\infty}^{\infty} = (P_{\mathcal{F}_k} \Gamma_{k-1} y_{k-1})_{k=-\infty}^{\infty} \\ &= P_{\mathcal{H}'} U' \vec{y}. \end{aligned} \tag{2.7}$$

Therefore, $P_{\mathcal{H}'}U'P_{\mathcal{H}'} = P_{\mathcal{H}'}U'$, and hence $\mathcal{H}'$ is co-invariant for U'. Now let T' on $\mathcal{H}'$ be the operator obtained by compressing U' to $\mathcal{H}'$, that is,

$$T' = P_{\mathcal{H}'}U'|\mathcal{H}' . \tag{2.8}$$

For $\vec{h} = (h_k)_{-\infty}^{\infty}$ in $\mathcal{H}$, using the definition of the operators in (2.5) and (2.6) along with $T_{k-1}^* Q_k T_{k-1} \geq Q_{k-1}$, it follows that

$$\begin{aligned}(QT\vec{h}, T\vec{h}) &= (Q(T_{k-1}h_{k-1})_{k=-\infty}^{\infty}, (T_{k-1}h_{k-1})_{k=-\infty}^{\infty}) \\ &= \sum_{k=-\infty}^{\infty} (Q_k T_{k-1}h_{k-1}, T_{k-1}h_{k-1}) \\ &\geq \sum_{k=-\infty}^{\infty} (Q_{k-1}h_{k-1}, h_{k-1}) = (Q\vec{h}, \vec{h}).\end{aligned}$$

This shows that we have the relation

$$T^*QT \geq Q. \tag{2.9}$$

Furthermore, since by (2.1) we have $A_k^* A_k \leq Q_k$, it follows that $A^*A \leq Q$. We will now show that A intertwines T with T'. Indeed, for $\vec{h} = (h_k)_{-\infty}^{\infty}$ in $\mathcal{H}$, we have

$$\begin{aligned}AT\vec{h} &= A((T_{k-1}h_{k-1})_{-\infty}^{\infty}) = (A_k T_{k-1}h_{k-1})_{-\infty}^{\infty} \\ &= (P_{\mathcal{F}_k}\Gamma_{k-1}A_{k-1}h_{k-1})_{-\infty}^{\infty} = P_{\mathcal{H}'}((\Gamma_{k-1}A_{k-1}h_{k-1})_{-\infty}^{\infty}) \\ &= P_{\mathcal{H}'}U'((A_k h_k)_{-\infty}^{\infty}) = P_{\mathcal{H}'}U'A\vec{h} = T'A\vec{h} .\end{aligned}$$

To obtain the third equality we used the second condition in (2.1). Thus, we may invoke Theorem 1.1 to conclude that there exists an operator B from $\mathcal{H}$ into $\mathcal{K}'$ satisfying

$$BT = U'B\,,\ P_{\mathcal{H}'}B = A \text{ and } B^*B \leq Q. \tag{2.10}$$

Let Π_k be the operator from $\mathcal{K}'$ into $\mathcal{K}_k$ which picks out the k-th coordinate of $\mathcal{K}'$, that is, $\Pi_k \vec{y} = y_k$ where $\vec{y} = (y_k)_{-\infty}^{\infty}$ is in $\mathcal{K}'$. To obtain a weighted interpolating sequence, let $\{B_k\}$ be the sequence obtained by extracting the diagonal entries from the matrix corresponding to B. To be precise, let B_k be the operator from $\mathcal{H}_k$ into $\mathcal{K}_k$ defined by

$$B_k h_k = \Pi_k B(\delta_{k,j}h_k)_{j=-\infty}^{\infty} \qquad (h_k \in \mathcal{H}_k) \tag{2.11}$$

where $\delta_{k,j}$ denotes the Kronecker delta. We will show that this $\{B_k\}$ is a weighted interpolating sequence for $\{A_k, T_k, \Gamma_k, Q_k\}$. Using (2.10) with $h_k \in \mathcal{H}_k$, we have

$$\|B_k h_k\|^2 \leq \|B(\delta_{k,j} h_k)_{j=-\infty}^{\infty}\|^2 \leq \|Q^{1/2}(\delta_{k,j} h_k)_{j=-\infty}^{\infty}\|^2 = \|Q_k^{1/2} h_k\|^2 .$$

Hence, $B_k^* B_k \leq Q_k$. Also, note that, for h_k in $\mathcal{H}_k$,

$$\begin{aligned} P_{\mathcal{F}_k} B_k h_k &= P_{\mathcal{F}_k} \Pi_k B(\delta_{k,j} h_k)_{j=-\infty}^{\infty} \\ &= \Pi_k P_{\mathcal{H}'} B(\delta_{k,j} h_k)_{j=-\infty}^{\infty} = \Pi_k A(\delta_{k,j} h_k)_{j=-\infty}^{\infty} = A_k h_k . \end{aligned}$$

Therefore, $P_{\mathcal{F}_k} B_k = A_k h_k$ for all k in $\mathbb{Z}$.

To complete the proof, it remains to show that $B_{k+1} T_k = \Gamma_k B_k$ holds for all k. By employing the intertwining relation $BT = U'B$ with h_k in $\mathcal{H}_k$, we obtain

$$\begin{aligned} B_{k+1} T_k h_k &= \Pi_{k+1} B(\delta_{k+1,j} T_k h_k)_{j=-\infty}^{\infty} \\ &= \Pi_{k+1} BT(\delta_{k,j} h_k)_{j=-\infty}^{\infty} = \Pi_{k+1} U' B(\delta_{k,j} h_k)_{j=-\infty}^{\infty} \\ &= \Gamma_k \Pi_k B(\delta_{k,j} h_k)_{j=-\infty}^{\infty} = \Gamma_k B_k h_k . \end{aligned}$$

This concludes the proof of the lemma. □

Next, it will be shown that the conditions in (2.1) are sufficient to obtain a weighted interpolating sequence for finitely many $\{A_k, T_k, \Gamma_k, Q_k\}$. More precisely, we have the following lemma.

LEMMA 2.3. *Let $\{A_k, T_k, \Gamma_k, Q_k\}$ be an admissible nonstationary set of data. If equation (2.1) holds, then for every integer $N \geq 1$ there exists a finite sequence of operators $\{B_k\}_{k=-N}^{N}$ such that*

$$P_{\mathcal{F}_k} B_k = A_k \text{ and } B_k^* B_k \leq Q_k, \qquad (-N \leq k \leq N), \tag{2.12}$$

and

$$B_k T_{k-1} = \Gamma_{k-1} B_{k-1} \qquad (-N+1 \leq k \leq N). \tag{2.13}$$

Proof. Define the sequence of spaces $\{\mathcal{H}_k^1\}$, $\{\mathcal{K}_k^1\}$ and $\{\mathcal{F}_k^1\}$ by

$$\mathcal{H}_k^1 = \mathcal{H}_k\,,\ \mathcal{K}_k^1 = \mathcal{K}_k\,,\ \mathcal{F}_k^1 = \mathcal{F}_k \qquad (\text{for } -N \leq k \leq N) \tag{2.14}$$

and

$$\mathcal{H}_k^1 = \mathcal{H}_{N+1}\,,\ \mathcal{K}_k^1 = \mathcal{K}_{N+1}\,,\ \mathcal{F}_k^1 = \mathcal{F}_{N+1} \qquad (\text{for } k \geq N+1). \tag{2.15}$$

For $k \leq -N-1$, the spaces $\mathcal{H}_k^1$, $\mathcal{K}_k^1$ and $\mathcal{F}_k^1$ are taken to consist only of zero vectors. The sequences of operators $\{A_k^{(1)}, T_k^{(1)}, \Gamma_k^{(1)}\}$ and the weights $\{Q_k^{(1)}\}$ are defined as follows:

$$A_k^{(1)} = A_k\,,\ T_k^{(1)} = T_k\,,\ \Gamma_k^{(1)} = \Gamma_k \text{ and } Q_k^{(1)} = Q_k \qquad (-N \leq k \leq N) \tag{2.16}$$

and

$$A_k^{(1)} = A_{N+1}\,,\ T_k^{(1)} = I\,,\ \Gamma_k^{(1)} = I \text{ and } Q_k^{(1)} = Q_{N+1} \qquad (\ k \geq N+1). \tag{2.17}$$

For $k \leq -N-1$, the operators $A_k^{(1)}$, $T_k^{(1)}$ and $\Gamma_k^{(1)}$ are taken to be zero operators and the weights $Q_k^{(1)}$ are also taken to be zero. It is now straightforward to verify that the conditions in (1.2) and (1.4) hold with $T_k^{(1)}$, $\Gamma_k^{(1)}$ and $Q_k^{(1)}$ in place of T_k, Γ_k and Q_k. Moreover, it is easy to see that the analogous condition in (2.1) holds and the condition (2.3) is satisfied. We can thus apply Lemma 2.2 to conclude that there exists a weighted interpolating sequence $\{B_k^{(1)}\}$ for $\{A_k^{(1)}, T_k^{(1)}, \Gamma_k^{(1)}, Q_k^{(1)}\}$. By choosing $B_k = B_k^{(1)}$ for $-N \leq k \leq N$ and recalling that (2.16) holds, it follows that the operators $\{B_k\}_{-N}^{N}$ satisfy (2.12) and (2.13). This completes the proof. □

We are now ready to prove the main result in this note.

THEOREM 2.1. *Let $\{A_k, T_k, \Gamma_k, Q_k\}$ be an admissible nonstationary set of data. Then there exists a weighted interpolating sequence $\{B_k\}$ for $\{A_k, T_k, \Gamma_k, Q_k\}$ if and only if*

$$A_k^* A_k \leq Q_k \text{ and } A_k T_{k-1} = P_{\mathcal{F}_k} \Gamma_{k-1} A_{k-1} \qquad (k \in \mathbb{Z}). \tag{2.18}$$

Proof. The necessity of this condition is already established in Lemma 2.1. So, assume that (2.18) holds. Using Lemma 2.3, for each integer $N \geq 1$, we can produce a finite sequence of operators $\{B_k^{(N)}\}_{k=-N}^{N}$ such that (2.12) and (2.13) hold with $B_k^{(N)}$ in place of B_k. For each k in $\mathbb{Z}$, we have $B_k^{(N)*}B_k^{(N)} \leq Q_k$ for all $N \geq |k|$. This implies that, by going through a proper subsequence if necessary, for each $k \in \mathbb{Z}$, we obtain $\lim_{N\to\infty} B_k^{(N)} = B_k$ exists weakly. We claim that $\{B_k\}$ is the desired interpolating sequence. First, note that since $B_k^{(N)*}B_k^{(N)} \leq Q_k$ for each k in $\mathbb{Z}$ with $N \geq |k|$, it follows that $B_k^* B_k \leq Q_k$. Moreover, by taking weak limits, we have

$$P_{\mathcal{F}_k} B_k = P_{\mathcal{F}_k} \lim_{N\to\infty} B_k^{(N)} = \lim_{N\to\infty} P_{\mathcal{F}_k} B_k^{(N)} = A_k \,.$$

The last equality follows by using the first relation in (2.12). The intertwining condition can be proven in a similar manner. Indeed, for all k in $\mathbb{Z}$,

$$B_k T_{k-1} = \lim_{N\to\infty} B_k^{(N)} T_{k-1} = \lim_{N\to\infty} \Gamma_{k-1} B_{k-1}^{(N)} = \Gamma_{k-1} B_{k-1}, \tag{2.19}$$

where all the limits above are weak limits. This concludes the proof of the theorem. □

REMARK 2.1. Let $\{A_k, T_k, \Gamma_k\}$ be a sequences of operators satisfying (1.1) and (1.2). Moreover, assume that $\Gamma_k^*\Gamma_k \geq I$ for all k in $\mathbb{Z}$. Then there exists a sequence of weights $\{Q_k\}$ such that $\{A_k, T_k, \Gamma_k, Q_k\}$ is an admissible set of nonstationary data if and only if there exists an interpolating sequence $\{B_k\}$ for $\{A_k, T_k, \Gamma_k\}$, that is, a sequence of operators $\{B_k\}$ such that

$$P_{\mathcal{F}_k} B_k = A_k \text{ and } B_{k+1} T_k = \Gamma_k B_k \qquad (k \in \mathbb{Z}). \tag{2.20}$$

Theorem 2.1 shows that if $\{A_k, T_k, \Gamma_k, Q_k\}$ is an admissible set of nonstationary data, then exists an interpolating sequence for this data. Now assume that $\{B_k\}$ is an interpolating sequence for $\{A_k, T_k, \Gamma_k\}$. Then we claim that $Q_k = B_k^* B_k$, for k in $\mathbb{Z}$, is a sequence of admissible weights and $\{B_k\}$ is the corresponding interpolating sequence with these weights. Indeed, due to the fact that $\Gamma_k^*\Gamma_k \geq I$ for all k, it follows that

$$\begin{aligned} T_{k-1}^* Q_k T_{k-1} &= T_{k-1}^* B_k^* B_k T_{k-1} \\ &= B_{k-1}^* \Gamma_{k-1}^* \Gamma_{k-1} B_{k-1} \geq B_{k-1}^* B_{k-1} = Q_{k-1}. \end{aligned}$$

Therefore, if $Q_k = B_k^* B_k$, then $\{A_k, T_k, \Gamma_k, Q_k\}$ is an admissible set of nonstationary data, which proves our claim.

□

REMARK 2.2. We claim that Theorems 1.1 and 2.1 are equivalent. To verify this, it is sufficient to derive the weighted commutant lifting Theorem 1.1 from Theorem 2.1. To this end, let A, T, U' and T' be the operators defined in Theorem 1.1. Then for each k in $\mathbb{Z}$, set $\mathcal{H}_k = \mathcal{H}$, $\mathcal{K}_k = \mathcal{K}'$, $\mathcal{F}_k = \mathcal{H}'$ and let

$$A_k = A\,,\ T_k = T\,,\ \Gamma_k = U' \text{ and } Q_k = Q.$$

It is easy to verify that all the hypotheses of Theorem 2.1 are satisfied. By employing this theorem, we obtain a weighted interpolating sequence $\{B_k\}$ for $\{A_k, T_k, \Gamma_k, Q_k\}$ satisfying

$$P_{\mathcal{H}'} B_k = A\,,\ B_k T = U' B_{k-1} \text{ and } B_k^* B_k \leq Q \qquad (k \in \mathbb{Z}). \tag{2.21}$$

Here each B_k maps $\mathcal{H}$ into $\mathcal{K}'$. Since $B_k^* B_k \leq Q$ for all k in $\mathbb{Z}$, the sequence $\{B_k\}$ is uniformly bounded in norm. Now let B be the operator from $\mathcal{H}$ into $\mathcal{K}'$ defined by

$$(Bh, y) = \text{LIM}_{k\to\infty}(B_k h, y) \qquad (h \in \mathcal{H} \text{ and } y \in \mathcal{K}') \tag{2.22}$$

where LIM denotes any generalized Banach limit on ℓ^∞. By consulting (2.21) it readily follows that

$$P_{\mathcal{H}'} B = A\,,\ BT = U'B \text{ and } B^* B \leq Q\,.$$

This completes the proof. □

3. Weighted three chains completions

In this section, we apply the geometric interpolation Theorem 2.1 to obtain a solution to the weighted three chains interpolation problem which is a direct generalization of the three chains interpolation problem considered in [3]. In order to state the problem, we need to establish some further notation. Let $\mathcal{X}$ and $\mathcal{Y}$ be two Hilbert spaces and for k in $\mathbb{Z}$, let $\mathcal{H}_k \subseteq \mathcal{X}$ and $\mathcal{M}_k \subseteq \mathcal{K}_k \subseteq \mathcal{Y}$ be ascending chains of Hilbert spaces satisfying the conditions

$$\mathcal{H}_k \subseteq \mathcal{H}_{k+1}\,,\ \mathcal{K}_k \subseteq \mathcal{K}_{k+1} \text{ and } \mathcal{M}_k \subseteq \mathcal{M}_{k+1} \qquad (k \in \mathbb{Z}). \tag{3.1}$$

The data for the weighted three chains interpolation problem is a sequence of operators $\{A_k, Q_k\}$ where A_k maps $\mathcal{H}_k$ into $\mathcal{K}_k \ominus \mathcal{M}_k$ and Q_k is a positive operator on $\mathcal{H}_k$ satisfying

$$P_{\mathcal{H}_k} Q_{k+1} | \mathcal{H}_k \geq Q_k \text{ and } \sup_{k\in\mathbb{Z}}\{\|Q_k\|\} < \infty. \tag{3.2}$$

An operator B is an *interpolant* for the sequence $\{A_k\}$ if B is an operator from $\mathcal{X}$ into $\mathcal{Y}$ with the following properties:

$$B\mathcal{H}_k \subseteq \mathcal{K}_k \text{ and } (I - P_{\mathcal{M}_k})B|\mathcal{H}_k = A_k \qquad (k \in \mathbb{Z}). \tag{3.3}$$

An operator B is called a *weighted three chains interpolant* for $\{A_k, Q_k\}$ if B is an interpolant for $\{A_k\}$ and

$$P_{\mathcal{H}_k} B^* B|\mathcal{H}_k \leq Q_k \qquad (k \in \mathbb{Z}). \tag{3.4}$$

The weighted three chains interpolation problem is to find necessary and sufficient conditions for the existence of a weighted interpolant of the sequence $\{A_k, Q_k\}$. If the weights $Q_k = \gamma I$ for all k in $\mathbb{Z}$ where $\gamma > 0$ is a constant, then the corresponding weighted three chains interpolation problem reduces to the three chains interpolation problem considered in [3]. The following result provides necessary and sufficient condition for the existence of a weighted interpolant.

THEOREM 3.1. *Let $\{A_k, Q_k\}$ be the data for the weighted three chains interpolation problem. There exists a weighted three chains interpolant B for the $\{A_k, Q_k\}$ if and only if*

$$(I - P_{\mathcal{M}_k})A_{k-1} = A_k|\mathcal{H}_{k-1} \text{ and } A_k^* A_k \leq Q_k \qquad (k \in \mathbb{Z}). \tag{3.5}$$

Proof. For every k in $\mathbb{Z}$, let T_k be the operator embedding $\mathcal{H}_k$ in $\mathcal{H}_{k+1}$ and Γ_k the operator embedding $\mathcal{K}_k$ in $\mathcal{K}_{k+1}$. The first condition in (3.2) shows that (1.4) holds for Q_k with $c_k = 1$. Let $\mathcal{F}_k$ be the subspaces defined by $\mathcal{F}_k = \mathcal{K}_k \ominus \mathcal{M}_k$. Using $\mathcal{M}_k \subseteq \mathcal{M}_{k+1}$ with $y_k \in \mathcal{K}_k$, we obtain

$$P_{\mathcal{F}_{k+1}} y_k = (P_{\mathcal{K}_{k+1}} - P_{\mathcal{M}_{k+1}})(P_{\mathcal{K}_{k+1}} - P_{\mathcal{M}_k})y_k = P_{\mathcal{F}_{k+1}} P_{\mathcal{F}_k} y_k \qquad (k \in \mathbb{Z}).$$

Since $\Gamma_k = I|\mathcal{K}_k$, this shows that (1.2) is satisfied. Therefore, $\{A_k, T_k, \Gamma_k, Q_k,\}$ is an admissible set of data for our previous nonstationary interpolation problem. Finally, it is noted that the sequence $\{A_k\}$ satisfies the relations in (2.18) if and only if it satisfies the relations in (3.5).

Now assume that there exists a weighted interpolant B for the three chains interpolation problem with data $\{A_k, Q_k\}$. Recall that $B\mathcal{H}_k \subseteq \mathcal{K}_k$ for all k in $\mathbb{Z}$. Let B_k be the operator from $\mathcal{H}_k$ into $\mathcal{K}_k$ defined by $B_k = B|\mathcal{H}_k$. We claim that $\{B_k\}$ is a weighted interpolating sequence for the data $\{A_k, T_k, \Gamma_k, Q_k\}$. Since B is a weighted three chains interpolant, $B_k^* B_k = P_{\mathcal{H}_k} B^* B|\mathcal{H}_k \leq Q_k$. Using $B\mathcal{H}_k \subseteq \mathcal{K}_k$ once again, we obtain

$$P_{\mathcal{F}_k} B_k = P_{\mathcal{F}_k} B|\mathcal{H}_k = (I - P_{\mathcal{M}_k})B|\mathcal{H}_k = A_k \qquad (k \in \mathbb{Z}).$$

From the definition of the operators B_k, it is clear that $B_{k+1}|\mathcal{H}_k = B|\mathcal{H}_k = B_k$. Because T_k and Γ_k are inclusion operators, we see that the second condition in (1.3) holds. Hence, $\{B_k\}$ is a weighted interpolating sequence for $\{A_k, T_k, \Gamma_k, Q_k\}$. By Theorem 2.1 it follows that (2.18) holds, and thus the necessity of (3.5) follows.

Conversely assume that (3.5) holds. This implies that the conditions in (2.18) are satisfied. By Theorem 2.1 there exists a weighted interpolating sequence $\{B_k\}$ for the nonstationary data $\{A_k, T_k, \Gamma_k, Q_k\}$. The second condition in (1.3), along with the definitions of the operators T_k and Γ_k, imply that $B_k|\mathcal{H}_{k-1} = B_{k-1}$ for all k in $\mathbb{Z}$. Since $B_k^* B_k \leq Q_k$ and $\sup_k\{\|Q_k\|\} < \infty$, it follows that B_k strongly approaches an operator B mapping $\mathcal{X}$ into $\mathcal{Y}$ as k tends to infinity. Notice that $B|\mathcal{H}_k = B_k$ for all k in $\mathbb{Z}$. Using this B, we have $P_{\mathcal{H}_k} B^* B|\mathcal{H}_k = B_k^* B_k \leq Q_k$ for all k. Moreover, it is easy to verify that (3.3) holds. So, B is a weighted three chains interpolant for the data $\{A_k, Q_k\}$. This completes the proof.

□

4. Weighted Carswell–Schubert Theorem

In Section XII.6 of [3], a proof of the Carswell–Schubert Theorem (see [2]) was obtained by using the three chains completion theorem. Motivated by these results, we will use the weighted three chains interpolation theorem in order to obtain and prove a weighted version of the Carswell–Schubert Theorem. Recall that an operator U on a space $\mathcal{K}$ is called a co-shift if U^* is an isometry and U^n approaches zero strongly as n tends to infinity. The following result is a weighted version of the Carswell–Schubert Theorem.

THEOREM 4.1. *Let V on $\mathcal{Y}$ be a co-shift and V' on $\mathcal{Y}'$ be a co-isometry. Let Q be a positive operator on $\mathcal{Y}$ satisfying $VQV^* \leq Q$. Let A be an operator from $\mathcal{Y}$ into $\mathcal{H}$ where $\mathcal{H}$ is a co-invariant subspace for V'. Finally, for every integers $n \geq 1$, let P_n be the orthogonal projection onto $\ker V^n = \mathcal{H}_n$ and P_n' be the orthogonal projection onto $\ker V'^n = \mathcal{K}_n$. Then there exists an operator B mapping $\mathcal{Y}$ into $\mathcal{Y}'$ satisfying*

$$P_{\mathcal{H}} B = A\,, \; BV = V'B \; \text{ and } \; B^*B \leq Q \tag{4.1}$$

if and only if the following conditions are fulfilled:

(a) *The intertwining condition $P_{\mathcal{H}} V' A = AV$ holds.*

(b) *For every integer $n \geq 1$, the relation*

$$A_n^* P_n' h = P_n A^* h \qquad (h \in \mathcal{H}) \tag{4.2}$$

defines a (bounded) operator A_n from $\mathcal{H}_n$ into $\overline{P_n' \mathcal{H}}$.

(c) *The inequality $A_n^* A_n \leq P_n Q|\mathcal{H}_n$ holds for all integers $n \geq 1$.*

REMARK 4.1. The conditions in (b) and (c) of Theorem 4.1 imply that there exists a sequence of operators $\{A_n\}_1^\infty$ such that $\|A_n\| \le \|Q\|^{1/2}$ and the factorization $P_\mathcal{H} A_n = A|\mathcal{H}_n$ holds for all integers $n \ge 1$. Moreover, $A^*A \le Q$. To see this, note that if (b) holds, then $A_n^* P_n' | \mathcal{H} = P_n A^*$. By taking the adjoint, this readily implies that $P_\mathcal{H} A_n = A|\mathcal{H}_n$ holds for all integers $n \ge 1$. The inequality $A_n^* A_n \le P_n Q | \mathcal{H}_n$ yields $\|A_n\| \le \|Q\|^{1/2}$. We will now show that $A^*A \le Q$. Because V is a co-shift, P_n is an increasing sequence of orthogonal projections which converge strongly to the identity operator. Using $P_\mathcal{H} A_n = A|\mathcal{H}_n$, it follows that

$$P_n A^* A P_n = P_n A_n^* P_\mathcal{H} A_n P_n \le P_n A_n^* A_n P_n \le P_n Q P_n \qquad (n \ge 1).$$

By letting n approach infinity, we obtain $A^*A \le Q$.

Following Section XII.6 in [3], the proof of Theorem 4.1 will be broken down into several lemmas. First, let us establish some further notation. Let $\mathcal{M}_n$ be the subspace of $\mathcal{Y}'$ defined by

$$\mathcal{M}_n = \mathcal{K}_n \cap (\mathcal{Y}' \ominus \mathcal{H}) \qquad (n \ge 1). \tag{4.3}$$

For any integer $n \le 0$, the spaces $\mathcal{H}_n$, $\mathcal{K}_n$ and $\mathcal{M}_n$ are set equal to zero. It is easy to show that

$$\mathcal{K}_n \ominus \mathcal{M}_n = \overline{P_n \mathcal{H}} \qquad (n \in \mathbb{Z}). \tag{4.4}$$

We are now ready to prove the necessary part of the Theorem 4.1.

LEMMA 4.1. *If there exists an operator B from $\mathcal{Y}$ into $\mathcal{Y}'$ satisfying* (4.1), *then the conditions* (a), (b) *and* (c) *in Theorem* 4.1 *hold.*

Proof. Assume that there exists an operator B satisfying all three conditions in (4.1). Since $\mathcal{H}$ is a co-invariant subspace for V', we have $P_\mathcal{H} V' P_\mathcal{H} = P_\mathcal{H} V'$. This observation along with the first two relations in (4.1) yield

$$\begin{aligned} P_\mathcal{H} V' A &= P_\mathcal{H} V' P_\mathcal{H} A = P_\mathcal{H} V' P_\mathcal{H} B \\ &= P_\mathcal{H} V' B = P_\mathcal{H} B V = AV. \end{aligned}$$

Hence, Part (a) holds. Since $B^*B \le Q$, we obtain

$$\|Ay\| = \|P_\mathcal{H} By\| \le \|By\| \le \|Q^{1/2} y\| \qquad (y \in \mathcal{Y}). \tag{4.5}$$

In other words, $A^*A \le Q$.

To prove Part (b), notice that the third condition in (4.1) yields $\|B\| \le \|Q\|^{1/2}$. Now observe that the intertwining relation $BV = V'B$ readily implies that B maps the kernel of V^n into kernel of ${V'}^n$, and consequently,

$$BP_n = P'_n B P_n \qquad (n \in \mathbb{Z}). \tag{4.6}$$

By taking adjoint of $P_{\mathcal{H}}B = A$, it follows that $B^*h = A^*h$ for all $h \in \mathcal{H}$. This fact along with (4.6) gives

$$\|P_n A^* h\| = \|P_n B^* h\| = \|P_n B^* P'_n h\| \le \|Q\|^{1/2} \|P'_n h\|.$$

So, there exists an operator A_n from $\mathcal{H}_n$ into $\overline{P'_n \mathcal{H}}$ satisfying $A_n^* P'_n h = P_n A^* h$ for all h in $\mathcal{H}$. This shows that the operators A_n in (4.2) are well defined and $\|A_n\| \le \|Q\|^{1/2}$. Therefore, Part (b) holds.

To show that $A_n^* A_n \le P_n Q|\mathcal{H}_n$ holds, let Q_n be the positive operator on $\mathcal{H}_n$ defined by $Q_n = P_n Q|\mathcal{H}_n$ for all n in $\mathbb{Z}$. Obviously, A_n and Q_n are zero for all integers $n \le 0$. Since (3.1) and (3.2) hold, the sequence of operators $\{A_n, Q_n\}$ can be viewed as the data for the weighted three chains interpolation problem. If B is any operator satisfying (4.1), then we claim that B is a weighted three chains interpolant for the data $\{A_n, Q_n\}$. Since $BP_n = P'_n B P_n$, we see that $B\mathcal{H}_n \subseteq \mathcal{K}_n$ for all n. So, the first equation in (3.3) holds. Hence, to prove that B is an interpolant, it suffices to show that the second condition in (3.3) holds for $n \ge 1$. To this end, recall that $P_{\mathcal{H}}B = A$. Using this along with the definition of A_n yields

$$P_n B^* P'_n h = P_n B^* h = P_n A^* h = A_n^* P'_n h \qquad (h \in \mathcal{H}).$$

Since $\mathcal{K}_n \ominus \mathcal{M}_n = \overline{P'_n \mathcal{H}}$, it follows that $P_n B^*|(\mathcal{K}_n \ominus \mathcal{M}_n) = A_n^*$. By taking the adjoint, $A_n = (I - P_{\mathcal{M}_n})B|\mathcal{H}_n$, that is, (3.3) holds. Moreover, because $B^*B \le Q$, it follows that $P_n B^* B|\mathcal{H}_n \le Q_n$. Therefore, B is a weighted three chains interpolant for the data $\{A_n, Q_n\}$.

To verify that Part (c) holds, let B_n be the operator from $\mathcal{H}_n$ into $\mathcal{K}_n$ defined by $B_n = B|\mathcal{H}_n$. Using $P_n B^* B|\mathcal{H}_n \le Q_n$, it follows that $B_n^* B_n \le Q_n = P_n Q|\mathcal{H}_n$ for all n in $\mathbb{Z}$. Since B is a weighted three chains interpolant for $\{A_n, Q_n\}$, we have $(I - P_{\mathcal{M}_n})B_n = A_n$. Hence, it follows that $A_n^* A_n \le P_n Q|\mathcal{H}_n$ for all n. This yields Part (c) and completes the proof of the lemma. □

The sufficiency of the conditions (a), (b) and (c) will be proven in the next lemma. We will only sketch the proof, for the details we refer to the proof of Theorem 6.2 in Section XII.6 in [3].

LEMMA 4.2. *If the conditions in* (a) (b) *and* (c) *in Theorem* 4.1 *hold, then there exists an operator* B *from* $\mathcal{Y}$ *into* $\mathcal{Y}'$ *satisfying* (4.1).

Proof. As before for each n in $\mathbb{Z}$, let Q_n be the positive operator on $\mathcal{H}_n$ defined by $Q_n = P_n Q|\mathcal{H}_n$. The condition (c) in the statement of Theorem 4.1 implies that $A_n^* A_n \leq Q_n$ for all n. The weights $\{Q_n\}$ clearly satisfy (3.2).

Now let us construct a weighted three chains interpolant B of $\{A_n, Q_n\}$ satisfying $BV = V'B$. Applying Theorem 3.1, it follows that there exists a weighted three chains interpolant B of $\{A_n, Q_n\}$. Because V is a co-shift, P_n is an increasing sequence of positive operators which strongly converge to I. Using this and the fact that $P_n B^* B|\mathcal{H}_n \leq P_n Q|\mathcal{H}_n$, we obtain $B^*B \leq Q$. By consulting formula (6.15) and its proof in Theorem XII.6.2 of [3], it follows that $V'BV^*$ is also a three chains interpolant of $\{A_n\}$. Moreover, using $V'^*V' = (I - P_1')$, we obtain

$$VB^*V'^*V'BV^* \leq VB^*BV^* \leq VQV^* \leq Q.$$

So, $V'BV^*$ is a weighted three chains interpolant for $\{A_n, Q_n\}$. This observation shows that the map $B \mapsto V'BV^*$ is a selfmap on the set of all weighted interpolants of $\{A_n, Q_n\}$. Hence, by the Tychonoff-Schauder fixed point theorem, we may assume that the weighted interpolant that we obtained by using Theorem 3.1 has the property

$$B = V'BV^*. \tag{4.7}$$

Note that since B is an interpolant, $B\mathcal{H}_1 \subset \mathcal{K}_1 = \ker V'$. Due to (4.7) this yields

$$BV = V'BV^*V = V'B - V'BP_1 = V'B.$$

This concludes the proof of the lemma and consequently Theorem 4.1 as well. □

5. An application

For a Hilbert space $\mathcal{E}$, let $\mathcal{L}(\mathcal{E})$ denote the algebra of all bounded linear operators on $\mathcal{E}$ and let $L^\infty(\mathcal{E}, \mathcal{E})$ be the space of all $\mathcal{L}(\mathcal{E})$-valued bounded, strongly measurable functions on the unit circle in the complex plane. Recall that $H^\infty(\mathcal{E}, \mathcal{E})$ is the Hardy subspace of $L^\infty(\mathcal{E}, \mathcal{E})$ consisting of the set of all uniformly bounded analytic functions in the open unit disc with values in $\mathcal{L}(\mathcal{E})$. The space of all measurable functions on the unit circle in the complex plane with values in $\mathcal{E}$ which are square integrable is denoted by $L^2(\mathcal{E})$. The Hardy subspace $H^2(\mathcal{E})$ of $L^2(\mathcal{E})$ is the Hilbert space consisting of the set of all analytic functions u in the open unit disc with values in $\mathcal{E}$ such that $\|u\|^2 = \sum_0^\infty \|u_n\|^2$ is finite where $u(\lambda) = \sum_0^\infty u_n \lambda^n$ is the power series expansion for u. If Ω is in $L^\infty(\mathcal{E},\mathcal{E})$, then T_Ω is the Toeplitz operator on $H^2(\mathcal{E})$ defined by $(T_\Omega u)(\lambda) = P_+\Omega(\lambda)u(\lambda)$ where u is in $H^2(\mathcal{E})$ and P_+ denotes the orthogonal projection from $L^2(\mathcal{E})$ onto $H^2(\mathcal{E})$. Note that if $\Omega \in H^2(\mathcal{E},\mathcal{E})$, then the projection P_+ in the above definition of a Toeplitz operator is redundant. We denote by S the canonical unilateral shift on $H^2(\mathcal{E})$, that is, $Su = \lambda u$. For more details, see Chapter V in [4] or Chapter I in [3].

Let F, Θ and M be functions in $H^\infty(\mathcal{E},\mathcal{E})$ such that M is an outer function, that is, T_M is with dense range in $H^2(\mathcal{E})$. In this setup, as we will show below, Theorem 4.1 provides necessary and sufficient conditions for the existence of a bounded analytic function G in $H^\infty(\mathcal{E},\mathcal{E})$ satisfying

$$G\Theta = F \quad \text{and} \quad G(e^{\imath t})G(e^{\imath t})^* \le M(e^{\imath t})^* M(e^{\imath t}) \text{ a.e.} \tag{5.1}$$

Notice that the existence of such a G is equivalent to the existence of a G satisfying

$$T_G T_\Theta = T_F \quad \text{and} \quad T_G = T_M^* T_Z \tag{5.2}$$

for some $Z \in L^\infty(\mathcal{E},\mathcal{E})$ with $\|Z\|_\infty \le 1$. Indeed, (5.2) readily follows from (5.1) while the first equalities in (5.1) and (5.2) are clearly equivalent. On the other hand, if the second equality in (5.2) is true, it follows that the function f_n, defined by $f_n(e^{\imath t}) = e^{\imath nt}[G(e^{\imath t}) - M(e^{\imath t})^* Z(e^{\imath t})]e_0$, belongs to $L^2(\mathcal{E})$ and is orthogonal to $H^2(\mathcal{E})$ for all $e_0 \in \mathcal{E}$ and integers $n \ge 1$. Therefore $f_0 = [G(e^{\imath t}) - M(e^{\imath t})^* Z(e^{\imath t})]e_0 = 0$. Hence the second equality in (5.1) is valid too.

Assume that there exists a function G in $H^\infty(\mathcal{E},\mathcal{E})$ such that (5.1) holds. Then, denoting $\gamma = \|T_G\|$ we have

$$\|T_F u\| \le \gamma \|T_\Theta u\| \qquad (u \in H^2(\mathcal{E}))\,. \tag{5.3}$$

We denote the Hilbert space $\mathcal{H} = \overline{T_\Theta H^2(\mathcal{E})}$ where the closure is taken in $H^2(\mathcal{E})$. Let A be the operator from $H^2(\mathcal{E})$ into $\mathcal{H}$ the adjoint of which is defined by

$$A^* T_\Theta u = T_F u \qquad (u \in H^\infty(\mathcal{E}))\,. \tag{5.4}$$

In case (5.3) is satisfied, then the operator A^*, and hence A, is well defined. Let $\mathcal{Y} = \mathcal{Y}' = H^2(\mathcal{E})$ and let the coisometries V on $\mathcal{Y}$ and V' on $\mathcal{Y}'$ both be the adjoint of the canonical unilateral shift S on $H^2(\mathcal{E})$. With these notations, we have the following lemma.

LEMMA 5.1. *Assume that the inequality in* (5.3) *is satisfied for some* $\gamma > 0$ *and let the operator* A *be as defined in* (5.4). *Then, there exists a* $G \in H^\infty(\mathcal{E},\mathcal{E})$ *satisfying* (5.1) *if and only if there exists a* B *satisfying* (4.1) *with* $Q = T_M^* T_M$.

Proof. First, assume that an operator B satisfying (4.1) exists. Then, note that the second relation in (4.1) implies that B commutes with V. Recalling $V = S^*$, it follows that B^* commutes with S. Consequently, $B^* = T_G$ for some G in $H^\infty(\mathcal{E},\mathcal{E})$. From the inequality $B^*B \le Q = T_M^* T_M$, it follows that $T_G^* = XT_M$ for some contraction X. By multiplying the last relation on the left by S^* and on the right by S and using the fact that $S^*T_G^*S = T_G^*$, we obtain

$$(S^*XS - X)T_M = 0.$$

Since M is assumed to be outer, this implies that X is Toeplitz and consequently $X = T_Z$, for some contractive $Z \in L^\infty(\mathcal{E},\mathcal{E})$. So, T_G admits the second factorization in (5.2). Moreover, since B also satisfies $P_{\mathcal{H}}B = A$, by taking adjoints, we get $B^*|\mathcal{H} = A^*$. Using the definition of the space $\mathcal{H}$ and (5.4), it follows that $T_G T_\Theta = T_F$. This proves the existence of a G in $H^\infty(\mathcal{E},\mathcal{E})$ satisfying (5.2), or equivalently, (5.1), provided there exists a B satisfying (4.1).

Conversely, if there exists a function G in $H^\infty(\mathcal{E},\mathcal{E})$ satisfying (5.1), then taking $B = T_G^*$ it is easy to check that B satisfies (4.1). This concludes the proof of the lemma.

□

For an integer $n \ge 1$, let $\mathcal{H}_n$ denote the closed subspace of $H^2(\mathcal{E})$ spanned by polynomials of degree less than or equal to n and note that $\mathcal{H}_n$ is precisely the kernel of S^{*n}. The orthogonal projection on $\mathcal{H}_n$ is denoted by P_n, that is, for $u = \sum_{k=0}^\infty u_k\lambda^k$ in $H^2(\mathcal{E})$, we have $P_n u = u_0 + u_1\lambda + \cdots + u_n\lambda^n$. Finally, let $\mathcal{F}_n = \overline{P_n\mathcal{H}}$ for $n \ge 1$. Recall that $\mathcal{H} = \overline{T_\Theta H^2(\mathcal{E})}$. The theorem below gives necessary and sufficient conditions for the existence of a bounded analytic function G in $H^\infty(\mathcal{E},\mathcal{E})$ satisfying (5.1).

THEOREM 5.1. *There exists a bounded analytic function* G *satisfying* (5.1) *if and only if there exist operators* E_n *from* $\mathcal{F}_n$ *to* $\mathcal{H}_n$ *satisfying*

$$E_n P_n T_\Theta u = P_n T_F u \qquad (n \ge 1,\ u \in H^2(\mathcal{E})) \tag{5.5}$$

and

$$E_n E_n^* \le P_n T_M^* T_M|\mathcal{H}_n \qquad (n \ge 1). \tag{5.6}$$

Proof. Assume that there exists a function G in $H^\infty(\mathcal{E},\mathcal{E})$ such that (5.1) or equivalently, (5.2) holds. Then, the inequality in (5.3) is satisfied (with $\gamma = \|T_G\|$) and consequently the operator A, defined via its adjoint in (5.4), is well defined. By Lemma 5.1, there exists a B satisfying (4.1) in Theorem 4.1. Invoking Theorem 4.1 in the setting of Lemma 5.1 taking $E_n = A_n^*$, we obtain the operators E_n satisfying (5.5) and (5.6).

Conversely, assume that there exist operators E_n from $\mathcal{F}_n$ to $\mathcal{H}_n$ satisfying (5.5) and (5.6). Notice that (5.6) implies that $\|E_n^*\| \leq \|T_M\|$ for all $n \geq 1$. This implies that $\|E_n\| \leq \gamma_1$ where $\gamma_1 = \|T_M\|$. From relation (5.5), it follows that

$$\|P_n T_F u\| \leq \gamma_1 \|P_n T_\Theta u\| \qquad (u \in H^2(\mathcal{E}),\ n \geq 1). \tag{5.7}$$

By letting n go to infinity in (5.7), and noting that the P_n's converge to the identity on $\mathcal{Y} = H^2(\mathcal{E})$, it follows that the inequality (5.3) holds with $\gamma = \gamma_1$. Thus, the operator A in (5.4) is well defined and taking $A_n = E_n^*$, we see that the two relations in (5.5) and (5.6) imply that the necessary and sufficient conditions for the existence of an operator B in Theorem 4.1 satisfying the relations in (4.1) are satisfied. Lemma 5.1 now implies the existence of a G satisfying (5.1). This completes the proof of the theorem.

□

ACKNOWLEDGEMENTS. This research was supported in part by National Science Foundation Grant Number DMS-9706903.

References

[1] A. BISWAS, C. FOIAS and A. E. FRAZHO, Weighted commutant lifting, *Acta Sci. Math. (Szeged)*, **65** (1999), 657–686.

[2] J. G. W. CARSWELL and C. F. SCHUBERT, Lifting of operators that commute with shifts, *Mich. Math. J.*, **22** (1975), 65–69.

[3] C. FOIAS, A. E. FRAZHO, I. GOHBERG and M. A. KAASHOEK, *Metric Constrained Interpolation, Commutant Lifting and Systems,* Operator Theory: Advances and Applications **100**, Birkhäuser, 1998.

[4] B. SZ.-NAGY and C. FOIAS, *Harmonic Analysis of Operators on Hilbert Space,* North Holland Publishing Co., Amsterdam – Budapest, 1970.

[5] S. TREIL and A. VOLBERG, A fixed point approach to Nehari's problem and its applications, Operator Theory: Advances and Applications **71**, Birkhäuser, 1994, 165–186.

A. BISWAS, Department of Mathematics, UNC-Charlotte, Charlotte, NC 28223-9998, U.S.A.; *e-mail*: abiswas@uncc.edu

C. FOIAS, Department of Mathematics, Indiana University, Bloomington, IN 47405-5701, U.S.A.; *e-mail*: foias@indiana.edu

A. E. FRAZHO, School of Aeronautics and Astronautics, Purdue University, West Lafayette, IN 47907-1282, U.S.A.; *e-mail*: frazho@ecn.purdue.edu

Operator Theory: Advances and Applications, Vol. 127, 145–162

Semigroups in finite von Neumann algebras

GILLES CASSIER

Dedicated to the memory of Professor Béla Szőkefalvi-Nagy

Let M be a finite von Neumann algebra. In the first part, we give asymptotic results about M-stable sequences of weak*-continuous mappings which are related with operators belonging to M. In the second part, we extend, by a shorter way, similarity results given in [CaFa2] to unbounded semigroups of operators contained in a finite von Neumann algebra.

I. Introduction and preliminaries

Let H be a separable complex Hilbert space and let $B(H)$ be the algebra of bounded linear operators acting on H. The ultra-weak topology of $B(H)$ is the weak* topology (in the sequel we will shorten weak* to w*) that comes from the well known duality $B(H) = (C_1(H))^*$, where $C_1(H)$ is the Banach space of trace class operators on H endowed with the trace norm (see [Dix]). A von Neumann algebra acting on H is by definition an ultra-weakly closed *-subalgebra of $B(H)$. Such a von Neumann algebra M is finite if it admits a faithful normal trace τ, which means that τ is an ultra-weakly continuous linear functional on M satisfying:

1) $\tau(AB) = \tau(BA)$ for any $A, B \in M$;
2) for any positive element A in M, we have $\tau(A) \geq 0$ and $\tau(A) = 0 \Longrightarrow A = 0$.

We denote by $\mathcal{T}(M)$ the set of all faithful normal traces acting on M. A good example of a finite von Neumann algebra is the w*-algebra generated by the left regular representation of a countable discrete group. We will denote by M_* the predual of M. For any subset $\mathcal{F}$ of M, we shall denote by $\mathcal{F}'$ the family of operators commuting with every element of $\mathcal{F}$.

AMS Subject Classification (1991): 47A65, 46L10, 47A25, 47C15.

Let $B(M)$ denote the algebra of bounded linear operators acting on M, and let $B_w(M)$ stand for the algebra of operators $T \in B(M)$ which are weak*-continuous. Recall that $\phi \in B_w(M)$ if and only if ϕ is the adjoint of a bounded linear operator acting on the Banach space M_* (see for instance [BCP]). For any $\phi \in B_w(M)$, let ϕ_* denote the uniquely determined operator whose (Banach space) adjoint is ϕ, that is $(\phi_*)^* = \phi$. For more details on von Neumann algebras, we refer the reader to [Dix] and [Sak].

As usual $[A_{i,j}]_{1 \leq i,j \leq n} \in \mathcal{M}_n(B(H))$ denotes the $n \times n$ matrix which acts on the orthogonal sum of n copies of H; its entries are operators acting on H. We remind the reader that $\mathcal{M}_n(M)$ inherits a unique structure of von Neumann algebra. Let ψ be a linear mapping from M into itself, we define $\psi_n : \mathcal{M}_n(M) \to \mathcal{M}_n(M)$ by $\psi_n([A_{i,j}]_{1 \leq i,j \leq n}) = [\psi(A_{i,j})]_{1 \leq i,j \leq n}$. We call ψ n-positive if ψ_n is positive (that is positive operators are transformed into positive ones) and we call ψ completely positive if ψ is n-positive for all n.

We proved in [CaFa2] that a power bounded operator T in a finite von Neumann algebra M is similar to a unitary element of M if and only if $T^n x \not\to 0$ for any $x \in H \backslash \{0\}$ (T is said to be of class $C_{1\cdot}$ in the Sz.-Nagy–Foias terminology). We will extend this result into two directions. On the one hand, we will consider general semigroups. On the other hand, we will work with operators which are not necessarily power bounded. To achieve this, we have to find a proper framework, which will allow short and well adapted methods.

In similarity problems, the idea of using limits in the sense of Banach comes from B. Sz.-Nagy [Nag]. In the sequel, we frequently use this idea. Recall that a Banach limit is a state, that is a linear functional L with $\|L\| = L(\mathbf{1}) = 1$, acting on the classical space ℓ^∞ of all complex bounded sequences and satisfying $L((u_{n+1})) = L((u_n))$. A bounded sequence $(u_n)_{n \geq 1}$ is said to be almost convergent to a complex number c if

$$\lim_{n \to \infty} \sup_{k \in \mathbf{N}} \left| \frac{1}{n+1} \sum_{i=k}^{k+n} u_i - c \right| = 0.$$

Lorentz proved in [Lor] that $(u_n)_{n \geq 1}$ is almost convergent to c if and only if for every Banach limit L we have $c = L((u_n))$. A sequence $(u_n)_{n \geq 1}$ is said to be strongly almost convergent to c if the sequence $(|u_n - c|)_{n \geq 1}$ is almost convergent to 0. We will say that a sequence $(\phi_n)_{n \geq 1}$ of operators in $B_w(M)$ is weakly almost convergent to $\phi \in B_w(M)$ if $[l, \phi_n(X) - \phi(X)]$ almost converges to 0 for any $(l, X) \in M_* \times M$.

DEFINITION 1.1. A mapping $p: \mathbf{N} \to (0, \infty)$ is called a *gauge* if there exists $c_p > 0$ such that the sequence $p(n+1)/p(n)$ is strongly almost convergent to c_p. Moreover, if in addition the sequence $c_p^n/p(n)$ strongly almost converges to 1, then we say that p is a *regular gauge*.

We will say that a sequence $(T_n)_{n \geq 1}$ of operators, acting on a Banach space, is *dominated* by a gauge p if $\|T_n\| \leq p(n)$ holds for every positive integer n. We follow [Ker] in saying that $(T_n)_{n \geq 1}$ is *compatible* with a gauge p if in addition the sequence $\|T_n\|/p(n)$ does not almost converge to 0. An operator T is dominated by (compatible with) p if the sequence $(T^n)_{n \geq 1}$ is dominated by (resp. compatible with) p. Finally, a family $\mathcal{F}$ of operators is called dominated by (compatible with) p if each operator in $\mathcal{F}$ is dominated by (resp. compatible with) p. For some recent contributions in this area, we refer the reader to [Ker], [Ker1], [Ker2], [Ker3], [Ker4], [Ker5] and [KeMü].

Assume that p is a gauge and $T \in M$ is dominated by p. Given a Banach limit L, let us introduce the (bounded, linear) operator $E_{L,T}$, acting on M, by setting

$$[l, E_{L,T}(X)] = L(\{[l, T^{*n} X T^n] p(n)^{-2}\}_{n \geq 1})$$

for any $(l, X) \in M_* \times M$. The following proposition summarizes some useful properties of the operator $E_{L,T}$.

PROPOSITION 1.2. *Let T be an element in a von Neumann algebra M acting on a separable Hilbert space H. Assume that T is dominated by a gauge p. Then, for any Banach limit L, we have*

(i) *$E_{L,T}$ is a completely positive mapping;*
(ii) *$E_{L,T}(T^* X T) = c_p^2 E_{L,T}(X)$ for any $X \in M$;*
(iii) *if $A, B \in M$ commute with T, then we have $E_{L,T}(A^* X B) = A^* E_{L,T}(X) B$ for any $X \in M$;*
(iv) *$T^* E_{L,T}(X) T = c_p^2 E_{L,T}(X)$ for any $X \in M$;*
(v) *there exists $\rho_L(p) \in [0,1]$ such that $E_{L,T} \circ E_{L,T} = \rho_L(p) E_{L,T}$;*
(vi) *moreover, if M is a finite von Neumann algebra, then the mapping $E_{L,T}$ belongs to $B_w(M)$.*

REMARK 1.3. If $T \in M$ is compatible with a gauge p, then the spectral radius $r(T)$ satisfies $r(T) = c_p$ (see [Ker)]).

Proof. (i) Let $[X_{i,j}]_{1\le i,j\le p}$ be a positive $p\times p$ matrix whose entries are operators in M and let $x_1,\dots,x_n$ be vectors in H. For any $(i,j)\in\{1,..,p\}^2$, we define the linear functional $l_{i,j}$ acting on M by setting $l_{i,j}(X)=\langle Xx_j,x_i\rangle$. It is obvious that $l_{i,j}\in M_*$, hence

$$\left\langle \begin{bmatrix} E_{L,T}(X_{1,1}) & \cdots & E_{L,T}(X_{1,p}) \\ \vdots & \ddots & \vdots \\ E_{L,T}(X_{p,1}) & \cdots & E_{L,T}(X_{p,p}) \end{bmatrix} \begin{bmatrix} x_1 \\ \vdots \\ x_p \end{bmatrix}, \begin{bmatrix} x_1 \\ \vdots \\ x_p \end{bmatrix} \right\rangle$$
$$= \sum_{i,j=1}^{p} \langle E_{L,T}(X_{i,j})x_j,x_i\rangle = L\left(\sum_{i,j=1}^{p}\{[l_{i,j},T^{*n}X_{i,j}T^n]p(n)^{-2}\}_{n\ge1}\right)$$
$$= L\left(\left\{\left\langle \begin{bmatrix} X_{1,1} & \cdots & X_{1,p} \\ \vdots & \ddots & \vdots \\ X_{p,1} & \cdots & X_{p,p} \end{bmatrix} \begin{bmatrix} T^n x_1 \\ \vdots \\ T^n x_p \end{bmatrix}, \begin{bmatrix} T^n x_1 \\ \vdots \\ T^n x_p \end{bmatrix} \right\rangle p(n)^{-2}\right\}_{n\ge1}\right) \ge 0.$$

The positivity of the last term follows from the positivity of the matrix $[X_{i,j}]_{1\le i,j\le p}$ and the positivity of the state L.

(ii) Given any l in M_*, we have

$$\begin{aligned}[l,E_{L,T}(T^*XT)] &= L(\{[l,T^{*n+1}XT^{n+1}]p(n)^{-2}\}_{n\ge1}) \\ &= L\left(\left\{\frac{[l,T^{*n+1}XT^{n+1}]}{p(n+1)^2}\frac{p(n+1)^2}{p(n)^2}\right\}_{n\ge1}\right).\end{aligned}$$

Since p is a gauge, we see that the sequence $(|p(n+1)/p(n)-c_p|)_{n\ge1}$ is almost convergent to 0. It follows that $(|p(n+1)^2/p(n)^2-c_p^2|)_{n\ge1}$ also almost converges to 0. By Lemma 1 from [Ker], we get

$$\begin{aligned}[l,E_{L,T}(T^*XT)] &= c_p^2L(\{[l,T^{*n+1}XT^{n+1}]p(n+1)^{-2}\}_{n\ge1}) \\ &= c_p^2L(\{[l,T^{*n}XT^n]p(n)^{-2}\}_{n\ge1}) = c_p^2[l,E_{L,T}(X)]\end{aligned}$$

and (ii) follows.

(iii) Let A, B be two operators in M commuting with T, we have

$$\begin{aligned}[l,E_{L,T}(A^*XB)] &= L(\{[l,T^{*n}A^*XBT^n]p(n)^{-2}\}_{n\ge1}) \\ &= L(\{[l,A^*T^{*n}XT^nB]p(n)^{-2}\}_{n\ge1}) = [l,A^*E_{L,T}(X)B].\end{aligned}$$

This establishes the formula.

(iv) follows immediately from (ii) and (iii).

(v) Let $l \in M_*$; using (iv), we get

$$\begin{aligned}[l, E_{L,T}(E_{L,T}(X))] &= L(\{[l, T^{*n}E_{L,T}(X)T^n]p(n)^{-2}\}_{n\geq 1}) \\ &= L\left(\left\{c_p^{2n}p(n)^{-2}\right\}_{n\geq 1}\right)[l, E_{L,T}(X)] = \rho_L(p)[l, E_{L,T}(X)],\end{aligned}$$

by setting $\rho_L(p) = L(\{c_p^{2n}p(n)^{-2}\}_{n\geq 1})$. From the formula $c_p = \inf\{p(n)^{1/n} : n \in \mathbf{N}\}$ (see [Ker] Proposition 1), we immediately deduce that $\rho_L(p) \in [0, 1]$.

(vi) It suffices to show that the linear functional $M \ni X \mapsto l(E_{L,T}(X))$ is ultra-weakly continuous for any $l \in M_*$. Let Z be in M, for clarity we will denote by l_Z the element in M_* given by $l_Z(X) = \tau(ZX)$ for any $X \in M$. Given $X, Y \in M$, we have

$$\begin{aligned}\tau(E_{L,T}(X)Y) &= [l_Y, E_{L,T}(X)] = L(\{[l_Y, T^{*n}XT^n]p(n)^{-2}\}_{n\geq 1}) \\ &= L(\{\tau(YT^{*n}XT^n)p(n)^{-2}\}_{n\geq 1}) = L(\{\tau(XT^nYT^{*n})p(n)^{-2}\}_{n\geq 1}) \\ &= [l_X, E_{L,T^*}(Y)] = \tau(XE_{L,T^*}(Y)),\end{aligned}$$

hence

$$\tau(E_{L,T}(X)Y) = \tau(XE_{L,T^*}(Y)).$$

Let Y be in M, we deduce from the last equation that $l_Y \circ E_{L,T}$ is ultra-weakly continuous. Since the linear functionals l_Y with $Y \in M$ are dense in M_*, it follows that $E_{L,T}$ is ultra-weakly continuous. This completes the proof. □

II. Convergence of τ-M-stable maps

Given any $X \in M$, the linear functional $l_X(Y) := \tau(XY)$ $(Y \in M)$ is weak*-continuous, and so $l_X \in M_*$. The mapping

$$\Psi_\tau : M \to M_*, \ X \mapsto l_X$$

is a bounded linear quasiaffinity; the linear manifold $\widehat{M_\tau} := \text{ran}\Psi$ is dense in M_*.

Let us consider the set

$$B_\tau(M) := \{\phi \in B_w(M) : \phi_*(\widehat{M_\tau}) \subset \widehat{M_\tau}\}$$

of τ-M-stable weak*-continuous operators. For any $\phi \in \widehat{B}_\tau(M)$, we can introduce the linear mapping

$$\widehat{\phi}_\tau := \Psi_\tau^{-1}\phi_*\Psi_\tau : M \to M.$$

For any $X, Y \in M$, we have

$$\tau(X\phi(Y)) = [l_X, \phi(Y)] = [\phi_*(l_X), Y] = [l_{\widehat{\phi}_\tau(X)}, Y] = \tau(\widehat{\phi}_\tau(X)Y).$$

An application of the Closed Graph Theorem yields that $\widehat{\phi}_\tau$ is bounded. In fact, we see that $\widehat{\phi}_\tau$ is also in $\widehat{B}_\tau(M)$ and we have $\widehat{\left(\widehat{\phi}_\tau\right)}_\tau = \phi$. We say that ϕ is M-stable if it is τ-M-stable for every $\tau \in \mathcal{T}(M)$. We will consider the set

$$\widehat{B}(M) = \bigcap_{\tau \in \mathcal{T}(M)} \widehat{B}_\tau(M)$$

of all M-stable operators.

REMARKS 2.1. **1.** Denote by $\mathcal{M}_p(\mathbf{C})$ the algebra of square matrices of order p, and consider the finite von Neumann algebra $M = \oplus_{p\geq 2}\mathcal{M}_p(\mathbf{C})$ acting on the Hilbert space $H = \oplus_{p\geq 2}\mathbf{C}^p$ in an obvious sense. We consider the faithful normal traces τ_1 and τ_2 defined by setting

$$\tau_1(\oplus_{p\geq 2}X_p) = \sum_{p\geq 2} \frac{1}{p^3}\mathrm{Tr}(X_p),$$

$$\tau_2(\oplus_{p\geq 2}X_p) = \sum_{p\geq 2} \alpha_p \mathrm{Tr}(X_p),$$

where $\mathrm{Tr}(\cdot)$ is the usual trace acting on $\mathcal{M}_p(\mathbf{C})$ and α_p is given by

$$\alpha_p = \begin{cases} \frac{1}{2^p} & \text{if} \quad p \notin 3^{\mathbf{N}} \\ \frac{p}{2^p} & \text{if} \quad p \in 3^{\mathbf{N}}. \end{cases}$$

Let us consider the mapping ϕ defined by

$$\phi(X_2, X_3, \ldots) = (0, X_2', X_3', \ldots)$$

where $X_p' \in \mathcal{M}_{p+1}(\mathbf{C})$ is given in an obvious sense by

$$X_p' = \begin{bmatrix} X_p & 0 \\ 0 & 0 \end{bmatrix}.$$

Then, we can check that $\phi \in \widehat{B}_{\tau_1}(M)$ but $\phi \notin \widehat{B}_{\tau_2}(M)$.

2. If M is a factor ($M \cap M' = \mathbf{C}I$), then all faithful normal traces are proportional (see [Dix, p. 249]). Consequently we have $\widehat{B}(M) = \widehat{B}_\tau(M)$ for every $\tau \in \mathcal{T}(M)$.

3. Let M be a finite von Neumann algebra and $A, B \in M$, then the mapping $\phi: X \mapsto AXB$ belongs to $\widehat{B}(M)$.

Let M be a finite von Neumann algebra. Recall that $\mathcal{M}_n(M)$ is also a finite von Neumann algebra with the faithful normal trace τ_n defined by setting

$$\tau_n([A_{i,j}]_{1 \le i,j \le n}) = \sum_{k=1}^{n} \tau(A_{k,k}).$$

We begin with some useful properties of the operators $\widehat{\phi}$ when ϕ is a τ-M-stable mapping.

Proposition 2.2. *Let M be a finite von Neumann algebra, τ a faithful normal trace on M and $\phi \in B(M)$. Then*

(i) *the mapping ϕ belongs to $\widehat{B}_\tau(M)$ if and only if there exists $\psi \in B(M)$ such that $\tau(\phi(X)Y) = \tau(X\psi(Y))$ for every $X, Y \in M$; and then $\psi = \widehat{\phi}_\tau$;*

(ii) *the set $\widehat{B}_\tau(M)$ is an algebra; the mapping $\phi \mapsto \widehat{\phi}_\tau$ is linear, involutive and $\widehat{(\phi_1\phi_2)}_\tau = (\widehat{\phi}_2)_\tau(\widehat{\phi}_1)_\tau$;*

(iii) *if $\phi \in \widehat{B}_\tau(M)$, then $\phi_n \in \widehat{B}_\tau(\mathcal{M}_n(M))$ and we have $(\widehat{\phi_n})_\tau = (\widehat{\phi}_\tau)_n$;*

(iv) *if ϕ is n-positive ($n \in \mathbf{N}$), then $\widehat{\phi}_\tau$ is also n-positive;*

(v) *if ϕ is completely positive, then $\widehat{\phi}_\tau$ is completely positive;*

(vi) *assume that ϕ is 2-positive, then the mappings ϕ and $\widehat{\phi}_\tau$ extend uniquely to bounded operators from $L^2(M, \tau)$ into itself; moreover, we have*

$$\|\phi\|_{B(L^2(M,\tau))} = \left\|\widehat{\phi}_\tau\right\|_{B(L^2(M,\tau))} \le (\|\phi(I)\|_M)^{1/2} \left(\left\|\widehat{\phi}_\tau(I)\right\|_M\right)^{1/2}.$$

Proof. (i) If $\phi \in \widehat{B}_\tau(M)$, it suffices to set $\psi = \widehat{\phi}_\tau$. Conversely, assume that there exists $\psi \in B(M)$ such that $\tau(\phi(X)Y) = \tau(X\psi(Y))$ for every $X, Y \in M$. We immediately deduce that the linear functional $X \mapsto \tau(\phi(X)Y)$ is ultra-weakly continuous for each $Y \in M$. Since $\widehat{M}_\tau$ is dense in M_*, we see that $\phi \in B_w(M)$. Moreover, we have

$$\phi_*(l_X) = l_{\psi(X)}$$

for any $X \in M$, thus we have $\phi_*(\widehat{M}_\tau) \subset \widehat{M}_\tau$. This gives $\phi \in \widehat{B}_\tau(M)$.

(ii) This statement follows clearly from the characterization of elements of $\widehat{B}_\tau(M)$ given in (i).

(iii) Assume that $\phi \in \widehat{B}_\tau(M)$. Let $A = [A_{i,j}]_{1\le i,j\le n}$ and $B = [B_{i,j}]_{1\le i,j\le n}$ be two elements in $\mathcal{M}_n(M)$, then

$$\begin{aligned}\tau_n((\widehat{\phi})_n(A)B) &= \tau_n([\widehat{\phi}(A_{i,j})]_{1\le i,j\le n}[B_{i,j}]_{1\le i,j\le n}) = \sum_{k=1}^{n}\tau\left(\sum_{l=1}^{n}\widehat{\phi}(A_{k,l})B_{l,k}\right)\\ &= \sum_{k=1}^{n}\sum_{l=1}^{n}\tau(\widehat{\phi}(A_{k,l})B_{l,k}) = \sum_{k=1}^{n}\sum_{l=1}^{n}\tau(A_{k,l}\phi(B_{l,k}))\\ &= \tau_n(A\phi_n(B)).\end{aligned}$$

It follows easily by (i) that $\phi_n \in \widehat{B}_\tau(\mathcal{M}_\tau(M))$ and we have $\widehat{\phi_n} = (\widehat{\phi})_n$.

(iv) Assume that $\phi \in \widehat{B}_\tau(M)$ is positive. Let A and B be two positive elements in M, we have

$$\tau(\widehat{\phi}_\tau(A)B) = \tau(A\phi(B)) = \tau(\sqrt{A}\phi(B)\sqrt{A}) \ge 0.$$

Hence, we derive easily the positivity of $\widehat{\phi}_\tau$ from the previous calculation. If ϕ is n-positive, the map ϕ_n is positive, thus $(\widehat{\phi_n})_\tau$ is positive and the formula $(\widehat{\phi_n})_\tau = (\widehat{\phi}_\tau)_n$ implies that $\widehat{\phi}_\tau$ is n-positive.

(v) It is clear from (iv) that $\widehat{\phi}$ is completely positive if ϕ is completely positive.

(vi) Let $\phi \in \widehat{B}_\tau(M)$ be 2-positive and $Y \in M$, then the matrix

$$\begin{bmatrix}\phi(I) & \phi(Y)\\ \phi(Y)^* & \phi(Y^*Y)\end{bmatrix}$$

is positive, a fact which implies that

$$\phi(Y)^*\phi(Y) \le \|\phi(I)\|\,\phi(Y^*Y).$$

Given a pair (X, Y) of elements of M, we deduce from the previous inequality that

$$\begin{aligned}\left|\tau(\widehat{\phi}_\tau(X)Y)\right| &= |\tau(X\phi(Y))|\\ &\le \sqrt{\tau(X^*X)}\sqrt{\tau(\phi(Y)^*\phi(Y))}\\ &\le \sqrt{\|\phi(I)\|}\sqrt{\tau(X^*X)}\sqrt{\tau(\phi(Y^*Y))} = \sqrt{\|\phi(I)\|}\|X\|_2\sqrt{\tau(Y^*Y\widehat{\phi}_\tau(I))}\\ &\le \sqrt{\|\phi(I)\|}\sqrt{\left\|\widehat{\phi}_\tau(I)\right\|}\|X\|_2\|Y\|_2.\end{aligned}$$

It follows that $\|\widehat{\phi}_\tau(X)\|_2 \leq \sqrt{\|\phi(I)\|}\sqrt{\left\|\widehat{\phi}_\tau(I)\right\|}\|X\|_2$. Using the density of M in $L^2(M,\tau)$, we see that the map $\widehat{\phi}_\tau$ extends uniquely to a bounded operator from $L^2(M,\tau)$ into itself. We also get

$$\left\|\widehat{\phi}\right\|_{B(L^2(M))} \leq \sqrt{\|\phi(I)\|_M} \cdot \sqrt{\left\|\widehat{\phi}_\tau(I)\right\|_M}.$$

Observe that the adjoint of ϕ in $L^2(M,\tau)$ is given by $\phi^*(X) = \widehat{\phi}_\tau(X^*)^*$ for any $X \in M$. The rest of the proof follows immediately.

□

REMARK 2.3. It follows immediately from (ii) that $\widehat{B}(M)$ is also an algebra.

Let $\Phi = (\phi_n)_{n\geq 1}$ be a sequence in $B(M)$ dominated by a gauge p. Given a Banach limit L, let us consider the limit operator $E_{\Phi,L} \in B(M)$, defined by

$$[l, E_{\Phi,L}(X)] = L(\{[l,\phi_n(X)]p(n)^{-1}\}_{n\geq 1})$$

for any $(l,X) \in M_* \times M$. Note that the previous formulas actually define $E_{\Phi,L}$ as an element of $B(M)$. We write $\gamma_L(p) = L(\{c_p^n p(n)^{-1}\}_n)$.

The following theorem seems to be of independent interest. It presents some fine properties of abelian sequences included in $\widehat{B}_\tau(M)$ which are compatible with a gauge p, where M is a finite von Neumann algebra and τ is a faithful normal trace on M.

THEOREM 2.4. *Let M be a finite von Neumann algebra, τ a faithful normal trace on M and $\Phi = (\phi_n)_{n\geq 1}$ a sequence in $\widehat{B}_\tau(M)$ dominated by a gauge p and such that $\widehat{\Phi}_\tau = ((\widehat{\phi_n})_\tau)_{n\geq 1}$ is also dominated by p.*

(i) *The operator $E_{\Phi,L}$ belongs to $\widehat{B}_\tau(M)$ and we have $(\widehat{E_{\Phi,L}})_\tau = E_{\widehat{\Phi}_\tau,L}$ for any Banach limit L.*
(ii) *Suppose $\Phi = (\phi_n)_{n\geq 1}$ is abelian, then the operators E_{Φ,L_1} and E_{Φ,L_2} commute for any pair (L_1,L_2) of Banach limits.*
(iii) *Assume that $\psi \in \widehat{B}_\tau(M)$ and that the sequences $\phi_n = \psi^n$ and $\widehat{\phi_n} = \widehat{\psi}^n$ are dominated by the gauge p. Then we have*

$$E_{\Phi,L_2} \circ E_{\Phi,L_1} = E_{\Phi,L_1} \circ E_{\Phi,L_2} = \gamma_{L_1}(p)E_{\Phi,L_2} = \gamma_{L_2}(p)E_{\Phi,L_1}$$

for any Banach limits L_1 and L_2. In particular, if $c_p^n/p(n)$ almost converges to a nonzero limit, then $(p(n)^{-1}\phi_n)_{n\geq 1}$ weakly almost converges to an operator ϕ belonging to $\widehat{B}_\tau(M)$.

Proof. (i) Given a pair (X, Y) of elements of M, we get

$$\begin{aligned}\tau(E_{\Phi,L}(X)Y) &= [l_Y, E_{\Phi,L}(X)] = L(\{[l_Y, \phi_n(X)]p(n)^{-1}\}_{n\geq 1}) \\ &= L(\{\tau(\phi_n(X)Y)p(n)^{-1}\}_{n\geq 1}) = L(\{\tau(X(\widehat{\phi_n})_\tau(Y))p(n)^{-1}\}_{n\geq 1}) \\ &= L(\{[l_X, (\widehat{\phi_n})_\tau(Y)]p(n)^{-1}\}_{n\geq 1}) = [l_X, E_{\hat{\Phi}_\tau,L}(Y)] = \tau(XE_{\hat{\Phi}_\tau,L}(Y))\end{aligned}$$

for every Banach limit L. Now Proposition 2.2.(i) implies the statement.

(ii) Assume that the sequence $(\phi_n)_{n\geq 1}$ is abelian. We have

$$\begin{aligned}[l_{(\widehat{\phi_m})_\tau(Y)}, \phi_n(X)] &= \tau(\phi_n(X)(\widehat{\phi_m})_\tau(Y)) = \tau(\phi_m(X)(\widehat{\phi_n})_\tau(Y)) \\ &= [l_{\phi_m(X)}, (\widehat{\phi_n})_\tau(Y)]\end{aligned}$$

for any pair (m, n) of positive integers and any pair $(X, Y) \in M^2$. By taking L_1-limit with respect to n we get

$$\begin{aligned}[l_{E_{\Phi,L_1}(X)}, (\widehat{\phi_m})_\tau(Y)] &= \tau(E_{\Phi,L_1}(X)(\widehat{\phi_m})_\tau(Y)) = [l_{(\widehat{\phi_m})_\tau(Y)}, E_{\Phi,L_1}(X)] \\ &= [l_{\phi_m(X)}, E_{\hat{\Phi}_\tau,L_1}(Y)] = \tau(\phi_m(X)E_{\hat{\Phi}_\tau,L_1}(Y)) \\ &= [l_{E_{\hat{\Phi}_\tau,L_1}(Y)}, \phi_m(X)].\end{aligned}$$

Now, taking L_2-limit with respect to m and using (i), we obtain that

$$\begin{aligned}\tau(E_{\Phi,L_2} \circ E_{\Phi,L_1}(X)Y) &= \tau(E_{\Phi,L_1}(X)(\widehat{E_{\Phi,L_2}})_\tau(Y)) = \tau(E_{\Phi,L_1}(X)E_{\hat{\Phi}_\tau,L_2}(Y)) \\ &= [l_{E_{\Phi,L_1}(X)}, E_{\hat{\Phi}_\tau,L_2}(Y)] = [l_{E_{\hat{\Phi},L_1}(Y)}, E_{\Phi,L_2}(X)] \\ &= \tau(E_{\hat{\Phi}_\tau,L_1}(Y)E_{\Phi,L_2}(X)) \\ &= \tau((\widehat{E_{\Phi,L_1}})_\tau(Y)E_{\Phi,L_2}(X)) = \tau(E_{\Phi,L_1} \circ E_{\Phi,L_2}(X)Y).\end{aligned}$$

We thus have $E_{\Phi,L_2} \circ E_{\Phi,L_1} = E_{\Phi,L_1} \circ E_{\Phi,L_2}$.

(iii) Let $\psi \in B_w(M)$ and (L_1, L_2) be a pair of Banach limits. For any $X, Y \in M$, we have

$$\begin{aligned}[l_{\phi_n(X)}, (\widehat{\phi_m})_\tau(Y)]p(n)^{-1}p(m)^{-1} &= \tau(\phi_n(X)(\widehat{\phi_m})_\tau(Y))p(n)^{-1}p(m)^{-1} \\ &= \tau(\psi^{m+n}(X)Y)p(n)^{-1}p(m)^{-1} \\ &= \frac{p(m+n)}{p(m)p(n)}[l_Y, \psi^{m+n}(X)]p(m+n)^{-1}.\end{aligned}$$

The sequence $p(m+n)/p(m) = p(m+1)/p(m) \cdots p(m+n)/p(m+n-1)$ is strongly almost convergent to c_p^n, when m goes to infinity. By taking L_1-limit with respect to m, we get

$$\begin{aligned}[l_{E_{\hat{\Phi}_\tau,L_1}(Y)}, \phi_n(X)]p(n)^{-1} &= \tau(\phi_n(X)E_{\hat{\Phi}_\tau,L_1}(Y))p(n)^{-1} \\ &= [l_{\phi_n(X)}, E_{\hat{\Phi}_\tau,L_1}(Y)]p(n)^{-1} = c_p^n p(n)^{-1}[l_Y, E_{\Phi,L_1}(X)].\end{aligned}$$

After taking L_2-limit with respect to n, it follows that

$$\begin{aligned}\tau(E_{\Phi,L_1}\circ E_{\Phi,L_2}(X)Y) &= [l_{E_{\widehat{\Phi}_\tau,L_1}(Y)}, E_{\Phi,L_2}(X)]\\ &= \gamma_{L_2}(p)[l_Y, E_{\Phi,L_1}(X)] = \gamma_{L_2}(p)\tau(E_{\Phi,L_1}(X)Y),\end{aligned}$$

whence $E_{\phi,L_1}\circ E_{\phi,L_2} = \gamma_{L_2}(p)E_{\phi,L_1}$. We deduce from (ii) that E_{ϕ,L_1} and E_{ϕ,L_2} commute. Interchanging the role of L_1 and L_2 we conclude that

$$E_{\Phi,L_2}\circ E_{\Phi,L_1} = E_{\Phi,L_1}\circ E_{\Phi,L_2} = \gamma_{L_1}(p)E_{\Phi,L_2} = \gamma_{L_2}(p)E_{\Phi,L_1}.$$

When the sequence $c_p^n/p(n)$ almost converges to a nonzero number, then we deduce that the limit of $(p(n)^{-1}\phi_n)_{n\geq 1}$ is independent of L. Applying Lorentz's result, we obtain that $(p(n)^{-1}\phi_n)_{n\geq 1}$ is weakly almost convergent to an operator in $\widehat{B}_\tau(M)$. The proof is now complete.

□

REMARKS 2.5. **1.** If $c_p^n p(n)^{-1}$ is almost convergent to a nonzero limit, then we see that the Cesaro means $(n+1)^{-1}(p(0)^{-1}\phi_0+\cdots+p(n)^{-1}\phi_n)$ weakly converge to an operator in $B_w(M)$ (which is obviously ultra-weakly continuous).

2. The assumption that $\Phi = (\phi_n)_{n\geq 1}$ is a sequence in $\widehat{B}_\tau(M)$ dominated by a gauge p does not imply that $\widehat{\Phi}_\tau = (\widehat{(\phi_n)_\tau})_{n\geq 1}$ is also dominated by p. Denote by $\mathcal{M}_p(\mathbf{C})$ the algebra of square matrices of order p, and consider the finite von Neumann algebra $M = \oplus_{p\geq 2}\mathcal{M}_p(\mathbf{C})$ acting on the Hilbert space $H = \oplus_{p\geq 2}\mathbf{C}^p$ in an obvious sense. We consider the faithful normal trace τ defined by

$$\tau(\oplus_{p\geq 2}X_p) = \sum_{p\geq 2}\frac{1}{p^3}\mathrm{Tr}(X_p),$$

where $\mathrm{Tr}(\cdot)$ is the usual trace on $\mathcal{M}_p(\mathbf{C})$. Fix a unit vector e_p in $\mathbf{C}^p$ and write P_p for the orthogonal projection onto $\mathbf{C}^p\ominus\mathbf{C}e_p$. For any $n\geq 2$, set $\phi_n(\oplus_{p\geq 2}X_p) = \langle X_n e_n, e_n\rangle P_n$. We can easily see that $\Phi = (\phi_n)_{n\geq 1}$ is a sequence in $\widehat{B}_\tau(M)$ such that $\|\phi_n\| = 1$ for every n. Hence Φ is dominated by the constant gauge p equal to 1, but $\widehat{\Phi}_\tau = (\widehat{(\phi_n)_\tau})_{n\geq 1}$ is not dominated by p , actually $\|\widehat{(\phi_n)_\tau}\| = n-1$.

Let $\mathcal{F}$ be an abelian set included in $\widehat{B}_\tau(M)$. We consider the (abelian) semigroup $\mathcal{E}(\mathcal{F})$ induced by $\mathcal{F}$, that is

$$\mathcal{E}(\mathcal{F}) = \{\phi_1\circ\cdots\circ\phi_n : \phi_1,\ldots,\phi_n\in\mathcal{F}, n\in\mathbf{N}\}.$$

We define a partial ordering on $\mathcal{E}(\mathcal{F})$ by setting $F\leq F'$ if there exists F'' in $\mathcal{E}(\mathcal{F})$ such that $F' = F''F$. (It is clear that $\mathcal{E}(\mathcal{F})$ is a directed set with this partial ordering, and it can be considered as a net (generalized sequence) indexed by itself.)

PROPOSITION 2.6. *Let M be a finite von Neumann algebra, τ a faithful normal trace on M, and let $\mathcal{F}$ be an abelian set of m-positive projections belonging to $\widehat{B}_\tau(M)$ ($m \geq 2$). Assume that $\mathcal{E}(\mathcal{F})$ and $\widehat{\mathcal{E}(\mathcal{F})}$ are bounded in $B(M)$. Then the net $\mathcal{E}(\mathcal{F})$ weakly converges to an m-positive projection $E \in \widehat{B}_\tau(M)$.*

Proof. Let us introduce the classical Hilbert space $L^2(M,\tau)$ equipped with the inner product $\langle X, Y\rangle := \tau(Y^*X)$ $(X, Y \in M)$. Since every $\phi \in \mathcal{E}(\mathcal{F})$ is 2-positive, Proposition 2.2.(vi) shows that ϕ extends uniquely to a bounded projection (still denoted by ϕ) from $L^2(M,\tau)$ into itself. Using again Proposition 2.2.(vi), we see that the set $\mathcal{E}(\mathcal{F})$ is bounded in $B(L^2(M,\tau))$, thus it is weakly relatively compact.

Choose two cofinal subnets $(E_i)_{i\in\mathcal{I}}$ and $(E_j)_{j\in\mathcal{J}}$ in $\mathcal{E}(\mathcal{F})$, which converge respectively to F and G in the weak operator topology of $B(L^2(M,\tau))$. Fix $i \in \mathcal{I}$ and consider the set $\mathcal{J}_i \doteq \{j \in \mathcal{J} : j \geq i\}$. Then we have

$$\begin{aligned}\langle E_i(X), E_j^*(Y)\rangle &= \tau(E_i(X)(\widehat{E_j})_\tau(Y^*)) = \tau(E_j \circ E_i(X)Y^*)\\ &= \tau(E_j(X)Y^*) = \langle E_j(X), Y\rangle\end{aligned}$$

for any $j \in \mathcal{J}_i$ and any pair $(X, Y) \in M^2$. Thus, taking limit with respect to the set $\mathcal{J}_i$, we obtain

$$\langle E_i(X), G^*(Y)\rangle = \langle G(X), Y\rangle.$$

Now, taking limit with respect to the directed set $\mathcal{I}$, we get

$$\langle G\circ F(X), Y\rangle = \langle F(X), G^*(Y)\rangle = \langle G(X), Y\rangle,$$

whence $G\circ F = G$ follows. Interchanging the role of F and G, we see that $F\circ G = F$. Since F and G are limit points of elements belonging to the commutative set $\mathcal{E}(\mathcal{F})$, they commute. Hence $F = F\circ G = G\circ F = G$, in particular $F = F\circ G = F\circ F$. We deduce that $\mathcal{E}(\mathcal{F})$ is weakly convergent in $B(L^2(M,\tau))$ to a projection E.

Now, we want to show that $E(M) \subset M$ and $E|M \in B(M)$. To this order let $X \in M$ be arbitrary, and let us consider the linear functional $\varphi(Y) := \langle E(X), Y^*\rangle$ $(Y \in M)$. Choosing a cofinal subnet $\{E_i\}_{i\in\mathcal{I}}$ in $\mathcal{E}(\mathcal{F})$, we have $\langle E(X), Y^*\rangle = \lim_i\langle E_i(X), Y^*\rangle = \lim_i \tau(E_i(X)Y)$. Thus

$$|\varphi(Y)| = \lim_i |\tau(E_i(X)Y)| \leq \liminf_i \|E_i(X)\|\|Y\|_1 \leq C\|X\|\|Y\|_1,$$

where $C = \sup\{\|F\| : F \in \mathcal{E}(\mathcal{F})\} < \infty$ and $\|Y\|_1 := \tau(|Y|) = \|l_Y\|$ (see [Dix] Section I.6.10). We deduce that there exists unique $\tilde{X} \in M$ such that

$$\langle E(X), Y^*\rangle = \varphi(Y) = (\varphi\circ\Psi_\tau^{-1})(l_Y) = [l_Y, \tilde{X}] = \tau(\tilde{X}Y) = \langle\tilde{X}, Y^*\rangle$$

holds for every $Y \in M$. Hence $E(X) = \tilde{X} \in M$ and $\|E(X)\| = \|\tilde{X}\| \leq C\|X\|$.

It is clear that

$$[l_Y, E(X)] = \tau(E(X)Y) = \langle E(X), Y^* \rangle = \lim_i \langle E_i(X), Y^* \rangle = \lim_i [l_Y, E_i(X)]$$

is true for every $X, Y \in M$. Since $\widehat{M}_\tau$ is dense in M_*, and $\mathcal{E}(\mathcal{F})$ is bounded, it follows that $\mathcal{E}(\mathcal{F})$ weakly converges to E.

We can prove in the same manner that $\widehat{\mathcal{E}(\mathcal{F})}$ converges weakly to an operator $F \in B(M)$. Taking into account that $\tau(E_i(X)Y) = \tau(X(\widehat{E_i})_\tau(Y))$, we obtain by passing to the limit that

$$\tau(E(X)Y) = \tau(XF(Y))$$

holds for every $X, Y \in M$. It follows by Proposition 2.2.(i) that E belongs to $B_w(M)$.

It remains to prove that E is m-positive. It is clear that every operator in $\mathcal{E}(\mathcal{F})$ is m-positive. Let $[X_{k,l}]_m \in \mathcal{M}_m(M)$ be a positive operator. Given any vector $x = x_1 \oplus \cdots \oplus x_m \in H^{(m)}$, we have

$$\begin{aligned}\left\langle [E(X_{k,l})]_m x, x \right\rangle &= \sum_{k,l=1}^{m} [l_{x_l, x_k}, E(X_{k,l})] \\ &= \lim_i [l_{x_l, x_k}, E_i(X_{k,l})] = \lim_i \left\langle [E_i(X_{k,l})]_m x, x \right\rangle \geq 0,\end{aligned}$$

and so E is an m-positive projection.

□

Let M be a finite von Neumann algebra and let T be an operator in M dominated by the regular gauge p. We know by Theorem 2.4.(iii) that the sequence of M-stable mappings $(\phi_{T,n})_{n\geq 0}$ defined by $\phi_{T,n}(X) = p(n)^{-2} T^{*n} X T^n$ is weakly almost convergent, we will denote its limit by E_T. Now Proposition 1.2 shows that E_T is a completely positive projection. Notice also that E_T is an ultra-weakly continuous M-stable operator with $\widehat{E}_T = E_{T^*}$ and that $\|E_T\| \leq 1$. Let $\mathcal{S}$ be an abelian subset of M, which is dominated by the regular gauge p. We consider the abelian semigroup $\mathcal{E}(\mathcal{S})$ induced by the (abelian) set $\{E_T : T \in \mathcal{S}\}$.

COROLLARY 2.7. *Let M be a finite von Neumann algebra and let $\mathcal{S}$ be an abelian subset of M. Assume that $\mathcal{S}$ is dominated by a regular gauge p. Then the net $\mathcal{E}(\mathcal{S})$ weakly converges to a completely positive M-stable projection E satisfying the following properties:*

(i) $E(T^*XT) = c_p^2 E(X)$ *for any* $T \in \mathcal{S}$ *and* $X \in M$;

(ii) $E(A^*XB) = A^*E(X)B$ *for any pair* $(A, B) \in (\mathcal{S}')^2$ *and* $X \in M$.

Proof. We apply Proposition 2.6 to the net $\mathcal{E}(\mathcal{S})$. We deduce that $\mathcal{E}(\mathcal{S})$ converges weakly to $E \in \widehat{B}_\tau(M)$, for any $\tau \in \mathcal{T}(M)$. Since properties (i) and (ii) are true for the operators E_R $(R \in \mathcal{S})$ by Proposition 1.2, we see that the same properties hold for E.

□

PROPOSITION 2.8. *Let M be a finite von Neumann algebra, τ a faithful normal trace on M, and let $\mathcal{S}$ be an abelian subset of M which is dominated by a regular gauge p. Let us assume that the limit projection E of the net $\mathcal{E}(\mathcal{S})$ is such that $E(I)$ is injective. Then there exists an abelian set $\mathcal{S}_1$ of unitaries belonging to M such that for any $T \in \mathcal{S}$ there exists $U_T \in \mathcal{S}_1$ satisfying $\sqrt{E(I)}T = c_p U_T \sqrt{E(I)}$. Moreover, if F denotes the limit of the net $\mathcal{E}(\mathcal{S}_1)$, then we have the following properties:*
(i) $E(\sqrt{E(I)}X\sqrt{E(I)}) = \sqrt{E(I)}F(X)\sqrt{E(I)}$ *for any* $X \in M$;
(ii) $\sqrt{E(I)}\widehat{E}_\tau(X)\sqrt{E(I)} = \widehat{F}_\tau(\sqrt{E(I)}X\sqrt{E(I)})$ *for any* $X \in M$.

Proof. (i) First of all, observe that the equation $T^*E(I)T = c_p^2 E(I)$ and the injectivity of $E(I)$ imply that $\sqrt{E(I)}T$ is also injective. Taking the polar decomposition of $\sqrt{E(I)}T$, we see that there exists a unique isometry U_T such that $\sqrt{E(I)}T = c_p U_T \sqrt{E(I)}$. Since $U_T \in M$ and M is finite, it follows that U_T is unitary. The previous intertwining relations readily imply that the set $\mathcal{S}_1 := \{U_T : T \in \mathcal{S}\}$ is abelian.

Given $T \in \mathcal{S}$ and $X \in M$, we have

$$p(n)^{-2}T^{*n}\sqrt{E(I)}X\sqrt{E(I)}T^n = \frac{c_p^{2n}}{p(n)^2}\sqrt{E(I)}U_T^{*n}XU_T^n\sqrt{E(I)}$$

for every positive integer n. Taking a Banach limit we get the relation

$$E_T(\sqrt{E(I)}X\sqrt{E(I)}) = \sqrt{E(I)}E_{U_T}(X)\sqrt{E(I)}.$$

Now taking limits in the nets $\mathcal{E}(\mathcal{S})$ and $\mathcal{E}(\mathcal{S}_1)$ we get (i).

(ii) Since E and F are τ-M-stable, we can now get (ii) by the following computation. For any $(X, Y) \in M^2$ we have

$$\begin{aligned}\tau(X\sqrt{E(I)}\widehat{E}_\tau(Y)\sqrt{E(I)}) &= \tau(\sqrt{E(I)}X\sqrt{E(I)}\widehat{E}_\tau(Y)) \\ &= \tau(E(\sqrt{E(I)}X\sqrt{E(I)})Y) = \tau(\sqrt{E(I)}F(X)\sqrt{E(I)}Y) \\ &= \tau(X\widehat{F}_\tau(\sqrt{E(I)}Y\sqrt{E(I)})).\end{aligned}$$

Hence, we have $\sqrt{E(I)}\widehat{E}_\tau(Y)\sqrt{E(I)} = \widehat{F}_\tau(\sqrt{E(I)}Y\sqrt{E(I)})$ for any $Y \in M$. This completes the proof.

□

III. Similarity

We say that an operator T is *asymptotically controlled* by a gauge p if T is compatible with p and satisfies the condition that $q'(\{\|T^n x\|^2/p(n)^2\}_n) > 0$, for every nonzero vector $x \in H$, where

$$q'(\xi) := \sup\left\{\liminf_k \frac{1}{m}\sum_{i=1}^{m} \xi(n_i + k) : m \in \mathbf{N}, n_1, \dots, n_m \in \mathbf{N}\right\}$$

for any bounded real sequence ξ (see [Ker] for the role of this functional in the study of Banach limits).

For any real sequence $\xi \in \ell^\infty(\mathbf{N}^n)$ $(n \in \mathbf{N}, n > 1)$, let $Q_n\xi := \eta \in \ell^\infty(\mathbf{N}^{n-1})$, where $\eta(j_1, \dots, j_{n-1}) := q'(\xi_{j_1,\dots,j_{n-1}})$ with $\xi_{j_1,\dots,j_{n-1}}(j) := \xi(j_1, \dots, j_{n-1}, j)$. Let $\widetilde{Q}_n := Q_1 \circ \dots \circ Q_{n-1} \circ Q_n$, where $Q_1 := q'$.

A set $\mathcal{F}$ of operators, acting on the Hilbert space H, is called asymptotically controlled by a gauge p, if every operator in $\mathcal{F}$ is compatible with p, and if for every nonzero vector $x \in H$ there exists $\rho(x) > 0$ such that

$$\widetilde{Q}_n\left(\left\{\frac{1}{p(j_1)^2 \cdots p(j_n)^2}\left\langle T_n^{*j_n} \cdots T_1^{*j_1} T_1^{j_1} \cdots T_n^{j_n} x, x\right\rangle\right\}_{j_1,\dots,j_n=1}^{\infty}\right) \geq \rho(x)$$

is true for every $n \in \mathbf{N}$ and $T_1, \dots, T_n \in \mathcal{F}$.

REMARK 3.1. Let T be an operator compatible with a gauge p. Assume that T satisfies

$$\inf\{\|T^n x\|/p(n) : n \in \mathbf{N}\} > 0$$

for any nonzero x in H; then T is asymptotically controlled by p. In particular, power bounded operators of class $C_{1\cdot}$ (in the Sz.-Nagy–Foias terminology) are exactly operators which are asymptotically controlled by constant gauges.

THEOREM 3.2. *Let $\mathcal{S}$ be an abelian set of operators which is contained in a finite von Neumann algebra. Assume that $\mathcal{S}$ is asymptotically controlled by a regular gauge p. Then, there exists an invertible operator A in M such that $r(T)^{-1}ATA^{-1}$ is a unitary operator for any $T \in \mathcal{S}$.*

Proof. Let τ be a faithful normal trace acting on M. Let E be the completely positive limit projection provided by Corollary 2.7. Since $\mathcal{S}$ is asymptotically controlled by the gauge p, we can infer by a short computation that, given any nonzero vector $x \in H$,

$$[l_{x,x}, E_{T_1} \circ E_{T_2} \circ \cdots \circ E_{T_n}(I)] \geq \rho(x)$$

is true for every choice of $T_1, \ldots, T_n \in \mathcal{S}, n \in \mathbf{N}$ with a $\rho(x) > 0$, whence

$$\langle E(I)x, x\rangle = [l_{x,x}, E(I)] \geq \rho(x) > 0.$$

Thus, the positive operator $E(I)$ is injective. Let us consider the associated set $\mathcal{S}_1$ and the corresponding limit operator F occurring in Proposition 2.8.

Set $X = E(I)$, $Y = \widehat{E}_\tau(I)$ and consider the positive operator $R = \sqrt{X}Y\sqrt{X} = \widehat{F}_\tau(X)$. Note that R commutes with U_T ($T \in \mathcal{S}$). Let P be a projection associated with the spectral decomposition of R (which still commutes with U_T). By the Cauchy–Schwarz Inequality, we get

$$\tau(PRP) = \tau(P\sqrt{X}Y\sqrt{X}P) \leq \sqrt{\tau(PXP)}\sqrt{\tau(P\sqrt{X}Y^2\sqrt{X}P)}. \tag{1}$$

Applying the properties of E and F described in Corollary 2.7 and Proposition 2.8, we infer that

$$\begin{aligned}\tau(PXP) &= \tau(XP) = \tau(XF(P)) = \tau(\sqrt{X}F(P)\sqrt{X}) = \tau(E(\sqrt{X}P\sqrt{X}))\\ &= \tau(\sqrt{X}P\sqrt{X}\widehat{E}_\tau(I)) = \tau(\sqrt{X}P\sqrt{X}Y) = \tau(PR) = \tau(PRP).\end{aligned}$$

Now, note that the operator $C = \sqrt{X}P\sqrt{X}Y$ commutes with T^*, because we have

$$\begin{aligned}T^*C &= T^*\sqrt{X}P\sqrt{X}Y = c_p\sqrt{X}U_T^*P\sqrt{X}Y = c_p\sqrt{X}PU_T^*\sqrt{X}Y\\ &= (1/c_p)\sqrt{X}PU_T^*\sqrt{X}TYT^* = \sqrt{X}PU_T^*U_T\sqrt{X}YT^* = \sqrt{X}P\sqrt{X}YT^*\\ &= CT^*.\end{aligned}$$

Using again Corollary 2.7 and Proposition 2.8 we obtain

$$\begin{aligned}\tau(P\sqrt{X}Y^2\sqrt{X}P) &= \tau(Y\sqrt{X}P\sqrt{X}Y) = \tau(YC) = \tau(\widehat{E}_\tau(I)C) = \tau(\widehat{E}_\tau(C))\\ &= \tau(\widehat{E}_\tau(C)E(I)) = \tau(\widehat{E}_\tau(I)CE(I)) = \tau(Y\sqrt{X}P\sqrt{X}YX)\\ &= \tau(P\sqrt{X}YXY\sqrt{X}) = \tau(PR^2) = \tau(PR^2P).\end{aligned}$$

Substituting these results into (1), we get

$$\tau(PRP) \leq \sqrt{\tau(PRP)}\sqrt{\tau(PR^2P)},$$

whence

$$\tau(PRP) \le \tau(PR^2P). \tag{2}$$

Let K be a compact set contained in the interval $(0,1)$, and denote by P the spectral projection associated to K by the functional calculus of R. We thus have

$$PRP \ge (PRP)^2 = PR^2P \ge 0. \tag{3}$$

Combining (2) with (3) yields

$$\tau(PRP - PR^2P) = 0.$$

The operator $PRP - PR^2P$ is positive, so it is necessarily equal to 0. Therefore $Q = PRP$ is an orthogonal projection. But K is compact and contained in $(0,1)$, thus there exists $\rho \in (0,1)$ such that $Q \le \rho I$. Consequently, we have $Q = 0$. It follows that

$$\sigma(R) \cap (0,1) = \emptyset.$$

The last step is devoted to show that $0 \notin \sigma(R)$. Let us denote by P the spectral projection associated to 0. We have $RP = 0$, thus

$$0 = \tau(RP) = \tau(\widehat{F}_\tau(X)P) = \tau(XF(P)) = \tau(XF(I)P) = \tau(XP) = \tau(PXP).$$

It follows that $PXP = 0$. Since X is injective, we deduce that $P = 0$. Finally, we see that R is invertible (actually, $\sigma(R) \subset [1,\infty)$), therefore X is also invertible. By Proposition 2.8 we know that $\sqrt{X}T = r(T)U_T\sqrt{X}$ ($c_p = r(T)$, see [Ker]). It follows that, for any $T \in \mathcal{S}$, $U_T = r(T)^{-1}ATA^{-1}$ is unitary, where $A = \sqrt{X}$ is invertible.

□

Acknowledgments. The author wishes to thank the organizers of the "Memorial Conference for Béla Szőkefalvi-Nagy", the organizers of "Journées d'Analyse Fonctionnelle de Lens" and the University of Besançon, where these results were presented. The author wishes to express his gratitude to László Kérchy for some stimulating questions.

References

[Bea] B. Beauzamy, *Introduction to Operator Theory and Invariant Subspaces*, North-Holland, Amsterdam, 1988.

[BCP] S. BROWN, B. CHEVREAU and C. PEARCY, Contractions with rich spectrum have invariant subspaces, *J. Operator Theory*, **1** (1979), 123–136.

[CaFa] G. CASSIER and T. FACK, Contractions in von Neumann algebras, *J. Funct. Anal.*, **135** (1996), 297–338.

[CaFa2] G. CASSIER and T. FACK, On power-bounded operators in finite von Neumann algebras, *J. Funct. Anal.*, **141** (1996), 133–158.

[Dix] J. DIXMIER, *Les algèbres d'opérateurs dans l'espace Hilbertien* (*Algèbre de von Neumann*), Gauthier-Villards, Paris, 1969.

[Dix2] J. DIXMIER, Les moyennes invariantes dans les semi-groupes et leurs applications, *Acta Sci. Math.* (*Szeged*), **12** (195[illegible]), 213–227.

[Lor] G. G. LORENTZ, A contribution to the theory of divergent sequences, *Acta. Math.*, **80** (1948), 167–190.

[Ker] L. KÉRCHY, Operators with regular norm sequences, *Acta Sci. Math.* (*Szeged*), **63** (1997), 571–605.

[Ker2] L. KÉRCHY, Criteria of regularity for norm sequences, *Integral Equations Operator Theory*, **34** (1999), 458–477.

[Ker3] L. KÉRCHY, Representations with regular norm-behavior of discrete abelian semigroups, *Acta Sci. Math.* (*Szeged*), **65** (1999), 702–726.

[Ker4] L. KÉRCHY, Hyperinvariant subspaces of operators with non-vanishing orbits, *Proc. Amer. Math. Soc.*, **127** (1999), 1363–1370.

[Ker5] L. KÉRCHY, Unbounded representations of discrete abelian semigroups, *Progress in Nonlinear Differential Equations and Their Applications*, **42** (2000), 141–150.

[KeMü] L. KÉRCHY and V. MÜLLER, Criteria of regularity for norm sequences. II, *Acta Sci. Math.* (*Szeged*), **65** (1999), 131–138.

[Sak] S. SAKAI, *C^* algebras and W^* algebras*, Ergebnisse der Mathematik und ihrer Grenzgebiete **60**, Springer-Verlag, Berlin – New York, 1971.

[Nag] B. SZ.-NAGY, On uniformly bounded linear transformations in Hilbert space, *Acta Sci. Math.* (*Szeged*), **11** (1947), 152–157.

G. CASSIER, Institut Girard Desargues, UPRES-A 5028 Mathématiques, Université Claude Bernard Lyon I, 69622 Villeurbanne Cedex, France;
e-mail: cassier@desargues.univ-lyon1.fr

Operator Theory: Advances and Applications, Vol. 127, 163–170

Singly generated algebras containing a compact operator

JOHN B. CONWAY and GABRIEL PRĂJITURĂ

Dedicated to the memory of Professor Béla Szőkefalvi-Nagy

For an operator T, consider $P(T)$, the norm closed algebra generated by T and the identity operator. The question is, "When does this algebra contain a compact operator?". The question remains unanswered. However it is shown that the set of all operators such that $P(T)$ contains a compact operator is dense in $\mathcal{B}(\mathcal{H})$ and the interior of this set is characterized. If the term "compact operator" is replaced by "finite rank operator" or if $P(T)$ is replaced by the weakly closed or weak-star closed algebra generated by T and the identity, the same results are obtained. Similar questions are raised and answered for other algebras associated with an operator.

For an operator T on a Hilbert space, let $W(T)$ be the weak operator topology closed algebra generated by T and the identity operator. This paper addresses the question, "When does $W(T)$ contain a non-zero, compact operator?". Variations on this question arise when the algebra $W(T)$ is replaced by $P(T)$, the norm closed algebra generated by T and 1; or $P^\infty(T)$, the weak* closed algebra generated by T and 1. Still other variations occur when we look for a non-zero, finite rank operator.

Definitive answers to such questions are not obtained. Instead, we will investigate the topological properties of the set of all operators having such properties. To phrase these questions, introduce the following notation. Let $\mathcal{B}_0(\mathcal{H})$, or $\mathcal{B}_0$, denote the set of all compact operators on the separable Hilbert space $\mathcal{H}$; let $\mathcal{B}_{00} = \mathcal{B}_{00}(\mathcal{H})$ be the finite rank operators. $\mathcal{B}(\mathcal{H})$ denotes the algebra of all bounded operators on

AMS Subject Classification (1991): 47A58, 47B07, 47C05.

$\mathcal{H}$. Now define the following classes:

1.*i* $\quad \mathcal{C}_{00} \equiv \{T \in \mathcal{B}(\mathcal{H}) : P(T) \cap \mathcal{B}_{00}(\mathcal{H}) \neq (0)\}$

1.*ii* $\quad \mathcal{C}_{00}^{\infty} \equiv \{T \in \mathcal{B}(\mathcal{H}) : P^{\infty}(T) \cap \mathcal{B}_{00}(\mathcal{H}) \neq (0)\}$

1.*iii* $\quad \mathcal{C}_{00}^{w} \equiv \{T \in \mathcal{B}(\mathcal{H}) : W(T) \cap \mathcal{B}_{00}(\mathcal{H}) \neq (0)\}$

1.*iv* $\quad \mathcal{C}_{0} \equiv \{T \in \mathcal{B}(\mathcal{H}) : P(T) \cap \mathcal{B}_{0}(\mathcal{H}) \neq (0)\}$

1.*v* $\quad \mathcal{C}_{0}^{\infty} \equiv \{T \in \mathcal{B}(\mathcal{H}) : P^{\infty}(T) \cap \mathcal{B}_{0}(\mathcal{H}) \neq (0)\}$

1.*vi* $\quad \mathcal{C}_{0}^{w} \equiv \{T \in \mathcal{B}(\mathcal{H}) : W(T) \cap \mathcal{B}_{0}(\mathcal{H}) \neq (0)\}.$

Later it is shown that each of these sets are norm dense in $\mathcal{B}(\mathcal{H})$. It is also shown that they all have the same interior and that interior can be characterized in terms of spectral properties alone.

In general, the notation will be that of [5]. The spectrum of an operator T is denoted by $\sigma(T)$. $\sigma_0(T)$ is the set of isolated eigenvalues of T having finite multiplicity, and $\sigma_e(T)$ is the essential spectrum of T. That is, $\sigma_e(T)$ is the spectrum of the image of T in the Calkin algebra $\mathcal{B}(\mathcal{H})/\mathcal{B}_0(\mathcal{H})$. Finally, for any set X in a metric space and $r > 0$, $X_r \equiv \{z : \text{dist}\,(z, X) < r\}$.

It is clear that each of these classes is invariant under similarity and $\mathcal{C}_{00} \subseteq \mathcal{C}_{00}^{\infty} \subseteq \mathcal{C}_{00}^{w}$; $\mathcal{C}_0 \subseteq \mathcal{C}_0^{\infty} \subseteq \mathcal{C}_0^{w}$; $\mathcal{C}_{00} \subseteq \mathcal{C}_0$, $\mathcal{C}_{00}^{\infty} \subseteq \mathcal{C}_0^{\infty}$ and $\mathcal{C}_{00}^{w} \subseteq \mathcal{C}_0^{w}$.

2. EXAMPLES. (a) If V is the Volterra operator, then $V \in \mathcal{C}_0$, but $V \notin \mathcal{C}_{00}$.

(b) Let U be the bilateral shift operating on K and let $T = U \oplus 1$ operating on $\mathcal{K} \oplus \mathbb{C}$. It is easy to check that $T \notin \mathcal{C}_0$, but a version of Szego's Theorem (see, for example, Proposition VI.4.9 in [6]) implies that $T \in \mathcal{C}_0^{w}$.

3. PROPOSITION. *For an operator T the following statements are equivalent.*

(a) *There is a finite rank operator in $P(T)$ that is not nilpotent.*

(b) *There is a finite rank idempotent in $P(T)$.*

(c) *T has an isolated eigenvalue of finite multiplicity that does not belong to the polynomially convex hull of $\sigma_e(T)$.*

Proof. If A is a finite rank, non-nilpotent operator in $P(T)$, let α be a non-zero eigenvalue of A and let $\{p_n\}$ be a sequence of polynomials such that $\|p_n(T) - A\| \to 0$. Let $r < 3^{-1}\text{dist}\,(\alpha, \sigma(A)\backslash\{\alpha\})$. If $\epsilon > 0$, then an easy argument involving spectral theory and the Riesz functional calculus shows that there is an n_0 such that for $n \geq n_0$, $\|p_n(T) - A\| < \epsilon$, $\sigma(p_n(T)) \subseteq B(\alpha; r) \cup [\sigma(A)\backslash\{\alpha\}]_r$ and $E_n \equiv E(p_n(T), B(\alpha; r))$, the Riesz idempotent for the operator $p_n(T)$ corresponding to the set $B(\alpha; r)$, has the same rank as $E(A; \alpha) \equiv E$. Let $n \geq n_0$. So if χ is the characteristic function of $B(\alpha; r)$, χ is analytic in a neighborhood of $\sigma(p_n(T))$ and $\chi(p_n(T)) = E_n$. By Runge's Theorem $E_n \in P(p_n(T)) \subseteq P(T)$. Clearly $\|E_n - E\| \to 0$, so that $E \in P(T)$.

Now assume there is a finite rank idempotent operator E that belongs to $P(T)$. Let $\{p_n\}$ be a sequence of polynomials such that $\|p_n(T) - E\| \to 0$. As in the preceding paragraph, for $0 < \epsilon < 1/2$, there is an n_0 such that for $n \geq n_0$, $p_n(\sigma(T)) = \sigma(p_n(T)) \subseteq B(0;\epsilon) \cup B(1;\epsilon)$ and the rank of $E(p_n(T), B(1;\epsilon))$ is the same as that of E. Fix $n \geq n_0$. Put $K_0 = p_n^{-1}(B(0;\epsilon)) \cap \sigma(T)$ and $K_1 = p_n^{-1}(B(1;\epsilon)) \cap \sigma(T)$. So $\sigma(T) = K_0 \cup K_1$, $|p_n(z)| < \epsilon$ for z in K_0 and $|p_n(z) - 1| < \epsilon$ for z in K_1. This implies that $\emptyset = K_1 \cap \widehat{K_0} \supseteq K_1 \cap \widehat{\sigma_e(T)}$. But since $E(p_n(T), B(1;\epsilon))$ has finite rank, $K_1 \subseteq \sigma_0(T)$, proving (c).

If we now assume that (c) holds and $\lambda \in \sigma_0(T)$ such that $\lambda \notin \widehat{\sigma_e(T)}$, then $E = E(T, \lambda)$ has finite rank. By Runge's Theorem there is a sequence of polynomials $\{p_n\}$ such that $p_n(z) \to 0$ in a neighborhood of $\sigma(T)\backslash\{\lambda\}$ and $p_n(z) \to 1$ in a neighborhood of λ. It follows that $\|p_n(T) - E\| \to 0$ so that $E \in P(T)$. □

4. EXAMPLE. Let A be the operator defined on $\mathbb{C}^2$ by the matrix

$$A = \begin{bmatrix} 0 & 1 \\ 0 & 0 \end{bmatrix}$$

and put $T = A \oplus 0$ on $\mathcal{H} = \mathbb{C}^2 \oplus \mathcal{K}$, where K is infinite dimensional. Clearly $T \in \mathcal{C}_{00}$, but $P(T)$ does not contain a non-nilpotent, finite rank operator.

5. THEOREM. *Each of the classes $\mathcal{C}_{00}$ and $\mathcal{C}_0$ is norm dense in $\mathcal{B}(\mathcal{H})$.*

Proof. Since $\mathcal{C}_{00} \subseteq \mathcal{C}_0$, it suffices to show that $\mathcal{C}_{00}$ is norm dense in $\mathcal{B}(\mathcal{H})$. So let T be an arbitrary operator and let $\lambda \in \partial\left[\widehat{\sigma_e(T)}\right]$. Since $\partial\left[\widehat{\sigma_e(T)}\right] \subseteq \partial\sigma(T)$, basic spectral theory implies that either $\lambda \in \sigma_0(T)$ or λ is an approximate eigenvalue. In the first case, we can write $T = T_1 + \lambda E$, where E is a finite rank idempotent and $T_1 E = E T_1 = 0$. If $\mu \notin \widehat{\sigma_e(T)}$ and $|\mu - \lambda| < \epsilon$, then $S = T_1 + \mu E \in \mathcal{C}_{00}$ by the preceding proposition and $\|T - S\| < \epsilon$.

Assume that λ is an approximate eigenvalue. The following argument is familiar to many but is repeated here for completeness. Let $\{e_n\}$ be an orthonormal sequence in H such that $\|(T - \lambda)e_n\| \to 0$. If $\epsilon > 0$, then, by passing to a subsequence if necessary, we may assume that $\sum_n \|(T - \lambda)e_n\|^2 < \epsilon^2$. Let M be the closed span of $\{e_n\}$ and consider the matrix decomposition of T with respect to $\mathcal{H} = \mathcal{M} \oplus \mathcal{M}^\perp$:

$$T = \begin{bmatrix} A & B \\ C & D \end{bmatrix}.$$

Since $\|A - \lambda\| < \epsilon$, if μ is as in the first part of this proof and

$$S = \begin{bmatrix} \mu & B \\ C & D \end{bmatrix},$$

then $\|T - S\| < 2\epsilon$ and $S \in \mathcal{C}_{00}$ by Proposition 1. □

Before characterizing the interior of the various sets, we need some background material. For a bounded open set Ω with smooth boundary, let $M(\partial\Omega)$ denote the normal operator defined as multiplication by the independent variable on $L^2(\partial\Omega)$, the space of square integrable functions with respect to arc length measure on $\partial\Omega$. $M(\partial\Omega)^{(\infty)}$ denotes the direct sum of $M(\partial\Omega)$ with itself an infinite number of times.

In this paper we will be especially interested in bounded open sets Ω that are of *Jordan type*; that is, $\Omega = \bigcup_{j=1}^n R_j$, where each R_j is the inside of a smooth Jordan loop γ_j and the sets $\{R_j\}$ have pairwise disjoint closures. Thus $\operatorname{cl} R_j = R_j \cup \gamma_j$ and $\operatorname{cl}\Omega = \bigcup_j \operatorname{cl} R_j$.

We will need the following result from [3]. The proof will be sketched as a small modification of this result is needed for this paper. The sketch of the proof will be helpful to see that the modification holds.

6. THEOREM. *If $T \in \mathcal{B}(\mathcal{H})$ and $\epsilon > 0$, then there is an open set of Jordan type Ω_ϵ and an operator T_ϵ such that $\|T - T_\epsilon\| < \epsilon$ and T_ϵ is unitarily equivalent to*

$$S_\epsilon = \begin{pmatrix} M(\partial\Omega_\epsilon)^{(\infty)} & A_\epsilon \\ 0 & B_\epsilon \end{pmatrix},$$

where

(i) $\sigma_e(B_\epsilon) \subseteq \sigma_e(T) \subseteq \Omega_\epsilon$;
(ii) $\sigma(B_\epsilon) \cap \partial\Omega_\epsilon = \emptyset$.

Proof. The only difference in this sketch from the proof that appears in [3] is that we begin by applying their Theorem 2.1 to obtain an open set of Jordan type, Ω_ϵ, such that $\sigma_e(T) \subseteq \Omega_\epsilon$, $\partial\Omega_\epsilon \subseteq [\partial\widehat{\sigma_e(T)}]_{\epsilon/3}$, and $\partial\widehat{\sigma_e(T)} \subseteq [\partial\Omega_\epsilon]_{\epsilon/3}$. Now apply the result of [1] to find a compact operator K_ϵ with $\|K_\epsilon\| < \epsilon/3$ and an operator T_ϵ that is unitarily equivalent to $T + K_\epsilon$ and such that

$$T_\epsilon = \begin{bmatrix} N_\epsilon & A_\epsilon \\ C & B_\epsilon \end{bmatrix}$$

with N_ϵ a diagonalizable normal operator satisfying $\sigma(N_\epsilon) = \sigma_e(N_\epsilon) = \partial\widehat{\sigma_e(T)}$. Now proceed as in [3] and note that the operator B_ϵ just defined is the same as the operator that appears in the final statement of the theorem.

□

The modification of the preceding theorem that is needed is as follows.

7. PROPOSITION. *With the notation of the preceding theorem, there is an $\epsilon_0 > 0$ such that for all $\epsilon < \epsilon_0$, the operator T_ϵ can be chosen with $\sigma(B_\epsilon) \subseteq \Omega_\epsilon$ if and only if $\sigma_0(T) \subseteq \widehat{\sigma_e(T)}$.*

Proof. First assume that such an $\epsilon_0 > 0$ exists, and, in order to arrive at a contradiction, suppose there is a λ in $\sigma_0(T)\backslash\widehat{\sigma_e(T)}$. Put $r = \frac{1}{2}\text{dist}\,(\lambda, \widehat{\sigma_e(T)})$.

Claim. $\partial\Omega_\epsilon \subseteq \sigma_e(T_\epsilon)$ for all $\epsilon < \epsilon_0$.

In fact, if $\alpha \in \partial\Omega_\epsilon$, then $B_\epsilon - \alpha$ is invertible by hypothesis and $M(\partial\,\Omega_\epsilon) - \alpha$ is not semi-Fredholm. By an elementary argument it follows that $S_\epsilon - \alpha$ is not semi-Fredholm, so that $\alpha \in \sigma_e(T_\epsilon)$.

Let G_0 be an open set of Jordan type with the property that $\widehat{\sigma_e(T)} \subseteq G_0 \subseteq \text{int}\,[\widehat{\sigma_e(T)}]_r$. By the upper-semicontinuity of the essential spectrum, there is an $\epsilon_1 < \epsilon_0$ such that $\sigma_e(T_\epsilon) \subseteq G_0$ for all $\epsilon < \epsilon_1$. Hence for such ϵ, $\text{cl}\,\Omega_\epsilon \subseteq G_0 \subseteq [\widehat{\sigma_e(T)}]_r$. Choose $\delta < r$ such that $B(\lambda;\delta) \cap \sigma(T) = \{\lambda\}$. Basic behavior of the spectrum implies that there is an $\epsilon_2 > 0$ such that if $\epsilon < \epsilon_2$ and $\|T - C\| < \epsilon$, then $\sigma_p(C) \cap B(\lambda;\delta) \neq \emptyset$. Letting $\epsilon < \min\{\epsilon_1, \epsilon_2\}$, it follows that there is a β in the point spectrum of T_ϵ, $\sigma_p(T_\epsilon)$, such that $\beta \notin \text{cl}\,\Omega_\epsilon$. Thus $\beta \in \sigma_p(S_\epsilon)$ and $M(\partial\,\Omega_\epsilon)^{(\infty)} - \beta$ is invertible. Again an elementary argument shows that it must be that $\beta \in \sigma_p(B_\epsilon)$, contradicting the hypothesis that $\sigma(B_\epsilon) \subseteq \Omega_\epsilon$ for all $\epsilon < \epsilon_0$.

For the converse, assume that $\sigma_0(T) \subseteq \sigma_e(T)$. Begin as in the proof of Theorem 6 by picking the open set of Jordan type Ω_ϵ such that $\sigma_e(T) \subseteq \Omega_\epsilon$. Since Ω_ϵ is of Jordan type, $\widehat{\sigma_e(T)} \subseteq \Omega_\epsilon$. By assumption, this implies that $\sigma_0(T) \subseteq \Omega_\epsilon$ so that $\sigma(T) \subseteq \Omega_\epsilon$. For any $\epsilon > 0$, there is a $\delta > 0$ such that $\|T - X\| < \delta$ implies that $\sigma(X) \subseteq \Omega_\epsilon$. Now choose the compact operator $K = K_\epsilon$ as in the proof of Theorem 6, but insist that $\|K\| < \min\{\delta, \epsilon/3\}$. We have the operator T_ϵ as in (6) and, as pointed out there, the operator B_ϵ is the same as appears in the final statement of the result.

By the construction of T_ϵ, $\sigma_e(B_\epsilon) \subseteq \Omega_\epsilon$. Let $\lambda \in \sigma_0(B_\epsilon)$; so $\bar{\lambda} \in \sigma_p(B_\epsilon^*)$. Since $N - \lambda$ is invertible, it follows that $\bar{\lambda} \in \sigma_p(T_\epsilon^*)$, so that $\lambda \in \sigma(T) \subseteq \Omega_\epsilon$.

□

8. THEOREM. *The interior of each of the six sets defined in* (1) *is*

$$\{T \in \mathcal{B}(\mathcal{H}) : \sigma_0(T) \setminus \widehat{\sigma_e(T)} \neq \emptyset\}$$

and this is the same as

$$\{T \in \mathcal{B}(\mathcal{H}) : P(T) \text{ contains a non-nilpotent, finite rank operator}\}.$$

Proof. Denote by I the first set of operators displayed in the statement of the theorem. The fact that I equals the second displayed set is immediate from Proposition 3. Using basic inclusions between the six sets, it suffices to show that the following statements are equivalent:

(a) $T \in \mathcal{I}$,
(b) $T \in \operatorname{int} \mathcal{C}_{00}$,
(c) $T \in \operatorname{int} \mathcal{C}_0^w$.

(a) *implies* (b). By Proposition 3, $\mathcal{I} \subseteq \mathcal{C}_{00}$. So it suffices to show that I is an open set of operators. Fix T in $\mathcal{I}$, and let $\lambda \in \sigma_0(T) \setminus \widehat{\sigma_e(T)}$. Let $\delta > 0$ be such that $B(\lambda; 2\delta) \cap \widehat{\sigma(T)} = \{\lambda\}$. Put $\Delta = \sigma(T) \setminus \{\lambda\}$. There is an $\epsilon > 0$ such that if $\|T - S\| < \epsilon$, then $\sigma(S)$ is contained in a δ-neighborhood of $\sigma(T)$. For such an S this implies that $\sigma(S) = \sigma_1 \cup \sigma_2$, where σ_1 and σ_2 are disjoint, non-empty compact sets, $\sigma_1 \subseteq B(\lambda; \delta)$, $\sigma_2 \subseteq \{z : \operatorname{dist}(z, \Delta) < \delta\}$, and $\sigma_e(S) \subseteq \sigma_2$. It follows that $\sigma_0(S) \setminus \widehat{\sigma_e(S)} \neq \emptyset$ so that $S \in \mathcal{I}$.

It is clear that (b) implies (c).

(c) *implies* (a). Let $T \in \operatorname{int} \mathcal{C}_0^w$ and let $\epsilon > 0$ be such that $X \in \mathcal{C}_0^w$ whenever $\|T - X\| < \epsilon$. Suppose that $T \notin \mathcal{I}$; so $\sigma_0(T) \subseteq \widehat{\sigma_e(T)}$. By Theorem 6 there is an operator T_1 with $\|T - T_1\| < \epsilon$ such that T_1 is unitarily equivalent to the operator S defined there that corresponds to the open set of Jordan type Ω. Thus any such operator T_1 belongs to $\mathcal{C}_0^w$. By Proposition 7, we can assume that $\sigma(B) \subseteq \Omega$.

As pointed out in [3], the fact that $\sigma(B) \cap \sigma(M(\partial\,\Omega))^{(\infty)} = \emptyset$ and the well known result of Rosenblum [4] imply that T_1 is similar to $D = M(\partial\,\Omega)^{(\infty)} \oplus B$. Thus $D \in \mathcal{C}_0^w$ and so there is a non-zero compact operator C and a net of polynomials $\{p_i\}$ such that $p_i(D) \to C$ (SOT).

This implies that $C = C_1 \oplus C_2$, and $p_i(M(\partial\,\Omega)^{(\infty)}) \to C_1$ and $p_i(B) \to C_2$ in the strong operator topology. But the operator $M(\partial\,\Omega)^{(\infty)}$ is normal with no isolated eigenvalues. Hence $C_1 = 0$. It follows that $p_i(M(\partial\,\Omega)) \to 0$ (SOT) in $\mathcal{B}(L^2(\partial\,\Omega))$.

Let $H^2(\Omega)$ denote the Hardy space for Ω consisting of the closure of the set of polynomials in $L^2(\partial\,\Omega)$, and denote by D_1 the restriction of $M(\partial\,\Omega)$ to $H^2(\Omega)$. It follows that $p_i(D_1) \to 0$ (SOT). Therefore $p_i = p_i(D_1)1 \to 0$ in $H^2(\Omega)$. But this implies that $p_i(\lambda) \to 0$ uniformly on compact subsets of Ω. Since $\sigma(B) \subseteq \Omega$, this implies that $\|p_i(B)\| \to 0$. That is, $C_2 = 0$ and therefore, $C = 0$, contradicting the fact that $D \in \mathcal{C}_0^w$.

□

9. EXAMPLE. Let T be the operator $T = A \oplus 0$ defined on $\mathbb{C}^2 \oplus \mathcal{K}$ as in Example 4. If S is the unilateral shift on $\mathcal{K}$, then $T_n = A \oplus n^{-1}S$ converges to T, but $T_n \notin \mathcal{C}_{00}$ for all $n \geq 1$. Thus $T \notin \operatorname{int} \mathcal{C}_{00}$.

It is possible to vary this theme by replacing the algebras $P(T)$, $P^\infty(T)$, and $W(T)$ by the algebras $R(T)$, the norm closure of all rational functions in T, the weak* closure of $R(T)$, the weak operator topology closure of $R(T)$, the double commutant of T, and the commutant of T, $\{T\}'$. If one considers the sets of operators such that each of these algebras contains either a non-zero finite rank operator or a non-zero compact operator, an argument similar to that used to prove Theorem 5 will show that these sets are norm dense in $\mathcal{B}(\mathcal{H})$.

The characterization of the interiors of these sets requires the use of the Apostol–Morrel model [2] rather than the results of [3], although each set has the same interior. This argument will be sketched for only the following two sets of operators. Let

10(*i*) $$\mathcal{C}_1 \equiv \{T \in \mathcal{B}(\mathcal{H}) : R(T) \cap \mathcal{B}_{00}(\mathcal{H}) \neq (0)\}$$

10(*ii*) $$\mathcal{C}_2 \equiv \{T \in \mathcal{B}(\mathcal{H}) : \{T\}' \cap \mathcal{B}_0(\mathcal{H}) \neq (0)\}.$$

11. EXAMPLE. Let A be the operator on $\mathbb{C}^2$ defined in Example 4, and put $T = A^{(\infty)}$. Clearly $T \in \mathcal{C}_2$, but $T \notin \mathcal{C}_1$.

The proof of the next proposition is similar to that of Proposition 3 and is left to the reader.

12. PROPOSITION. *If $\sigma_0(T) \neq \emptyset$, then $T \in \mathcal{C}_1$. If $R(T)$ contains a finite rank operator that is non-nilpotent, then $\sigma_0(T) \neq \emptyset$.*

Note that $\mathcal{C}_{00} \subseteq \mathcal{C}_1$, so $\mathcal{C}_1$ is dense. The density of $\mathcal{C}_2$ and other similar classes of operators follows from the density of $\mathcal{C}_1$.

13. THEOREM. *$\operatorname{int}\mathcal{C}_1 = \operatorname{int}\mathcal{C}_2 = \{T \in \mathcal{B}(\mathcal{H}) : \sigma_0(T) \neq \emptyset\}$ and this is the same as*

$$\{T \in \mathcal{B}(\mathcal{H}) : R(T) \ \textit{contains a non-nilpotent, finite rank operator.}\}$$

It is easy to see that the operator T defined in Example 11 does not belong to the interior of $\mathcal{C}_2$.

References

[1] C. APOSTOL, C. FOIAS and D. VOICULESCU, Some results on non-quasitriangular operators. IV, *Revue Roum. Math. Pures et Appl.*, **18** (1973), 487–514.

[2] C. APOSTOL and B. B. MORREL, On uniform approximation of operators by simple models, *Indiana Univ. Math. J.*, **26** (1977), 427–442.

[3] A. BROWN and C. PEARCY, Jordan loops and decomposition of operators, *Can. J. Math.*, **29** (1977), 1112–1119.

[4] M. ROSENBLUM, On the operator equation $BX - XA = Q$, *Duke Math. J.*, **23** (1956), 263–269.

[5] J. B. CONWAY, *A Course in Functional Analysis*, Springer–Verlag, New York, 1990.

[6] J. B. CONWAY, *The Theory of Subnormal Operators,* Amer. Math. Soc. Surveys and Monographs **36**, Amer. Math. Soc., Providence, 1991.

J. B. CONWAY, University of Tennessee, Knoxville, TN 37996-1300, USA; *e-mail*: conway@math.utk.edu

G. PRĂJITURĂ, Bucknell University, Lewisburg, PA 17837, USA; *e-mail*: gprajitu@bucknell.edu

Operator Theory: Advances and Applications, Vol. 127, 171–202

Analytic extension of vector valued functions

CLAUDIO D'ANTONI and LÁSZLÓ ZSIDÓ

Dedicated to the memory of Professor Béla Szőkefalvi-Nagy

Let D be a plane domain and X a complex Banach space. It is well known that the analyticity of some $F: D \to X$ with respect to the norm of X follows from the analyticity of the compositions of F with all functionals belonging to a closed linear subspace $\mathcal{F}$ of the dual of X, which determines the norm of X. Thus, roughly speaking, scalar analyticity implies vector valued analyticity.

We investigate here, whether in reasonable situations scalar analytic extendibility implies vector valued analytic extendibility. More precisely, we describe situations in which an X-valued map on a Jordan arc Γ in the boundary of D, whose compositions with all functionals belonging to an $\mathcal{F}$ as above have analytic extensions on D satisfying weak continuity properties at Γ, has X-valued analytic extension on D, continuous at Γ.

Introduction

The following situation frequently occurs in problems of Functional Analysis:

Given a simply connected plane domain D, a Jordan arc Γ on the boundary of D, no part of which is slice in D, and a map F from Γ in a complex Banach space X, such that all compositions of F with the functionals belonging to a linear subspace $\mathcal{F}$ of the dual of X, which determines the norm of X, have continuous extension to $D \cup \Gamma$, analytic in the interior, we look for the existence of an extension of F to a $\sigma(X, \mathcal{F})$-continuous map $\tilde{F}: D \cup \Gamma \to X$, which is analytic in the interior.

AMS Subject Classification (1991): 47A56, 46G10.

Supported by M.U.R.S.T., C.N.R. and E.U.

In certain cases the above problem can be easily settled. For example, let H be a complex Hilbert space, $B(H)$ the Banach space of all bounded linear operators on H, and $F: (-a,a) \to B(H)$ a map such that, for every $\xi, \eta \in H$, the functions $(a,b) \ni t \mapsto (F(t)\xi \mid \eta) \in \mathbb{C}$ have continuous extensions $F_{\xi,\eta}$ on a rectangle $(-a,a) + i\,[0,b)$, which are analytic in the interior. Then there exists a weak operator continuous extension $\tilde{F}: (-a,a) + i\,[0,b) \to B(H)$ of F, which is analytic in the interior. Indeed, replacing a, b by a', b' with $a < a' < b' < b$, we can assume withlout loss of generality that the functions $F_{\xi,\eta}$ are bounded. Then the closed graph theorem entails the existence of a constant $c > 0$ such that $\|F_{\xi,\eta}\| \leq c\,\|\xi\|\,\|\eta\|$ and, for every $z \in (-a,a) + i\,[0,b)$, $\tilde{F}(z)$ is defined by the sesqui-linear form $(\xi, \eta) \mapsto F_{\xi,\eta}(z)$.

Not every case is so easy. For example, if $A \subset B(H)$ is a C^*-algebra, ξ a cyclic vector for A, and $\mathbb{R} \ni t \mapsto \alpha_t$ a strongly continuous one-parameter group of $*$-automorphisms of A, such that the "two-point functions" $\mathbb{R} \ni t \mapsto (x\,\alpha_t(y)\,\xi \mid \xi) \in \mathbb{C}$, $x, y \in A$ have bounded, continuous extensions on the strip $\{z \in \mathbb{C}\,;\, 0 \leq \text{ Im } < 1\}$, which are analytic in the interior, then the derivation of the corresponding analytic extendibility properties of the vector valued maps $\mathbb{R} \ni t \mapsto \alpha_t(x)\xi \in H$, of interest for the investigation of the relativistic KMS-condition, is much more involved (see [D-Z 2]).

Our approach of the general problem has the following strategy. We approximate F with analytically extendable mollifications relative to the arc length measure of Γ, for which we need the local rectifiability of Γ and a Pettis integrability property of F, and then show that the values in any $z \in D$ of the analytic extensions of these mollifications are norm-convergent. In the case of Krein completeness of $\sigma(X, \mathcal{F})$ (the $\sigma(X, \mathcal{F})$-closed convex hull of any $\sigma(X, \mathcal{F})$-compact set is $\sigma(X, \mathcal{F})$-compact), the Pettis integrability of $\sigma(X, \mathcal{F})$-continuous maps is automatic, so in this situation the case of arbitrary Γ can be reduced to the rectifiable case by using conformal representations (which, in general, does not preserve rectifiability). The main results of the paper are the analytic extension criteria Theorems 5.1 and 5.4, supplemented by the continuity criterion at the boundary Theorem 4.3.

The authors wish to thank H. Helson for the reference [He], which enabled them to replace a weaker first version of Theorem 5.4 with the present one. They are grateful also to W. Arendt for calling their attention on the reference [Are-Ni].

1. Preliminaries on weak topologies

We call *dual pair of normed vector spaces* a pair $(X, \mathcal{F})$ of complex normed vector spaces together with a bilinear functional

$$X \times \mathcal{F} \ni (x, \varphi) \mapsto \langle x, \varphi >\in \mathbb{C}$$

satisfying

$$\|x\| = \sup\{|\langle x, \varphi\rangle| \,;\ \varphi \in \mathcal{F},\ \|\varphi\| \le 1\}, \quad x \in X,$$
$$\|\varphi\| = \sup\{|\langle x, \varphi\rangle| \,;\ x \in \mathcal{F},\ \|x\| \le 1\}, \quad \varphi \in \mathcal{F}.$$

We denote, as usual, by $\sigma(X, \mathcal{F})$ the weak topology defined on X by the linear functionals $X \ni x \mapsto \langle x, \varphi\rangle \in \mathbb{C}$, $\varphi \in \mathcal{F}$. We recall that a map $F\colon \Omega \to X$, defined on a locally compact Hausdorff topological space Ω, is called Pettis integrable with respect to a Radon measure μ on Ω and the topology $\sigma(X, \mathcal{F})$ if all functions $\Omega \ni \omega \mapsto \langle F(\omega), \varphi\rangle \in \mathbb{C}$, $\varphi \in \mathcal{F}$ belong to $\mathcal{L}^1(\mu)$ and there exists a (necessarily unique) $x_{F,\mu} \in X$ such that

$$\int_\Omega \langle F(\omega), \varphi\rangle \, d\mu(\omega) = \langle x_{F,\mu}, \varphi\rangle \quad \text{for all } \varphi \in \mathcal{F}.$$

In this case $x_{F,\mu}$ is denoted by

$$\sigma(X, \mathcal{F}) - \int_\Omega F(\omega)\, d\mu(\omega)$$

and it is called the Pettis integral of F with respect to μ and $\sigma(X, \mathcal{F})$.

Relevant examples of dual pairs of normed spaces will be given in 1.5.

We say that a dual pair of normed vector spaces $(X, \mathcal{F})$ has the

- *Banach–Steinhaus Property* if for $S \subset X$,
  $\sup\{|\langle x, \varphi\rangle| \,;\ x \in S\} < +\infty$ for all $\varphi \in \mathcal{F}$ implies $\sup\{\|x\| \,;\ x \in S\} < +\infty$;
- *Krein Property* if the $\sigma(X, \mathcal{F})$-closed convex hull of every $\sigma(X, \mathcal{F})$-compact subset of X is $\sigma(X, \mathcal{F})$-compact.

Let us recall the basic criterion for the Banach–Steinhaus property and formulate an immediate consequence of it:

1.1. PROPOSITION. *Let $(X, \mathcal{F})$ be a dual pair of normed spaces.*

(i) *If the norm of $\mathcal{F}$ is complete, then $(X, \mathcal{F})$ has the Banach–Steinhaus Property.*

(ii) *If $(X, \mathcal{F})$ has the Banach–Steinhaus Property, $\Omega \subset \mathbb{R}^n$ is open and $F\colon \Omega \to X$ is such that $\Omega \ni \mathbf{t} \mapsto \langle F(\mathbf{t}), \varphi\rangle$ has locally bounded partial derivatives for all $\varphi \in \mathcal{F}$, then F is locally Lipschitz (in particular, norm-continuous).*

Proof. (i) is the classical uniform boundedness principle (see [Hi-P], Theorem 2.8.6).

For (ii) let $\mathbf{a} \in \Omega$ and $\delta > 0$ be such that Ω contains the (compact) closure of $U_\delta(\mathbf{a}) = \{\mathbf{t} \in \Omega\,;\, \|\mathbf{t} - \mathbf{a}\| < \delta\}$. By the assumption on F, the function $U_\delta(\mathbf{a}) \ni \mathbf{t} \mapsto F_\varphi(\mathbf{t}) = \langle F(\mathbf{t})\,,\, \varphi\rangle$ has bounded partial derivatives for all $\varphi \in \mathcal{F}$.

Now, denoting by $\mathbf{e_1}, \ldots, \mathbf{e_n}$ the canonical basis of $\mathbb{R}^n$ and using the mean value theorem, we get for all $\mathbf{t} \neq \mathbf{t}'$ in $U_\delta(\mathbf{a})$ and all $\varphi \in \mathcal{F}$:

$$\begin{aligned}
&\Big| \operatorname{Re} \Big\langle \frac{1}{\|\mathbf{t}-\mathbf{t}'\|} (F(\mathbf{t}) - F(\mathbf{t}')) , \varphi \Big\rangle \Big| \\
&= \frac{1}{\|\mathbf{t}-\mathbf{t}'\|} \Big| \operatorname{Re} \, (F_\varphi(\mathbf{t}) - F_\varphi(\mathbf{t}')) \Big| \\
&= \frac{1}{\|\mathbf{t}-\mathbf{t}'\|} \Big| \operatorname{Re} \sum_{k=1}^{n} \Big(F_\varphi\Big(\mathbf{t} + \sum_{j=1}^{k-1} (t'_j - t_j)\mathbf{e_j}\Big) - F_\varphi\Big(\mathbf{t} + \sum_{j=1}^{k} (t'_j - t_j)\mathbf{e_j}\Big)\Big) \Big| \\
&= \frac{1}{\|\mathbf{t}-\mathbf{t}'\|} \Big| \sum_{k=1}^{n} (t_k - t'_k) \operatorname{Re} \frac{\partial F_\varphi}{\partial t_k} \Big(\sum_{j=1}^{k-1} t'_j \mathbf{e_j} + \tau_k \mathbf{e_k} + \sum_{j=k+1}^{n} t_j \mathbf{e_j} \Big) \Big| \\
&\leq \Big(\sum_{k=1}^{n} \Big\| \frac{\partial F_\varphi}{\partial t_k} \Big\|_\infty \Big)^{1/2},
\end{aligned}$$

where τ_k is an appropriate real between t_k and t'_k, and $\|\frac{\partial F_\varphi}{\partial t_k}\|_\infty$ stands for the uniform norm of $\frac{\partial F_\varphi}{\partial t_k}$ on $U_\delta(\mathbf{a})$. A similar inequality holds for the imaginary part, so

$$\Big| \Big\langle \frac{1}{\|\mathbf{t}-\mathbf{t}'\|} (F(\mathbf{t}) - F(\mathbf{t}')), \varphi \Big\rangle \Big| \leq 2 \Big(\sum_{k=1}^{n} \Big\| \frac{\partial F_\varphi}{\partial t_k} \Big\|_\infty \Big)^{1/2}.$$

By the assumed Banach–Steinhaus Property it follows that

$$\sup \Big\{ \frac{1}{\|\mathbf{t}-\mathbf{t}'\|} \|F(\mathbf{t}) - F(\mathbf{t}'\|\,;\, \mathbf{t}, \mathbf{t}' \in U_\delta(\mathbf{a})\,, \mathbf{t} \neq \mathbf{t}' \Big\} < +\infty. \qquad \square$$

We notice that far reaching extensions of the classical uniform boundedness principle were given by H. Helson in [He].

Let $(X, \mathcal{F})$ be a dual pair of normed spaces having the Banach–Steinhaus Property and let us assume that the norm of X is complete. If $\Omega \subset \mathbb{C}$ is open and $U : \Omega \to X$ is such that $\Omega \ni z \mapsto \langle U(z)\,,\, \varphi\rangle$ is harmonic for all $\varphi \in \mathcal{F}$, then Proposition 1.1 (ii) yields the norm-continuity of U. Thus, if $a \in \Omega$ and $\delta > 0$ are such that the closure of $U_\delta(a) = \{z \in \mathbb{C};\, |z-a| < \delta\}$ is contained in Ω, then for every $z \in U_\delta(a)$ the Poisson integral

$$\frac{1}{2\pi} \int_{-\pi}^{\pi} \frac{\delta^2 - |z-a|^2}{|\delta e^{i\theta} - (z-a)|^2}\, U(a + \delta e^{i\theta})\, d\theta$$

exists in the Riemann sense with respect to the norm of X and, taking into account that by the harmonicity assumption on U

$$\frac{1}{2\pi} \int_{-\pi}^{\pi} \frac{\delta^2 - |z-a|^2}{|\delta e^{i\theta} - (z-a)|^2} \langle U(a + \delta e^{i\theta}),\, \varphi\rangle\, d\theta = \langle U(z),\, \varphi\rangle \quad \text{for all } \varphi \in \mathcal{F},$$

it is equal to $U(z)$. Consequently, U is harmonic with respect to the norm of X: it is locally expressible as a norm-convergent power series in the real part and the imaginary part of z and its Laplacian vanishes identically.

Similarly, if $(X, \mathcal{F})$, $\Omega \subset \mathbb{C}$ are as above and $F\colon \Omega \to X$ is such that $\Omega \ni z \mapsto \langle F(z),\, \varphi\rangle$ is analytic for all $\varphi \in \mathcal{F}$, then F is norm-continuous and, for every $a \in \Omega$, $\delta > 0$ with the closure of $U_\delta(a)$ contained in Ω, we have

$$F(z) = \frac{1}{2\pi i} \int_{\partial U_\delta(a)} \frac{1}{z - \zeta}\, F(\zeta)\, d\zeta,$$

where $\partial U_\delta(a)$ denotes the counterclockwise oriented boundary of $U_\delta(a)$ and the Cauchy integral exists in the Riemann sense with respect to the norm of X. Consequently, F is analytic with respect to the norm of X: it is locally expressible as a norm-convergent power series in z.

Thus we got the following consequence of the Banach–Steinhaus Property (see [Hi-P], Theorem 3.10.1):

1.2. COROLLARY. *Let $(X, \mathcal{F})$ be a dual pair of normed spaces having the Banach–Steinhaus Property and let us assume that the norm of X is complete. If $\Omega \subset \mathbb{C}$ is open and $U\colon \Omega \to X$ (respectively $F\colon \Omega \to X$) is such that $\Omega \ni z \mapsto \langle U(z),\, \varphi\rangle$ is harmonic (respectively $\Omega \ni z \mapsto \langle U(z),\, \varphi\rangle$ is analytic) for all $\varphi \in \mathcal{F}$, then U is harmonic (respectively F is analytic) with respect to the norm of X. Thus we can speak simply about harmonic and analytic X-valued maps.*

Clearly, the statement of Corollary 1.2 holds, with similar proof, for the harmonicity (respectively analiticity) of X-valued maps of more than two real variables (respectively more than one complex variable).

We notice also that a recent result of W. Arendt and N. Nikolski entails that, assuming the completeness of the norm of X, the conclusion of Corollary 1.2 implies the Banach–Steinhaus Property for $(X, \mathcal{F})$ (see [Are-Ni], Theorem 1.5).

The Krein Property of dual pairs of normed spaces is intimately related with the Pettis integrability of X-valued maps with respect to $\sigma(X, \mathcal{F})$:

1.3. PROPOSITION. *Let $(X, \mathcal{F})$ be a dual pair of normed spaces. Then $(X, \mathcal{F})$ has the Krein Property if and only if every $\sigma(X, \mathcal{F})$ -continuous map from a compact Hausdorff topological space Ω in X is Pettis integrable with respect to any Radon measure on Ω and $\sigma(X, \mathcal{F})$. In particular, the Krein Property for $(X, \mathcal{F})$ implies the completeness of the norm of X .*

Proof. The necessity is well known (see e.g. [R], Theorem 3.27, [Sch], Exercise IV.39(a), [Arv], Proposition 1.2): if $(X, \mathcal{F})$ has the Krein Property, Ω is a compact Hausdorff topological space, $F: \Omega \to X$ is $\sigma(X, \mathcal{F})$-continuous and μ is a Radon measure on Ω, then the $\sigma(X, \mathcal{F})$-closed convex hull $\mathcal{K}$ of the $\sigma(X, \mathcal{F})$-compact set $F(\Omega)$ is $\sigma(X, \mathcal{F})$-compact and

$$\left| \int_{\Omega} \langle F(\omega) , \varphi \rangle \, d\mu(\omega) \right| \leq \mu(\Omega) \sup\{|\langle x, \varphi \rangle| \, ; \, x \in \mathcal{K}\} \text{ for all } \varphi \in \mathcal{F} ,$$

so the linear functional $\mathcal{F} \ni \varphi \mapsto \int_{\Omega} \langle F(\omega) , \varphi \rangle \, d\mu(\omega)$ is continuous with respect to the Mackey topology associated to $\sigma(X, \mathcal{F})$ and it follows that it is defined by some $x_{F,\mu} \in X$.

Now let us assume that every $\sigma(X, \mathcal{F})$ -continuous map from a compact Hausdorff topological space Ω in X is Pettis integrable with respect to any Radon measure on Ω and $\sigma(X, \mathcal{F})$. Let $\mathcal{K}$ be any $\sigma(X, \mathcal{F})$-compact subset of X and let $\mathcal{P}(\mathcal{K})$ denote the weak*-compact convex set of all probability Radon measures on $\mathcal{K}$. By our assumption we can define the affine map

$$\mathcal{P}(\mathcal{K}) \ni \mu \mapsto \Phi(\mu) = \sigma(X, \mathcal{F}) - \int_{\mathcal{K}} x \, d\mu(x) \in X,$$

and it is easy to see that Φ is continuous with respect to the weak* topology on $\mathcal{P}(\mathcal{K})$ and $\sigma(X, \mathcal{F})$ on X. Consequently, $\Phi(\mathcal{P}(\mathcal{K}))$ is a $\sigma(X, \mathcal{F})$-compact convex subset of X, containing

$$\{\Phi(\delta_{\varphi}) \, ; \; \delta_{\varphi} \text{ the Dirac measure in } \varphi \in \mathcal{K}\} = \mathcal{K} .$$

Finally, let us assume that the above equivalent conditions are satisfied. For the completeness of the norm of X it is enough to prove that, for every sequence $(x_n)_{n \geq 1}$ in X with $\|x_n\| \leq 4^{-n}$ for all $n \geq 1$, the series $\sum_{n \geq 1} x_n$ is $\sigma(X, \mathcal{F})$-convergent. Denoting by Ω the compact set $\left\{0, 1, \frac{1}{2}, \frac{1}{3}, \cdots \right\} \subset [0, 1]$, the function $F: \Omega \to X$ defined by

$$F(0) = 0 \text{ and } F\left(\frac{1}{n}\right) = 2^n \, x_n \text{ for } n \geq 1$$

is norm-continuous. On the other hand, the formulas

$$\mu(0)=0 \quad \text{and} \quad \mu\Big(\Big\{\frac{1}{n}\Big\}\Big)=2^{-n} \quad \text{for} \quad n\geq 1$$

define a Radon measure on Ω. Then F is Pettis integrable with respect to μ and $\sigma(X,\mathcal{F})$ and the partial sums of the series $\sum_{n\geq 1} x_n$ are clearly $\sigma(X,\mathcal{F})$-convergent to the Pettis integral of F with respect to μ and $\sigma(X,\mathcal{F})$.

□

For Ω a locally compact Hausdorff topological space and μ a Radon measure on it, we call a map F from Ω into a topological space X μ-measurable in the Lusin sense if for every compact $K\subset\Omega$ and $\varepsilon>0$ there exists a compact $L\subset K$ such that $\mu(K\backslash L)<\varepsilon$ and $F|L$ is continuous. Clearly, if $F:\Omega\to X$ is μ-measurable in the Lusin sense then the same holds for $G\circ F$ whenever G is a continuous map from X into a topological space. By the classical Lusin Theorem, a scalar valued function on Ω is μ-measurable if and only if it is μ-measurable in the Lusin sense. We notice that the same holds for maps from Ω into arbitrary metric spaces, according to a deep result of D. H. Fremlin (see [F] or [Ku-Pr]).

Proposition 1.3 implies the following Pettis integrability criterion (see [Arv], Proposition 1.2, [C-Z], Proposition 1.4):

1.4. COROLLARY. *Let $(X,\mathcal{F})$ be a dual pair of normed spaces having the Krein Property. Let further Ω be a locally compact Hausdorff topological space, μ a Radon measure on it, and $F:\Omega\to X$. If F is μ-measurable in the Lusin sense and $\Omega\ni\omega\mapsto\|F(\omega)\|$ is majorized by some function from $\mathcal{L}^1(\mu)$, then F is Pettis integrable.*

Proof. Let $f\in\mathcal{L}^1(\mu)$ be such that $\|F(\omega)\|\leq f(\omega)$ for all $\omega\in\Omega$, choose compact sets $K_1\subset K_2\subset\ldots\subset\Omega$ with

$$\int_{\Omega\backslash K_n} f(\omega)\,d\mu(\omega)\leq\frac{1}{n},$$

and then compact sets $L_n\subset K_n$ such that $\mu(K_n\backslash L_n)$ is small enough to have

$$\int_{K_n\backslash L_n} f(\omega)\,d\mu(\omega)\leq\frac{1}{n}$$

and $F|L_n$ is $\sigma(X,\mathcal{F})$-continuous. By Proposition 1.3 there exist

$$x_n=\sigma(X,\mathcal{F})-\int_{L_n} F(\omega)\,d\mu(\omega)\in X$$

and we have, for every $\varphi \in \mathcal{F}$,

$$\begin{aligned}|\langle x_j - x_k , \varphi\rangle| &= \left| \int_{L_j} \langle F(\omega) , \varphi\rangle \, d\mu(\omega) - \int_{L_k} \langle F(\omega) , \varphi\rangle \, d\mu(\omega) \right| \\ &= \left| \int_{(L_k \setminus L_j)\cup(L_j \setminus L_k)} \langle F(\omega) , \varphi\rangle \, d\mu(\omega) \right| \\ &\leq \|\varphi\| \int_{(\Omega \setminus L_j)\cup(\Omega \setminus L_k)} f(\omega) \, d\mu(\omega) \\ &\leq \|\varphi\| \left(\frac{2}{j} + \frac{2}{k} \right) .\end{aligned}$$

Then $\|x_j - x_k\| \leq \left(\frac{2}{j} + \frac{2}{k}\right)$, so $(x_n)_{n\geq 1}$ is a Cauchy sequence in X. Now it is easy to verify that the limit of this sequence is the Pettis integral of F with respect to μ and $\sigma(X, \mathcal{F})$.

□

1.5. EXAMPLES. Let X be a complex normed vector space. According to the Hahn–Banach Theorem, the natural pairing between X and its dual X^* gives rise to the dual pairs (X, X^*), (X^*, X). By the uniform boundedness principle

- (X, X^*) has always the Banach–Steinhaus Property,
- (X^*, X) has the Banach–Steinhaus Property whenever X is complete.

By the Krein Theorem, respectively the uniform boundedness principle and the Alaoglu theorem,

- (X, X^*) and (X^*, X) have the Krein Property whenever X is complete.

Now let A be a C^*-algebra and $M(A)$ its multiplier algebra, considered imbedded in the second dual A^{**} (see [Ped], 3.12 or [WO], Chapter 2). Then the natural pairing between A^* and A^{**} induces a pairing between A^* and $M(A)$, which makes $(A^*, M(A))$ and $(M(A), A^*)$ dual pairs of Banach spaces. We notice that $M(A)$ is the dual of A^* only in very particular cases and A^* is the dual of $M(A)$ only in the trivial case $M(A) = A$. By the completeness of the norms of A^* and $M(A)$, $(M(A), A^*)$ and $(A^*, M(A))$ have the Banach–Steinhaus Property. While the Krein Property of $(M(A), A^*)$ follows from Grothendieck's extension of Krein's theorem, the Krein Property of $(A^*, M(A))$ was proved only recently (see [D-Z 1], Theorem 1.1).

Further examples can be obtained from the above ones by passing "from vectors to operators". More precisely, let $(X, \mathcal{F})$ be any dual pair of normed spaces, and Y a complex normed vector space. Let us denote by $B(Y, X)$ the vector space of all bounded linear operators from Y into X, endowed with the operator norm.

Putting

$$\left\langle T, \sum_{j=1}^{n} y_j \otimes \varphi_j \right\rangle = \sum_{j=1}^{n} \langle T y_j, \varphi_j \rangle \quad \text{for } T \in B(Y, X),\ y_j \in Y,\ \varphi_j \in \mathcal{F},$$

it is easy to verify that

$$\sum_{j=1}^{n} y_j \otimes \varphi_j \neq 0 \Rightarrow \left\langle T, \sum_{j=1}^{n} y_j \otimes \varphi_j \right\rangle \neq 0 \quad \text{for some } T \in B(Y, X).$$

Therefore the formula

$$\left\| \sum_{j=1}^{n} y_j \otimes \varphi_j \right\| = \sup \left\{ \left| \left\langle T, \sum_{j=1}^{n} y_j \otimes \varphi_j \right\rangle \right| ; T \in B(Y, X), \|T\| \leq 1 \right\}$$

defines a norm on $Y \otimes \mathcal{F}$, whith respect to which $(B(Y, X), Y \otimes \mathcal{F})$ becomes a dual pair of normed spaces.

We notice that in general the norm of $Y \otimes \mathcal{F}$ is not complete. In the case $\mathcal{F} = X^*$, $\sigma(B(Y, X), Y \otimes \mathcal{F})$ is the weak operator topology on $B(Y, X)$.

1.6. PROPOSITION. *Let $(X, \mathcal{F})$ be a dual pair of normed spaces, and Y a complex normed vector space.*

(i) *If $(X, \mathcal{F})$ has the Banach–Steinhaus Property and the norm of Y is complete, then $(B(Y, X), Y \otimes \mathcal{F})$ has the Banach–Steinhaus Property.*

(ii) *If $(X, \mathcal{F})$ has the Krein Property and $(B(Y, X), Y \otimes \mathcal{F})$ has the Banach–Steinhaus Property, then $(B(Y, X), Y \otimes \mathcal{F})$ has the Krein Property.*

Proof. (i) follows immediately by using the Banach–Steinhaus Property of $(X, \mathcal{F})$ and the classical uniform boundedness principle in $B(Y, X)$.

Taking into account Proposition 1.3, for (ii) it is enough to prove that, under the assumptions from (ii), every $\sigma(B(Y, X), Y \otimes \mathcal{F})$-continuous map F from a compact Hausdorff topological space Ω into $B(Y, X)$ is Pettis integrable with respect to any Radon measure μ on Ω and $\sigma(B(Y, X), Y \otimes \mathcal{F})$. But then F is norm-bounded and the Pettis integrals

$$\sigma(X, \mathcal{F}) - \int_{\Omega} F(\omega) y \, d\mu(\omega) \in X,\ y \in Y$$

exist. Denoting by $\|F\|_\infty$ the uniform norm of F, we have for every $\varphi \in \mathcal{F}$

$$\left| \int_{\Omega} \langle F(\omega) y, \varphi \rangle d\mu(\omega) \right| \leq \int_{\Omega} |\langle F(\omega) y, \varphi \rangle| d\mu(\omega) \leq \mu(\Omega) \|F\|_\infty \|y\| \|\varphi\|,$$

so

$$Y \ni y \mapsto \sigma(X,\mathcal{F}) - \int_\Omega F(\omega) y d\mu(\omega) \in X$$

is a linear operator T of norm $\leq \mu(\Omega)\|F\|_\infty$. Clearly, T is the Pettis integral of F with respect to μ and $\sigma(B(Y,X), Y \otimes \mathcal{F})$.

□

2. Scalar analytic extendibility versus vector valued analytic extendibility

In all this section $(X,\mathcal{F})$ will denote a dual pair of normed spaces. In the case of a complete norm on X, harmonicity and analyticity of X-valued maps will be understood with respect to the norm of X.

For any $S \subset \mathbb{C}$, ∂S will stand for the boundary of S, and $d(z,\partial S)$ for the distance of $z \in \mathbb{C}$ to ∂S. For $a \in \mathbb{C}$ and $r > 0$, $U_r(a)$ will denote the disk $\{z \in \mathbb{C}, |z-a| < r\}$. By complex integrals on $\partial U_r(0)$ we shall always use the counterclockwise orientation of $\partial U_r(0)$.

For any set S, normed vector space Y and $F: S \to Y$, $\|F\|_\infty$ will stand for the uniform norm $\sup\{\|F(s)\|, s \in S\} \in [0,+\infty]$. The Banach space of all bounded, continuous complex functions on $S \subset \mathbb{C}$, endowed with the uniform norm, will be denoted by $\mathcal{C}_b(S)$. If $S \subset \mathbb{C}$ and $f: S \to \mathbb{C}$, we call f *regular* whenever it is continuous and its restriction to the interior of S is analytic. Clearly, the set of all bounded, regular complex functions on $S \subset \mathbb{C}$, endowed with the uniform norm, is a complex Banach space $\mathcal{R}_b(S)$. We shall also denote:

$$\mathbf{S}_\beta = \{z \in \mathbb{C}, |\text{ Im } z| < |\beta|, \beta \cdot (\text{ Im } z) > 0\}, \quad \beta \in \mathbb{R}.$$

The aim of this section is to find reasonable conditions, under which a map $F: \mathbb{R} \to X$ such that $\mathbb{R} \ni t \mapsto \langle F(t), \varphi \rangle \in \mathbb{C}$ has a (necessarily unique) extension to some $F_\varphi \in \mathcal{R}_b(\mathbf{S}_1 \cup \mathbb{R})$ for all $\varphi \in \mathcal{F}$, can be extended to a bounded, $\sigma(X,\mathcal{F})$ -continuous map $\mathbf{S}_1 \cup \mathbb{R} \to X$, which is analytic in the interior. Using conformal representations, we shall see in Section 5 that the obtained result still holds if we replace $\mathbf{S}_1$ with more general simply connected domains, and $\mathbb{R}$ with a suitable Jordan arc on the boundary of the domain. We have chosen $\mathbf{S}_1$ and $\mathbb{R}$ for model because of the easy way to integrate on $\mathbb{R}$.

Let us first prove an elementary inequality for bounded, regular functions on $\mathbf{S}_1 \cup \mathbb{R}$:

2.1. Lemma. *For every $f \in \mathcal{R}_b(\mathbf{S}_1 \cup \mathbb{R})$ we have*

$$\left|\sqrt{\frac{n}{\pi}}\int_{-\infty}^{+\infty} e^{-n(t-z)^2} f(t)dt - f(z)\right| \leq \frac{4\|f\|_\infty}{d(z,\partial\mathbf{S}_1)\cdot\sqrt{\pi n}}, \quad z \in \mathbf{S}_1,\ n > 0.$$

Proof. For arbitrary $0 < \delta < r < d(z,\partial\mathbf{S}_1)$ we have

$$\begin{aligned}&\left|\sqrt{\frac{n}{\pi}}\int_{-\infty}^{+\infty} e^{-n(t-z)^2} f(t)dt - f(z)\right| \\ &\quad = \left|\sqrt{\frac{n}{\pi}}\int_{-\infty}^{+\infty} e^{-nt^2}(f(t+z)-f(z))dt\right| \\ &\quad \leq \sqrt{\frac{n}{\pi}}\int_{|t|\geq\delta} e^{-nt^2}2\|f\|_\infty dt + \sqrt{\frac{n}{\pi}}\left|\int_{-\delta}^{+\delta} e^{-nt^2}(f(t+z)-f(z))dt\right|.\end{aligned}$$

But

$$\sqrt{n}\int_{|t|\geq\delta} e^{-nt^2}dt = \int_{|s|\geq\delta\sqrt{n}} e^{-s^2}ds \leq \frac{2}{\delta\sqrt{n}}\int_{\delta\sqrt{n}}^{+\infty} e^{-s^2}sds = \frac{e^{-n\delta^2}}{\delta\sqrt{n}}$$

implies

$$\sqrt{\frac{n}{\pi}}\int_{|t|\geq\delta} e^{-nt^2}2\|f\|_\infty dt \leq \frac{\|f\|_\infty}{\delta\cdot\sqrt{\pi n}}2e^{-n\delta^2}$$

and, using the Cauchy integral formula, we get also

$$\begin{aligned}&\sqrt{\frac{n}{\pi}}\left|\int_{-\delta}^{+\delta} e^{-nt^2}(f(t+z)-f(t))dt\right| \\ &= \sqrt{\frac{n}{\pi}}\left|\int_{-\delta}^{+\delta} e^{-nt^2}\left(\frac{1}{2\pi i}\int_{\partial U_r(z)} f(\zeta)\left(\frac{1}{\zeta-z-t}-\frac{1}{\zeta-z}\right)d\zeta\right)dt\right| \\ &= \sqrt{\frac{n}{\pi}}\frac{1}{2\pi}\left|\int_{\partial U_r(z)} f(\zeta)\left(\int_{-\delta}^{+\delta}\frac{e^{-nt^2}\cdot t}{(\zeta-z-t)\cdot(\zeta-z)}dt\right)d\zeta\right| \\ &\leq \sqrt{\frac{n}{\pi}}\frac{1}{2\pi}2r\pi\|f\|_\infty\frac{1}{(r-\delta)\delta}\int_{-\delta}^{+\delta} e^{-nt^2}\cdot tdt = \frac{\|f\|_\infty}{\delta\cdot\sqrt{\pi n}}\frac{r}{r-\delta}(1-e^{-n\delta^2}).\end{aligned}$$

Consequently,

$$\begin{aligned}&\left|\sqrt{\frac{n}{\pi}}\int_{-\infty}^{+\infty} e^{-n(t-z)^2} f(t)dt - f(z)\right| \\ &\quad \leq \frac{\|f\|_\infty}{\delta\cdot\sqrt{\pi n}}\left(2e^{-n\delta^2}+\frac{r}{r-\delta}(1-e^{-n\delta^2})\right) = \frac{\|f\|_\infty}{\delta\cdot\sqrt{\pi n}}\left(\frac{r}{r-\delta}+\frac{r-2\delta}{r-\delta}e^{-n\delta^2}\right).\end{aligned}$$

Now, putting $\delta = r/2$ and taking into account that r can be chosen arbitrarily close to $d(z, \partial\mathbf{S}_1)$, the desired inequality follows. □

Now we prove that, roughly speaking, vector valued analytic extendibility means scalar analytic extendibility together with a boundedness property. We shall see later that this boundedness property we get often for free.

2.2. PROPOSITION. *Let us assume that the norm of X is complete and let $F: \mathbb{R} \to X$ be norm-bounded, Pettis integrable with respect to all bounded, absolutely continuous Radon measures on $\mathbb{R}$ and $\sigma(X, \mathcal{F})$, and such that*

- *$\mathbb{R} \ni t \mapsto \langle F(t), \varphi\rangle \in \mathbb{C}$ has an extension to some $F_\varphi \in \mathcal{R}_b(\mathbf{S}_1 \cup \mathbb{R})$ for all $\varphi \in \mathcal{F}$.*

Then

- *F has a norm-bounded, $\sigma(X, \mathcal{F})$ -continuous extension $\tilde{F}: \mathbf{S}_1 \cup \mathbb{R} \to X$, which is analytic in the interior,*

if and only if

- *the linear map $\mathcal{F} \ni \varphi \mapsto F_\varphi \in \mathcal{R}_b(\mathbf{S}_1 \cup \mathbb{R})$ is bounded,*

in which case the norm of the above linear map is equal to $\|\tilde{F}\|_\infty$.

Proof. The necessity is clear: the norm of $\mathcal{F} \ni \varphi \mapsto F_\varphi \in \mathcal{R}_b(\mathbf{S}_1 \cup \mathbb{R})$ will be equal to $\|\tilde{F}\|_\infty$, hence finite.

Now let us assume that the norm c of $\mathcal{F} \ni \varphi \mapsto F_\varphi \in \mathcal{R}_b(\mathbf{S}_1 \cup \mathbb{R})$ is finite and define the analytic maps $F_n: \mathbb{C} \to X, n \geq 1$ by

$$F_n(z) = \sqrt{\frac{n}{\pi}}\, \sigma(X, \mathcal{F}) - \int_{-\infty}^{+\infty} e^{-n(t-z)^2} F(t) dt.$$

Let $z \in \mathbf{S}_1$ and $n, k \geq 1$ be arbitrary. By Lemma 2.1 we have for every $\varphi \in \mathcal{F}$

$$\begin{aligned} \|\langle F_n(z) - F_k(z), \varphi\rangle\| &\leq \|\langle F_n(z), \varphi\rangle - F_\varphi(z)\| + \|\langle F_k(z), \varphi\rangle - F_\varphi(z)\| \\ &\leq \frac{4c}{d(z, \partial\mathbf{S}_1) \cdot \sqrt{\pi}} \left(\frac{1}{\sqrt{n}} + \frac{1}{\sqrt{k}}\right) \cdot \|\varphi\|, \end{aligned}$$

so

$$\|F_n(z) - F_k(z)\| \leq \frac{4c}{d(z, \partial\mathbf{S}_1) \cdot \sqrt{\pi}} \left(\frac{1}{\sqrt{n}} + \frac{1}{\sqrt{k}}\right).$$

Therefore the sequence $(F_n)_{n \geq 1}$ is uniformly Cauchy on every compact subset of $\mathbf{S}_1$, so it converges uniformly on the compacts in $\mathbf{S}_1$ to some analytic map

$F_\infty : \mathbf{S}_1 \to X$. Since

$$\begin{aligned}\langle F_n(z), \varphi\rangle &= \sqrt{\frac{n}{\pi}} \int_{-\infty}^{+\infty} e^{-n(t-z)^2} \langle F(t), \varphi\rangle dt \\ &= \sqrt{\frac{n}{\pi}} \int_{-\infty}^{+\infty} e^{-nt^2} F_\varphi(t+z) dt \longrightarrow F_\varphi(z), \quad \varphi \in \mathcal{F},\end{aligned}$$

we have

$$\langle F_\infty(z), \varphi\rangle = F_\varphi(z) \text{ for all } \varphi \in \mathcal{F}.$$

In particular, $\|F_\infty\|_\infty \leq c$.

We conclude that the map

$$\mathbf{S}_1 \cup \mathbb{R} \ni z \longmapsto \tilde{F}(z) = \begin{cases} F(z) & \text{if } z \in \mathbb{R}, \\ F_\infty(z) & \text{if } z \in \mathbf{S}_1 \end{cases}$$

is $\sigma(X, \mathcal{F})$ -continuous and norm-bounded, and $\tilde{F}|\mathbf{S}_1 = F_\infty$ is analytic.

□

Proposition 2.2 makes reasonable to say that $(X, \mathcal{F})$ has the

- *Dominated Analytic Extension Property* if, for any norm-bounded $F: \mathbb{R} \to X$ which is Pettis integrable with respect to all bounded, absolutely continuous Radon measures on $\mathbb{R}$ and $\sigma(X, \mathcal{F})$, and such that $\mathbb{R} \ni t \mapsto \langle F(t), \varphi\rangle \in \mathbb{C}$ has an extension to some $F_\varphi \in \mathcal{R}_b(\mathbf{S}_1 \cup \mathbb{R})$ for all $\varphi \in \mathcal{F}$, the linear map $\mathcal{F} \ni \varphi \mapsto F_\varphi \in \mathcal{R}_b(\mathbf{S}_1 \cup \mathbb{R})$ is bounded.

Using the classical closed graph theorem, it is easily seen that $(X, \mathcal{F})$ has the Dominated Analytic Extension Property whenever the norm of $\mathcal{F}$ is complete.

Proposition 2.2 implies the following supplement to Proposition 1.6:

2.3. Corollary. *Let us assume that $(X, \mathcal{F})$ has the Dominated Analytic Extension Property and the norm of X is complete. Then, for every complex Banach space Y, $(B(Y, X), Y \otimes \mathcal{F})$ has the Dominated Analytic Extension Property.*

Proof. Let $F: \mathbb{R} \to B(Y, X)$ be norm-bounded, Pettis integrable with respect to all bounded, absolutely continuous Radon measures on $\mathbb{R}$ and $\sigma(B(Y, X), Y \otimes \mathcal{F})$, and such that $\mathbb{R} \ni t \mapsto \langle F(t)y, \varphi\rangle$ has an extension to some $F_{y,\varphi} \in \mathcal{R}_b(\mathbf{S}_1 \cup \mathbb{R})$ for all $y \in Y$ and $\varphi \in \mathcal{F}$.

The complex vector space $\mathcal{R}_b(\mathbf{S}_1 \cup \mathbb{R}; X)$ of all norm-bounded, $\sigma(X, \mathcal{F})$ - continuous maps $\mathbf{S}_1 \cup \mathbb{R} \to X$ with analytic restriction to $\mathbf{S}_1$, endowed with the uniform norm, is a Banach space and Proposition 2.2 implies, for every $y \in Y$, the existence of an extension $F_y \in \mathcal{R}_b(\mathbf{S}_1 \cup \mathbb{R}; X)$ of $\mathbb{R} \ni t \mapsto F(t)y \in X$.

It turns out that the graph of the linear operator $Y \ni y \mapsto F_y \in \mathcal{R}_b(\mathbf{S}_1 \cup \mathbb{R}; X)$ is closed:

$$\|y_n - y\| \to 0 \text{ and } \|F_{y_n} - f\| \to 0$$

imply for every $\varphi \in \mathcal{F}$ and $t \in \mathbb{R}$

$$\begin{aligned}|\langle (f - F_y)(t), \varphi\rangle| &\leq |\langle (f - F_{y_n}(t)), \varphi\rangle| + |\langle (F_{y_n} - F_y)(t), \varphi\rangle| \\ &\leq \|f - F_{y_n}\|_\infty \cdot \|\varphi\| + \|F(t)\| \cdot \|y_n - y\| \cdot \|\varphi\| \longrightarrow 0.\end{aligned}$$

Thus the closed graph theorem yields the existence of some constant $c > 0$ with

$$\|F_y\|_\infty \leq c \cdot \|y\|, \quad y \in Y.$$

It follows that the norms of the linear operators

$$\tilde{F}(z): Y \ni y \mapsto F_y(z) \in X, \quad z \in \mathbf{S}_1 \cup \mathbb{R}$$

are $\leq c$, so we got a map $\tilde{F}: \mathbf{S}_1 \cup \mathbb{R} \to B(Y, X)$ with $\|\tilde{F}\|_\infty \leq c$. For every $y \in Y$ and $\varphi \in \mathcal{F}$,

$$F_{y,\varphi}(t) = \langle F(t)y, \varphi\rangle = \langle F_y(t), \varphi\rangle, \quad t \in \mathbb{R}$$

implies

$$F_{y,\varphi}(z) = \langle F_y(z), \varphi\rangle = \langle \tilde{F}(z)y, \varphi\rangle, \quad z \in \mathbf{S}_1 \cup \mathbb{R}.$$

Therefore we have for a generic element $\sum_{j=1}^n y_j \otimes \varphi_j \in Y \otimes \mathcal{F}$ and every $z \in \mathbf{S}_1 \cup \mathbb{R}$:

$$\begin{aligned}\Big|\sum_{j=1}^n F_{y_j,\varphi_j}(z)\Big| &= \Big|\sum_{j=1}^n \langle \tilde{F}(z)y_j, \varphi_j\rangle\Big| = \Big|\Big\langle \tilde{F}(z), \sum_{j=1}^n y_j \otimes \varphi_j\Big\rangle\Big| \\ &\leq c \cdot \Big\|\sum_{j=1}^n y_j \otimes \varphi_j\Big\|.\end{aligned}$$

□

In particular, Corollary 2.3 supplies us with examples of dual pairs $(X, \mathcal{F})$ of normed spaces having the Dominated Analytic Extension Property, but with a non-complete norm on $\mathcal{F}$.

3. Preliminaries on conformal maps

Assuming that a conformal representation of $U_1(0)$ onto a simply connected domain D can be extended continuously to $\overline{U_1(0)}$, we investigate here, in which way this extension parametrizes arcs in the boundary of D.

A *Jordan arc* J is the range of an injective, continuous map $f: [0,1] \to \mathbb{C}$ and then $f(0), f(1)$ are called the endpoints of J, while all other points of J are called inner points of J. Then we say that f parametrizes J and we denote the set of all inner points of J by $\mathring{J}$. An *open Jordan arc* is the range of an injective, continuous map $(0,1) \to \mathbb{C}$. It should not necessarily have endpoints.

An *endcut* of a domain $D \subset \mathbb{C}$ is a Jordan arc, one endpoint of which belongs to the boundary of D, and all other points to D. We say that $f: D \to \mathbb{C}$ has the *asymptotic value* $w \in \mathbb{C}$ at $z \in \partial D$ if there exists an endcut C of D with one endpoint equal to z, such that

$$\lim_{\mathring{C} \ni \zeta \to z} f(\zeta) = w.$$

A *crosscut* of a domain $D \subset \mathbb{C}$ is a Jordan arc, whose both endpoints belong to the boundary of D and whose inner points are in D.

A *Jordan curve* is the range of an injective, continuous map $\partial U_1(0) \to \mathbb{C}$. If J is a Jordan curve then, according to the Jordan Curve Theorem (see e.g. [D], Theorem XVII.5.4 or [N], Theorem V.10.2), $\mathbb{C} \backslash J$ has exactly two connected components, one of which is bounded and simply connected, and the common boundary of the two components is J. The bounded component is called the *Jordan domain* defined by J.

By the well known Riemann Mapping Theorem, for every simply connected domain $\mathbb{C} \neq D \subset \mathbb{C}$ there exists a conformal map Φ of $U_1(0)$ onto D. Moreover, according to [Pom], Theorem 2.1, if the boundary $\hat{\partial} D$ of D in the Riemann sphere $\hat{\mathbb{C}} = \mathbb{C} \cup \{\infty\}$ is locally connected, then Φ has a continuous extension to a map from $\overline{U_1(0)}$ in $\hat{\mathbb{C}}$, which we shall still denote by Φ. Clearly, Φ will map $U_1(0)$ conformally onto D, and $\partial U_1(0)$ onto $\hat{\partial} D$.

Let us first prove a lemma concerning continuous extensions of conformal maps up to the boundary:

3.1. Lemma. *Let $D \subset \mathbb{C}$ be a simply connected domain such that $\hat{\partial} D$ is locally connected, and let Φ denote an arbitrary conformal map of $U_1(0)$ onto D. Let further $J \subset \partial D = \mathbb{C} \cap \hat{\partial} D$ be a Jordan arc, parametrized by $f: [0,1] \to J$, such that $\mathring{J} = f((0,1))$ is an open subset of ∂D, hence also of $\hat{\partial} D$. Then either*

(1) *there exist reals $\theta_1 < \theta_2 < \theta_1 + 2\pi$ such that Φ maps $U_1(0) \cup \{e^{i\theta}; \theta \in [\theta_1, \theta_2]\}$ homeomorphically onto $D \cup J$, and $\{e^{i\theta}; \theta \in [\theta_2, \theta_1 + 2\pi]\}$ onto $\hat{\partial} D \backslash \mathring{J}$,*

or

(2) *there are reals* $\theta_1 < \theta_2 \leq \theta_1' < \theta_2' \leq \theta_1 + 2\pi$ *such that* $\Phi(\{e^{i\theta}, \theta \in (\theta_1, \theta_2)\}) = \Phi(\{e^{i\theta}, \theta \in (\theta_1', \theta_2')\}) = \overset{\circ}{J}$, *the function* $[\theta_1, \theta_2] \ni \theta \mapsto f^{-1}(\Phi(e^{i\theta})) \in [0,1]$ *is strictly increasing, while the function* $[\theta_1', \theta_2'] \ni \theta \mapsto f^{-1}(\Phi(e^{i\theta})) \in [0,1]$ *is strictly decreasing, and* Φ *maps* $\{e^{i\theta}; \theta \in [\theta_2, \theta_1'] \cup [\theta_2', \theta_1 + 2\pi]\}$ *onto* $\hat{\partial} D \backslash \overset{\circ}{J}$.

Moreover, (1) *can only occur if* $\overset{\circ}{J}$ *is disjoint to the interior of the closure of* D, *and* (2) *only if* $\overset{\circ}{J}$ *is contained in the interior of the closure of* D.

Proof. According to [Pom], Proposition 2.5, Φ maps the connected components of $\partial U_1(0) \backslash \Phi^{-1}(\{f(0)\})$ exactly onto the connected components of $\hat{\partial} D \backslash \{f(0)\}$. Therefore there are reals $\varphi_1 < \varphi_2 \leq \varphi_1 + 2\pi$ such that

$$\Phi(\{e^{i\theta}; \varphi_1 < \theta < \varphi_2\}) \text{ is the component of } \hat{\partial} D \backslash \{f(0)\} \text{ containing } f((0,1]),$$

$$\Phi(\{e^{i\theta}; \varphi_2 \leq \theta \leq \varphi_1 + 2\pi\}) \text{ is disjoint to } f((0,1]).$$

It is similarly seen that there exist reals $\varphi_1 \leq \psi_1 < \varphi_1 + 2\pi$ and $\psi_1 < \psi_2 \leq \psi_1 + 2\pi$ such that

$$\Phi(\{e^{i\theta}; \psi_1 < \theta < \psi_2\}) \text{ is the component of } \hat{\partial} D \backslash \{f(1)\} \text{ containing } f([0,1)),$$

$$\Phi(\{e^{i\theta}; \psi_2 \leq \theta \leq \psi_1 + 2\pi\}) \text{ is disjoint to } f([0,1)).$$

Then $\Phi(\{e^{i\theta}; \varphi_2 \leq \theta \leq \varphi_1 + 2\pi \text{ or } \psi_1 < \theta \leq \psi_1 + 2\pi\})$ is disjoint to $f((0,1))$, so

$$\Phi(\{e^{i\theta}; \varphi_1 < \theta < \varphi_2\} \cap \{e^{i\theta}; \psi_1 < \theta < \psi_2\}) \text{ contains } f((0,1)).$$

But a short inspection shows that

$$\{e^{i\theta}; \varphi_1 < \theta < \varphi_2\} \cap \{e^{i\theta}; \psi_1 < \theta < \psi_2\} = \{e^{i\theta}; \theta \in I_1 \cup I_2\},$$

where $I_1 = (\varphi_1, \ \min(\varphi_2, \psi_2 - 2\pi))$ and $I_2 = (\psi_1, \ \min(\varphi_2, \psi_2))$, hence

$$(*) \qquad \Phi(\{e^{i\theta}; \theta \in I_1\}) \cup \Phi(\{e^{i\theta}; \theta \in I_2\}) \supset f((0,1)),$$

$$(**) \qquad \Phi(\{e^{i\theta}; \theta \in (\varphi_1, \varphi_1 + 2\pi], \theta \notin I_1 \cup I_2\}) \text{ is disjoint to } f((0,1)).$$

We notice that if $I \subset \mathbb{R}$ is any (necessarily bounded) open interval such that $\Phi(\{e^{i\theta}; \theta \in I\})$ intersects $\overset{\circ}{J}$, but does not contain the endpoints of J, then

$$\Phi(\{e^{i\theta}; \theta \in I\}) \subset \overset{\circ}{J}.$$

Indeed, since $\overset{\circ}{J}$ is open in $\hat{\partial}D$, $\{\theta \in I, \Phi(e^{i\theta}) \in \overset{\circ}{J}\}$ is an open subset of I. On the other hand, $\{\theta \in I, \Phi(e^{i\theta}) \in \overset{\circ}{J}\} = \{\theta \in I, \Phi(e^{i\theta}) \in J\}$ is also closed in I and not empty by assumption, so it coincides with I by the connectedness of this.

Moreover, assuming additionally that $\theta \mapsto \Phi(e^{i\theta})$ carries the endpoints of $I = (\alpha, \beta)$ in the endpoints of J (we notice that the equality $\Phi(\{e^{i\theta}; \theta \in I\}) = \overset{\circ}{J}$ would imply this), the function $[\alpha, \beta] \ni \theta \mapsto g(\theta) = f^{-1}(\Phi(e^{i\theta})) \in [0,1]$ follows to be strictly increasing if $g(\alpha) = 0$, $g(\beta) = 1$, and strictly decreasing if $g(\alpha) = 1$, $g(\beta) = 0$. Let us consider only the first case, the treatment of the second case being completely similar.

If g would not be increasing, then it would exist $\theta' < \theta''$ in (α, β) such that $g(\theta') > g(\theta'')$. By the intermediate value theorem it would then follow that every value $t \in (g(\theta''), g(\theta'))$ is taken at least once in each one of the intervals $(\alpha, \theta'), (\theta', \theta''), (\theta'', \beta)$, that is three times in (α, β). But having uncountably many such values t, this would contradict [Pom], Corollary 2.19. Consequently g must be increasing. Now if it would not be strictly increasing, then it would be constant on some non-empty subinterval of I and by the Privalov Uniqueness Theorem (see e.g. [K], III:D.2 or [Pom], Corollary 6.14) it would follow that Φ is constant, what is clearly not true.

Now, going back to $(*)$, it implies that either only one of the sets $\Phi(\{e^{i\theta}; \theta \in I_1\})$ and $\Phi(\{e^{i\theta}; \theta \in I_2\})$ intersects $f((0,1))$, or both of them.

Assuming that only $\Phi(\{e^{i\theta}; \theta \in I_1\})$ intersects $f((0,1))$, $(*)$ implies that it contains $f((0,1))$ and then the above remarks yield first that $\Phi(\{e^{i\theta}; \theta \in I_1\}) = f((0,1))$, and then that $\theta \mapsto \Phi(e^{i\theta})$ maps the closure of I_1 homeomorphically onto J. Taking into account also $(**)$, it is easily seen that (1) holds with $\theta_1 = \varphi_1$ and $\theta_2 = \min(\varphi_2, \psi_2 - 2\pi)$. Similarly, if only $\Phi(\{e^{i\theta}; \theta \in I_2\})$ intersects $f((0,1))$, then (1) holds with $\theta_1 = \psi_1$ and $\theta_2 = \min(\varphi_2, \psi_2)$.

Let us now assume that both sets $\Phi(\{e^{i\theta}; \theta \in I_1\})$ and $\Phi(\{e^{i\theta}; \theta \in I_2\})$ intersect $f((0,1))$. By the first one of the above remarks and by $(*)$, their union must then be equal to $f((0,1))$. But

$$\begin{aligned} I_1 \neq \emptyset \quad &\Rightarrow \quad \varphi_1 < \psi_2 - 2\pi \quad \Rightarrow \quad \varphi_2 \leq \varphi_1 + 2\pi < \psi_2, \\ I_1 \neq \emptyset \quad &\Rightarrow \quad \psi_1 < \varphi_2 \quad \Rightarrow \quad \psi_2 - 2\pi \leq \psi_1 < \varphi_2, \end{aligned}$$

so $I_1 = (\varphi_1, \psi_2 - 2\pi)$ and $I_2 = (\psi_1, \varphi_2)$. Thus $\theta \mapsto \Phi(e^{i\theta})$ acts on the endpoints of I_1 and I_2 as follows:

$$\begin{aligned} \Phi(e^{i\varphi_1}) = f(0), \quad &\Phi(e^{i(\psi_2 - 2\pi)}) = f(1), \\ \Phi(e^{i\psi_1}) = f(1), \quad &\qquad \Phi(e^{i\varphi_1}) = f(0). \end{aligned}$$

Using the second one of the above remarks and (∗∗), we conclude that (2) holds with $\theta_1 = \varphi_1, \theta_2 = \psi_2 - 2\pi, \theta_1{}' = \psi_1, \theta_2{}' = \varphi_2$.

Next we show that (1) implies that $\overset{\circ}{J}$ is disjoint to the interior of $\overline{D}$. Indeed, assuming the contrary, it would exist some $\theta_\circ \in (\theta_1, \theta_2)$ and $0 < r <$ distance $(\Phi(e^{i\theta_\circ}), \partial D \setminus \overset{\circ}{J})$ with $U_r(\Phi(e^{i\theta_\circ})) \subset \overline{D}$ and it would follow

$$U_r(\Phi(e^{i\theta_\circ})) \subset \overline{D} \setminus (\partial D \setminus \overset{\circ}{J}) = D \cup \overset{\circ}{J}.$$

But the restriction Ψ of Φ to $U_1(0) \cup \{e^{i\theta}; \theta_1 < \theta < \theta_2\}$ is a homeomorphism onto $D \cup \overset{\circ}{J}$ and Ψ^{-1} would then map the open subset $U_r(\Phi(e^{i\theta_\circ}))$ of $\mathbb{C}$ homeomorphically onto some subset of $\overline{U_1(0)}$ containing $e^{i\theta_\circ}$, which can not be open. This would contradict Brouwer's theorem on the invariance of open sets (see e.g. [D], Theorem XVII.3.1 or [N], Theorem V.13.2).

To complete the proof, it remains only to show that (2) implies that $\overset{\circ}{J}$ is contained in the interior of $\overline{D}$. If C (respectively C') is the chord joining $e^{i\theta_1}$ and $e^{i\theta_2}$ (respectively $e^{i\theta_1{}'}$ and $e^{i\theta_2{}'}$) then $\Phi(C) \cup J$ (respectively $\Phi(C') \cup J$) is a Jordan curve and let D_1 (respectively D_2) denote the corresponding Jordan domain. Since Φ is injective on the region U_1 between C and the arc $\{e^{i\theta}, \theta \in [\theta_1, \theta_2]\}$ and it maps ∂U_1 homeomorphically onto $\Phi(C) \cup J$, it follows that $\Phi(U_1) = D_1$ (see e.g. [Pom], Theorem 1.9). Similarly, Φ maps the region U_2 between C' and the arc $\{e^{i\theta}, \theta \in [\theta_1{}', \theta_2{}']\}$ onto D_2. In particular, $D_1 \cup D_2 \subset D$.

Now let G denote the Jordan domain defined by the Jordan curve $\Phi(C) \cup \Phi(C')$ and choose some $z_\circ$ in the intersection of the unbounded components of the sets $\mathbb{C} \setminus (\Phi(C) \cup J)$, $\mathbb{C} \setminus (\Phi(C') \cup J)$, $\mathbb{C} \setminus (\Phi(C) \cup \Phi(C'))$. We prove that then

$$(***) \qquad G \subset D_1 \cup D_2 \cup \overset{\circ}{J}.$$

For let us assume that there exists some $z \in G \setminus (D_1 \cup D_2 \cup \overset{\circ}{J})$. Then z belongs to the unbounded component $\mathbb{C} \setminus (\Phi(C) \cup J \cup D_1)$ of $\mathbb{C} \setminus (\Phi(C) \cup J)$, so z and $z_\circ$ are not separated by $\Phi(C) \cup J$. Similarly, they are not separated also by $\Phi(C') \cup J$. Since $(\Phi(C) \cup J) \cup (\Phi(C') \cup J) = J$ is connected, the Janiszewski Theorem (see e.g. [D], Theorem XVII.5.2 or [N], Theorem V.9.2) implies that z and $z_\circ$ are not separated by $\Phi(C) \cup J \cup \Phi(C')$. But this is false, because they were chosen in different components of $\mathbb{C} \setminus (\Phi(C) \cup \Phi(C'))$.

Finally we prove that actually $D_1 \cup D_2 \cup \overset{\circ}{J} = G$, what implies first $\overline{G} = \overline{D_1 \cup D_2} \subset \overline{D}$, and then $\overset{\circ}{J} \subset G \subset$ the interior of $\overline{D}$. For we notice that, the intersection of the connected sets $D_1 \cup \overset{\circ}{J}$ and $D_2 \cup \overset{\circ}{J}$ being not empty, their union $D_1 \cup D_2 \cup \overset{\circ}{J}$ is connected. On the other hand, since G is not contained in $\overset{\circ}{J}$, there exists $z \in G \backslash \overset{\circ}{J}$ and by $(***)$ it follows that $z \in (D_1 \cup D_2 \cup \overset{\circ}{J}) \cap G$. Therefore the union $D_1 \cup D_2 \cup \overset{\circ}{J} \cup G \subset \mathbb{C} \backslash (\Phi(C) \cup \Phi(C'))$ is connected, hence it is equal to the connected component G and it follows that $D_1 \cup D_2 \cup \overset{\circ}{J} \subset G$.

□

The following result will allow to reduce the investigation of the boundary behaviour of maps defined on rather general simply connected domains in $\mathbb{C}$, as well as of extension possibilities from the boundary of such domains, to the case when the domain is a Jordan domain:

3.2. PROPOSITION. *Let $D \subset \mathbb{C}$ be a simply connected domain such that $\hat{\partial} D$ is locally connected. Let further $\Gamma \subset \partial D = \mathbb{C} \cap \hat{\partial} D$ be an open Jordan arc, which is open in ∂D and not contained in the interior of the closure of D.*

(i) *For any conformal map Φ of $U_1(0)$ onto D there exist reals $\theta_1 < \theta_2 \leq \theta_1 + 2\pi$ such that Φ maps $U_1(0) \cup \{e^{i\theta}, \theta \in (\theta_1, \theta_2)\}$ homeomorphically onto $D \cup \Gamma$, and $\{e^{i\theta}; \theta \in [\theta_2, \theta_1 + 2\pi]\}$ onto $\hat{\partial} D \backslash \Gamma$. Moreover, if Γ is locally rectifiable, then the homeomorphism $\Phi | \{e^{i\theta}, \theta \in (\theta_1, \theta_2)\}$ onto Γ is absolutely bicontinuous relative to the arc length measures.*

(ii) *Assuming that C is a crosscut of D with endpoints $a(C), b(C)$ belonging to Γ, let J_C denote the Jordan arc on Γ between $a(C)$ and $b(C)$, and D_C the Jordan domain defined by the Jordan curve $C \cup J_C$. Then $D_C \subset D$ and $D_C \cup \overset{\circ}{J_C}$ is open in $\overline{D}$, so every point of $\overset{\circ}{J_C}$ has a neighborhood N in $\mathbb{C}$ with $N \cap D \subset D_C$.*

(iii) *For every compact $K \subset D \cup \Gamma$ there exists a rectifiable crosscut C of D with endpoints on $\Gamma \backslash K$, such that $K \backslash \Gamma$ is contained in the Jordan domain defined by the union of C with the Jordan subarc of Γ having the same endpoints as C, and K is contained in the closure of the above Jordan domain.*

(iv) *There exists a sequence $C_1, C_2, \ldots$ of rectifiable crosscuts of D with endpoints on Γ, such that, denoting by J_n the Jordan subarc of Γ having the same endpoints as C_n, and by D_n the Jordan domain defined by $C_n \cup J_n$,*

$$J_n \subset \overset{\circ}{J}_{n+1} \text{ and } D_n \cup \overset{\circ}{C}_n \subset D_{n+1} \text{ for } n \geq 1, \quad \bigcup_{n \geq 1} J_n = \Gamma, \quad \bigcup_{n \geq 1} D_n = D.$$

Proof. Let us first notice that, according to the last remarks in Lemma 3.1, Γ is actually disjoint to the interior of $\overline{D}$.

Applying now Lemma 3.1 to an arbitrary conformal representation $\Phi: U_1(0) \to D$ and to the Jordan subarcs of Γ, it is easily seen that the first statement in (i) holds. In the remaining part of the proof Φ, θ_1, θ_2 will be as in (i).

If C is as in (ii), then there are uniquely defined $\theta_1 < \varphi_1 \neq \varphi_2 < \theta_2$ such that $\Phi(e^{i\varphi_1}) = a(C)$ and $\Phi(e^{i\varphi_2}) = b(C)$. Without loss of generality we can assume that $\varphi_1 < \varphi_2$. Then the restriction of Φ to $U_1(0) \cup e^{i[\varphi_1,\varphi_2]}$ is a homeomorphism Ψ onto $D \cup J_C$ and Φ maps $e^{i[\varphi_2,\varphi_1+2\pi]}$ onto $\hat{\partial} D \backslash \overset{\circ}{J}_C$.

Let U_1 denote the Jordan domain defined by the Jordan curve $C_1 = \Psi^{-1}(C) \cup e^{i[\varphi_1,\varphi_2]}$, and U_2 that one defined by $C_2 = \Psi^{-1}(C) \cup e^{i[\varphi_2,\varphi_1+2\pi]}$. Clearly, U_1 and U_2 are contained in $U_1(0)$. Actually holds

$$U_1(0) = U_1 \cup U_2 \cup \Psi^{-1}(\overset{\circ}{C}), \text{ so } D = \Phi(U_1) \cup \Phi(U_2) \cup \overset{\circ}{C}.$$

Indeed, if z would belong to $U_1(0) \backslash (U_1 \cup U_2 \cup \Psi^{-1}(\overset{\circ}{C}))$, then it would not be separated from 2 by anyone of C_1 and C_2, so the Janiszewski Theorem (see again [D], Theorem XVII.5.2 or [N], Theorem V.9.2) would imply that z and 2 are not separated by $C_1 \cup C_2 = \partial U_1(0) \cup \Psi^{-1}(C)$, what is clearly false. On the other hand, since Φ is injective on U_1 and maps C_1 homeomorphically onto $C \cup J_C$, it maps U_1 onto D_C (see e.g. [Pom], Theorem 1.9), so

$$D = D_C \cup \Phi(U_2) \cup \overset{\circ}{C}.$$

By the aboves we have

$$D \cup \hat{\partial} D = D_C \cup \Phi(U_2) \cup \overset{\circ}{C} \cup \hat{\partial} D,$$

$$\Phi(\overline{U_2}) = \Phi\left(U_2 \cup \Psi^{-1}(\overset{\circ}{C}) \cup e^{i[\varphi_2,\varphi_1+2\pi]}\right) = \Phi(U_2) \cup \overset{\circ}{C} \cup (\hat{\partial} D \backslash \overset{\circ}{J}_C),$$

where D_C, $\Phi(U_2)$, $\overset{\circ}{C}$, $\hat{\partial} D$ are mutually disjoint. Therefore

$$(D \cup \hat{\partial} D) \backslash \Phi(\overline{U_2}) = D_C \cup \overset{\circ}{J}_C$$

and it follows that $D_C \cup \overset{\circ}{J}_C$ is open in $D \cup \hat{\partial} D \supset D \cup \partial D$.

For (iii) we notice that $\Phi^{-1}(K)$ is a compact subset of $U_1(0) \cup e^{i(\theta_1,\theta_2)}$, so there exists $0 < \delta < \frac{1}{2}(\theta_2 - \theta_1)$ such that

$$\{re^{i\theta}; r \geq 1-\delta, \theta_2 - \delta \leq \theta \leq \theta_1 + 2\pi + \delta\} \cap \Phi^{-1}(K) = \emptyset.$$

Indeed, if we define the function g on $[0,1] \times [\theta_1, \theta_1 + 2\pi]$ by putting

$$g(r,\theta) = \max\left(1 - r, \min(\theta_2 - \theta, \theta - \theta_1)\right),$$

then g takes a minimum value $\delta_\circ > 0$ on the compact set $\{(r,\theta) \in [0,1] \times [\theta_1, \theta_1 + 2\pi]; re^{i\theta} \in \Phi^{-1}(K)\}$ and we can choose for δ any number in $(0, \delta_\circ)$, which is $< \frac{1}{2}(\theta_2 - \theta_1)$.

According to [Pom], Corollary 4.18, there exist $\theta_1 < \alpha_1 < \theta_1 + \delta$ and $\theta_2 - \delta < \alpha_2 < \theta_2$ such that

$$\text{length}(\Phi([0,1] \cdot e^{i\alpha_1})) \leq \text{distance}(\Phi(0), \partial D) \cdot \exp\frac{A}{\delta} < +\infty,$$
$$\text{length}(\Phi([0,1] \cdot e^{i\alpha_2})) \leq \text{distance}(\Phi(0), \partial D) \cdot \exp\frac{A}{\delta} < +\infty,$$

where A is an absolute constant. Therefore the crosscut

$$C = \Phi\left(([1-\delta,1] \cdot e^{i\alpha_1}) \cup ((1-\delta) \cdot e^{i[\alpha_1,\alpha_2]}) \cup ([1-\delta,1] \cdot e^{i\alpha_2})\right)$$

of D is rectifiable. Moreover, since $\Phi^{-1}(K \backslash \Gamma)$ is contained in the Jordan domain $U_{1-\delta}(0) \cup ([1-\delta, 1) \cdot e^{i(\alpha_1,\alpha_2)})$ and this is mapped by Φ onto the Jordan domain defined by $C \cup \Phi(e^{i[\alpha_1,\alpha_2]})$ (see e.g. [Pom], Theorem 1.9), $K \backslash \Gamma$ is contained in the Jordan domain defined by $C \cup \Phi(e^{i[\alpha_1,\alpha_2]})$. Similarly, K is contained in the closure of the above Jordan domain.

For (iv) we have just to choose compact sets $L_1 \subset L_2 \subset \cdots \subset D \cup \Gamma$ such that $\bigcup_{n\geq 1} L_n = D \cup \Gamma$, and to apply (iii) first with $K = L_1$ to get C_1, and then with $K = L_n \cup \overline{\text{the Jordan domain defined by } C_n \cup J_n}$ to get C_{n+1}.

Finally, in order to prove the second statement in (i), let us assume that Γ is locally rectifiable. Let J be any Jordan arc on Γ. Reproducing the proof of statement (iii) in the case $K = J$, we get $\theta_1 < \beta_1 < \beta_2 < \theta_2$ and a rectifiable crosscut B of $U_1(0)$ with endpoints $e^{i\beta_1}, e^{i\beta_2}$, such that $\Phi(B)$ is rectifiable and $J \subset \Phi(e^{i(\beta_1,\beta_2)})$. Since Φ maps the Jordan curve $B \cup e^{i[\beta_1,\beta_2]}$ onto the Jordan curve $\Phi(B) \cup \Phi(e^{i[\beta_1,\beta_2]})$, it maps the Jordan domain defined by $B \cup e^{i[\beta_1,\beta_2]}$ onto the Jordan domain defined by $\Phi(B) \cup \Phi(e^{i[\beta_1,\beta_2]})$ (see e.g. [Pom], Theorem 1.9). But the two Jordan curves being rectifiable, the Riemann Mapping Theorem, the

Carathéodory Theorem (see e.g. [K], I:C.1 or [Pom], Theorem 2.6) and the F. and M. Riesz Theorem (see e.g. [K], II:D.2 or [Pom], Theorem 6.8) imply that the homeomorphism $\Phi|B \cup e^{i[\beta_1,\beta_2]}$ onto $\Phi(B) \cup \Phi(e^{i[\beta_1,\beta_2]})$ is absolutely bicontinuous relative to the arc length measures. In particular, the homeomorphism $\Phi|e^{i[\beta_1,\beta_2]}$ onto $\Phi(e^{i[\beta_1,\beta_2]}) \supset J$ is absolutely bicontinuous relative to the arc length measures. Since the Jordan arc $J \subset \Gamma$ was arbitrary, this yields the desired statement.

□

4. Continuity properties of vector valued analytic maps at the boundary

The goal of this section is to give conditions under which harmonic or analytic maps taking values in a Banach space and having certain weak continuity properties at the boundary, allow norm-continuous extensions on the closure.

In all this section $(X, \mathcal{F})$ will denote a dual pair of normed spaces, and τ a locally convex vector space topology between $\sigma(X, \mathcal{F})$ and the norm-topology of X, such that the closed unit ball of X is τ-complete (the existence of such τ implies the completeness of the norm of X). Trivial example for such τ is the norm-topology of X in the case of a complete norm on X. But there are also other relevant examples:

- the strict topology on $M(A)$ for $(M(A), A^*)$, where A is a C^*-algebra;
- the strong operator topology on $B(Z,Y)$ for $(B(Z,Y), Z \otimes \mathcal{G})$, where $(Y, \mathcal{G})$ is a dual pair of normed spaces with the norm of Y complete.

"Almost everywhere" (abreviated: a.e.) in a rectifiable curve in $\mathbb{C}$ (e.g. in a segment or arc) will always be understood with respect to the arc length measure.

The following result concerning the Dirichlet problem is "folklore", but we sketch the proof for the convenience of the reader:

4.1. Lemma. *Every norm-bounded, τ-continuous $F\colon \partial U_1(0) \to X$ has a τ-continuous extension $\tilde{F}\colon \overline{U_1(0)} \to X$, which is harmonic in the interior.*

Proof. Standard arguments show that the Poisson integrals

$$P_F(z) = \frac{1}{2\pi} \int_{-\pi}^{\pi} \frac{1-|z|^2}{|e^{i\theta} - z|^2} F(e^{i\theta})d\theta \in X, \quad z \in U_1(0)$$

exist in the Riemann sense with respect to τ. Define $\tilde{F}\colon \overline{U_1(0)} \to X$ by

$$\tilde{F}(z) = \begin{cases} P_F(z), & \text{for } z \in U_1(0) \\ F(z), & \text{for } z \in \partial U_1(0). \end{cases}$$

By the classical theory of the Dirichlet problem, the composition of $\tilde{F}$ with every bounded linear functional on X is continuous on $\overline{U_1(0)}$ and harmonic on $U_1(0)$, so, taking into account Corollary 1.2, it remains only to show that

$$\lim_{U_1(0)\ni z\to e^{i\theta}} p(P_F(z)-F(e^{i\theta}))=0 \text{ for every } \tau\text{-continuous seminorm } p \text{ on } X$$
$$\text{and for every } 0\leq\theta<2\pi.$$

But this can be done similarly as in the case of scalar valued functions, using the properties of the Poisson kernel and the τ-continuity of F (see e.g. [A-B-R], proof of Theorem 1.14).

□

Next we prove a criterion for the τ-continuity at the boundary of harmonic maps from the unit disc in X:

4.2. Lemma. *Let* $U:\overline{U_1(0)}\to X$ *be a map such that*

- $U|U_1(0)$ *is harmonic and norm-bounded,*
- $U|\partial U_1(0)$ *is* τ*-continuous,*
- *for every* $\varphi\in\mathcal{F}$ *we have* $\lim_{1>r\to 1}\langle U(re^{i\theta}),\varphi\rangle=\langle U(e^{i\theta}),\varphi\rangle$ *for a.e.* $0\leq\theta<2\pi$.

Then U *is* τ*-continuous.*

Proof. By Lemma 4.1, the norm-bounded, τ- continuous map $F=U|\partial U_1(0)$ has a τ- continuous extension $\tilde{F}:\overline{U_1(0)}\to X$, which is harmonic in the interior. Then $V=U-\tilde{F}:\overline{U_1(0)}\to X$ is

- harmonic and norm-bounded on $U|U_1(0)$,
- vanishing on $U|\partial U_1(0)$,
- such that, for every $\varphi\in\mathcal{F}$, the function $U_1(0)\ni z\mapsto V_\varphi(z)=\langle V(z),\varphi\rangle$ has zero radial limit in a.e. point of $\partial U_1(0)$.

But bounded harmonic functions on the open unit disc, which have zero radial limits a.e. on the unit circle, must vanish identically (see e.g. [A-B-R], Theorems 6.12 (b) and 6.31, or [K], the second theorem in I:C and the fourth theorem in I:D.2), so it follows that V_φ vanishes identically for every $\varphi\in\mathcal{F}$. Consequently V is vanishing also on $U_1(0)$ and we conclude that $U=\tilde{F}$ is τ-continuous.

□

Now we are ready to prove a general result concerning the τ-continuity at the boundary of analytic maps with values in X:

4.3. THEOREM. *Let $D \subset \mathbb{C}$ be a simply connected domain such that $\hat{\partial} D$ is locally connected, $\Gamma \subset \partial D$ a locally rectifiable open Jordan arc, which is open in ∂D and not contained in the interior of the closure of D, and $F: D \cup \Gamma \to X$ a map with the properties:*

- *$F|D$ is analytic and norm-bounded,*
- *$F|\Gamma$ is τ-continuous,*
- *for every $\varphi \in \mathcal{F}$ there exists a set $N_\varphi \subset \Gamma$ of zero arc length measure such that $D \ni \zeta \mapsto \langle F(\zeta), \varphi \rangle$ has the asymptotic value $\langle F(z), \varphi \rangle$ at z for every $z \in \Gamma \backslash N_\varphi$.*

Then F is τ-continuous.

Proof. Taking into account statements (ii) and (iii) in Proposition 3.2, we can assume without loss of generality that $\Gamma = \overset{\circ}{J}$ for some rectifiable Jordan arc $J \subset \partial D$ and that there exists a rectifiable crosscut C of D with the same endpoints a, b as J, being enough to prove the τ-continuity of Φ on $D_C \cup \overset{\circ}{J}$, where $D_C \subset D$ denotes the Jordan domain defined by $C \cup J$.

Let $\Phi: U_1(0) \to D_C$ be a conformal representation. By the Carathéodory Theorem (see e.g. [K], II:C.1 or [Pom], Theorem 2.6) Φ has an extension to a homeomorphism of $\overline{U_1(0)}$ onto $\overline{D_C}$, which will be still denoted by Φ. Moreover, by the F. and M. Riesz Theorem (see e.g. [K], II:D.2 or [Pom], Theorem 6.8) a subset S of $\partial U_1(0)$ has arc length measure zero if and only if $\Phi(S)$ has arc length measure zero on ∂D_C. Therefore, if we define $U: \overline{U_1(0)} \to X$ by putting

$$U(z) = (\Phi(z) - a)(\Phi(z) - b) F(\Phi(z)) \text{ for } z \in \overline{U_1(0)} \backslash \{\Phi^{-1}(a), \Phi^{-1}(b)\},$$
$$U(\Phi^{-1}(a)) = U(\Phi^{-1}(b)) = 0,$$

then

- $U|U_1(0)$ is analytic and norm-bounded,
- $U|\partial U_1(0)$ is τ-continuous,
- $U|U_1(0) \cup (\partial U_1(0) \backslash \Phi^{-1}(J))$ is τ-continuous,
- for every $\varphi \in \mathcal{F}$ and for every $z \in \Phi^{-1}(\overset{\circ}{J})$, not belonging to the set $\Phi^{-1}(N_\varphi)$ of zero arc length measure, $U_1(0) \ni \zeta \mapsto \langle U(\zeta), \varphi \rangle$ has the asymptotic value $\langle \Phi(z), \varphi \rangle$ at z.

But, according to a theorem of Lindelöf (see e.g. [Col-L] Theorem 2.3 or [Con], Theorem 13.5.4 or, for a more general result, [Pom], Theorem 4.3), if a bounded analytic function on $U_1(0)$ has the asymptotic value w at $z \in \partial U_1(0)$ then it has the angular limit w at z. Consequently, the last two properties of U imply:

- for every $\varphi \in \mathcal{F}$ we have $\lim\limits_{1 > r \to 1} \langle U(re^{i\theta}), \varphi \rangle = \langle U(e^{i\theta}), \varphi \rangle$ for a.e. $0 \leq \theta < 2\pi$.

Now we have only to apply Lemma 4.2 to conclude the τ-continuity of U, which clearly implies the τ-continuity of $F|D_C \cup \overset{\circ}{J}$.

□

We notice that the proof of Theorem 4.3 does not work if we assume only the harmonicity of $F|D$, because in contrast to analyticity, harmonicity is not preserved by multiplication with analytic functions, needed in the passage from F to U.

Taking into account statement (i) in Proposition 3.2, Theorem 4.3 implies immediately:

4.4. COROLLARY. *Let $D \subset \mathbb{C}$ be a simply connected domain such that $\hat{\partial} D$ is locally connected, $\Gamma \subset \partial D$ an open Jordan arc, which is open in ∂D and not contained in the interior of the closure of D, and $F: D \cup \Gamma \to X$ a map with the properties:*

- *$F|D$ is analytic and norm-bounded,*
- *$F|\Gamma$ is τ-continuous,*
- *for every $\varphi \in \mathcal{F}$ and every $z \in \Gamma$, $D \ni \zeta \mapsto \langle F(\zeta), \varphi \rangle$ has the asymptotic value $\langle F(z), \varphi \rangle$ at z.*

Then F is τ-continuous.

We recall that $f \in L^\infty(\partial U_1(0))$ is a.e. angular limit of a bounded analytic function $\tilde{f}$ on $U_1(0)$ if and only if $\int_0^{2\pi} f(e^{i\theta}) e^{in\theta} d\theta = 0$ for all integers $n \geq 1$, in which case $\tilde{f}$ is the Poisson integral of f and $\|\tilde{f}\|_\infty$ is equal to the norm $\|f\|_\infty$ of f in $L^\infty(\partial U_1(0))$ (see e.g. in [Ho] the discussion at the end of Chapter 3 or [Con], Theorem 20.1.5). Therefore the set of all these f is a weak*closed linear subspace $\mathcal{H}^\infty(\partial U_1(0))$ of $L^\infty(\partial U_1(0))$. In particular, the closed unit ball of $\mathcal{H}^\infty(\partial U_1(0))$ is weak*compact. Furthermore, using the Poisson formula, it is easily seen that the map $\{f \in \mathcal{H}^\infty(\partial U_1(0)); \|f\|_\infty \leq 1\} \ni f \mapsto \tilde{f}$ is continuous with respect to the weak*topology on the closed unit ball of $\mathcal{H}^\infty(\partial U_1(0))$ and the topology of the uniform convergence on the compact subsets of $U_1(0)$.

Now let $\theta_1 < \theta_2 \leq \theta_1 + 2\pi$ be arbitrary. Since the restriction map $\mathcal{H}^\infty(\partial U_1(0)) \ni f \mapsto f|e^{i(\theta_1,\theta_2)} \in L^\infty(e^{i(\theta_1,\theta_2)})$ is weak*continuous and injective, it maps the closed unit ball of $\mathcal{H}^\infty(\partial U_1(0))$ bijectively onto the weak*compact set

$$\{f|e^{i(\theta_1,\theta_2)}; f \in \mathcal{H}^\infty(\partial U_1(0)), \|f\|_\infty \leq 1\} \subset L^\infty(e^{i(\theta_1,\theta_2)}),$$

and this bijection is a homeomorphism with respect to the weak*topologies. It follows that the intersection of the above set with $\mathcal{C}_b(e^{i(\theta_1,\theta_2)})$

$$\left\{ f \in \mathcal{C}_b(e^{i(\theta_1,\theta_2)}); \begin{array}{l} f \text{ is a.e. angular boundary value of some} \\ \text{analytic } \tilde{f}: U_1(0) \to \mathbb{C} \text{ with } \|\tilde{f}\|_\infty \leq 1 \end{array} \right\},$$

which is by Theorem 4.3 (applied to $X = \mathbb{C}$) equal to

$$\mathcal{C}_{\theta_1,\theta_2} = \left\{ f \in \mathcal{C}_b(e^{i(\theta_1,\theta_2)});\ \begin{matrix} f \text{ has an extension } \tilde{f} \in \mathcal{R}_b(U_1(0) \cup e^{i(\theta_1,\theta_2)}) \\ \text{with } \|\tilde{f}\|_\infty \le 1 \end{matrix} \right\},$$

is weak*closed in $\mathcal{C}_b(e^{i(\theta_1,\theta_2)})$ and the map $\mathcal{C}_{\theta_1,\theta_2} \ni f \mapsto \tilde{f}|U_1(0)$ is continuous with respect to the weak*topology on $\mathcal{C}_{\theta_1,\theta_2}$ and the topology of the uniform convergence on the compact subsets of $U_1(0)$.

Thus, using statement (i) in Proposition 3.2, we get:

4.5. COROLLARY. *Let $D \subset \mathbb{C}$ be a simply connected domain such that $\hat{\partial} D$ is locally connected, and $\Gamma \subset \partial D$ an open Jordan arc, which is open in ∂D and not contained in the interior of the closure of D. Then the set*

$$\mathcal{C}_\Gamma = \{f \in \mathcal{C}_b(\Gamma); f \text{ has an extension } \tilde{f} \in \mathcal{R}_b(D \cup \Gamma) \text{ with } \|\tilde{f}\|_\infty \le 1\}$$

is uniformly closed in $\mathcal{C}_b(\Gamma)$ and the map $\mathcal{C}_\Gamma \ni f \mapsto \tilde{f}|D$ is continuous with respect to the topology of the uniform convergence on $\mathcal{C}_\Gamma$ and the topology of the uniform convergence on the compact subsets of D. Moreover, if Γ is locally rectifiable and λ denotes the arc length measure on Γ then the above statements hold with the topology of the uniform convergence on $\mathcal{C}_b(\Gamma)$ replaced by $\sigma(L^\infty(\lambda), L^1(\lambda))$.

5. General analytic extension criteria for vector valued functions

This section contains the main results of the paper. In it $(X, \mathcal{F})$ will always denote a dual pair of normed spaces, $D \subset \mathbb{C}$ be a simply connected domain such that $\hat{\partial} D$ is locally connected, and $\Gamma \subset \partial D$ an open Jordan arc, which is open in ∂D and not contained in the interior of the closure of D.

5.1. THEOREM (ON ANALYTIC EXTENSION TO A FIXED DOMAIN). *Let us assume that $(X, \mathcal{F})$ has the Banach-Steinhaus Property and the Dominated Analytic Extension Property, the norm of X is complete, Γ is locally rectifiable, and $F: \Gamma \to X$ is locally Pettis integrable with respect to any Radon measure on Γ, which is absolutely continuous relative to the arc length measure and $\sigma(X, \mathcal{F})$. If*

- *$\Gamma \ni z \mapsto \langle F(z), \varphi \rangle \in \mathbb{C}$ has a regular extension on $D \cup \Gamma$ for all $\varphi \in \mathcal{F}$,*

then

- *F has a $\sigma(X,\mathcal{F})$-continuous extension $\tilde{F}: D \cup \Gamma \to X$, which is analytic in the interior.*

Moreover:

- *If the regular extension of $\Gamma \ni z \mapsto \langle F(z), \varphi\rangle \in \mathbb{C}$ on $D \cup \Gamma$ is bounded for every $\varphi \in \mathcal{F}$, then $\tilde{F}$ is bounded.*
- *If F is continuous with respect to a locally convex vector space topology τ between $\sigma(X,\mathcal{F})$ and the norm-topology of X, which is complete on the closed unit ball of X, then $\tilde{F}$ is τ-continuous.*

Proof. Let C be any rectifiable crosscut of D with endpoints a, b on Γ, J_C the Jordan arc on Γ between a and b, and D_C the Jordan domain defined by $C \cup J_C$. We recall that $D_C \subset D$ by statement (ii) of Proposition 3.2. By the Banach–Steinhaus Property of $(X,\mathcal{F})$ and the integrability assumption on F, the restriction $F|J_C$ is norm-bounded and Pettis integrable with respect to any Radon measure on J_C, which is absolutely continuous relative to the arc length measure and $\sigma(X,\mathcal{F})$.

Let $\Phi_C: U_1(0) \to D_C$ be a conformal representation. By the Carathéodory Theorem (see e.g. [K], II:C.1 or [Pom], Theorem 2.6) Φ_C has an extension to a homeomorphism of $\overline{U_1(0)}$ onto $\overline{D_C}$, which will be still denoted by Φ_C. By the F. and M. Riesz Theorem (see e.g. [K], II:D.2 or [Pom], Theorem 6.8) $\Phi_C|\partial U_1(0)$ is an absolutely bicontinuous homeomorphism between $\partial U_1(0)$ and $C \cup J_C$ relative to the arc length measures. Let $\theta_1 < \theta_2 < \theta_1 + 2\pi$ be reals such that Φ_C maps $e^{i[\theta_1,\theta_2]}$ onto J_C.

Now let Ψ_{θ_1,θ_2} be a conformal representation of $\mathbf{S}_1$ onto $U_1(0)$ which maps $\mathbb{R}$ onto $e^{i(\theta_1,\theta_2)}$. For example, we can take

$$\Psi_{\theta_1,\theta_2}: \zeta \longmapsto e^{i\theta_2} \frac{e^{\pi\zeta} + e^{i\frac{\theta_1-\theta_2}{2}}}{e^{\pi\zeta} + e^{-i\frac{\theta_1-\theta_2}{2}}}.$$

Clearly, $\Psi_{\theta_1,\theta_2}|\mathbb{R}$ is an absolutely bicontinuous homeomorphism between $\mathbb{R}$ and $e^{i(\theta_1,\theta_2)}$ relative to the arc length measures. Therefore $\Phi_C \circ \Psi_{\theta_1,\theta_2}$ carries the bounded Radon measures on $\mathbb{R}$, which are absolutely continuous relative to the Lebesgue measure, onto the bounded Radon measures on $\mathring{J}_C$, which are absolutely continuous relative to the arc length measure. It follows that $F \circ \Phi_C \circ \Psi_{\theta_1,\theta_2}: \mathbb{R} \to X$ is norm-bounded and Pettis integrable with respect to all bounded, absolutely continuous Radon measures on $\mathbb{R}$ and $\sigma(X,\mathcal{F})$.

By our assumptions on $(X,\mathcal{F})$ and F we can apply Proposition 2.2 to

$$F_C: \mathbb{R} \ni t \longmapsto F \circ \Phi_C \circ \Psi_{\theta_1,\theta_2}(t) \in X$$

and deduce the existence of a bounded, $\sigma(X, \mathcal{F})$-continuous extension $\tilde{F}_C: \mathbf{S}_1 \cup \mathbb{R} \to X$ of F, which is analytic in the interior. The map

$$D_C \cup \overset{\circ}{J}_C \ni z \longmapsto \tilde{F}_C \circ \Psi_{\theta_1,\theta_2}^{-1} \circ \Phi_C^{-1}(z) \in X$$

will be a $\sigma(X, \mathcal{F})$-continuous extension of

$$\overset{\circ}{J}_C \ni z \longmapsto F(z) \in X,$$

which is analytic in the interior.

Using statement (iv) in Proposition 3.2 and the above part of the proof, it follows the existence of a $\sigma(X, \mathcal{F})$-continuous extension of F, which is analytic in the interior.

Let next assume that the regular extensions of the functions $\Gamma \ni z \mapsto \langle F(z), \varphi \rangle \in \mathbb{C}$ on $D \cup \Gamma$ are bounded. We notice that then F must be norm-bounded. For any conformal representation $\Phi: U_1(0) \to D$, let θ_1, θ_2 be as in statement (i) of Proposition 3.2. Choosing some $\theta_1 < \alpha < \beta < \theta_2$, let $\Psi: \mathbf{S}_1 \to U_1(0)$ be a conformal representation mapping $\mathbb{R}$ onto $e^{i(\alpha,\beta)}$. Then $\Phi \circ \Psi|\mathbb{R}$ is an absolutely bicontinuous homeomorphism with respect to the arc length measures, hence it carries the bounded, absolutely continuous Radon measures on $\mathbb{R}$ onto the bounded Radon measures on its image, which are absolutely continuous relative to the arc length measure. It follows that the norm-bounded map $\mathbb{R} \ni t \longmapsto F \circ \Phi \circ \Psi(t) \in X$ is Pettis integrable with respect to all bounded, absolutely continuous Radon measures on $\mathbb{R}$. Now, using the Dominated Analytic Extension Property of $(X, \mathcal{F})$, it is easy to verify that $\tilde{F} \circ \Phi \circ \Psi$, hence also $\tilde{F}$ is norm-bounded.

Finally, the last statement follows easily using statement (i) in Proposition 3.2 and Corollary 4.4.

□

We notice that, assuming that $(X, \mathcal{F})$ has also the Krein property, the statement of the above theorem holds without requiring the local rectifiability of Γ and the local Pettis integrability property of F. Indeed, according to statement (i) in Proposition 3.2, we can assume in the proof without loss of generality that $D = U_1(0)$. But then Γ is rectifiable and the local Pettis integrability property of F follows from its $\sigma(X, \mathcal{F})$-continuity, the Banach–Steinhaus Property of $(X, \mathcal{F})$ and the Krein Property of $(X, \mathcal{F})$.

Next we prepare the proof of a criterion on analytic extendibility to a not previously fixed domain. First we prove a result of interest on its own:

5.2. PROPOSITION. *Let $F:\Gamma \to X$ be a norm-bounded, $\sigma(X,\mathcal{F})$ -continuous map. Then the sets*

$$\mathcal{F}_{F,c} = \left\{ \varphi \in \mathcal{F};\ \begin{array}{l} \Gamma \ni \zeta \mapsto \langle F(\zeta),\varphi\rangle \text{ has an extension} \\ F_\varphi \in \mathcal{R}_b(D\cup\Gamma) \text{ with } \|F_\varphi\|_\infty \le c \end{array} \right\}, \quad c>0$$

are norm closed in $\mathcal{F}$ and each map $\mathcal{F}_{F,c} \ni \varphi \mapsto F_\varphi|D$ is continuous with respect to the norm topology on $\mathcal{F}$ and the topology of the uniform convergence on the compact subsets of D. In particular,

$$\left\{ \varphi \in \mathcal{F};\ \begin{array}{l} \Gamma \ni \zeta \mapsto \langle F(\zeta),\varphi\rangle \text{ has a} \\ \text{bounded, regular extension on } D\cup\Gamma \end{array} \right\} = \bigcup_{n\ge 1} \mathcal{F}_{F,n}$$

is a linear subspace of $\mathcal{F}$, which is a Borel set with respect to the Borel structure defined by the norm of $\mathcal{F}$. Moreover, assuming that Γ is locally rectifiable with arc length measure λ and F is Pettis integrable with respect to any bounded Radon measure on Γ, which is absolutely continuous relative to λ and $\sigma(X,\mathcal{F})$, the sets $\mathcal{F}_{F,c}$ are even $\sigma(\mathcal{F},X)$-closed in $\mathcal{F}$ and the maps $\mathcal{F}_{F,c} \ni \varphi \mapsto F_\varphi|D$ are continuous with respect to $\sigma(\mathcal{F},X)$ on $\mathcal{F}$ and the topology of the uniform convergence on the compact subsets of D.

Proof. The first statement is an immediate consequence of Corollary 4.5.

Also the second statement follows from Corollary 4.5, once we prove the continuity of the linear map $T:\mathcal{F} \to \mathcal{C}_b(\Gamma)$, defined by $T(\varphi)(\zeta) = \langle F(\zeta),\varphi\rangle$, with respect to the topologies $\sigma(\mathcal{F},X)$ and $\sigma(L^\infty(\lambda),L^1(\lambda))$, that is the $\sigma(\mathcal{F},X)$-continuity of the linear functionals

$$\mathcal{F} \ni \varphi \mapsto \int_\Gamma g(\zeta)\langle F(\zeta),\varphi\rangle d\lambda(\zeta)\ ,\ g\in L^1(\Gamma)\,.$$

But this follows plainly from the equality

$$\int_\Gamma g(\zeta)\langle F(\zeta),\varphi\rangle d\lambda(\zeta) = \langle \sigma(X,\mathcal{F}) - \int_\Gamma g(\zeta)F(\zeta)d\lambda(\zeta),\varphi\rangle,$$

valid for every $g\in L^1(\Gamma)$ and every $\varphi\in\mathcal{F}$.

□

We recall that, according to a result due to H. Helson (see [He], (i)), for any sequence $Y_1 \subset Y_2 \subset \dots$ of linear subspaces of a Banach space Y, each one of which is Borel set with respect to the Borel structure defined by the norm of Y and such that $\bigcup_{n\ge 1} Y_n = Y$, there exists $n_\circ \ge 1$ such that $Y_{n_\circ} = Y$. The proof of the next statement uses this result and is inspired from the proof of a theorem of D. A. Herrero given in [He], (iii).

5.3. LEMMA. *Let $J \subset \Gamma$ be a compact, connected set (that is a Jordan arc or a singleton) and $F\colon \Gamma \to X$ a norm-bounded, $\sigma(X, \mathcal{F})$ -continuous map. If the norm of $\mathcal{F}$ is complete and*

- *for every $\varphi \in \mathcal{F}$ there exists an open set $J \subset G_\varphi \subset \mathbb{C}$ such that $\Gamma \cap G_\varphi \ni \zeta \mapsto \langle F(\zeta), \varphi\rangle$ has a bounded, regular extension on $(D \cup \Gamma) \cap G_\varphi$,*

then there exists an open set $J \subset G \subset \mathbb{C}$ such that

- *$\Gamma \cap G \ni \zeta \mapsto \langle F(\zeta), \varphi\rangle$ has an extension to some $F_\varphi \in \mathcal{R}_b((D \cup \Gamma) \cap G)$ for all $\varphi \in \mathcal{F}$.*

Proof. By statement (i) in Proposition 3.2 we can assume without loss of generality that $D = U_1(0)$, $\Gamma = e^{i(\theta_1, \theta_2)}$ and $J = e^{i[\varphi_1, \varphi_2]}$ for some $\theta_1 < \varphi_1 \leq \varphi_2 < \theta_2 \leq \theta_1 + 2\pi$. Choosing an integer $n_\circ \geq 1$ with $\frac{1}{n_\circ} < \min(\varphi_1 - \theta_1, \theta_2 - \varphi_2)$, let us consider the open sets

$$G_n = \left\{ re^{i\theta} ; r \in \left(1 - \frac{1}{n}, 1 + \frac{1}{n}\right), \theta \in \left(\varphi_1 - \frac{1}{n}, \varphi_2 + \frac{1}{n}\right)\right\}, \quad n \geq n_\circ .$$

Then every open subset of $\mathbb{C}$ containing J contains some G_n.

Now, according to Proposition 5.2, each one of the linear subspaces

$$\mathcal{F}_n = \left\{ \varphi \in \mathcal{F} ; \begin{array}{l} \Gamma \cap G_n \ni \zeta \mapsto \langle F(\zeta), \varphi\rangle \text{ has a} \\ \text{bounded, regular extension on } (D \cup \Gamma) \cap G_n \end{array} \right\} \subset \mathcal{F}, \quad n \geq n_\circ$$

is a Borel set with respect to the Borel structure defined by the norm of $\mathcal{F}$. But, by the assumptions of the lemma, $\bigcup_{n \geq n_\circ} \mathcal{F}_n = \mathcal{F}$, so the above quoted result of H. Helson implies that $\mathcal{F} = \mathcal{F}_n$ for some $n \geq n_\circ$. Put $G = G_n$ for this n.

□

Now, Lemma 5.3 and Theorem 5.1 imply immediately:

5.4. THEOREM (ON ANALYTIC EXTENSION TO AN UNDETERMINED DOMAIN). *Let us assume that $(X, \mathcal{F})$ has the Banach–Steinhaus Property and the Dominated Analytic Extension Property, the norm of X is complete, Γ is locally rectifiable, $J \subset \Gamma$ is a compact, connected set, and $F\colon \Gamma \to X$ is locally Pettis integrable with respect to any Radon measure on Γ, which is absolutely continuous relative to the arc length measure, and $\sigma(X, \mathcal{F})$. If*

- *for every $\varphi \in \mathcal{F}$ there exists an open set $J \subset G_\varphi \subset \mathbb{C}$ such that $\Gamma \cap G_\varphi \ni \zeta \mapsto \langle F(\zeta), \varphi\rangle$ has a bounded, regular extension on $(D \cup \Gamma) \cap G_\varphi$,*

then there exists an open set $J \subset G \subset \mathbb{C}$ such that

- *$\Gamma \cap G \ni z \mapsto F(z) \in X$ has a bounded, $\sigma(X, \mathcal{F})$-continuous extension on $(D \cup \Gamma) \cap G$, which is analytic in the interior.*

Similarly to Theorem 5.1, also the statement of Theorem 5.4 holds without assuming the local rectifiability of Γ and the local Pettis integrability property of F whenever $(X, \mathcal{F})$ has additionally the Krein Property.

References

[Are-Ni] W. Arendt and N. Nikolski, Vector-valued holomorphic functions revisited, *Math. Z.*, **234** (2000), 777–805.

[Arv] W. Arveson, On groups of automorphisms of operator algebras, *J. Funct. Analysis*, **15** (1974), 217–243.

[A-B-R] S. Axler, P. Bourdon and W. Ramey, *Harmonic Function Theory*, Springer–Verlag, 1992.

[C-Z] I. Ciorănescu and L. Zsidó, Analytic generators for one-parameter groups, *Tôhoku Math. J.*, **28** (1976), 327–362.

[Col-L] E. F. Collingwood and A. J. Lohwater, *The Theory of Cluster Sets*, Cambridge University Press, 1966.

[Con] J. B. Conway, *Functions of One Complex Variable. II*, Springer–Verlag, 1995.

[D-Z 1] C. D'Antoni and L. Zsidó, Groups of linear isometries on multiplier C^*-algebras, *Pacific J. Math.*, **193** (2000), 279–306.

[D-Z 2] C. D'Antoni and L. Zsidó, Hardy states for C^*-dynamical systems, to appear.

[D] J. Dugundji, *Topology*, Allyn and Bacon, 1968.

[F] D. H. Fremlin, Measurable functions and almost continuous functions, *Manuscripta Math.*, **33** (1981), 387–405.

[He] H. Helson, Boundedness from measure theory. II, *Operator Theory: Advances and Applications* **11**, Birkhäuser Verlag, 1983, 179–190.

[Hi-P] E. Hille and R. S. Phillips, *Functional Analysis and Semi-Groups,* Amer. Math. Soc. Colloquium Publ. **31**, Amer. Math. Soc., 1981.

[Ho] K. Hoffman, *Banach Spaces of Analytic Functions*, Prentice-Hall, 1962.

[K] P. Koosis, *Introduction to H_p-Spaces*, Cambridge University Press, 1998.

[Ku-Pr] J. Kupka and K. Prikry, The measurability of uncountable unions, *Amer. Math. Monthly*, **91** (1984), 85–97.

[N] M. H. A. Newman, *Elements of the Topology of Plane Sets of Points*, Cambridge University Press, 1951.

[Ped] G. K. Pedersen, *C^*-Algebras and their Automorphism Groups*, Academic Press, 1979.

[Pom] Ch. Pommerenke, *Boundary Behaviour of Conformal Maps*, Springer–Verlag, 1992.

[R] W. Rudin, *Functional Analysis*, McGraw-Hill, 1973.

[Sch] H. H. Schaefer, *Topological Vector Spaces*, Springer–Verlag, 1980.

[WO] N. E. WEGGE-OLSEN, *K-theory and C^*-algebras*, Oxford University Press, 1993.

C. D'ANTONI, Dipartimento di Matematica, Università di Roma "Tor Vergata", Via della Ricerca Scientifica, 00133 Roma, Italia; *e-mail*: dantoni@axp.mat.uniroma2.it

L. ZSIDÓ, Dipartimento di Matematica, Università di Roma "Tor Vergata", Via della Ricerca Scientifica, 00133 Roma, Italia; *e-mail*: zsido@axp.mat.uniroma2.it

Operator Theory: Advances and Applications, Vol. 127, 203–209

On quotient modules

Ronald G. Douglas and Gadadhar Misra*

In this paper we obtain an extension of one of the main results in [5] relating the fundamental class of the zero set defining a quotient Hilbert module $\mathcal{M}_q$, the curvatures of the two modules $\mathcal{M}$, $\mathcal{M}_0$ and the map X in a topologically exact sequence

$$0 \longrightarrow \mathcal{M}_0 \xrightarrow{X} \mathcal{M} \longrightarrow \mathcal{Q} \longrightarrow 0.$$

Let Ω be a bounded open connected set in $\mathbb{C}^m$. Let $\mathcal{A}(\Omega)$ denote the closure of the algebra of holomorphic functions in some neighborhood of $\bar{\Omega}$ with respect to the supremum norm. Let $\mathcal{M}$ be a Hilbert module, consisting of holomorphic functions on Ω over the function algebra $\mathcal{A}(\Omega)$. In an earlier paper [5], we considered the problem of finding invariants for $\mathcal{Q}$ using the resolution

$$0 \longrightarrow \mathcal{M}_0 \xrightarrow{X} \mathcal{M} \longrightarrow \mathcal{Q} \longrightarrow 0, \tag{1}$$

where $X\colon \mathcal{M}_0 \to \mathcal{M}$ is the inclusion map and $\mathcal{M}_0$ is the submodule of all functions vanishing on a hypersurface $\mathcal{Z} \subseteq \Omega$. In this paper we reconsider this problem extending our earlier results.

Let $\mathbb{C}_w$ be the one dimensional module over $\mathcal{A}(\Omega)$, where the module map is given by evaluation at $w \in \Omega$, that is, $(f, \lambda) \to f(w)\lambda$, $f \in \mathcal{A}(\Omega)$, $\lambda \in \mathbb{C}_w$. Let

$$X \otimes_{\mathcal{A}(\Omega)} 1\colon \mathcal{M} \otimes_{\mathcal{A}(\Omega)} \mathbb{C}_w \to \mathcal{N} \otimes_{\mathcal{A}(\Omega)} \mathbb{C}_w$$

be the map obtained by localising a module map $X\colon \mathcal{M} \to \mathcal{N}$ between any two Hilbert modules $\mathcal{M}$ and $\mathcal{N}$ over a function algebra $\mathcal{A}(\Omega)$ (cf. [6, p. 114–115]). We

AMS Subject Classification (1991): 47B38, 53C07, 58A25.

* Research was partly carried out during a visit to Texas A&M University in August, 1999. Also supported by a grant from the National Board for Higher Mathematics, India.

let $X(w)$ denote the map $X \otimes_{\mathcal{A}(\Omega)} 1$. Finally, let

$$\mathcal{K}_X(w) \overset{\text{def}}{=} \sum_{i,j=1}^{m} \frac{\partial^2}{\partial w_i \partial \bar{w}_j} \log \|X(w)\|^2 dw_i \wedge d\bar{w}_j. \tag{2}$$

Note that $\|X(w)\|^2$ vanishes on $\mathcal{Z}$ and that the right hand side in the above definition is thought of as a (1,1) form with distributional coefficients.

Let $\mathcal{K}$ and $\mathcal{K}_0$ be the curvatures associated with the module $\mathcal{M}$ and the submodule $\mathcal{M}_0$ of functions vanishing on the hypersurface $\mathcal{Z}$, respectively. In the paper [5], we proved that if Ω is a bounded domain in $\mathbb{C}^m$ for which the second Cousin problem is solvable, $X\colon \mathcal{M} \to \mathcal{M}_0$ is the inclusion map, then

$$\mathcal{K}_X(w) - \mathcal{K}_0(w) + \mathcal{K}(w)$$

represents the fundamental class $[\mathcal{Z}]$ of the hypersurface $\mathcal{Z}$.

In this note, we show that this alternating sum represents the fundamental class $[\mathcal{Z}]$ not just for the inclusion map but for any injective map $X\colon \tilde{\mathcal{M}} \to \mathcal{M}$ which has dense range in $\mathcal{M}_0 \subseteq \mathcal{M}$.

We begin by describing the exact hypothesis on the domain Ω and the module $\mathcal{M}$. Let $\mathcal{Z}$ be an irreducible analytic hypersurface in Ω (complex submanifold of dimension $m-1$) in the sense of [8, Definition 8, p. 17]. We assume that the second Cousin problem is solvable on the domain Ω. Consequently, as pointed out in the remark preceding Corollary 3 in [8, p. 34], there exists a global defining function φ for the hypersurface $\mathcal{Z}$. We also assume that Ω is polynomially convex. Then the algebra $\mathcal{A}(\Omega)$ equals the uniform limits of polynomials with respect to the supremum norm on Ω. We assume that $\mathcal{M}$ is a complex separable Hilbert space of holomorphic functions on Ω and that the evaluation functionals on $\mathcal{M}$ are bounded. Consequently, recall from [1] that $\mathcal{M}$ admits a reproducing kernel K. The reproducing kernel $K\colon \Omega \times \Omega \to \mathbb{C}$ is holomorphic in the first variable and anti-holomorphic in the second variable. Further, $K(\cdot, w) \in \mathcal{M}$ for each fixed $w \in \Omega$ and $K(z,w) = \overline{K(w,z)}$. Finally, K has the reproducing property

$$\langle h, K(\cdot, w)\rangle = h(w) \text{ for } w \in \Omega, \quad h \in \mathcal{M}.$$

Since $K(w,w) = \langle K(\cdot,w), K(\cdot,w)\rangle$, it follows that $K(w,w) \neq 0$ for $w \in \Omega$ whenever $K(\cdot, w)$ is a nonzero vector in $\mathcal{M}$. Assume that $\mathcal{M}$ is a bounded module over $\mathcal{A}(\Omega)$, in particular, the tuple $\mathbf{M}^* \overset{\text{def}}{=} (M_1^*, \dots, M_m^*)$ is bounded. Here M_k denotes the multiplication operator on $\mathcal{M}$ defined by $(M_k h)(w) = w_k h(w)$ for $h \in \mathcal{M}$ and

$w \in \Omega$. Finally, we assume that the tuple $\mathbf{M}^*$ is in the class $B_1(\Omega)$ introduced in [2] and [3]. As shown in [3], in this case the curvature of the module $\mathcal{M}$

$$\mathcal{K}_{\mathcal{M}}(w) = \sum_{i,j=1}^{m} \frac{\partial^2}{\partial w_i \partial \bar{w}_j} \log K(w,w) dw_i \wedge d\bar{w}_j$$

is a complete unitary invariant.

Lemma. *Let $\mathcal{M}$ and $\mathcal{N}$ be two modules satisfying the hypotheses stated in the preceding paragraph. Let $L: \mathcal{M} \to \mathcal{N}$ be a module map with dense range. Then*

$$\mathcal{K}_L(w) = \mathcal{K}_{\mathcal{N}}(w) - \mathcal{K}_{\mathcal{M}}(w).$$

Proof. Assuming that the tuple $\mathbf{M}^*$ is in $B_1(\Omega)$ ensures, on the one hand, the existence of eigenvectors

$$\{\gamma(w) : M_f^* \gamma(w) = \overline{f(w)} \gamma(w), \quad \gamma(w) \in \mathcal{M}, \ w \in \Omega\}$$

which span $\mathcal{M}$, and on the other hand, it also ensures that $\mathcal{M} \otimes_{\mathcal{A}(\Omega)} \mathbb{C}_w$ is the one dimensional module spanned by $\gamma(w)$. Similarly, $\mathcal{N} \otimes_{\mathcal{A}(\Omega)} \mathbb{C}_w$ is spanned by the eigenvector $\tilde{\gamma}(w)$. Furthermore, γ, $\tilde{\gamma}: \Omega \to \mathcal{M}_k \otimes_{\mathcal{A}(\Omega)} \mathbb{C}_w$ may be chosen to be anti-holomorphic. Indeed if the Hilbert module $\mathcal{M}$ consists of holomorphic functions on Ω such that all of them do not vanish at any point in Ω, then $\gamma(w)$ may be taken to be the reproducing kernel at $w \in \Omega$, that is, $\gamma(w) = K(\cdot, w)$. In this case, $K(\cdot, w) \neq 0$ for each $w \in \Omega$. Hence $\|\gamma(w)\|^2 = K(w,w) \neq 0$. If all the functions vanish on a common zero set $\mathcal{Z} \subseteq \Omega$, then it can be shown (cf. [5, p. 91]) that the reproducing kernel factors as $K(z,w) = \varphi(z)\chi(z,w)\overline{\varphi(w)}$, where $\chi(w,w)$ does not vanish for any $w \in \Omega$ and φ is holomorphic. In this case, we may use $\chi(w,w)$ for calculating the curvature $\mathcal{K}_{\mathcal{M}}$. Consequently, without loss of generality, we shall assume that all reproducing kernels are nonvanishing. Of course, similar considerations apply to the module $\mathcal{N}$ as well.

The fact that L is a module map implies $L \otimes_{\mathcal{A}(\Omega)} 1(\mathcal{M} \otimes_{\mathcal{A}(\Omega)} \mathbb{C}_w) \subseteq \mathcal{N} \otimes_{\mathcal{A}(\Omega)} \mathbb{C}_w$. However this map cannot be zero since the range of L is dense. Hence $L \otimes_{\mathcal{A}(\Omega)} 1\gamma(w) = a(w)\tilde{\gamma}(w)$, where $a(w)$ is a non-vanishing anti-holomorphic function. Therefore

$$\|L \otimes_{\mathcal{A}(\Omega)} 1\| = \|\tilde{\gamma}(w)\| |a(w)| / \|\gamma(w)\|.$$

Taking logarithm on both sides and differentiating verifies the claim. □

Let $\mathcal{N}$, $\mathcal{N}_k$, $k = 1, 2$ be Hilbert modules. We say that

$$\mathcal{N}_1 \xrightarrow{X_1} \mathcal{N} \xrightarrow{X_2} \mathcal{N}_2$$

is *topologically exact* at $\mathcal{N}$ if the clos(ran X_1) $= \ker X_2$. This differs from the usual notion of exactness in that the range of the module map X_1 is not assumed to be closed.

Let $\mathcal{M}$, $\tilde{\mathcal{M}}$ and $\mathcal{N}$, $\tilde{\mathcal{N}}$, be modules over the function algebra $\mathcal{A}(\Omega)$ satisfying the hypotheses stated in the paragraph preceding the Lemma. Suppose that $L\colon \mathcal{M} \to \mathcal{N}$ is a bijective module map. In other words, the two Hilbert spaces $\mathcal{M}$ and $\mathcal{N}$ are isomorphic. In this case, it is easy to see that both $\mathcal{M}$ and $\mathcal{N}$ consist of the same set of functions on Ω. To see this, first note that for each $w \in \Omega$, we must have $LK_{\mathcal{M}}(\cdot, w) = c(w)K_{\mathcal{N}}(\cdot, w)$, for some scalar $c(w) \in \mathbb{C}$. The fact that L is a bounded invertible transform is equivalent to saying that there are positive constants a, b such that

$$aK_{\mathcal{M}}(z, w) \leq c(z)\overline{c(w)}K_{\mathcal{N}}(z, w) \leq bK_{\mathcal{M}}(z, w).$$

As usual, these are to be interpreted as inequalities involving the positive matrices $K_{\mathcal{M}}(w_j, w_i)$, $1 \leq i, j \leq n$ (respectively, $K_{\mathcal{N}}(w_j, w_i)$, $1 \leq i, j \leq n$) for all finite subsets $\{w_1, \dots w_n\} \subseteq \Omega$. From these inequalities, it also follows that $a' \leq |c(z)| \leq b'$ for all $z \in \Omega$ and some positive constants a', b'. This implies, in view of [1, Corollary IV$_3$, page 383], that the module $\mathcal{M}$ coincides with the module whose reproducing kernel is given by the positive definite kernel $c(z)\overline{c(w)}K_{\mathcal{N}}(z, w)$. Clearly, this latter module coincides with $\mathcal{N}$.

THEOREM. *Let $L\colon \mathcal{M} \to \mathcal{N}$ be a bijective module map with $L(\text{ran } X) \subseteq \text{ran } Y$. Assume that the diagram*

$$\begin{array}{ccccccccc} 0 & \longrightarrow & \tilde{\mathcal{M}} & \xrightarrow{X} & \mathcal{M} & \xrightarrow{p} & \mathcal{Q} & \longrightarrow & 0 \\ & & \downarrow & & \downarrow L & & \| & & \\ 0 & \longrightarrow & \tilde{\mathcal{N}} & \xrightarrow{Y} & \mathcal{N} & \xrightarrow{q} & \mathcal{Q} & \longrightarrow & 0 \end{array}$$

is topologically exact, the range of the module map Y is closed and that $L(\ker p) = \ker q$. Then we have

$$\mathcal{K}_X(w) - \mathcal{K}_{\tilde{\mathcal{M}}}(w) + \mathcal{K}_{\mathcal{M}}(w) = \mathcal{K}_Y(w) - \mathcal{K}_{\tilde{\mathcal{N}}}(w) + \mathcal{K}_{\mathcal{N}}(w).$$

Proof. Notice that the assumption of topological exactness at $\mathcal{M}$ together with the fact that ran Y is closed implies

$$\operatorname{ran} Y = \ker q = L(\ker p) = L(\operatorname{clos}(\operatorname{ran} X)).$$

Furthermore, from the exactness at $\mathcal{Q}$, it follows that $\mathcal{Q}$ may be identified with the quotient module $\mathcal{N}/\operatorname{ran} Y = L(\mathcal{M})/L(\operatorname{ran} \mathrm{X})$. Define a map $Z: \tilde{\mathcal{M}} \to \tilde{\mathcal{N}}$ by setting

$$Zh = Y^{-1}LX(h), \quad h \in \tilde{\mathcal{M}}.$$

This definition of the map Z makes the square on the left of our diagram commutative. Clearly, the map Z is a module map which has dense range in $\tilde{\mathcal{N}}$. Tensoring the entire diagram given in the statement of the theorem with the module $\mathbb{C}_w$, $w \in \Omega$, we obtain a new diagram of one dimensional Hilbert modules in which the map

$$Z \otimes_{\mathcal{A}(\Omega)} 1: \tilde{\mathcal{M}} \otimes_{\mathcal{A}(\Omega)} \mathbb{C}_w \to \tilde{\mathcal{N}} \otimes_{\mathcal{A}(\Omega)} \mathbb{C}_w$$

is surjective. Also, the square of Hilbert modules remains commutative even after the localisation, that is,

$$Z \otimes_{\mathcal{A}(\Omega)} 1 = (Y \otimes_{\mathcal{A}(\Omega)} 1)^{-1}(L \otimes_{\mathcal{A}(\Omega)} 1)(X \otimes_{\mathcal{A}(\Omega)} 1).$$

However, since all these operators act on one dimensional Hilbert spaces, it follows that $\|Z \otimes_{\mathcal{A}(\Omega)} 1\| = \|X \otimes_{\mathcal{A}(\Omega)} 1\| \; \|(Y \otimes_{\mathcal{A}(\Omega)} 1)\|^{-1} \; \|L \otimes_{\mathcal{A}(\Omega)} 1)\|$. Hence in the notation of (2), we have

$$\mathcal{K}_Z(w) = \mathcal{K}_X(w) - \mathcal{K}_Y(w) + \mathcal{K}_L(w). \tag{3}$$

The Lemma shows that $\mathcal{K}_Z(w) = \mathcal{K}_{\tilde{\mathcal{N}}}(w) - \mathcal{K}_{\tilde{\mathcal{M}}}(w)$. Similarly, $\mathcal{K}_L(w) = \mathcal{K}_{\mathcal{N}}(w) - \mathcal{K}_{\mathcal{M}}(w)$. Going back to the equation (3), we arrive at the desired conclusion when we substitute the values for $\mathcal{K}_Z(w)$ and $\mathcal{K}_L(w)$.

□

Recall that $\mathcal{M}_0$ is the sub-module of functions in $\mathcal{M}$ which vanish on the hypersurface $\mathcal{Z}$. Let $\mathcal{Q}$ be the quotient $\mathcal{M}/\mathcal{M}_0$. Let $0 \to \tilde{\mathcal{M}} \xrightarrow{X} \mathcal{M} \to \mathcal{Q} \to 0$ be a topologically exact resolution. Then the range of X must be dense in $\mathcal{M}_0$ and conversely. If $Y: \tilde{\mathcal{N}} \to \mathcal{N}$ is any injective module map such that the range of Y coincides with $L(\operatorname{clos}(\operatorname{ran} X))$ then we can apply the theorem. Specialise to the case, where $\tilde{\mathcal{N}} = \mathcal{M}_0$, $\mathcal{N} = \mathcal{M}$ and $Y: \mathcal{M}_0 \to \mathcal{M}$ is the inclusion map. Note that in this case, the alternating sum

$$\mathcal{K}_Y(w) - \mathcal{K}_{\tilde{\mathcal{N}}}(w) + \mathcal{K}_{\mathcal{N}}(w) = \sum_{i,j=1}^{m} \frac{\partial^2}{\partial w_i \partial \bar{w}_j} \log |\varphi(w)|^2 dw_i \wedge d\bar{w}_j \tag{4}$$

represents the fundamental class of the hypersurface $\mathcal{Z}$ [5, Theorem 1.4]. Having established the equation (4) for a module map with closed range, we can apply the theorem to arrive at the same conclusion for the alternating sum

$$\mathcal{K}_X(w) - \mathcal{K}_{\tilde{\mathcal{M}}}(w) + \mathcal{K}_{\mathcal{M}}(w)$$

for an injective module map $X: \tilde{\mathcal{M}} \to \mathcal{M}$ which has dense but not necessarily closed range in $\mathcal{M}_0 \subseteq \mathcal{M}$. We have therefore proved

COROLLARY. *Let* $0 \longrightarrow \tilde{\mathcal{M}} \xrightarrow{X} \mathcal{M} \longrightarrow \mathcal{Q} \longrightarrow 0$ *be a short topologically exact sequence of Hilbert modules. Suppose that* $\operatorname{ran} X$ *is dense in* $\mathcal{M}_0$. *Then the alternating sum*

$$\mathcal{K}_X(w) - \mathcal{K}_{\tilde{\mathcal{M}}}(w) + \mathcal{K}_{\mathcal{M}}(w)$$

represents the fundamental class of the hypersurface $\mathcal{Z}$.

One way to view this result is that it provides an invariant for any topologically exact resolution of quotient modules of the form $\mathcal{M}/\mathcal{M}_0$. Such resolutions give an analogue of the Sz.-Nagy–Foias model for contraction operators to the multivariate case (cf. the Introduction in [5]).

We now give some examples of maps $X: \tilde{\mathcal{M}} \to \mathcal{M}$ which satisfy the hypothesis of the Corollary. Let $\Omega_1 \times \Omega_2 \subseteq \mathbb{C}$ be a product domain containing $(0,0)$. Let $\mathcal{M}$ be a module over the function algebra $\mathcal{A}(\Omega_1 \times \Omega_2)$. Assume that the reproducing kernel K of the module is of the form

$$K(z,w) = K_1(z_1,w_1)K_2(z_2,w_2), \quad z_i, w_i \in \Omega_i, \quad i = 1,2,$$

and that K_1, K_2 possess diagonal expansion about the origin, that is,

$$K_i(z_i,w_i) = \sum_{n=0}^{\infty} a_n(i) z_i^n \bar{w}_i^n, \quad i = 1,2.$$

As shown in [5, Proposition 2.4], the reproducing kernel for the submodule $\mathcal{M}_0$ of functions vanishing on the set $\{(0,z_2) \in \Omega_1 \times \Omega_2\}$ is of the form

$$K_0(z,w) = z_1\bar{w}_1\Big(\sum_{n=1}^{\infty} a_n(1) z_1^{n-1}\bar{w}_1^{n-1}\Big)K_2(z_2,w_2).$$

Consequently, $\mathcal{M}_0 = \{z_1 f : f \in \mathcal{M}\}$. If $X: \mathcal{M} \to \mathcal{M}$ is the multiplication operator defined by $(Xf)(z_1,z_2) = z_1 f(z_1,z_2)$ then we see that the range of this operator coincides with $\mathcal{M}_0$.

We point out that similar considerations as in the paragraph preceding the Corollary apply to the commutative square on the right of the diagram in our theorem. This yields the relation

$$\mathcal{K}_{\mathcal{M}}(w) - \mathcal{K}_{\mathcal{Q}}(w) + \mathcal{K}_p(w) = \mathcal{K}_{\mathcal{N}}(w) - \mathcal{K}_{\mathcal{Q}}(w) + \mathcal{K}_q(w) \tag{5}$$

for $w \in \mathcal{Z}$. This relationship is obtained by restricting p to $\mathcal{M}/(\text{ran } X)$ and then writing the identity map on $\mathcal{Q}$ as qLp^{-1}. In particular, if we take $p\colon \mathcal{M} \to \mathcal{Q}$ to be the quotient map, it is easy to verify that

$$p \otimes_{\mathcal{A}(\Omega)} 1\colon \mathcal{M}/(\text{ran } X) \otimes_{\mathcal{A}(\Omega)} \mathbb{C}_w \to \mathcal{Q} \otimes_{\mathcal{A}(\Omega)} \mathbb{C}_w$$

is the constant map identically equal to 1. Now if we further assume that $X(\tilde{\mathcal{M}})$ is dense in $\mathcal{M}_0$, then $\mathcal{Q}$ equals the quotient $\mathcal{M}_0^{\perp}$. It was shown, in this case, in [5] that $\mathcal{K}_{\mathcal{Q}}(w) = \mathcal{K}_{\mathcal{M}}(w)$ for $w \in \mathcal{Z}$. Therefore we obtain the equality $\mathcal{K}_{\mathcal{M}}(w) - \mathcal{K}_{\mathcal{Q}}(w) + \mathcal{K}_p(w) = 0$ for $w \in \Omega$. Hence in view of equation (5), this alternating sum is seen to be 0 even if p is not necessarily the quotient map.

References

[1] N. Aronszajn, Theory of Reproducing kernels, *Trans. Amer. Math. Soc.*, **68** (1950), 337–404.

[2] M. Cowen and R. G. Douglas, Complex geometry and operator theory, *Acta Math.*, **141** (1978), 187–261.

[3] M. Cowen and R. G. Douglas, *Operators possessing an open set of eigenvalues,* Proc. Fejér–Riesz Conf., Budapest, 1980.

[4] R. E. Curto and N. Salinas, Generalised Bergman kernels and the Cowen–Douglas theory, *Amer J. Math.*, **106** (1984), 447–488.

[5] R. G. Douglas and G. Misra, Geometric invariants for resolutions of Hilbert modules, *Operator Theory: Advances and Applications,* Birkhäuser Verlag, **104** (1998), 83–112.

[6] R. G. Douglas and V. I. Paulsen, *Hilbert Modules over Function Algebras,* Longman Research Notes **217**, 1989.

[7] P. Griffiths and J. Harris, *Principles of Algebraic Geometry,* John Wiley & Sons, 1978.

[8] R. C. Gunning and H. Rossi, *Analytic functions of Several Complex Variables,* Prentice Hall, 1965.

R. G. Douglas, Texas A & M University, College Station, Texas 77843; *e-mail:* rgd@tamu.edu

G. Misra, Indian Statistical Institute, R. V. College Post, Bangalore 560 059; *e-mail:* gm@isibang.ac.in

Operator Theory: Advances and Applications, Vol. 127, 211–242

On the structure of spherical contractions

JÖRG ESCHMEIER

We show that a spherical contraction T with a spherical dilation and isometric H^∞-functional calculus on the unit ball belongs to the class $\mathbb{A}_{1,\aleph_0}$ if and only if it possesses an analytic invariant subspace, or if and only if, there is a compression of T to a semi-invariant subspace which is of type $C_{\cdot 0}$ and possesses dominating Harte spectrum in the unit ball. If any of these conditions holds, then T is proved to be reflexive. As a byproduct we obtain that the dual algebra generated by a spherical isometry has property $(\mathbb{A}_{1,\aleph_0})$. Applied to the particular case of subnormal tuples our results lead to a new reflexivity proof for subnormal spherical contractions with rich Taylor spectrum in the unit ball.

Let $T \in L(H)$ be a contraction on a complex Hilbert space H. A classical result of Brown, Chevreau and Pearcy [5] from 1979 shows that T has a non-trivial invariant subspace if the spectrum of T is dominating in the open unit disc. Since then this result has been improved considerably. The richness condition on the spectrum of T occurring above implies that the unit circle is contained in the spectrum of T and that the absolutely continuous part of T has an isometric H^∞-functional calculus. In 1988 Brown, Chevreau and Pearcy [6] proved that each contraction $T \in L(H)$ whose spectrum contains the unit circle possesses a non-trivial invariant subspace. In the same year Brown and Chevreau [4] showed that each contraction $T \in L(H)$ whose absolutely continuous part has an isometric H^∞-functional calculus is reflexive. An important intermediate step for the proof of the last observation was a result of Chevreau, Exner and Pearcy [9] saying that each contraction $T \in L(H)$ in the class $\mathbb{A}_{1,\aleph_0}$ is reflexive.

Let $T = (T_1, \ldots, T_n) \in L(H)^n$ be a spherical contraction, that is, a commuting tuple of operators on H such that $\sum_{i=1}^n \|T_i x\|^2 \leq \|x\|^2$ holds for all vectors $x \in H$. In [14] we proved that T has a non-trivial joint invariant subspace if T possesses

---

AMS Subject Classification (1991): 47A15, 47D27, 47A13.

a spherical dilation and if the Harte spectrum of T is dominating in the open Euclidean unit ball B in $\mathbb{C}^n$. In the present paper we give various characterizations of spherical contractions in the class $\mathbb{A}_{1,\aleph_0}$. In particular, we show that each spherical contraction in the class $\mathbb{A}_{1,\aleph_0}$ is reflexive. As consequences we prove that the dual algebra generated by an arbitrary spherical isometry has property $(\mathbb{A}_{1,\aleph_0})$, and we obtain a new proof of a result from [15] saying that each subnormal spherical contraction with rich spectrum in the unit ball is reflexive.

By a spherical contraction of class $\mathbb{A}$ we mean a spherical contraction $T \in L(H)^n$ which possesses a spherical dilation and an isometric w^*-continuous H^∞-functional calculus over the Euclidean unit ball B in $\mathbb{C}^n$. We prove that a spherical contraction $T \in L(H)^n$ of class $\mathbb{A}$ is in the class $\mathbb{A}_{1,\aleph_0}$ if and only if T possesses an analytic invariant subspace, or if and only if, there is a semi-invariant subspace L for T such that the compression of T to L is of type $C_{\cdot 0}$ and possesses dominating Harte spectrum in B. In this case, the set of vectors in H generating an analytic invariant subspace for T is dense in H, and the tuple T is reflexive. These observations form part of a list of equivalent characterizations of the class $\mathbb{A}_{1,\aleph_0}$ which establishes the complete analogue of corresponding one-variable results from [9].

In more detail, in Section 1 we make systematic use of the observation that each spherical co-isometry decomposes into a $C_{0\cdot}$-part and a spherical unitary. We show that the unitary part of the minimal spherical co-isometric extension of a given spherical contraction (provided that it is non-trivial) allows a certain $\mathbb{A}_{1,\aleph_0}$-factorization. As a byproduct we obtain that the dual algebra $\mathfrak{A}_T$ generated by an arbitrary spherical isometry $T \in L(H)^n$ has property $(\mathbb{A}_{1,\aleph_0})$. In Section 2 we show that each spherical contraction with property $F^r_{\theta,\gamma}$, for some $0 \leq \theta < \gamma \leq 1$, is in the class $\mathbb{A}_{1,\aleph_0}$. Finally, in Section 3 we obtain the equivalent characterizations of spherical contractions in the class $\mathbb{A}_{1,\aleph_0}$ referred to above. In particular, we show that spherical contractions in the class $\mathbb{A}_{1,\aleph_0}$ are reflexive. As an immediate consequence it follows that each subnormal tuple $S \in L(H)^n$ whose Taylor spectrum is contained in the closed unit ball $\overline{B}$, and is dominating in the open unit ball B, is reflexive. Thus the results of the present paper generalize corresponding parts from [15].

Preliminaries

Let H be a complex Hilbert space, and let $T = (T_1, \ldots, T_n) \in L(H)^n$ be an n-tuple of commuting continuous linear operators on H. We denote by $\sigma(T)$ the Taylor spectrum of T (see Chapter 2 in [16]). The Harte spectrum $\sigma^H(T)$ of T is the set of those points $\lambda \in \mathbb{C}^n$ for which the map

$$H \to H^n, \quad x \mapsto \Big((\lambda_i - T_i)x\Big)_{i=1}^n$$

is not bounded from below or for which the map

$$H^n \to H, \quad (x_i)_{i=1}^n \mapsto \sum_{i=1}^n (\lambda_i - T_i)x_i$$

is not surjective. The point spectrum $\sigma_p(T)$ of T consists of those points $\lambda \in \mathbb{C}^n$ for which the first of the above two maps is not injective.

Let K be a compact subset of $\mathbb{C}^n$. We write $M(K)$ for the Banach space of all complex regular Borel measures on K. We denote by $M^+(K)$ and $M_1^+(K)$ the subsets of $M(K)$ consisting of all positive measures and probability measures, respectively. For $\mu \in M^+(K)$ and $1 \leq q < \infty$, we define $P^q(\mu)$ as the closure of the polynomials in $L^q(\mu)$, while $P^\infty(\mu)$ stands for the w^*-closure of the polynomials in $L^\infty(\mu)$. The space $P^\infty(\mu)$ is regarded as the norm dual of the quotient Banach space $Q(\mu) = L^1(\mu)/{}^\perp P^\infty(\mu)$.

We denote by $A(B)$ the Banach algebra of all continuous complex-valued functions on $\overline{B}$ which are analytic on B, equipped with the supremum norm. The Banach algebra $H^\infty(B)$ of all bounded analytic functions on B, equipped with the supremum norm $\|f\| = \sup\{|f(z)|;\ z \in B\}$, is a w^*-closed subspace of $L^\infty(B)$ regarded as the dual space of $L^1(B)$. Here $L^1(B)$ and $L^\infty(B)$ are formed with respect to the $2n$-dimensional Lebesgue measure. We identify $H^\infty(B)$ with the norm dual of the Banach space $Q = L^1(B)/{}^\perp H^\infty(B)$.

A sequence (f_k) in $H^\infty(B)$ is a w^*-zero sequence if and only if (f_k) is norm-bounded and converges to zero pointwise (or equivalently, uniformly on all compact subsets of B). For $\lambda \in B$ and $k \in \mathbb{N}^n$, the w^*-continuous linear functionals

$$\mathcal{E}_\lambda \colon H^\infty(B) \to \mathbb{C}, \quad f \mapsto f(\lambda),$$
$$\mathcal{E}_\lambda^{(k)} \colon H^\infty(B) \to \mathbb{C}, \quad f \mapsto f^{(k)}(\lambda)$$

are regarded as elements in Q. A subset M of $\mathbb{C}^n$ is called dominating in B if $\|f\| = \sup\{|f(z)|;\ z \in M \cap B\}$ holds for all functions $f \in H^\infty(B)$, or equivalently, if the closed absolutely convex hull of the set of all point evaluations $\mathcal{E}_\lambda$ ($\lambda \in M \cap B$) is the closed unit ball in Q.

A Montel sequence (f_k) is a sequence (f_k) in $A(B)$ that is a w^*-zero sequence in $H^\infty(B)$. A measure $\mu \in M(S)$ is a Henkin measure (also called A-measure) if

$$\lim_{k \to \infty} \int_S f_k d\mu = 0$$

for each Montel sequence (f_k). For each Henkin measure $\mu \in M(S)$, there is a unique w^*-continuous algebra homomorphism $r = r_\mu: H^\infty(B) \to P^\infty(\mu)$ extending the restriction map

$$A(B) \to C(S), \quad f \mapsto f|S$$

from $A(B)$ into the Banach algebra of all continuous complex-valued functions on S. For details on Henkin measures we refer the reader to Chapter 9 in [18] and to §11.4 in [10].

The Banach space $L(H)$ of all continuous linear operators on H is the norm dual of the Banach space $\mathcal{C}^1(H)$ of all trace-class operators on H via the duality

$$\mathcal{C}^1(H) \times L(H) \to \mathbb{C}, \quad (A, B) \mapsto \mathrm{Tr}(AB).$$

Let $T = (T_1, \ldots, T_n) \in L(H)^n$ be a commuting system. The smallest unital w^*-closed subalgebra $\mathfrak{A}_T$ of $L(H)$ containing $T_1, \ldots, T_n$ is regarded as the norm dual of the Banach space $Q_T = \mathcal{C}^1(H)/{}^\perp\mathfrak{A}_T$. For $x, y \in H$, we denote by $[x \otimes y] \in Q_T$ the equivalence class of the rank-one operator $H \to H$, $\xi \mapsto \langle \xi, y\rangle x$. Let p, q be any cardinal numbers with $1 \leq p, q \leq \aleph_0$. The algebra $\mathfrak{A}_T$ possesses property $(\mathbb{A}_{p,q})$ if, for each matrix (L_{ij}) of functionals $L_{ij} \in Q_T$ $(0 \leq i < p,\ 0 \leq j < q)$, there are vectors $(x_i)_{0 \leq i < p}$ and $(y_j)_{0 \leq j < q}$ in H solving the equations

$$L_{ij} = [x_i \otimes y_j] \quad (0 \leq i < p,\ 0 \leq j < q).$$

If $p = q$, then we write $\mathbb{A}_p$ instead of $\mathbb{A}_{p,q}$.

Suppose that there is a w^*-continuous algebra homomorphism $\Phi: H^\infty(B) \to L(H)$ extending the polynomial functional calculus of T. For $x, y \in H$,

$$x \otimes y: H^\infty(B) \to \mathbb{C}, \quad f \mapsto \langle \Phi(f)x, y\rangle$$

defines an element in Q with $\Phi_*([x \otimes y]) = x \otimes y$, where $\Phi_*: Q_T \to Q$ is the predual of $\Phi: H^\infty(B) \to \mathfrak{A}_T$. If Φ is isometric, then the latter map is a dual algebra isomorphism, that is, an isometric isomorphism of Banach algebras and a w^*-homeomorphism. In this case the properties $(\mathbb{A}_{p,q})$ are equivalent to corresponding factorization properties of the elements of the predual Q of $H^\infty(B)$.

For a subset $M \subset L(H)$, we denote by $\mathrm{Lat}(M)$ the set of all closed linear subspaces that are invariant for each operator $T \in M$. The same notation is used for the invariant subspace lattice of a single operator or a commuting tuple of operators on H. A subalgebra $\mathfrak{A} \subset L(H)$ is reflexive if $\mathfrak{A} = \mathrm{Alg\,Lat}(\mathfrak{A})$, where the space on the right consists of all operators $T \in L(H)$ with $\mathrm{Lat}(\mathfrak{A}) \subset \mathrm{Lat}(T)$. A commuting n-tuple $T = (T_1, \ldots, T_n) \in L(H)^n$ of continuous linear operators on H

is called reflexive if the smallest WOT-closed subalgebra W_T of $L(H)$ containing $T_1, \dots, T_n$ and the identity operator on H is reflexive.

1. Spherical dilations and co-isometric extensions

Let $T \in L(H)^n$ be a commuting system of Hilbert space operators. A spherical dilation of T is a commuting system $N \in L(K)^n$ of normal operators on a Hilbert space K containing H such that $\sigma(N) \subset S$ and such that

$$T^\alpha = PN^\alpha | H \quad (\alpha \in \mathbb{N}^n),$$

where P is the orthogonal projection from K onto H. Each commuting system $T \in L(H)^n$ with a spherical dilation is a spherical contraction, that is,

$$\sum_{i=1}^n \|T_i x\|^2 \leq \|x\|^2 \quad (x \in H).$$

Conversely, for $n \geq 2$, not each spherical contraction $T \in L(H)^n$ possesses a spherical dilation (see [12]). But it follows from the work of Athavale [1], Vasilescu [20], Müller and Vasilescu [17] that each spherical contraction $T \in L(H)^n$ that satisfies certain natural positivity conditions possesses a spherical dilation.

More precisely, define $M_T: L(H) \to L(H)$, $X \mapsto \sum_{i=1}^n T_i^* X T_i$, and consider the operators

$$\Delta_T^{(m)} = (1 - M_T)^m (1_H) \quad (m = 0, 1, 2, \dots).$$

The system T is a spherical contraction if and only if $\Delta_T^{(1)} \geq 0$. Each commuting system $T \in L(H)^n$ that satisfies the positivity conditions $\Delta_T^{(1)} \geq 0$ and $\Delta_T^{(n)} \geq 0$ possesses a spherical dilation [17].

A commuting system $T \in L(H)^n$ is a spherical isometry if

$$\sum_{i=1}^n \|T_i x\|^2 = \|x\|^2 \quad (x \in H).$$

By a result of Athavale [1] each spherical isometry $T \in L(H)^n$ is subnormal, and its minimal normal extension is a spherical unitary, that is, a commuting tuple of normal operators with Taylor spectrum contained in the unit sphere. We call a tuple $T \in L(H)^n$ a spherical co-isometry if the adjoint system $T^* = (T_1^*, \dots, T_n^*) \in L(H)^n$ is a spherical isometry.

Each commuting tuple $T \in L(H)^n$ with a spherical dilation possesses a spherical co-isometric extension, that is, there is a spherical co-isometry $C \in L(K)^n$ on a

Hilbert space K containing H such that T is the restriction of C to one of its closed invariant subspaces (Lemma 3.1 in [14]). Using the above cited result of Athavale [1] on spherical isometries one easily obtains that, conversely, each commuting system $T \in L(H)^n$ with a spherical co-isometric extension possesses a spherical dilation.

Suppose that $T \in L(H)^n$ possesses a spherical dilation. Then T satisfies von Neumann's inequality over the unit ball, that is,

$$\|p(T)\| \leq \|p\|_{\infty,B} \quad (p \in \mathbb{C}[z]),$$

and hence T possesses a contractive functional calculus over the ball algebra $\Psi: A(B) \to L(H)$. We say that T is of class $C_{0\cdot}$ if $\left(\Psi(f_k)\right)$ tends to zero strongly for each Montel sequence (f_k) in $A(B)$. By definition T is of class $C_{\cdot 0}$ if the adjoint tuple T^* is of class $C_{0\cdot}$, or equivalently, if $\left(\Psi(f_k)^*\right)$ tends to zero strongly for each Montel sequence (f_k) in $A(B)$. For single contractions these definitions yield the classical notions of contractions of class $C_{0\cdot}$ and $C_{\cdot 0}$, respectively. The tuple T is said to be absolutely continuous if Ψ extends to a w^*-continuous algebra homomorphism

$$\Phi: H^\infty(B) \to L(H).$$

If T is of class $C_{0\cdot}$, then T is absolutely continuous and its w^*-continuous $H^\infty(B)$-functional calculus Φ is a $C_{0\cdot}$-representation, that is, $\left(\Phi(f_k)\right)$ tends to zero strongly for each w^*-zero sequence (f_k) in $H^\infty(B)$ (see Lemma 1.1 and the beginning of §2 in [14]). A similar remark holds in the case when T is $C_{\cdot 0}$. We shall say that T is a spherical contraction of class $\mathbb{A}$ if T is absolutely continuous and its unique w^*-continuous $H^\infty(B)$-functional calculus Φ is isometric.

A spherical co-isometric extension $C \in L(K)^n$ of a spherical contraction $T \in L(H)^n$ is called minimal if there is no reducing subspace $M \neq K$ for C that contains H. In [14] (Lemma 3.2) it was shown that a minimal spherical dilation of an absolutely continuous spherical contraction T is absolutely continuous again. In exactly the same way one obtains the following result.

LEMMA 1.1. *Let $C \in L(K)^n$ be a minimal spherical co-isometric extension of T. If T is absolutely continuous, then C is absolutely continuous.*

In the remainder of Section 1 we denote by $T \in L(H)^n$ a commuting tuple with a spherical dilation and a fixed minimal spherical co-isometric extension $C \in L(K)^n$.

There is a unique orthogonal decomposition

$$K = \mathcal{S} \oplus \mathcal{R}$$

with reducing subspaces $\mathcal{S}, \mathcal{R} \in \mathrm{Lat}(C)$ for C such that $C|\mathcal{S}$ is completely non-unitary and $C|\mathcal{R}$ is a spherical unitary (see [14]). As a completely non-unitary spherical isometry the tuple $S = C^*|\mathcal{S}$ is of class $C_{\cdot 0}$ (Corollary 2.4 in [14]). Therefore the spherical co-isometry C is of the form

$$C = S^* \oplus R \in L(\mathcal{S} \oplus \mathcal{R})^n,$$

where S^* is of class $C_{0\cdot}$ and R is a spherical unitary. The $A(B)$-functional calculi of T, S^* and R satisfy

$$\Psi(f) = \Big(\Psi_{S^*}(f) \oplus \Psi_R(f)\Big)|H \quad (f \in A(B)).$$

We use the notation

$$\Psi_R: L^\infty(S, \mu) \to W^*(R)$$

also for the isomorphism of von Neumann algebras associated with the spherical unitary R. We denote by $A \in L(K)$ and $Q \in L(K)$ the orthogonal projections of K onto $\mathcal{R}$ and $\mathcal{S}$, respectively.

As in the one-variable case [9] our invariant-subspace constructions will be based on the following observation concerning the structure of the unitary part of the minimal co-isometric extension.

LEMMA 1.2. *Suppose that $\mathcal{R} \neq \{0\}$. Then there is a probability measure μ in $M_1^+(S)$ such that $\mathcal{R}$ contains a reducing subspace $\mathcal{R}_0$ for R with*

(a) *$R_0 = R|\mathcal{R}_0$ is unitarily equivalent to the tuple M_z given by the multiplication with the coordinate functions on $L^2(\mu)$,*

(b) *the subspace $\mathcal{R}_0^+$ of $\mathcal{R}_0$ corresponding to $P^2(\mu)$ under the above unitary equivalence satisfies $\mathcal{R}_0^+ \subset \overline{AH}$.*

If T is absolutely continuous, then μ is necessarily a Henkin measure.

Proof. Obviously, $\overline{AH} \in \mathrm{Lat}(R)$. The minimality assumption on C implies that R is the minimal normal extension of $R|\overline{AH}$.

Choose a separating unit vector e for $W^*(R)$ in $\overline{AH}$ (cf. Proposition V.17.14 in [11]) and denote by $\mu \in M_1^+(S)$ the associated scalar spectral measure for R, that is,

$$\mu(A) = \Big\langle E\Big(A \cap \sigma(R)\Big)e, e\Big\rangle \quad (A \in \mathcal{B}(S)),$$

where E is the projection-valued spectral measure of R. The space $\mathcal{R}_0 = \vee(R^k R^{*m} e;\ k, m \in \mathbb{N}^n)$ is reducing for R, and the unitary operator $W: L^2(\mu) \to \mathcal{R}_0$ satisfying

$$Wp(z, \overline{z}) = p(R, R^*)e$$

for all polynomials p in $2n$ variables intertwines $M_z \in L(L^2(\mu))^n$ and $R|\mathcal{R}_0$.

Suppose that T is absolutely continuous. By Lemma 1.1 the tuple R is absolutely continuous, hence also M_z on $L^2(\mu)$. Then, for each Montel sequence (f_k) in $A(B)$,

$$\int_S f_k d\mu = \langle f_k, 1\rangle_{L^2(\mu)} \xrightarrow{k} 0,$$

and the proof is complete.

□

Let us fix a probability measure $\mu \in M_1^+(S)$ as in the preceding proof. For any pair of vectors $x, y \in \mathcal{R}$, we write $x \cdot y \in L^1(\mu)$ for the Radon–Nikodym derivative of the measure

$$\mu_{x,y}: \mathcal{B}(S) \to \mathbb{C}, \quad \mu_{x,y}(A) = \langle E\Big(A \cap \sigma(R)\Big)x, y\rangle$$

with respect to μ, and we denote by

$$x \odot y = [x \cdot y]$$

the induced equivalence in $Q(\mu) = L^1(\mu)/{}^{\perp}P^\infty(\mu)$. For $x \in \mathcal{R}_0$, let $\{x\}$ be the function in $L^2(\mu)$ which corresponds to x under the unitary operator $W: L^2(\mu) \to \mathcal{R}_0$. An elementary exercise shows that $x \cdot y = \{x\}\overline{\{y\}}$ for all vectors $x, y \in \mathcal{R}_0$.

Our factorization results are based on the following measure-theoretic observation.

LEMMA 1.3. *Let $\rho \in M_1^+(X)$ be a probability measure on a compact set $X \subset \mathbb{C}^n$. For each $L \in Q(\rho)$ and each $\varepsilon > 0$, there are functions $f, g \in P^2(\rho)$ with $\|f\|$, $\|g\| \leq \|L\|^{1/2}$ and*

$$\|L - [f\overline{g}]\| < \varepsilon.$$

Proof. Fix a real number $r > \|L\|^{1/2}$. It suffices to find $f, g \in P^2(\rho)$ with $\|f\|$, $\|g\| < r$ and $\|L - [f\overline{g}]\| < \varepsilon$.

Since $L^\infty(\rho) \subset L^1(\rho)$ is dense, we may suppose that $L = [\ell]$, where $\ell \in L^\infty(\rho)$ satisfies $\|\ell\|_1 < r^2$. Because $\|\ell\|_1 = \lim_{q \to 1} \|\ell\|_q$ there is an odd integer $p \geq 3$ such that $\|\ell\|_q < r^2$ holds for the conjugate exponent q of p. By Hahn–Banach there is a function $h \in L^q(\rho)$ with $\ell - h \in P^p(\rho)^-$ and

$$\|h\|_q = \sup\Big\{\Big|\int_X fhd\rho\Big|;\ f \in P^p(\rho) \text{ with } \|f\|_p \leq 1\Big\}.$$

Exactly as in the one-variable case (Lemma V.4.3 in [11]) one can show that there is a function $f \in P^2(\rho)$ with $|h| = |f|^2$ ρ-almost everywhere. Note that $[h] = [\ell] = L$ in $Q(\rho)$ and that

$$\|h\|_1 \leq \|h\|_q \leq \|\ell\|_q < r^2.$$

The measurable function $G: X \to \mathbb{C}$, defined by

$$G(z) = \overline{h(z)}/\overline{f(z)} \quad \text{if } f(z) \neq 0, \quad G(z) = 0 \text{ otherwise},$$

satisfies $|G|^2 = |h|$ and $h = f\overline{G}$ ρ-almost everywhere. Define $g \in P^2(\rho)$ as the image of $G \in L^2(\rho)$ under the orthogonal projection from $L^2(\rho)$ onto $P^2(\rho)$. To complete the proof observe that

$$[f\overline{g}] = [f\overline{G}] = [h] = L.$$

□

The following result on the possible boundary values of functions in $A(B)$ follows easily from Chapter 15 in [19] (see Lemma 1.6 in [15]).

LEMMA 1.4. *Let $\kappa: S \to \mathbb{R}$ be a Borel measurable function with $c \leq \kappa \leq d$, where $c, d > 0$ are real numbers. For any finite positive Borel measure ρ on S and any real number $\varepsilon > 0$, there is a function $f \in A(B)$ with $|f| \leq d$ on B and*

$$\rho(\{z \in S;\ \kappa(z) \neq |f(z)|\}) < \varepsilon.$$

In the remainder of Section 1, we shall always suppose that the unitary part R of the spherical co-isometry C is non-trivial. We shall apply the previous two results to the scalar spectral measure $\mu \in M_1^+(S)$ of the spherical unitary R fixed before. In the setting of Lemma 1.2, let $A_0 \in L(K)$ and $A_1 \in L(K)$ be the orthogonal projections from K onto $\mathcal{R}_0$ and $\mathcal{R} \ominus \mathcal{R}_0$, respectively.

PROPOSITION 1.5. *Let $L \in Q(\mu)$ and let $0 < \rho < 1$ be a real number. For any given vectors $a \in H$, $b \in \mathcal{R}$, and any $\varepsilon > 0$, there are vectors $x \in H$, $c \in \mathcal{R}$, and a Borel set $Z \subset S$ with $\mu(Z) < \varepsilon$, $c - b \in \mathcal{R}_0$, and*

$$\|L - A(a+x) \odot c\| < \varepsilon, \quad \|x\| \leq 2\|L - Aa \odot b\|^{1/2},$$
$$\|Qx\| < \varepsilon, \quad \|A_1 x\| < \varepsilon, \quad \|c\| \leq (1/\rho)(\|b\| + \|L - Aa \odot b\|^{1/2}),$$
$$|\{A_0(a+x)\}| \geq \rho|\{A_0 a\}| \quad \mu\text{-a.e. on } S \setminus Z.$$

If $a = 0$ and $b = 0$, then one can achieve that even

$$\|x\| \leq \|L\|^{1/2}, \quad \|c\| \leq \|L\|^{1/2}.$$

Proof. Let $\delta > 0$ be arbitrary. Set $d = \|L - Aa \odot b\|^{1/2}$. By Lemma 1.3 there are $\alpha, \beta \in \mathcal{R}_0^+$ with $\|\alpha\|$, $\|\beta\| \leq d$ and

$$\|L - Aa \odot b - \alpha \odot \beta\| < \delta.$$

Fix $x' \in H$ with $\|\alpha - Ax'\| < \delta$ and choose a Montel sequence (θ_k) in the closed unit ball of $A(B)$ with $(|\theta_k|) \xrightarrow{k} 1$ μ-almost everywhere on S (Theorem 3.5 in [19]). Since

$$\Psi_R(\theta_k)\alpha \cdot \Psi_R(\theta_k)\beta = |\theta_k|^2\{\alpha\}\overline{\{\beta\}} \xrightarrow{k} \{\alpha\}\overline{\{\beta\}}$$

in $L^1(\mu)$, we can choose a natural number k such that the vectors defined by

$$u = \Psi_R(\theta_k)\alpha \in \mathcal{R}_0^+,\ v = \Psi_R(\theta_k)\beta \in \mathcal{R}_0^+,\ \tilde{x} = \Psi(\theta_k)x' \in H$$

satisfy the estimates $\|Q\tilde{x}\| = \|\Psi_{S^*}(\theta_k)Qx'\| < \delta/2$ and

$$\|L - Aa \odot b - u \odot v\| < \delta.$$

Note that $\|u\|$, $\|v\| \leq d$ and $\|u - A\tilde{x}\| < \delta$.

Let us choose a real number $\eta > 0$ such that

$$\int_Z \left(|\{A_0a\}\{A_0b\}| - |\{A_0\tilde{x}\}\{v\}|\right) d\mu < \delta$$

for each Borel set $Z \subset S$ with $\mu(Z) < \eta$. By Lemma 1.4 there are a Borel set $Z \subset S$ with $\mu(Z) < \eta$ and a function $f \in A(B)$ with $\|f\| \leq 1 + \rho$ such that, for $z \in S \setminus Z$,

$$|f(z)| = \begin{cases} 1+\rho; & |\{A_0a\}(z)| \leq |\{A_0\tilde{x}\}(z)| \\ 1-\rho; & \text{otherwise.} \end{cases}$$

Here as in the following we write $\{A_0a\}$, $\{A_0\tilde{x}\}$,..., when we really mean some fixed representatives of the corresponding equivalence classes in $L^2(\mu)$. Define $x = \Psi(f)\tilde{x} \in H$. Note that $\|Qx\| < \delta$, $\|x\| \leq (1+\rho)\|\tilde{x}\|$, and that

$$\begin{aligned} \{A_0x\} &= \{\Psi_R(f)A_0\tilde{x}\} = f\{A_0\tilde{x}\}, \\ \|A_1x\| &= \|\Psi_R(f)A_1\tilde{x}\| \leq (1+\rho)\|A_1(A\tilde{x} - u)\| < 2\delta. \end{aligned}$$

For $z \in S \setminus Z$ with $|\{A_0a\}(z)| \leq |\{A_0\tilde{x}\}(z)|$, we have

$$|\{A_0(a+x)\}(z)| \geq |(f\{A_0\tilde{x}\})(z)| - |\{A_0\tilde{x}\}(z)| = \rho|\{A_0\tilde{x}\}(z)|.$$

For the remaining points z in $S \setminus Z$, we obtain

$$|\{A_0(a+x)\}(z)| \geq |\{A_0a\}(z)| - |(f\{A_0\tilde{x}\})(z)| \geq \rho|\{A_0a\}(z)|.$$

Combining these two estimates we obtain on $S \setminus Z$

$$|\{A_0(a+x)\}| \geq \rho \max(|\{A_0\tilde{x}\}|, |\{A_0 a\}|).$$

Set $h_0 = (A_0 a) \cdot (A_0 b) + (A_0 \tilde{x}) \cdot v \in L^1(\mu)$. Define $h: S \to \mathbb{C}$ by

$$\overline{h(z)} = h_0(z)/\{A_0(a+x)\}(z)$$

if $z \in S \setminus Z$ with $\{A_0(a+x)\}(z) \neq 0$, and $h(z) = 0$ otherwise. Then $h \in L^2(\mu)$ and there is a vector $c_1 \in \mathcal{R}_0$ with $h = \{c_1\}$ and

$$\|c_1\| = \|h\| \leq (\|A_0 b\| + \|v\|)/\rho.$$

Our choices guarantee that

$$\|A_0(a+x) \cdot c_1 - h_0\|_{L^1(\mu)} < \delta.$$

Define $c = c_1 + A_1 b \in \mathcal{R}$. Since $r \cdot s = 0$ for each pair of vectors $r \in \mathcal{R}_0$, $s \in \mathcal{R} \ominus \mathcal{R}_0$, we obtain that

$$\begin{aligned} &\|L - A(a+x) \odot c\| \\ &\qquad \leq \|L - Aa \odot b - A\tilde{x} \odot v\| + \\ &\qquad\qquad + \|Aa \odot b + A_0\tilde{x} \odot v - A_0(a+x) \odot c_1 - A_1(a+x) \odot A_1 b\|. \end{aligned}$$

We abbreviate the first term on the right-hand side by r and the second by r'. Then

$$\begin{aligned} r &\leq \|L - Aa \odot b - u \odot v\| + \|(u - A\tilde{x}) \odot v\| \leq \delta(1+d), \\ r' &\leq \|[h_0] - A_0(a+x) \odot c_1\| + \|A_1 x \odot A_1 b\| \leq \delta(1 + 2\|A_1 b\|). \end{aligned}$$

Furthermore, $c - b = c_1 - b + A_1 b = c_1 - A_0 b \in \mathcal{R}_0$ and

$$\begin{aligned} \|x\| &\leq (1+\rho)\|u - (u - A\tilde{x}) + Q\tilde{x}\| \leq (1+\rho)(d + 2\delta), \\ \|c\|^2 &= \|c_1\|^2 + \|A_1 b\|^2 \leq (1/\rho)^2(\|b\| + d)^2. \end{aligned}$$

A careful analysis of the above proof shows that in the case where $a = 0$ and $b = 0$ we can achieve that

$$\|c\| = \|c_1\| \leq \|v\|/(1+\rho).$$

This clearly suffices to prove the extra assertion in this case.

□

An iterative application of Proposition 1.5 allows an almost factorization of finite systems in $Q(\mu)$.

PROPOSITION 1.6. *Let $N \in \mathbb{N}$, let $L_1, \ldots, L_N \in Q(\mu)$, and let $\varepsilon > 0$, $\mu_1, \ldots, \mu_N > 0$ be given real numbers. Suppose that $a \in H$ and $b_1, \ldots, b_N \in \mathcal{R}$ satisfy*

$$\|L_k - Aa \odot b_k\| < \mu_k \quad (k = 1, \ldots, N).$$

Then there are vectors $x \in H$ and $y_1, \ldots, y_N \in \mathcal{R}$ with

$$\|L_k - A(a+x) \odot y_k\| < \varepsilon, \quad \|Qx\| < \varepsilon, \quad \|A_1 x\| < \varepsilon, \quad \|x\| < 2\sum_{i=1}^{N} \mu_i^{1/2},$$

$$\|y_k\| < \|b_k\| + \mu_k^{1/2}, \quad y_k - b_k \in \mathcal{R}_0$$

for $k = 1, \ldots, N$.

Proof. Fix positive real numbers $\eta < \varepsilon$, $\rho < 1$. Apply Proposition 1.5 to obtain $x_1 \in H$, $c_1 \in \mathcal{R}$, and a Borel set $Z_1 \subset S$ with $\mu(Z_1) < \eta$ and

$$\|L_1 - A(a+x_1) \odot c_1\| < \varepsilon, \quad \|x_1\| < 2\mu_1^{1/2}, \quad \|Qx_1\| < \varepsilon, \quad c_1 - b_1 \in \mathcal{R}_0,$$
$$\|c_1\| < \|b_1\| + \mu_1^{1/2}, \quad \|A_1 x_1\| < \eta, \quad |\{A_0(a+x_1)\}| \geq \rho|\{A_0 a\}| \quad \mu\text{-a.e. on } S \setminus Z_1.$$

Define $f_1 \in L^\infty(S, \mu)$ by setting

$$\overline{f_1} = \{A_0 a\}/\{A_0(a+x_1)\}$$

on $S \setminus Z_1$ (with $t/0 = 0$ for all $t \in \mathbb{C}$) and $f_1|Z_1 = 0$. Set

$$b_k(1) = A_1 b_k + \Psi_R(f_1) A_0 b_k \in \mathcal{R} \quad (k = 1, \ldots, N).$$

If η is chosen small enough, then one obtains the estimates

$$\begin{aligned}
&\|L_k - A(a+x_1) \odot b_k(1)\| \\
&\quad \leq \|L_k - A_1 a \odot A_1 b_k - A_0(a+x_1) \odot \Psi_R(f_1) A_0 b_k\| + \eta \|A_1 b_k\| \\
&\quad \leq \|L_k - Aa \odot b_k\| + \int_{Z_1} |\{A_0 a\}\{A_0 b_k\}| \, d\mu + \eta \|A_1 b_k\| < \mu_k
\end{aligned}$$

for $k = 1, \ldots, N$.

In the next step we repeat the above constructions, but this time with L_1, a, b_1 replaced by $L_2, a + x_1, b_2(1)$, and $b_1, \ldots, b_N$ replaced by $b_1(1), \ldots, b_N(1)$. Inductively we obtain vectors $x_1, \ldots, x_N \in H$, $c_1, \ldots, c_N \in \mathcal{R}$, and Borel sets $Z_1, \ldots, Z_N \subset S$ with

$$\Big\|L_k - A\Big(a + \sum_{i=1}^{k} x_i\Big) \odot c_k\Big\| < \varepsilon, \quad \|x_k\| < 2\mu_k^{1/2}, \quad \|Qx_k\| < \varepsilon,$$

$$c_k - b_k(k-1) \in \mathcal{R}_0, \quad \|c_k\| < \|b_k(k-1)\| + \mu_k^{1/2}, \quad \|A_1 x_k\| < \varepsilon,$$

$$\Big|\Big\{A_0\Big(a + \sum_{i=1}^{k} x_i\Big)\Big\}| \geq \rho\Big|\Big\{A_0\Big(a + \sum_{i=1}^{k-1} x_i\Big)\Big\}\Big| \quad \mu\text{-a.e. on } S \setminus Z_k$$

for $k = 1, \ldots, N$. Here

$$b_k(k-1) = A_1 b_k + \Psi_R(f_{k-1} \cdot \ldots \cdot f_1) A_0 b_k \quad (= b_1 \text{ for } k = 1),$$

where $f_1, \ldots, f_N \in L^\infty(S, \mu)$ are defined by setting

$$\overline{f_k} = \Big\{A_0\Big(a + \sum_{i=1}^{k-1} x_i\Big)\Big\} \Big/ \Big\{A_0\Big(a + \sum_{i=1}^{k} x_i\Big)\Big\}$$

on $S \setminus Z_k$ and $f_k = 0$ on Z_k. The sets $Z_1, \ldots, Z_N$ are chosen in such a way that

$$\int_{Z_k} \Big|\Big\{A_0\Big(a + \sum_{i=1}^{j} x_i\Big)\Big\}\{A_0 c_j\}\Big| \, d\mu < \varepsilon / N$$

for $k = 2, \ldots, N$ and $j = 1, \ldots, k-1$.

Let us define $x = \sum_{i=1}^{N} x_i \in H$, $y_N = c_N \in \mathcal{R}$, and

$$y_k = A_1 c_k + \Psi_R(f_{k+1} \cdots f_N) A_0 c_k \in \mathcal{R} \quad (1 \leq k < N).$$

Then we obtain the estimates

$$\|y_k\| \leq (1/\rho)^{N-1}(\|b_k\| + \mu_k^{1/2})$$

for $k = 1, \ldots, N$. For $k = 1, \ldots, N-1$, it follows that

$$\begin{aligned}
&|L_k - A(a+x) \odot y_k\| \\
&= \Big\|L_k - A_1\Big(a + \sum_{i=1}^{k} x_i\Big) \odot A_1 c_k - A_1\Big(\sum_{i=k+1}^{N} x_i\Big) \odot A_1 c_k - \\
&\quad - A_0(a+x) \odot \Psi_R(f_{k+1} \cdots f_N) A_0 c_k\Big\| \\
&\leq \Big\|L_k - A_1\Big(a + \sum_{i=1}^{k} x_i\Big) \odot A_1 c_k - A_0\Big(a + \sum_{i=1}^{k} x_i\Big) \odot A_0 c_k\Big\| + \\
&\quad + \sum_{i=k+1}^{N} \|A_1 x_i\| \|A_1 c_k\| + \sum_{j=k+1}^{N} \int_{Z_j} \Big|\Big\{A_0\Big(a + \sum_{i=1}^{k} x_i\Big)\Big\}\{A_0 c_k\}\Big| \, d\mu \\
&< \varepsilon(2 + N\|b_k\|),
\end{aligned}$$

while, for $k = N$, we have the inequality

$$\|L_N - A(a+x) \odot y_N\| < \varepsilon.$$

Furthermore, our choices guarantee that $y_k - b_k \in \mathcal{R}_0$ for $k = 1, \ldots, N$.

By starting the proof with a sufficiently small ε and with suitable positive real numbers $\mu_i' < \mu_i$, $\rho < 1$, one easily obtains vectors $x \in H$ and $y_1, \ldots, y_N \in \mathcal{R}$ satisfying the estimates as stated in Proposition 1.6.

□

As in the one-variable case (§2 in [9]) one can easily improve the last result to an exact form.

COROLLARY 1.7. *Let $N \in \mathbb{N}$, let $L_1, \ldots, L_N \in Q(\mu)$, and let $\varepsilon > 0$, $\mu_1, \ldots, \mu_N > 0$ be real numbers. For given vectors $a \in H$ and $b_1, \ldots, b_N \in \mathcal{R}$ with*

$$\|L_k - Aa \odot b_k\| < \mu_k \quad (k = 1, \ldots, N),$$

there are vectors $x \in H$ and $c_1, \ldots, c_N \in \mathcal{R}$ with

$$L_k = A(a+x) \odot c_k, \quad \|Qx\| < \varepsilon, \quad \|A_1 x\| < \varepsilon, \quad \|x\| < 2\sum_{i=1}^{N} \mu_i^{1/2},$$

$$\|c_k\| < \|b_k\| + \mu_k^{1/2}, \quad c_k - b_k \in \mathcal{R}_0$$

for $k = 1, \ldots, N$.

Proof. Choose a sequence $(\varepsilon_j)_{j \geq 1}$ of positive real numbers $\varepsilon_j < 1$ such that

$$\sum_{j=1}^{\infty} \varepsilon_j^{1/2} < \varepsilon.$$

An inductive application of Proposition 1.6 yields sequences $(x_j)_{j \geq 1}$ in H and $(b_{jk})_{j \geq 1}$ in $\mathcal{R}$ $(k = 1, \ldots, N)$ with

$$\|x_1\| < 2\sum_{i=1}^{N} \mu_i^{1/2}, \quad \|b_{1k}\| < \|b_k\| + \mu_k^{1/2}, \quad b_{1k} - b_k \in \mathcal{R}_0$$

and

$$\Big\|L_k - A\Big(a + \sum_{i=1}^{j} x_i\Big) \odot b_{jk}\Big\| < \varepsilon_j, \quad \|Qx_j\| < \varepsilon_j, \quad \|A_1 x_j\| < \varepsilon_j, \quad \|x_{j+1}\| < 2N\varepsilon_j^{1/2},$$

$$\|b_{j+1,k}\| < \|b_{jk}\| + \varepsilon_j^{1/2}, \quad b_{j+1,k} - b_{jk} \in \mathcal{R}_0$$

for $j \geq 1$ and $k = 1, \dots, N$.

Note that $x = \sum_{j=1}^{\infty} x_j \in H$ exists and satisfies

$$\|x\| < 2\Big(\sum_{i=1}^{N} \mu_i^{1/2} + N\varepsilon\Big)$$

as well as $\|Qx\| < \varepsilon$, $\|A_1 x\| < \varepsilon$. For $j \geq 1$ and $k = 1, \dots, N$,

$$\|b_{jk}\| < \|b_k\| + \mu_k^{1/2} + \varepsilon.$$

After dropping to suitable subsequences we may therefore suppose that the weak limits

$$c_k = w - \lim_{j \to \infty} b_{jk} \in \mathcal{R} \quad (k = 1, \dots, N)$$

exist. Clearly, $\|c_k\| \leq \|b_k\| + \mu_k^{1/2} + \varepsilon$, $c_k - b_k \in \mathcal{R}_0$, and

$$A(a + x) \odot c_k = L_k$$

for $k = 1, \dots, N$.

Again, by starting the proof with suitable real numbers $\mu_k' < \mu_k$ and with a sufficiently small ε, one can prove the stated version of Corollary 1.7. □

In the next step we obtain simultaneous factorizations of infinitely many functionals in $Q(\mu)$.

Proposition 1.8. *Let $(L_k)_{k \geq 1}$ be a sequence in $Q(\mu)$ such that there are real numbers $\delta_k > \|L_k\|$ with $\delta = \sum_{k=1}^{\infty} \delta_k^{1/2} < \infty$. For any given vector $a \in H$, there are vectors $x \in H$ and $z_k \in \mathcal{R}$ $(k \geq 1)$ such that $\|x - a\| < 2\delta$ and*

$$L_k = Ax \odot z_k, \quad \|z_k\| < \delta_k^{1/2} \quad (k \geq 1).$$

Proof. Choose real numbers μ_k with

$$\|L_k\| < \mu_k < \delta_k \quad (k \geq 1).$$

Define a sequence of positive real numbers c_k $(k \geq 1)$ by

$$c_k^{1/2} = \min(1/2^k,\ \delta_k^{1/2} - \mu_k^{1/2})$$

and set

$$\varepsilon_{kj} = c_k c_j \quad (k > j \geq 1).$$

We obtain the inequalities

$$\sum_{k=j+1}^{\infty} \varepsilon_{kj}^{1/2} < \delta_j^{1/2} - \mu_j^{1/2} \quad (j \geq 1), \qquad \sum_{j=1}^{k-1} \varepsilon_{kj}^{1/2} < \delta_k^{1/2} - \mu_k^{1/2} \quad (k \geq 1).$$

An iterative application of Corollary 1.7 allows us to choose vectors $a_k \in H$, $b_{kj} \in \mathcal{R}$ $(k \geq j \geq 1)$ with

$$L_j = Aa_k \odot b_{kj} \quad (k \geq j \geq 1)$$

and such that (with $a_0 = a$)

$$\|a_k - a_{k-1}\| < 2\Big(\mu_k^{1/2} + \sum_{j=1}^{k-1} \varepsilon_{kj}^{1/2}\Big) \quad (k \geq 1),$$

$$\|b_{kj}\| < \|b_{k-1,j}\| + \varepsilon_{kj}^{1/2} \quad (k > j \geq 1),$$

$$\|b_{jj}\| < \mu_j^{1/2} \quad (j \geq 1).$$

Note that the sequence $(a_k)_{k\geq 1}$ converges and that its limit x satisfies

$$\|x - a\| < 2 \sum_{k=1}^{\infty} \delta_k^{1/2} = 2\delta.$$

Since

$$\|b_{kj}\| < \mu_j^{1/2} + \sum_{i=j+1}^{k} \varepsilon_{ij}^{1/2} < \delta_j^{1/2}$$

for $k \geq j \geq 1$, we can choose, for each $j \geq 1$, a weakly convergent subsequence of the sequence $(b_{kj})_{k\geq j}$. The limit z_j of this sequence satisfies $\|z_j\| < \delta_j^{1/2}$ and $L_j = Ax \odot z_j$.

□

The following version of Proposition 1.8 is more suitable for applications.

COROLLARY 1.9. *Let $\varepsilon > 0$ be given. There is a constant $c = c(\varepsilon) > 0$ such that, for each sequence $(L_k)_{k\geq 1}$ in $Q(\mu)$ and each vector $a \in H$, there are elements $x \in H$, $z_k \in \mathcal{R}$ $(k \geq 1)$ with $\|x - a\| < \varepsilon$ and*

$$L_k = Ax \odot z_k, \quad \|z_k\| \leq ck^2 \|L_k\| \quad (k \geq 1).$$

Proof. Set $d_k = \|L_k\|$ if $L_k \neq 0$ and $d_k = 1$ otherwise. Define $\beta = 2\sum_{j=1}^{\infty}(1/j^2)$ and

$$\delta_k = \varepsilon^2/\beta^2 k^4, \quad M_k = \delta_k\,(L_k/2d_k) \quad (k \geq 1).$$

Then $\|M_k\| < \delta_k$ for all $k \geq 1$, and by Proposition 1.8, there are vectors $x \in H$ and $b_k \in \mathcal{R}$ with $\|x - a\| < \varepsilon$ and

$$M_k = Ax \odot b_k, \quad \|b_k\| < \delta_k^{1/2} \quad (k \geq 1).$$

Then $L_k = Ax \odot z_k$ with $z_k = (2d_k/\delta_k)b_k$ and $\|z_k\| < (2/\delta_k^{1/2})d_k$ for $k \geq 1$. Hence we can choose $c(\varepsilon) = c/\varepsilon$ with a suitable universal constant $c > 0$.

□

Let us consider the particular case that $T \in L(H)^n$ is a spherical isometry. By the above cited result of Athavale [1] each such tuple is subnormal and its minimal normal extension is a spherical unitary. Thus in all the previous results we may assume that $K = \mathcal{R}$ and that $C = R \in L(\mathcal{R})^n$ is the minimal normal extension of T. Let $\mu \in M_1^+(S)$ be the scalar spectral measure for R chosen as in Lemma 1.2.

The w^*-continuous isomorphism of von Neumann algebras

$$\Psi_R\colon L^\infty(\mu) \to W^*(R)$$

induces a dual algebra isomorphism

$$\gamma\colon P^\infty(\mu) \to \mathfrak{A}_T.$$

We denote by $\gamma_*\colon Q_T \to Q(\mu)$ the predual of this map. It is easy to check that

$$\gamma_*([x \otimes y]) = x \odot y$$

for all $x, y \in H$, where $[x \otimes y]$ is the equivalence class of the rank-one operator

$$x \otimes y\colon H \to H, \quad \xi \mapsto \langle \xi, y\rangle x$$

in $Q_T = \mathcal{C}^1(H)/{}^\perp\mathfrak{A}_T$.

COROLLARY 1.10. *Let $T \in L(H)^n$ be a spherical isometry. Then the dual algebra $\mathfrak{A}_T$ possesses property $(\mathbb{A}_{1,\aleph_0})$. Furthermore, there is a universal constant $c > 0$ such that, for each $\varepsilon > 0$ and for each sequence $(L_k)_{k\geq 1}$ in Q_T, there are $x \in H$, $y_k \in H$ $(k \geq 1)$ with $\|x - a\| < \varepsilon$ and*

$$L_k = [x \otimes y_k], \quad \|y_k\| \leq (c/\varepsilon)k^2\|L_k\| \quad (k \geq 1).$$

Proof. By Corollary 1.9 there are vectors $x \in H$ and $z_k \in \mathcal{R}$ with

$$\gamma_*(L_k) = x \odot z_k \quad (k \geq 1)$$

satisfying the right growth conditions. Since

$$x \odot z_k = x \odot Pz_k \quad (k \geq 1),$$

where P denotes the orthogonal projection from K onto H, it suffices to replace the vectors z_k by the vectors $y_k = Pz_k$.

□

By a result of Bercovici [3] each commuting family of isometries on a Hilbert space is reflexive. Corollary 1.10 suggests that the same result should hold for spherical isometries. The answer is positive if $\sigma(T)$ is dominating in B or if T possesses an isometric $H^\infty(B)$-functional calculus [15]. But in the general case the question remains open.

2. Factorizations of type $(\mathbb{A}_{1,\aleph_0})$

Recall that by a spherical contraction of class $\mathbb{A}$ we mean a commuting tuple $T \in L(H)^n$ of Hilbert space operators such that T possesses a spherical dilation and a w^*-continuous isometric H^∞-functional calculus $\Phi: H^\infty(B) \to L(H)$. We write $\mathbb{A}_{1,\aleph_0}$ for the set of all spherical contractions $T \in L(H)^n$ of class $\mathbb{A}$ for which the dual algebra $\mathfrak{A}_T$ possesses property $(\mathbb{A}_{1,\aleph_0})$. Throughout this section $T \in L(H)^n$ will be a fixed spherical contraction of class $\mathbb{A}$ with minimal co-isometric extension

$$C = S^* \oplus R \in L(\mathcal{S} \oplus \mathcal{R})^n$$

as explained in Section 1. By Lemma 1.[illegible] the tuple C is absolutely continuous. The H^∞-functional calculi of C, S^*, and R satisfy

$$\Phi(f) = \Phi_C(f)|H = [\Phi_{S^*}(f) \oplus \Phi_R(f)]|H \quad (f \in H^\infty(B)).$$

For $x, y \in K = \mathcal{S} \oplus \mathcal{R}$, we regard the functional

$$x \otimes y: H^\infty(B) \to \mathbb{C}, \quad f \mapsto \langle \Phi_C(f)x, y \rangle$$

as an element in the predual $Q = L^1(B)/{}^\perp H^\infty(B)$ of $H^\infty(B)$. As before we denote by $Q \in L(K)$ and $A \in L(K)$ the orthogonal projections from K onto $\mathcal{S}$ and $\mathcal{R}$, respectively. We shall use that, for $x \in H$ and $w \in \mathcal{S}$, $b \in \mathcal{R}$,

$$x \otimes w = x \otimes Pw, \quad x \otimes b = x \otimes Pb,$$

where $P \in L(K)$ is the orthogonal projection from K onto H.

In the one-variable case the following definitions can be found in [9].

Definition 2.1. Let $T \in L(H)^n$ be a spherical contraction of class $\mathbb{A}$. For $0 \leq \theta < 1$, we define $\mathcal{E}_\theta^r(T)$ as the set of all elements $L \in Q$ for which there are sequences $(x_k)_{k\geq 0}$ and $(y_k)_{k\geq 0}$ in the closed unit ball of H with

(a) $\overline{\lim}_{k\to\infty}\|L - x_k \otimes y_k\| \leq \theta$,

(b) $(x_k \otimes z) \xrightarrow{k} 0$ for all $z \in H$,

(c) $(y_k) \xrightarrow{k} 0$ weakly.

A standard argument (cf. Remark 3.1 in [9]) shows that condition (c) in Definition 2.1 is actually superfluous.

In the case when $\mathcal{R} \neq \{0\}$ we fix a Henkin probability measure $\mu \in M_1^+(S)$ as explained in Lemma 1.2, and we denote by

$$r_*: Q(\mu) \longrightarrow Q$$

the predual of the canonical w^*-continuous algebra homomorphism

$$r: H^\infty(B) \longrightarrow P^\infty(\mu)$$

associated with μ. For $x, y \in \mathcal{R}$, let $x \odot y \in Q(\mu)$ be defined as explained in the section following Lemma 1.2. It is an elementary exercise to show that $r_*(x \odot y) = x \otimes y$ for all $x, y \in \mathcal{R}$. If $\mathcal{R} = \{0\}$, then we set $Q(\mu) = \{0\} = P^\infty(\mu)$.

Definition 2.2. Let $T \in L(H)^n$ be a spherical contraction of class $\mathbb{A}$ with minimal co-isometric extension $C = S^* \oplus R$, and let μ be a scalar spectral measure for R as before. Let $0 \leq \theta < \gamma \leq 1$ be real numbers. We say that T possesses property $F_{\theta,\gamma}^r$ if

$$\overline{\Gamma}\Big(\mathcal{E}_\theta^r(T) \bigcup r_*\{L \in Q(\mu);\ \|L\| \leq 1\}\Big)$$

contains the closed ball $\{L \in Q;\ \|L\| \leq \gamma\}$. If this ball is contained in the smaller set $\overline{\Gamma}\Big(\mathcal{E}_\theta^r(T)\Big)$, then T will be said to have property $E_{\theta,\gamma}^r$.

It will become clear in Section 3 (Corollary 3.7) that property $F_{\theta,\gamma}^r$ does not depend on the choice of the minimal co-isometric extension $C = S^* \oplus R$ of T. Our aim in this section is to show that spherical contractions with property $F_{\theta,\gamma}^r$ belong to the class $\mathbb{A}_{1,\aleph_0}$. The proof follows closely the corresponding one-variable ideas (cf. Proposition 3.4 in [9]).

PROPOSITION 2.3. *Suppose that T has property $F^r_{\theta,\gamma}$ for some $0 < \theta < \gamma \leq 1$. Let $L_1, \ldots, L_N \in Q$ and $\mu_1, \ldots, \mu_N > 0$ be given. For any vectors $a \in H$, $w_1, \ldots, w_N \in \mathcal{S}$, and $b_1, \ldots, b_N \in \mathcal{R}$ with*

$$\|L_j - a \otimes (w_j + b_j)\| < \mu_j \quad (j = 1, \ldots, N),$$

there are elements $a' \in H$, $w'_1, \ldots, w'_N \in \mathcal{S}$, and $b'_1, \ldots, b'_N \in \mathcal{R}$ such that

$$\|L_j - a' \otimes (w'_j + b'_j)\| < (\theta/\gamma)\mu_j, \quad \|a' - a\| < (3/\gamma^{1/2}) \sum_{i=1}^{N} \mu_i^{1/2},$$

$$\|w'_j - w_j\| < (\mu_j/\gamma)^{1/2}, \quad \|b'_j\| < \|b_j\| + (\mu_j/\gamma)^{1/2}$$

for $j = 1, \ldots, N$.

Proof. For $j = 1, \ldots, N$, define

$$L'_j = L_j - a \otimes (w_j + b_j), \quad d_j = \max\left(\|L'_j\|, \frac{\mu_j}{2}\right).$$

Choose $\varepsilon > 0$ such that

$$(\theta/\gamma)d_j + 1\varepsilon < (\theta/\gamma)\mu_j$$

and define $s_j = (\theta/\gamma)d_j + \varepsilon$ for all j. The functionals $(\gamma/d_j)L'_j$ $(j = 1, \ldots, N)$ can be approximated by absolute convex combinations of elements in $\mathcal{E}^r_\theta(T)$ and $r_*\Big(\{L \in Q(\mu); \|L\| \leq 1\}\Big)$. Hence there are integers $0 = k_0 < k_1 < \ldots < k_N$, elements $K_i \in \mathcal{E}^r_\theta(T)$, $\alpha_i \in \mathbb{C}$ $(i = 1, \ldots, k_N)$, and functions $f_j \in L^1(\mu)$ $(j = 1, \ldots, N)$ with

$$\Big\|L'_j - r_*([f_j]) - \sum_{i \in I_j} \alpha_i K_i\Big\| < \varepsilon/2$$

and

$$\|f_j\|_{L^1(S,\mu)} + \sum_{i \in I_j} |\alpha_i| < d_j/\gamma$$

for $j = 1, \ldots, N$ where $I_j = \{k_{j-1} + 1, \ldots, k_j\}$.

Since $K_i \in \mathcal{E}^r_\theta(T)$, one can choose sequences $(x^i_k)_{k \geq 0}$ and $(y^i_k)_{k \geq 0}$ $(i = 1, \ldots, k_N)$ in the closed unit ball of H such that, for $i = 1, \ldots, k_N$,

$$(x^i_k \otimes z) \xrightarrow{k} 0 \quad (z \in F), \qquad (y^i_k) \xrightarrow{k} 0 \quad \text{weakly}$$

and such that, for each tuple $\nu = (n_1, \ldots, n_{k_N}) \in \mathbb{N}^{k_N}$ and each $j = 1, \ldots, N$,

$$\Big\|L'_j - r_*([f_j]) - \sum_{i \in I_j} \alpha_i x^i_{n_i} \otimes y^i_{n_i}\Big\| < s_j.$$

For ν as above and $j = 1, \dots, N$, we define

$$A^j(\nu) = a \otimes b_j + r_*([f_j]) + \sum_{i \in I_j} \alpha_i A x^i_{n_i} \otimes y^i_{n_i},$$

$$Q^j(\nu) = a \otimes w_j + \sum_{i \in I_j} \alpha_i Q x^i_{n_i} \otimes y^i_{n_i}.$$

Then, for $j = 1, \dots, N$ and $\nu = (n_1, \dots, n_{k_N})$,

$$\|L_j - Q^j(\nu) - A^j(\nu)\| < s_j.$$

Choose complex numbers β_i $(i = 1, \dots, k_N)$ with $\beta_i^2 = \alpha_i$ and define

$$u_\nu = \sum_{i=1}^{k_N} \beta_i x^i_{n_i}, \quad v_{\nu,j} = \sum_{i \in I_j} \overline{\beta_i} Q y^i_{n_i}$$

for $j = 1, \dots, N$ and ν arbitrary.

Let $\eta > 0$ be arbitrary. Since S^* is of type $C_{0\cdot}$, one can choose inductively, for each index $i = 1, \dots, k_N$, a natural number $n_i \geq 1$ such that

$$\|Qx^i_{n_i} \otimes y^\ell_{n_\ell}\| < \eta, \quad \|x^i_{n_i} \otimes w_j\| < \eta, \quad \|Qa \otimes y^i_{n_i}\| < \eta,$$
$$\|x^i_{n_i} \otimes b_j\| < \eta, \quad |\langle x^i_{n_i}, x^\ell_{n_\ell}\rangle| < \eta, \quad |\langle Qy^i_{n_i}, y^\ell_{n_\ell}\rangle| < \eta$$

for all $i, \ell = 1, \dots, k_N$ with $i \neq \ell$ and each $j = 1, \dots, N$. By making η small enough, we can achieve that $\|u_\nu\|^2 < \sum_{j=1}^N \mu_j/\gamma$, $\|v_{\nu,j}\|^2 < \mu_j/\gamma$, and that $\|u_\nu \otimes b_j\| < \varepsilon$,

$$\|Q^j(\nu) - (a + u_\nu) \otimes (w_j + v_{\nu,j})\| < \varepsilon.$$

We fix a tuple $\nu = (n_1, \dots, n_{k_N})$ such that the above estimates hold, and define

$$a_1 = a + u_\nu \in H, \quad w'_j = w_j + v_{\nu,j} \in \mathcal{S}, \quad x^i = x^i_{n_i} \in H, \quad y^i = y^i_{n_i} \in H,$$
$$h_j = f_j + \sum_{i \in I_j} \alpha_i A x^i \cdot A y^i \in L^1(\mu)$$

for $j = 1, \dots, N$ and $i = 1, \dots, k_N$.

Since $\|h_j\| < d_j/\gamma$, Corollary 1.7 allows us to choose $x \in H$ and $b'_1, \dots, b'_N \in \mathcal{R}$ with

$$Aa_1 \odot b_j + [h_j] = A(a_1 + x) \odot b'_j, \quad \|Qx\| < \varepsilon/(\|w'_j\| + 1),$$

$$\|x\| < 2\sum_{i=1}^N (d_i/\gamma)^{1/2}, \quad \|b'_j\| < \|b_j\| + (d_j/\gamma)^{1/2}$$

for $j = 1, \ldots, N$. By comparing the definitions of $A^j(\nu)$ and h_j, we obtain that

$$(a_1 + x) \otimes b'_j = A^j(\nu) + u_\nu \otimes b_j.$$

Summarizing the previous estimates we conclude that

$$\begin{aligned} &\|L_j - (a_1 + x) \otimes (w'_j + b'_j)\| \\ &\quad = \|L_j - (a + u_\nu) \otimes (w_j + v_{\nu,j}) - (a_1 + x) \otimes b'_j - x \otimes w'_j\| \\ &\quad < \|L_j - (a + u_\nu) \otimes (w_j + v_{\nu,j}) - (a_1 + x) \otimes b'_j + u_\nu \otimes b_j\| + 2\varepsilon \\ &\quad < \|L_j - Q^j(\nu) - A^j(\nu)\| + 3\varepsilon < s_j + 3\varepsilon < (\theta/\gamma)\mu_j. \end{aligned}$$

Since

$$\|u_\nu + x\| \leq \|u_\nu\| + \|x\| < (3/\gamma^{1/2}) \sum_{i=1}^{N} \mu_i^{1/2},$$

and since, for $j = 1, \ldots, N$,

$$\|w'_j - w_j\| = \|v_{\nu,j}\| < (\mu_j/\gamma)^{1/2},$$

it suffices to define $a' = a + u_\nu + x \in H$ to complete the proof.

□

A standard procedure (cf. §3 in [9]) very similar to the constructions leading from Proposition 1.6 to Corollary 1.9 in Section 1, can be used to prove property $(\mathbb{A}_{1,\aleph_0})$. We indicate the main steps.

COROLLARY 2.4. *Suppose that T has property $F^r_{\theta,\gamma}$ for some $0 \leq \theta < \gamma \leq 1$. Let $L_1, \ldots, L_N \in Q$ and let $\mu_1, \ldots, \mu_N > 0$ be given real numbers. Then, for any given vectors $a \in H$, $w_1, \ldots, w_N \in \mathcal{S}$, and $b_1, \ldots, b_N \in \mathcal{R}$ with*

$$\|L_j - a \otimes (w_j + b_j)\| < \mu_j \quad (j = 1, \ldots, N),$$

there are vectors $a' \in H$, $w'_1, \ldots, w'_N \in \mathcal{S}$, and $b'_1, \ldots, b'_N \in \mathcal{R}$ such that

$$L_j = a' \otimes (w'_j + b'_j),$$

$$\|a' - a\| < 3\alpha \sum_{i=1}^{N} \mu_i^{1/2}, \quad \|w'_j - w_j\| < \alpha\mu_j^{1/2}, \quad \|b'_j\| < \|b_j\| + \alpha\mu_j^{1/2}$$

for $j = 1, \ldots, N$, where $\alpha = 1/(\gamma^{1/2} - \theta^{1/2})$.

Proof. We may of course assume that $\theta > 0$. Otherwise one can replace the numbers μ_j by numbers $\mu'_j < \mu_j$, and then replace $\theta = 0$ by a sufficiently small $\theta' > 0$.

An inductive application of Proposition 2.3 allows us to choose sequences $(a_k)_{k\geq 1}$ in H, $(w_{kj})_{k\geq 1}$ in $\mathcal{S}$, and $(b_{kj})_{k\geq 1}$ in $\mathcal{R}$ for $j = 1, \ldots, N$ such that (with $a_0 = a$, $w_{0j} = w_j$, and $b_{0j} = b_j$)

$$\|L_j - a_k \otimes (w_{kj} + b_{kj})\| < (\theta/\gamma)^k \mu_j,$$

$$\|a_k - a_{k-1}\| < (3/\gamma^{1/2})\Big(\sum_{i=1}^{N} \mu_i^{1/2}\Big)(\theta/\gamma)^{(k-1)/2},$$

$$\|w_{kj} - w_{k-1,j}\| < (\mu_j/\gamma)^{1/2}(\theta/\gamma)^{(k-1)/2},$$

$$\|b_{kj}\| < \|b_{k-1,j}\| + (\mu_j/\gamma)^{1/2}(\theta/\gamma)^{(k-1)/2}$$

for $j = 1, \ldots, N$ and $k \geq 1$.

The sequences $(a_k)_{k\geq 1}$ and $(w_{kj})_{k\geq 1}$ $(j = 1, \ldots, N)$ are Cauchy sequences and their limits a' and w'_j satisfy the right estimates. Since

$$\|b_{kj}\| < \|b_j\| + (\mu_j/\gamma)^{1/2} \sum_{i=0}^{k-1} (\theta/\gamma)^{i/2}$$

for all $j = 1, \ldots, N$ and all $k \geq 1$, we may suppose after dropping to suitable subsequences that the weak limits

$$b'_j = w - \lim_{k\to\infty} b_{kj} \quad (j = 1, \ldots, N)$$

exist. Then $\|b'_j\| \leq \|b_j\| + \alpha\mu_j^{1/2}$ and $a' \otimes (w'_j + b'_j) = L_j$ for $j = 1, \ldots, N$. □

Exactly as Proposition 1.8 was deduced from Corollary 1.7, the next result can be deduced from Corollary 2.4.

THEOREM 2.5. *Suppose that T satisfies property $F^r_{\theta,\gamma}$ for some $0 \leq \theta < \gamma \leq 1$. Let $(L_k)_{k\geq 1}$ be a sequence in Q and let $\delta_k > \|L_k\|$ be real numbers with $\delta = \sum_{k=1}^{\infty} \delta_k^{1/2} < \infty$. For any given vector $a \in H$, there are vectors $x \in H$, $y_k \in \mathcal{S}$ and $z_k \in \mathcal{R}$ $(k \geq 1)$ such that $\|x - a\| < 3\alpha\delta$ and*

$$L_k = x \otimes (y_k + z_k), \quad \|y_k\| < \alpha\delta_k^{1/2}, \quad \|z_k\| < \alpha\delta_k^{1/2}$$

for all $k \geq 1$. Here as before $\alpha = 1/(\gamma^{1/2} - \theta^{1/2})$.

The preceding result shows that each spherical contraction with property $F^r_{\theta,\gamma}$ satisfies property $(\mathbb{A}_{1,\aleph_0})$. In the same way as Corollary 1.9 was obtained from Proposition 1.8 one can prove the following corollary.

COROLLARY 2.6. *Suppose that T satisfies property $F^r_{\theta,\gamma}$ for some real numbers $0 \le \theta < \gamma \le 1$. For each $\varepsilon > 0$, there is a constant $c = c(\varepsilon, \theta, \gamma) > 0$ such that, for each sequence $(L_k)_{k \ge 1}$ in Q and each vector $a \in H$, there are elements $x, y_k \in H$ with $\|x - a\| < \varepsilon$ and*

$$L_k = x \otimes y_k, \quad \|y_k\| \le c k^2 \|L_k\| \quad (k \ge 1).$$

In [15] it was shown that the quantitative version of property $(\mathbb{A}_{1,\aleph_0})$ obtained in Corollary 2.6 is sufficient to deduce the existence of a rich supply of analytic invariant subspaces for T and to prove the reflexivity of T. We shall return to these questions at the end of the next section.

3. Spherical contractions of class $\mathbb{A}_{1,\aleph_0}$ and reflexivity

Let $T \in L(H)$ be a single contraction of class $\mathbb{A}$. It was shown in [9] that $T \in \mathbb{A}_{1,\aleph_0}$ if and only if $\mathfrak{A}_T$ has property $E^r_{0,1}$ and that, in this case, T is reflexive. In the present section we prove the same results for spherical contractions of class $\mathbb{A}$.

DEFINITION 3.1. Let $T \in L(H)^n$ be a spherical contraction in the class $\mathbb{A}$, and let $\rho > 0$. We say that T has the *ρ-almost factorization property* if, for each $L \in Q$ and each $\varepsilon > 0$, there are vectors $x, y \in H$ with $\|x\|, \|y\| \le \rho \|L\|^{1/2}$ and

$$\|L - x \otimes y\| < \varepsilon.$$

If these conditions hold with $\rho = 1$, then T is said to possess the *almost factorization property*.

We fix, for each point $\mu \in B$, a conformal mapping $\varphi_\mu : B \to B$ with $\varphi_\mu(\mu) = 0$ and such that φ_μ extends to a holomorphic $\mathbb{C}^n$-valued map on a neighbourhood of $\overline{B}$ which induces a homeomorphism $\varphi_\mu : \overline{B} \to \overline{B}$ (cf. §2.2 in [18]).

PROPOSITION 3.2. *Let $T \in L(H)^n$ be a spherical contraction of class $\mathbb{A}$. Suppose that T is of type $C_{\cdot 0}$ and that T satisfies the ρ-almost factorization property. Then T has property $E^r_{0,\gamma}$ with $\gamma = 1/\rho^2$.*

Proof. We show that $\gamma\mathcal{E}_\mu \in \mathcal{E}_0^r(T)$ for all $\mu \in B$.

Let $\mu = 0$. Fix $i \in \{1, \ldots, n\}$ arbitrary. Since the functionals

$$L_k\colon H^\infty(B) \to \mathbb{C}, \quad f \mapsto \frac{\partial^{2k} f}{\partial z_i^{2k}}(0)/(2k)! \quad (k \in \mathbb{N})$$

are w^*-continuous with $\|L_k\| \leq 1$, there are sequences $(u_k)_{k\geq 0}$, $(v_k)_{k\geq 0}$ in H with $\|u_k\|$, $\|v_k\| \leq \rho$ for all k and

$$\lim_{k\to\infty} (L_k - u_k \otimes v_k) = 0.$$

Define $x_k = T_i^k u_k$ and $y_k = T_i^{*k} v_k$. Then $\|x_k\|, \|y_k\| \leq \rho$, and for each $z \in H$,

$$(x_k \otimes z) = (u_k \otimes T_i^{*k} z) \xrightarrow{k} 0.$$

For each $k \geq 0$, choose a function h_k in the closed unit ball of $H^\infty(B)$ with $\|L_0 - x_k \otimes y_k\| = \langle L_0 - x_k \otimes y_k, h_k\rangle$. Then we obtain

$$\begin{aligned}\|L_0 - x_k \otimes y_k\| &\leq |\langle L_0, h_k\rangle - \langle L_k, z_i^{2k} h_k\rangle| + |\langle L_k - u_k \otimes v_k, z_i^{2k} h_k\rangle| \\ &\leq \|L_k - u_k \otimes v_k\| \xrightarrow{k} 0.\end{aligned}$$

Let $\mu \in B$ be arbitrary, and let $\varphi_\mu\colon B \to B$ be a conformal mapping as above. Denote by $\gamma_*\colon Q \to Q$ the predual of the dual algebra isomorphism

$$\gamma = \gamma_\mu\colon H^\infty(B) \to H^\infty(B), \quad f \mapsto f \circ \varphi_\mu.$$

Then $\Phi_\mu = \Phi \circ \gamma_\mu$ is the w^*-continuous H^∞-functional calculus of the tuple $T_\mu = \Phi(\varphi_\mu)$. For $x, y \in H$, write

$$x \otimes_\mu y\colon H^\infty(B) \to \mathbb{C}, \quad f \mapsto \langle \Phi_\mu(f)x, y\rangle.$$

Note that $\gamma_*(x \otimes y) = x \otimes_\mu y$. Since T_μ satisfies the same hypotheses as the original tuple T, the first part of the proof yields sequences $(x_k)_{k\geq 0}$ and $(y_k)_{k\geq 0}$ with $\|x_k\|, \|y_k\| \leq \rho$ such that $(x_k \otimes_\mu z) \xrightarrow{k} 0$ for each $z \in H$ and

$$\mathcal{E}_0 = \lim_{k\to\infty} x_k \otimes_\mu y_k.$$

Then $(x_k \otimes z) = \left(\gamma_*^{-1}(x_k \otimes_\mu z)\right) \xrightarrow{k} 0$ for each $z \in H$ and

$$\mathcal{E}_\mu = \gamma_*^{-1}(\mathcal{E}_0) = \lim_{k\to\infty} x_k \otimes y_k.$$

□

The above proof still works if the hypothesis that T is of type $C_{\cdot 0}$ is replaced by the condition that there is a family of conformal mappings $\varphi_\mu: B \to B$ $(\mu \in B)$ with $\varphi_\mu(\mu) = 0$ and an index $i \in \{1, \ldots, n\}$ such that the ith components of T and $\Phi(\varphi_\mu)$ $(\mu \in B)$ are of type $C_{\cdot 0}$.

In the one-variable case Bercovici [2] and Chevreau [8] proved that each single contraction $T \in L(H)$ in the class $\mathbb{A}$ has property $(\mathbb{A}_1)$. Essential for the proof of this result were certain L^1-factorization results of Bercovici which imply in particular that each contraction $T \in L(H)$ of class $\mathbb{A}$ has the almost factorization property. We do not know whether the latter result remains true for spherical contractions in the class $\mathbb{A}$. If the answer to this question is positive, then the results of this paper can be improved considerably.

THEOREM 3.3. *Let $T \in L(H)^n$ be a spherical contraction of class $\mathbb{A}$. If T is of class $\mathbb{A}_{1,\aleph_0}$, then T satisfies property $E_{0,1}^r$.*

Proof. We claim that $\mathcal{E}_\mu \in \mathcal{E}_0^r(T)$ for all $\mu \in B$. As in the previous proof it suffices to prove this assertion for $\mu = 0$.

Choose vectors $x, y_k \in H$ with

$$\mathcal{E}_0^{(k)}/k! = x \otimes y_k \quad (k \in \mathbb{N}^n).$$

Define $M = \vee(T^k x;\ k \in \mathbb{N}^n) \in \operatorname{Lat}(T)$. We may of course assume that $y_k \in M$ for all k. Since, for all $j, k \in \mathbb{N}^n$ and each $i = 1, \ldots, n$, we have

$$\langle T^j x, (T_i|M)^* y_k \rangle = \langle T^j x, y_{k-e_i} \rangle$$

(with $y_{k-e_i} = 0$ if $k_i = 0$), it follows that

$$(T_i|M)^* y_k = y_{k-e_i} \quad (k \in \mathbb{N}^n,\ i = 1, \ldots, n).$$

Using the Gram–Schmidt orthogonalization process and a suitable enumeration of $\mathbb{N}^n$, we obtain an orthonormal family $(x_k)_{k \in \mathbb{N}^n}$ in M such that

$$\mathrm{LH}\{y_j;\ |j| \le r\} = \mathrm{LH}\{x_j;\ |j| \le r\}$$

for all $r \in \mathbb{N}$. Denoting these spaces by M_r, we obtain that

$$(T_i|M)^* M_r \subset M_{r-1} \quad (M_{-1} = \{0\})$$

for $i = 1, \ldots, n$ and $r \in \mathbb{N}$.

We claim that $\mathcal{E}_0 = x_k \otimes x_k$ for all k. To verify this assertion choose, for $f \in H^\infty(B)$, functions $f_1, \dots, f_n \in H^\infty(B)$ with $f - f(0) = \sum_{i=1}^n z_i f_i$ and observe that

$$x_k \otimes x_k(f) = f(0) + \sum_{i=1}^n \langle x_k, (T_i|M)^*(\Phi(f_i)|M)^* x_k \rangle = f(0).$$

To complete the proof let us fix a minimal co-isometric extension

$$C_M = S_M^* \oplus R_M \in L(\mathcal{S}_M \oplus \mathcal{R}_M)^n$$

of the tuple $(T|M)^*$, where as usual S_M^* is the $C_{0\cdot}$-part and R_M is the (spherical) unitary part of C_M. An elementary induction on $|k|$ shows that $x_k \in \mathcal{S}_M$ for all $k \in \mathbb{N}^n$.

Let Q_M be the orthogonal projection from $\mathcal{S}_M \oplus \mathcal{R}_M$ onto $\mathcal{S}_M$. For $f \in H^\infty(B)$ and $z \in M$, we obtain

$$x_k \otimes z(f) = \langle x_k, \Phi_{(T|M)^*}(\tilde{f})z \rangle = \langle x_k, \Phi_{S_M^*}(\tilde{f}) Q_M z \rangle,$$

where $\Phi_{(T|M)^*}$ and $\Phi_{S_M^*}$ denote the H^∞-functional calculi of $(T|M)^*$ and S_M^*, respectively, and $\tilde{f} \in H^\infty(B)$ is defined by $\tilde{f}(\lambda) = \overline{f(\bar{\lambda})}$. Since S_M^* is of type $C_{0\cdot}$, any injective map $k: \mathbb{N} \to \mathbb{N}^n$ yields a weak zero sequence $(\tilde{x}_j)_{j \in \mathbb{N}} = (x_{k(j)})_{j \in \mathbb{N}}$ in the closed unit ball of H with

$$\mathcal{E}_0 = \tilde{x}_j \otimes \tilde{x}_j \quad (j \in \mathbb{N})$$

and the property that $(\tilde{x}_j \otimes x) \xrightarrow{j} 0$ for each vector $x \in H$.

□

Let $T \in L(H)^n$ be a spherical contraction of class $\mathbb{A}$. Combining Corollary 2.6 and Theorem 3.3 we see that the factorization properties $F_{\theta,\gamma}^r$ (for arbitrary $0 \le \theta < \gamma \le 1$) and $E_{0,1}^r$ are equivalent, and that T satisfies these properties if and only if $T \in \mathbb{A}_{1,\aleph_0}$.

For each natural number $N \ge 1$, we denote by $T^{(N)}$ the tuple

$$T^{(N)} = (T_1^{(N)}, \dots, T_n^{(N)}) \in L(H^N)^n,$$

where each $T_i^{(N)}$ is the direct sum of N copies of the operator T_i on the Hilbert space direct sum H^N. If T is of class $\mathbb{A}$, then $T^{(N)}$ is of class $\mathbb{A}$ again with H^∞-functional calculus

$$\Phi^N: H^\infty(B) \to L(H^N), \quad f \mapsto \Phi(f)^{(N)}.$$

For $x, y \in H^N$, denote by $x \otimes_N y \in Q$ the functionals defined by $x \otimes_N y(f) = \langle \Phi^N(f)x, y \rangle$.

LEMMA 3.4. *Let $T \in L(H)^n$ be a spherical contraction of class $\mathbb{A}$. Suppose that, for some integer $N \geq 1$, the tuple $T^{(N)}$ is of class $\mathbb{A}_{1,\aleph_0}$. Then $T \in \mathbb{A}_{1,\aleph_0}$.*

Proof. By Theorem 3.3 the tuple $T^{(N)}$ possesses property $E^r_{0,1}$. The technique used in the proof of Theorem 3.8 in [7] shows that T satisfies property $E^r_{(N-1/N),1}$. But then Corollary 2.6 implies that $T \in \mathbb{A}_{1,\aleph_0}$.

□

The following result describes a concrete spectral condition that is sufficient to imply the factorization property $(\mathbb{A}_{1,\aleph_0})$.

THEOREM 3.5. *Let $T \in L(H)^n$ be a spherical contraction of class $\mathbb{A}$. Suppose that $\sigma^H(T)$ is dominating in B and that T is of type $C_{\cdot 0}$. Then T satisfies property $E^r_{0,1}$.*

Proof. Exactly as in the first part of the proof of Theorem 2.11 in [13], it follows that, for each absolute convex combination $L = \sum_{\lambda \in J} c_\lambda \mathcal{E}_\lambda$ of point evaluations $\mathcal{E}_\lambda$ with $\lambda \in \sigma^H(T) \cap B$ and each $\varepsilon > 0$, there are vectors $x, y \in H^{16}$ with $\|x\|, \|y\| \leq 8$ and

$$\|L - x \otimes_{16} y\| < \varepsilon.$$

By Proposition 3.2 the tuple $T^{(16)}$ has property $E^r_{0,\gamma}$ with $\gamma = 1/64$. The remarks following Theorem 3.3 together with Lemma 3.4 show that T has property $E^r_{0,1}$.

□

As usual we call a closed linear subspace $L \subset H$ semi-invariant for $T \in L(H)^n$ if there are $M, N \in \operatorname{Lat}(T)$ such that $L = M \ominus N$. In this case, we denote by T_L the compression of T onto L. To conclude that a spherical contraction $T \in L(H)^n$ of class $\mathbb{A}$ satisfies property $E^r_{0,1}$, it suffices to know that there is a semi-invariant subspace L for T such that T_L is of type $C_{\cdot 0}$ and has dominating Harte spectrum $\sigma^H(T_L)$ in B. Indeed, by Theorem 3.5, $T_L \in \mathbb{A}_{1,\aleph_0}$ and since L is semi-invariant for T, it follows that $T \in \mathbb{A}_{1,\aleph_0}$.

To prove that spherical contractions of class $\mathbb{A}_{1,\aleph_0}$ are reflexive, one introduces the notion of analytic invariant subspaces. A space $M \in \operatorname{Lat}(T)$ is an analytic invariant subspace for T if there is a non-zero conjugate analytic function $e \colon B \to M$ such that

$$(\lambda_i - T_i|M)^* e(\lambda) = 0$$

for $\lambda \in B$ and $i = 1, \ldots, n$.

LEMMA 3.6. *Let $T \in L(H)^n$ be a spherical contraction of class $\mathbb{A}$. If T possesses an analytic invariant subspace M, then T satisfies property $E^r_{0,1}$.*

Proof. Let $e: B \to M$ be a conjugate analytic function as above. Then the set $\sigma = \{\lambda \in B; e(\lambda) \neq 0\}$ is dominating in B and

$$L = \bigvee \Big(\bigcap_{i=1}^{n} \operatorname{Ker}(\lambda_i - T_i|M)^*;\ \lambda = (\lambda_i) \in \sigma \Big) \in \operatorname{Lat}\Big((T|M)^* \Big)$$

is a semi-invariant subspace for T.

Since by definition $\{\overline{\lambda};\ \lambda \in \sigma\} \subset \sigma_p(T_L^*)$, it is clear that σ is contained in $\sigma^H(T_L)$. To see that $T_L^* = (T|M)^*|L$ is of type $C_{0\cdot}$, it suffices to observe that, for each function $f \in H^\infty(B)$ and each $\lambda \in \sigma$, the restriction of $\Phi_{(T|M)^*}(f)$ to the space

$$L(\lambda) = \bigcap_{i=1}^{n} \operatorname{Ker}(\lambda_i - T_i|M)^*$$

coincides with the operator $f(\overline{\lambda})1_{L(\lambda)}$.

Thus we have shown that T_L is of type $C_{\cdot 0}$ and has dominating Harte spectrum in B. By the remarks preceding the lemma the tuple T has property $E^r_{0,1}$. □

A standard argument carried out in detail in [15] (Theorem 3.1 and Lemma 3.2) shows that conversely each spherical contraction of class $\mathbb{A}$ with property $E^r_{0,1}$ possesses a rich supply of analytic invariant subspaces. More precisely, let us say that a vector $x \in H$ generates an analytic invariant subspace for T if the space

$$M_x = \bigvee (T^k x;\ k \in \mathbb{N}^n) \in \operatorname{Lat}(T)$$

is an analytic invariant subspace for T. The cited results in [15] show that, for a spherical contraction with property $E^r_{0,1}$, the set of all vectors in H generating an analytic invariant subspace is dense in H.

The following corollary summarizes the results obtained so far.

COROLLARY 3.7. *For a spherical contraction $T \in L(H)^n$ of class $\mathbb{A}$, the following conditions are equivalent:*

(i) *T has property $E^r_{0,1}$;*
(ii) *T has property $E^r_{\theta,\gamma}$ for some $0 \leq \theta < \gamma \leq 1$;*
(iii) *T has property $F^r_{\theta,\gamma}$ for some $0 \leq \theta < \gamma \leq 1$;*
(iv) *$T \in \mathbb{A}_{1,\aleph_0}$;*

(v) *the set of those vectors $x \in H$ generating an analytic invariant subspace for T is dense in H;*
(vi) *T has an analytic invariant subspace;*
(vii) *$T^{(N)} \in \mathbb{A}_{1,\aleph_0}$ for some natural number $N \geq 1$;*
(viii) *there is a compression T_L of T to a semi-invariant subspace L of T such that T_L is of type $C_{\cdot 0}$ with dominating Harte spectrum in B.*

If any of these conditions is satisfied, then T is reflexive.

The reflexivity of T in Corollary 3.7 can be proved in exactly the same way as in the case of subnormal spherical contractions of class $\mathbb{A}$ (see the proof of Theorem 3.7 in [15]). In (viii) the condition that the Harte spectrum of T_L is dominating in B, can be replaced by the condition that T_L satisfies the ρ-almost factorization property for some $\rho > 0$. In particular, each spherical contraction T of class $\mathbb{A}$ which is of type $C_{\cdot 0}$ and possesses the ρ-almost factorization property, for some $\rho > 0$, has property $(\mathbb{A}_{1,\aleph_0})$ and is reflexive.

As mentioned before, in the one-dimensional case $n = 1$, each single contraction T of class $\mathbb{A}$ automatically satisfies the almost factorization property. It is an open question whether the same result holds true in the multivariable case.

If T satisfies one and hence all of the conditions in Corollary 3.7, then T admits the factorization properties described in Corollary 2.4, Theorem 2.5, and Corollary 2.6 with $\alpha = 1$. In particular, it follows that $T \in \mathbb{A}_1(r)$ with $r = 3\sqrt{2}$. Here $\mathbb{A}_1(r)$ denotes the set of all spherical contractions $T \in L(H)^n$ for which the dual algebra $\mathfrak{A}_T$ satisfies property $(\mathbb{A}_1(r))$ in the sense of [7] (Definition 2.01).

Using Proposition 2.09 from [7] one obtains the following consequence.

COROLLARY 3.8. *Let $T \in L(H)^n$ be a spherical contraction of class $\mathbb{A}$. If T satisfies one and hence all of the conditions in Corollary* 3.7, *then* Alg Lat(T) = $\mathfrak{A}_T = W_T$ *and the* weak* *and weak operator topologies on this algebra coincide.*

By modifying the proofs in Section 1 and 2, one can show that each spherical contraction $T \in L(H)^n$ satisfying the equivalent conditions of Corollary 3.7 belongs to the class $\mathbb{A}_{1,\aleph_0}(r)$ for some $r \geq 1$. Here $\mathbb{A}_{1,\aleph_0}(r)$ consists of all spherical contractions of class $\mathbb{A}$ for which the dual algebra $\mathfrak{A}_T$ satisfies property $\left(\mathbb{A}_{1,\aleph_0}(r)\right)$ in the sense of [9].

We end this paper by indicating how one of the main results from [15] can be obtained as an application of Corollary 3.7. Let $T \in L(H)^n$ be a subnormal tuple with $\sigma(T) \subset \overline{B}$ such that $\sigma(T)$ is dominating in B. In [15] it was shown that T is reflexive. We indicate very briefly without aiming at completeness how to deduce this result from Corollary 3.7.

Let T be as above. It is well known that each subnormal tuple with Taylor spectrum contained in $\overline{B}$ satisfies the positivity conditions $\Delta_T^{(1)} \geq 0$ and $\Delta_T^{(n)} \geq 0$. Hence T possesses a spherical dilation. Let

$$T = T_0 \oplus T_1 \in L(H_0 \oplus H_1)^n$$

be the decomposition of T into its completely non-unitary part $T_0 \in L(H_0)^n$ and its (spherical) unitary part $T_1 \in L(H_1)^n$. Then the spectrum $\sigma(T_0)$ of T_0 is still dominating in B. By Corollary 2.4 in [14] the tuple T_0 is of class $C_{\cdot 0}$. It follows that the completely non-unitary part T_0 of T is a spherical contraction of class $\mathbb{A}$. By a standard argument (see e.g. the last part of the proof of Theorem VII.8.5 in [11]) to prove the reflexivity of T it suffices to show that T_0 and T_1 are reflexive and that the dual algebras $\mathfrak{A}_{T_0}$ and $\mathfrak{A}_{T_1}$ generated by these tuples satisfy property $(\mathbb{A}_1(r))$ for some $r \geq 1$. It is well known that the normal tuple T_1 has these properties (Proposition 2.04 and Proposition 2.05 in [7]).

To complete the argument it suffices to show that T_0 satisfies one, or equivalently, all the conditions in Corollary 3.7. By the remark immediately following Corollary 3.7, we only need to verify that T_0 has the almost factorization property. By applying the spectral theorem for normal tuples to the minimal normal extension of T_0 one can reduce this assertion to the result formulated in Lemma 1.3. The reader will easily fill in the missing details to obtain a complete proof of the following result from [15] along these lines.

COROLLARY 3.9. *Let $T \in L(H)^n$ be a subnormal tuple with $\sigma(T) \subset \overline{B}$ such that $\sigma(T)$ is dominating in B. Then T is reflexive.*

References

[1] A. ATHAVALE, On the intertwining of joint isometries, *J. Operator Theory*, **23** (1990), 330–350.

[2] H. BERCOVICI, Factorization theorems and the structure of operators on Hilbert space, *Ann. of Math.*, **128** (1988), 399–413.

[3] H. BERCOVICI, A factorization theorem with applications to invariant subspaces and the reflexivity of isometries, *Mathematical Research Letters*, **1** (1994), 511–518.

[4] S. BROWN and B. CHEVREAU, Toute contraction à calcul fonctionnel isométrique est réflexive, *C.R. Acad. Sc. Paris*, **307** (1988), 185–188.

[5] S. BROWN, B. CHEVREAU and C. PEARCY, Contractions with rich spectrum have invariant subspaces, *J. Operator Theory*, **1** (1979), 123–136.

[6] S. BROWN, B. CHEVREAU and C. PEARCY, On the structure of contraction operators. II, *J. Funct. Anal.*, **76** (1988), 30–55.

[7] H. BERCOVICI, C. FOIAS and C. PEARCY, *Dual algebras with applications to invariant subspaces and dilation theory,* CBMS Regional Conference Series in Math. **56**, AMS, Providence, Rhode Island, 1985.

[8] B. CHEVREAU, Sur les contractions à calcul fonctionnel isométrique. II, *J. Operator Theory,* **20** (1988), 269–293.

[9] B. CHEVREAU, G. EXNER and C. PEARCY, On the structure of contraction operators. III, *Michigan Math. J.*, **36** (1989), 29–62.

[10] U. CEGRELL, *Capacities in complex analysis,* Aspects of Mathematics **14**, Vieweg, Braunschweig, 1988.

[11] J. B. CONWAY, *The theory of subnormal operators,* Math. Surveys and Monographs **38**, AMS, Providence, Rhode Island, 1991.

[12] S. W. DRURY, A generalization of von Neumann's inequality to complex ball, *Proc. Amer. Math. Soc.*, **68** (1978), 300–304.

[13] J. ESCHMEIER, C_{00}-representations with dominating Harte spectrum, *Proceedings of the 13th Conference on Banach algebras, Blaubeuren,* Banach algebras '97, editors: E. Albrecht and M. Mathieu, Walter de Gruyter, Berlin, 1998.

[14] J. ESCHMEIER, Invariant subspaces for spherical contractions, *Proc. London Math. Soc.*, **75** (1997), 157–176.

[15] J. ESCHMEIER, Algebras of subnormal operators on the unit ball, *J. Operator Theory,* **42** (1999), 37–76.

[16] J. ESCHMEIER and M. PUTINAR, *Spectral decompositions and analytic sheaves,* LMS Monograph Series, Oxford University Press, Oxford, 1996.

[17] V. MÜLLER and F.-H. VASILESCU, Standard models for some commuting multioperators, *Proc. Amer. Math. Soc.*, **117** (1993), 979–989.

[18] W. RUDIN, *Function theory in the unit ball of* $\mathbb{C}^n$, Springer, Heidelberg, 1980.

[19] W. RUDIN, *New constructions of functions holomorphic in the unit ball of* $\mathbb{C}^n$, CBMS Regional Conference Series in Math. **63**, AMS, Providence, Rhode Island, 1986.

[20] F.-H. VASILESCU, An operator valued Poisson kernel, *J. Funct. Anal.*, **110** (1992), 47–72.

J. ESCHMEIER, Fachbereich Mathematik, Universität des Saarlandes, Postfach 15 11 50, D-66041 Saarbrücken, Germany; *e-mail:* eschmei@math.uni-sb.de

Operator Theory: Advances and Applications, Vol. 127, 243–266

Apostol's bilateral weighted shifts are hyper-reflexive

JEAN ESTERLE

C. Apostol showed in 1984 that when a weight ω on $\mathbb{Z}$ has sufficiently "irregular" behavior then the shift operator S on $\ell^2_\omega(\mathbb{Z})$ has nontrivial hyperinvariant subspaces.

We reprove this result by making explicitely use of an approximation scheme which goes back to S. Brown. We also prove that some of these spaces are "analytic translation invariant subspaces" in a sense analogous to the notion of "analytic invariant subspaces" introduced by Exner, Chevreau and Pearcy for contractions of the class $\mathbb{A}_{1,\aleph_0}$, and we also show that Apostol's weighted shifts are in fact "hyper-reflexive": if a bounded operator T on $\ell^2_\omega(\mathbb{Z})$ satisfies $T(M) \subset M$ for every closed subspace of $\ell^2_\omega(\mathbb{Z})$ hyperinvariant for S, then T belongs to the commutant of S.

1. Introduction

Let $\omega: \mathbb{Z} \to (0, \infty)$ and assume that $0 < \inf_{n\in\mathbb{Z}} \frac{\omega(n+1)}{\omega(n)} \leq \sup_{n\in\mathbb{Z}} \frac{\omega(n+1)}{\omega(n)} < +\infty$, so that the usual shift $S: (u_n)_{n\in\mathbb{Z}} \to (u_{n-1})_{n\in\mathbb{Z}}$ is bounded and invertible on the Hilbert space $\ell^2_\omega(\mathbb{Z}) := \left\{u = (u_n)_{n\in\mathbb{Z}} \middle| \|u\|_\omega = \left[\sum_{n\in\mathbb{Z}} |u_n|^2\omega^2(n)\right]^{1/2} < +\infty\right\}$.

A closed subspace M of $\ell^2_\omega(\mathbb{Z})$ is said to be translation invariant if $S(M) = M$. The translation invariant subspaces of $\ell^2_\omega(\mathbb{Z})$ are exactly those which are hyperinvariant for S, i.e. satisfy the condition $T(M) \subset M$ for every bounded operator T which commutes with S, see [Sh]. The existence of a nontrivial translation invariant subspace of $\ell^2_\omega(\mathbb{Z})$ is a long-standing open problem.

For recent progress on this question see [Atz], [Atz–So], [Do], [Es], [Es–Vo$_1$], [Es–Vo$_2$], [Es–Vo$_3$].

In this paper we have a fresh look at a breakthrough performed in 1984 by Apostol [Ap]. Denote by $\rho(S)$ and $\rho(S^{-1})$ the spectral radii of S and S^{-1}. Apostol

AMS Subject Classification (1991): 47A15, 47D27, 47A65.

showed that if the set

$$\left\{\overline{\lim_{n\to\infty}}\,\omega(n)^{1/n},\ \underline{\lim}_{n\to\infty}\,\omega(n)^{1/n},\ \overline{\lim_{n\to-\infty}}\,\omega(n)^{1/n},\ \underline{\lim}_{n\to-\infty}\,\omega(n)^{1/n}\right\}$$

is not a singleton, or if

$$\begin{aligned}\frac{1}{\rho(S^{-1})} &< \inf\left(\underline{\lim}_{n\to\infty}\,\omega(n)^{1/n},\ \underline{\lim}_{n\to-\infty}\,\omega(n)^{1/n}\right)\\ &\le \sup\left(\overline{\lim_{n\to\infty}}\,\omega(n)^{1/n},\ \overline{\lim_{n\to-\infty}}\,\omega(n)^{1/n}\right) < \rho(S),\end{aligned}$$

then $\ell^2_\omega(\mathbb{Z})$ does have nontrivial translation invariant subspaces. This result was recently extended to the spaces $\ell^p_\omega(\mathbb{Z})$, $1 \le p < +\infty$ by G. Amir [Am]. Apostol's proof is based on ingenious computations, which involve some auxiliary weights $\sigma\colon \mathbb{Z} \to (0,\infty)$.

In fact, the existence of a nontrivial translation invariant subspace of $\ell^2_\omega(\mathbb{Z})$ is equivalent to the existence of $u \in \ell^2_\omega(\mathbb{Z})\backslash\{0\}$ and $v \in \ell^2_{\omega^*}(\mathbb{Z})\backslash\{0\}$ such that $u * v = 0$, where $\omega^*\colon \mathbb{Z} \to (0,\infty)$ is the "dual weight" defined by the formula $\omega^*(n) = \omega^{-1}(-n)$. Our first observation is that the auxiliary weights σ introduced by Apostol enjoy a "$A_{\aleph_0}$-type property": given any family $(w_{i,j})_{i\ge 0,\ j\ge 0}$ in $\ell^1_\sigma(\mathbb{Z})$ there exists $(u_i)_{i\ge 0}$ in $\ell^2_\omega(\mathbb{Z})$ and $(v_j)_{j\ge 0}$ in $\ell^2_{\omega^*}(\mathbb{Z})$ such that $u_i * v_j = w_{i,j}$. This follows from a general result of Bercovici, Foias and Pearcy [BFP] (and the first technical computation of Apostol is a special case of the "superdominancy property" also discussed in [BFP]). The existence of a nontrivial translation invariant subspace of $\ell^2_\omega(\mathbb{Z})$ follows immediately from this, since there exists $u_1, u_2 \in \ell^2_\omega(\mathbb{Z})$, $v_1, v_2 \in \ell^2_{\omega^*}(\mathbb{Z})$ satisfying $u_1 * v_1 = u_2 * v_2 = 0$, $u_1 * v_2 \neq 0$, $u_2 * v_1 \neq 0$.

These considerations also show that, in all the situations considered by Apostol, there exists an open annulus Ω contained in the spectrum of S such that for every $\lambda \in \Omega$ there exists $u \in \ell^2_\omega(\mathbb{Z})$ and $v \in \ell^2_{\omega^*}(\mathbb{Z})$ such that $u * v = (\lambda^{-n})_{n\in\mathbb{Z}}$. This gives a "concrete" way to obtain nontrivial translation invariant subspaces of $\ell^2_\omega(\mathbb{Z})$, since in this situation $u \neq 0$ and $u \notin \bigvee_{n\in\mathbb{Z}} S^n(S-\lambda I)u$, so that either u is an eigenvector of S or $\bigvee_{n\in\mathbb{Z}} S^n(S-\lambda I)u$ is a nontrivial translation invariant subspace of $\ell^2_\omega(\mathbb{Z})$. These results are given in Section 3. In Section 4 we prove much more in this direction. Adapting a notion from [CEP], we will say that a translation invariant subspace $M \neq \{0\}$ of $\ell^2_\omega(\mathbb{Z})$ is full holomorphic on a connected open subset Ω of $\mathbb{C}$ if there is a nonzero conjugate-holomorphic map $g\colon \Omega \to M$ such that $\left(S_{|M} - \lambda I\right)^* g(\lambda) = 0$ for every $\lambda \in \mathbb{C}$ and such that the closed linear span of $g(\Omega)$ equals M. Set $\widetilde{u}(\lambda) = [u, g(\lambda)]$ for $\lambda \in \Omega$, where we denote by $[.,.]$ the hermitian product on $\ell^2_\omega(\mathbb{Z})$. Then the map $u \to \widetilde{u}$ is a one-to-one linear map from M into

$\mathcal{H}(\Omega)$, and $\widetilde{Su}(\lambda) = \lambda\widetilde{u}(\lambda)$ for $\lambda \in \Omega$. It is then easy to see that $(ST - TS)u = 0$ for every $u \in M$ if T is a bounded operator on $\ell^2_\omega(\mathbb{Z})$ such that $T(N) \subset N$ for every translation invariant subspace N of $\ell^2_\omega(\mathbb{Z})$.

Using some precise estimates concerning the solutions of a finite system of equations given in [BFP] we show then that if the point spectrum of S is empty, and if ω satisfies Apostol's condition then there exists an open annulus $\Omega \subset \operatorname{Spec} S$ centered at the origin such that the set of all $u \in \ell^2_\omega(\mathbb{Z})$ generating a full holomorphic translation invariant subspace of $\ell^2_\omega(\mathbb{Z})$ is dense in $\ell^2_\omega(\mathbb{Z})$.

Let R be a bounded operator on a Hilbert space H. Denote by $\mathcal{A}(R)$ the ultraweakly closed unital subalgebra of $\mathcal{B}(H)$ generated by R, denote by $\mathcal{C}(R)$ the commutant of R and denote by $\operatorname{Lat} R$ the lattice of closed subspaces of H which are invariant for R. Recall that R is said to be reflexive if every bounded operator T such that $\operatorname{Lat} R \subset \operatorname{Lat} T$ belongs to $\mathcal{A}(R)$. Sarason showed long ago that every normal operator is reflexive [Sa]. More generally, every subnormal operator is reflexive [Ol–Th]. Also every absolutely continuous contraction with isometric H^∞-functional calculus is reflexive [Br–Ch]. As in [BFN], we will say that an operator R is hyper-reflexive if every bounded operator T on H such that $T(M) \subset M$ for every closed subspace M of H which is hyperinvariant for R commutes with R. Corollary 3.7 shows that the shift operator S on $\ell^2_\omega(\mathbb{Z})$ is in fact hyper-reflexive (so that S and S^{-1} are both reflexive) in all the situations considered by Apostol.

These methods are not sufficient to prove the existence of nontrivial translation invariant subspaces of $\ell^2_\omega(\mathbb{Z})$ for all weights ω for which $1/\rho(S^{-1}) < \rho(S)$, i.e. for all weighs ω for which $\operatorname{Spec} S$ has nonempty interior but we used only in Section 4 properties "of type $\mathbb{A}_{1,\aleph_0}$", while the results of Section 3 give a stronger property "of type $\mathbb{A}_{\aleph_0}$" for the auxiliary weights σ introduced by Apostol. The author believes that a modification of the Brown approximation scheme used in the present paper can solve the translation invariant subspace problem for all weights ω for which the spectrum of the shift operator has nonempty interior.

2. Preliminaries

A map $\omega\colon \mathbb{Z} \to (0, \infty)$ will be called a weight if

$$0 < \inf_{n\in\mathbb{Z}} \frac{\omega(n+1)}{\omega(n)} \le \sup_{n\in\mathbb{Z}} \frac{\omega(n+1)}{\omega(n)} < +\infty.$$

If ω is a weight on $\mathbb{Z}$, we set

$$\widetilde{\omega}(n) = \sup_{p\in\mathbb{Z}} \frac{\omega(n+p)}{\omega(p)}, \quad \omega^*(n) = \frac{1}{\omega(-n)} \qquad (n \in \mathbb{Z}),$$

then

$$\ell^2_\omega(\mathbb{Z}) = \Big\{u = (u_n)_{n\in\mathbb{Z}} \mid \|u\|_\omega = \Big[\sum_{n\in\mathbb{Z}} |u_n|^2\, \omega^2(n)\Big]^{1/2} < +\infty\Big\},$$

$$\ell^1_\omega(\mathbb{Z}) = \Big\{u = (u_n)_{n\in\mathbb{Z}} \mid \|u\|^1_\omega = \sum_{n\in\mathbb{Z}} |u_n|\, \omega(n) < +\infty\Big\},$$

$$\ell^\infty_\omega(\mathbb{Z}) = \Big\{u = (u_n)_{n\in\mathbb{Z}} \mid \|u\|^\infty_\omega = \sup_{n\in\mathbb{Z}} |u_n|\, \omega(n) < +\infty\Big\}.$$

We will denote by $[x, y] = \sum_{n\in\mathbb{Z}} x_n \overline{y}_n \omega^2(n)$ the standard hermitian product on $\ell^2_\omega(\mathbb{Z})$. We can identify the dual of $\ell^2_\omega(\mathbb{Z})$ to $\ell^2_{\omega^*}(\mathbb{Z})$, the duality being implemented by the formula

$$\langle u, v\rangle = \sum_{n\in\mathbb{Z}} u_n v_{-n} \qquad (u = (u_n)_{n\in\mathbb{Z}} \in \ell^2_\omega(\mathbb{Z}),\ v = (v_n)_{n\in\mathbb{Z}} \in \ell^2_{\omega^*}(\mathbb{Z})). \tag{2.1}$$

Similarly we can use formula (2.1) to identify the dual of $\ell^1_\omega(\mathbb{Z})$ to $\ell^\infty_{\omega^*}(\mathbb{Z})$. We will denote by $S\colon (u_n)_{n\in\mathbb{Z}} \to (u_{n-1})_{n\in\mathbb{Z}}$ the usual shift operator on $\ell^2_\omega(\mathbb{Z})$. The shift operator is invertible, and we have

$$\|S^n\| = \widetilde{\omega}(n) \qquad (n \in \mathbb{Z}). \tag{2.2}$$

In particular, the spectral radii of S and S^{-1} are given by the formulae $\rho(S) = \lim_{n\to\infty} \widetilde{\omega}(n)^{1/n}$, $\rho(S^{-1}) = \lim_{n\to-\infty} \widetilde{\omega}(n)^{1/|n|}$.

For $u \in \ell^2_\omega(\mathbb{Z})$, $v \in \ell^2_{\omega^*}(\mathbb{Z})$ we define $u * v$ by the formula

$$(u * v)_n = \sum_{p\in\mathbb{Z}} u_p v_{n-p} = \langle S^{-n} u, v\rangle = \langle u, S^{-n} v\rangle \qquad (n \in \mathbb{Z}).$$

A closed subspace M of $\ell^2_\omega(\mathbb{Z})$ will be said to be translation invariant if $SM \cup S^{-1}M \subset M$ (or, equivalently, if $SM = M$). The existence of a nontrivial translation invariant subspace of $\ell^2_\omega(\mathbb{Z})$ is clearly equivalent to the existence of $u \in \ell^2_\omega(\mathbb{Z})\setminus\{0\}$, $v \in \ell^2_{\omega^*}(\mathbb{Z})\setminus\{0\}$ such that $u * v = 0$.

Denote by $\mathcal{U}_\omega = \{R \in \mathcal{B}(\ell^2_\omega(\mathbb{Z})) \mid RS = SR\}$ the commutant of S. Set $e_m = (\delta_{m,n})_{n\in\mathbb{Z}}$ for $m \in \mathbb{Z}$, where we denote by $\delta_{n,m}$ the usual Kronecker symbol. For $R \in \mathcal{U}_\omega$ set $\widehat{R}(n) = \langle \mathrm{Re}_o, e_{-n}\rangle = \langle \mathrm{Re}_o, S^{-n} e_o\rangle$, so that $\mathrm{Re}_o = (\widehat{R}(n))_{n\in\mathbb{Z}}$.

Let $R \in \mathcal{U}_\omega$, and set

$$s_n(R) = \sum_{|p|\le n} \widehat{R}(p) S^p, \qquad \sigma_n(R) = \frac{1}{n+1} \sum_{m=0}^{n} s_m(R).$$

It follows from [Sh, Th.12] that $\|Ru - \sigma_n(R)u\|_\omega \to 0$ $(n \to \infty)$ for $u \in \ell^2_\omega(\mathbb{Z})$ and so $RM \subset M$ for every translation invariant subspace M of $\ell^2_\omega(\mathbb{Z})$, and every translation invariant subspace of $\ell^2_\omega(\mathbb{Z})$ is in fact hyperinvariant for S. Denote by $\mathcal{M}_\omega$ the set of all sequences $\varphi = (\varphi_n)_{n\in\mathbb{Z}}$ such that $(\varphi * u)_n = \lim\limits_{m\to\infty} \sum_{|p|\le m} \varphi_p . u_{n-p}$ exists for every $n \in \mathbb{Z}$ and every $u \in \ell^2_\omega(\mathbb{Z})$ and such that $\varphi * u \in \ell^2_\omega(\mathbb{Z})$ for every $u \in \ell^2_\omega(\mathbb{Z})$.

It follows from [Sh. Th. 3] that the map $R \to \widehat{R}$ is a bijection from $\mathcal{U}_\omega$ onto $\mathcal{M}_\omega$. We will equip $\mathcal{U}_\omega$ with the usual operator norm $\|.\|_{op}$ and set $\left\|\widehat{R}\right\|_{op} = \|R\|_{op}$ for $R \in \mathcal{U}_\omega$, so that $\|\varphi\|_{op} = \sup\{\|\varphi * u\|_\omega \mid u \in \ell^2_\omega(\mathbb{Z}), \|u\|_\omega \le 1\}$ for $\varphi \in \mathcal{M}_\omega$. Clearly, $\mathcal{M}_\omega \subset \ell^2_\omega(\mathbb{Z})$ and $(\mathcal{M}_\omega, \|.\|_{op})$ is a Banach algebra with respect to convolution. For $R \in \mathcal{U}_\omega$ define ${}^tR\colon \ell^2_{\omega^*}(\mathbb{Z}) \to \ell^2_{\omega^*}(\mathbb{Z})$ by the formula

$$\langle u, {}^t Rv\rangle = \langle Ru, v\rangle \qquad (u \in \ell^2_\omega(\mathbb{Z}),\ v \in \ell^2_{\omega^*}(\mathbb{Z})).$$

Clearly, ${}^tR \in \mathcal{U}_{\omega^*}$ and ${}^t({}^tR) = R \in \mathcal{U}_\omega$, so that $\mathcal{U}_{\omega^*} = \{{}^tR\}_{R\in\mathcal{U}_\omega}$. For $n \in \mathbb{Z}$, we have

$$\begin{aligned}\widehat{{}^tR}(n) &= \langle e_{-n}, {}^t R e_o\rangle = \langle Re_{-n}, e_o\rangle = \langle RS^{-n}e_o, e_o\rangle \\ &= \langle S^{-n}Re_o, e_o\rangle = \langle Re_o, e_{-n}\rangle = \widehat{R}(n).\end{aligned}$$

Hence $\widehat{{}^tR} = \widehat{R}$ and $\mathcal{M}_\omega = \mathcal{M}_{\omega^*}$.

Now let $R \in \mathcal{U}_\omega$, $u \in \ell^2_\omega(\mathbb{Z})$, $v \in \ell^2_{\omega^*}(\mathbb{Z})$.

We have

$$\sum_{|k|\le p} \widehat{R}(k)(u * v)_{n-k} = \sum_{|k|\le p} \widehat{R}(k)\langle S^{k-n}u, v\rangle = \langle s_p(R)S^{-n}u, v\rangle.$$

We thus see that the sequence $\sum_{|k|\le p} \widehat{R}(k)(u * v)_{n-k}$ converges in the sense of Cesaro to $\langle RS^{-n}u, v\rangle = \langle S^{-n}Ru, v\rangle = (Ru * v)_n$. So if we define $(\widehat{R} * (u * v))_n$ to be the limit in the sense of Cesaro of the sequence $\left(\sum_{|k|\le p} \widehat{R}(k)(u*v)_{n-k}\right)_{p\ge1}$ we see that $\varphi * (u * v)$ is well defined for $\varphi \in \mathcal{M}_\omega$, $u \in \ell^2_\omega(\mathbb{Z})$, $v \in \ell^2_{\omega^*}(\mathbb{Z})$ and that we have

$$\varphi * (u * v) = (\varphi * u) * v = u * (\varphi * v) \qquad (u \in \ell^2_\omega(\mathbb{Z}),\ v \in \ell^2_{\omega^*}(\mathbb{Z}), \varphi \in \mathcal{M}_\omega = \mathcal{M}_{\omega^*}). \tag{2.3}$$

Recall that if H is a separable Hilbert space we can identify $\mathcal{B}(H)$ to the dual of the space $\mathcal{C}^1(H)$ of trace-class operators, the duality being implemented by the formula $\langle N, T\rangle = \mathrm{Tr}(TN)$ for $N \in \mathcal{C}^1(H)$, $T \in \mathcal{B}(H)$. This allows to consider the

w^* (or ultraweak) topology on $\mathcal{B}(H)$. Now $\mathcal{U}_\omega$ is w^*-closed in $\mathcal{B}(\ell^2_\omega(\mathbb{Z}))$, and so we can identify $\mathcal{U}_\omega$ to the dual of $\mathcal{U}_{\omega,*} = \mathcal{C}^1(\ell^2_\omega(\mathbb{Z}))/\mathcal{U}_\omega^\perp$, where

$$\mathcal{U}_\omega^\perp = \bigcap_{T\in\mathcal{U}_\omega} \{N \in \mathcal{C}^1(H) \mid \mathrm{Tr}(TN) = 0\}.$$

We will denote by $\|.\|^*_{op}$ the quotient norm on $\mathcal{U}_{\omega,*}$. The map $v \to v_* = \left(\frac{\overline{v_{-n}}}{\omega^2(n)}\right)_{n\in\mathbb{Z}}$ is an antilinear isometry from $\ell^2_{\omega^*}(\mathbb{Z})$ onto $\ell^2_\omega(\mathbb{Z})$. Now let $u \otimes v_*$ be the rank one operator $w \to [w, v_*]u$ and let $[u * v_*]$ be the image of $u * v_*$ in $\mathcal{U}_{\omega,*}$. We have, for $R \in \mathcal{U}_{\omega*}$

$$\langle R, [u \otimes v_*]\rangle = [Ru, v_*] = \langle Ru, v\rangle. \tag{2.4}$$

For $\theta \in \mathcal{U}_{\omega,*}$ set

$$\widehat{\theta}(n) = \langle S^{-n}, \theta\rangle \qquad (n \in \mathbb{Z}). \tag{2.5}$$

Then $[u \widehat{\otimes} v_*](n) = \langle S^{-n}u, v\rangle = (u * v)_n$ and we have

$$[u \widehat{\otimes} v_*] = u * v \qquad (u \in \ell^2_\omega(\mathbb{Z}),\ v \in \ell^2_{\omega^*}(\mathbb{Z})). \tag{2.6}$$

For $R \in \mathcal{U}_\omega, \theta \in \mathcal{U}_{\omega,*}$ set

$$\langle \widehat{R}, \widehat{\theta}\rangle = \langle R, \theta\rangle. \tag{2.7}$$

Also set $\left\|\widehat{\theta}\right\|^*_{op} = \|\theta\|^*_{op}$ for $\theta \in \mathcal{U}_{\omega,*}$. By using formula (2.7), we can identify $\mathcal{M}_\omega$ to the dual of $(\mathcal{M}_{\omega,*}, \|.\|^*_{op}) := (\widehat{\mathcal{U}_{\omega,*}}, \|.\|^*_{op})$. If $u \in \ell^2_\omega(\mathbb{Z})$, $v \in \ell^2_{\omega^*}(\mathbb{Z})$, then $u * v = [u \widehat{\otimes} v_*] \in \mathcal{M}_{\omega,*}$ and we have

$$\langle \varphi, u * v\rangle = \langle \varphi * u, v\rangle \qquad (\varphi \in \mathcal{M}_\omega). \tag{2.8}$$

In fact $\|u * v\|^*_{op} = \inf \sum_{p=1}^{\infty} \left\|u^{(p)}\right\|_\omega \left\|v^{(p)}\right\|_{\omega^*}$, the infimum being taken over all families $(u^{(p)}, v^{(p)})_{p\geq 1}$ of elements of $\ell^2_\omega(\mathbb{Z}) \times \ell^2_{\omega^*}(\mathbb{Z})$ such that $\sum_{p=1}^{\infty} \left\|u^{(p)}\right\|_\omega \left\|v^{(p)}\right\|_{\omega^*} < +\infty$ and such that $(u * v)_n = \sum_{p=1}^{\infty} (u^{(p)} * v^{(p)})_n$ for every $n \in \mathbb{Z}$.

Set again $e_m = (\delta_{m,n})_{n\in\mathbb{Z}}$ for $m \in \mathbb{Z}$. Since $e_m = e_{p+m} * e_{-p}$ for $p \in \mathbb{Z}$ we obtain

$$\|e_m\|^*_{op} \leq \inf_{p\in\mathbb{Z}} \frac{\omega(p+m)}{\omega(p)} = \frac{1}{\widetilde{\omega}(-m)} \qquad (m \in \mathbb{Z}). \tag{2.9}$$

It follows from the definition of $(\mathcal{M}_{\omega^*}, \|.\|_{\omega,*})$ that the closed linear span of the convolution products $u * v$, where $u \in \ell^2_\omega(\mathbb{Z})$, $v \in \ell^2_{\omega^*}(\mathbb{Z})$, is norm-dense in $\mathcal{M}_{\omega,*}$. Let $\mathcal{E}$ be the set if sequences $w = (w_n)_{n\in\mathbb{Z}}$ such that $\operatorname{Supp} w = \{n \in \mathbb{Z} \mid w_n \neq 0\}$ is finite. Since $\mathcal{E} * \mathcal{E} \subset \mathcal{E}$, and since $\mathcal{E} * \mathcal{E}$ is dense in $\ell^2_\omega(\mathbb{Z}) \times \ell^2_{\omega^*}(\mathbb{Z})$, we see that $\mathcal{E}$ is norm dense in $\mathcal{M}_{\omega,*}$.

Now for $m \geq 0$, $w = (w_n)_{n\in\mathbb{Z}} \in \mathcal{M}_{\omega,*}$, set

$$s_m(w) = \sum_{|k|\leq m} w_k e_k, \qquad \sigma_m(w) = \frac{1}{m+1}\sum_{p=0}^{m} s_p(w).$$

For $R \in \mathcal{U}_\omega$, $\theta \in \mathcal{U}_{\omega,*}$, $p \geq 0$ we have

$$\langle s_p(R), \theta\rangle = \sum_{|k|\leq p} \widehat{R}(k)\langle S^k, \theta\rangle = \sum_{|k|\leq p} \widehat{R}(k)\widehat{\theta}(-k) = \langle \widehat{R}, s_p(\widehat{\theta})\rangle.$$

Hence $\langle \sigma_m(R), \theta\rangle = \langle \widehat{R}, \sigma_m(\widehat{\theta})\rangle$ for $R \in \mathcal{U}_\omega, \theta \in \mathcal{U}_{\omega,*}$. Since $\|\sigma_m(R)\|_{op} \leq \|R\|_{op}$ for $m \geq 0$, this shows that $\|\sigma_m(w)\|^*_{op} \leq \|w\|^*_{op}$ for $w \in \mathcal{M}_{\omega,*}$.

Since $\sigma_m(w) = w$ for $w \in \mathcal{E}$ when m is sufficiently large, we obtain

$$\|\sigma_m(w) - w\|^*_{op} \to 0 \qquad (m \to \infty,\ w \in \mathcal{M}_{\omega,*}). \tag{2.10}$$

Now denote by $\mathcal{W}$ the set of all sequences $w = (w_n)_{n\in\mathbb{Z}}$ such that $\overline{\lim}_{|n|\to\infty} |w_n|^{1/|n|} < +\infty$. For $w \in \mathcal{W}$, consider the formal Laurent series $\widetilde{w}(z) = \sum_{n\in\mathbb{Z}} w_n z^n$. Notice that $\sum_{n=0}^{\infty} w_n z^n$ converges for $|z| < 1/\overline{\lim}_{n\to\infty} |w_n|^{1/n}$, and that $\sum_{n<0} w_n z^n$ converges for $|z| > \overline{\lim}_{n\to-\infty} |w_n|^{1/|n|}$. Now set

$$R_+ = \overline{\lim_{n\to\infty}}\, \omega(n)^{1/n}, \qquad r_+ = \underline{\lim_{n\to\infty}}\, \omega(n)^{1/n},$$

$$R_- = \overline{\lim_{n\to-\infty}}\, \omega(n)^{1/n}, \quad r_- = \underline{\lim_{n\to-\infty}}\, \omega(n)^{1/n},$$

so that $\sup(R_+, R_-) \leq \rho(S)$, $\inf(r_+, r_-) \geq 1/\rho(S^{-1})$. If $u \in \ell^2_\omega(\mathbb{Z})$, the series $\sum_{n=0}^{\infty} u_n z^n$ converges for $|z| < r_+$, and the series $\sum_{n<0} u_n z^n$ converges for $|z| > R_-$. Also if $v \in \ell^2_{\omega^*}(\mathbb{Z})$ the series $\sum_{n=0}^{\infty} v_n z^n$ converges for $|z| < r_-$, and the series $\sum_{n<0} v_n z^n$ converges for $|z| > R_+$.

It is well-known [Sh, Th. 10'] that if $\varphi \in \mathcal{M}_\omega$ then the series $\widetilde{\varphi}(z) = \sum_{n\in\mathbb{Z}} \varphi_n z^n$ is absolutely convergent for $1/\rho(S^{-1}) < |z| < \rho(S)$. In fact we have the following result.

PROPOSITION 2.1. *Assume that* $\frac{1}{\rho(S^{-1})} < \rho(S)$. *Then the map* $\varphi \to \widetilde{\varphi}(\lambda)$ *is a* w^**-continuous character of* $\mathcal{M}_\omega$ *for* $\frac{1}{\rho(S^{-1})} < |\lambda| < \rho(S)$.

Proof. We have $\varphi_n = \langle \varphi, e_{-n} \rangle$ for $n \in \mathbb{Z}$, and $\|e_{-n}\|_{op}^* \leq \frac{1}{\widetilde{\omega}(n)}$ for $n \in \mathbb{Z}$. Hence $\sum\limits_{n\in\mathbb{Z}} |\lambda|^n \|e_{-n}\|_{op}^* \leq \sum\limits_{n\in\mathbb{Z}} \frac{|\lambda|^n}{\widetilde{\omega}(n)} < +\infty$, since

$$\rho(S) = \lim_{n\to\infty} \widetilde{\omega}(n)^{1/n}, \qquad \frac{1}{\rho(S^{-1})} = \lim_{n\to-\infty} \widetilde{\omega}(n)^{1/n}.$$

Hence $\widetilde{\varphi}(\lambda) = \langle \varphi, \sum\limits_{n\in\mathbb{Z}} \lambda^n e_{-n} \rangle$ and the map $\lambda \to \widetilde{\varphi}(\lambda)$ is w^*-continuous. □

Notice that, defining if possible the product of two formal Laurent series by convolution of their coefficients, we have for $\lambda \neq 0$ the trivial equation

$$(z - \lambda) \cdot \Big(\sum_{n\in\mathbb{Z}} \lambda^{-n} z^n \Big) = 0. \tag{2.11}$$

In other terms

$$(e_1 - \lambda e_0) * (\lambda^{-n})_{n\in\mathbb{Z}} = 0. \tag{2.12}$$

Conversely if $w = (w_n)_{n\in\mathbb{Z}}$ is any sequence of complex numbers we can define the convolution product $(e_1 - \lambda e_0) * w = (w_{n-1} - \lambda w_n)_{n\in\mathbb{Z}}$ and we see that $(e_1 - \lambda e_0) * w = 0$ if and only if $w = w_0 \cdot (\lambda^{-n})_{n\in\mathbb{Z}}$. We obtain

PROPOSITION 2.2. *Let* $u \in \ell^2_\omega(\mathbb{Z})$ *and* $\lambda \in \mathbb{C}\setminus\{0\}$. *Then* $u \notin \bigvee\limits_{n\in\mathbb{Z}} S^n(S - \lambda I)u$ *if and only if there exists* $v \in \ell^2_{\omega^*}(\mathbb{Z})$ *such that* $u * v = (\lambda^{-n})_{n\in\mathbb{Z}}$.

Proof. Assume that $u * v = (\lambda^{-n})_{n\in\mathbb{Z}}$ Then $(S - \lambda)u * v = [(e_1 - \lambda e_0) * u] * v = (e_1 - \lambda e_0) * (u * v) = 0$, by (2.3) and (2.12). Hence $v \perp \bigvee\limits_{n\in\mathbb{Z}} S^n(S - \lambda I)u$, and $\langle u, v \rangle = (u * v)_0 = 1$, so that $u \notin \bigvee\limits_{n\in\mathbb{Z}} S^n(S - \lambda I)u$.

Conversely assume that $u \notin \bigvee\limits_{n\in\mathbb{Z}} S^n(S - \lambda I)u$, and let $v \in \ell^2_{\omega^*}(\mathbb{Z})$ such that $\langle u, v \rangle = 1$ and $\langle S^n(S - \lambda I)u, v \rangle = 0$ for $n \in \mathbb{Z}$. Then $(S - \lambda I)u * v = 0$. Again

$$(S - \lambda I)u * v = [(e_1 - \lambda e_0) * u] * v = (e_1 - \lambda e_0) * (u * v)$$

and

$$u * v = \langle u, v \rangle (\lambda^{-n})_{n\in\mathbb{Z}} = (\lambda^{-n})_{n\in\mathbb{Z}}.$$

□

3. Apostol's results

Apostol [Ap] showed that if $\{r_+, r_-, R_+, R_-\}$ is not a singleton, or if

$$r_+ = r_- = R_+ = R_- \in \left(\frac{1}{\rho(S^{-1})}, \rho(S)\right),$$

then $\ell^2_\omega(\mathbb{Z})$ possesses a nontrivial translation invariant subspace. His results are based on the use of auxiliary weights, related as follows to a given weight ω.

DEFINITION 3.1. Let ω be a weight on $\mathbb{Z}$. We denote by $\mathcal{A}(\omega)$, (resp. $A^+(\omega)$, resp. $\mathcal{A}^-(\omega)$) the set of weights σ on $\mathbb{Z}$ such that for every $n \in \mathbb{Z}$ there exists a sequence $(k_p)_{\,p\geq 1}$ of integers (resp. positive integers, resp. negative integers) which satisfies the two following conditions

(i) $\omega(k_p)\sigma(-k_p) + \frac{\sigma(k_p)}{\omega(k_p)} \to 0 \ (p \to \infty)$,

(ii) $\overline{\lim\limits_{p\to\infty}} \frac{\omega(k_p)}{\omega(k_{p+n})} \geq \frac{1}{\sigma(n)}$.

Denote by $\widetilde{S}$ the image of S in the Calkin algebra. Then

$$\left\|\widetilde{S}^{-n}\right\| = \overline{\lim_{|k|\to\infty}} \frac{\omega(k-n)}{\omega(k)} = \overline{\lim_{|k|\to\infty}} \frac{\omega(k)}{\omega(k+n)}.$$

So if $\sigma \in \mathcal{A}(\omega)$ then $\frac{1}{\sigma(n)} \leq \left\|\widetilde{S}^{-n}\right\|$, and

$$\sigma(n)\widetilde{\omega}(-n) = \sigma(n)\left\|S^{-n}\right\|_{op} \geq \sigma(n)\left\|\widetilde{S}^{-n}\right\| \geq 1.$$

Set again $e_m = (\delta_{m,n})_{n\in\mathbb{Z}}$ for $m \in \mathbb{Z}$. It follows from (2.8) that

$$\|e_n\|^*_{op} \leq \frac{1}{\widetilde{\omega}(-n)} \leq \sigma(n).$$

So if $w \in \ell^1_\sigma(\mathbb{Z})$ then $\sum\limits_{n\in\mathbb{Z}} |w_n|\,\|e_n\|^*_{op} < +\infty$ and $w = \sum\limits_{n\in\mathbb{Z}} w_n e_n \in \mathcal{M}_{\omega,*}$, which shows that $\ell^1_\sigma(\mathbb{Z}) \subset \mathcal{M}_{\omega,*}$.

By using elementary computations in $\ell^1_\sigma(\mathbb{Z})$, Apostol showed that $\ell^2_\omega(\mathbb{Z})$ has a nontrivial translation invariant subspace if $\mathcal{A}(\omega) \neq \phi$. His arguments show also that if $\mathcal{A}^+(\omega) \neq \phi$ then there exists $u \in \ell^2_\sigma(\mathbb{Z}^+)\backslash\{0\}$, $v \in \ell^2_{\omega^*}(\mathbb{Z}^-)\backslash\{0\}$ such that $u * v = 0$, where

$$\ell^2_\omega(\mathbb{Z}^+) = \{u \in \ell^2_\omega(\mathbb{Z}) \mid u_n = 0 \ \ (n < 0)\},$$
$$\ell^2_{\omega^*}(\mathbb{Z}^-) = \{v \in \ell^2_{\omega^*}(\mathbb{Z}) \mid v_n = 0 \ \ (n \geq 0)\}.$$

Similarly if $\mathcal{A}^-(\omega) \neq \phi$ there exists $u \in \ell^2_\omega(\mathbb{Z}^-)\backslash\{0\}$ and $v \in \ell^2_\omega(\mathbb{Z}^+)\backslash\{0\}$ such that $u * v = 0$.

Our aim is to show that Apostol's elementary computations are a special case of "Banach spaces methods in operator theory" presented by Bercovici, Foias and Pearcy in [BFP]. Recall that a subset E of a complex Banach space X is said to be balanced if $\lambda E \subset E$ for $|\lambda| \leq 1$, and E is said to be absolutely convex if it is convex and balanced.

Also if $E \subset X, F \subset X$ we set $\mathrm{Dist}(E, F) = \sup_{x\in E}[\inf_{y\in F} \|x - y\|]$. We will denote by $\mathrm{aco}(B)$ the absolutely convex hull of a subset B of X. Recall that the Radon–Nikodym property for a Banach space X means that every bounded harmonic function $\theta: \mathbb{D} \to X$ has nontangential limits almost everywhere on $\mathbb{T}$. Every separable Banach space X which is a dual space (i.e. $X = Y^*$ for some Banach space Y) has the Radon–Nikodym property. In particular, $\ell^1_\sigma(\mathbb{Z})$ has the Radon–Nikodym property for every weight σ.

The first part of the paper [BFP] is devoted to the theory of "super-dominating sets". They obtain in particular the following result.

PROPOSITION 3.2. *Let X be a separable Banach space having the Radon–Nikodym property, let C be a closed absolutely convex subset of X and let B be a balanced subset of C. Assume that for every $f \in X^*$, every $\varepsilon > 0$ and every finite subset F of B there exists $x \in B$ saytisfying the two following conditions*

(i) $|f(x)| + \varepsilon > \sup\{|f(y)| : y \in C\}$,

(ii) $\mathrm{Dist}(\mathrm{aco}(x, b), B) < \varepsilon$ *for every* $b \in F$.

Then B is norm-dense in C.

Now let ω and σ be two weights on $\mathbb{Z}$. The bilinear map $\tau: (u, v) \to u * v$ is partially defined from $\ell^2_\omega(\mathbb{Z}) \times \ell^2_{\omega^*}(\mathbb{Z})$ into $\ell^1_\sigma(\mathbb{Z})$. Moreover if we denote by $\mathcal{E}$ the space of complex sequences with finite support we have $\mathcal{E} * \mathcal{E} \subset \mathcal{E}$, so that $\mathcal{E} * \mathcal{E} \subset \mathcal{D}(\tau)$. Set again $e_m = (\delta_{m,n})_{n\in\mathbb{Z}}$ for $m \in \mathbb{Z}$. Then $u * e_0 = u$ for $u \in \ell^2_\omega(\mathbb{Z})$, and so $u = 0$ if $u * \mathcal{E} = 0$. Similarly $v = 0$ if $\mathcal{E} * v = 0$.

Also since the maps $(u, v) \to (u * v)_n$ and $w \to w_n$ are continuous on $\ell^2_\omega(\mathbb{Z}) \times \ell^2_{\omega^*}(\mathbb{Z})$, and on $\ell^1_\sigma(\mathbb{Z})$ for every $n \in \mathbb{Z}$, we see that the graph

$$\mathcal{G} = \{(u, v, w) \mid (u, v) \in \mathcal{D}(\tau), w \in \ell^1_\sigma(\mathbb{Z}), w = u * v\}$$

is closed in $\ell^2_\omega(\mathbb{Z}) \times \ell^2_{\omega^*}(\mathbb{Z}) \times \ell^1_\sigma(\mathbb{Z})$.

The following result follows then immediately from [BFP, Th. 6.6 and Th. 8.3].

LEMMA 3.3. *Let ω and σ be two weights on $\mathbb{Z}$. Assume that for every $w \in \ell^1_\sigma(\mathbb{Z})$, for every finite subset F of $\mathcal{E}$ and every $\varepsilon > 0$ there exists $u \in \ell^2_\omega(\mathbb{Z})$ and $v \in \ell^2_{\omega^*}(\mathbb{Z})$ satisfying the following properties*
(i) $(\|u\|_\omega)^2 \le \|w\|^1_\sigma (1+\varepsilon)$, $(\|v\|_\omega)^2 \le \|w\|^1_\sigma (1+\varepsilon)$,
(ii) $\|w - u * v\|^1_\sigma < \varepsilon$,
(iii) $\sum_{x \in F} \left[\|x * v\|^1_\sigma + \|u * x\|^1_\sigma\right] < \varepsilon$.
Then for every family $(w_{i,j})_{i\ge1, j\ge1}$ of elements of $\ell^1_\sigma(\mathbb{Z})$ there exists a sequence $(u_i)_{i\ge1}$ of elements of $\ell^2_\omega(\mathbb{Z})$ and a sequence $(v_j)_{j\ge1}$ of elements of $\ell^2_{\omega^*}(\mathbb{Z})$ such that $w_{i,j} = u_i * v_j \quad (i \ge 1,\ j \ge 1)$.

REMARK 1. If conditions (i), (ii) and (iii) are satisfied with $u \in \ell^2_\omega(\mathbb{Z}^+)$, $v \in \ell^2_{\omega^*}(\mathbb{Z}^-)$ then the factorizations of Lemma 3.3 can be performed with $(u_i)_{i\ge1} \subset \ell^2_\omega(\mathbb{Z}^+)$ and $(v_j)_{j\ge1} \subset \ell^2_{\omega^*}(\mathbb{Z}^-)$. A similar property holds if conditions (i), (ii) and (iii) are satisfied with $u \in \ell^2_\omega(\mathbb{Z}^-)$, $v \in \ell^2_{\omega^*}(\mathbb{Z}^+)$.

Using Proposition 3.2 and Lemma 3.3, we obtain the following result.

THEOREM 3.4. *Let ω and σ be two weights on $\mathbb{Z}$, and assume that $\sigma \in \mathcal{A}(\omega)$. Then for every family $(w_{i,j})_{i\ge1,j\ge1}$ of elements of $\ell^1_\sigma(\mathbb{Z})$ there exists a sequence $(u_i)_{i\ge1}$ of elements of $\ell^2_\omega(\mathbb{Z})$ and a sequence $(v_j)_{j\ge1}$ of elements of $\ell^2_{\omega^*}(\mathbb{Z})$ such that*

$$w_{i,j} = u_i * v_j \qquad (i \ge 1, j \ge 1).$$

If $\sigma \in \mathcal{A}^+(\omega)$, the same properties hold for $\ell^2_\omega(\mathbb{Z}^+)$ and $\ell^2_{\omega^*}(\mathbb{Z}^-)$, and if $\sigma \in \mathcal{A}^-(\omega)$ the same properties hold for $\ell^2_\omega(\mathbb{Z}^-)$ and $\ell^2_{\omega^*}(\mathbb{Z}^+)$.

Proof. Let $\sigma \in \mathcal{A}(\omega)$. It is possible to show by elementary computations analogous to the proof of [Ap, Prop.1] that σ satisfies conditions (i), (ii), (iii) of Lemma 3.3, but we prefer to use Proposition 3.2. Let $\varepsilon > 0$, and let $F = \{x_1, \dots, x_k\}$ be a finite subset of $\mathcal{E}$.

Set

$$B = \Big\{u * v \;\Big|\; u \in \mathcal{E},\ v \in \mathcal{E},\ \|u\|_\omega < 1+\varepsilon,\ \|v\|_{\omega^*} < 1+\varepsilon,\ \|u * v\|^1_\sigma \le 1, \\ \sum_{x \in F} (\|x * v\|^1_\sigma + \|u * x\|^1_\sigma) < \varepsilon\Big\},$$

and let C be the closed unit ball of $\ell^1_\sigma(\mathbb{Z})$. In order to show that σ satisfies conditions (i), (ii) and (iii) of Lemma 3.3 it is sufficient to show that the balanced set B is dense in C.

We can identify the dual of $\ell^1_\sigma(\mathbb{Z})$ to $\ell^\infty_{\sigma^*}(\mathbb{Z})$ by using formula (2.1). Fix $\eta > 0$. Let $h = (h_n)_{n\in\mathbb{Z}} \in \ell^\infty_{\sigma^*}(\mathbb{Z})\backslash\{0\}$, and let $n \in \mathbb{Z}$ be such that $\frac{|h-n|}{\sigma(n)} > \|h\|^\infty_{\sigma^*}(1-\eta)$.

Since $\sigma \in \mathcal{A}(\omega)$, there exists a sequence $(k_p)_{p\geq 1}$ of integers such that $\omega(k_p)\sigma(-k_p)+\frac{\sigma(k_p)}{\omega(k_p)}\to 0$ $(p\to\infty)$ and such that $\frac{\omega(k_p)\sigma(n)}{\omega(k_{p+n})}\to\ell$ $(p\to\infty)$, where $\ell\geq 1$. Set

$$f_p = \frac{e_{k_p+n}}{\omega(k_p+n)}, \qquad g_p = \omega(k_p)e_{-k_p},$$

so that $\|f_p\|_\omega = \|g_p\|_{\omega^*} = 1$, and $\|f_p * g_p\|^1_\sigma = \frac{\omega(k_p)}{\omega(k_p+n)}\sigma(n)$. Set

$$\delta_p = \max(1, \|f_p * g_p\|^1_\sigma),$$

so that $\delta_p\to\ell$ and $\delta_p^{-1}\|f_p * g_p\|^1_\sigma\to 1$ $(p\to\infty)$, with $\delta_p^{-1}\|f_p * g_p\|^1_\sigma \leq 1$ for $p\geq 1$. Also set $u_p = \delta_p^{-1/2}\cdot f_p, v_p = \delta_p^{-1/2}\cdot g_p$. For $m\in\mathbb{Z}$ we have with $p\to\infty$

$$\|u_p * e_m\|^1_\sigma = \frac{\delta_p^{-1/2}\sigma(k_p+m+n)}{\omega(k_p+n)} \leq \delta_p^{-1/2}\widetilde{\sigma}(n+m)\widetilde{\omega}(-n)\frac{\sigma(k_p)}{\omega(k_p)}\to 0.$$

Also with $p\to\infty$

$$\|e_m * v_p\|^1_\sigma = \delta_p^{-1/2}\omega(k_p)\sigma(m-k_p) \leq \delta_p^{-1/2}\widetilde{\sigma}(m)\omega(k_p)\sigma(-k_p)\to 0.$$

Hence $\|x * v_p\|^1_\sigma + \|u_p * x\|^1_\sigma \to 0$ $(p\to\infty)$ for every $x\in\mathcal{E}$.

Since $\|u_p\|_\omega \leq 1, \|v_p\|_{\omega^*}\leq 1$ and $\|u_p * v_p\|^1_\sigma \leq 1$, we see that $u_p * v_p \in B$ when p is sufficiently large. We have

$$|\langle u_p * v_p, h\rangle| = \delta_p^{-1}\frac{\omega(k_p)}{\omega(k_p+n)}|\langle e_n, h\rangle| = \delta_p^{-1}\frac{\omega(k_p)}{\omega(k_p+n)}|h_{-n}|$$

so that $|\langle u_p * v_p, h\rangle| \to |h_{-n}|/\sigma(n)$ $(p\to\infty)$, and $|\langle u_p * v_p, h\rangle| > (1-\eta)\|h\|^\infty_{\sigma^*}$ when p is sufficiently large.

In order to apply Proposition 3.2 it suffices to show that $\mathrm{Dist}(\mathrm{aco}(u_p*v_p, y), B)\to 0$ $(p\to\infty)$ for every $y = u*v\in B$. Let $\lambda, \mu\in\mathbb{C}$, with $|\lambda|+|\mu|\leq 1$, and let $\alpha,\beta\in\mathbb{C}$ such that $\alpha^2=\lambda$, $\beta^2=\mu$. We have

$$\lambda y + \mu(u_p * v_p) = (\alpha u + \beta u_p) * (\alpha v + \beta v_p) - \alpha\beta(u * v_p + u_p * v).$$

Set $\varepsilon_p = \|u * v_p\|^1_\sigma + \|u_p * v\|^1_\sigma$. Then $\varepsilon_p \to 0$ $(p\to\infty)$ and $\|\alpha\beta(u*v_p+u_p*v)\|^1_\sigma \leq \varepsilon_p$ for $p\geq 1$. We have

$$\|\alpha u + \beta u_p\|^2_\omega \leq |\lambda|\,\|u\|^2_\omega + |\mu|\,\|u_p\|^2_\omega + 2\left|[u, u_p]_\omega\right| < (1+\varepsilon)^2 + 2\left|[u, u_p]_\omega\right|$$

and similarly $\|\alpha v + \beta v_p\|^2_{\omega^*} < (1+\varepsilon)^2 + 2\left|[v, v_p]_{\omega^*}\right|$ where we denote by $[.,.]_\omega$ and $[.,.]_{\omega^*}$ the hermitian products on $\ell^2_\omega(\mathbb{Z})$ and $\ell^2_{\omega^*}(\mathbb{Z})$. Also, for $i \leq k$,

$$\begin{aligned}\|(\alpha u + \beta u_p) * x_i\|^1_\sigma &+ \|x_i * (\alpha v + \beta v_p)\|^1_\sigma \\ &\leq \|u * x_i\|^1_\sigma + \|u_p * x_i\|^1_\sigma + \|x_i * v\|^1_\sigma + \|x_i * v_p\|^1_\sigma .\end{aligned}$$

Hence there exists $p_1 \geq 1$, which does not depend on λ and μ, such that $\|\alpha u + \beta u_p\|_\omega < 1 + \varepsilon$, $\|\alpha u + \beta v_p\|_{\omega^*} < 1 + \varepsilon$ and

$$\sum_{i=1}^{k} \|(\alpha u + \beta u_p) * x_i\|^1_\sigma + \|x_i * (\alpha v + \beta v_p)\|^1_\sigma < \varepsilon$$

for $p \geq p_1$.

Also $\|(\alpha u + \beta u_p) * (\alpha v + \beta v_p)\|^1_\sigma \leq 1 + \varepsilon_p$. We thus see that $\frac{1}{1+\varepsilon_p}(\alpha u + \beta u_p) * (\alpha v + \beta v_p) \in B$, so that $\mathrm{Dist}\left[(\alpha u + \beta u_p) * (\alpha v + \beta v_p), B\right] \leq \varepsilon_p$ for $p \geq p_1$.

Since $\|\lambda y + \mu(u_p * v_p) - (\alpha u + \beta v_p) * (\alpha v + \beta v_p)\|^1_\sigma < \varepsilon_p$ for $p \geq 1$, we see that $\mathrm{Dist}(\mathrm{aco}(u_p * v_p, y), B) \to 0$ $(p \to \infty)$ for every $y \in B$. It follows then from Proposition 3.2 that B is dense in C, and the Theorem follows then immediately from Lemma 3.3 in the case where $\sigma \in \mathcal{A}(\omega)$. Using Remark 1, we see that the same proof works in the case where $\sigma \in \mathcal{A}^+(\omega)$ or $\sigma \in \mathcal{A}^-(\omega)$.

□

Using a standard argument from the theory of dual algebras, we obtain Apostol's result

COROLLARY 3.5. *Let ω be a weight. If $\mathcal{A}(\omega) \neq \phi$, then $\ell^2_\omega(\mathbb{Z})$ has a nontrivial invariant subspace.*

Proof. Let $\sigma \in \mathcal{A}(\omega)$ and let w be a nonzero element of $\ell^1_\sigma(\mathbb{Z})$. It follows from the Theorem that there exists $u_1, u_2 \in \ell^2_\omega(\mathbb{Z})$ and $v_1, v_2 \in \ell^2_{\omega^*}(\mathbb{Z})$ such that $u_1 * v_1 = u_2 * v_2 = 0$ and such that $u_1 * v_2 = u_2 * v_1 = w \neq 0$.

□

Now let ω be a weight. Set again

$$R_+ = \overline{\lim_{n\to\infty}}\, \sigma(n)^{1/n}, \qquad r_+ = \underline{\lim_{n\to\infty}}\, \sigma(n)^{1/n},$$

$$R_- = \overline{\lim_{n\to-\infty}}\, \sigma(n)^{1/n}, \quad r_- = \underline{\lim_{n\to\infty}}\, \sigma(n)^{1/n}.$$

Lemmas 1, 2, 3 from [Ap] are implicitely based on the following elementary observation.

LEMMA 3.6. *Let $(t_n)_{n\geq 0}$ be a sequence of real numbers and assume that $t_{n+1} - t_n \to 0$ $(n \to \infty)$. Let $\gamma \in \left(\underline{\lim}_{n\to\infty} t_n, \overline{\lim}_{n\to\infty} t_n\right)$. Then for every $m \in \mathbb{Z}\backslash\{0\}$ there exists a strictly increasing sequence $(k_p)_{p\geq 1}$ of integers such that $t_{k_p+m} \geq \gamma$, $t_{k_p} \leq \gamma$ for $p \geq 1$, and $\lim_{p\to\infty} t_{k_p} = \lim_{p\to\infty} t_{k_p+m} = \gamma$.*

Proof. Let $\varepsilon > 0$ be such that $\gamma + \varepsilon < \overline{\lim}_{n\to\infty} t_n$. Let $m \geq 1$ and $N \geq 1$. There exists $q > N$ such that $t_q < \gamma$. Since $\sup_{|r|\leq m-1} |t_n - t_{n+r}| \to 0$ $(n \to \infty)$ there exists $N_1 \geq q$ such that $|t_n - t_{n+r}| < \varepsilon$ for $n \geq N_1$, $|r| \leq m$. Let $n \geq N_1 + m$ such that $t_n > \gamma + \varepsilon$. There exists $k \geq 1$ such that $0 \leq n - q - km \leq m-1$ and so $t_{q+km} > \gamma$. If $k \geq 1$ is the smallest integer having this property we have $t_{q+(k-1)m} \leq \gamma < t_{q+km}$.

We can thus construct by induction a strictly increasing sequence $(k_p)_{p\geq 1}$ of integers such that $t_{k_p} \leq \gamma$, $t_{k_p+m} > \gamma$ and since $t_{k_p+m} - t_{k_p} \to 0$ $(p \to \infty)$ we have $\lim_{p\to\infty} t_{k_p} = \lim_{p\to\infty} t_{k_p+m} = \gamma$. The case $m \leq -1$ follows from the same argument applied to the sequence $(-t_n)_{n\geq 1}$.

□

Notice that if $(s_n)_{n\geq 1}$ is a sequence of positive real numbers such that $0 < \inf_{n\geq 0} \frac{s_{n+1}}{s_n} \leq \sup_{n\geq 0} \frac{s_{n+1}}{s_n} < +\infty$, we have $s_{n+1}^{1/n+1} - s_n^{1/n} \to 0$ $(n \to \infty)$. To see this set

$$\delta = \inf_{n\geq 0} \frac{s_{n+1}}{s_n}, \qquad \rho = \sup_{n\geq 0} \frac{s_{n+1}}{s_n}.$$

Then $\delta^{1/n} \leq \left(\frac{s_{n+1}}{s_n}\right)^{1/n} \leq \rho^{1/n}$, and so $\left(\frac{s_{n+1}}{s_n}\right)^{1/n} \to 1$ $(n \to \infty)$. Also

$$\delta^n s_0 \leq s_n \leq \rho^n s_0, \qquad \delta \leq \liminf_{n\to\infty} s_n^{1/n} \leq \limsup_{n\to\infty} s_n^{1/n} \leq \rho$$

and

$$\log s_{n+1}^{\frac{1}{n+1}-\frac{1}{n}} = -\frac{1}{n(n+1)} \log s_{n+1} \to 0 \quad (n \to \infty).$$

Hence $s_n^{-1/n} \cdot s_{n+1}^{1/n+1} \to 1$ $(n \to \infty)$ and

$$s_{n+1}^{1/n+1} - s_n^{1/n} = s_n^{1/n}\left(s_n^{-1/n} \cdot s_{n+1}^{1/n+1} - 1\right) \to 0 \quad (n \to \infty).$$

PROPOSITION 3.7. *Let ω be a weight on $\mathbb{Z}$.*

(i) *If $r_+ < R_+$ and if $\sigma(n) = r_+^n$ for $n \geq 0$, $\sigma(n) = R_+^n$ for $n < 0$, then $\sigma \in \mathcal{A}^+(\omega)$.*

(ii) *If $r_- < R_-$ and if $\sigma(n) = r_-^n$ for $n \geq 0$, $\sigma(n) = R_-^n$ for $n < 0$, then $\sigma \in \mathcal{A}^-(\omega)$.*

(iii) *If $\frac{1}{\rho(S^{-1})} < \inf(r_+, r_-), \rho(S) > \sup(R_+, R_-)$ and if $\sigma(n) = \frac{1}{\rho(S^{-1})^n}$ for $n \geq 0$, $\sigma(n) = \rho(S)^n$ for $n < 0$ then $\sigma \in \mathcal{A}(\omega)$.*

Proof. This argument is essentially the same as Apostol original proof from [Ap], but the use of Lemma 3.6 seems to bring some clarification in the proof of (i) and (ii).

(i) It follows from Lemma 3.6 that for every $\gamma \in (r_+, R_+)$ and for every $n \neq 0$ there exists a strictly increasing sequence $(k_p)_{p \geq 1}$ of positive integers, with $k_1 > |n|$ such that $\omega(k_p)^{1/k_p} \leq \gamma$, $\omega(k_p - n)^{1/k_p - n} \geq \gamma$. Set $q_p = k_p - n$. We have

$$\frac{\sigma(q_p)}{\omega(q_p)} = \frac{\sigma(k_p - n)}{\omega(k_p - n)} \leq \left(\frac{r^+}{\gamma}\right)^{k_p - n}$$

and

$$\omega(q_p)\sigma(-q_p) = \omega(k_p - n)\sigma(-k_p + n) \leq \widetilde{\omega}(-n)\widetilde{\sigma}(n)\omega(k_p)\sigma(-k_p)$$
$$\leq \widetilde{\omega}(-n)\widetilde{\sigma}(n)\left(\frac{\gamma}{R^+}\right)^{k_p}.$$

We thus see that $\frac{\sigma(q_p)}{\omega(q_p)} + \omega(q_p)\sigma(-q_p) \to 0$ $(p \to \infty)$. Also $\frac{\omega(q_p)}{\omega(q_p+n)}\sigma(n) = \frac{\omega(k_p-n)}{\omega(k_p)}\sigma(n) \geq \gamma^{-n}\sigma(n)$. If $n > 0$ then $\gamma^{-n}\sigma(n) = \left(\frac{r_+}{\gamma}\right)^n$, and if $n < 0$ then $\gamma^{-n}\sigma(n) = \left(\frac{\gamma}{R_+}\right)^{|n|}$. We thus see that for every $\varepsilon > 0$ and every $N \geq 1$ there exists $k \geq N$ such that

$$\omega(k)\sigma(-k) + \frac{\sigma(k)}{\omega(k)} < \varepsilon, \qquad \frac{\omega(k)}{\omega(k+n)}\sigma(n) > 1 - \varepsilon.$$

Since $\sigma(0) = 1$, wee see that $\sigma \in \mathcal{A}^+(\omega)$.

(ii) We have $R_- = \overline{\lim}_{n \to \infty} \omega^*(n)^{1/n}$, $r_- = \underline{\lim}_{n \to \infty} \omega^*(n)^{1/n}$. Hence $\sigma \in \mathcal{A}^+(\omega^*) = \mathcal{A}^-(\omega)$.

(iii) Set $R = \rho(S), r = \frac{1}{\rho(S)}$. Denote again by $\widetilde{S}$ the image of S in the Calkin algebra. We have $\left\|\widetilde{S}^n\right\| = \overline{\lim}_{|k| \to \infty} \frac{\omega(k+n)}{\omega(k)} = \overline{\lim}_{|k| \to \infty} \frac{\omega(k)}{\omega(k-n)}$. Also $\rho(S) = \rho(\widetilde{S})$ and $r = \frac{1}{\rho(S^{-1})} = \frac{1}{\rho(\widetilde{S^{-1}})}$. We obtain

$$\overline{\lim}_{|k| \to \infty} \frac{\omega(k)}{\omega(k+n)} = \left\|\widetilde{S}^{-n}\right\| \geq R^{-n} = \frac{1}{\sigma(n)}$$

for $n < 0$,

$$\overline{\lim_{|k|\to\infty}} \frac{\omega(k)}{\omega(k+n)} = \left\|\widetilde{S}^{-n}\right\| \geq \rho(\widetilde{S}^{-1})^n = \frac{1}{\sigma(n)}$$

for $n \geq 0$. Also

$$\omega(k)\sigma(-k) = \frac{\omega(k)}{R^k}, \qquad \frac{\sigma(k)}{\omega(k)} = \frac{r^k}{\omega(k)} \quad \text{for } k \geq 0,$$

and since $R_+ < R$ and $r_+ > r$ we have

$$\omega(k)\sigma(-k) + \frac{\sigma(k)}{\omega(k)} \to 0 \quad (k \to \infty).$$

Similarly $\omega(k)\sigma(-k) = \omega(k)r^{-k}$ and $\frac{\sigma(k)}{\omega(k)} = \frac{R^k}{\omega(k)}$ for $k \leq 0$, and since $R_- < R$ and $r_- > r$ we have $\omega(k)\sigma(-k) + \frac{\sigma(k)}{\omega(k)} \to 0$ $(k \to -\infty)$. This shows that $\sigma \in \mathcal{A}(\omega)$. □

It follows in particular from Theorem 3.4 that if $\sigma \in \mathcal{A}(\omega)$, and if $\overline{\lim_{n\to\infty}} \sigma(n)^{1/n} < |\lambda| < \underline{\lim}_{n\to-\infty} \sigma(n)^{1/n}$, then there exists $u \in \ell^2_\omega(\mathbb{Z})$ and $v \in \ell^2_{\omega^*}(\mathbb{Z})$ such that $u * v = (\lambda^{-n})_{n\in\mathbb{Z}}$, and it follows then from Proposition 2.2 that $u \notin \bigvee_{n\in\mathbb{Z}} S^n(S-\lambda I)u$.

It follows then from Proposition 3.7 that this property holds for $r_+ < |\lambda| < R_+$ (with $u \in \ell^2_\omega(\mathbb{Z}^+)$) if $r_+ < R_+$, and for $r_- < |\lambda| < R_-$ (with $u \in \ell^2_\omega(\mathbb{Z}^-)$) if $r_- < R_-$. If

$$\frac{1}{\rho(S^{-1})} < \inf(r_+, r_-) \leq \sup(R_+, R_-) < \rho(S)$$

then this property holds for $\frac{1}{\rho(S^{-1})} < |\lambda| < \rho(S)$, i.e. for all interior points of $\operatorname{Spec} S$. We will see in the next section that in these situations we can choose $u \notin \bigvee_{n\in\mathbb{Z}} S^n(S-\lambda I)u$ which does not depend on λ.

If $R_+ < r_-$ then $u_\lambda = (\lambda^{-n})_{n\in\mathbb{Z}} \in \ell^2_\omega(\mathbb{Z})$ for $R_+ < |\lambda| < r_-$. In this case $(S-\lambda I)u_\lambda = 0$ and, trivially, $u_\lambda \notin \bigvee_{n\in\mathbb{Z}} S^n(S-\lambda I)u_\lambda$. If $R_- < r_+$ then the series $\sum_{n\in\mathbb{Z}} u_n\lambda^n$ converges absolutely for $R_- < |\lambda| < r_+$ for every $u \in \ell^2_\omega(\mathbb{Z})$. Set $\widetilde{u}(\lambda) = \sum_{n\in\mathbb{Z}} u_n\lambda^n$. The map $u \to \widetilde{u}$ is then a one to one continuous linear map from $\ell^2_\omega(\mathbb{Z})$ into $\mathcal{H}(\Omega)$, where $\Omega = \{\lambda \in \mathbb{C} \mid R_- < |\lambda| < r_+\}$. In this situation $u \notin \bigvee_{n\in\mathbb{Z}} S^n(S-\lambda I)u$ for every $u \in \ell^2_\omega(\mathbb{Z}) \setminus \{0\}$. We thus see that if (r_+, r_-, R_+, R_-) is not a singleton, or if $\frac{1}{\rho(S^{-1})} < \inf(r_+, r_-) \leq \sup(R_+, R_-) < \rho(S)$ then there always exists $\lambda \in \mathbb{C}$, with $\frac{1}{\rho(S^{-1})} < |\lambda| < \rho(S)$ such that $u \notin \bigvee_{n\in\mathbb{Z}} S^n(S-\lambda I)u$ for some $u \in \ell^2_\omega(\mathbb{Z})$.

4. Reflexivity properties of Apostol's weighted shifts

We now want to show that in all the situations considered by Apostol the lattice of translation invariant subspaces of $\ell^2_\omega(\mathbb{Z})$ is very rich: if T is a bounded operator on $\ell^2_\omega(\mathbb{Z})$, and if $T(M) \subset M$ for every translation invariant subspace M of $\ell^2_\omega(\mathbb{Z})$ then T belongs to the commutant of S.

The following notion is an adaptation to our context of [CEP, Definition 4.1].

DEFINITION 4.1. A translation invariant subspace M of $\ell^2_\omega(\mathbb{Z})$ is said to be *holomorphic* on a connected nonempty subset Ω of $\mathbb{C}$ if there exists a holomorphic function $f: \Omega \to \ell^2_{\omega^*}(\mathbb{Z})$ such that $M * f(\Omega) \neq \{0\}, (S - \lambda I)(M) * f(\lambda) = 0$ for $\lambda \in \Omega$. If, further, $M \cap f(\Omega)^\perp = \{0\}$, then M is said to be *full holomorphic* on Ω.

For $v \in \ell^2_\omega(\mathbb{Z})$ denote by v^* the unique element of M satisfying $[u, v^*] = \langle u, v\rangle$ for every $u \in M$, where we denote again by $[.,.]$ the usual scalar product on $\ell^2_\omega(\mathbb{Z})$. Clearly, $(S - \lambda I)(M) * f(\lambda) = 0$ for $\lambda \in \Omega$ if and only if $S^*_{|M} \cdot f^*(\lambda) = \overline{\lambda} f^*(\lambda)$ for every $\lambda \in \Omega$, and so M is holomorphic if and only if there exists a nonzero conjugate holomorphic function $g: \Omega \to M$ such that $g(\lambda) \in \operatorname{Ker}\left[(S - \lambda I)_{|M}\right]^*$ for every $\lambda \in \Omega$. Also M is a full holomorphic translation invariant subspace of $\ell^2_\omega(\mathbb{Z})$ if and only if $\bigvee_{\lambda \in \Omega} f^*(\lambda) = M$. Clearly, $\operatorname{Spec}(S_{|M}) \supset \Omega$ if M is holomorphic on Ω, and so

$$\Omega \subset \left\{z \in \mathbb{C} \mid \frac{1}{\rho(S^{-1})} < |z| < \rho(S)\right\}.$$

Notice also that if M is a holomorphic translation invariant subspace of $\ell^2_\omega(\mathbb{Z})$ then the function $\widetilde{u}: \lambda \to \langle u, f(\lambda)\rangle$ is holomorphic on Ω for every $u \in M$, and $\widetilde{Su}(\lambda) = \lambda \widetilde{u}(\lambda)$ and $\widetilde{S^{-1}u}(\lambda) = \lambda^{-1}\widetilde{u}(\lambda)$ for $u \in M$, $\lambda \in \Omega$. As in the proof of Proposition 1.2 we see that $u * f(\lambda) = \widetilde{u}(\lambda)(\lambda^{-n})_{n\in\mathbb{Z}}$ for $u \in M$, $\lambda \in \Omega$. Also the map $u \to \widetilde{u}$ is one to one if and only if M is full holomorphic.

We mentioned at the end of Section 3 the fact that Apostol's results produce for some λ in the interior of $\operatorname{Spec}(S)$ elements u of $\ell^2_\omega(\mathbb{Z})$ such that $u \notin \bigvee_{n\in\mathbb{Z}} S^n(S - \lambda I)u$. In the case of full holomorphic translation invariant subspaces, the choice of u does not depent on $\lambda \in \Omega$, as shown by the following proposition.

PROPOSITION 4.2. *If a translation invariant subspace M of $\ell^2_\omega(\mathbb{Z})$ is full holomorphic on a connected open set Ω, then $u \notin \bigvee_{n\in\mathbb{Z}} S^n(S - \lambda I)u$ for every $u \in M \setminus \{0\}$ and every $\lambda \in \Omega$.*

Proof. We have $\langle S^n(S-\lambda I)u, f(\lambda)\rangle = 0$ for $n \in \mathbb{Z}$, and so $S^p u \notin \bigvee_{n\in\mathbb{Z}} S^n(S-\lambda I)u$ for some $p \in \mathbb{Z}$, which implies that

$$u \notin \bigvee_{n\in\mathbb{Z}} S^n(S-\lambda I)u$$

if $u \in M$, $\lambda \in \Omega$ and if $u * f(\lambda) \neq 0$. Now assume that $u \in M\setminus\{0\}$ and that $u * f(\lambda_o) = 0$ for some $\lambda_o \in \Omega$. Since M is full holomorphic on Ω and since Ω is connected, there exists $k \geq 1$ such that $u * f^{(k)}(\lambda_o) \neq 0$ and such that $u * f^{(p)}(\lambda_o) = 0$ for $0 \leq p < k$. Since $(S-\lambda I)u * f(\lambda) = 0$ for $\lambda \in \Omega$ it follows from the Leibnitz rule that $(S-\lambda_o I)u * f^{(k)}(\lambda_o) = ku * f^{(k-1)}(\lambda_o) = 0$, and so $S^p u \notin \bigvee_{n\in\mathbb{Z}} S^n(S-\lambda_o I)u$ for some $p \in \mathbb{Z}$, which means again that

$$u \notin \bigvee_{n\in\mathbb{Z}} S^n(S-\lambda_o I)\,u.$$

□

In what follows we will say that a translation invariant subspace M of $\ell^2_\omega(\mathbb{Z})$ is full holomorphic if M is full holomorphic with respect to some nonempty connected open subset Ω of $\mathbb{C}$. The relevance of this notion to reflexivity is given by the following simple observation.

PROPOSITION 4.3. *Let M be a full holomorphic translation invariant subspace of $\ell^2_\omega(\mathbb{Z})$, and let T be a bounded operator on $\ell^2_\omega(\mathbb{Z})$ such that $T(N) \subset N$ for every translation invariant subspace of $\ell^2_\omega(\mathbb{Z})$. Then $S_{|M} \circ T_{|M} = T_{|M} \circ S_{|M}$.*

Proof. Let Ω be a connected nonempty open set and let $f: \Omega \to \ell^2_{\omega^*}(\mathbb{Z})$ be a holomorphic map satisfying the conditions of Definition 4.1 with respect to M.

For $\lambda \in \Omega$ set $M(\lambda) = \{u \in M \mid u * f(\lambda) = 0\}$. Then $T(M) \subset M$, and $T(M(\lambda)) \subset M(\lambda)$ for every $\lambda \in \Omega$, since $M(\lambda)$ is translation invariant. Let $u \in M$. Then $(S-\lambda I)u \in M(\lambda)$ and $(S-\lambda I)Tu \in M(\lambda)$, so that $TSu - \lambda Tu \in M(\lambda)$ and $STu - \lambda Tu \in M(\lambda)$ for every $\lambda \in \Omega$. Hence $(TS-ST)u \in M \cap f(\Omega)^\perp = \{0\}$.

□

In order to construct full holomorphic translation invariant subspaces, we need to use the full strenghth of the estimates of [BFP]. With the notations of Section 3, the proofs of Theorems 6.6 and 7.3 of [BFP] give immediately the following result, which we state as a lemma.

LEMMA 4.4. *Let ω be a weight and assume that $\sigma \in \mathcal{A}(\omega)$. Let $(a_0, \dots, a_{n-1})$, $(b_o, \dots, b_{n-1}) \in \mathcal{E}$ and let $(c_{i,j})_{0\le i<n, 0\le j<n}$ be a family of elements of $\ell^1_\sigma(\mathbb{Z})$ such that*

$$\|c_{i,j} - a_i * b_j\|^1_\sigma < \alpha \qquad (0 \le i,j < n).$$

Then for every $\varepsilon > 0$ there exists $(u_0, \dots, u_{n-1}) \in \mathcal{E}$, $(v_0, \dots, v_{n-1}) \in \mathcal{E}$ such that
(i) $\|u_i * v_j - c_{i,j}\|^1_\sigma < \varepsilon \quad (0 \le i,j < n)$,
(ii) $\|u_i - a_i\|_\omega \le n\alpha^{1/2}$, $\|v_j - b_j\|_{\omega^*} \le n\alpha^{1/2} \quad (0 < i,j \le n)$.

We can now use a strategy inspired by the proof of [CEP, Theorem 5.4]. We deduce from Lemma 4.4 the following technical result.

LEMMA 4.5. *Let ω be a weight and let $\sigma \in \mathcal{A}(\omega)$. Then for every $a \in \ell^2_\omega(\mathbb{Z})$ and every $\varepsilon > 0$ there exists a dense subset $\mathcal{F}$ of $\ell^2_{\omega^*}(\mathbb{Z})$, an element u of $\ell^2_\omega(\mathbb{Z})$ and a sequence $(v_n)_{n\in\mathbb{Z}}$ of elements of $\ell^2_{\omega^*}(\mathbb{Z})$ satisfying the following conditions*
(1) $\|a - u\|_\omega < \varepsilon$,
(2) $u * v_n = e_n \ (n \in \mathbb{Z})$,
(3) $\overline{\lim}_{n\to\infty} \frac{\|v_n\|_{\omega^*}}{(|n|+1)^5\sigma(n)} < +\infty$,
(4) $u * \mathcal{F} \subset \mathcal{E}$.

Proof. We can assume without loss of generality that $a \in \mathcal{E}$. Let $(r_n)_{n\ge1}$ be a sequence of elements of $\mathcal{E}$ such that $(r_n)_{n\ge m}$ is dense in $\ell^2_{\omega^*}(\mathbb{Z})$ for every $m \ge 1$. We construct by induction a sequence $(a_p)_{p\ge0}$ of elements of $\mathcal{E}$, a sequence $(d_p)_{p\ge1}$ of elements of $\mathcal{E}$ and two families $(b_{p,n})_{n\in\mathbb{Z}, p\ge|n|}$ and $(c_{p,n})_{n\ge1, p\ge n}$ of elements of $\mathcal{E}$ satisfying the following conditions
(1) $\|a - a_0\|_\omega < \frac{\varepsilon}{2}$
(2) $\|b_{n,n}\|_{\omega^*} < \frac{\varepsilon}{2(n+1)^2} \quad (n \ge 0)$
(3) $\|b_{n,-n}\|_{\omega^*} < \frac{\varepsilon}{2(n+1)^2} \quad (n \ge 0)$
(4) $\|c_{n,n} - r_n\|_{\omega^*} < \frac{\varepsilon}{2(n+1)^2} \quad (n \ge 0)$
(5) $\|a_{p+1} - a_p\|_\omega < \frac{\varepsilon}{2(p+1)^2} \quad (p \ge 0)$
(6) $\|b_{p,n} - b_{p+1,n}\|_{\omega^*} < \frac{\varepsilon}{2(p+1)^2} \quad (p \ge |n| \ge 0)$
(7) $\|c_{p,n} - c_{p+1,n}\|_{\omega^*} < \frac{\varepsilon}{2(p+1)^2} \quad (p \ge n \ge 1)$
(8) $\left\|a_p * b_{p,n} - \frac{\varepsilon^2 e_n}{40(n+1)^6\sigma(n)}\right\|^1_\sigma < \frac{\varepsilon}{2(p+2)^2} \quad (p \ge |n| \ge 0)$
(9) $\|a_p * c_{p,n} - d_n\|^1_\sigma < \frac{\varepsilon}{2(p+2)^2} \quad (p \ge n \ge 1)$.

Since $\left\|\frac{\varepsilon^2 e_0}{40\sigma(0)} - a * 0\right\|^1_\sigma = \left\|\frac{\varepsilon^2 e_0}{40\sigma(0)}\right\|^1_\sigma < \left(\frac{\varepsilon}{2}\right)^2$ it follows from Lemma 4.4 that there exists $a_0 \in \mathcal{E}$ and $b_{0,0} \in \mathcal{E}$ such that $\|a - a_0\|_\omega < \frac{\varepsilon}{2}$, $\|b_{0,0}\|_{\omega^*} < \frac{\varepsilon}{2}$ and $a_0 * b_{0,0} = \frac{\varepsilon^2 e_0}{40\sigma(0)}$.

Now let $k \geq 1$ and assume that $(a_p)_{0\leq p<k}$, $(d_p)_{0\leq p<k}$, $(b_{p,n})_{0\leq |n|\leq p<k}$ and $(c_{p,n})_{1\leq n\leq p<k}$ satisfying the conditions above have been constructed.

Set $d_k = a_{k-1} * r_k$. We have

$$\left\| a_{k-1} * b_{k-1,n} - \frac{\varepsilon^2 e_n}{40(n+1)^6\sigma(n)} \right\|_\sigma^1 < \frac{\varepsilon^2}{40(k+1)^6}$$

for $0 \leq |n| \leq k-1$, and

$$\|a_{k-1} * c_{k-1,n} - d_n\|_\sigma^1 < \frac{\varepsilon^2}{40(k+1)^6}$$

for $1 \leq n \leq k-1$ if $k \geq 2$. Also

$$\left\| \frac{\varepsilon^2 e_k}{40(k+1)^6\sigma(k)} \right\|_\sigma^1 = \left\| \frac{\varepsilon^2 e_{-k}}{40(k+1)^6\sigma(-k)} \right\|_\sigma^1 = \frac{\varepsilon^2}{40(k+1)^6}.$$

Applying Lemma 4.4 with $\alpha = \frac{\varepsilon^2}{40(k+1)^6}$, $r = 3k+1$ and noticing that $\frac{3k+1}{\sqrt{40}(k+1)} < \frac{1}{2}$, we see that there exists a_k, $(b_{k,n})_{0\leq|n|\leq k}$ and $(c_{k,n})_{1\leq n\leq k}$ satisfying the required conditions. This shows that the construction can be carried out by induction.

It follows from (5), (6), (7) that the sequences $(a_p)_{p\geq 1}$, $(b_{p,n})_{p\geq |n|}$ and $(c_{p,n})_{p\geq n}$ are convergent. Set $u = \lim\limits_{p\to\infty} a_p$, set $v_n = \frac{40}{\varepsilon^2}(|n|+1)^6\sigma(n) \lim\limits_{p\to\infty} b_{p,n}$ for $n \in \mathbb{Z}$ and set $s_n = \lim\limits_{p\to\infty} c_{p,n}$ for $n \geq 1$. It follows from (1) and (5) that $\|a-u\|_\omega < \frac{\varepsilon}{2}\sum\limits_{p=1}^{\infty}\frac{1}{p^2} < \varepsilon$. Also it follows from (8) and (9) that $u*v_n = e_n$ for $n \in \mathbb{Z}$ and $u*s_n = d_n \in \mathcal{E}$ for $n \geq 1$. It follows from (4) and (7) that $\|s_n - r_n\|_{\omega^*} < \frac{\varepsilon}{2}\sum\limits_{k=n+1}^{\infty}\frac{1}{k^2}$, and so $\|s_n - r_n\|_{\omega^*} \to 0$ $(\omega^* n \to \infty)$ and $\mathcal{F} = \{s_n\}_{n\geq 1}$ is dense in $\ell^2_{\omega^*}(\mathbb{Z})$ and satisfies $u * \mathcal{F} \subset \mathcal{E}$. It follows also from (2), (3) and (6) that $\|v_n\|_{\omega^*} \leq \frac{20}{\varepsilon}(|n|+1)^6\sigma(n)\sum\limits_{k=|n|+1}^{\infty}\frac{1}{k^2}$. Since $\sum\limits_{k=n+1}^{\infty}\frac{1}{k^2} = 0(\frac{1}{n})$ as $n \to \infty$, this concludes the proof of the lemma. □

The following result is analogous to Proposition 5.3 of [CEP].

LEMMA 4.6. *Let ω be a weight, and let $\sigma \in \mathcal{A}(\omega)$.*
Assume that $\overline{\lim\limits_{n\to\infty}}\, \sigma(n)^{1/n} < \underline{\lim\limits_{n\to-\infty}}\, \sigma(n)^{1/n}$, and set

$$\Omega = \left\{ \lambda \in \mathbb{C} \mid \overline{\lim_{n\to\infty}}\, \sigma(n)^{1/n} < |\lambda| < \underline{\lim_{n\to-\infty}}\, \sigma(n)^{1/n} \right\}.$$

Let $u \in \ell^2_\omega(\mathbb{Z})$. If there exists a sequence $(v_n)_{n\in\mathbb{Z}}$ of elements of $\ell^2_{\omega^*}(\mathbb{Z})$ and a dense subset $\mathcal{F}$ of $\ell^2_{\omega^*}(\mathbb{Z})$ satisfying conditions (2), (3), (4) of Lemma 4.5 with respect to u, then $\bigvee\limits_{n\in\mathbb{Z}} S^n u$ is full holomorphic on Ω.

Proof. It follows from condition (3) of Lemma 4.5 that the series $\sum_{n\in\mathbb{Z}} |\lambda|^{-n}\|v_n\|_{\omega^*}$ is convergent for $\lambda \in \Omega$. Set $f(\lambda) = \sum_{n\in\mathbb{Z}} \lambda^{-n}v_n$, so that $f\colon\Omega \to \ell^2_{\omega^*}(\mathbb{Z})$ is holomorphic. We have $u * f(\lambda) = (\lambda^{-n})_{n\in\mathbb{Z}}$ and so $(S-\lambda I)u * f(\lambda) = 0$ for $\lambda \in \Omega$. Set $M = \bigvee_{n\in\mathbb{Z}} S^n u$. Since the set $\{a \in \ell^2_\omega(\mathbb{Z}) \mid (S-\lambda I)a * f(\lambda) = 0\}$ is translation invariant, $(S-\lambda I)(M) * f(\Omega) = 0$, and so M is holomorphic on Ω.

Let $y \in \mathcal{F}$. Since $u * y \in \mathcal{E}$, there exists $p \geq 1$ and a family $(\alpha_k)_{|k|\leq p}$ of complex numbers such that

$$u * y = \sum_{|k|\leq p} \alpha_k e_k = \sum_{|k|\leq p} \alpha_k (u * v_k).$$

Hence $u * (y - \sum_{|k|\leq p} \alpha_k v_k) = 0$, and so $a * (y - \sum_{|k|\leq p} a_k v_k) = 0$ for every $a \in M$. Now assume that $a \in M \cap f(\Omega)^\perp$.

Since $v_k r^k = \frac{1}{2\pi}\int_0^{2\pi} f(re^{it})e^{ikt}\,dt$ for $\overline{\lim}_{n\to\infty} \sigma(n)^{1/n} < r < \underline{\lim}_{n\to-\infty} \sigma(n)^{1/n}$, we have $\langle a, v_k\rangle = 0$ for every $k \in \mathbb{Z}$, and so $\langle a, y\rangle = 0$ for every $y \in \mathcal{F}$. Hence $M \cap f(\Omega)^\perp = \{0\}$, and M is full holomorphic on Ω.

□

For $0 < \rho_1 < \rho_2$, denote by $\mathcal{C}_\omega(\rho_1, \rho_2)$ the set of all $u \in \ell^2_\omega(\mathbb{Z})$ such that $\bigvee_{n\in\mathbb{Z}} S^n u$ is full holomorphic on $\Omega(\rho_1, \rho_2) := \{\lambda \in \mathbb{C} \mid \rho_1 < |\lambda| < \rho_2\}$.

We now deduce immediately from Proposition 3.7, Lemma 4.5 and Lemma 4.6 the following result

THEOREM 4.7. *Let ω be a weight.*

(i) *If $r_+ < R_+$, then $\mathcal{C}_\omega(r_+, R_+)$ is dense in $\ell^2_\omega(\mathbb{Z})$.*

(ii) *If $r_- < R_-$, then $\mathcal{C}_\omega(r_-, R_-)$ is dense in $\ell^2_\omega(\mathbb{Z})$.*

(iii) *If*

$$\frac{1}{\rho(S^{-1})} < \inf(r_+, r_-) \leq \sup(R_+, R_-) < \rho(S)$$

then $\mathcal{C}_\omega(\frac{1}{\rho(S^{-1})}, \rho(S))$ is dense in $\ell^2_\omega(\mathbb{Z})$.

Notice that if $r_+ < R_+$ we can choose $u \in \ell^2_\omega(\mathbb{Z}^+)$ and $(v_n)_{n\geq 1} \cup \mathcal{F} \subset \ell^2_{\omega^*}(\mathbb{Z}^-)$ in Lemma 4.5. We thus obtain elements u of $\ell^2_\omega(\mathbb{Z}^+)$ such that $u \notin \bigvee_{n\in\mathbb{Z}} S^n(S-\lambda I)u$ for every $\lambda \in \Omega(r_+, R_+)$, and these elements do not depend on $\omega_{|\mathbb{Z}^-}$. Similarly if $r_- < R_-$ we obtain elements u of $\ell^2_\omega(\mathbb{Z}^-)$ such that $u \notin \bigvee_{n\in\mathbb{Z}} S^n(S-\lambda I)u$ for every $\lambda \in \Omega(r_-, R_-)$, and these elements do not depend on $\omega_{|\mathbb{Z}^+}$.

Notice also that if $\frac{1}{\rho(S^{-1})} < \inf(r_+, r_-) \le \sup(R_+, R_-) < \rho(S)$ we obtain elements u of $\ell^2_\omega(\mathbb{Z})$ such that $u \notin \bigvee_{n\in\mathbb{Z}} S^n(S-\lambda I)u$ for all λ in the interior of Spec(S).

As mentioned in the introduction, we will say as in [BFN] that a bounded operator T on a Hilbert space H is hyper-reflexive if every bounded operator R on H such that $R(M) \subset M$ for every closed subspace M of H hyperinvariant for T actually commutes with T. We now obtain the following strenghthening of Apostol's results.

COROLLARY 4.8. *Let ω be a weight on $\mathbb{Z}$. If $\{r_-, r_+, R_-, R_+\}$ is not a singleton, or if*

$$\frac{1}{\rho(S^{-1})} < \inf(r_+, r_-) \le \sup(R_+, R_-) < \rho(S),$$

then the shift operator on $\ell^2_\omega(\mathbb{Z})$ is hyper-reflexive.

Proof. If $r_+ < R_+$, or if $r_- < R_-$, or if

$$\frac{1}{\rho(S^{-1})} < \inf(r_+, r_-) \le \sup(R_+, R_-) < \rho(S),$$

the result is an immediate consequence of Proposition 4.2 and Theorem 4.7. If $R_- < r_+$ then the series $\sum_{n\in\mathbb{Z}} |\lambda|^n |u_n|$ converges for $u \in \ell^2_\omega(\mathbb{Z})$, $\lambda \in \Omega(R_-, r_+)$. Set $\widetilde{u}(\lambda) = \sum_{n\in\mathbb{Z}} \lambda^n u_n$ for $\lambda \in \Omega(R_-, r_+)$, $u \in \ell^2_\omega(\mathbb{Z})$ and set $f(\lambda) = (\lambda^{-n})_{n\in\mathbb{Z}}$ for $\lambda \in \Omega(R_-, r_+)$. Then $f: \Omega(R_-, r_+) \to \ell^2_{\omega^*}(\mathbb{Z})$ is holomorphic, and $u * f(\lambda) = \widetilde{u}(\lambda)(\lambda^{-n})_{n\in\mathbb{Z}}$, so that $(S-\lambda I)u * f(\lambda) = 0$ for every $u \in \ell^2_\omega(\mathbb{Z})$. Since the map $u \to \widetilde{u}$ is one-to-one we see that in this situation $\ell^2_\omega(\mathbb{Z})$ is itself full holomorphic on $\Omega(R_-, r_+)$, and the result follows from Proposition 4.2. Now if $R_+ < r_-$ set $f(\lambda) = (\lambda^{-n})_{n\in\mathbb{Z}}$ for $\lambda \in \Omega(R_+, r_-)$. Then f is holomorphic, and we see as above that $\ell^2_{\omega^*}(\mathbb{Z})$ is full holomorphic on $\Omega(R_+, r_-)$. So the shift operator on $\ell^2_{\omega^*}(\mathbb{Z})$ is hypereflexive, which implies immediately that the shift operator on $\ell^2_\omega(\mathbb{Z})$ is hyper-reflexive (one can also observe directly in this situation that $f(\lambda)$ is an eigenvector of S for $\lambda \in \Omega(R_+, r_-)$, and that $\bigvee_{R_+<|\lambda|<r_-} f(\lambda) = \ell^2_\omega(\mathbb{Z})$), which implies immediately that S is hyper-reflexive).

□

REMARK. As mentioned in the introduction, the hyper-reflexivity of S implies that S and S^{-1} are both reflexive. To see this assume that S is hyper-reflexive, and let R be a bounded operator on $\ell^2_\omega(\mathbb{Z})$ such that Lat $S \subset$ Lat R. Then R belongs to the commutant of S, and $Re_0 \in \ell^2_\omega(\mathbb{Z}^+)$. Hence $\widehat{R}(n) = 0$ for $n < 0$, and $Ru = \lim_{n\to\infty}\left[\frac{1}{n+1}\sum_{p=0}^{n}\widehat{R}(p)S^p u\right]$ in the sense of Cesàro for $u \in \ell^2_\omega(\mathbb{Z})$, so that $R \in \mathcal{A}(S)$. The same argument shows that S^{-1} is reflexive.

References

[Am] G. Amir, Translation invariant subspaces of $\ell^p_\omega(\mathbb{Z})$, *preprint.*

[Ap] C. Apostol, Hyperinvariant subspaces for bilateral weighted shifts, *J. Int. Eq. and Op. Theory*, **7** (1984), 1–9.

[Atz] A. Atzmon, Entire functions, invariant subspaces and Fourier transforms, *Israel Math. Conf. Proceedings*, **11** (1977), 37–52.

[Atz–So] A. Atzmon and M. Sodin, Completely indecomposable operators and a uniqueness theorem of Cartwright–Levinson type, *J. Funct. Anal.*, **169** (1999), 164–188.

[BFN] H. Bercovici, C. Foias and B. Sz-Nagy, Reflexive and hyper-reflexive operators of class C_0, *Acta Sci. Math. (Szeged)*, **43** (1981), 5–14.

[BFP] H. Bercovici, C. Foias and C. Pearcy, Two Banach spaces methods and dual operator algebras, *J. Funct. Anal.*, **78** (1988), 306–345.

[Br–Ch] S. Brown and B. Chevreau, Toute contraction à calcul isométrique est réflexive, *C. R. Acad. Sci. Paris*, **307** (1988), 185–188.

[CEP] B. Chevreau, G. Exner and C. Pearcy, On the structure of contraction operators. III, *Michigan Math.*, **36** (1989), 29–62.

[Do] I. Domar, Entire functions of order ≤ 1, with bounds on both axes, *Ann. Acad. Sci. Fenn.*, **22** (1997), 339–348.

[Es] J. Esterle, Singular inner functions and biinvariant subspaces for dissymetric weighted shifts, *J. Funct. Anal.*, **44** (1997), 64–104.

[Es–Vo$_1$] J. Esterle and A. Volberg, Sous-espaces invariants par translation de certains espaces de Hilbert de suites quasianalytiquement pondérées, *C.R. Acad. Sci. Paris t.*, **326** (1998), 295–300.

[Es–Vo$_2$] J. Esterle and A. Volberg, Analytic left-invariant subspaces of weighted Hilbert spaces of sequences, *J. Op. Theory*, to appear.

[Es–Vo$_3$] J. Esterle and A. Volberg, Asymptotically holomorphic functions and translation invariant subspaces of weighted Hilbert spaces of sequences, *Ann. Sci. Ec. Norm. Sup.*, to appear.

[Ol–Th] R. F. Olin and J. E. Thomson, Algebras of subnormal operators, *J. Funct. Anal.*, **37** (1980), 271–301.

[Sa] D. Sarason, Invariant subspaces and unstarred operator algebras, *Pac. J. Math.*, **17** (1966), 511–517.

[Sh] A. SHIELDS, Weighted shift operators and analytic function theory, *Topics in Operator Theory* (Ed. by C. Pearcy), Amer. Math. Soc. Math. Surveys, **13** (1974), 49–128.

J. ESTERLE, Laboratoire de Mathématiques Pures, Upresa 5467, Université Bordeaux 1, 351, Cours de la Libération, 33405 Talence Cedex; *e-mail*: esterle@math.u-bordeaux.fr

Operator Theory: Advances and Applications, Vol. 127, 267–277

Wielandt type extensions of the Heinz–Kato–Furuta inequality

MASATOSHI FUJII and YUKI SEO

The Wielandt inequality asserts that if a positive operator A on a Hilbert space H satisfies $0 < m \leq A \leq M$ for some $0 < m < M$, then

$$|(Ax, y)|^2 \leq \left(\frac{M-m}{M+m}\right)^2 (Ax, x)(Ay, y)$$

for every orthogonal pair x and y. In this paper, we show Wielandt type extensions of the Heinz–Kato–Furuta inequality, which is based on some generalizations of the Wielandt inequality by Fujii–Katayama–Nakamoto and Bauer–Householder. The obtained inequalities are simultaneous extensions of the Heinz–Kato–Furuta and the Wielandt inequalities. Related to our extensions, we discuss some applications of the Furuta inequality and the grand Furuta inequality.

1. Introduction

In what follows, a capital letter means a bounded linear operator on a complex Hilbert space H. An operator T is said to be positive (in symbol: $T \geq 0$) if $(Tx, x) \geq 0$ for all $x \in H$. The order $A \geq B$ means that A and B are selfadjoint and $A - B$ is positive. Furuta [9] extended the Heinz–Kato inequality as follows:

THE HEINZ–KATO–FURUTA INEQUALITY. *Let T be an operator on a Hilbert space H. If A and B are positive operators such that $\|Tx\| \leq \|Ax\|$ and $\|T^*y\| \leq \|By\|$ for all $x, y \in H$, then*

$$|(T|T|^{\alpha+\beta-1}x, y)| \leq \|A^{\alpha}x\|\|B^{\beta}y\| \tag{1}$$

AMS Subject Classification (1991): 47A30, 47A63.

for all $\alpha, \beta \in [0,1]$ *with* $\alpha + \beta \geq 1$ *and* $x, y \in H$.

As noted in [9], the Heinz–Kato inequality is just the case of $\alpha + \beta = 1$, and the assumption $\alpha + \beta \geq 1$ in above can be deleted if T is invertible. For simplicity, we call it the HKF inequality in the sequel. Moreover, Furuta [12] showed the following generalization of the HKF inequality as an application of the grand Furuta inequality.

THEOREM A. *Let* T *be an invertible operator on a Hilbert space* H. *If* A *and* B *are positive operators such that* $\|Tx\| \leq \|Ax\|$ *and* $\|T^*y\| \leq \|By\|$ *for all* $x, y \in H$, *then, for each* $r_1,\ r_2 \geq 0$ *and* $t_1,\ t_2 \in [0,1]$ *with* $r_j \geq t_j$ $(j = 1, 2)$,

$$\begin{aligned} &|(T|T|^{(1-t_1+r_1)\alpha+(1-t_2+r_2)\beta-1}x, y)|^2 \\ &\leq (\{|T|^{r_1}(|T|^{-t_1}A^{2p_1}|T|^{-t_1})^{s_1}|T|^{r_1}\}^{\frac{(1-t_1+r_1)\alpha}{(p_1-t_1)s_1+r_1}}x, x)\times \\ &\quad \times (\{|T^*|^{r_2}(|T^*|^{-t_2}B^{2p_2}|T^*|^{-t_2})^{s_2}|T^*|^{r_2}\}^{\frac{(1-t_2+r_2)\beta}{(p_2-t_2)s_2+r_2}}y, y) \end{aligned}$$

holds for all $p_1, p_2, s_1, s_2 \geq 1$, $\alpha, \beta \in [0,1]$ *and* $x, y \in H$.

We now pay our attention to the Wielandt inequality which is regarded as an improvement of the Schwarz inequality.

THE WIELANDT INEQUALITY. *If* $0 < m \leq A \leq M$, *then*

$$|(Ay, x)|^2 \leq \left(\frac{M-m}{M+m}\right)^2 (Ax, x)(Ay, y)$$

for every orthogonal pair x *and* y.

Recently the Wielandt inequality was considered without assumption on the orthogonality of vectors in [5]:

THEOREM B. *If* $0 < m \leq A \leq M$, *then*

$$|(Ay, x)| \leq \frac{M-m}{M+m}(Ax, x)^{1/2}(Ay, y)^{1/2} + \frac{2Mm}{M+m}|(x, y)|$$

for all $x, y \in H$.

In this paper, we show Wielandt type extensions of the HKF inequality, which are simultaneous extensions of the HKF inequality and the Wielandt inequality. It is based on Theorem B and the Bauer–Householder Theorem [2]. Moreover, as an application of the grand Furuta inequality, we have a Wielandt type extension of Theorem A. We note that it is equivalent to the grand Furuta inequality.

2. Wielandt type HKF inequalities

In this section, we discuss Wielandt type variations of the HKF inequality.

THEOREM 1. *Let T be an operator on H satisfying $0 < m \leq |T| \leq M$ for some $0 < m \leq M$. Then, for each $\gamma > 0$,*

$$|(T|T|^{\alpha+\beta-1}x, y)| \leq \frac{M^\gamma - m^\gamma}{M^\gamma + m^\gamma}\||T|^\alpha x\|\||T^*|^\beta y\| + \frac{2M^\gamma m^\gamma}{M^\gamma + m^\gamma}|(T|T|^{\alpha+\beta-\gamma-1}x, y)|$$

for all $\alpha, \beta \in \mathbb{R}$ and $x, y \in H$.

Proof. Let $T = U|T|$ be the polar decomposition of T. For given $x, y \in H$, we put $x_1 = |T|^{\alpha-\frac{\gamma}{2}}x$ and $y_1 = |T|^{\beta-\frac{\gamma}{2}}U^*y$. Thus we use Theorem B for x_1, y_1 and $A = |T|^\gamma$. Since $0 < m^\gamma \leq |T|^\gamma \leq M^\gamma$ and $U|T|^\beta U^* = |T^*|^\beta$, we have the desired inequality.

□

In particular, we take $\gamma = \alpha + \beta$ in above:

COROLLARY 2. *Let T be an operator on H satisfying $0 < m \leq |T| \leq M$ for some $0 < m \leq M$. Then*

$$\begin{aligned}&|(T|T|^{\alpha+\beta-1}x, y)| \\ &\leq \frac{M^{\alpha+\beta} - m^{\alpha+\beta}}{M^{\alpha+\beta} + m^{\alpha+\beta}}\||T|^\alpha x\|\||T^*|^\beta y\| + \frac{2M^{\alpha+\beta}m^{\alpha+\beta}}{M^{\alpha+\beta} + m^{\alpha+\beta}}|(T|T|^{-1}x, y)|\end{aligned}$$

for all $\alpha, \beta \in \mathbb{R}$ with $\alpha + \beta > 0$ and $x, y \in H$.

Next we obtain the following Wielandt type HKF inequalities by the Löwner–Heinz inequality, i.e., for $\alpha \in [0, 1]$, $A \geq B \geq 0$ implies $A^\alpha \geq B^\alpha$.

THEOREM 3. *Let T be an operator on H satisfying $0 < m \leq |T| \leq M$. If A and B are positive operators such that $\|Tx\| \leq \|Ax\|$ and $\|T^*y\| \leq \|By\|$ for all $x, y \in H$, then for each $\gamma > 0$*

$$|(T|T|^{\alpha+\beta-1}x, y)| \leq \frac{M^\gamma - m^\gamma}{M^\gamma + m^\gamma}\|A^\alpha x\|\|B^\beta y\| + \frac{2M^\gamma m^\gamma}{M^\gamma + m^\gamma}|(T|T|^{\alpha+\beta-\gamma-1}x, y)|$$

for all $\alpha, \beta \in [0,1]$ and $x, y \in H$.

In particular,

$$|(T|T|^{\alpha+\beta-1}x, y)| \leq \frac{M^{\alpha+\beta} - m^{\alpha+\beta}}{M^{\alpha+\beta} + m^{\alpha+\beta}}\|A^\alpha x\|\|B^\beta y\| + \frac{2M^{\alpha+\beta}m^{\alpha+\beta}}{M^{\alpha+\beta} + m^{\alpha+\beta}}|(T|T|^{-1}x, y)|$$

for all $\alpha, \beta \in [0,1]$, and $x, y \in H$.

COROLLARY 4. *Let A, B and T be as in Theorem 3, and $\alpha, \beta \in [0,1]$ and $\gamma > 0$ be given. If $x, y \in H$ satisfy*

$$T|T|^{\alpha+\beta-\gamma-1}x \neq 0, \quad (T|T|^{\alpha+\beta-\gamma-1}x, y) = 0,$$

then

$$|(T|T|^{\alpha+\beta-1}x, y)| \leq \left(\frac{M^\gamma - m^\gamma}{M^\gamma + m^\gamma}\right)\|A^\alpha x\|\|B^\beta y\|.$$

REMARK. We note that Theorem 3 is a simultaneous extension of the HKF inequality and the Wielandt inequality. As a matter of fact, we take $\gamma = 1$, $\alpha = \beta = \frac{1}{2}$ and $T = A = B \geq 0$ in Theorem 3. Then we have Theorem B. On the other hand, we have also the HKF inequality by tending m to 0.

3. Applications of the grand Furuta inequality

We start this section by introducing the following order preserving operator inequalities.

THE FURUTA INEQUALITY. *If $A \geq B \geq 0$, then for each $r \geq 0$,*

(i) $(B^{r/2}A^pB^{r/2})^{1/q} \geq (B^{r/2}B^pB^{r/2})^{1/q}$ *and*

(ii) $(A^{r/2}A^pA^{r/2})^{1/q} \geq (A^{r/2}B^pA^{r/2})^{1/q}$

hold for $p \geq 0$ and $q \geq 1$ with $(1+r)q \geq p + r$.

The original proof of the Furuta inequality is in [7], mean theoretic proofs are in [3], [13], and a one-page proof is in [8]. The domain drawn for p, q and r in Figure 1 is the best possible one for (i) and (ii) of the Furuta inequality in [14]. Therefore it can be rewritten as follows by Löwner–Heinz Theorem.

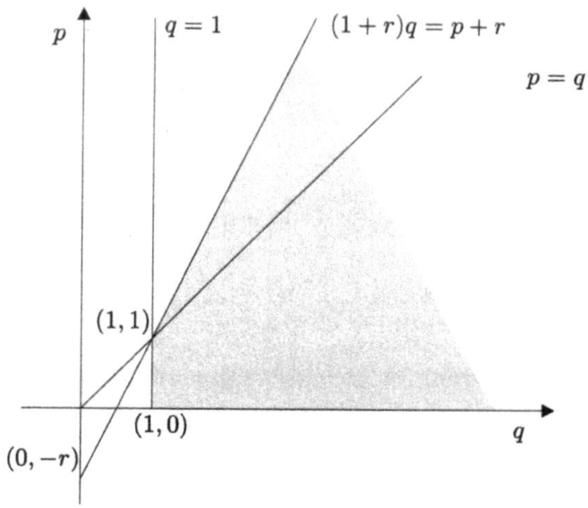

Figure 1

THEOREM F′. *If $A \geq B \geq 0$, then for each $r \geq 0$*

$$\left(B^{\frac{r}{2}} A^p B^{\frac{r}{2}}\right)^{\frac{(1+r)\alpha}{p+r}} \geq B^{(1+r)\alpha}$$

and

$$A^{(1+r)\alpha} \geq \left(A^{\frac{r}{2}} B^p A^{\frac{r}{2}}\right)^{\frac{(1+r)\alpha}{p+r}}$$

hold for all $p \geq 1$ and $\alpha \in [0,1]$.

The following Theorem GFI interpolates the Furuta inequality itself and a useful inequality equivalent to the main theorem of log majorization by Ando–Hiai [1].

THEOREM GFI. *If $A \geq B > 0$, then for each $t \in [0,1]$*

$$\{B^{\frac{r}{2}}(B^{-\frac{t}{2}} A^p B^{-\frac{t}{2}})^s B^{\frac{r}{2}}\}^{\frac{1-t+r}{(p-t)s+r}} \geq B^{1-t+r}$$

and

$$A^{1-t+r} \geq \{A^{\frac{r}{2}}(A^{-\frac{t}{2}} B^p A^{-\frac{t}{2}})^s A^{\frac{r}{2}}\}^{\frac{1-t+r}{(p-t)s+r}}$$

hold for all $p \geq 1$, $r \geq t$ and $s \geq 1$.

The original proof of Theorem GFI is in [10], a mean theoretic one is in [4], and a one-page proof can be found in [11]. An excellent and tough proof of the best possibility of the power $\frac{1-t+r}{(p-t)s+r}$ is in [15], and simple proofs are in [16] and [6]. Therefore Theorem GFI can be rewritten as follows by the Löwner–Heinz Theorem for the sake of convenience in this paper.

Theorem GFI′. *If $A \geq B > 0$, then for each $t \in [0,1]$*

$$\{B^{\frac{r}{2}}(B^{-\frac{t}{2}}A^pB^{-\frac{t}{2}})^sB^{\frac{r}{2}}\}^{\frac{(1-t+r)\alpha}{(p-t)s+r}} \geq B^{(1-t+r)\alpha}$$

and

$$A^{(1-t+r)\alpha} \geq \{A^{\frac{r}{2}}(A^{-\frac{t}{2}}B^pA^{-\frac{t}{2}})^sA^{\frac{r}{2}}\}^{\frac{(1-t+r)\alpha}{(p-t)s+r}}$$

hold for all $p \geq 1$, $r \geq t$ and $s \geq 1$.

Now the following theorem is an application of the grand Furuta inequality, whose base is Theorem 1

Theorem 5. *Let T be an operator on H satisfying $0 < mI \leq |T| \leq MI$. If A and B are strictly positive operators such that $\|Tx\| \leq \|Ax\|$ and $\|T^*y\| \leq \|By\|$ for all $x, y \in H$, then, for each $r_1,\ r_2 \geq 0$, $t_1,\ t_2 \in [0,1]$ with $r_j \geq t_j$ $(j = 1,2)$ and $\gamma > 0$,*

$$\begin{aligned}
&|(T|T|^{(1-t_1+r_1)\alpha+(1-t_2+r_2)\beta-1}x, y)| \\
&\leq \frac{M^\gamma - m^\gamma}{M^\gamma + m^\gamma}(\{|T|^{r_1}(|T|^{-t_1}A^{2p_1}|T|^{-t_1})^{s_1}|T|^{r_1}\}^{\frac{(1-t_1+r_1)\alpha}{(p_1-t_1)s_1+r_1}}x, x)^{1/2}\times \\
&\qquad \times (\{|T^*|^{r_2}(|T^*|^{-t_2}B^{2p_2}|T^*|^{-t_2})^{s_2}|T^*|^{r_2}\}^{\frac{(1-t_2+r_2)\beta}{(p_2-t_2)s_2+r_2}}y, y)^{1/2} + \\
&\quad + \frac{2M^\gamma m^\gamma}{M^\gamma + m^\gamma}|(T|T|^{(1-t_1+r_1)\alpha-(1-t_2+r_2)\beta-\gamma-1}x, y)|
\end{aligned}$$

holds for all $p_1, p_2, s_1, s_2 \geq 1$, $\alpha, \beta \in [0,1]$ and $x, y \in H$.

Remark. We note that Theorem 5 is a simultaneous extension of Theorem A and the Wielandt inequality. Since Theorem 5 is an extension of Theorem 3, it is an extension of the Wielandt inequality automatically. On the other hand, we have Theorem A by tending m to 0 in Theorem 5. In other words, Theorem 5 is an alternative expression of the grand Furuta inequality because Theorem A is equivalent to the grand Furuta inequality.

Corollary 6. *Let A, B and T be as in Theorem 5, and $r_1,\ r_2 \geq 0$, $t_1,\ t_2 \in [0,1]$ with $r_j \geq t_j$ $(j = 1,2)$, $\alpha, \beta \in [0,1]$ and $\gamma > 0$ be given. If $x, y \in H$ satisfy*

$$T|T|^{(1-t_1+r_1)\alpha+(1-t_2+r_2)\beta-\gamma-1}x \neq 0, \quad (T|T|^{(1-t_1+r_1)\alpha+(1-t_2+r_2)\beta-\gamma-1}x, y) = 0,$$

then

$$|(T|T|^{(1-t_1+r_1)\alpha+(1-t_2+r_2)\beta-1}x,y)|^2$$
$$\leq K(\{|T|^{r_1}(|T|^{-t_1}A^{2p_1}|T|^{-t_1})^{s_1}|T|^{r_1}\}^{\frac{(1-t_1+r_1)\alpha}{(p_1-t_1)s_1+r_1}}x,x)\times$$
$$\times(\{|T^*|^{r_2}(|T^*|^{-t_2}B^{2p_2}|T^*|^{-t_2})^{s_2}|T^*|^{r_2}\}^{\frac{(1-t_2+r_2)\beta}{(p_2-t_2)s_2+r_2}}x,x)$$

holds for all $p_1,p_2,s_1,s_2\geq 1$, *where* $K=K_\gamma=\left(\frac{M^\gamma-m^\gamma}{M^\gamma+m^\gamma}\right)^2$.

Proof of Theorem 5. Let $T=U|T|$ be the polar decomposition of T. We put $a=(1-t_1+r_1)\alpha$ and $b=(1-t_2+r_2)\beta$.

We may assume $b>0$. Then it follows from Theorem 1 that

$$|(T|T|^{a+b-1}x,y)|\leq\left(\frac{M^\gamma-m^\gamma}{M^\gamma+m^\gamma}\right)\||T|^a x\|\||T^*|^b y\|+\frac{2M^\gamma m^\gamma}{M^\gamma+m^\gamma}|(T|T|^{a+b-\gamma-1}x,y)|.$$

On the other hand, since $|T|^2\leq A^2$ and $|T^*|^2\leq B^2$ by the assumption, the grand Furuta inequality ensures that

$$|T|^{2a}=|T|^{2(1-t_1+r_1)\alpha}\leq\{|T|^{r_1}(|T|^{-t_1}A^{2p_1}|T|^{-t_1})^{s_1}|T|^{r_1}\}^{\frac{(1-t_1+r_1)\alpha}{(p_1-t_1)s_1+r_1}}$$

and

$$|T|^{2b}=|T^*|^{2(1-t_2+r_2)\beta}\leq\{|T^*|^{r_2}(|T^*|^{-t_2}B^{2p_2}|T^*|^{-t_2})^{s_2}|T^*|^{r_2}\}^{\frac{(1-t_2+r_2)\beta}{(p_2-t_2)s_2+r_2}}.$$

Combining these three inequalities, we have the desired inequality.

□

The next corollaries are also applications of the Furuta inequality. That is, they can be obtained by combining the Furuta inequality and Theorem 1.

Corollary 7. *Let A,B and T be as in Theorem* 5 *and $r,s\geq 0$, $\alpha,\beta\in[0,1]$ and $\gamma>0$ are given. Then*

$$|(T|T|^{(1+r)\alpha+(1+s)\beta-1}x,y)|$$
$$\leq\frac{M^\gamma-m^\gamma}{M^\gamma+m^\gamma}(\{|T|^r A^{2p}|T|^r\}^{\frac{(1+r)\alpha}{p+r}}x,x)^{1/2}(\{|T^*|^s B^{2q}|T^*|^s\}^{\frac{(1+s)\beta}{q+s}}y,y)^{1/2}+$$
$$+\frac{2M^\gamma m^\gamma}{M^\gamma+m^\gamma}|(T|T|^{(1+r)\alpha+(1+s)\beta-\gamma-1}x,y)|$$

holds for all $p,q\geq 1$ and $x,y\in H$.

In particular, if $x, y \in H$ satisfy

$$T|T|^{(1+r)\alpha+(1+s)\beta-\gamma-1}x \neq 0, \quad (T|T|^{(1+r)\alpha+(1+s)\beta-\gamma-1}x, y) = 0,$$

then

$$|(T|T|^{(1+r)\alpha+(1+s)\beta-1}x, y)|^2 \le \left(\frac{M^\gamma - m^\gamma}{M^\gamma + m^\gamma}\right)^2 (\{|T|^r A^{2p}|T|^r\}^{\frac{(1+r)\alpha}{p+r}} x, x)(\{|T^*|^s B^{2q}|T^*|^s\}^{\frac{(1+s)\beta}{q+s}} y, y)$$

holds for all $p, q \ge 1$.

Proof. In Theorem 5 and Corollary 6 we take $t_1 = t_2 = 0$, $s_1 = s_2 = 1$ and $p_1 = p$, $p_2 = q$, $r_1 = r$, $r_2 = s$. Then we obtain the desired inequalities.

It is easily seen that Corollary 7 can be proved in a similar way as Theorem 5 by the use of the Furuta inequality instead of the grand Furuta inequality. □

4. Bauer–Householder type extensions

For convenience, we cite the Bauer–Householder Theorem [2] which is also a generalization of the Wielandt inequality, see [5].

Theorem C. *If a positive operator C on H satisfies $0 < m \le C \le M$ for some $0 < m < M$, then*

$$|(Cx, y)|^2 \le \frac{\tilde{M} - \tilde{m}}{\tilde{M} + \tilde{m}}(Cx, x)(Cy, y)$$

for all unit vectors $x, y \in H$, where $\tilde{M} = M(1 + |(x, y)|)$ and $\tilde{m} = m(1 - |(x, y)|)$.

Theorem C gives us another Wielandt type inequality, by which we obtain an alternative extension of the HKF inequality via the Löwner–Heinz inequality:

Theorem 8. *Let T be an operator on H satisfying $0 < m \le |T| \le M$ for some $0 < m \le M$. Then, for each $\gamma > 0$,*

$$|(T|T|^{\alpha+\beta-1}x, y)| \le \frac{\tilde{M}_\gamma - \tilde{m}_\gamma}{\tilde{M}_\gamma + \tilde{m}_\gamma} \||T|^\alpha x\| \||T^*|^\beta y\|$$

for all $\alpha, \beta \in \mathbb{R}$ and $x, y \in H$, where

$$\tilde{M}_\gamma = M^\gamma(\||T|^{\alpha-\frac{\gamma}{2}}x\| \||T^*|^{\beta-\frac{\gamma}{2}}y\| + |(T|T|^{\alpha+\beta-\gamma-1}x, y)|),$$

$$\tilde{m}_\gamma = m^\gamma(\||T|^{\alpha-\frac{\gamma}{2}}x\| \||T^*|^{\beta-\frac{\gamma}{2}}y\| - |(T|T|^{\alpha+\beta-\gamma-1}x, y)|).$$

Proof. First of all, we have to generalize Theorem C as follows: If $0 < m \le C \le M$, then

$$|(Cx,y)|^2 \le \frac{M_1 - m_1}{M_1 + m_1}(Cx,x)(Cy,y)$$

for all $0 \ne x, y \in H$, where $M_1 = M(\|x\|\|y\| + |(x,y)|)$ and $m_1 = m(\|x\|\|y\| - |(x,y)|)$. This is obtained by putting $x' = x/\|x\|$ and $y' = y/\|y\|$ in Theorem C.

Now, as in the proof of Theorem 1, we put $x_1 = |T|^{\alpha-\gamma/2}x$ and $y_1 = |T|^{\beta-\gamma/2}U^*y$. Thus we use the above for x_1, y_1 and $C = |T|^\gamma$ and so we have the desired inequality.

□

COROLLARY 9. *Let A, B and T be as in Theorem 3. Then, for each $\gamma > 0$,*

$$|(T|T|^{\alpha+\beta-1}x,y)| \le \frac{\tilde{M}_\gamma - \tilde{m}_\gamma}{\tilde{M}_\gamma + \tilde{m}_\gamma}\|A^\alpha x\|\||B|^\beta y\|$$

for all $\alpha, \beta \in [0,1]$ and $x, y \in H$, where $\tilde{M}_\gamma$ and $\tilde{m}_\gamma$ are as in Theorem 8.

The grand Furuta inequality gives us the following Bauer–Householder type extension of the HKF inequality:

THEOREM 10. *Let T be an operator on H satisfying $0 < mI \le |T| \le MI$. If A and B are strictly positive operators such that $\|Tx\| \le \|Ax\|$ and $\|T^*y\| \le \|By\|$ for all $x, y \in H$, then, for each r_1, $r_2 \ge 0$, t_1, $t_2 \in [0,1]$ with $r_j \ge t_j$ $(j = 1,2)$, $\alpha, \beta \in [0,1]$ and $\gamma > 0$,*

$$\begin{aligned}
&|(T|T|^{(1-t_1+r_1)\alpha+(1-t_2+r_2)\beta-1}x,y)|^2 \\
&\quad \le \tilde{K}(\{|T|^{r_1}(|T|^{-t_1}A^{2p_1}|T|^{-t_1})^{s_1}|T|^{r_1}\}^{\frac{(1-t_1+r_1)\alpha}{(p_1-t_1)s_1+r_1}}x,x)\times \\
&\qquad \times (\{|T^*|^{r_2}(|T^*|^{-t_2}B^{2p_2}|T^*|^{-t_2})^{s_2}|T^*|^{r_2}\}^{\frac{(1-t_2+r_2)\beta}{(p_2-t_2)s_2+r_2}}y,y)
\end{aligned}$$

holds for all $p_1, p_2, s_1, s_2 \ge 1$ and $x, y \in H$, where $\tilde{K} = \tilde{K}_\gamma = \left(\frac{\tilde{M}_\gamma - \tilde{m}_\gamma}{\tilde{M}_\gamma + \tilde{m}_\gamma}\right)^2$ and $\tilde{M}_\gamma$ and $\tilde{m}_\gamma$ are as in Theorem 8.

Proof. Let $T = U|T|$ be the polar decomposition of T. We put $a = (1-t_1+r_1)\alpha$ and $b = (1-t_2+r_2)\beta$. Then it follows from Theorem 8 that

$$|(T|T|^{a+b-1}x, y)|^2 \le \tilde{K}(|T|^{2a}x, x)(|T^*|^{2b}y, y)$$

As in the proof of Theorem 5, we use the grand Furuta inequality. That is, the right hand side of the above is not less than

$$\tilde{K}(\{|T|^{r_1}(|T|^{-t_1}A^{2p_1}|T|^{-t_1})^{s_1}|T|^{r_1}\}^{\frac{(1-t_1+r_1)\alpha}{(p_1-t_1)s_1+r_1}}x, x)\times$$
$$\times(\{|T^*|^{r_2}(|T^*|^{-t_2}B^{2p_2}|T^*|^{-t_2})^{s_2}|T^*|^{r_2}\}^{\frac{(1-t_2+r_2)\beta}{(p_2-t_2)s_2+r_2}}y, y).$$

□

REMARK. Theorem 10 is also a simultaneous extension of Theorem A and the Wielandt inequality. Theorem A is obtained by tending m to 0 in Theorem 10 because $\tilde{m}_\gamma \to 0$.

5. Concluding remarks

Finally we mention some corollaries of Theorem 1 which are generalizations of the Wielandt inequality. Let $T = U|T|$ be the polar decomposition.

COROLLARY 11. *Let $T = U|T|$ be an operator on H such that $0 < mI \le |T| \le MI$. Then*

$$|(Tx, y)| \le \frac{M-m}{M+m}(|T|x, x)^{1/2}(|T^*|y, y)^{1/2} + \frac{2Mm}{M+m}|(Ux, y)|$$

holds for all $x, y \in H$.

COROLLARY 12. *Let $T = U|T|$ be an operator on H such that $0 < mI \le |T| \le MI$. Then*

$$|(Tx, y)|^2 \le \left(\frac{M-m}{M+m}\right)^2(|T|x, x)(|T^*|y, y)$$

holds for all $x, y \in H$ with $(Ux, y) = 0$.

References

[1] T. Ando and F. Hiai, Log majorization and complementary Golden–Thompson type inequalities, *Linear Alg. and Appl.*, **197/198** (1994), 113–131.

[2] F. L. Bauer and A. S. Householder, Some inequalities involving the euclidean condition of a matrix, *Numerische Math.*, **2** (1960), 308–311.

[3] M. Fujii, Furuta's inequality and its mean theoretic approach, *J. Operator Theory*, **23** (1990), 67–72.

[4] M. Fujii and E. Kamei, Mean theoretic approach to the grand Furuta inequality, *Proc. Amer. Math. Soc.*, **124** (1996), 2751–2756.

[5] M. Fujii, Y. Katayama and R. Nakamoto, Generalizations of the Wielandt theorem, *Math. Japon.*, **49** (1999), 217–222.

[6] M. Fujii, A. Matsumoto and R. Nakamoto, A short proof of the best possibility for the grand Furuta inequality, *J. Inequal. Appl.*, **4** (1999), 339–344.

[7] T. Furuta, $A \geq B \geq 0$ assures $(B^r A^p B^r)^{1/q} \geq B^{(p+2r)/q}$ for $r \geq 0$, $p \geq 0$, $q \geq 1$ with $(1+2r)q \geq p+2r$, *Proc. Amer. Math. Soc.*, **101** (1987), 85–88.

[8] T. Furuta, Elementary proof of an order preserving inequality, *Proc. Japan Acad.*, **65** (1989), 126.

[9] T. Furuta, An extension of the Heinz–Kato Theorem, *Proc. Amer. Math. Soc.*, **120** (1994), 785–787.

[10] T. Furuta, Extension of the Furuta inequality and Ando–Hiai log-majorization, *Linear Alg. and Appl.*, **219** (1995), 139–155.

[11] T. Furuta, Simplifed proof of an order preserving inequality, *Proc. Japan. Acad. Ser. A*, **75** (1998), 114.

[12] T. Furuta, Equivalence relations among Reid, Löwner–Heinz and Heinz–Kato inequalities, and extensions of these inequalities, *Integral Equations Operator Theory*, **29** (1997), 1–9.

[13] E. Kamei, A satellite to Furuta's inequality, *Math. Japon.*, **33** (1988), 883–886.

[14] K. Tanahashi, Best possibility of the Furuta inequality, *Proc. Amer. Math. Soc.*, **124** (1996), 141–146.

[15] K. Tanahashi, The best possibility of the grand Furuta inequality, *Proc. Amer. Math. Soc.*, **128** (2000), 511–519.

[16] T. Yamazaki, Simplified proof of Tanahashi's result on the best possibility of generalized Furuta inequality, *Math. Inequal. Appl.*, **2** (1999), 473–477.

M. Fujii, Department of Mathematics, Osaka Kyoiku University, Asahigaoka, Kashiwara, Osaka 582-8582, Japan; *e-mail*: mfujii@cc.osaka-kyoiku.ac.jp

Y. Seo, Tennoji Branch, Senior Highschool, Osaka Kyoiku University, Tennoji, Osaka 543-0054, Japan; *e-mail*: yukis@cc.osaka-kyoiku.ac.jp

Operator Theory: Advances and Applications, Vol. 127, 279–290

Logarithmic order and dual logarithmic order

TAKAYUKI FURUTA

Dedicated to the memory of Professor Béla Szőkefalvi-Nagy in deep sorrow

We shall define the following four orders for strictly positive operators A and B on a Hilbert space H.
Strictly logarithmic order (denoted by $A \succ_{sl} B$) is defined by $\frac{A-I}{\log A} > \frac{B-I}{\log B}$.
Logarithmic order (denoted by $A \succ_{l} B$) is defined by $\frac{A-I}{\log A} \geq \frac{B-I}{\log B}$.
Strictly dual logarithmic order (denoted by $A \succ_{sdl} B$) is defined by $\frac{A \log A}{A-I} > \frac{B \log B}{B-I}$.
Dual Logarithmic order (denoted by $A \succ_{dl} B$) is defined by $\frac{A \log A}{A-I} \geq \frac{B \log B}{B-I}$.

Firstly we shall show direct and simplified proofs of operator monotonicity of the logarithmic function $f(t) = \frac{t-1}{\log t}$ and the dual logarithmic function $f^*(t) = \frac{t \log t}{t-1}$. In what follows, let A and B be strictly positive operators on a Hilbert space H.

Secondary we shall show the following:

($\star$) $\log A > \log B \Longrightarrow$ there exists $\beta \in (0,1]$ such that $A^\alpha \succ_{sl} B^\alpha$ holds for all $\alpha \in (0, \beta)$;

($\ddagger$) $\log A > \log B \Longrightarrow$ there exists $\beta \in (0,1]$ such that $A^\alpha \succ_{sdl} B^\alpha$ holds for all $\alpha \in (0, \beta)$.

By using these two results ($\star$) and ($\ddagger$), we summarize the following interesting contrast among $A > B > 0$, $A \geq B > 0$, $\log A > \log B$ and $\log A \geq \log B$.

(l-i) $A > B > 0 \Longrightarrow$ there exists $\beta \in (0,1]$ such that $A^\alpha \succ_{sl} B^\alpha$ holds for all $\alpha \in (0, \beta)$.

(l-ii) $A \geq B > 0 \Longrightarrow A^\alpha \succ_{l} B^\alpha$ for all $\alpha \in (0,1]$.

(l-iii) $\log A \geq \log B \Longrightarrow$ for any $\delta \in (0,1]$, there exists $\beta = \beta_\delta \in (0,1]$ such that $(e^\delta A)^\alpha \succ_{sl} B^\alpha$ holds for all $\alpha \in (0, \beta)$.

(l-iv) $\log A \geq \log B \Longrightarrow$ for any $p \geq 0$, there exists $K_p > 1$ such that $K_p \to 1$ as $p \to +0$ and $(K_p A)^{p\alpha} \succ_{l} B^{p\alpha}$ for all $\alpha \in (0,1]$.

AMS Subject Classification (1991): 47A63.

(dl-i) $A > B > 0 \Longrightarrow$ there exists $\beta \in (0,1]$ such that $A^\alpha \succ_{sdl} B^\alpha$ holds for all $\alpha \in (0,\beta)$.

(dl-ii) $A \geq B > 0 \Longrightarrow A^\alpha \succ_{dl} B^\alpha$ for all $\alpha \in (0,1]$.

(dl-iii) $\log A \geq \log B \Longrightarrow$ for any $\delta \in (0,1]$, there exists $\beta = \beta_\delta \in (0,1]$ such that $(e^\delta A)^\alpha \succ_{sdl} B^\alpha$ holds for all $\alpha \in (0,\beta)$.

(dl-iv) $\log A \geq \log B \Longrightarrow$ for any $p \geq 0$, there exists $K_p > 1$ such that $K_p \to 1$ as $p \to +0$ and $(K_p)^{p\alpha} \succ_{dl} B^{p\alpha}$ for all $\alpha \in (0,1]$.

Finally we cite a counterexample related to (l-iii) and (dl-iii).

1. Introduction

A capital letter means a bounded linear operator on a complex Hilbert space H. An operator T is said to be positive (denoted by $T \geq 0$) if $(Tx,x) \geq 0$ for all $x \in H$, and also an operator T is said to be strictly positive (denoted by $T > 0$) if T is positive and invertible. *The strictly chaotic order is defined by* $\log A > \log B$ *for strictly positive operators* A *and* B.

It is well known that the usual order $A \geq B > 0$ ensures the chaotic order $\log A \geq \log B$ since $\log t$ is operator monotone function.

Also it is known by [6, Theorem] and [5, Example 5.1.12 and Corollary 5.1.11] that $A \geq B > 0$ ensures $\frac{A-I}{\log A} \geq \frac{B-I}{\log B}$ and $A \geq B > 0$ ensures $\frac{A \log A}{A-I} \geq \frac{B \log B}{B-I}$ since $f(t) = \frac{t-1}{\log t}$ $(t > 0)$ and $f^*(t) = \frac{t \log t}{t-1}$ $(t > 0)$ are both operator monotone functions (see Theorem A below). The function $f(t) = \frac{t-1}{\log t}$ $(t > 0)$ is said to be "*logarithmic function*" which is widely used in the theory of heat transfer of the heat engineering and fluid mechanics. Also the function $f^*(t) = \frac{t \log t}{t-1}$ $(t > 0)$ is said to be "*dual logarithmic function*". Related to these two operator inequalities, we shall define the following four orders for strictly positive operators A and B.

DEFINITION 1. Let A and B be strictly positive operators on a Hilbert space H.

(d1) *Strictly logarithmic order* (denoted by $A \succ_{sl} B$) is defined by $\frac{A-I}{\log A} > \frac{B-I}{\log B}$.

(d2) *Logarithmic order* (denoted by $A \succ_l B$) is defined by $\frac{A-I}{\log A} \geq \frac{B-I}{\log B}$.

(d3) *Strictly dual logarithmic order* (denoted by $A \succ_{sdl} B$) is defined by $\frac{A \log A}{A-I} > \frac{B \log B}{B-I}$.

(d4) *Dual Logarithmic order* (denoted by $A \succ_{dl} B$) is defined by $\frac{A \log A}{A-I} \geq \frac{B \log B}{B-I}$.

2. Simplified proofs of operator monotonicity of logarithmic function and dual logarithmic function

We shall show a direct and simplified proof of the results [6, Theorem] and [5, Example 5.1.12 and Corollary 5.1.11] without use of Löwner's general result.

THEOREM A. *The functions f and f^* given by*

$$f(t)=\begin{cases}\frac{t-1}{\log t} & (t>0, t\neq 1)\\ 1 & (t=1)\\ 0 & (t=0)\end{cases}$$

and

$$f^*(t)=\begin{cases}\frac{t\log t}{t-1} & (t>0, t\neq 1)\\ 1 & (t=1)\\ 0 & (t=0)\end{cases}$$

are operator monotone functions satisfying the symmetry condition:

$$f(t)=tf\Big(\frac{1}{t}\Big) \quad \textit{and} \quad f^*(t)=tf^*\Big(\frac{1}{t}\Big).$$

Proof. Let A and B be strictly positive operators. We have only to show the following (i) and (ii) since the latter half is obvious.

(i) If $A\geq B$, then $\frac{A-I}{\log A}\geq\frac{B-I}{\log B}$.

(ii) If $A\geq B$, then $\frac{A\log A}{A-I}\geq\frac{B\log B}{B-I}$.

First of all, we cite the following obvious result:

(1) $T-I=(T^{\frac{1}{n}}-I)(T^{1-\frac{1}{n}}+T^{1-\frac{2}{n}}+\cdots+T^{\frac{1}{n}}+I)$ for $T\geq 0$ and for any natural number n;

(2) $\lim_{n\to\infty}n(T^{\frac{1}{n}}-I)=\log T$ holds for any $T>0$;

(3) if $A\geq B\geq 0$, then $A^{\alpha}\geq B^{\alpha}$ holds for any $\alpha\in[0,1]$ (Löwner–Heinz inequality).

(i):

$$\begin{aligned}\frac{A-I}{n(A^{\frac{1}{n}}-I)}&=\frac{1}{n}(A^{1-\frac{1}{n}}+A^{1-\frac{2}{n}}+\cdots+A^{\frac{1}{n}}+I) \text{ by (1) for any natural number } n\\ &\geq\frac{1}{n}(B^{1-\frac{1}{n}}+B^{1-\frac{2}{n}}+\cdots+B^{\frac{1}{n}}+I) \text{ by (3) for any natural number } n\\ &=\frac{B-I}{n(B^{\frac{1}{n}}-I)} \text{ for any natural number } n \text{ by (1) tending } n \text{ to } \infty,\end{aligned}$$

so we obtain (i) by (2).
(ii):

$$\begin{aligned}\frac{n(A^{\frac{1}{n}}-I)A}{A-I} &= \frac{n}{(A^{-\frac{1}{n}}+A^{-\frac{2}{n}}+\cdots+A^{-1})} \text{ by (1) for any natural number } n\\ &\geq \frac{n}{(B^{-\frac{1}{n}}+B^{-\frac{2}{n}}+\cdots+B^{-1})} \text{ by (3) for any natural number } n\\ &= \frac{n(B^{\frac{1}{n}}-I)B}{B-I} \text{ by (1) tending } n \text{ to } \infty,\end{aligned}$$

so we obtain (ii) by (2).

□

REMARK 1. It is well known that (i) is equivalent to (ii) in Theorem A. Alternative proof of (i) in the proof of Theorem A is cited in [5]. Related to Theorem A, we remark the following result; see [4, Corollary 2.6], [7, Theorem 2] and [5, Corollary 5.1.11]. Let $g(t)$ be a continuous positive function such that $(0,\infty)\to(0,\infty)$. Then $g(t)$ is operator monotone function if and only if $g^*(t)=\frac{t}{g(t)}$ is operator monotone function. Actually, $f(t)$ and $f^*(t)$ in Theorem A satisfy this condition: $f^*(t)=\frac{t}{f(t)}$.

3. Strictly logarithmic order $A \succ_{sl} B$ and logarithmic order $A \succ_l B$

Let A and B be strictly positive operators. Firstly we shall give Theorem 1 asserting the following

$$(\star)\quad \begin{array}{l}\log A > \log B \Longrightarrow \text{there exists } \beta \in (0,1] \text{ such that } A^\alpha \succ_{sl} B^\alpha\\ \text{holds for all } \alpha \in (0,\beta).\end{array}$$

Secondary, we shall give Corollary 2 showing that there exists an interesting contrast between $A \geq B > 0$ and $A > B > 0$ related to $A \succ_{sl} B$ and $A \succ_l B$. Thirdly, we shall give some applications of two characterizations (Theorem A and Theorem B under below) of chaotic order to $A \succ_{sl} B$ and $A \succ_l B$ in Corollary 3.

LEMMA 1. *Let A and B be selfadjoint operators on a Hilbert space H. If $A > B$, then there exists $\beta \in (0,1]$ such that the following inequality holds for all $\alpha \in (0,\beta)$:* $\frac{e^{\alpha A}-I}{\alpha A} > \frac{e^{\alpha B}-I}{\alpha B}$ *, i.e.,* $e^{\alpha A} \succ_{sl} e^{\alpha B}$.

Proof. There exists ε such that $A - B \geq \varepsilon > 0$. Choose α and β such that

$$(4) \qquad 0 < \alpha < \mathrm{Min}\left\{\frac{\varepsilon}{2}\left(\frac{e^{||A||}}{||A||} + \frac{e^{||B||}}{||B||}\right)^{-1}, 1\right\} = \beta.$$

By an easy calculation, we obtain

$$\begin{aligned}
\frac{e^{\alpha A} - I}{A} - \frac{e^{\alpha B} - I}{B} &= \sum_{n=1}^{\infty} \frac{\alpha^n}{n!} A^{n-1} - \sum_{n=1}^{\infty} \frac{\alpha^n}{n!} B^{n-1} \\
&= \sum_{n=2}^{\infty} \frac{\alpha^n}{n!} (A^{n-1} - B^{n-1}) \\
&= \frac{\alpha^2}{2!}(A - B) + \sum_{n=3}^{\infty} \frac{\alpha^n}{n!}(A^{n-1} - B^{n-1}) \\
&\geq \frac{\alpha^2}{2!}\varepsilon - \alpha^3\Big[\sum_{n=3}^{\infty} \frac{1}{n!}(||A||^{n-1} + ||B||^{n-1})\Big] \\
&\geq \alpha^2\Big[\frac{\varepsilon}{2!} - \alpha\Big(\frac{e^{||A||}}{||A||} + \frac{e^{||B||}}{||B||}\Big)\Big] > 0
\end{aligned}$$

by (4), so that $\frac{e^{\alpha A} - I}{\alpha A} - \frac{e^{\alpha B} - I}{\alpha B}$ holds, i.e., there exists $\beta \in (0, 1]$ such that $e^{\alpha A} \succ_{sl} e^{\alpha B}$ holds for all $\alpha \in (0, \beta)$ and the proof is complete.

□

THEOREM 1. *Let A and B be strictly positive operators. If $\log A > \log B$, then there exists $\beta \in (0, 1]$ such that $A^\alpha \succ_{sl} B^\alpha$ holds for all $\alpha \in (0, \beta)$.*

Proof. We have only to replace A by $\log A$ and also B by $\log B$, respectively in Lemma 1.

□

COROLLARY 2. *Let A and B be strictly positive operators. Then*

(i) *If $A > B > 0$, then there exists $\beta \in (0, 1]$ such that $A^\alpha \succ_{sl} B^\alpha$ holds for all $\alpha \in (0, \beta)$.*

(ii) *If $A \geq B > 0$, then $A^\alpha \succ_l B^\alpha$ holds for all $\alpha \in (0, 1]$.*

In Corollary 2, it is interesting to point out the contrast between $A > B > 0$ and $A \geq B > 0$.

Proof of Corollary 2. (i) We cite the following obvious and fundamental result

$$\text{(5)} \qquad \text{if } A > B > 0, \text{ then } \log A > \log B.$$

In fact if $A > B > 0$, then $A \geq B + \varepsilon > B$ for some $\varepsilon > 0$, so that $\log A \geq \log(B + \varepsilon) > \log B$, that is, (5) holds. (i) follows by (5) and Theorem 1.

(ii) If $A \geq B > 0$, then $A^\alpha \geq B^\alpha$ for all $\alpha \in (0, 1]$ by Löwner–Heinz inequality and (ii) follows by the result that the function $f(t) = \frac{t-1}{\log t}$ $(t > 0, t \neq 1)$ is an operator monotone function by Theorem A, i.e., $f(A^\alpha) \geq f(B^\alpha)$ for all $\alpha \in (0, 1]$, so we have (ii).

□

COROLLARY 3. *Let A and B be strictly positive operators such that* $\log A \geq \log B$. *Then*

(i) *For any* $\delta \in (0, 1]$ *there exists* $\beta = \beta_\delta \in (0, 1]$ *such that* $(e^\delta A)^\alpha \succ_{sl} B^\alpha$ *holds for all* $\alpha \in (0, \beta)$.

(ii) *For any* $p \geq 0$ *there exists* $K_p \geq 1$ *such that* $K_p \to 1$ *as* $p \to +0$ *and* $(K_p A)^{p\alpha} \succ_l B^{p\alpha}$ *for all* $\alpha \in (0, 1]$.

We cite the following two results in order to give a proof of Corollary 3.

THEOREM A. ([1], [3]). *Let A and B be invertible positive operators on a Hilbert space H. Then* $\log A \geq \log B$ *holds if and only if for any* $\delta \in (0, 1]$ *there exists* $\alpha = \alpha_\delta > 0$ *such that* $(e^\delta A)^\alpha > B^\alpha$.

THEOREM B. ([8]). *Let A and B be invertible positive operators on a Hilbert space H. Then* $\log A \geq \log B$ *if and only if for any* $p \geq 0$ *there exists a* $K_p > 1$ *such that* $K_p \to 1$ *as* $p \to +0$ *and* $(K_p A)^p \geq B^p$.

Proof of Corollary 3. (i) As $\log A \geq \log B$ holds, then for any $\delta \in (0, 1]$, there exists $\alpha' = \alpha'_\delta > 0$ such that $(e^\delta A)^{\alpha'} > B^{\alpha'}$ by Theorem A. Then $\log e^\delta A > \log B$ by (5), so that there exists $\beta = \beta_\delta \in (0, 1]$ such that $(e^\delta A)^\alpha \succ_{sl} B^\alpha$ holds for all $\alpha \in (0, \beta)$ by Theorem 1.

(ii) As $\log A \geq \log B$ holds, then for any $p \geq 0$ there exists a $K_p > 1$ such that $K_p \to 1$ as $p \to +0$ and $(K_p A)^p \geq B^p$ by Theoren B, so we have $(K_p A)^{p\alpha} \geq B^{p\alpha}$ for all $\alpha \in (0, 1]$ by (ii) of Corollary 2.

4. Strictly dual logarithmic order $A \succ_{sdl} B$ and dual logarithmic order $A \succ_{dl} B$

Let A and B be strictly positive operators. Firstly we shall give Theorem 4 asserting the following

(‡) $\log A > \log B \Longrightarrow$ there exists $\beta \in (0,1]$ such that $A^\alpha \succ_{sdl} B^\alpha$ holds for all $\alpha \in (0,\beta)$.

Secondary, we shall give Corollary 5 showing that there exists an interesting contrast between $A \geq B > 0$ and $A > B > 0$ related to $A \succ_{sdl} B$ and $A \succ_{dl} B$. Thirdly, we shall give some applications of Theorem A and Theorem B to $A \succ_{sdl} B$ and $A \succ_{dl} B$ in Corollary 6.

LEMMA 2. *Let A and B be selfadjoint operators on a Hilbert space H. If $A > B$, then there exists $\beta \in (0,1]$ such that the following inequality holds for all $\alpha \in (0,\beta)$: $\frac{\alpha A e^{\alpha A}}{e^{\alpha A}-I} > \frac{\alpha B e^{\alpha B}}{e^{\alpha B}-I}$, i.e., $e^{\alpha A} \succ_{sdl} e^{\alpha B}$.*

Proof. As $-B > -A$ holds, by applying Lemma 1, there exists $\beta \in (0,1]$ such that

$$\frac{e^{-\alpha B}-I}{-\alpha B} > \frac{e^{-\alpha A}-I}{-\alpha A}$$

holds for all $\alpha \in (0,\beta)$. That is, $\frac{e^{\alpha B}-I}{\alpha B e^{\alpha B}} > \frac{e^{\alpha A}-I}{\alpha A e^{\alpha A}}$ holds iff

$$\frac{\alpha A e^{\alpha A}}{e^{\alpha A}-I} > \frac{\alpha B e^{\alpha B}}{e^{\alpha B}-I}$$

holds, i.e., there exists $\beta \in (0,1]$ such that $e^{\alpha A} \succ_{sdl} e^{\alpha B}$ holds for all $\alpha \in (0,\beta)$ and the proof is complete. □

THEOREM 4. *Let A and B be strictly positive operators. If $\log A > \log B$, then there exists $\beta \in (0,1]$ such that $A^\alpha \succ_{sdl} B^\alpha$ holds for all $\alpha \in (0,\beta)$.*

Proof. We have only to replace A by $\log A$ and also B by $\log B$, respectively in Lemma 2. □

COROLLARY 5. *Let A and B be strictly positive operators. Then*

(i) *If $A > B > 0$, then there exists $\beta \in (0,1]$ such that $A^\alpha \succ_{sdl} B^\alpha$ holds for all $\alpha \in (0,\beta)$.*

(ii) *If $A \geq B > 0$, then $A^\alpha \succ_{dl} B^\alpha$ for all $\alpha \in (0,1]$.*

In Corollary 5, it is interesting to point out the contrast between $A > B > 0$ and $A \geq B > 0$.

Proof of Corollary 5. By the same way as in the proof of Corollary 2, we shall give the following proofs of (i) and (ii).

(i) If $A > B > 0$, then $\log A > \log B$ holds by (5), so that (i) follows by Theorem 4.

(ii) If $A \geq B > 0$, then $A^\alpha \geq B^\alpha$ for all $\alpha \in (0,1]$ by Löwner–Heinz inequality. The function $f^*(t) = \frac{t \log t}{t-1}$ $(t > 0)$ is also an operator monotone function by Theorem A, so that $f^*(A^\alpha) \geq f^*(B^\alpha)$ for all $\alpha \in (0,1]$, so we have (ii).

□

COROLLARY 6. *Let A and B be strictly positive operators such that $\log A \geq \log B$. Then*

(i) *For any $\delta \in (0,1]$ there exists $\beta = \beta_\delta \in (0,1]$ such that $(e^\delta A)^\alpha \succ_{sdl} B^\alpha$ holds for all $\alpha \in (0,\beta)$.*

(ii) *For any $p \geq 0$ there exists $K_p > 1$ such that $K_p \to 1$ as $p \to +0$ and $(K_p A)^{p\alpha} \succ_{dl} B^{p\alpha}$ for all $\alpha \in (0,1]$*

Proof of Corollary 6. We shall obtain Corollary 6 by the same way as Corollary 3.

(i) As $\log A \geq \log B$ holds, then for any $\delta \in (0,1]$, there exists $\alpha' = \alpha'_\delta > 0$ such that $(e^\delta A)^{\alpha'} > B^{\alpha'}$ by Theorem A. Then $\log e^\delta A > \log B$ by (5), so that there exists $\beta = \beta_\delta \in (0,1]$ such that $(e^\delta A)^\alpha \succ_{sdl} B^\alpha$ holds for all $\alpha \in (0,\beta)$ by Theorem 4.

(ii) As $\log A \geq \log B$ holds, then for any $p \geq 0$ there exists a $K_p > 1$ such that $K_p \to 1$ as $p \to +0$ and $(K_p A)^p \geq B^p$ by Theorem B, so that $(K_p A)^{p\alpha} \succ_{dl} B^{p\alpha}$ for all $\alpha \in (0,1]$ by (ii) of Corollary 5.

□

5. An example related to strictly logarithmic order $A \succ_{sl} B$ and strictly dual logarithmic order $A \succ_{sdl} B$

Related to (i) of Corollary 3, we consider the following problem:

(Q1) "Does $\log A \geq \log B$ ensure that there exists an $\alpha > 0$ such that $A^\alpha \succ_l B^\alpha$?"

Also related to (i) of Corollary 6, we consider the following problem too:

(Q2) "Does $\log A \geq \log B$ ensure that there exists an $\alpha > 0$ such that $A^\alpha \succ_{dl} B^\alpha$?"

In fact, we cite a counterexample to (Q1) and (Q2) as follows.

EXAMPLE 1. Take A and B as follows: $\log A = \begin{pmatrix} 2 & 2 \\ 2 & -1 \end{pmatrix}$ and $\log B = \begin{pmatrix} 1 & 0 \\ 0 & -5 \end{pmatrix}$. Then $\log A \geq \log B$ holds, but

(i) $A^\alpha \succ_l B^\alpha$ does not hold for any $\alpha > 0$,

(ii) $A^\alpha \succ_{dl} B^\alpha$ does not hold for any $\alpha > 0$,

(iii) $A^\alpha \geq B^\alpha$ does not hold for any $\alpha > 0$.

In fact, $\log A$ is diagonalized by $U = \frac{1}{\sqrt{5}} \begin{pmatrix} -1 & 2 \\ 2 & 1 \end{pmatrix}$ as follows: $U(\log A)U = \begin{pmatrix} -2 & 0 \\ 0 & 3 \end{pmatrix}$, and $UAU = \begin{pmatrix} e^{-2} & 0 \\ 0 & e^3 \end{pmatrix}$, so that we have $A^\alpha = U \begin{pmatrix} e^{-2\alpha} & 0 \\ 0 & e^{3\alpha} \end{pmatrix} U$ and $B^\alpha = \begin{pmatrix} e^\alpha & 0 \\ 0 & e^{-5\alpha} \end{pmatrix}$. Then $x = e^\alpha > 1$ since $\alpha > 0$. At first we show (i). By a slightly elaborate calculation, we have

$$\begin{aligned}
\det\Big(\frac{A^\alpha - I}{\log A} - \frac{B^\alpha - I}{\log B}\Big) &= \begin{vmatrix} \frac{5}{6} - \frac{1}{10x^2} - x + \frac{4x^3}{15} & -\frac{1}{3} + \frac{1}{5x^2} + \frac{2x^3}{15} \\ -\frac{1}{3} + \frac{1}{5x^2} + \frac{2x^3}{15} & \frac{2}{15} + \frac{1}{5x^5} - \frac{2}{5x^2} + \frac{x^3}{15} \end{vmatrix} \\
&= \frac{-1}{50x^7} + \frac{1}{6x^5} - \frac{1}{5x^4} - \frac{4}{25x^2} + \frac{2}{5x} - \frac{3x}{10} + \frac{9x^3}{50} - \frac{x^4}{15} \\
&= \frac{-(x-1)^6(10x^5 + 33x^4 + 48x^3 + 38x^2 + 18x + 3)}{150x^7} \\
&< 0 \quad \text{since } x > 1.
\end{aligned}$$

Whence $A^\alpha \succ_l B^\alpha$ does not hold for any $\alpha > 0$, so the proof of (i) is complete.

Next we show (ii). By more elaborate calculation than (i), we obtain

$$\det\left(\frac{A^\alpha \log A}{A^\alpha - I} - \frac{B^\alpha \log B}{B^\alpha - I}\right)$$
$$= \begin{vmatrix} \frac{7x^3+9x^2+x-2}{5(x^3+2x^2+2x+1)} & \frac{2(3x^3+6x^2+4x+2)}{5(x^3+2x^2+2x+1)} \\ \frac{2(3x^3+6x^2+4x+2)}{5(x^3+2x^2+2x+1)} & \frac{3x^4+3x^3+8x^2+8x+8}{5(x^4+x^3-x-1)} + \frac{5(x-1)}{-x^6+x^5+x-1} \end{vmatrix}$$
$$= \frac{7x^3+9x^2+x-2}{5(x^3+2x^2+2x+1)}\left(\frac{3x^4+3x^3+8x^2+8x+8}{5(x^4+x^3-x-1)} + \frac{5(x-1)}{-x^6+x^5+x-1}\right) - \left(\frac{2(3x^3+6x^2+4x+2)}{5(x^3+2x^2+2x+1)}\right)^2$$
$$= \frac{-(x-1)^2(3x^5+18x^4+38x^3+48x^2+33x+10)}{5(x+1)(x^2+x+1)(x^4+x^3+x^2+x+1)}$$
$$< 0 \quad \text{since } x > 1.$$

Whence $A^\alpha \succ_{dl} B^\alpha$ does not hold for any $\alpha > 0$, so the proof of (ii) is complete.

Incidentally, we remark that this example also shows that $\log A \geq \log B$ does not ensure $A^\alpha \geq B^\alpha$ for any $\alpha > 0$. Actually we have

$$\det(A^\alpha - B^\alpha) = \begin{vmatrix} \frac{1}{5x^2} - x + \frac{4x^3}{5} & \frac{2(x^5-1)}{5x^2} \\ \frac{2(x^5-1)}{5x^2} & \frac{x^8+4x^3-5}{5x^5} \end{vmatrix}$$
$$= \frac{-1}{5x^7} + x^{-4} - \frac{4}{5x^2} - \frac{4}{5x} + x - \frac{x^4}{5}$$
$$= \frac{-(x-1)^4(x+1)(x^2+x+1)(x^4+2x^3+4x^2+2x+1)}{5x^7}$$
$$< 0 \quad \text{since } x > 1,$$

that is, $A^\alpha \geq B^\alpha$ does not hold for any $\alpha > 0$, so (iii) is shown. In [2], there is another nice example that $\log A \geq \log B$ does not ensure $A^\alpha \geq B^\alpha$ for any $\alpha > 0$. In fact, we constructed Example 1 inspired by an excellent method in [2].

6. Concluding remarks

Let A and B be strictly positive operators. We can obtain the following interesting contrast among $A > B > 0$, $A \geq B > 0$, $\log A > \log B$ and $\log A \geq \log B$ by summarizing our results in this paper.

($\star$) $\log A > \log B \Longrightarrow$ there exists $\beta \in (0,1]$ such that $A^\alpha \succ_{sl} B^\alpha$ holds for all $\alpha \in (0,\beta)$.

($\ddagger$) $\log A > \log B \Longrightarrow$ there exists $\beta \in (0,1]$ such that $A^\alpha \succ_{sdl} B^\alpha$ holds for all $\alpha \in (0,\beta)$.

(l-i) $A > B > 0 \Longrightarrow$ there exists $\beta \in (0,1]$ such that $A^\alpha \succ_{sl} B^\alpha$ holds for all $\alpha \in (0,\beta)$.

(l-ii) $A \geq B > 0 \Longrightarrow A^\alpha \succ_l B^\alpha$ for all $\alpha \in (0,1]$.

(l-iii) $\log A \geq \log B \Longrightarrow$ for any $\delta \in (0,1]$, there exists $\beta = \beta_\delta \in (0,1]$ such that $(e^\delta A)^\alpha \succ_{sl} B^\alpha$ holds for all $\alpha \in (0,\beta)$.

(l-iv) $\log A \geq \log B \Longrightarrow$ for any $p \geq 0$, there exists $K_p > 1$ such that $K_p \to 1$ as $p \to +0$ and $(K_pA)^{p\alpha} \succ_l B^{p\alpha}$ for all $\alpha \in (0,1]$.

(dl-i) $A > B > 0 \Longrightarrow$ there exists $\beta \in (0,1]$ such that $A^\alpha \succ_{sdl} B^\alpha$ holds for all $\alpha \in (0,\beta)$.

(dl-ii) $A \geq B > 0 \Longrightarrow A^\alpha \succ_{dl} B^\alpha$ for all $\alpha \in (0,1]$.

(dl-iii) $\log A \geq \log B \Longrightarrow$ for any $\delta \in (0,1]$, there exists $\beta = \beta_\delta \in (0,1]$ such that $(e^\delta A)^\alpha \succ_{sdl} B^\alpha$ holds for all $\alpha \in (0,\beta)$.

(dl-iv) $\log A \geq \log B \Longrightarrow$ for any $p \geq 0$, there exists $K_p > 1$ such that $K_p \to 1$ as $p \to +0$ and $(K_p)^{p\alpha} \succ_{dl} B^{p\alpha}$ for all $\alpha \in (0,1]$.

ACKNOWLEDGMENT. We would like to express our cordial thanks to the referee for nice advice to our first version.

References

[1] M. FUJII, J. F. JIANG and E. KAMEI, Characterization of chaotic order and its application to Furuta inequality, *Proc. Amer. Math. Soc.*, **125** (1997), 3655–3658.

[2] M. FUJII, J. F. JIANG, E. KAMEI and K. TANAHASHI, A characterization of chaotic order and a problem, *J. Inequality Appl.*, **2** (1998), 149–156.

[3] M. FUJII, J. F. JIANG and E. KAMEI, Characterization of chaotic order and its application to Furuta's type operator inequalities, *Linear Multilinear Algebra*, **43** (1998), 339–349.

[4] F. HANSEN and G. K. PEDERSEN, Jensen's inequality for operators and Löwner's theorem, *Math. Ann.*, **258** (1982), 229–241.

[5] F. HIAI and K. YANAGI, *Hilbert space and linear operators*, Makinoshoten, 1995 (in Japanese).

[6] F. KUBO, On logarithmic operator means, *Tenth Symposium on Applied Functional Analysis* 1987, 47–61.

[7] M. K. KWONG, Some results on matrix monotone functions, *Linear Alg. Appl.*, **118** (1989), 129–153.

[8] T. YAMAZAKI and M. YANAGIDA, Characterizations of chaotic order associated with Kantorovich inequality, *Scienticae Mathematicae*, **2** (1999), 37-50.

T. FURUTA, Department of Applied Mathematics, Faculty of Science, Science University of Tokyo, 1-3 Kagurazaka, Shinjuku-ku, Tokyo, 162-8601, Japan; *e-mail*: furuta@rs.kagu.sut.ac.jp

Operator Theory: Advances and Applications, Vol. 127, 291–304

On the generalized von Neumann inequality

DUMITRU GAŞPAR and NICOLAE SUCIU

To the memory of Béla Szőkefalvi-Nagy

The classical von Neumann inequality states

$$\|p(T)\| \leq \sup_{|z|=1} |p(z)|,$$

for any contraction T on a Hilbert space $\mathcal{H}$ and any (scalar) polynomial p. We prove a generalized von Neumann inequality for a special class of commuting multi-contractions $T = (T_1, \ldots, T_N)$ on $\mathcal{H}$ and for any polynomial p of N independent complex variables, such that the coefficients of p are operators on $\mathcal{H}$ that doubly commute with each T_i, $i = \overline{1, N}$. This is made by means of a construction of a kind of partially doubly commuting isometric dilation for our class of multi-contractions. For similar generalizations of the von Neumann inequality, in the commutative context, we refer to [11] (for the case $N = 1$) and to [14] (for $N = 2$).

1. Introduction

In this paper we are interested in the validity of the von Neumann inequality as

$$\|P(T_1, \ldots, T_N)\| \leq \sup_{\substack{|z_i|=1 \\ 1\leq i\leq N}} \|P(z_1, \ldots, z_N)\|, \tag{1.1}$$

where N is a positive integer, $T_1, \ldots, T_N$ are pairwise commuting contractions on a Hilbert space $\mathcal{H}$ and P is any polynomial of the form

$$P(z_1, \ldots, z_N) = \sum_{0\leq|n|\leq n_0} A_n z^n$$

AMS Subject Classification (1991): 47A63, 47A20.

such that the coefficients A_n are bounded linear operators on $\mathcal{H}$, that doubly commute with each T_i, while $z^n = z_1^{n_1} z_2^{n_2} \cdots z_N^{n_N}$ for $n = (n_1, \ldots, n_N) \in \mathbb{Z}_+^N$ and $|n| = n_1 + n_2 + \cdots + n_N$.

The inequality (1.1) was originally discussed by S. Parrott in [10], where it was proved for $N = 1$. The validity of (1.1) for any pair of commuting contractions, i.e. the case $N = 2$, was proved by the second author in [14]. In general, by the well known dilation theorem of W. Arveson ([4]), the inequality (1.1) might hold relative to any polynomial only for commuting contractions, having common isometric dilations. Because of the example of S. Parrott in [10], the inequality (1.1) fails for $N \geq 3$.

We notice that W. Mlak in [9] and J. Janas in [8] have obtained some inequalities, which somehow generalize (1.1) in the context of semi-spectral measures. Let us also mention that some variants of the von Neumann inequality for commuting non-contractions are contained in the recent paper [1].

In this paper we show that the inequality (1.1) holds for a special class of N-tuples of commuting contractions. For our purpose we need some investigations on the regular dilations of multi-contractions.

2. Preliminaries

Let $\mathcal{H}$ be a complex Hilbert space and $\mathcal{B}(\mathcal{H})$ the Banach algebra of all bounded linear operators on $\mathcal{H}$.

If $A \in \mathcal{B}(\mathcal{H})$ and $B \in \mathcal{B}(\mathcal{K})$, with $\mathcal{K}$ a Hilbert space containing $\mathcal{H}$, then B is called a *lifting* of A if $P_{\mathcal{H}} B = A P_{\mathcal{H}}$, where $P_{\mathcal{H}}$ is the orthogonal projection from $\mathcal{K}$ onto $\mathcal{H}$. It is easy to see that $P_{\mathcal{H}}^*$ is just the natural embedding $J_{\mathcal{H}}$ of $\mathcal{H}$ into $\mathcal{K}$. The operator B is called an *extension* of A if $B J_{\mathcal{H}} = J_{\mathcal{H}} A$. In this case $\mathcal{H}$ is invariant to B and A will be called the *restriction* of B to $\mathcal{H}$. We shall denote this by $A = B|_{\mathcal{H}}$. If B and B^* are extensions of A and A^* respectively, we may say that B is a $*$-*extension* of A.

An N-tuple $T = (T_1, \ldots, T_N) \in \mathcal{B}(\mathcal{H})^N$ is called a (*commuting*) *multi-contraction* on $\mathcal{H}$, if T_i are (pairwise commuting) contractions on $\mathcal{H}$, $i = \overline{1, N}$. A commuting pair (T_1, T_2) of contractions on $\mathcal{H}$ is briefly called a *bi-contraction* on $\mathcal{H}$ and denoted $[\mathcal{H}, (T_1, T_2)]$.

For a commuting multi-contraction $T = (T_1, \ldots, T_N)$, we shall use the multi-index notations: $T^* := (T_1^*, \ldots, T_N^*)$, T_i^* being the adjoint of T_i, $i = \overline{1, N}$, and $T^n := T_1^{n_1} \cdots T_N^{n_N}$ for $n = (n_1, \ldots, n_N) \in \mathbb{Z}_+^N$, $\mathbb{Z}$ ($\mathbb{Z}_+$) being the set of all (positive) integers.

An *isometric* (*unitary*) *dilation* of T is a commuting multi-contraction $V = (V_1, \dots, V_N)$ on a Hilbert space $\mathcal{K} \supset \mathcal{H}$, such that V_i are isometric (unitary) operators on $\mathcal{K}$ ($i = \overline{1, N}$) and

$$T^n = P_{\mathcal{H}} V^n J_{\mathcal{H}} \qquad (n \in \mathbb{Z}_+^N). \tag{2.1}$$

It is known from the Parrott example [10], that the existence of isometric dilations is not certain for any commuting N-tuple of contractions for $N \geq 3$. For $N = 2$ by theorem of Ando [3] (or by the commutant lifting theorem of B. Sz.-Nagy and C. Foiaş [16]) such dilations exist.

The isometric dilation is called *minimal* if

$$\mathcal{K} = \bigvee_{n \in \mathbb{Z}_+^N} V^n \mathcal{H}. \tag{2.2}$$

If the minimality condition (2.2) is satisfied, then the condition (2.1) is equivalent to the fact that V_i are liftings of T_i, or equivalently V_i^* are extensions of T_i^*, $i = \overline{1, N}$.

We denote by $[\mathcal{K}, V]$ a minimal isometric dilation of the multi-contraction $[\mathcal{H}, T]$. $[\mathcal{K}, V]$ is called *regular* or $*$-*regular* if it satisfies

$$T^{*n_-} T^{n_+} = P_{\mathcal{H}} V^{*n_-} V^{n_+} J_{\mathcal{H}} \qquad (n \in \mathbb{Z}^N), \tag{2.3}$$

respectively

$$T^{n_+} T^{*n_-} = P_{\mathcal{H}} V^{*n_-} V^{n_+} J_{\mathcal{H}} \qquad (n \in \mathbb{Z}^N), \tag{2.3'}$$

where for $n = (n_1, \dots, n_N) \in \mathbb{Z}^N$, we denote $n_+ := (n_1^+, \dots, n_N^+)$, $n_- := (n_1^-, \dots, n_N^-)$ with $n_i^+ := \max\{n_i, 0\}$ and $n_i^- := \max\{-n_i, 0\}$, $i = \overline{1, N}$.

A minimal regular (or $*$-regular) isometric dilation of a multi-contraction, if it exists, is uniquely determined up to unitary equivalence (see [16]). In general this is not true for any minimal isometric dilation of a multi-contraction.

The regular dilations were orginally introduced by S. Brehmer [5] and by B. Sz.-Nagy [15] and during recent decades were studied by many authors. Regarding (2.3) and (2.3′) we remark that T has a regular isometric dilation iff T^* has a $*$-regular isometric dilation. We also recall ([7], [17]) that a minimal isometric dilation $V = (V_1, \dots, V_N)$ of T is $*$-regular iff V_i are doubly commuting isometries (i.e. $V_i V_j^* = V_j^* V_i; i, j = \overline{1, N}, i \neq j$).

Let us mention that in our earlier paper [7], for a three-contraction $T = (T_1, T_2, T_3)$ such that the pairs (T_1, T_2) and (T_1, T_3) have regular (or $*$-regular) isometric dilations, we have constructed an isometric dilation $V = (V_1, V_2, V_3)$ such that (V_1, V_2) and (V_1, V_3) are doubly commuting. Further, the second author used in [14] this fact to prove an (1.1)-type von Neumann inequality for (commuting) bi-contractions.

In what follows we use these facts to prove, for an arbitrary N, the existence of isometric dilations for a particular class of commuting multi-contractions, which contains the multi-contactions having regular dilations. This fact allows us to prove the generalized von Neumann inequality (1.1) for such multi-contractions. As a consequence, in the case $N = 2$, we obtain the inequality (1.1) for any bi-contraction on $\mathcal{H}$.

The authors are grateful to Ciprian Foiaş for very helpful discussions about the subject, in August 1999, during the B. Sz.-Nagy Memorial Conference in Szeged and his visit at the University of the West of Timişoara. Thanks are also due to the referees for bringing to our attention the papers [1], [2], [13] as well as for valuable suggestions regarding improvements of the presentation of the paper.

3. Multi-contractions with isometric dilations

We first prove

THEOREM 3.1. *Let $T = (T_1, \dots, T_N)$ be a commuting multi-contraction on $\mathcal{H}$ such that both $T' = (T_1, \dots, T_{N-2}, T_{N-1})$ and $T'' = (T_1, \dots, T_{N-2}, T_N)$ have regular, or $*$-regular, isometric dilations. Then there exists an isometric dilation $V = (V_1, \dots, V_N)$ of T on a space $\mathcal{K} \supset \mathcal{H}$, such that V_i doubly commute with V_j, for $i = 1, \dots, N-2$, $j = 1, \dots, N$, $i \neq j$. Moreover, $V_1, \dots, V_{N-2}$ are unitary operators on $\mathcal{K}$ if T' and T'' have regular dilations, while the dilation $[\mathcal{K}, V]$ can be chosen minimal if T' and T'' have $*$-regular dilations.*

Proof. Suppose T' and T'' have regular dilations. Let $[\mathcal{M}, (\tilde{S}_1, \dots, \tilde{S}_{N-2}, \tilde{S}_{N-1})]$ be the minimal $*$-regular isometric dilation of T'^* and let $\mathfrak{M}$ be the subspace of $\mathcal{M}$ defined by

$$\mathfrak{M} := \bigvee_{\substack{n_i \in \mathbb{Z}_+ \\ 1 \le i \le N-2}} \tilde{S}_1^{n_1} \cdots \tilde{S}_{N-2}^{n_{N-2}} \mathcal{H}.$$

Since the isometries $\tilde{S}_i$ doubly commute and $\tilde{S}_i^*$ are extensions of T_i, we have that $\mathfrak{M}$ is a reducing subspace for $\tilde{S}_i$, $i = \overline{1, N-2}$. Putting $S_i := \tilde{S}_i|_{\mathfrak{M}}$, we have $S_i^* = \tilde{S}_i^*|_{\mathfrak{M}}$, while S_i are doubly commuting isometries on $\mathfrak{M}$, $i = \overline{1, N-2}$. Hence $[\mathfrak{M}, (S_1, \dots, S_{N-2})]$ is a minimal $*$-regular isometric dilation of $(T_1^*, \dots, T_{N-2}^*)$.

On the other hand, since $\tilde{S}^*_{N-1}$ commutes with $\tilde{S}_i$ $(1 \le i \le N-2)$ and $\tilde{S}^*_{N-1}$ is an extension of T_{N-1}, we infer that $\mathfrak{M}$ is an invariant subspace for $\tilde{S}^*_{N-1}$. Therefore $S_{N-1} := \tilde{S}^*_{N-1}|_{\mathfrak{M}}$ is a contractive extension to $\mathfrak{M}$ of T_{N-1}. Also, it is clear that S_{N-1} doubly commutes with S_i, $i = \overline{1, N-2}$.

Now, let $[\mathcal{N}, (\hat{S}_1, \dots, \hat{S}_{N-2}, \hat{S}_N)]$ be the minimal $*$-regular isometric dilation of T''^*. Then the subspace

$$\mathfrak{N} := \bigvee_{\substack{n_i \in \mathbb{Z}_+ \\ 1 \le i \le N-2}} \hat{S}_1^{n_1} \cdots \hat{S}_{N-2}^{n_{N-2}} \mathcal{H}$$

is reducing for $\hat{S}_i$ $(1 \le i \le N-2)$ and if we put $S'_i := \hat{S}_i|_{\mathfrak{N}}$, then as above $[\mathfrak{N}, (S'_1, \dots, S'_{N-2})]$ is a minimal $*$-regular isometric dilation of $(T^*_1, \dots, T^*_{N-2})$. Also $\mathfrak{N}$ is an invariant subspace for $\hat{S}^*_N$ and $S'_N := \hat{S}^*_N|_{\mathfrak{N}}$ is a contractive extension to $\mathfrak{N}$ of T_N, such that S'_N doubly commutes with S'_i, $i = \overline{1, N-2}$. In virtue of the uniqueness of the ($*$-) regular dilation (see [16], [7]), the dilations $[\mathfrak{M}, (S_1, \dots, S_{N-2})]$ and $[\mathfrak{N}, (S'_1, \dots, S'_{N-2})]$ are unitarily equivalent, i.e there exists a unitary operator W from $\mathfrak{M}$ onto $\mathfrak{N}$ satisfying $WS_i = S'_i W$, $i = 1, \dots, N-2$ and $Wh = h$, $h \in \mathcal{H}$. We put $S_N := W^* S'_N W$. Obviously, S_N is a contractive extension to $\mathfrak{M}$ of T_N. Moreover, S_N doubly commutes with S_i $(1 \le i \le N-2)$ and also, S_N comutes with S_{N-1}. Indeed, on one hand we have

$$S_N S_i = W^* S'_N W S_i = W^* S'_N S'_i W = W^* S'_i S'_N W = S_i W^* S'_N W = S_i S_N,$$

$$S^*_N S_i = W^* S'^*_N W S_i = W^* S'^*_N S'_i W = W^* S'_i S'^*_N W = S_i W^* S'^*_N W = S_i S^*_N.$$

On the other hand, for any element $g \in \mathfrak{M}$ of the form $g = \prod_{i=1}^{N-2} S_i^{n_i} h$, where $n_i \in \mathbb{Z}_+$ and $h \in \mathcal{H}$, we have

$$S_N S_{N-1} g = \prod_{i=1}^{N-2} S_i^{n_i} T_N T_{N-1} h = \prod_{i=1}^{N-2} S_i^{n_i} T_{N-1} T_N h = S_{N-1} S_N g,$$

and by the minimality condition of the space $\mathfrak{M}$ we infer $S_N S_{N-1} = S_{N-1} S_N$ on $\mathfrak{M}$. Since S_i doubly commute with S_{N-1} and S_N for $i = 1, \dots, N-2$, by Theorem 1 of [14] there exists a minimal isometric dilation $[\widetilde{\mathfrak{M}}, (W_{N-1}, W_N)]$ of the bi-contraction (S_{N-1}, S_N), such that each co-isometry S^*_i has a co-isometric $*$-extension W_i to $\widetilde{\mathfrak{M}}$ and W_i doubly commutes with both W_{N-1} and W_N. Using the minimality condition of $\widetilde{\mathfrak{M}}$, we can show that $W_i W_j = W_j W_i$, $i, j = 1, \dots, N-2$. Therefore $(W_1, \dots, W_{N-2})$ is a commuting multi-contraction on $\widetilde{\mathfrak{M}}$ having a (unique) minimal $*$-regular isometric dilation $[\mathcal{K}, (V_1, \dots, V_{N-2})]$.

Since W_i are co-isometries, V_i are unitary operators on $\mathcal{K}$. Now let V_{N-1}, V_N be the isometric $*$-extension on $\mathcal{K}$ of W_{N-1}, W_N respectively, which doubly commute with V_i, $i = 1, \ldots, N-2$. Using the minimality condition of $\mathcal{K}$, we get that V_{N-1} and V_N commute as well.

Finally we observe that $V = (V_1, \ldots, V_N)$ is an isometric dilation of $S = (S_1^*, \ldots, S_{N-2}^*, S_{N-1}, S_N)$, while S is an extension of T, hence V is an isometric dilation of T and V has the desired properties.

Suppose now that T' and T'' have $*$-regular isometric dilations. Using the above argument, we can obtain that, if $[\mathcal{M}, (S_1, \ldots, S_{N-2})]$ is the minimal $*$-regular isometric dilation of $(T_1, \ldots, T_{N-2})$, then there are contractive liftings S_{N-1} and S_N of T_{N-1} and T_N respectively, such that $S_{N-1}S_N = S_N S_{N-1}$ and $S_i S_j = S_j S_i$, $S_i S_j^* = S_j^* S_i$, for $i = 1, \ldots, N-2$, $j = N-1, N$. Using Theorem 1 of [14], we get a minimal isometric dilation $[\mathcal{K}, (V_{N-1}, V_N)]$ of (S_{N-1}, S_N) and an isometric $*$-extension V_i to $\mathcal{K}$ of S_i, such that V_i doubly commutes with both V_{N-1} and V_N, $i = 1, \ldots, N-2$. Because S_i and S_j doubly commute for $i \neq j$, it follows that V_i and V_j doubly commute as well, if $i, j = 1, \ldots, N-2$, $i \neq j$. Finally, the space $\mathcal{K}$ verifies the minimality condition relative to $V = (V_1, \ldots, V_N)$, so $[\mathcal{K}, V]$ is a minimal isometric dilation of T, which has the desired properties.

□

REMARK 1. It is clear that the hypothesis on T in the previous theorem is a necessary but not sufficient condition as that T admits regular (or $*$-regular) isometric dilations.

In the case of T having regular (or $*$-regular) isometric dilations, we can obtain from the proof above even the minimal regular (or $*$-regular) isometric dilation $[\mathcal{K}, V]$ of T.

REMARK 2. Theorem 3.1 generalizes and completes Theorem 12 of [7], which was obtained for three commuting contractions. In Theorem 3.1 we remark a special feature of the dilation V of T, namely that some isometries of V doubly commute with all the others and we will use this feature of V for obtaining

THEOREM 3.2. *Let $T = (T_1, \ldots, T_N)$ be a commuting multi-contraction on $\mathcal{H}$ such that both $(T_1, \ldots, T_{N-2}, T_{N-1})$ and $(T_1, \ldots, T_{N-2}, T_N)$ have regular, or $*$-regular isometric dilations and let $[\mathcal{K}, (V_1, \ldots, V_N)]$ be the isometric dilation of T given by Theorem 3.1. Then for any operator $A \in \mathcal{B}(\mathcal{H})$ which doubly commutes with T_i $(1 \leq i \leq N)$, there exists a unique operator $B \in \mathcal{B}(\mathcal{K})$, such that B is a $*$-extension of A and B doubly commutes with V_i $(1 \leq i \leq N)$. Moreover $\|B\| = \|A\|$.*

Proof. Let $A \in \mathcal{B}(\mathcal{H})$ be an operator satisfying $AT_i = T_iA$, $AT_i^* = T_i^*A$, for $i = 1, \ldots, N$. Suppose first that $(T_1, \ldots, T_{N-2}, T_{N-1})$ and $(T_1, \ldots, T_{N-2}, T_N)$ have regular isometric dilations. As in the previous proof, let $[\mathfrak{M}, (S_1, \ldots, S_{N-2})]$ be the minimal $*$-regular isometric dilation of $(T_1^*, \ldots, T_{N-2}^*)$. We identify in this dilation the minimal isometric dilation $[\mathcal{M}_1, M_1]$ of T_1^*, where

$$\mathcal{M}_1 := \bigvee_{m \in \mathbb{Z}_+} S_1^m \mathcal{H}, \quad M_1 := S_1|_{\mathcal{M}_1}.$$

Because for $i = 2, \ldots, N-2$, S_i^* commutes with S_1 and $S_i^*|_{\mathcal{H}} = T_i$, it follows that $\mathcal{M}_1$ is invariant to S_i^*. Then $S_{*i} := S_i^*|_{\mathcal{M}_1}$ is a contractive extension of T_i and S_{*i} commutes with M_1. But $\mathcal{M}_1$ is a reducing subspace for S_1, therefore $M_1^* = S_1^*|_{\mathcal{M}_1}$ and it results that S_{*i} commutes with M_1^*. Hence S_{*i} doubly commutes with M_1. We also remark that $S_{*i}S_{*j} = S_{*j}S_{*i}$, for any i, j. Now since A doubly commutes with T_1^*, there exists a $*$-extension $A_1 \in \mathcal{B}(\mathcal{M}_1)$ of A such that A_1 doubly commutes with M_1 and $\|A_1\| = \|A\|$. The minimality condition on $\mathfrak{M}$ also assures the double commutativity of A_1 with S_{*i}, because S_{*i} is an extension of T_i and T_i doubly comutes with A, $i = 2, \ldots, N-2$.

Notice that $[\mathfrak{M}, (S_2, \ldots, S_{N-2})]$ is a (possible non-minimal) $*$-regular isometric dilation of $(S_{*2}^*, \ldots, S_{*N-2}^*)$. We identify in this dilation the minimal $*$-regular isometric dilation $[\mathfrak{M}', (S_2', \ldots, S_{N-2}')]$ of $(S_{*2}^*, \ldots, S_{*N-2}^*)$, where

$$\mathfrak{M}' := \bigvee_{\substack{n_i \in \mathbb{Z}_+ \\ 2 \le i \le N-2}} S_2^{n_2} \cdots S_{N-2}^{n_{N-2}} \mathcal{H}, \qquad S_i' := S_i|_{\mathfrak{M}'}.$$

Also let $[\mathcal{M}_2, M_2]$ be the minimal isometric dilation of S_{*2}^* given by

$$\mathcal{M}_2 := \bigvee_{m \in \mathbb{Z}_+} S_2'^m \mathcal{M}_1, \qquad M_2 := S_2'|_{\mathcal{M}_2}.$$

Obviously, M_2 is an isometric lifting on $\mathcal{M}_2$ of T_2^*.

Now using an analogous argument as above, we have for $i = 3, \ldots, N-2$, that $\mathcal{M}_2$ is invariant to $S_i'^*$ and $S_{*i}' := S_i'^*|_{\mathcal{M}_2}$ is a contractive extension of S_{*i} such that S_{*i}' doubly commutes with M_2. We also have $S_{*i}'S_{*j}' = S_{*j}'S_{*i}'$ for any i, j. On the other hand, since A_1 and M_1 doubly commute with S_{*2}^*, there exist the $*$-extensions A_2 of A_1 and M_1' of M_1 on $\mathcal{M}_2$, such that A_2 and M_1' doubly commute with M_2, $\|A_2\| = \|A_1\|$, while M_1' is an isometry on $\mathcal{M}_2$. We infer that A_2 is a $*$-extension of A, $\|A_2\| = \|A\|$ and that M_1' is an isometric lifting of T_1^*. Next, since A_1 and M_1 doubly commute with S_{*i}, using the minimality condition of the space $\mathcal{M}_2$ we obtain that A_2 and M_1' doubly commute with S_{*i}', $i = 3, \ldots, N-2$. Let us also remark that $[\mathfrak{M}, (S_3, \dots, S_{N-2})]$ is a (non-minimal) $*$-regular isometric dilation of $(S'^*_{*3}, \dots, S'^*_{*N-2})$. As above, we get eventually a Hilbert space $\mathcal{R}$, the doubly commuting isometries R_i on $\mathcal{R}$ $(1 \leq i \leq N-2)$ and an operator $\tilde{A} = A_{N-2} \in \mathcal{B}(\mathcal{R})$ such that R_i is a lifting of T_i^*, $\tilde{A}$ is a $*$-extension of A, $\|\tilde{A}\| = \|A\|$ and $\tilde{A}$ doubly commutes with R_i. Moreover, we have

$$\mathcal{R} = \mathcal{M}_{N-2} = \bigvee_{m \in \mathbb{Z}_+} R_{N-2}^m \mathcal{M}_{N-3} = \bigvee_{\substack{n_i \in \mathbb{Z}_+ \\ 2 \leq i \leq N-2}} R_{N-2}^{n_{N-2}} \cdots R_2^{N_2} \mathcal{M}_1$$

$$= \bigvee_{\substack{n_i \in \mathbb{Z}_+ \\ 1 \leq i \leq N-2}} R_1^{n_1} \cdots R_{N-2}^{n_{N-2}} \mathcal{H}.$$

Therefore $[\mathcal{R}, (R_1, \dots, R_{N-2})]$ is a minimal $*$-regular isometric dilation of $(T_1^*, \dots, T_{N-2}^*)$. This dilation will be equivalent to $[\mathfrak{M}, (S_1, \dots, S_{N-2})]$, that is, there exists a unitary operator Q from $\mathfrak{M}$ onto $\mathcal{R}$, such that $Qh = h$ and $QS_i = R_iQ$, $i = 1, \dots, N-2$.

Define now the operator $A_0 := Q^*\tilde{A}Q \in \mathcal{B}(\mathfrak{M})$. Then A_0 is a $*$-extension of A, therefore $\|A_0\| \geq \|A\|$. Since $\|A_0\| \leq \|\tilde{A}\| = \|A\|$, we infer that $\|A_0\| = \|A\|$. Moreover A_0 doubly commutes with S_i, $i = 1, \dots, N-2$. If S_{N-1}, S_N are the contractive extensions to $\mathfrak{M}$ of T_{N-1}, T_N respectively, defined as in the proof of Theorem 3.1, then, since A doubly commutes with T_{N-1} and T_N, using the minimalilty condition on $\mathfrak{M}$, we get that A_0 doubly commutes with S_{N-1} and S_N. Let $[\tilde{\mathfrak{M}}, (W_{N-1}, W_N)]$ be the minimal isometric dilation of (S_{N-1}, S_N) given in the proof of Theorem 3.1 . By Theorem 1 of [14] there exists a $*$-extension $B_0 \in \mathcal{B}(\tilde{\mathfrak{M}})$ of A_0, such that $\|B_0\| = \|A_0\| = \|A\|$ and B_0 doubly commutes with W_{N-1} and W_N.

Since A_0 doubly commutes with S_i^* we deduce that B_0 doubly commutes with W_i, where W_i is the $*$-extension of S_i^* on $\tilde{\mathfrak{M}}$ which doubly commutes with W_{N-1} and W_N, $i = 1, \dots, N-2$. In the same way, we can obtain a $*$-extension B of B_0 on the space $\mathcal{K}$ of the minimal $*$-regular isometric dilation $(V_1, \dots, V_{N-2})$ of $(W_1, \dots, W_{N-2})$ such that $\|B\| = \|B_0\|$ and B doubly commutes with V_i, $i = 1, \dots, N-2$. Finally, B also doubly commutes with V_j for $j = N-1, N$, where V_j is the isometric $*$-extension to $\mathcal{K}$ of W_j, which doubly commutes with V_i, $i = 1, \dots, N-2$. We conclude that B is a $*$-extension of A, $\|B\| = \|A\|$ and B doubly commutes with the dilation $V = (V_1, \dots, V_N)$ on $\mathcal{K}$ of T given by Theorem 3.1.

For the uniqueness of B, let $B' \in \mathcal{B}(\mathcal{K})$ be another $*$-extension of A which doubly commutes with V. We remark firstly that

$$\mathcal{K} = \bigvee_{n \in \mathbb{Z}_+^N}' V^n \mathfrak{M}$$

and $V_i^*|_{\mathfrak{M}} = W_i^*|_{\mathfrak{M}} = S_i$ for $i = 1, \ldots, N-2$. Then for any element $g \in \mathfrak{M}$ of the form

$$g = \prod_{i=1}^{N-2} S_i^{n_i} h = \prod_{i=1}^{N-2} V_i^{*n_i} h,$$

where $h \in \mathcal{H}$, we have

$$B'g = \prod_{i=1}^{N-2} V_i^{*n_i} B'h = \prod_{i=1}^{N-2} V_i^{*n_i} Ah = \prod_{i=1}^{N-2} V_i^{*n_i} Bh = Bg.$$

Using now the minimality condition on $\mathfrak{M}$, we obtain that $B' = B$ on $\mathfrak{M}$ and consequently, the above minimality condition of $\mathcal{K}$ implies $B' = B$ on $\mathcal{K}$.

In the second case, we suppose that $(T_1, \ldots, T_{N-2}, T_{N-1})$ and $(T_1, \ldots, T_{N-2}, T_N)$ have $*$-regular isometric dilations. As in the proof of Theorem 3.1, we denote by $[\mathfrak{M}, (S_1, \ldots, S_{N-2})]$ the minimal $*$-regular isometric dilation of $(T_1, \ldots, T_{N-2})$. Considering the above argument, there exists a $*$-extension $A_0 \in \mathcal{B}(\mathfrak{M})$ of A such that $\|A_0\| = \|A\|$ and A_0 doubly commutes with each S_i, $i = 1, \ldots, N-2$. Also by the minimality condition on $\mathfrak{M}$, A_0 doubly commutes with S_j $(j = N-1, N)$, S_j^* being the contractive extension to $\mathfrak{M}$ of T_j^* which doubly commutes with each S_i. Now, if $[\mathcal{K}, (V_{N-1}, V_N)]$ is a minimal isometric dilation of (S_{N-1}, S_N) obtained in the proof of Theorem 3.1 by applying Theorem 1 of [14], then there is a $*$-extension $B \in \mathcal{B}(\mathcal{K})$ of A_0, such that $\|B\| = \|A_0\|$ and B doubly commutes with V_j, $j = N-1, N$. If V_i $(1 \leq i \leq N-2)$ is the isometric $*$-extension of S_i on $\mathcal{K}$, which doubly commutes with V_{N-1} and V_N, then necessarily V_i and B doubly commute for any i. We conclude that B is a $*$-extension of A, $\|B\| = \|A\|$ and B doubly commutes with $V = (V_1, \ldots, V_N)$, where $[\mathcal{K}, V]$ is just the minimal isometric dilation of T given by Theorem 3.1 . The uniqueness of B is assured in this case by the minimality condition of $\mathcal{K}$.

□

In virtue of Remark 1, Theorem 3.2 implies the following

COROLLARY 3.3. *Let T be a commuting multi-contraction on $\mathcal{H}$ having regular, or $*$-regular, isometric dilation. Then for any operator $A \in \mathcal{B}(\mathcal{H})$ which doubly commutes with T, there exists a unique $*$-extension B of A on the space $\mathcal{K}$ of the minimal regular, respectively $*$-regular, isometric dilation V of T, such that B doubly commutes with V and moreover B preserves the norm of A.*

4. The generalized von Neumann inequality

We are now able to prove the generalized von Neumann inequality (1.1) for the multi-contractions described above.

Theorem 4.1. *Let $T = (T_1, \ldots, T_N)$ be a commuting multi-contraction on $\mathcal{H}$ such that both $(T_1, \ldots, T_{N-2}, T_{N-2})$ and $(T_1, \ldots, T_{N-2}, T_N)$ have regular, or $*$-regular, isometric dilations. Then the inequality (1.1) holds for T and for any polynomial P of N complex variables, where the coefficients of P are bounded linear operators on $\mathcal{H}$, which doubly commute with each of the contractions T_i, $i = 1, \ldots, N$.*

Proof. Let $P(z_1, \ldots, z_N) = \sum\limits_{0 \le |n| \le n_0} A_n z^n$ be a polynomial of independent variables $z_1, \ldots, z_N$, where $z = (z_1, \ldots, z_N)$ and $n \in \mathbb{Z}_+^N$ with $|n| \le n_0$, $A_n \in \mathcal{B}(\mathcal{H})$ such that A_n doubly commutes with T_i, $i = 1, \ldots, N$.

Let $V = (V_1, \ldots, V_N)$ be the isometric dilation of T on the space $\mathcal{K}$ given by Theorem 3.1. Then Theorem 3.2 implies that there exist operators $B_n \in \mathcal{B}(\mathcal{K})$ such that B_n is a $*$-extension of A_n and B_n doubly commutes with V_i, $i = 1, \ldots, N$. Since for any $z \in \mathbb{C}^n$, $P(z)$ doubly commutes with T and $\widetilde{P}(z) := \sum\limits_{0 \le |n| \le n_0} B_n z^n$ is a $*$-extension to $\mathcal{K}$ of $P(z)$ which doubly commutes with V, by Theorem 3.2 we have

$$\|\widetilde{P}(z)\| = \|P(z)\| \qquad (z \in \mathbb{C}^N).$$

Now let F be the $\mathcal{B}(\mathcal{K})$-valued semispectral measure on the N-torus $\mathbb{T}^N$ of the multi-isometry V. Because B_n is a $*$-extension of A_n, B_n will be a lifting of A_n and we have on the one hand

$$\begin{aligned} P(T) &= \sum A_n T^n = \sum A_n P_{\mathcal{H}} V^n J_{\mathcal{H}} = \sum P_{\mathcal{H}} B_n \Big(\int_{\mathbb{T}^N} z^n \, dF(z) \Big) J_{\mathcal{H}} \\ &= \sum P_{\mathcal{H}} \Big(\int_{\mathbb{T}^N} B_n z^n \, dF(z) \Big) J_{\mathcal{H}} = P_{\mathcal{H}} \Big(\int_{\mathbb{T}^N} \widetilde{P}(z) \, dF(z) \Big) J_{\mathcal{H}}. \end{aligned}$$

On the other hand, since B_n and V doubly commute, it follows easily that B_n commutes with the values of F (which are positive operators on $\mathcal{K}$). For this, let $[\widetilde{\mathcal{K}}, \tilde{V}]$ be the minimal unitary extension of V (i.e. $[\widetilde{\mathcal{K}}, \tilde{V}^*]$ is the minimal $*$-regular isometric dilation of V^*) and let $\widetilde{B} \in \mathcal{B}(\widetilde{\mathcal{K}})$ be the $*$-extension of B, such that $\widetilde{B}$ commutes with $\tilde{V}$ (given for instance by Corollary 3.3). If E is the $\mathcal{B}(\widetilde{\mathcal{K}})$-valued spectral measure on $\mathbb{T}^N$ of $\tilde{V}$, then $F(\sigma) = P_{\mathcal{H}}E(\sigma)J_{\mathcal{H}}$ for any Borel set $\sigma \subset \mathbb{T}^N$, and for $m \in \mathbb{Z}^N$ and $n \in \mathbb{Z}_+^N$ with $|n| \leq n_0$ and $h \in \mathcal{H}$ we have

$$\begin{aligned}\int_{\mathbb{T}^N} z^m \, d\langle F(z)B_n h, h\rangle &= \int_{\mathbb{T}^N} z^m \, d\langle E(z)\widetilde{B}_n h, h\rangle = \langle \widetilde{V}^m h, \widetilde{B}_n^* h\rangle \\ &= \int_{\mathbb{T}^N} z^m \, d\langle E(z)h, B_n^* h\rangle = \int_{\mathbb{T}^N} z^m \, d\langle F(z)h, B_n^* h\rangle.\end{aligned}$$

Then $B_n F(\sigma) = F(\sigma)B_n$ for any n and σ as above, for the trigonometric polynomials on $\mathbb{T}^N$ are dense in the set of all continuous functions on $\mathbb{T}^N$ ([6], p.73).

Finally, we can apply the theorem from [9] or Theorem 1.2 of [8] to conclude

$$\begin{aligned}\|P(T)\| &= \Big\|P_{\mathcal{H}}\Big(\int_{\mathbb{T}^N} \widetilde{P}(z)\, dF(z)\Big)J_{\mathcal{H}}\Big\| \leq \Big\|\int_{\mathbb{T}^N} \widetilde{P}(z)\, dF(z)\Big\| \\ &\leq \sup_{z\in\mathbb{T}^N} \|\widetilde{P}(z)\| = \sup_{z\in\mathbb{T}^N} \|P(z)\|,\end{aligned}$$

which is nothing else than the inequality (1.1) for the multi-contraction T and the polynomial P.

□

Let us remark that in the case $N = 2$ the preceding theorem yields

COROLLARY 4.2. *For any (commuting) bi-contraction* (T_1, T_2) *on* $\mathcal{H}$ *and any polynomial* $P(z_1, z_2)$ *of two complex variables, with coefficients in* $\mathcal{B}(\mathcal{H})$*, which doubly commute with both* T_1 *and* T_2*, the following inequality holds*

$$\|P(T_1, T_2)\| \leq \sup_{|z_1|=|z_2|=1} \|P(z_1, z_2)\|.$$

Proof. This corollary was also proved in [14], but we will give here another proof, that matches our actual context. The three-contraction (I, T_1, T_2) on $\mathcal{H}$ (I being the identity operator) satisfies the hypothesis of Theorem 4.1, for both bi-contractions (I, T_1) and (I, T_2) have regular isometric dialtions.

□

In virtue of Theorem 4.1, the generalized von Neumann inequality (1.1) is true for any commuting multi-contraction which has regular (or $*$-regular) isometric dilations. But as noticed in [9], for such multi-contractions an even more general von Neumann inequality holds, in the sense that it relates to any trigonometric polynomial on $\mathbb{T}^N$, with coefficients operators which doubly commute with the multi-contraction (see also [14] for the case $N = 2$).

A special class of sequences of contractions on $\mathcal{H}$, which admit $*$-regular dilations is the class of sequences of contractions having the property (P) in the sense of Gelu Popescu in [12], [13].

More precisely, let $N, n_1, \ldots, n_N \geq 1$ be fixed integers and let $\{T_{ij}\}_{\substack{i=1,\ldots,N \\ j=1,\ldots,n_i}} \subset \mathcal{B}(\mathcal{H})$ be a sequence of contractions such that

$$\sum_{j=1}^{n_i} T_{ij}T_{ij}^* \leq I_{\mathcal{H}} \text{ and } T_{ij}T_{pq} = T_{pq}T_{ij}, \tag{4.1}$$

for any i, $p = 1, \ldots, N$, $p \neq i$ and $j = 1, \ldots, n_i$, $q = 1, \ldots, n_p$. For $i = 1, \ldots, N$ let $\mathbb{F}_{n_i}^+$ be the unital free semigroup on n_i generators $s_1, \ldots, s_{n_i}$. If $\alpha_i = s_{j_1} \cdots s_{j_m} \in \mathbb{F}_{n_i}^+$ we define $T_{i,\alpha_i} := T_{ij_1} \cdots T_{ij_m}$ and $|\alpha_i| = m$, while if $\alpha_i = e_i$ is the neutral element in $\mathbb{F}_{n_i}^+$ we put $T_{i,e_i} := I_{\mathcal{H}}$ and $|e_i| = 0$. For $\{T_{ij}\}$ as above we define the self-adjoint operator

$$\Delta(\{T_{ij}\}) := \sum_{\substack{1 \leq i \leq N \\ \alpha_i \in \mathbb{F}_{n_i}^+, |\alpha_i| \leq 1}} (-1)^{|\alpha_1| - \cdots + |\alpha_N|} T_{1,\alpha_1} \cdots T_{N,\alpha_N} T_{N,\alpha_N}^* \cdots T_{1,\alpha_1}^*. \tag{4.2}$$

According to [12] Section 3, we may say that the sequence of contractions $\{T_{ij}\}_{\substack{i=1,\ldots,N \\ j=1,\ldots,n_i}} \subset \mathcal{B}(\mathcal{H})$ verifying (4.1) has property (P), if there exists $\rho \in [0, 1)$ such that for any $r \in (\rho, 1)$ the operator $\Delta(\{rT_{ij}\})$ is positive.

It is shown in [12] (see Corollary 5.2) that any sequence of contractions $\{T_{ij}\}$ on $\mathcal{H}$ with property (P) has a minimal doubly commuting (that is $*$-regular) isometric dilation. Thus from Theorem 4.1 we infer (see also [13])

COROLLARY 4.3. *Let $\{T_{ij}\}_{\substack{i=1,\ldots,N \\ j=1,\ldots,n_i}}$ be a sequence of contractions on $\mathcal{H}$ such that $\{T_{ij}\}$ or $\{T_{ij}^*\}$ have the property (P). Then for any polynomial $P(\{z_{ij}\})$ of $m = n_1 + \cdots + n_N$ complex variables $\{z_{ij}\}_{\substack{i=1,\ldots,N \\ j=1,\ldots,n_i}}$ such that the coefficients of P are operators in $\mathcal{B}(\mathcal{H})$ which doubly commute with each of the contractions T_{ij}, the following inequality holds*

$$\|P(\{T_{ij}\})\| \leq \sup_{\substack{|z_{ij}|=1 \\ 1 \leq i \leq N \\ 1 \leq j \leq n_i}} \|P(\{z_{ij}\})\|.$$

It is clear that any commuting multi-contraction $T = (T_1, \ldots, T_N)$ on $\mathcal{H}$ can be identified as a sequence $\{T_{ij}\}_{\substack{i=1,\ldots,N \\ j=1,\ldots,n_i}} \subset \mathcal{B}(\mathcal{H})$ verifying conditions (4.1), in the case $n_i = 1$ for $i = 1, \ldots, N$, where $T_{i1} = T_i$. In this case we have

THEOREM 4.4. *A commuting multi-contraction $T = (T_1, \ldots, T_N)$ has property (P) iff T has a $*$-regular isometric dilation.*

Proof. The "if" part results from a remark above. Conversely, suppose T has $*$-regular isometric dilation. Then for any $r \in (0,1)$ the commuting multi-contraction $T_r = (rT_1, rT_2, \ldots, rT_N)$ has also $*$-regular isometric dilation. This can be proved as follows. Because T has $*$-regular isometric dilation, from the argument used in the proof of Theorem 3.1 we infer that T_1 has a contractive lifting T_1' on the space of the minimal $*$-regular isometric dilation $(W_2', \ldots W_N')$ of $(T_2, \ldots, T_N)$, such that T_1' doubly commutes with each isometry W_i'. Then rT_1' is a contractive lifting of rT_1, which doubly commutes with W_i' as well $(i = 2, \ldots, N)$. Now, let $[\mathcal{K}', V_1']$ be the minimal isometric dilation of T_1' and for $i = 2, \ldots, N$ let V_i' be the $*$-extension of W_i' to $\mathcal{K}'$ such that W_i' and V_1' doubly commute. Since W_i' doubly commutes with W_j', it follows that V_i' doubly commutes with V_j' as well $(i, j = 2, \ldots, N;\ i \neq j)$. Thus $[\mathcal{K}', (V_1', \ldots, V_N')]$ is a minimal $*$-regular isometric dilation of $(rT_1, T_2, \ldots, T_N)$. Analoguously, we can show that $(rT_1, rT_2, T_3, \ldots, T_N)$ has a $*$-regular isometric dilation and eventually we obtain that $rT = (rT_1, rT_2, \ldots, rT_N)$ has a $*$-regular isometric dilation. According to (4.2), we have for our setting that

$$\Delta(\{T\}) = \sum_{\substack{\varepsilon_i \in \{0,1\} \\ 1 \leq i \leq N}} (-1)^{\varepsilon_1 + \cdots + \varepsilon_N} T_1^{\varepsilon_1} \cdots T_N^{\varepsilon_N} T_N^{*\varepsilon_N} \cdots T_1^{*\varepsilon_1}.$$

Because $rT^* = (rT_1^*, \ldots, rT_N^*)$ has a regular isometric dilation, we have (see [5], [15], [16]) that $\Delta(\{rT\}) \geq 0$, for any $r \in (0,1)$. Hence the commuting multi-contraction T has property (P).

□

It is known (see [12], Corollary 9.1), that any commuting multi-contraction $T = (T_1, \ldots, T_N)$ on $\mathcal{H}$ which is a spherical contraction, i.e. it verifies

$$T_1 T_1^* + \cdots + T_N T_N^* \leq I,$$

has property (P), which by Theorem 4.4 is equivalent to the fact that T has a $*$-regular isometric dilation. Consequently, Theorem 4.1 implies the following special result, which was also obtained by Gelu Popescu in [13].

COROLLARY 4.5. *The generalized von Neumann inequality* (1.1) *holds for any commuting multi-contraction T, if T or T^* is a spherical contraction.*

References

[1] C. AMBROZIE, M. ENGLIS and V. MÜLLER, Operator tuples and analytic models over general domains in $\mathbb{C}^n$, *J. Operator Theory*, to appear.

[2] T. ANDO, Unitary dilation for a triple of commuting contractions, *Bull. Acad. Polon. Sci. Math.*, **XXIV, 10** (1976), 851–853.

[3] T. ANDO, On a pair of commutative contractions, *Acta Sci. Math. (Szeged)*, **24** (1963), 88–90.

[4] W. B. ARVESON, Subalgebras of C^*-algebras. II, *Acta Math.*, **128** (1972), 271–308.

[5] S. BREHMER, Über vertauschbare Kontraktionen der hilbertschen Räume, *Acta Sci. Math. (Szeged)*, **22** (1961), 106–111.

[6] T. W. GAMELIN, *Uniform algebras*, Englewood Cliffs, N. J., 1969.

[7] D. GAŞPAR and N. SUCIU, On the intertwinings of regular dilations, *Ann. Polonici Math.*, **66** (1997), 105–121.

[8] J. JANAS, Semi-spectral integrals and related mappings, *Studia Math. T.*, **46** (1973), 301–313.

[9] W. MLAK, Note on the von Neumann Inequality, *Bull. Acad. Pol. Sci. Math. XIX*, **9** (1971), 843–848.

[10] S. PARROTT, Unitary dilations for commuting contractions, *Pacific J. Math.*, **34** (1970), 481–490.

[11] S. PARROTT, A new look at the theorem of von Neumann, *Preprint Univ. Mass.*, Boston, **21** (1969), 1–14.

[12] G. POPESCU, Poisson transforms on some C^*-algebras generated by isometries, *J. Funct. Anal.*, **161** (1999), 27–61.

[13] G. POPESCU, Commutant lifting, tensor algebras and functional calculus, *Proc. Edinburgh Math. Soc.*, to appear.

[14] N. SUCIU, On the von Neumann inequality for operator-valued polynomials of two variables, *Anal. Univ. Timişoara, Ser. Matem. Inform.*, **37, 1** (1999), 133–142.

[15] B. SZŐKEFALVI-NAGY, Bemerkungen zur vorstehenden Arbeit des Herrn S. Brehmer, *Acta Sci. Math. (Szeged)*, **22** (1961), 112–114.

[16] B. SZŐKEFALVI-NAGY and C. FOIAS, *Harmonic Analysis of Operators on Hilbert Space*, North Holland, Amsterdam - Budapest, 1970.

[17] D. TIMOTIN, Regular dilations and models for multi-contractions, *Indiana Univ. Math. J.*, **47** (1998), 671–684.

D. GAŞPAR, University of the West, Timisoara, Department of Mathematics, Bul. V. Parvan 4, Timisoara, Romania; *e-mail*: gaspar@math.uvt.ro

N. SUCIU, University of the West, Timisoara, Department of Mathematics, Bul. V. Parvan 4, Timisoara, Romania; *e-mail*: suciu@tim1.math.uvt.ro

Operator Theory: Advances and Applications, Vol. 127, 305–326

Ultraproducts of C*-algebras

LIMING GE and DON HADWIN

Dedicated to the memory of Professor Béla Szőkefalvi-Nagy, a great mathematician and a great human being

We initiate a general study of ultraproducts of C*-algebras, including topics on representations, homomorphisms, isomorphisms, positive linear maps and their ultraproducts. We partially settle a question of D. McDuff by proving, for a finite von Neumann algebra with a separable predual, that the continuum hypothesis implies the isomorphism of all of its tracial ultrapowers with respect to different free ultrafilters on the natural numbers. The analog for C*-ultrapowers of separable C*-algebras is equivalent to the continuum hypothesis. We also prove a finite local reflexivity theorem for operator spaces that implies that the second dual of a C*-algebra can be embedded in some ultrapower of the algebra using a completely positive completely isometric map.

1. Introduction

Ultraproducts were first introduced in model theory and played a fundamental role in understanding elementary equivalence and constructing nonstandard models. They were then applied to many areas of mathematics. For example, H. G. Dales and W. H. Woodin [DW] used ultraproducts to prove the independence from the axioms of set theory (ZFC) of the assertion that every algebra homomorphism from a commutative C*-algebra into a Banach algebra is continuous. A version of ultraproducts for Banach spaces was defined and used to construct important examples and settle a number of important questions (see [He]). The ultrapower construction for finite von Neumann algebras was introduced by D. McDuff [Mc]

AMS Subject Classification (1991): 46L05, 46B08, 46M07, 03C20.
Both authors were supported in part by NSF research grants.

and G. Janssen [J] independently. McDuff proved that if the relative commutant of a factor of type II_1 in its ultrapower is noncommutative, then the factor is isomorphic to the tensor product of the factor with the hyperfinite II_1 factor. Elements in the relative commutant of an algebra in its ultrapower correspond to central sequences in the algebra. Central sequences and ultraproducts played a very important role in Connes's work on the classification of injective von Neumann algebras and their automorphisms (see, e.g., [C]). Some aspects of Connes's results were extended to C*-algebras by Phillips [Ph]. Akemann and Pedersen [AP] proved that a separable C*-algebra has continuous trace if and only if it has no nontrivial central sequences. Ultraproducts of C*-algebras are closely related to approximation properties of elements in the algebras. Several people have used ultraproducts of C*-algebras as technical tools to study problems in C*-algebras (see, e.g., [Ha]). In this paper, we initiate a general study of ultraproducts of C*-algebras.

In Section 2, we define a generalized notion of ultraproduct that contains the classical notions and those used in Banach spaces and von Neumann algebras as special cases. The generalized notion of ultraproducts has uses in K-theory, since the K-groups of a C*-ultraproduct are rarely the algebraic ultraproduct of the K-groups, but they can often be expressed as generalized ultraproducts. We also summarize some of the basic properties of ultraproducts of C*-algebras. Two isomorphism problems on ultrapowers are studied in Section 3. First, assuming the continuum hypothesis, we answer affirmatively the problem whether ultrapowers of a separable C*-algebra (or a finite von Neumann algebra with a separable predual) on all free ultrafilters on $\mathbb{N}$ are all $*$-isomorphic. Then we study the problem of C*-algebras with isomorphic ultrapowers, which is related to embeddings of C*-algebras (or finite von Neumann algebras) into certain ultrapowers. In Section 4, we prove a version of local reflexivity for operator spaces and extend a classical result for Banach spaces by showing that the second dual $\mathfrak{A}^{\#\#}$ of a C*-algebra $\mathfrak{A}$ can be isometrically embedded into an ultrapower of $\mathfrak{A}$ where the embedding is completely positive (but not usually a $*$-homomorphism) such that the range of the embedding is the range of a completely positive idempotent on the ultrapower. Exact sequences of C*-algebras are preserved by taking ultraproducts. We also show that nontrivial ultraproducts are usually highly nonseparable. In Section 5, spatial ultraproducts of C*-algebras, ultraproducts of states and representations, and simplicity of ultraproducts are discussed. We prove a generalization of Kadison's transitivity theorem for certain ultraproducts with finite-dimensional spaces replaced by separable ones. We show that an ultraproduct of irreducible representations (respectively, pure states, primitive C*-algebras) is irreducible (respectively, pure, primitive), and that ultraproducts of simple C*-algebras need not be simple, but in certain cases they are. The last section contains discussions on ultraproducts with respect to ultrafilters closed under countable intersections. Strange things happen. This leads to an ultraproduct characterization of the existence of measurable cardinals.

2. Definitions and basic properties

Suppose $\mathbb{I}$ is an infinite set. An *ultrafilter* α on $\mathbb{I}$ is a collection of subsets of $\mathbb{I}$ such that the empty set $\emptyset \notin \alpha$, α is closed under finite intersections, and, for each subset A of $\mathbb{I}$, either $A \in \alpha$ or $\mathbb{I} \setminus A \in \alpha$. One example of an ultrafilter is obtained by choosing an ι in $\mathbb{I}$ and letting α be the collection of all subsets of $\mathbb{I}$ that contain ι. Such an ultrafilter is called a *principal* ultrafilter; ultrafilters not of this form are called *free.*

In general, there are two different types of free ultrafilters we must consider. We will call a free ultrafilter α *countably cofinal* if there is a sequence $\{A_n\}$ in α whose intersection is empty. Otherwise, α is called $\aleph_1$-*complete* [CN]. Free ultrafilters on a countable set are always countably cofinal. The existence of free $\aleph_1$-complete ultrafilters is an open problem (equivalent to the existence of measurable cardinals [CN]). We will call an ultrafilter *nontrivial* if it is both free and countably cofinal. We leave some of the discussions on ultraproducts with respect to $\aleph_1$-complete ultrafilters to the last section (Section 6).

Nontrivial ultrafilters on a countable set $\mathbb{I}$ can be identified as points in $\beta(\mathbb{I}) \setminus \mathbb{I}$, where $\beta(\mathbb{I})$ is the Stone–Céch compactification of $\mathbb{I}$ (when $\mathbb{I}$ is endowed with the discrete topology). One may also view $\beta(\mathbb{I})$ as the compact Hausdorff space so that $l^\infty(\mathbb{I})$ (uniformly bounded complex-valued functions on $\mathbb{I}$) is $*$-isomorphic to $C(\beta(\mathbb{I}))$ as C*-algebras. One important property of $\beta(\mathbb{I})$ is that every bounded function on $\mathbb{I}$ extends to a continuous function on $\beta(\mathbb{I})$.

Suppose $\mathcal{X}$ is another set, $f\colon \mathbb{I} \to \mathcal{X}$ is a mapping and $E \subseteq \mathcal{X}$. We say that $f(\iota)$ is *eventually in* E along α if $f^{-1}(E) = \{\iota \in \mathbb{I} : f(\iota) \in E\} \in \alpha$. If $\mathcal{X}$ is a topological space, we say that $f(\iota)$ converges to x (in $\mathcal{X}$) along α, denoted by $\lim_{\iota\to\alpha} f(\iota) = x$, if $f(\iota)$ is eventually in each neighborhood of x. It is a well-known fact about ultrafilters that if $\mathcal{X}$ is a compact Hausdorff space, then $\lim_{\iota\to\alpha} f(\iota)$ always exists in $\mathcal{X}$ for every $f\colon \mathbb{I} \to \mathcal{X}$ and every ultrafilter α on $\mathbb{I}$.

Now we review, briefly, some of the existing constructions of ultraproducts of sets, groups, algebras, Banach spaces, C*-algebras, etc.

First, let $\mathcal{Y}_\iota$, $\iota \in \mathbb{I}$, be sets, $\prod_{\iota\in\mathbb{I}} \mathcal{Y}_\iota$ or $\prod_\iota \mathcal{Y}_\iota$ be the Cartesian product of $\mathcal{Y}_\iota$'s. Elements in $\prod_{\iota\in\mathbb{I}} \mathcal{Y}_\iota$ are denoted by $\{Y_\iota\}$ for Y_ι in $\mathcal{Y}_\iota$. Let α be a nontrivial ultrafilter on $\mathbb{I}$. We define an equivalence relation $\sim$ on $\prod_{\iota\in\mathbb{I}} \mathcal{Y}_\iota$ by $\{Y_\iota\} \sim \{Z_\iota\}$ if and only if $Y_\iota = Z_\iota$ eventually along α. Let $\prod^\alpha \mathcal{Y}_\iota$ be the equivalence classes of elements $\{Y_\iota\}$ in $\prod_{\iota\in\mathbb{I}}\mathcal{Y}_\iota$ with respect to $\sim$ and α. In this case, $\prod^\alpha \mathcal{Y}_\iota$ is the original ultraproduct defined in model theory [CK], which we call the *classical ultraproduct* of the $\mathcal{Y}_\iota$'s. The equivalence class of an element $\{Y_\iota\}$ in $\prod_\iota \mathcal{Y}_\iota$ (with respect to $\sim$ and α) is denoted by $[\{Y_\iota\}]$ (or $\{Y_\iota\}_\alpha$). Note that if the $\mathcal{Y}_\iota$'s have certain additional structure, e.g., they are groups or partially ordered sets, then $\prod^\alpha \mathcal{Y}_\iota$ inherits this structure in the obvious way, i.e., the Cartesian product $\prod_\iota \mathcal{Y}_\iota$ has the same structure and $\sim$ is a congruence on $\prod_\iota \mathcal{Y}_\iota$.

Next, we assume that each $\mathcal{Y}_\iota$ is a Banach space. In this case, the ultraproduct $\prod^\alpha \mathcal{Y}_\iota$ of the $\mathcal{Y}_\iota$'s as Banach spaces is the l^∞-product of the $\mathcal{Y}_\iota$'s modulo the closed subspace $\sum^\alpha \mathcal{Y}_\iota$ of all elements $\{\xi_\iota\}$ that converge in norm to 0 along α. This ultraproduct is the ultraproduct of Banach spaces discussed in [He]. Note that $\|[\{u_\iota\}]\| = \lim_{\iota\to\alpha}\|u_\iota\|$ defines a norm on $\prod^\alpha \mathcal{Y}_\iota$. If each $\mathcal{Y}_\iota$ is a Banach algebra or a C*-algebra, then so is $\prod^\alpha \mathcal{Y}_\iota$. More interestingly, if each $\mathcal{Y}_\iota$ is a Hilbert space, then so is $\prod^\alpha \mathcal{Y}_\iota$ with the inner product $\langle[\{u_\iota\}],[\{v_\iota\}]\rangle = \lim_{\iota\to\alpha}\langle u_\iota, v_\iota\rangle$.

Now, suppose that each $\mathcal{Y}_\iota$ is a finite von Neumann algebra with a faithful normal tracial state τ_ι. Let $\prod_\iota \mathcal{Y}_\iota$ be the l^∞-product of the $\mathcal{Y}_\iota$'s. Then $\prod_\iota \mathcal{Y}_\iota$ is a von Neumann algebra (with pointwise multiplication). If $\{X_\iota\}$ and $\{Y_\iota\}$ are two elements in $\prod_\iota \mathcal{Y}_\iota$, then we define $\{X_\iota\}\sim\{Y_\iota\}$ when $\lim_{\iota\to\alpha}\|X_\iota - Y_\iota\|_2 = 0$, where $\|X\|_2 = \tau_\iota(X^*X)^{1/2}$ is the trace norm given by τ_ι. Then $\prod^\alpha \mathcal{Y}_\iota$ is $\prod_\iota \mathcal{Y}_\iota$ modulo the equivalence relation $\sim$. We will call such ultraproducts *tracial ultraproducts*.

When $\mathcal{Y}_\iota = \mathcal{Y}$ for all ι, then $\prod^\alpha \mathcal{Y}_\iota$ is called the *ultrapower* of $\mathcal{Y}$, denoted by $\mathcal{Y}^\alpha$. For example, McDuff [Mc] considered certain tracial ultrapowers of factors of type II_1 along free ultrafilters on $\mathbb{N}$. A tracial ultrapower of a finite von Neumann algebra with respect to a faithful normal trace is again a von Neumann algebra [Sa].

In the following, we introduce a general construction of ultraproducts. In order to avoid getting bogged down in logic and model theory, we consider the ultraproduct construction at the level of sets and elementary algebraic structures. For the reader interested in the logical aspect of ultraproducts, we refer the reader to [CK] and [CN].

Suppose, for each ι in $\mathbb{I}$, $\mathcal{Y}_\iota$ is a nonempty set. Suppose $\mathcal{E}$ is a collection of mappings on $\mathbb{I}$, such that, for each ι in $\mathbb{I}$, and each E in $\mathcal{E}$, $E(\iota)$ defines a nonnegative function on $\mathcal{Y}_\iota$, i.e., $E(\iota)\colon \mathcal{Y}_\iota \to [0,\infty]$. Also suppose $\mathcal{F}$ is a family of mappings on $\mathbb{I}$ such that, for each ι in $\mathbb{I}$, and each F in $\mathcal{F}$, $F(\iota)\colon \mathcal{Y}_\iota\times\mathcal{Y}_\iota \to [0,\infty]$ is a semi-metric on $\mathcal{Y}_\iota$. We define the $\mathcal{E}$-bounded Cartesian product of $\{\mathcal{Y}_\iota : \iota\in\mathbb{I}\}$, denoted by $\prod_\mathcal{E} \mathcal{Y}_\iota$, to be the set of $Y = \{Y_\iota\}_{\iota\in\mathbb{I}}$ in the Cartesian product of the $\mathcal{Y}_\iota$'s such that, for each E in $\mathcal{E}$, $\{E(\iota)(Y_\iota)\}$ is bounded on $\mathbb{I}$. Let α be an ultrafilter on $\mathbb{I}$. We define an equivalence relation $\sim_{\mathcal{F}(\alpha)}$ on $\prod_\mathcal{E} \mathcal{Y}_\iota$: $\{Y_\iota\}\sim_{\mathcal{F}(\alpha)}\{Z_\iota\}$ if and only if, for each F in $\mathcal{F}$, $\lim_{\iota\to\alpha} F(\iota)(Y_\iota, Z_\iota) = 0$. We define the *ultraproduct* of the $\mathcal{Y}_\iota$'s along α with respect to $(\mathcal{E}, \mathcal{F})$, denoted by $\prod_{\mathcal{E},\mathcal{F}}^{\alpha} \mathcal{Y}_\iota$, to be $(\prod_{\mathcal{E}} \mathcal{Y}_\iota)/\sim_{\mathcal{F}(\alpha)}$. If we have that each $\mathcal{Y}_\iota = \mathcal{Y}$, we call the ultraproduct $\prod_{\mathcal{E},\mathcal{F}}^{\alpha} \mathcal{Y}_\iota$ an *ultrapower* of $\mathcal{Y}$. This general notion of ultraproduct contains the previously mentioned examples.

For example, assume each $\mathcal{Y}_\iota$ is a Banach space, and suppose $\mathcal{E} = \{E\}$ and $\mathcal{F} = \{F\}$, such that, for each ι in $\mathbb{I}$ and each u, v in $\mathcal{Y}_\iota$, $E(\iota)(u) = \|u\|$ and $F(\iota)(u,v) = \|u - v\|$. In this case, $\prod_{\mathcal{E},\mathcal{F}}^{\alpha} \mathcal{Y}_\iota$ is the l^∞-product of the $\mathcal{Y}_\iota$'s modulo the closed subspace of all elements $\{\xi_\iota\}$ that converge in norm to 0 along α, thus it agrees with the above ultraproduct $\prod^{\alpha} \mathcal{Y}_\iota$ of $\mathcal{Y}_\iota$'s as Banach spaces.

Sometimes, it is useful to have this general notion of ultraproducts. It helps to describe objects that cannot be expressed by classical ultraproducts. For example, suppose each $\mathcal{Y}_\iota$ is an abelian group, $\mathcal{E} = \{E\}$ and $\mathcal{F} = \{F\}$, where $E(\iota)$ is the constant 1 function on $\mathcal{Y}_\iota$, and $F(\iota)$ is the discrete metric on $\mathcal{Y}_\iota$ (i.e., $F(\iota)(X,Y) = 1$ if $X \neq Y$ and $F(\iota)(X,Y) = 0$ if $X = Y$). Then $\prod_{\mathcal{E},\mathcal{F}}^{\alpha} \mathcal{Y}_\iota$ is the classical ultraproduct of $\mathcal{Y}_\iota$'s. If $\mathcal{E}_1 = \mathcal{E} \cup \{H\}$, where, for each ι in $\mathbb{I}$ and each Y in $\mathcal{Y}_\iota$, $H(\iota)(Y)$ is the order of Y in the group $\mathcal{Y}_\iota$, then $\prod_{\mathcal{E}_1,\mathcal{F}}^{\alpha} \mathcal{Y}_\iota$ is the torsion subgroup of $\prod_{\mathcal{E},\mathcal{F}}^{\alpha} \mathcal{Y}_\iota$.

Another more pertinent example comes from C*-algebraic K-theory. For each positive integer n, it is well-known that $K_0(M_n) = \mathbb{Z}$, where M_n is the C*-algebra of all $n \times n$ matrices. It is not difficult to show, for any nontrivial ultrafilter α on $\mathbb{N}$, that $K_0(\prod^{\alpha} M_n)$ is not the algebraic ultraproduct $\prod^{\alpha} \mathbb{Z}$. However, it is the generalized ultraproduct $\prod_{\mathcal{E},\mathcal{F}}^{\alpha} \mathbb{Z}$, where $\mathcal{E} = \{E\}$, $\mathcal{F} = \{F\}$ with $E(n)(z) = |z/n|$ and $F(n)$ is the discrete metric on $\mathbb{Z}$. We will discuss the K-theory of ultraproducts more fully in a later paper.

We end this section by summarizing some of the well-known, and useful facts concerning elements in ultraproducts of C*-algebras and their representatives.

PROPOSITION 2.1. *Let $\mathfrak{A}_\iota$, $\iota \in \mathbb{I}$, be unital C*-algebras and α an ultrafilter on $\mathbb{I}$. Then*

1. *if A is a self-adjoint operator in $\prod^{\alpha} \mathfrak{A}_\iota$, then there are self-adjoint operators A_ι in $\mathfrak{A}_\iota$ such that $A = [\{A_\iota\}]$;*
2. *if P is a projection in $\prod^{\alpha} \mathfrak{A}_\iota$, then there are projections P_ι in $\mathfrak{A}_\iota$ such that $P = [\{P_\iota\}]$;*
3. *if U is a unitary operator in $\prod^{\alpha} \mathfrak{A}_\iota$, then there are unitary operators U_ι in $\mathfrak{A}_\iota$ such that $U = [\{U_\iota\}]$;*
4. *if $P = \{P_\iota\}_\alpha$, $Q = \{Q_\iota\}_\alpha$ are in $\prod^{\alpha} \mathfrak{A}_\iota$ and each P_ι, Q_ι are projections and if $V \in \prod^{\alpha} \mathfrak{A}_\iota$ is a partial isometry with $V^*V = P$ and $VV^* = Q$, then there are V_ι in $\mathfrak{A}_\iota$ such that, eventually along α, $V_\iota^* V_\iota = P_\iota$ and $V_\iota V_\iota^* = Q_\iota$;*
5. *if $P = \{P_\iota\}_\alpha \in \prod^{\alpha} \mathfrak{A}_\iota$ and each P_ι is a projection, and if Q is a projection in $\prod^{\alpha} \mathfrak{A}_\iota$ such that $Q \leq P$, then there are projections $Q_\iota \in \mathfrak{A}_\iota$ with $Q_\iota \leq P_\iota$, such that $Q = \{Q_\iota\}_\alpha$.*

Proof. 1. If $A = [\{A_\iota\}]$ and A is selfadjoint, then $A = [\{(A_\iota + A_\iota^*)/2\}]$.

2. Let $\{A_\iota\}$ be a representative for the projection P. From 1, we may assume that each A_ι is selfadjoint. From $P^2 = P$, we have that $\lim_{\iota\to\alpha} \|A_\iota^2 - A_\iota\| = 0$. Let f be a continuous nondecreasing function on $\mathbb{R}$ such that $f(t) = 0$ when $t \leq \frac{1}{4}$ and $f(t) = 1$ when $t \geq \frac{3}{4}$. Then we know that when $\|A_\iota^2 - A_\iota\| \leq \frac{1}{8}$, the spectrum of A_ι lies in $[-\frac{1}{4}, \frac{1}{4}] \cup [\frac{3}{4}, \frac{5}{4}]$ and so $f(A_\iota)$ is a projection. It is easy to see that $\lim_{\iota\to\alpha} \|A_\iota - f(A_\iota)\| = 0$. Replacing A_ι by I when $\|A_\iota^2 - A_\iota\| > \frac{1}{8}$ and by $f(A_\iota)$ when $\|A_\iota^2 - A_\iota\| \leq \frac{1}{8}$, we have that $P = [\{A_\iota\}]$ and each A_ι is a projection.

3. Similar to (2). We leave it as an exercise.

4. First write $V = \{A_\iota\}_\alpha$. Since $V = QVP$, we can assume $A_\iota = Q_\iota V_\iota P_\iota$ for every ι. Let $g\colon \mathbb{R} \to \mathbb{R}$ be continuous with $g(t) = 0$ when $0 \leq t \leq 1/4$, and $g(t) = 1/\sqrt{t}$ when $3/4 \leq t \leq 5/4$, and for each ι, let $V_\iota = g(A_\iota A_\iota^*)A_\iota$. Since $V = g(VV^*)V$, it follows that $V = \{V_\iota\}_\alpha$. Note that we still have $V_\iota = Q_\iota V_\iota P_\iota$. However, as in the proof of (2), we have, eventually along α, that $V_\iota V_\iota^*$ is a projection, $V_\iota V_\iota^* \leq Q_\iota$, and $\|V_\iota V_\iota^* - Q_\iota\| < 1$, which implies that $V_\iota V_\iota^* = Q_\iota$. Similarly, $V_\iota^* V_\iota$ is eventually a projection dominated by P_ι but close to P_ι, which implies that, eventually along α, $V_\iota^* V_\iota = P_\iota$.

5. Write $Q = \{B_\iota\}_\alpha$ with $B_\iota = B_\iota^*$. Since $Q = PQP$, we can assume that $B_\iota = P_\iota B_\iota P_\iota$ for each ι. If we apply the technique of the proof of (2) above, letting Q_ι be $f(P_\iota B_\iota P_\iota)$, we see that, eventually along α, Q_ι is a projection and $Q_\iota \leq P_\iota$.

□

3. Isomorphisms of ultrapowers

In this section, we study isomorphism problems of ultrapowers. First, a natural question about an ultrapower of a C*-algebra is whether it depends on the choice of the nontrivial ultrafilter. In the case $I = \mathbb{N}$, we have the following result of H. J. Keisler [CK; 4.3] concerning classical ultrapowers.

THEOREM 3.1. *Assume that the continuum hypothesis holds. Suppose $\mathcal{R}$ is a universal algebra with cardinality at most $2^{\aleph_0}$ and having finitely many operations. Suppose also $\mathcal{S}$ is a countable subset of $\mathcal{R}$. If α, β are nontrivial ultrafilters on $\mathbb{N}$, then there is an isomorphism $\pi\colon \mathcal{R}^\alpha \to \mathcal{R}^\beta$ such that $\pi|\mathcal{S}$ is the identity (i.e., for every $S \in \mathcal{S}$, $\pi(\{(S, S, \ldots)\}_\alpha) = \{(S, S, \ldots)\}_\beta$).*

We can use Keisler's theorem to prove analogs results for more general ultrapowers.

THEOREM 3.2. *Assume the continuum hypothesis. Suppose $\mathcal{M}$ is a finite von Neumann algebra with a faithful finite normal trace τ. If $\mathcal{M}$ is trace-norm separable, then $\mathcal{M}^\alpha$ and $\mathcal{M}^\beta$ are $*$-isomorphic von Neumann algebras for any free ultrafilters α and β over $\mathbb{N}$, where $\mathcal{M}^\alpha$ and $\mathcal{M}^\beta$ are tracial ultrapowers of $\mathcal{M}$. Moreover, the relative commutant of $\mathcal{M}$ in $\mathcal{M}^\alpha$ is $*$-isomorphic to that of $\mathcal{M}$ in $\mathcal{M}^\beta$.*

Proof. We let $\mathcal{R}$ be the universal algebra $\mathcal{M}$ with the operations $+, \cdot, *$, and we define predicates "$\|x\| < 1$", "$\|x\|_2 < 1$" and "$x = iy$". Choose a countable trace-norm dense subset $\mathcal{S}$ of $\mathcal{M}$.

It follows from Theorem 3.1 that there is an isomorphism $\pi: \mathcal{R}^\alpha \to \mathcal{R}^\beta$ that fixes the elements of $\mathcal{S}$ and preserves the predicates. In particular, π maps $\{X \in \mathcal{R}^\alpha : \|X\| < 1\}$ and $\{X \in \mathcal{R}^\alpha : \|X\|_2 < 1\}$ onto $\{X \in \mathcal{R}^\beta : \|X\| < 1\}$ and $\{X \in \mathcal{R}^\beta : \|X\|_2 < 1\}$, respectively.

Note that the predicate $\|[\{X_n\}]\| < 1$ holds in $\mathcal{R}^\alpha$ (respectively, $\mathcal{R}^\beta$) if and only if, eventually along α (respectively, β), $\|X_n\| < 1$. We define

$$\mathcal{R}_b^\alpha = \{\{X_n\}_\alpha \in \mathcal{R}^\alpha : \lim_{n\to\alpha} \|X_n\| < \infty\}.$$

We define $\mathcal{R}_b^\beta$ similarly. Note that $\mathcal{R}_b^\alpha$ is the additive subgroup of $\mathcal{R}^\alpha$ generated by $\{X \in \mathcal{R}^\alpha : \|X\| < 1\}$. It follows that $\pi(\mathcal{R}_b^\alpha) = \mathcal{R}_b^\beta$.

Now, let $\mathcal{R}_0^\alpha = \{\{X_n\}_\alpha \in \mathcal{R}^\alpha : \lim_{n\to\alpha} \|X_n\|_2 = 0\} \cap \mathcal{R}_b^\alpha$ and $\mathcal{R}_0^\beta = \{\{X_n\}_\beta \in \mathcal{R}^\beta : \lim_{n\to\beta} \|X_n\|_2 = 0\} \cap \mathcal{R}_b^\beta$. It is clear that $\mathcal{R}_0^\alpha$ is the unique maximal additive subgroup contained in $\{X \in \mathcal{R}^\alpha : \|X\|_2 < 1\} \cap \mathcal{R}_b^\alpha$. It follows that $\pi(\mathcal{R}_0^\alpha) = \mathcal{R}_0^\beta$. Hence π induces an isomorphism $\rho: \mathcal{R}_b^\alpha/\mathcal{R}_0^\alpha \to \mathcal{R}_b^\beta/\mathcal{R}_0^\beta$.

But $\mathcal{R}_b^\alpha/\mathcal{R}_0^\alpha = \mathcal{M}^\alpha$ and $\mathcal{R}_b^\beta/\mathcal{R}_0^\beta = \mathcal{M}^\beta$. It is clear that ρ is bijective, additive, multiplicative $*$-mapping. Moreover, $\rho(iX) = i\rho(X)$ since π preserves the predicate "$Y = iX$". Also an element X in $\mathcal{M}^\alpha$ satisfies $\|X\| < 1$ (or $\|X\|_2 < 1$) if and only if X has a pre-image W in $\mathcal{R}_b^\alpha$ such that $\|W\| < 1$ (or $\|W\|_2 < 1$, respectively). Hence, for every X in $\mathcal{M}^\alpha$, $\|X\| < 1$ (or $\|X\|_2 < 1$) if and only if $\|\rho(X)\| < 1$ (or $\|\rho(X)\|_2 < 1$, respectively), which implies that ρ is an isometry and preserves trace(-norm). Since every additive isometry on a Banach space is linear over $\mathbb{R}$, it follows that ρ is linear over $\mathbb{C}$. Therefore ρ is a trace preserving $*$-isomorphism.

Since π fixes the elements of $\mathcal{S}$, which implies ρ fixes the points in (a trace-norm dense subset of) $\mathcal{M}$.

□

REMARK 3.3. The above theorem answers a question of D. McDuff [Mc]. The ideas in the proof of Theorem 3.2 can be used to prove analogs for Banach space ultrapowers or C*-ultrapowers.

COROLLARY 3.4. *Assume the continuum hypothesis. If* $\mathfrak{A}$ *is a* C^**-algebra with cardinality* $2^{\aleph_0}$*, then* $\mathfrak{A}^\alpha$ *and* $\mathfrak{A}^\beta$ *are* $*$*-isomorphic. Moreover, if* $\mathfrak{A}$ *is separable, there is a* $*$*-isomorphism from* $\mathfrak{A}^\alpha$ *to* $\mathfrak{A}^\beta$ *that fixes the elements of* $\mathfrak{A}$*. Thus the relative commutants of* $\mathfrak{A}$ *in* $\mathfrak{A}^\alpha$ *and* $\mathfrak{A}^\beta$ *are* $*$*-isomorphic.*

If the continuum hypothesis fails, the conclusion for Keisler's theorem (Theorem 3.1) no longer holds. However, P. Eklof [E] has shown that all nontrivial ultrapowers over $\mathbb{N}$ of the additive group $(\mathbb{Z}, +)$ of the integers are isomorphic. On the other hand, if we view $\mathbb{Z}$ as an ordered structure with the usual ordering, A. Dow [D] and S. Shelah [S2] have independently proved that nonisomorphic ultrapowers exist:

THEOREM 3.5. *Assume the continuum hypothesis fails. Suppose* $(X, \leq)$ *is a partially ordered set containing an infinite chain. Then there are nontrivial ultrapowers* α*,* β *on* $\mathbb{N}$ *such that the classical ultrapowers* $(X, \leq)^\alpha$ *and* $(X, \leq)^\beta$ *are not order-isomorphic.*

To prove analogs of the preceding theorem we need to consider ordered structures associated with C*-algebras that are preserved under ultraproducts. The usual ordering on a C*-algebra fails in this regard, since the ordered structure of a C*-ultraproduct is not the classical ultraproduct of the ordered structures of the C*-algebras. One natural choice is the ordered K_0-group, but this too, except in certain cases, does not behave nicely with respect to ultraproducts; also the K_0-group throws away some of the natural structure on the projections in the algebra. A simpler choice comes from the partial ordering on the projections in the algebra induced by Murry–von Neumann equivalence.

Suppose $\mathfrak{A}$ is a C*-algebra and $P, Q \in \mathfrak{A}$ are projections. We say P and Q are Murry–von Neumann equivalent if there is a V in $\mathfrak{A}$ with $VV^* = P$ and $V^*V = Q$. We will say that two projections in $\mathfrak{A}$ are *equivalent* if each is Murry–von Neumann equivalent to a subprojection of the other. We let $\mathcal{P}(\mathfrak{A})$ denote the set of equivalence classes of projections in $\mathfrak{A}$, and we define a partial ordering $\leq$ on $\mathcal{P}(\mathfrak{A})$: $[P] \leq [Q]$ if and only if P is equivalent to a subprojection of Q. The fact that the functor $\mathcal{P}$ from the category of C*-algebras to the category of partially ordered sets behaves nicely with respect to ultraproducts is contained in the following lemma. To avoid confusion, we use our alternate notation $\{Y_\iota\}_\alpha$ for the usual equivalence class of an element $\{Y_\iota\}$ in an ultraproduct.

LEMMA 3.6. *Suppose* $\{\mathfrak{A}_\iota : \iota \in \mathbb{I}\}$ *is a family of* C^**-algebras and* α *is a nontrivial ultrafilter on* $\mathbb{I}$*. Then* $\mathcal{P}(\prod^\alpha \mathfrak{A}_\iota)$ *is order-isomorphic to the classical ultraproduct* $\prod^\alpha \mathcal{P}(\mathfrak{A}_\iota)$*.*

Proof. This all follows from Proposition 2.1. Projections P and Q in $\prod^\alpha \mathfrak{A}_\iota$ can be written as $P = \{P_\iota\}_\alpha$ and $Q = \{Q_\iota\}_\alpha$, where each P_ι and Q_ι are projections. Then $[P] \leq [Q]$ if and only if, eventually along α, $[P_\iota] \leq [Q_\iota]$. Thus the map $\sigma: \mathcal{P}(\prod^\alpha \mathfrak{A}_\iota) \to \prod^\alpha \mathcal{P}(\mathfrak{A}_\iota)$ defined by $\sigma(\{P_\iota\}_\alpha) = \{[P_\iota]\}_\alpha$ is an order-isomorphism. □

EXAMPLES. 1. Suppose $\mathcal{K}$ is the C*-algebra of all compact operators on a separable infinite-dimensional Hilbert space. All the projections in $\mathcal{K}$ have finite rank, and two projections are equivalent if and only if they have the same rank. Moreover $[P] \leq [Q]$ if and only if $\mathrm{rank}(P) \leq \mathrm{rank}(Q)$. Hence $\mathcal{P}(\mathcal{K})$ is order-isomorphic to $\{0, 1, 2, \ldots\}$ with the usual ordering.

2. Suppose $\mathfrak{A}$ is a Glimm algebra with supernatural number m. Since two projections in $\mathfrak{A}$ are equivalent if and only if they have the same trace, $\mathcal{P}(\mathfrak{A})$ is order-isomorphic to $\{\frac{k}{n} : k, n \in \mathbb{Z}, 0 \leq k \leq n, n \neq 0, n|m\}$ with the usual ordering.

3. Suppose $\mathfrak{A}_\theta$ is an irrational rotation algebra, where θ is the associated irrational number. Then $\mathcal{P}(\mathfrak{A}_\theta)$ is order-isomorphic to $\{m + n\theta \in [0,1] : m, n \in \mathbb{Z}\}$.

4. $\mathcal{P}(\mathcal{B}(\mathcal{H}))$ is order-isomorphic to $\mathbb{N} \cup \{\infty\}$, with the usual ordering, where $\mathcal{H}$ is the separable infinite-dimensional Hilbert space. Note that $K_0(\mathcal{B}(\mathcal{H})) = \{0\}$.

5. If X is a compact Hausdorff space, then $\mathcal{P}(C(X))$ is order-isomorphic to $(\{A \subset X : A \text{ is clopen}\}, \subset)$.

6. If $\mathfrak{A}$ is a von Neumann algebra, $\mathcal{P}(\mathfrak{A})$ is the set of Murray–von Neumann equivalence classes of projections with the usual ordering.

The following theorem is an immediate consequence of Theorem 3.5 and Lemma 3.6. Note that there is no assumption on the cardinality of $\mathfrak{A}$.

THEOREM 3.7. *Suppose $\mathfrak{A}$ is a C*-algebra and $\mathcal{P}(\mathfrak{A})$ contains an infinite chain. If the continuum hypothesis fails, there are nontrivial ultrafilters α, β on $\mathbb{N}$ such that $\mathfrak{A}^\alpha$ and $\mathfrak{A}^\beta$ are not $*$-isomorphic.*

COROLLARY 3.8. *Suppose $\mathfrak{A}$ is a C*-algebra with cardinality $2^{\aleph_0}$ such that $\mathcal{P}(\mathfrak{A})$ contains an infinite chain. The following are equivalent:*

1. *The continuum hypothesis holds.*
2. *For all nontrivial ultrafilters α, β on $\mathbb{N}$, $\mathfrak{A}^\alpha$ and $\mathfrak{A}^\beta$ are $*$-isomorphic.*

4. Embeddings into ultraproducts and finite local reflexivity

In the classical theory of ultraproducts, a theorem of S. Shelah [S1] states that two models $\mathfrak{A}$ and $\mathcal{B}$ have isomorphic ultrapowers if and only if they are elementarily equivalent (i.e., certain first-order statements are true in $\mathfrak{A}$ if and only if they are true in $\mathcal{B}$). A version of this result for isometric isomorphism of ultrapowers of Banach spaces was obtained by C. W. Henson [Hn1,2], and a version for homeomorphic linear isomorphism of ultrapowers of Banach spaces was obtained by S. Heinrich and C. W. Henson [HH]. The techniques of Henson [Hn1, 2] can be used to prove a similar result for $*$-isomorphism of ultrapowers of C*-algebras, but it is beyond the scope of this paper. However, the following result, which is a variant of Theorem 1.7 in [Hn2] is somewhat interesting. The proof, which is similar to that of [He; 6.3], is omitted. The gist of the following result is that the C*-algebra $\mathfrak{A}$ can be embedded in an ultrapower of $\mathcal{B}$ if and only if each finite "part" of $\mathfrak{A}$ can be approximately embedded in $\mathcal{B}$.

PROPOSITION 4.1. *Suppose $\mathfrak{A}$ and $\mathcal{B}$ are C^*-algebras. The following are equivalent:*

1. *$\mathfrak{A}$ is $*$-isomorphic to a subalgebra of an ultrapower of $\mathcal{B}$.*
2. *For any $X_1, \ldots, X_n$ in $\mathfrak{A}$, $*$-polynomials $p_1, \ldots, p_n$ in n variables and positive numbers $r_1, s_1, \ldots, r_n, s_n$ such that $r_k < \|p_k(X_1, \ldots, X_n)\| < s_k$ for $1 \le k \le n$, there are $Y_1, \ldots, Y_n$ in $\mathcal{B}$ such that $r_k < \|p_k(Y_1, \ldots, Y_n)\| < s_k$, for $1 \le k \le n$.*
3. *There is a net $\{\varphi_\lambda\}$ of maps whose domains form an increasingly directed collection of subsets of $\mathfrak{A}$ whose union $\mathcal{D}$ is a dense $*$-subalgebra (over the field $\mathbb{Q}+i\mathbb{Q}$) of $\mathfrak{A}$, and whose ranges are contained in $\mathcal{B}$ such that for all X, Y in $\mathcal{D}$ and all r, s in $\mathbb{Q}+i\mathbb{Q}$:*

$$\text{a. } \lim_\lambda \|\varphi_\lambda(rX + sY^*) - (r\varphi_\lambda(X) + s\varphi_\lambda(Y)^*)\| = 0,$$

$$\text{b. } \lim_\lambda \|\varphi_\lambda(XY) - \varphi_\lambda(X)\varphi_\lambda(Y)\| = 0,$$

$$\text{c. } \liminf_\lambda \|\varphi_\lambda(X)\| \ge \|X\|/2.$$

If $\mathfrak{A}$ and $\mathcal{B}$ in the preceding proposition are unital C*-algebras, and we want the $*$-isomorphism of $\mathfrak{A}$ into an ultrapower of $\mathcal{B}$ in (1) to be unital, then we change statement (2) by replacing X_1 by 1, and statement (3) by adding the conditions that $1 \in \mathcal{D}$ and $\|1 - \varphi_\lambda(1)\| \to 0$. In statement (2), it is sufficient to choose the X_k's from some generating set of $\mathfrak{A}$.

Whether every finite von Neumann algebra with a separable predual can be embedded into the ultrapower of the hyperfinite II_1 factor with respect to a free ultrafilter on $\mathbb{N}$ is an outstanding open question [C]. It has many deep connections with norms on tensor products (see [Ki]) and free entropy in free probablity theory [V]. As a consequence of the above proposition, we have the following corollary.

COROLLARY 4.2. *Suppose $\mathcal{M}$ and $\mathcal{N}$ are finite von Neumann algebras with faithful normal traces τ and σ, respectively. The following are equivalent:*

1. *There is a normal embedding π of $\mathcal{M}$ into some tracial ultrapower of $\mathcal{N}$ such that $\pi \circ \sigma = \tau$.*

2. *For every $n, N \in \mathbb{N}$, $\varepsilon > 0$, and every $A_1, \ldots, A_n$ in the unit ball of $\mathcal{M}$, there are elements $B_1, \ldots, B_n$ in the unit ball of $\mathcal{N}$ such that*

$$|\tau(A_{k_1} \cdots A_{k_s}) - \sigma(B_{k_1} \cdots B_{k_s})| < \varepsilon$$

for $1 \leq k_1, \ldots, k_s \leq n$ and $1 \leq s \leq N$.

It was proved by C. W. Henson, L. C. Moore, Jr. [HM] and J. Stern [St] (see [He; 6.7]) that if $\mathcal{Y}$ is a Banach space, there is an ultrapower $\mathcal{Y}^\alpha$ of $\mathcal{Y}$ and an isometric embedding π of the second dual $\mathcal{Y}^{\#\#}$ of $\mathcal{Y}$ into $\mathcal{Y}^\alpha$ so that π restricted to the natural embedding of $\mathcal{Y}$ in $\mathcal{Y}^{\#\#}$ induces the natural embedding of $\mathcal{Y}$ into $\mathcal{Y}^\alpha$. Moreover, there is a norm 1 projection mapping from $\mathcal{Y}^\alpha$ onto $\pi(\mathcal{Y}^{\#\#})$. This result is based on the principle of *local reflexivity* [LR], [JRZ], which says that if $\mathcal{Y}$ is a Banach space, $\mathcal{M}$ is a finite-dimensional subspace of $\mathcal{Y}^{\#\#}$ and $\mathcal{L}$ is a finite-dimensional subspace of $\mathcal{Y}^{\#}$, and $\varepsilon > 0$, then there is an operator $T: \mathcal{M} \to \mathcal{Y}$ that is a $(1+\varepsilon)$-isomorphism onto its range such that $T|\mathcal{M} \cap \mathcal{Y}$ is the identity map and, for each x in $\mathcal{M}$ and each f in $\mathcal{L}$, $f(Tx) = x(f)$.

If $\mathfrak{A}$ is a C*-algebra, then $\mathfrak{A}^{\#\#}$ is a von Neumann algebra, so it is natural to ask if the maps above can be chosen to be $*$-homomorphisms. However, this is not the case in general. For example, $C[0,1]$ is a commutative C*-algebra with no nontrivial projections, and thus any ultrapower $C[0,1]^\alpha$ has no nontrivial projections (see Proposition 2.1). But $C[0,1]^{\#\#}$ has many nontrivial projections. Hence there is no $*$-isomorphism of $C[0,1]^{\#\#}$ into an ultrapower of $C[0,1]$. In spite of the fact that the above embeddings cannot be made to be $*$-homomorphisms in the C*-algebra case, they can be chosen to be completely positive.

One conceivable way to prove the existence of a completely positive embedding of $\mathfrak{A}^{\#\#}$ into an ultrapower of $\mathfrak{A}$ would be to prove a completely positive version of local reflexivity for C*-algebras. However, Effros and Haagerup [EH] (see also [ER]) have shown that such a result is not generally true. Arveson [A] has shown that completely contractive unital maps are completely positive, so it follows that there is no completely contractive version of local reflexivity for operator spaces. However, we prove that there is a finitely contractive version.

An operator space is a closed linear subspace $\mathcal{S}$ of $\mathcal{B}(\mathcal{H})$ for some Hilbert space $\mathcal{H}$, along with the family $\{\| \ \|_n\}_{n=1}^{\infty}$ of norms, where each $\| \ \|_n$ is the norm on $M_n(\mathcal{S})$ as a subspace of $M_n(\mathcal{B}(\mathcal{H}))$. If $\mathcal{S}$ and $\mathcal{T}$ are operator spaces and $T: \mathcal{S} \to \mathcal{T}$ is linear, then $T_n: M_n(\mathcal{S}) \to M_n(\mathcal{T})$ is the map $T_n((S_{ij})) = (T(S_{ij}))$. We say T is *completely bounded* if $\|T\|_{\text{cb}} = \sup_{n\geq 1} \|T_n\|_n < \infty$. We say T is *completely contractive* if $\|T\|_{\text{cb}} \leq 1$. For background material on operator spaces we refer the reader to [Pi].

THEOREM 4.3. *Suppose $\mathcal{S}$ is an operator space, $\mathcal{M}$ is a finite-dimensional subspace of $\mathcal{S}^{\#\#}$, $\mathcal{L}$ is a finite-dimensional subspace of $\mathcal{S}^{\#}$, $\varepsilon > 0$, and n is a positive integer. Then there is a map $T: \mathcal{M} \to \mathcal{S}$ such that:*

1. *$T|(\mathcal{M} \cap \mathcal{S})$ is the identity map;*
2. *for each X in $\mathcal{M}$ and f in $\mathcal{L}$, $f(TX) = X(f)$;*
3. *$\|T_n\|, \|T_n^{-1}\| \leq 1 + \varepsilon$.*

Proof. Let $\delta > 0$. By replacing $\mathcal{L}$ with a larger finite-dimensional subspace, we can assume that, for every A in $M_n(\mathcal{M})$ with $\|A\| = 1$,

$$\sup\{|\langle A, F\rangle| : F \in M_n(\mathcal{L}), \|F\| = 1\} \geq 1 - \delta.$$

Using the fact that the Banach space $M_n(\mathcal{S})$ is locally reflexive, we obtain a linear map $\rho: M_n(\mathcal{M}) \to M_n(\mathcal{S})$ that is a $(1+\delta)$-isomorphism such that $\rho|M_n(\mathcal{M} \cap \mathcal{S})$ is the identity and such that, for each X in $M_n(\mathcal{M})$ and each f in $M_n(\mathcal{L})$, $f(\rho(X)) = X(f)$.

Let $\mathcal{G}$ be the group of all unitary $n \times n$ complex matrices such that each row has exactly one nonzero entry, and each nonzero entry is either 1 or -1. Define a mapping $\sigma: M_n(\mathcal{M}) \to M_n(\mathcal{S})$ by

$$\sigma(A) = \frac{1}{|\mathcal{G}|^2} \sum_{U,V \in \mathcal{G}} U^{-1} \rho(UAV) V^{-1}.$$

It is clear that $\|\sigma\| \leq \|\rho\| \leq 1 + \delta$, $\sigma|M_n(\mathcal{S} \cap \mathcal{M})$ is the identity map, and, for each X in $M_n(\mathcal{M})$ and each f in $M_n(\mathcal{L})$, $f(\sigma(X)) = X(f)$. It follows from the choice of $\mathcal{L}$ that, for each A in $M_n(\mathcal{M})$ with $\|A\| = 1$,

$$\|\sigma(A)\| \geq \sup\{|\langle A, F\rangle| : F \in M_n(\mathcal{L}), \|F\| = 1\} \geq 1 - \delta.$$

Hence $\|\sigma^{-1}\| \leq \frac{1}{1-\delta}$. We now specify δ so that $1+\delta, \frac{1}{1-\delta} \leq 1+\varepsilon$. Then $\|\sigma\|, \|\sigma^{-1}\| \leq 1 + \varepsilon$.

It follows from the definition of σ that $U^{-1}\sigma(UAV)V^{-1} = \sigma(A)$ for every A in $M_n(\mathcal{M})$ and every U, V in $\mathcal{G}$. Thus there must be a mapping $T: \mathcal{M} \to \mathcal{S}$ such that, for every matrix $(X_{ij}) \in M_n(\mathcal{M})$, $\sigma((X_{ij})) = (T(X_{ij}))$. In other words, $\sigma = T_n$ where $T_n: M_n(\mathcal{M}) \to M_n(\mathcal{S})$ is the canonical extension of T. It is clear that T satisfies (1) through (3) above.

□

COROLLARY 4.4. *If $\mathcal{S}$ is an operator space, then there are an infinite set $\mathbb{I}$ and a nontrivial ultrafilter α on $\mathbb{I}$, a completely isometric mapping $T: \mathcal{S}^{\#\#} \to \mathcal{S}^\alpha$, and a completely contractive surjective projection $E: \mathcal{S}^\alpha \to T(\mathcal{S}^{\#\#})$ such that $T|\mathcal{S}$ is the natural embedding. Moreover, if $\mathcal{S}$ is a C*-algebra, T and E are completely positive.*

Proof. Let $\mathbb{I}$ be the set of all 4-tuples $(\mathcal{M}, \mathcal{L}, \varepsilon, n)$ with $0 < \varepsilon < 1$, $n \in \mathbb{N}$, $\mathcal{M}$ a finite-dimensional subspace of $\mathcal{S}^{\#\#}$, and $\mathcal{L}$ a finite-dimensional subspace of $\mathcal{S}^\#$, and order $\mathbb{I}$ by $(\subset, \subset, >, <)$. For each $\iota = (\mathcal{M}, \mathcal{L}, \varepsilon, n)$ in $\mathbb{I}$, choose $T_\iota: \mathcal{M} \to \mathcal{S}$ satisfying (1)–(3) in Theorem 4.3.

Note that the collection of sets of the form $\{\iota \in \mathbb{I} : \iota \geq \iota_0\}$ (with ι_0 in $\mathbb{I}$), form a filter. Let α be an ultrafilter containing this filter. For each $\iota = (\mathcal{M}, \mathcal{L}, \varepsilon, n)$ in $\mathbb{I}$, define $\rho_\iota: \mathcal{S}^{\#\#} \to \mathcal{S}$ by $\rho_\iota|\mathcal{M} = T_\iota$, and $\rho|(\mathcal{S}^{\#\#} \setminus \mathcal{M}) = 0$. We define $T: \mathcal{S}^{\#\#} \to \mathcal{S}^\alpha$ by $T(x) = \{\rho_\iota(x)\}_\alpha$, we define $S: \mathcal{S}^\alpha \to T(\mathcal{S}^{\#\#})$ by $S(\{s_\iota\}_\alpha) = \lim_{\iota \to \alpha} s_\iota$ (limit is taken in weak $*$-topology), and we define $E = TS$. It is clear from the choice of the T_ι's that S is a left-inverse for T. The proof that T and E satisfy the required properties is an easy exercise from this point.

If $\mathcal{S}$ is a unital C*-algebra, then T, S, E are unital completely contractive maps, and thus, by Arveson's theorem [A], completely positive.

If $\mathfrak{A}$ is a nonunital C*-algebra, let $\mathcal{S}$ be the unitization of $\mathfrak{A}$. There is a unique state f on $\mathcal{S}$ whose kernel is $\mathfrak{A}$. We have $\mathfrak{A}^{\#\#}$ can be identified with $\{X \in \mathcal{S}^{\#\#} : X(f) = 0\}$. If, in the definition of $\mathbb{I}$, we assume, whenever $\iota = (\mathcal{M}, \mathcal{L}, \varepsilon, n)$, that $f \in \mathcal{L}$, then we have, for each $\iota \in \mathbb{I}$, $T_\iota(\mathcal{M} \cap \mathfrak{A}^{\#\#}) \subset \mathfrak{A}$. It follows that if T is defined for $\mathcal{S}$ as above, then $T(\mathfrak{A}^{\#\#}) \subset \mathfrak{A}$, and hence $E(\mathfrak{A}^\alpha) = T(\mathfrak{A}^{\#\#})$, so $T|\mathfrak{A}^{\#\#}$ and $E|\mathfrak{A}^\alpha$ are the required maps.

□

5. Maps on ultraproducts

Various maps on C*-algebras extend naturally to maps on ultraproducts.

Suppose $\{\mathfrak{A}_\iota : \iota \in \mathbb{I}\}$ and $\{\mathcal{B}_\iota : \iota \in \mathbb{I}\}$ are families of C*-algebras and, for each ι in $\mathbb{I}$, $\pi_\iota: \mathfrak{A}_\iota \to \mathcal{B}_\iota$ is a $*$-homomorphism, we define the ultraproduct $\pi = \prod^\alpha \pi_\iota$ of the π_ι's from $\prod^\alpha \mathfrak{A}_\iota$ into $\prod^\alpha \mathcal{B}_\iota$ by

$$\pi([\{A_\iota\}]) = [\{\pi_\iota(A_\iota)\}].$$

If each π_ι is a representation, i.e., $\mathcal{B}_\iota = \mathcal{B}(\mathcal{H}_\iota)$ for each ι in $\mathbb{I}$, then the ultraproduct π is a representation from $\prod^\alpha \mathfrak{A}_\iota$ into $\prod^\alpha \mathcal{B}(\mathcal{H}_\iota) \subseteq \mathcal{B}(\prod^\alpha \mathcal{H}_\iota)$.

If, for each ι in $\mathbb{I}$, $\varphi_\iota: \mathfrak{A}_\iota \to \mathcal{B}_\iota$ is a completely positive map, we define the ultraproduct $\varphi = \prod^\alpha \varphi_\iota$ in the same manner as $*$-homomorphisms. In particular, if each φ_ι is a state, then so is φ. The following proposition is an obvious consequence of the fact that if π is a $*$-homomorphism on a C*-algebra, then $\|\pi(X)\| = \operatorname{dist}(X, \ker\pi)$.

PROPOSITION 5.1. *Suppose $\{\mathfrak{A}_\iota : \iota \in \mathbb{I}\}$, $\{\mathcal{B}_\iota : \iota \in \mathbb{I}\}$, $\{\mathcal{J}_\iota : \iota \in \mathbb{I}\}$ are families of C*-algebras such that, for each ι in $\mathbb{I}$, $\mathcal{J}_\iota$ is a closed ideal in $\mathfrak{A}_\iota$, and $\pi_\iota: \mathfrak{A}_\iota \to \mathcal{B}_\iota$ is a $*$-homomorphism. Then, for any ultrafilter α on $\mathbb{I}$, $\ker(\prod^\alpha \pi_\iota) = \prod^\alpha \ker \pi_\iota$, and $\prod^\alpha \mathfrak{A}_\iota / \prod^\alpha \mathcal{J}_\iota = \prod^\alpha (\mathfrak{A}_\iota / \mathcal{J}_\iota)$.*

One of the consequences of the above proposition is that exact sequences of C*-algebras are preserved under ultraproduct construction.

The following theorem shows that ultraproducts with respect to nontrivial ultrafilters on $\mathbb{N}$ are usually highly nonseparable. Recall that $\prod_{n\in\mathbb{N}} \mathfrak{A}_n$, or simply $\prod \mathfrak{A}_n$, denotes the cartesian product of $\mathfrak{A}_n$'s, $\sum_{n\in\mathbb{N}} \mathfrak{A}_n$, or $\sum \mathfrak{A}_n$, denotes the subspace of $\prod \mathfrak{A}_n$ consisting of all sequences converging in norm to zero as n tends to infinity. We say that two elements A and B in a C*-algebra are *orthogonal* if $A^*B = AB^* = 0$.

THEOREM 5.2. *Suppose $\{\mathfrak{A}_n\}$ is a sequence of C*-algebras such that α and $\lim_{n\to\infty} \dim \mathfrak{A}_n = \infty$ is a nontrivial ultrafilter on $\mathbb{N}$. Then*

1. *there is a family $\{A_t : t \in [0,1]\}$ of positive elements in $\prod \mathfrak{A}_n$, $A_t = \{A_t(n)\}$, with $\|A_t(n)\| = 1$ for t in $[0,1]$ and n in $\mathbb{N}$, whose image in $\prod \mathfrak{A}_n / \sum \mathfrak{A}_n$ (and hence in $\prod^\alpha \mathfrak{A}_n$) is an orthogonal family;*
2. *there is no one-to-one continuous linear map from $\prod \mathfrak{A}_n / \sum \mathfrak{A}_n$ (or $\prod^\alpha \mathfrak{A}_n$) into $l^\infty(\mathbb{N})$;*
3. *$\prod \mathfrak{A}_n / \sum \mathfrak{A}_n$ and $\prod^\alpha \mathfrak{A}_n$ have no faithful continuous linear maps into $\mathcal{B}(l^2(\mathbb{N}))$;*
4. *if each $\mathfrak{A}_n$ is separable, then $\sum^\alpha \mathfrak{A}_n$ and $\sum \mathfrak{A}_n$ are not complemented in $\prod \mathfrak{A}_n$;*
5. *if each $\mathfrak{A}_n$ is unital and there is a positive integer N such that, for each n in $\mathbb{N}$, and each P in $\mathfrak{A}_n$, there are elements U_k V_k $(1 \le k \le N)$ in $\mathfrak{A}_n$ with norm less than N such that $\sum_{k=1}^N U_k P V_k = 1$. Then every separable representation of $\prod \mathfrak{A}_n$ is a*

direct sum of representations that factor through the coordinate homomorphisms. In particular, $\prod \mathfrak{A}_n / \sum \mathfrak{A}_n$ *and* $\prod^\alpha \mathfrak{A}_n$ *have no non-zero separable representations;*

6. *if* $\mathfrak{A}$ *is a purely infinite* C^**-algebra, then* $\mathfrak{A}^\alpha$ *has no separable unital representation.*

Proof. 1. For each n in $\mathbb{N}$, we can choose a finite orthogonal set $\mathbb{E}_n$ of positive norm one elements in $\mathfrak{A}_n$ so that the limit of the cardinalities of the $\mathbb{E}_n$'s approaches ∞ as $n \to \infty$. In fact, if $\mathfrak{A}_n$ is infinite-dimensional, then $\mathfrak{A}_n$ must contain a hermitian element with infinite spectrum, and if $\mathcal{X}$ is a compact Hausdorff space with infinitely many points, $C(\mathcal{X})$ contains an infinite orthogonal set of norm-one positive elements. We can assume that each $\mathbb{E}_n$ is the finite set $\{0, 1, \ldots, k_n\}$ (each number corresponds to a norm one positive operator) and that $\lim_{n\to\infty} k_n = \infty$. For each t in $[0,1]$ and each n in $\mathbb{N}$, we define $A_t(n) = [tk_n]$ (where $[\,]$ denotes the greatest integer function). Since $0 \le tk_n - [tk_n] < 1$, it follows, for every t in $[0,1]$, that $\lim_{n\to\infty} A_t(n)/k_n = t$. It is now clear that $\{n \in \mathbb{N} : A_s(n) = A_t(n)\}$ is finite whenever $0 \le s < t \le 1$. This implies the image of $\{A_t : t \in [0,1]\}$ in $\prod \mathfrak{A}_n / \sum \mathfrak{A}_n$ is orthogonal.

2. The linear span of the image of $\{A_t : t \in [0,1]\}$ in both $\prod \mathfrak{A}_n / \sum \mathfrak{A}_n$ and $\prod^\alpha \mathfrak{A}_n$ is isometrically isomorphic to the space $\mathcal{Y}$ of complex functions on $[0,1]$ having finite support with the supremum norm on $\mathcal{Y}$. It suffices to show that there is no continuous linear map $\varphi : \mathcal{Y} \to l^\infty$. If such a φ exists, there must be $k, n \in \mathbb{N}$ such that $\mathbb{E} = \{t \in [0,1] : |\varphi(\chi_{\{t\}})(n)| \ge 1/k\}$ is infinite. For each $t \in \mathbb{E}$, choose λ_t with $|\lambda_t| = 1$ so that $\varphi(\lambda_t \chi_{\{t\}}) \ge 1/k$. Any finite sum of distinct $\lambda_t \chi_{\{t\}}$'s has norm 1, while the image under φ of such sums is unbounded, violating the boundedness of φ.

Statements (3)–(6) follow from (2).

□

From the ultraproduct construction, we saw that the ultraproduct $\prod^\alpha \mathcal{H}_\iota$ of a family $\{\mathcal{H}_\iota : \iota \in \mathbb{I}\}$ of Hilbert spaces is a Hilbert space. There is a natural way in which the C*-ultraproduct $\prod^\alpha \mathcal{B}(\mathcal{H}_\iota)$ can be viewed as a subalgebra of $\mathcal{B}(\prod^\alpha \mathcal{H}_\iota)$, defined by $[\{T_\iota\}][\{h_\iota\}] = [\{T_\iota h_\iota\}]$.

If x, y are vectors in a Hilbert space $\mathcal{H}$, we let $\omega_{x,y}$ denote the rank-one operator defined by $\omega_{x,y}(h) = \langle h, y\rangle x$ for all h in $\mathcal{H}$. It is clear that if $x, y \in \prod^\alpha \mathcal{H}_\iota$, $x = [\{x_\iota\}]$, $y = [\{y_\iota\}]$, then $\omega_{x,y} = [\{\omega_{x_\iota, y_\iota}\}] \in \prod^\alpha \mathcal{B}(\mathcal{H}_\iota)$. We let $\mathcal{K}(\mathcal{H})$ denote the algebra of compact operators on $\mathcal{H}$.

LEMMA 5.3. *If* $\{\mathcal{H}_\iota\}_{\iota\in\mathbb{I}}$ *is a family of Hilbert spaces and* α *is a nontrivial ultrafilter on* $\mathbb{I}$, *then* $\prod^\alpha \mathcal{K}(\mathcal{H}_\iota) \supseteq \mathcal{K}(\prod^\alpha \mathcal{H}_\iota)$.

Proof. Since $\prod^\alpha \mathcal{K}(\mathcal{H}_\iota)$ is a C*-algebra containing all of the rank-one operators on $\prod^\alpha \mathcal{H}_\iota$, the lemma is proved.

□

We list some more basic properties of ultraproducts in the following theorem. The first result is an analogue of Kadison's transitivity theorem [K] for ultraproducts.

THEOREM 5.4. *Assume α is a nontrivial ultrafilter on $\mathbb{I}$.*

1. *Suppose, for each ι in $\mathbb{I}$, $\mathfrak{A}_\iota$ is a C^*-subalgebra of $\mathcal{B}(\mathcal{H}_\iota)$ and T in $\mathcal{B}(\prod^\alpha \mathcal{H}_\iota)$ is in the weak-operator closure of $\prod^\alpha \mathfrak{A}_\iota$, and suppose $\mathcal{H}_0$ is a separable subspace of $\prod^\alpha \mathcal{H}_\iota$. Then there is an A in $\prod^\alpha \mathfrak{A}_\iota$ such that $\|A\| \le \|T\|$, $A|\mathcal{H}_0 = T|\mathcal{H}_0$, and $A^*|\mathcal{H}_0 = T^*|\mathcal{H}_0$. Moreover, if T is hermitian, positive, or unitary, then A can be chosen to be the same.*
2. *Suppose, for each ι in $\mathbb{I}$, $\pi_\iota: \mathfrak{A}_\iota \to \mathcal{B}(\mathcal{H}_\iota)$ is a $*$-representation. Then $\prod^\alpha \pi_\iota$ is irreducible if and only if the π_ι's are eventually irreducible along α.*
3. *An ultraproduct of primitive C^*-algebras is primitive.*
4. *An ultraproduct of pure states is a pure state.*
5. *An ultraproduct of tracial states is a tracial state.*

Proof. 1. There is no harm in assuming that $\mathcal{H}_0$ reduces T, since the smallest subspace reducing T that contains $\mathcal{H}_0$ is also separable. Let $\{e_0, e_1, e_2, \ldots\}$ be an orthonormal basis for $\mathcal{H}_0$, and for each $n \ge 0$, write $e_n = [\{e_n(\iota)\}]$. We want to show that the $e_n(\iota)$'s can be chosen so that, for each ι in $\mathbb{I}$,

a. $\{e_n(\iota) : 0 \le n < \dim \mathcal{H}_\iota\}$ is orthonormal, and

b. $e_n(\iota) = 0$ when $n \ge \dim \mathcal{H}_\iota$.

There is no harm in assuming that $e_0(\iota) \ne 0$ for each ι in $\mathbb{I}$ for which $\dim \mathcal{H}_\iota > 0$. Since $\|e_0\| = 1$, e_0 is not affected if we replace each $e_0(\iota)$ with $e_n(\iota)/\|e_n(\iota)\|$. Thus we can assume that $e_0(\iota), \ldots, e_n(\iota)$ have been chosen for each ι in $\mathbb{I}$ in accordance with (a), (b) above. For each ι in $\mathbb{I}$, choose a vector $f_{n+1}(\iota)$ in $\mathcal{H}_\iota$ so that $f_{n+1}(\iota)$ is the normalized projection of $e_{n+1}(\iota)$ onto the orthogonal complement of the linear span of $\{e_0(\iota), e_1(\iota), \ldots, e_n(\iota)\}$, $f_{n+1}(\iota) = 0$ if $\dim \mathcal{H}_\iota = n+1$, and $f_{n+1}(\iota)$ is any unit vector perpendicular to $\{e_0(\iota), e_1(\iota), \ldots, e_n(\iota)\}$ otherwise. It is clear from the orthonormality of $\{e_1, e_2, \ldots, e_{n+1}\}$ that $e_{n+1} = [\{f_{n+1}(\iota)\}]$.

Let a_{jk} be $\langle Te_k, e_j\rangle$ for $0 \le j, k < \infty$. For each ι in $\mathbb{I}$, and each A in $\mathfrak{A}_\iota$, define

$$\delta_\iota(A) = \sum_{k=1}^{\infty} \frac{1}{2^k}\left(\left\|Ae_k(\iota) - \sum_{j=1}^{\infty} a_{jk}e_j(\iota)\right\| + \left\|A^*e_k(\iota) - \sum_{j=1}^{\infty} \bar{a}_{kj}e_j(\iota)\right\|\right),$$

and define $\delta_\iota = \inf\{\delta_\iota(A) : A \in \mathfrak{A}_\iota, \|A\| \le \|T\|\}$. It follows from the fact that T is in the $*$-strong closure of $\{S \in \prod^\alpha \mathfrak{A}_\iota : \|S\| \le \|T\|\}$, that $\lim_{\iota \to \alpha} \delta_\iota = 0$.

Next choose a decreasing sequence $\{\mathbb{I}_n\}$ of sets in α whose intersection is $\emptyset$, with $\mathbb{I}_1 = \mathbb{I}$. For each $n \geq 1$, and each ι in $\mathbb{I}_n \setminus \mathbb{I}_{n+1}$, choose an A_ι in $\mathfrak{A}_\iota$ such that $\delta_\iota(A_\iota) < \delta_\iota + \frac{1}{n}$. Let A be $[\{A_\iota\}]$ in $\prod^\alpha \mathfrak{A}_\iota$. Then it follows that $A|\mathcal{H}_0 = T|\mathcal{H}_0$ and $A^*|\mathcal{H}_0 = T^*|\mathcal{H}_0$.

2. Since α is an ultrafilter, either the π_ι's are eventually irreducible or eventually reducible. In the latter case we can choose, for each ι in $\mathbb{I}$, a projection P_ι commuting with π_ι such that $0 \neq P_\iota \neq 1$ whenever π_ι is reducible. It follows that $P = [\{P_\iota\}]$ is a nontrivial projection commuting with $\prod^\alpha \pi_\iota$.

Suppose that the π_ι's are eventually irreducible along α, and suppose e, f are unit vectors in $\prod^\alpha \mathcal{H}_\iota$, with $e = [\{e_\iota\}]$ and $f = [\{f_\iota\}]$. We can assume that π_ι is irreducible and $\|e_\iota\| = \|f_\iota\|$ for each ι in $\mathbb{I}$. It follows from Kadison's transitivity theorem [K] that, for each ι in $\mathbb{I}$, there is an A_ι in the closed unit ball of the range of π_ι such that $A_\iota e_\iota = f_\iota$. Choose $B = [\{B_\iota\}]$ in $\prod^\alpha \mathfrak{A}_\iota$ so that $(\prod^\alpha \pi_\iota)(B) = [\{A_\iota\}]$, which maps e to f. Hence $\prod^\alpha \pi_\iota$ is irreducible.

Statements (3) and (4) follow immediately from (2), and statement (5) is elementary.

□

Corollary 5.5. *An ultraproduct of simple C*-algebras need not be simple.*

Proof. Suppose $\mathfrak{A}$ is a simple infinite-dimensional unital C*-algebra with a faithful tracial state τ (e.g., the reduced free group C*-algebras $C^*_r(F_n)$). Let α be any nontrivial ultrafilter on $\mathbb{N}$ and choose a sequence $\{A_n\}$ of norm 1 positive elements of $\mathfrak{A}$ such that $\tau(A_n) \to 0$ as $n \to \infty$. Then $A = [\{A_n\}]$ is a nonzero positive element in $\mathfrak{A}^\alpha$ whose trace $\tau^\alpha(A) = 0$. Hence $\mathfrak{A}^\alpha$, having a non-faithful tracial state, is not simple.

□

There are some obvious conditions under which ultraproducts of simple algebras are simple.

Proposition 5.6. *Suppose $\{\mathfrak{A}_\iota : \iota \in \mathbb{I}\}$ is a family of unital C*-algebras and N is a positive integer such that, for every ι in $\mathbb{I}$ and every A in $\mathfrak{A}_\iota$ with $\|A\| = 1$, there are elements $B_1, C_1, \ldots, B_N, C_N$ in $\mathfrak{A}_\iota$ with norm at most N, such that $1 = \sum_{k=1}^N B_k A C_k$. Then $\prod^\alpha \mathfrak{A}_\iota$ is simple.*

Corollary 5.7. *If $\mathfrak{A}$ is the Calkin algebra (i.e., $\mathfrak{A} = \mathcal{B}(l^2(\mathbb{N}))/\mathcal{K}(l^2(\mathbb{N}))$), then for any ultrafilter α, $\mathfrak{A}^\alpha$ is simple.*

Proof. Suppose $A \in \mathfrak{A}$ and $\|A\| = 1$, and suppose $\varepsilon > 0$. We will show that there are elements X, Y in $\mathfrak{A}$ with $\|X\|, \|Y\| < 1 + \varepsilon$ such that $XAY = 1$. Replacing A with A^*A, we can assume that $0 \leq A \leq 1$. We can lift A to an operator, denoted again by A, in $\mathcal{B}(l^2(\mathbb{N}))$ with $0 \leq A \leq 1$. We can choose an isometry V whose range is the spectral projection of A of the interval $[\frac{1}{1+\varepsilon}, 1]$, such that V^*AV is an invertible operator whose norm is at most $1 + \varepsilon$. Let Y be the image in $\mathfrak{A}$ of V and X be the image in $\mathfrak{A}$ of $(V^*AV)^{-1}V^*$. □

6. Ultraproducts on $\aleph_1$-complete ultrafilters

We end this paper with some remarks and results concerning different types of ultrafilters.

Recall that an ultrafilter α is *countably cofinal* if there is a sequence $\{A_n\}$ in α whose intersection is empty. Otherwise, α is called $\aleph_1$*-complete* [CN]. An infinite cardinal m is *Ulam measurable* if there is a nontrivial $\aleph_1$-complete ultrafilter on a set with cardinality m. Clearly, $\aleph_0$ is not Ulam measurable. More generally, if m is an infinite cardinal, an ultrafilter α is m-complete if α is closed under intersections of subcollections with cardinality less than m. An uncountable cardinal m is *measurable* if there is an ultrafilter α on a set with cardinality m, such that α is m-complete. Clearly, a measurable cardinal is Ulam measurable. On the other hand, S. Ulam proved (see [CN; 8.31]) that a cardinal is Ulam measurable if and only if it is not less than the smallest measurable cardinal. Thus the existence of nontrivial $\aleph_1$-complete ultrafilters is equivalent to the existence of Ulam measurable cardinals, which, in turn, is equivalent to the existence of measurable cardinals. It is possible that the existence of measurable cardinals is independent of the usual axioms of set theory (ZFC), but it has been shown that the proof of the consistency of the existence of measurable cardinals can not be accomplished using only (ZFC). For a nice discussion of these ideas see [CN].

Ultraproducts with respect to $\aleph_1$-complete ultrafilters are strange, and when the factors are countable or separable they are often trivial. We give some examples in the next proposition.

PROPOSITION 6.1. *Suppose α is a nontrivial ultrafilter on the set $\mathbb{I}$. Then*

1. *α is $\aleph_1$-complete if and only if there is no funtion $f: \mathbb{I} \to (0, \infty)$ such that* $\lim_{\iota \to \alpha} f(\iota) = 0$;

2. *if m is an uncountable cardinal, α is m-complete and $\mathcal{X}$ is a Banach space with a separating family $\mathcal{F}$ of linear functionals such that* $\mathrm{card}(\mathcal{F}) < m$, *then the natural inclusion of $\mathcal{X}$ into $\mathcal{X}^\alpha$ is onto;*

3. *if α and m are as in (2), and k is a cardinal less than m, $\{\mathcal{X}_\iota : \iota \in \mathbb{I}\}$ is a family of separable Banach spaces such that each $\mathcal{X}_\iota$ has a dense subset $\mathcal{D}_\iota$ such that $\operatorname{card}(\mathcal{D}_\iota) \leq k$, then, eventually along α, all of the $\mathcal{X}_\iota$'s are isometrically isomorphisc to some Banach space $\mathcal{X}$, so that $\prod^\alpha \mathcal{X}_\iota$ is the same as $\mathcal{X}^\alpha$;*

4. *if α is $\aleph_1$-complete and $\{\mathcal{X}_\iota : \iota \in \mathbb{I}\}$ is a family of Banach spaces, then $\prod^\alpha \mathcal{X}_\iota$ is the same as the classical ultraproduct of the $\mathcal{X}_\iota$'s.*

Proof. 1. Suppose α is *not* $\aleph_1$-complete, then we can choose a countable decreasing sequence $\{\mathcal{A}_n\}$ in α whose intersection is empty. If we define $f: \mathbb{I} \to (0, \infty)$ by $f|(\mathcal{A}_n \setminus \mathcal{A}_{n+1}) = \frac{1}{n}$, then $\lim_{\iota \to \alpha} f(\iota) = 0$. On the other hand, suppose α is $\aleph_1$-complete, and suppose $f: \mathbb{I} \to (0, \infty)$ and $\lim_{\iota \to \alpha} f(\iota) = 0$. Then, for each positive integer n, $f^{-1}((0, \frac{1}{n}]) \in \alpha$. Since α is closed under countable intersections, $\emptyset = \cap_n f^{-1}((0, \frac{1}{n}]) \in \alpha$, a contradiction.

2. Suppose, for each ι in $\mathbb{I}$, $x_\iota \in \mathcal{X}$. It follows from (1) that, for each f in $\mathcal{F}$, $f(x_\iota)$ is eventually constant along α, and since $\operatorname{card}(\mathcal{F}) < m$, and α is m-complete, there is a set $\mathcal{S}$ in α, such that for every $\iota, \jmath$ in $\mathcal{S}$, and every f in $\mathcal{F}$, $f(x_\iota) = f(x_\jmath)$, which implies that $x_\iota = x_\jmath$. Hence $\{x_\iota\}$ is eventually constant along α.

3. Let $\mathcal{K}$ be a set with $\operatorname{card}(\mathcal{K}) = k$. For each ι in $\mathbb{I}$, we can find a bounded surjective linear mapping $T_\iota: l^1(\mathcal{K}) \to \mathcal{X}_\iota$ (i.e., map $\{\chi_{\{x\}} : x \in \mathcal{K}\}$ onto a dense subset of the closed unit ball of $\mathcal{X}_\iota$ and extend by linearity and continuity). Let $\mathcal{D}$ be the linear span of $\{\chi_{\{x\}} : x \in \mathcal{K}\}$ over the field $\mathbb{Q} + i\mathbb{Q}$. Then $\operatorname{card}(\mathcal{D}) = k$, and, for each x in $\mathcal{D}$, it follows from (1) that $\|T_\iota x\|$ is eventually constant along α, and since $\operatorname{card}(\mathcal{D}) < m$, there is a set $\mathcal{S}$ in α such that, for every $\iota, \jmath$ in $\mathcal{S}$, $\|T_\iota x\| = \|T_\jmath x\|$. It follows that, for every $\iota, \jmath$ in $\mathcal{S}$ and every x in $l^1(\mathcal{K})$, $\|T_\iota x\| = \|T_\jmath x\|$. Hence, for $\iota, \jmath$ in $\mathcal{S}$, the mapping $T_\iota x \to T_\jmath x$ defines an isometric linear isomorphism from $\mathcal{X}_\iota$ to $\mathcal{X}_\jmath$.

4. Suppose $\{x_\iota\}$ is in the Cartesian product of the $\mathcal{X}_\iota$'s (i.e., the x_ι's are not necessarily uniformly bounded). It follows from (1) that $\lim_{\iota \to \alpha} \|x_\iota\| \neq \infty$. Hence there is a bounded family $\{y_\iota\}$ such that $x_\iota = y_\iota$ is eventually along α. Furthermore, if $\|x_\iota\| \to 0$, then, by (1), $x_\iota = 0$ eventually along α. This proves (4).

□

REMARKS 6.2. 1. It is clear that the Banach space results in the preceding proposition also hold for C*-algebras. In the proof of (3), we can replace $l^1(\mathcal{K})$ with the universal C*-algebra with generators $\mathcal{K}$ subject to the conditions that $\|x\| \leq 1$ for every x in $\mathcal{K}$. Then the T_ι's are $*$-homomorphisms and the $\mathcal{X}_\iota$'s are eventually $*$-isomorphic along α.

2. By modifying the proofs of (2) and (3) in the preceding proposition, it can be shown that if α is $\aleph_m$-complete, $\mathcal{G}$ is a group with $\operatorname{card}(\mathcal{G}) < m$, and $\{\mathcal{G}_\iota : \iota \in \mathbb{I}\}$

is a family of groups with $\sup\{\mathrm{card}(\mathcal{G}_\iota) : \iota \in \mathbb{I}\} < m$, then the natural embedding of $\mathcal{G}$ into the classical ultrapower $\mathcal{G}^\alpha$ is onto, and the $\mathcal{G}_\iota$'s are eventually, along α, isomorphic to one another. The role of $l^1(\mathcal{K})$ in the proof is played by the free group with k generators.

We have seen (Theorem 5.2) that ultrapowers of a separable infinite-dimensional C*-algebra (or, an infinite-dimensional Banach space) with respect to countably cofinal ultrafilters are always highly non-separable. Thus we can express the existence of measurable cardinals in terms of ultrapowers of C*-algebras.

COROLLLARY 6.3. *The following are equivalent:*
1. *There exists a measurable cardinal.*
2. *There is a free ultrafilter α and a separable infinite-dimensional C^*-algebra $\mathfrak{A}$ such that $\mathfrak{A}^\alpha$ is separable.*

ACKNOWLEDGEMENTS. We wish to thank David Feldman for many informative discussions on ultraproducts. The path to Theorem 3.7 involved the help of many eminent and friendly logicians, including Robert Solovay, Paul Eklof, Ken Kunen, Hugh Woodin, H. Garth Dales, H. Jerome Keisler and Alan Dow. We are particularly indebted to Robert Solovay, who gave us many suggestions and pointed us in the right direction, and to Alan Dow, who not only knew the answers to the logic questions we asked, but patiently explained them. We also like to thank referees for corrections and helpful suggestions.

References

[AP] C. A. AKEMANN and G. K. PEDERSEN, Central sequences and inner derivations of separable C*-algebras, *Amer. J. Math.*, **101** (1979), 1047–1061.

[A] W. ARVESON, Subalgebras of C^*-algebras, *Acta Math.*, **123** (1969), 141–224.

[CK] C. C. CHANG and H. J. KEISLER, *Model Theory*, North-Holland, Amsterdam, 1978.

[C] A. CONNES, Classification of injective factors, Cases II_1, II_∞, III_λ, $\lambda \neq 1$, *Ann. Math.*, **104** (1976), 73–115.

[CN] W. W. COMFORT and S. NEGREPONTIS, *The Theory of Ultrafilters*, Springer-Verlag, New York, 1974.

[DW] H. G. DALES and W. H. WOODIN, *An Introduction to Independence for Analysists*, London Math. Soc. Lec. Notes, No. **115**, Cambridge Univ. Press, Cambridge, 1987.

[D] A. DOW, On the ultrapowers of Banach algebras, *Topology Proceedings*, **9** (1984), 269–291.

[E] P. Eklof, The structure of ultraproducts of abelian groups, *Pacific J. Math.*, **47** (1973), 69–79.

[ER] E. Effros and Z. Ruan, Mapping spaces and liftings for operator spaces, *Proc. London Math. Soc.*, **69** (1994), 171–197.

[EH] E. Effros and U. Haagerup, Lifting problems and local reflexivity for C*-algebras, *Duke Math. J.*, **52** (1985), 103–128.

[Ha] U. Haagerup, Solution of the similarity problem for cyclic representations of C*-algebras, *Ann. Math.*, **118** (1983), 215–240.

[He] S. Heinrich, Ultraproducts in Banach space theory, *J. Reine Angew. Math.*, **27** (1978), 72–104.

[HH] S. Heinrich and C. W. Henson, Banach space model theory. II. Isomorphic equivalence, *Math. Nachr.*, **125** (1986), 301–317.

[Hn1] C. W. Henson, When do two Banach spaces have isometrically isomorphic nonstandard hulls?, *Israel J. Math.*, **22** (1975), 57–67.

[Hn2] C. W. Henson, Nonstandard hulls of Banach spaces, *Israel J. Math.*, **25** (1976), 108–144.

[HM] C. W. Henson and L. C. Moore, Jr., Subspaces of the nonstandard hull of a normed space, *Trans. Amer. Math. Soc.*, **197** (1974), 131–143.

[J] G. Janssen, Restricted ultraproducts of finite von Neumann algebras, *Studies in Logic and Found. Math.*, Vol. **69**, North-Holland, Amsterdam, 1972, 101–114.

[JRZ] W. B. Johnson, H. P. Rosenthal and M. Zippin, On bases, finite-dimensional decompositions and weaker structures in Banach spaces, *Israel J. Math.*, **9** (1971), 488–506.

[K] R. V. Kadison, Irreducible operator algebras, *Proc. Nat. Acad. Sci. USA*, **43** (1957), 273–276.

[Ki] E. Kirchberg, On nonsemisplit extensions, tensor products and exactness of group C*-algebras, *Invent. Math.*, **112** (1993), 449–489.

[LR] J. Lindenstrauss and H. P. Rosenthal, The $\mathcal{L}_p$ spaces, *Israel J. Math.*, **7** (1969), 325–349.

[Mc] D. McDuff, Central sequences and the hyperfinite factor, *Proc. London Math. Soc.*, **21** (1970), 443–461.

[Ph] J. Phillips, Central sequences and automorphisms of C*-algebras, *Amer. J. Math.*, **110** (1988), 1095–1118.

[Pi] G. Pisier, *Espaces de Banach quantiques: une introduction à la thèorie des espaces d'opèrateurs*, SMF Journ. Annu., 1994; Soc. Math. France, Paris, 1994.

[Sa] S. Sakai, *The Theory of W* Algebras,* Lecture notes, Yale University, 1962.

[S1] S. Shelah, Every two elementarily equivalent models have isomorphic ultrapowers, *Israel J. Math.*, **10** (1971), 224–233.

[S2] S. Shelah, *Classification Theory and the Number of Nonisomorphic Models,* Studies in Logic and the Foundations of Mathematics **92**, North-Holland Publishing Co., Amsterdam, 1990.

[St] J. Stern, Ultrapowers and local properties of Banach spaces, *Trans. Amer. Math. Soc.*, **240** (1978), 231–252.

[V] D. VOICULESCU, The analogue of entropy and of Fisher's information measure in free probability theory. II, *Invent. Math.*, **118** (1994), 411–440.

L. GE, Department of Mathematics, University of New Hampshire, Durham, NH 03824; *e-mail*: liming@math.unh.edu

D. HADWIN, Department of Mathematics, University of New Hampshire, Durham, NH 03824; *e-mail*: don@math.unh.edu

Operator Theory: Advances and Applications, Vol. 127, 327–347

Intertwining extensions and a two-sided corona problem

CAIXING GU and RADU I. TEODORESCU

In the memory of Professor Béla Sz.-Nagy, one of the builders of Operator Theory

In this paper we discuss two seemingly unrelated problems: a two-sided corona problem and an extension problem for an intertwining operator with conservation of the intertwining property. We show that these two problems are intimately connected. Furthermore, we find situations when the two-sided corona problem is solvable.

1. Introduction

Let $\mathcal{E}$ and $\mathcal{F}$ be two complex separable Hilbert spaces. As usual $B(\mathcal{E},\mathcal{F})$ will be the set of all linear bounded operators from $\mathcal{E}$ into $\mathcal{F}$ and we write $B(\mathcal{E})$ instead of B($\mathcal{E}$,$\mathcal{E}$). If $\mathcal{G} \subset \mathcal{E}$ is a closed subspace then $P_{\mathcal{G}}$ will be the orthogonal projection onto $\mathcal{G}$. We recall that the Hardy space $H^2(\mathcal{E})$ is the set of all $\mathcal{E}$-valued analytic functions in the open unit disc $D = \{z \in C : |z| < 1\}$ whose Fourier coefficients are square summable; the space $H^\infty(\mathcal{E},\mathcal{F})$ is the set of all bounded analytic functions in D with values in $B(\mathcal{E},\mathcal{F})$; finally the function $\Theta \in H^\infty(\mathcal{E},\mathcal{F})$ is called *contractive* if $\|\Theta(z)\| \leq 1$ for $z \in D$ (see [7] and [14] for all definitions and related notations). The two-sided corona problem we are going to study is the following:

TWO-SIDED CORONA PROBLEM. (TSCP). *Let $\Theta_1 \in H^\infty(\mathcal{E}_1,\mathcal{F}_1)$ and $\Theta_2 \in H^\infty(\mathcal{E}_2,\mathcal{F}_2)$ be two contractive analytic functions. Find the condition(s) guaranteeing that for any bounded analytic function $\Theta \in H^\infty(\mathcal{E}_1,\mathcal{F}_2)$, there exist bounded analytic functions $\Phi \in H^\infty(\mathcal{F}_1,\mathcal{F}_2)$ and $\Psi \in H^\infty(\mathcal{E}_1,\mathcal{E}_2)$ such that*

$$\Phi(z)\Theta_1(z) + \Theta_2(z)\Psi(z) = \Theta(z) \quad \text{for} \quad z \in D. \tag{1}$$

AMS Subject Classification (1991): 47A20, 47A45.

This problem is inspired by the classical corona problem and its generalizations (see [2], [11], [8], [12]), as well as problems in pure operator theory (see [1], [3], [4]). In this note we will study an unexpected relationship of (TSCP) with the following problem about intertwining operators:

General Intertwining Extension Problem. (GIEP). *Let $T' \in B(\mathcal{K}')$ and $T \in B(\mathcal{H})$ be two contractions, $\mathcal{G}' \subset \mathcal{K}'$ an invariant subspace of T', and T_0' the restriction of T' to $\mathcal{G}'$, i.e. $T_0' = T'|\mathcal{G}'$. Let also $A \in B(\mathcal{G}', \mathcal{H})$ be an operator intertwining T and T_0', i.e. $TA = AT_0'$. Is there an operator $B \in B(\mathcal{K}', \mathcal{H})$ such that $TB = BT'$ and $B|\mathcal{G}' = A$?*

We call such an operator B an intertwining extension of A. If there exists an intertwining extension of A, we set

$$\mu(A, T, T', \mathcal{G}') = \inf\{\|B\| : B \in B(\mathcal{K}', \mathcal{H}), BT' = TB, B|\mathcal{G}' = A\},$$

and if A has no intertwining extension, we set $\mu(A, T, T', \mathcal{G}') = \infty$.

We remark that the classical commutant lifting theorem [14] solves a special case of (GIEP). To see this we first recall the following terminology: for $T \in B(\mathcal{E})$ the operator $S \in B(\mathcal{F})$ is a dilation of T, if $\mathcal{F} \supset \mathcal{E}$ and $T^n = P_{\mathcal{E}} S^n|\mathcal{E}$ for any $n \geq 0$. If T is a contraction, it is well known that T has a minimal isometric dilation; moreover this minimal isometric dilation of T is unique in the sense that any two minimal isometric dilations of the same contraction are isomorphic (see [13]).

The Commutant Lifting Theorem. (CLT). *Let S be an isometry of the complex Hilbert space $\mathcal{F}$, $T \in B(\mathcal{E})$ a contraction, and U on $\mathcal{K}(\supset \mathcal{E})$ the minimal isometric dilation of T. Let $A \in B(\mathcal{F}, \mathcal{E})$ be an operator intertwining T and S, i.e. $AS = TA$. Then there exists an operator $B \in B(\mathcal{F}, \mathcal{K})$ satisfying*

$$BS = UB, \ P_{\mathcal{E}} B = A \quad \text{and} \quad \|B\| = \|A\|.$$

The operator $B \in B(\mathcal{F}, \mathcal{K})$ satisfying $BS = UB$, $P_{\mathcal{E}} B = A$ is called an intertwining lifting of A. To see the connection between (CLT) and (GIEP), we note that if A is as in the (CLT) and B is an intertwining lifting of A, then $S^* A^* = A^* T^*$, $S^* B = B^* U^*$ and $B^*|\mathcal{E} = A^*$. That is, B^* is an extension of A^* in the sense of (GIEP). See also [5], [6], [9] for other special cases of (GIEP), for their connections with a nonlinear robust control problem and some problems concerning Toeplitz and Hankel operators.

We recall the following classical result: if S is the multiplication by e^{it} on $H^2(\mathcal{F})$ and U is the multiplication by e^{it} on $H^2(\mathcal{E})$, then the general form of an operator $A: H^2(\mathcal{E}) \longrightarrow H^2(\mathcal{F})$ intertwining S and U is the multiplication by a bounded analytic function $\Theta \in H^\infty(\mathcal{E},\mathcal{F})$.

Let $\Theta \in H^\infty(\mathcal{E},\mathcal{F})$ be a contractive analytic function. As in [14], we say that Θ is *purely contractive* if $\|\Theta(0)e\| < \|e\|$ for $e \in \mathcal{E}, e \neq 0$. It is a standard result that if Θ is a contractive analytic function, then there is a unique decomposition $\mathcal{E} = \mathcal{E}_p \oplus \mathcal{E}_u$, $\mathcal{F} = \mathcal{F}_p \oplus \mathcal{F}_u$ and $\Theta(z) = \Theta_p(z) \oplus U$ such that U is a unitary constant mapping $\mathcal{E}_u$ onto $\mathcal{F}_u$ and $\Theta_p(z)$ is a purely contractive analytic function mapping $\mathcal{E}_p$ into $\mathcal{F}_p$. Θ_p is called the *purely contractive part* of Θ. Now for $\Theta \in H^\infty(\mathcal{E},\mathcal{F})$ a purely contractive analytic function, we recall the Sz.-Nagy–Foias functional model associated to Θ, namely the operator T defined by

$$T(u \oplus v) = P_{\mathcal{H}}(e^{it}u(e^{it}) \oplus e^{it}v(t)) \quad \text{for} \quad u \oplus v \in \mathcal{H} \tag{2}$$

on the space

$$\mathcal{H} = \mathcal{K} \ominus \mathcal{G} \quad \text{where} \quad \mathcal{K} = H^2(\mathcal{F}) \oplus \overline{\Delta L^2(\mathcal{E})},$$

$$\mathcal{G} = \{\Theta(e^{it})w(e^{it}) \oplus \Delta(t)w(e^{it}); w \in H^2(\mathcal{E})\}$$

and $\Delta(t) = [I - \Theta^*(e^{it})\Theta(e^{it})]^{1/2}$. It is known that the operator T as above is a completely non-unitary contraction and its minimal isometric dilation is the operator S of multiplication by e^{it} on $\mathcal{K}$.

The Sz.-Nagy–Foias functional model theory states that the operator T defined by (2) is a model for *any* completely non-unitary contraction. To be more precise, if $T \in B(\mathcal{H})$ is a completely non-unitary contraction, then the function $\Theta_T \in H^\infty(\mathcal{D}_T, \mathcal{D}_{T^*})$,

$$\Theta_T(z) = [-T + zD_{T^*}(I - zT^*)^{-1}D_T]|\mathcal{D}_T,$$

called the *characteristic function* of T, is a purely contractive analytic function (here $D_T = (I - T^*T)^{1/2}$, $D_{T^*} = (I - TT^*)^{1/2}$, $\mathcal{D}_T = \overline{D_T\mathcal{H}}$ and $\mathcal{D}_{T^*} = \overline{D_{T^*}\mathcal{H}}$); moreover the Sz.-Nagy–Foias functional model associated with $\Theta_T \in H^\infty(\mathcal{D}_T, \mathcal{D}_{T^*})$ is unitary equivalent to the initial contraction T on $\mathcal{H}$. Now we can state our main theorem as follows:

MAIN THEOREM. *Let $T' \in B(\mathcal{K}')$ and $T \in B(\mathcal{H})$ be two contractions, $\mathcal{G}' \subset \mathcal{K}'$ an invariant subspace of T', T_0' and T_1' the restriction of T' to $\mathcal{G}'$ respectively the compression of T' to $\mathcal{H}' = \mathcal{K}' \ominus \mathcal{G}'$ (i.e. $T_0' = T'|\mathcal{G}'$ and $T_1' = P_{\mathcal{H}'}T'|\mathcal{H}'$).*

(1) *If the characteristic functions of T_1' and T satisfy* (TSCP), *then any operator $A \in B(\mathcal{G}', \mathcal{H})$ intertwining T and T_0' has an intertwining extension.*

(2) *If any operator* $A \in B(\mathcal{G}', \mathcal{H})$ *intertwining* T *and* T_0' *has an intertwining extension and the operator* T' *is an isometry then the characteristic functions of* T_1' *and* T *satisfy* (TSCP).

Let us outline the plan of this paper. In Section 2 we show that (TSCP) and (GIEP) can be reduced to special cases which are amenable to the functional model theory. Section 3 is dedicated to the connection between (TSCP) and (GIEP). Let A, T_1' and T be given as in (GIEP); we associate operators A, T_1' and T with bounded analytic functions Θ, Θ_1 and Θ_2, and prove that the existence of an intertwining extension B of A is equivalent to the existence of bounded analytic functions Ψ and Φ satisfying (1). When the operator A has an intertwining extension we obtain also estimations for $\mu(A, T, T', \mathcal{G}')$ as function of $\|\Psi\|_\infty$ and $\|\Phi\|_\infty$. In Section 4 we use the results from previous sections to obtain conditions for (TSCP) to be solvable.

Acknowledgment. We thank professor Ciprian Foias for initiating our collaboration and for his support during the preparation of this paper.

2. Reduction of (TSCP) and (GIEP)

In order to use the Sz.-Nagy–Foias functional models for contractions we shall first reduce both (TSCP) and (GIEP) to the following special cases:

(1) (TSCP) to the situation when both functions $\Theta_1 \in H^\infty(\mathcal{E}_1, \mathcal{F}_1)$ and $\Theta_2 \in H^\infty(\mathcal{E}_2, \mathcal{F}_2)$ are *pure* analytic contractive functions;

(2) (GIEP) to the situation when T' is an isometry and, if $\mathcal{G}' \subset \mathcal{K}'$ is the invariant subspace to T', then T' is exactly the minimal isometric dilation of the compression T_1' of T' to $\mathcal{H}' = \mathcal{K}' \ominus \mathcal{G}'$, and T_1' as well as T are completely non-unitary contractions.

Let us start with the following

Proposition 1. *In* (TSCP) *without loss of generality we can assume that both* $\Theta_1 \in H^\infty(\mathcal{E}_1, \mathcal{F}_1)$ *and* $\Theta_2 \in H^\infty(\mathcal{E}_2, \mathcal{F}_2)$ *are pure contractive analytic functions.*

Proof. To see this, let

$$\Theta_1 = \begin{bmatrix} \Theta_{1p} & 0 \\ 0 & U_1 \end{bmatrix} : \mathcal{E}_{1p} \oplus \mathcal{E}_{1u} \to \mathcal{F}_{1p} \oplus \mathcal{F}_{1u}$$

and

$$\Theta_2 = \begin{bmatrix} \Theta_{2p} & 0 \\ 0 & U_2 \end{bmatrix} : \mathcal{E}_{2p} \oplus \mathcal{E}_{2u} \to \mathcal{F}_{2p} \oplus \mathcal{F}_{2u}$$

be the decompositions of Θ_1 and Θ_2 into their purely contractive and unitary parts respectively. Let us consider the decomposition

$$\Theta = \begin{bmatrix} \Theta_{11} & \Theta_{12} \\ \Theta_{21} & \Theta_{22} \end{bmatrix} : \mathcal{E}_{1p} \oplus \mathcal{E}_{1u} \to \mathcal{F}_{2p} \oplus \mathcal{F}_{2u}$$

and similar decompositions for Ψ and Φ with respect to the corresponding direct sums.

Then equation (1) becomes

$$\begin{bmatrix} \Phi_1 & \Phi_2 \\ \Phi_3 & \Phi_4 \end{bmatrix} \begin{bmatrix} \Theta_{1p} & 0 \\ 0 & U_1 \end{bmatrix} + \begin{bmatrix} \Theta_{2p} & 0 \\ 0 & U_2 \end{bmatrix} \begin{bmatrix} \Psi_1 & \Psi_2 \\ \Psi_3 & \Psi_4 \end{bmatrix} = \begin{bmatrix} \Theta_1 & \Theta_2 \\ \Theta_3 & \Theta_4 \end{bmatrix},$$

equivalently

$$\begin{aligned} \Phi_1\Theta_{1p} + \Theta_{2p}\Psi_1 &= \Theta_{11} \\ \Phi_2 U_1 + \Theta_{2p}\Psi_2 &= \Theta_{12} \\ \Phi_3\Theta_{1p} + U_2\Psi_3 &= \Theta_{21} \\ \Phi_4 U_1 \ \ + U_2\Psi_4 &= \Theta_{22}. \end{aligned}$$

The second, the third, and the fourth equations of this system are always solvable. For example, to solve the second equation we can take $\Psi_2 = 0$ and $\Phi_2 = \Theta_{12}U_2^*$. Thus the equation (1) has a solution if and only if the first equation of our system has a solution; so we can assume that Θ_1 and Θ_2 in (TSCP) are purely contractive analytic functions.

□

Concerning (GIEP), we shall prove the following

PROPOSITION 2. (GIEP) *is equivalent to the situation when* T' *is an isometry and, if* $\mathcal{G}' \subset \mathcal{K}'$ *is the invariant subspace to* T', *then* T' *is exactly the minimal isometric dilation of the compression* T_1' *of* T' *to* $\mathcal{H}' = \mathcal{K}' \ominus \mathcal{G}'$, *and* T_1' *as well as* T *are completely non-unitary contractions.*

Proof. The reduction of (GIEP) will be done in four steps:

Step 1. *In* (GIEP) *we can assume that the contraction T' is an isometry* (see [5] and [6]).

Indeed if U' on $\mathcal{M}$ is the minimal isometric dilation of T', let us set

$$\mathcal{K}_0' = (\mathcal{M} \ominus \mathcal{K}') \oplus \mathcal{G}', \; U_0' = U'|\mathcal{K}_0' \text{ and } A' = AP_{\mathcal{G}'}|\mathcal{K}_0'.$$

It is easy to see that $\mathcal{K}_0'$ is invariant to U_0', $TA' = A'U_0'$ and $\mu(A,T,T',\mathcal{G}') = \mu(A',T,U',\mathcal{K}_0')$.

So from now on we will asumme that $T' \in B(\mathcal{K}')$ is an isometry. Set

$$\mathcal{H}' = \mathcal{K}' \ominus \mathcal{G}', \quad T_1' = P_{\mathcal{H}'}T'|\mathcal{H}'. \tag{3}$$

Step 2. *We can assume T_1' is a completely non-unitary contraction.*

To prove this, let $T_1' = U \oplus T''$ on $\mathcal{H} = \mathcal{H}_u' \oplus \mathcal{H}_p'$ be the decomposition of T_1' into its unitary and completely non-unitary parts. It is easy to see that $\mathcal{K}' \ominus \mathcal{H}_u' = \mathcal{H}_p' \oplus \mathcal{G}'$ is a reducing subspace for T'. We will show that A has an intertwining extension to the whole $\mathcal{K}'$ if and only if it has an intertwining extension to $\mathcal{K}' \ominus \mathcal{H}_u'$. Indeed if B on $\mathcal{K}'$ is an intertwining extension of A then $B_1 = B|\mathcal{K}' \ominus \mathcal{G}_u'$ is an extension of A to $\mathcal{K}' \ominus \mathcal{H}_u'$, and $TB_1 = B_1T'|(\mathcal{K}' \ominus \mathcal{G}_u')$. Conversely, if B_1 on $\mathcal{K}' \ominus \mathcal{G}_u'$ is an intertwining extension of A, then $B = B_1 P_{\mathcal{K}' \ominus \mathcal{G}_u'}$ is an extension of A to the whole $\mathcal{K}'$ such that $TB = BT'$.

So from now on we will assume that $T' \in B(\mathcal{K}')$ is an isometry with $\mathcal{G}' \subset \mathcal{K}'$ an invariant subspace for T'. Moreover if we set $\mathcal{H}' = \mathcal{K}' \ominus \mathcal{G}'$, then the operator $T_1' = P_{\mathcal{H}'}T'|\mathcal{H}'$ is a completely non-unitary contraction.

Step 3. *We can assume that T' is a minimal isometric dilation of T_1'.*

Let us consider

$$\mathcal{G}'' = \bigvee_{n=0}^{\infty} T'^n \mathcal{H}, \quad T'' = T'|\mathcal{G}''.$$

It is known that T'' is a minimal isometric dilation of T_1'. Set

$$\mathcal{G}_p = \mathcal{G}'' \ominus \mathcal{H}', \quad \mathcal{G}_u = \mathcal{K}' \ominus \mathcal{G}''.$$

It is easy to see that T' takes the following form:

$$T' = \begin{bmatrix} T_1' & 0 & 0 \\ T_2' & T_p' & 0 \\ 0 & 0 & T_u' \end{bmatrix} : \mathcal{H}' \in \mathcal{G}_p \oplus \mathcal{G}_u \longmapsto \mathcal{H}' \oplus \mathcal{G}_p \oplus \mathcal{G}_u$$

where T_p' and T_u' are isometries. The operators T_0' and T'' become

$$T_0' = \begin{bmatrix} T_p' & 0 \\ 0 & T_u' \end{bmatrix}, \quad T'' = \begin{bmatrix} T_1' & 0 \\ T_2' & T_p' \end{bmatrix}.$$

Let us write $A = [A_p, A_u]$ from $\mathcal{G}' = \mathcal{G}_p \oplus \mathcal{G}_u$ into $\mathcal{H}$. Then the relationship $TA = AT_0'$ becomes $TA_p = A_pT_p'$ and $TA_u = A_uT_u'$. We see that B is an extension of A if and only if $B_1 = B|\mathcal{H}' \oplus \mathcal{G}_p$ is an extension of A_p, i.e. $TB_1 = B_1T''$.

Step 4. We can assume that T is a completely non-unitary contraction.

To prove this, let $T = U \oplus T_p$ on $\mathcal{H} = \mathcal{H}_u \oplus \mathcal{H}_p$ be the decomposition of T into its unitary and completely non-unitary parts. We shall write the operators A from $\mathcal{G}$ into $\mathcal{H}_u \oplus \mathcal{H}_p$ and B from $\mathcal{K}$ into $\mathcal{H}_u \oplus \mathcal{H}_p$ as column operators:

$$A = \begin{bmatrix} A_u \\ A_p \end{bmatrix}, \quad B = \begin{bmatrix} B_u \\ B_p \end{bmatrix}.$$

With this notation the intertwining property

$$TA = AT_0' \text{ becomes } UA_u = A_uT_0' \text{ and } T_pA_p = A_pT_0',$$

and similarly

$$TB = BT' \text{ becomes } UB_u = B_uT' \text{ and } T_pB_p = B_pT'.$$

Therefore B is an intertwining extension of A if and only if B_p is an intertwining extension of A_p and B_u is an intertwining extension of A_u. Since U is a unitary operator, there always exists an intertwining extension B_u of A_u with $\|B_u\| = \|A_u\|$. (See [3], [4], [7, p.207, Theorem 3.5], or a more general version in [9, Theorem 3].) So it follows that A has an intertwining extension if and only if its completely non-unitary part A_p has an intertwining extension. But in $T_pA_p = A_pT_0'$ the factor T_p is a completely non-unitary contraction and so the existence of an intertwining extension of A should be proved only in the situation when T is a completely non-unitary contraction.

□

From the above analysis we conclude that there will be no significant loss of generality in considering the special cases of (GIEP) where both T and T_1' are completely non-unitary contractions and T' is the minimal isometric dilation of T_1'. This puts us exactly in the situation where we can replace T and T_1' by their functional models. For notational convienence we will write T_1 instead of T_1' and will assume that the characteristic function of T_1 is $\Theta_1 \in H^\infty(\mathcal{E}_1, \mathcal{F}_1)$. Thus the operator T_1 is given by

$$T_1(u \oplus v) = P_{\mathcal{H}_1}(e^{it}u(e^{it}) \oplus e^{it}v(t)) \text{ for } u \oplus v \in \mathcal{H}_1 \tag{4}$$

on the space

$$\mathcal{H}_1 = \mathcal{K}_1 \ominus \mathcal{G}_1 \quad \text{where} \quad \mathcal{K}_1 = H^2(\mathcal{F}_1) \oplus \overline{\Delta_1 L^2(\mathcal{E}_1)},$$

$$\mathcal{G}_1 = \{\Theta_1(e^{it})w(e^{it}) \oplus \Delta_1(t)w(e^{it}); w \in H^2(\mathcal{E}_1)\}$$

and $\Delta_1(t) = [I - \Theta_1^*(e^{it})\Theta_1(e^{it})]^{1/2}$. The minimal isometric dilation of T_1 is the operator S (previously called T') of multiplication by e^{it} on $\mathcal{K}_1$ and the invariant subspace to S is $\mathcal{G}_1$.

Similarly, we write T_2 instead of T and assume that $\Theta_2 \in H^\infty(\mathcal{E}_2, \mathcal{F}_2)$ is the characteristic function of T_2 which has the representation

$$T_2(u \oplus v) = P_{\mathcal{H}_2}(e^{it}u(e^{it}) \oplus e^{it}v(t)) \quad \text{for} \quad u \oplus v \in \mathcal{H}_2 \tag{5}$$

on the space

$$\mathcal{H}_2 = \mathcal{K}_2 \ominus \mathcal{G}_2 \quad \text{where} \quad \mathcal{K}_2 = H^2(\mathcal{F}_2) \oplus \overline{\Delta_2 L^2(\mathcal{E}_2)},$$

$$\mathcal{G}_2 = \{\Theta_2(e^{it})w(e^{it}) \oplus \Delta_2(t)w(e^{it}); w \in H^2(\mathcal{E}_2)\}$$

and $\Delta_2(t) = [I - \Theta_2^*(e^{it})\Theta_2(e^{it})]^{1/2}$. The minimal isometric dilation of T_2 is denoted by U, the multiplication by e^{it} on $\mathcal{K}_2$.

In this way we have obtained the following *reduced* (GIEP):

Let $T_i \in B(\mathcal{H}_i), i = 1, 2$ be defined by (4) and (5), S the multiplication by e^{it} on $\mathcal{K}_1$, $\mathcal{G}_1(\subset \mathcal{K}_1)$ an invariant subspace for S and S_1 the restriction of S to $\mathcal{G}_1$. Let also $A \in B(\mathcal{G}_1, \mathcal{H}_2)$ be an operator which intertwines S_1 with T_2, i.e. $T_2 A = AS_1$. Is there an operator $B \in B(\mathcal{K}_1, \mathcal{H}_2)$ such that $T_2 B = BS$ and $A = B|\mathcal{G}_1$?

Let us conclude this section by noticing the following: the (GIEP) and the reduced (GIEP) are equivalent in the sense that in (GIEP) the operator $A \in B(\mathcal{G}', \mathcal{H})$ has an intertwining extension if and only if in the reduced (GIEP) the corresponding operator $A \in B(\mathcal{G}_1, \mathcal{H}_2)$ has an intertwining extension. However the characteristic functions of $T_i \in B(\mathcal{H}_i)$, $i = 1, 2$ in the reduced (GIEP) may satisfy (TSCP), while the characteristic functions of $T \in B(\mathcal{H})$ and $T_1' = P_{\mathcal{H}'}T'|\mathcal{H}'$ in (GIEP) may not satisfy (TSCP). This can happen because passing from $T_1' = P_{\mathcal{H}'}T'|\mathcal{H}'$ in (GIEP), to $T_1 \in B(\mathcal{H}_1)$ in the reduced (GIEP), the characteristic function is changing.

3. The Connection Between (TSCP) and (GIEP)

The central result of this section is the following fact: in the reduced (GIEP) any operator $A \in B(\mathcal{G}_1, \mathcal{H}_2)$ intertwining S_1 and T_2 has an intertwining extension if and only if the characteristic function of T_1 and T_2 satisfy (TSCP). The main ingredients of the proof are Proposition 3 and Theorem 4 below. First we shall prove Proposition 3 showing that the solvability of (GIEP) is equivalent to the solvability of a commutator equation.

Recall that S is the multiplication by e^{it} on $\mathcal{K}_1$ and let us set

$$S_1' = P_{\mathcal{G}_1} S|\mathcal{H}_1.$$

With these notations the operator S can be expressed as

$$S = \begin{bmatrix} T_1 & 0 \\ S_1' & S_1 \end{bmatrix} : \mathcal{K}_1 = \mathcal{H}_1 \oplus \mathcal{G}_1 \to \mathcal{H}_1 \oplus \mathcal{G}_1.$$

PROPOSITION 3. *Let $A \in B(\mathcal{G}_1, \mathcal{H}_2)$ be an operator such that $T_2 A = A S_1$ and $A_0 \in B(\mathcal{H}_1, \mathcal{H}_2)$ given by $A_0 = A S_1'$. The operator A has an intertwining extension (i.e. $\mu(A, T_2, S, \mathcal{G}_1) < \infty$) if and only if the equation*

$$T_2 B_1 - B_1 T_1 = A_0 \tag{6}$$

has at least one solution $B_1 \in B(\mathcal{H}_1, \mathcal{H}_2)$. Moreover, if this is the case then

$$\begin{aligned} \mu(A; T_2, S; \mathcal{G}_1) : &= \inf\{\|B\| : \ B \text{ is an extension of } A \ \} \\ &= \inf\{\|[B_1 \ A]\| : \ B_1 \text{ is a solution of } (6) \ \}. \end{aligned}$$

Proof. Let $B \in B(\mathcal{K}_1, \mathcal{H}_2)$. B is an intertwining extension of A if and only if $B = [B_1, A]$ for some $B_1 \in B(\mathcal{H}_1, \mathcal{H}_2)$ and $BS = T_2 B$. That is

$$[B_1 \ A] \begin{bmatrix} T_1 & 0 \\ S_1' & S_1 \end{bmatrix} = T_2 [B_1 \ A]$$

which is equivalent to $T_2 B_1 - B_1 T_1 = A S_1' = A_0$.

□

Now we are ready for the following connection between (TSCP) and the commutator equation (6) in Proposition 3 above.

THEOREM 4. *Let $\Theta_1 \in H^\infty(\mathcal{E}_1, \mathcal{F}_1)$ and $\Theta_2 \in H^\infty(\mathcal{E}_2, \mathcal{F}_2)$ be two purely contractive analytic functions and $T_i \in B(\mathcal{H}_i)$, $i = 1, 2$, their associated model operators defined by (4) and (5). For any bounded analytic function $\Theta \in H^\infty(\mathcal{E}_1, \mathcal{F}_2)$ and bounded measurable function $\Lambda(t)\colon L^2(\mathcal{E}_1) \to \overline{\Delta_2 L^2(\mathcal{E}_2)}$, we define $A \in B(\mathcal{G}_1, \mathcal{H}_2)$ by*

$$A(\Theta_1 f \oplus \Delta_1 f) = P_{\mathcal{H}_2}(\Theta f \oplus \Lambda f), \quad f \in H^2(\mathcal{E}_1) \tag{7}$$

and $A_0 = AS_1'$. *There exist bounded analytic functions* $\Phi \in H^\infty(\mathcal{F}_1, \mathcal{F}_2)$ *and* $\Psi \in H^\infty(\mathcal{E}_1, \mathcal{E}_2)$ *such that*

$$\Phi(z)\Theta_1(z) + \Theta_2(z)\Psi(z) = \Theta(z) \quad \textit{for } z \in D \tag{8}$$

if and only if there exists an operator $B_1 \in B(\mathcal{H}_1, \mathcal{H}_2)$ *such that*

$$T_2 B_1 - B_1 T_1 = A_0. \tag{9}$$

Proof. We note that the operator A given by (7) satisfies $AS_1 = T_2 A$. Let us assume that (8) holds, and let us consider the operator $\tilde{B}: \mathcal{K}_1 \to \mathcal{K}_2$ given by multiplication by the matrix

$$\tilde{B} = \begin{bmatrix} \Phi & 0 \\ \Lambda\Theta_1^* - \Delta_2\Psi\Theta_1^* & \Lambda\Delta_1 - \Delta_2\Psi\Delta_1 \end{bmatrix} \tag{10}$$

and set $B = P_{\mathcal{H}_2}\tilde{B}$. We claim that B is an extension of A. To prove this we first prove that $BS = T_2 B$. Indeed it is obvious that $\tilde{B}S = U\tilde{B}$ (U is the multiplication by e^{it} on $\mathcal{K}_2$) and therefore

$$\begin{aligned} BS = P_{\mathcal{H}_2}\tilde{B}S = P_{\mathcal{H}_2}U\tilde{B} = P_{\mathcal{H}_2}UP_{\mathcal{H}_2}\tilde{B} + P_{\mathcal{H}_2}U(I - P_{\mathcal{H}_2})\tilde{B} = \\ = P_{\mathcal{H}_2}UP_{\mathcal{H}_2}\tilde{B} = T_2 B. \end{aligned}$$

Now to see that $B|\mathcal{G}_1 = A$, note that for $f \in H^2(\mathcal{E}_1)$

$$\begin{aligned} B(\Theta_1 f \oplus \Delta_1 f) &= P_{\mathcal{H}_2}(\Phi\Theta_1 f \oplus (\Lambda\Theta_1^*\Theta_1 f - \Delta_2\Psi\Theta_1^*\Theta_1 f + \Lambda\Delta_1^2 f - \Delta_2\Psi\Delta_1^2 f)) \\ &= P_{\mathcal{H}_2}(\Phi\Theta_1 f \oplus (\Lambda f - \Delta_2\Psi f)) \\ &= P_{\mathcal{H}_2}((\Theta f - \Theta_2\Psi f) \oplus (\Lambda f - \Delta_2\Psi f)) \\ &= P_{\mathcal{H}_2}(\Theta f \oplus \Lambda f - \Theta_1\Psi f \oplus \Delta_1\Psi f) \\ &= P_{\mathcal{H}_2}(\Theta f \oplus \Lambda f) \\ &= A(\Theta_1 f \oplus \Delta_1 f), \end{aligned}$$

where the third equality follows from (8). Setting $B_1 = B|\mathcal{H}_1$, it is clear by Proposition 1 that B_1 satisfies equation (9).

In order to prove the second half of the theorem, let us suppose that there is an operator $B_1: \mathcal{H}_1 \to \mathcal{H}_2$ satisfying $T_2 B_1 - B_1 T_1 = A_0$ and define the operator $B: \mathcal{K}_1 \to \mathcal{H}_2$ by $B = [B_1, A]$. By Proposition 1, it is clear that B is an intertwining extension of A; hence $BS = T_2 B$. By the commutant lifting theorem there is an operator

$$\tilde{B}: H^2(\mathcal{F}_1) \oplus \overline{\Delta_1 L^2(\mathcal{E}_1)} \to H^2(\mathcal{F}_2) \oplus \overline{\Delta_2 L^2(\mathcal{F}_2)}$$

such that $\tilde{B}S = U\tilde{B}$, $B = P_{\mathcal{H}_2}\tilde{B}$, and $\|B\| = \|\tilde{B}\|$. The operator $\tilde{B}$ has to have the following structure:

$$\tilde{B} = \begin{bmatrix} C_{11} & C_{12} \\ C_{21} & C_{22} \end{bmatrix}$$

where

$$C_{11}: H^2(\mathcal{F}_1) \to H^2(\mathcal{F}_2), \quad C_{12}: \overline{\Delta_1 L^2(\mathcal{E}_1)} \to H^2(\mathcal{F}_2)$$
$$C_{21}: H^2(\mathcal{F}_1) \to \overline{\Delta_2 L^2(\mathcal{E}_2)}, \text{ and } C_{22}: \overline{\Delta_2 L^2(\mathcal{E}_1)} \to \overline{\Delta_2 L^2(\mathcal{E}_2)}.$$

The operators C_{ik}, $(i,k = 1,2)$ commute with the multiplication by e^{it} on appropriate spaces and so $C_{12} = 0$ and C_{11} is M_Φ, multiplication by some bounded analytic function $\Phi \in H^\infty(\mathcal{F}_1, \mathcal{F}_2)$. Thus for the moment

$$\tilde{B} = \begin{bmatrix} M_\Phi & 0 \\ C_{21} & C_{22} \end{bmatrix}. \tag{11}$$

Now if $f \in H^2(\mathcal{E}_1)$ then $\Theta_1 f \oplus \Delta_1 f \in \mathcal{G}_1$ and

$$P_{\mathcal{H}_2}\tilde{B}(\Theta_1 f \oplus \Delta_1 f) = P_{\mathcal{H}_2}[\Phi\Theta_1 f \oplus (C_{21}\Theta_1 f + C_{22}\Delta_1 f)].$$

On the other hand,

$$P_{\mathcal{H}_2}\tilde{B}(\Theta_1 f \oplus \Delta_1 f) = B(\Theta_1 f \oplus \Delta_1 f) = A(\Theta_1 f \oplus \Delta_1 f) = P_{\mathcal{H}_2}(\Theta f \oplus \Lambda f).$$

Therefore $P_{\mathcal{H}_2}[(\Phi\Theta_1 f - \Theta f) \oplus (C_{21}\Theta_1 f + C_{22}\Delta_1 f - \Lambda f)] = 0$ and hence

$$\Phi\Theta_1 f - \Theta f = \Theta_2 w \tag{12}$$

$$C_{21}\Theta_1 f + C_{22}\Delta_1 f - \Lambda f = \Delta_2 w. \tag{13}$$

The application $f \longmapsto w$ of $H^2(\mathcal{E}_1)$ into $H^2(\mathcal{E}_2)$ is linear, continuous and commutes with multiplication by e^{it}, so there is a bounded analytic function $\Psi \in H^\infty(\mathcal{E}_1, \mathcal{E}_2)$ such that $w = M_{-\Psi} f = -\Psi f$. The first relation in the previous system becomes $\Phi\Theta_1 f - \Theta f = -\Theta_2 \Psi f$ and hence $\Phi\Theta_2 + \Theta_1\Psi = \Theta$. The proof of Theorem 4 is now complete.

□

Remark. If in the previous theorem the equation $T_2 B_1 - B_1 T_1 = A_0$ has a solution for an operator $A_0 = AS_1'$ where A is given by

$$A(\Theta_1 f \oplus \Delta_1 f) = P_{\mathcal{H}_2}(\Theta f \oplus \Lambda_0 f), \quad f \in H^2(\mathcal{E}_1)$$

for a particular measurable bounded function $\Lambda_0(t): L^2(\mathcal{E}_1) \to \overline{\Delta_2 L^2(\mathcal{E}_2)}$, then the equation $T_2 B_1 - B_1 T_1 = A_0$ has a solution for any measurable bounded function $\Lambda(t)$ in the second component of the operator A.

To see this, it is enough to notice that the equation (9) is equivalent to (8) which does not depend on $\Lambda(t)$.

Corollary 5. *Let $T_1 \in B(\mathcal{H}_1)$ and $T_2 \in B(\mathcal{H}_2)$ be as in the previous theorem. There is an operator $B_1 \in B(\mathcal{H}_1, \mathcal{H}_2)$ intertwining T_2 and T_1, i.e. $T_2B_1 = B_1T_1$, if and only if there are bounded analytic functions $\Phi \in H^\infty(\mathcal{F}_1, \mathcal{F}_2)$ and $\Psi \in H^\infty(\mathcal{E}_1, \mathcal{E}_2)$ such that $\Phi(z)\Theta_1(z) = \Theta_2(z)\Psi(z)$ for any $z \in D$.*

Proof. It is enough to make $\Theta(z) = 0$ and $\Lambda(t) = 0$ in the previous theorem. Notice that in this situation the operator A_0 is zero, therefore equation (9) becomes $B_1T_1 = T_2B_1$, and equation (8) becomes $\Phi(z)\Theta_1(z) + \Theta_2(z)\Psi(z) = 0$. The proof will be completed by replacing Ψ with $-\Psi$.

□

Next we study the structure of the operator $A \in B(\mathcal{G}_1, \mathcal{H}_2)$ such that $T_2A = AS_1$, which appears in (GIEP).

Theorem 6. *Let $A \in B(\mathcal{G}_1, \mathcal{H}_2)$ be such that $T_2A = AS_1$. Then there exists a bounded analytic function $\Theta_A \in H^\infty(\mathcal{E}_1, \mathcal{F}_2)$ and a bounded measurable function $\Lambda_A(t): L^2(\mathcal{E}_1) \to \overline{\Delta_2 L^2(\mathcal{E}_2)}$ such that the operator A has the representation*

$$A(\Theta_1 f \oplus \Delta_1 f) = P_{\mathcal{H}_2}(\Theta_A f \oplus \Lambda_A f), \quad f \in H^2(\mathcal{E}_1); \tag{14}$$

moreover $\|A\| = \left\| \begin{bmatrix} \Theta_A \\ \Lambda_A \end{bmatrix} \right\|_\infty$.

Proof. By the commutant lifting theorem there exists an intertwining lifting B of A; that is, there exists an operator $B: \mathcal{G}_1 \to \mathcal{K}_2$ such that $BS_1 = UB$, $A = P_{\mathcal{H}_2}B$ and $\|B\| = \|A\|$. Let us set

$$B(\Theta_1 f \oplus \Delta_1 f) = u \oplus v, \quad u \oplus v \in H^2(\mathcal{F}_2) \oplus \overline{\Delta_2 L^2(\mathcal{E}_2)}.$$

Since S_1 and U are multiplication by e^{it} on $\mathcal{G}_1$ and $\mathcal{K}_2$ respectively, it is easy to see that the operators $f \longmapsto u$ and $f \longmapsto v$ are linear, continuous and commute with multiplication by e^{it}. Therefore there exists a bounded analytic function $\Theta_A \in H^\infty(\mathcal{E}_1, \mathcal{F}_2)$ and a bounded measurable function $\Lambda_A(t): L^2(\mathcal{E}_1) \to \overline{\Delta_2 L^2(\mathcal{E}_2)}$ such that

$$B(\Theta_1 f \oplus \Delta_1 f) = \Theta_A f \oplus \Lambda_A f, \quad f \in H^2(\mathcal{E}_1).$$

The representation formula (6) is a consequence of $A = P_{\mathcal{H}_2}B$. It is also clear that

$$\|A\| = \|B\| = \left\| \begin{bmatrix} \Theta_A \\ \Lambda_A \end{bmatrix} \right\|_\infty.$$

The proof of the theorem is now complete.

□

Before going any further we mention that the pair $\{\Theta_A, \Lambda_A\}$ corresponding to the operator A is far from being unique. Indeed, for any bounded analytic function $F \in H^\infty(\mathcal{E}_1, \mathcal{E}_2)$ the pair given by

$$\Theta = \Theta_A + \Theta_2 F$$
$$\Lambda = \Lambda_A + \Delta_2 F$$

will generate the same operator A.

Combining Proposition 3, Theorem 4 and Theorem 6, the equivalence of (TSCP) and (GIEP) can be stated as follows:

THEOREM 7. *Let $\Theta_1 \in H^\infty(\mathcal{E}_1, \mathcal{F}_1)$ and $\Theta_2 \in H^\infty(\mathcal{E}_2, \mathcal{F}_2)$ be purely contractive analytic functions and $T_i \in B(\mathcal{H}_i)$, $i = 1, 2$ their functional models.*

1. *For any bounded analytic function $\Theta \in H^\infty(\mathcal{E}_1, \mathcal{F}_2)$, we define the operator $A_\Theta \in B(\mathcal{G}_1, \mathcal{H}_2)$ by*

$$A_\Theta(\Theta_1 f \oplus \Delta_1 f) = P_{\mathcal{H}_2}(\Theta f \oplus 0), \quad f \in H^2(\mathcal{E}_1).$$

There exist bounded analytic functions $\Phi \in H^\infty(\mathcal{F}_1, \mathcal{F}_2)$ and $\Psi \in H^\infty(\mathcal{E}_1, \mathcal{E}_2)$ such that

$$\Phi(z)\Theta_1(z) + \Theta_2(z)\Psi(z) = \Theta(z) \;\; \textit{for} \;\; z \in D$$

if and only if A_Θ has an intertwining extension.

2. *For any operator $A \in B(\mathcal{G}_1, \mathcal{H}_2)$ such that $T_2 A = A S_1$, let $\Theta_A \in H^\infty(\mathcal{E}_1, \mathcal{F}_2)$ be the analytic component from a representation of A as in Theorem 6. The operator A has an intertwining extension if and only if there exist bounded analytic functions $\Phi \in H^\infty(\mathcal{F}_1, \mathcal{F}_2)$ and $\Psi \in H^\infty(\mathcal{E}_1, \mathcal{E}_2)$ such that*

$$\Phi(z)\Theta_1(z) + \Theta_2(z)\Psi(z) = \Theta_A(z) \;\; \textit{for all} \;\; z \in D. \tag{15}$$

Proof. For the first part of the theorem it is enough to combine Proposition 3 and Theorem 4, and for the second part use Theorem 6 and then Proposition 3 and Theorem 4.

□

The two parts of the previous theorem can be combined into the following

Corollary 8. *In the reduced General Intertwining Extension Problem any operator $A \in B(\mathcal{G}_1, \mathcal{H}_2)$ intertwining S_1 and T_2 has an intertwining extension if and only if the characteristic functions of T_1 and T_2 satisfy the Two-Sided Corona Problem.*

The Main Theorem in the Introduction is now a consequence of the previous theorem and the Proposition 2.

Before ending this section we mention that according to a previous remark the operator A given by (6) has an intertwining extension if and only if its analytic component satisfies (15). If A has an intertwining extension, its measurable component is important in the size of $\mu(A, T_2, S, \mathcal{G}_1)$. We end this section with a result concerning estimations of $\mu(A, T_2, S, \mathcal{G}_1)$. Namely we shall prove the following result.

Theorem 9. *Let $A \in B(\mathcal{G}_1, \mathcal{H}_2)$ be an operator such that $T_2 A = A S_1$ and let us consider*

$$A(\Theta_1 f \oplus \Delta_1 f) = P_{\mathcal{H}_1}(\Theta_A f \oplus \Lambda_A f) \tag{16}$$

with $\|A\| = \left\| \begin{bmatrix} \Theta_A \\ \Lambda_A \end{bmatrix} \right\|_\infty$, a representation of the operator A given by Theorem 6. Let us also suppose that the analytic component $\Theta_A \in H^\infty(\mathcal{E}_1, \mathcal{F}_2)$ satisfies (15). If we set

$$\nu(\Theta_A, \Theta_1, \Theta_2) = \inf \left\{ \sqrt{\|\Phi\|_\infty^2 + \|\Psi\|_\infty^2}, \quad \Phi,\ \Psi \text{ satisfying } (15) \right\},$$

then the following inequalities hold:

$$\begin{aligned} &\inf \left\{ \max \left\{ \|\Phi\|_\infty, \frac{\|\Psi\|_\infty}{2}, \|\Lambda_A\|_\infty \right\},\ \Phi,\Psi \text{ satisfying } (15) \right\} \\ &\qquad \le \mu(A, T_2, S, \mathcal{G}_1) \\ &\qquad \le \inf \left\{ \sqrt{\|\Phi\|_\infty^2 + \|\Lambda_A - \Delta_2 \Psi\|_\infty^2},\ \Psi,\Psi \text{ satisfying } (15) \right\} \end{aligned} \tag{17}$$

and also

$$\begin{aligned} &\frac{1}{\sqrt{12}}(\nu(\Theta_A, \Theta_1, \Theta_2) + \|\Lambda_A\|_\infty) \le \mu(A, T_2, S, \mathcal{G}_1) \\ &\qquad\qquad \le \sqrt{2}(\nu(\Theta_A, \Theta_1, \Theta_2) + \|\Lambda_A\|_\infty). \end{aligned} \tag{18}$$

Proof. The first inequality in (17) can be proven in the following way. Let $B \in B(\mathcal{K}_1, \mathcal{H}_2)$ be an intertwining extension of A. Following the procedure from Theorem 4 we get $\tilde{B} \in B(\mathcal{K}_1, \mathcal{K}_2)$ such that $B = P_{\mathcal{H}_2}\tilde{B}$, $\|B\| = \|\tilde{B}\|$ and $\tilde{B}$ is given by (11). From (11) it is easy to see that $\|\tilde{B}\| \geq \|\Phi_A\|_\infty$. It is also clear that $\|\tilde{B}\| = \|B\| \geq \|A\| \geq \|\Lambda_A\|$. Now we show that

$$\|\tilde{B}\| \geq \frac{\|\Psi\|_\infty}{2}.$$

By substituting $w = M_{-\Psi}f = -\Psi f$ into (12) and (13) we have

$$\begin{aligned} \Phi\Theta_1 f - \Theta_A f &= -\Theta_2 \Psi f \\ C_{21}(\Theta_1 f) + C_{22}(\Delta_1 f) - \Lambda_A f &= -\Delta_2 \Psi f. \end{aligned}$$

Therefore

$$\begin{aligned} \|\Psi f\|^2 &= \|\Theta_2 \Psi f\|^2 + \|\Delta_2 \Psi f\|^2 \\ &= \|\Phi\Theta_1 f - \Theta_A f\|^2 + \|C_{21}(\Theta_1 f) + C_{22}(\Delta_1 f) - \Lambda_A f\|^2 \\ &\leq 2(\|\Phi\Theta_1 f\|^2 + \|\Theta_A f\|^2) + 2(\|C_{21}(\Theta_1 f) + C_{22}(\Delta_1 f)\|^2 + \|\Lambda_A f\|^2) \\ &= 2\left\| \begin{bmatrix} \Theta_A \\ \Lambda_A \end{bmatrix} f \right\|^2 + 2\|\tilde{B}(\Theta_1 f \oplus \Delta_1 f)\|^2 \\ &\leq 2(\|A\|^2 + \|\tilde{B}\|^2)\|f\|^2 = 4\|B\|^2\|f\|^2 \end{aligned}$$

and so the first inequality in (17) is proved. For the second inequality in (17) let us first notice that

$$\|\Theta_1^* u + \Delta_1 v\|^2 \leq \|u\|^2 + \|v\|^2 \quad \text{for any} \quad u \oplus v \in \mathcal{K}_2.$$

Now for $\tilde{B}$ given by (10) (with Θ and Λ replaced by Θ_A and Λ_A respectively), we have the following estimates:

$$\begin{aligned} \|\tilde{B}(u \oplus v)\|^2 &= \|\Phi u\|^2 + \|(\Lambda_A - \Delta_2\Psi)(\Theta_1^* u + \Delta_1 v)\|^2 \\ &\leq \|\Phi u\|^2 + \|\Lambda_A - \Delta_2\Psi\|_\infty^2 \|\Theta_1^* u + \Delta_1 v\|^2 \\ &\leq (\|\Phi\|_\infty^2 + \|\Lambda_A - \Delta_2\Psi\|_\infty^2)(\|u \oplus v\|^2). \end{aligned}$$

Thus we get

$$\|B\| \leq \|\tilde{B}\| \leq \sqrt{\|\Phi\|_\infty^2 + \|\Lambda_A - \Delta_2\Psi\|_\infty^2}.$$

This completes the proof of (17). For (18) let us notice that for any Φ and Ψ satisfying (15), the operator $B = P_{\mathcal{H}_2}\tilde{B}$ for $\tilde{B}$ as in (10) is an extension of A. Therefore

$$\begin{aligned} \|B\| \leq \|\tilde{B}\| &\leq \sqrt{\|\Phi\|_\infty^2 + \|\Lambda_A - \Delta_2\Psi\|_\infty^2} \leq \sqrt{\|\Phi\|_\infty^2 + 2\|\Psi\|_\infty^2 + 2\|\Lambda_A\|_\infty^2} \\ &\leq \sqrt{2}\left(\sqrt{\|\Phi\|_\infty^2 + \|\Psi\|_\infty^2} + \|\Lambda_A\|_\infty\right) \end{aligned}$$

and from here the second inequality in (18) is obvious. The first inequality (18) follows from the following elementary inequalities

$$\max\left\{\|\Phi\|_\infty^2, \frac{\|\Psi\|_\infty^2}{4}, \|\Lambda_A\|^2\right\} \geq \frac{1}{6}(\|\Phi\|_\infty^2 + \|\Psi\|_\infty^2 + \|\Lambda_A\|_\infty^2)$$
$$\geq \frac{1}{12}\left(\sqrt{\|\Phi\|_\infty^2 + \|\Psi\|_\infty^2} + \|\Lambda_A\|_\infty\right)^2.$$

This completes the proof of our theorem.

□

Remark. (i) If the measurable component Λ_A of the operator A is 0, the second inequality in (18) can be strengthened as

$$\mu(A, T_2, S, \mathcal{G}_1) \leq \nu(\Theta_A, \Theta_1, \Theta_2).$$

(ii) If furthermore Θ_2 is an inner function, i.e.,

$$\Delta_2^2 = I - \Theta_2^*(e^{it})\Theta_2(e^{it}) = 0,$$

then the first inequality in (17) becomes an equality.

4. Solvability of (TSCP)

In this section we study the solvability of (TSCP) by exploiting its connection with the solvability of the commutator equation $T_2B_1 - B_1T_1 = A_0$. We remark that similar results can also be obtained for the solvability of (GIEP).

We define the so called generalized derivation $\delta_{T_1T_2}$ on $B(\mathcal{H}_1, \mathcal{H}_2)$ by

$$\delta_{T_1T_2}(X) = T_2X - XT_1, \quad X \in B(\mathcal{H}_1, \mathcal{H}_2).$$

Recall that for a given bounded analytic function $\Theta \in H^\infty(\mathcal{E}_1, \mathcal{F}_2)$, the operator $A_\Theta \in B(\mathcal{G}_1, \mathcal{H}_2)$ associated with Θ is defined by

$$A_\Theta(\Theta_1 f \oplus \Delta_1 f) = P_{\mathcal{H}_2}(\Theta f \oplus 0), \quad f \in H^2(\mathcal{E}_1).$$

Let

$$\mathcal{R} = \{A_0 : \ A_0 = A_\Theta S_1' \text{ for some } \Theta \in H^\infty(\mathcal{E}_1, \mathcal{F}_2)\} \subset B(\mathcal{H}_1, \mathcal{H}_2);$$

by the Main Theorem (TSCP) is solvable if and only if Range$(\delta_{T_1T_2}) \supset \mathcal{R}$ where Range$(\delta_{T_1T_2})$ is the range of the operator $\delta_{T_1T_2}$.

There are natural conditions which guarantee that the operator $\delta_{T_1T_2}$ is onto. We recall such results now. For an operator T, let $\sigma(T)$, $\sigma_l(T)$ and $\sigma_r(T)$ be the spectrum, the left spectrum and the right spectrum of T, respectively.

ROSENBLUM'S THEOREM. [10]. *Let $\mathcal{E}$ and $\mathcal{F}$ be two separable complex Hilbert spaces and let $T \in B(\mathcal{E})$, $S \in B(\mathcal{E}_*)$ be two operators. For any $X \in B(\mathcal{E}_*, \mathcal{E})$ we define the operator $\delta_{TS}: B(\mathcal{E}_*, \mathcal{E}) \longrightarrow B(\mathcal{E}_*, \mathcal{E})$ by $\delta_{TS}(X) = TX - XS$. The right spectrum of the operator δ_{TS} is given by*

$$\sigma_r(\delta_{TS}) = \{\lambda_1 - \lambda_2;\ \lambda_1 \in \sigma_r(T), \lambda_2 \in \sigma_l(S)\}.$$

In particular, if $\sigma_r(T) \cap \sigma_l(S)$ is empty, then for any $Y \in B(\mathcal{E}_*, \mathcal{E})$ there exists $X \in B(\mathcal{E}_*, \mathcal{E})$ such that $TX - XS = Y$.

REMARK. There are similar results for the spectrum and left spectrum of δ_{TS}; see [10] for details.

The following result gives a sufficient condition for the solvability of (TSCP) and follows immediately from the above analysis.

THEOREM 10. *Let $T_1 \in B(\mathcal{H}_1)$ and $T_2 \in B(\mathcal{H}_2)$ be as in Theorem 2 and suppose that $\sigma_r(T_2) \cap \sigma_l(T_1) = \emptyset$. Then for any bounded analytic function $\Theta \in H^\infty(\mathcal{E}_1, \mathcal{F}_2)$ there exist bounded analytic functions $\Phi \in H^\infty(\mathcal{F}_1, \mathcal{F}_2)$ and $\Psi \in H^\infty(\mathcal{E}_1, \mathcal{E}_2)$ such that*

$$\Phi(z)\Theta_1(z) + \Theta_2(z)\Psi(z) = \Theta(z).$$

It seems that $\sigma_r(T_2) \cap \sigma_l(T_1) = \emptyset$ is a condition too strong for the solvability of (TSCP). Indeed the condition $\sigma_r(T_2) \cap \sigma_l(T_1) = \emptyset$ implies that $\delta_{T_1T_2}$ is onto while the solvability of (TSCP) only requires that the range of $\delta_{T_1T_2}$ contains $\mathcal{R}$. Nevertheless we will show that $\sigma_r(T_2) \cap \sigma_l(T_1) \cap D = \emptyset$ is a necessary condition for the solvability of (TSCP).

We recall the results concerning the spectrum of a completely non-unitary contraction T in relation to its characteristic function. If C_1 is the unit circle, let S_T be the set of points $\mu \in C_1$ not lying on any of the open arcs of C_1 on which $\Theta_T(z)$ is a unitary operator. Then the following equalities hold (see [14]):

$$\sigma(T) \cap C_1 = S_T$$

$$\sigma(T) \cap D = \{\lambda \in D; |\lambda| < 1,\ \Theta_T(\lambda) \text{ is not invertible}\}$$

$$\sigma_l(T) \cap D = \{\lambda \in D; |\lambda| < 1,\ \Theta_T(\lambda) \text{ is not left invertible}\} \tag{19}$$

$$\sigma_r(T) \cap D = \{\lambda \in D; |\lambda| < 1,\ \Theta_T(\lambda) \text{ is not right invertible}\}. \tag{20}$$

We need a couple of technical results.

Lemma 11. *The operator $A \in B(\mathcal{E}_1, \mathcal{E}_2)$ is not left invertible if and only if one of the following statements hold:*

(i) *There is a unit vector $f \in \mathcal{E}_1$ such that $Af = 0$.*

(ii) *There exists a sequence of orthogonal unit vectors $f_n \in \mathcal{E}_1$ such that $\|Af_n\| \to 0$.*

Lemma 12. *Let $A \in B(\mathcal{H}_1, \mathcal{H}_2)$ and $B \in B(\mathcal{K}_1, \mathcal{K}_2)$. If A is not left invertible and B is not right invertible, then there exists an operator $Z \in B(\mathcal{H}_1, \mathcal{K}_2)$ such that there exist no $X \in B(\mathcal{H}_2, \mathcal{K}_2)$ and no $Y \in B(\mathcal{H}_1, \mathcal{K}_1)$ satisfying $XA + BY = Z$.*

Proof. Recall that an operator is not right invertible if and only if its adjoint is not left invertible. Using the previous lemma, we get the following situations:

Case 1. There exists a unit vector $f \in \mathcal{H}_1$ such that $Af = 0$ and there exists a unit vector $g \in \mathcal{K}_2$ that is not in the range of B. Let $Z = g \otimes f$ be the rank one operator defined by $Zh = \langle h, f \rangle g$ for $h \in \mathcal{H}_1$. We claim that there exist no X and no Y such that

$$XA + BY = Z.$$

Indeed, in the contrary situation the above equation would imply that

$$XAf + BYf = \langle f, f \rangle g = g,$$

and therefore $BYf = g$, which is a contradiction, since g is not in the range of B.

Case 2. There exists a unit vector g such that $B^*g = 0$ and there exists a unit vector f such that f is not in the range of A^*. Let $Z = g \otimes f$. We claim that there exist no X and Y such that

$$XA + BY = Z.$$

Since otherwise the above equation implies that

$$A^*X^*g + Y^*B^*g = Z^*g = \langle g, g \rangle f = f.$$

That is $A^*X^*g = f$. This is a contradiction since f is not in the range of A^*.

Case 3. There exists a sequence of orthogonal unit vectors $f_n \in \mathcal{H}_1$ such that $\|Af_n\| \to 0$, and there exists a sequence of orthogonal unit vectors $g_n \in \mathcal{K}_2$ such that $\|B^*g_n\| \to 0$. Let Z be the operator given by $Z = \sum_{n=1}^{\infty} g_n \otimes f_n$. Since both sequences $\{f_n\}_{n=1}^{\infty}$ and $\{g_n\}_{n=1}^{\infty}$ are sequences of orthogonal unit vectors, Z is a linear bounded operator; in fact $\|Z\| = 1$ but this is not important now. We claim that there exist no X and Y such that

$$XA + BY = Z.$$

Indeed, in the contrary situation the above equation would imply that

$$\langle XAf_n, g_n\rangle + \langle BYf_n, g_n\rangle = \langle\langle f_n, f_n\rangle g_n, g_n\rangle = 1,$$

and taking the absolute value both sides we get:

$$|\langle XAf_n, g_n\rangle + \langle BYf_n, g_n\rangle| \le \|X\| \cdot \|Af_n\| + \|Y\| \cdot \|B^*g_n\| \to 0$$

which is a contradiction because the left hand side is 1.

□

Now we are ready to prove the following theorem, which gives a *necessary* condition for (TSCP) to be solvable.

THEOREM 13. *If* (TSCP) *is solvable, then* $\sigma_r(T_2) \cap \sigma_l(T_1) \cap D$ *is the empty set.*

Proof. Let us suppose that $z \in \sigma_r(T_2) \cap \sigma_l(T_1) \cap D$. By (20) and (19) we get that $\Theta_2(z)$ is not right invertible and $\Theta_1(z)$ is not left invertible; now with the previous lemma, there exists $Z \in B(\mathcal{E}_1, \mathcal{F}_2)$, such that there exist no X and Y satisfying $X\Theta_1(z) + \Theta_2(z)Y = Z$. Therefore, for the constant function $\Theta = Z$ there are no $\Phi(z)$ and $\Psi(z)$ such that $\Phi\Theta_1 + \Theta_2\Psi = \Theta$. The proof is complete.

□

By Θ_1 being unitary, we mean that $\Theta_1(z)$ is a unitary operator for any z on the unit circle. If either Θ_1 or Θ_2 is unitary, we have the following necessary and sufficient condition for (TSCP) to be solvable.

COROLLARY 14. *Assume either* Θ_1 *or* Θ_2 *is unitary. Then* (TSCP) *is solvable if and only of* $\sigma_r(T_2) \cap \sigma_l(T_1) \cap D$ *is the empty set.*

For completeness, we end the paper with a solvability criteria for (GIEP) (see the concluding remarks in [5]). Recall that the dual space of $B_1(\mathcal{H}_2, \mathcal{H}_1)$, the Banach space of trace class operators from $\mathcal{H}_2$ into $\mathcal{H}_1$ endowed with the trace norm, is $B(\mathcal{H}_1, \mathcal{H}_2)$ endowed with the operator norm. Thus by using Hahn–Banach theorem, we give the following characterization of the solvability of (GIEP).

PROPOSITION 15. *With the above notations* $\mu(A; T_2, S; \mathcal{G}_1) < \infty$ *if and only if there exists a constant* M *such that*

$$|\operatorname{Trace}(A_0Y)| \le M\|T_1Y - YT_2\|_1 \quad \textit{for all} \quad Y \in B_1(\mathcal{H}_2, \mathcal{H}_1), \tag{21}$$

where Trace$(\cdot)$ *denotes the trace of an operator and* $\|\cdot\|_1$ *denotes the trace norm of an operator.*

Proof. Let $\mathcal{L}$ be a Banach space. Recall the following easy consequence of the Hahn–Banach theorem: If $\Xi\colon \mathcal{L} \to \mathcal{L}$ is a bounded linear operator, then ξ^* is in the range of Ξ^*, the adjoint of Ξ, if and only if there exists a constant M such that

$$|\xi^*(\eta)| \leq M\|\Xi\eta\| \text{ for all } \eta \in \mathcal{L}. \tag{22}$$

Now apply this to the case

$$\mathcal{L} = B_1(\mathcal{H}_2, \mathcal{H}_1) \quad \text{and} \quad \Xi(Y) = T_1Y - YT_2, Y \in B_1(\mathcal{H}_2, \mathcal{H}_1).$$

Note that $\Xi^*(X) = XT_1 - T_2X$ for $X \in B(\mathcal{H}_1, \mathcal{H}_2)$. A_0 can be identified as a linear functional on $B_1(\mathcal{H}_2, \mathcal{H}_1)$ by its action on any $Y \in B_1(\mathcal{H}_2, \mathcal{H}_1)$ to be $\mathrm{Trace}(A_0Y)$. Therefore in this case (22) becomes (21). The proof is complete. □

References

[1] W. B. Arveson, Interpolation problems in nest algebras, *J. Funct. Anal.*, **20** (1975), 208–233.

[2] L. Carlson, Interpolation by bounded analytic functions and the Corona problem, *Ann. of Math.*, **76** (1962), 547–559.

[3] J. G. W. Carswell and C. F. Schubert, Lifting of operators that commute with shifts, *Michigan Math. J.*, **22** (1975), 65–69.

[4] Z. Ceausescu, Lifting of a contraction intertwining two isometries, *Michigan Math. J.*, **26** (1979), 1471–1482.

[5] C. Foias, On the extension of intertwining operators, *Contemporary Mathematics*, **189** (1995), 227–235.

[6] C. Foias, C. Gu and A. Tannenbaum, Intertwining dilations, intertwining extensions and causality, *Acta Sci. Math. (Szeged)*, **57** (1993), 103–125.

[7] C. Foias and A. E. Frazho, *The Commutant Lifting Approach to Interpolation Problems,* Operator Theory: Advances and Applications **44**, Birkhäuser, Boston, 1990.

[8] P. A. Fuhrmann, On the Corona theorem and its applications to spectral problems in Hilbert spaces, *Trans. Amer. Math. Soc.*, **132** (1968), 55–66.

[9] C. Gu, A generalization of Cowen's characterization of hyponormal Toeplitz operators, *J. Funct. Anal.*, **124** (1994), 135–148.

[10] M. Rosenblum, On the operator equation $BX - XA = Q$, *Duke Math. J.*, **23** (1956), 263–270.

[11] M. Rosenblum, A Corona theorem for countably many functions, *Integral Equations Operator Theory*, **3** (1980), 125–137.

[12] C. F. Schubert, A Corona theorem as an operator theorem, *Proc. Amer. Math. Soc.*, **69** (1978), 73–76.

[13] B. Sz.-Nagy, Transformations de l'espace de Hilbert, fonctions de type positif sur un groupe, *Acta Sci. Math. (Szeged)*, **15** (1954), 104–114.

[14] B. Sz.-Nagy and C. Foias, *Harmonic Analysis of Operators on Hilbert Space*, North Holland Publishing Co., Amsterdam – Budapest, 1970.

C. Gu, Department of Mathematics, California Polytechnic State University, San Luis Obispo, CA 93407

R. I. Teodorescu, Department of Mathematics and Statistics, Western Michigan University, Kalamazoo, MI 49008-5152

Operator Theory: Advances and Applications, Vol. 127, 349–374

On self-polar Hilbertian norms on (indefinite) inner product spaces

Gerald Hofmann

Considering a non-degenerate (possibly indefinite) inner product space with complete Hilbertian majorant, the self-polar Hilbertian majorants, which are suitable generalizations of the canonical norm on a pre-Hilbert space, are characterized by an operator equation involving the Gram operator. Among the self-polar Hilbertian norms, both the interesting class of pseudo-decomposition norms and the generic decomposition majorant are distinguished and investigated. Furthermore, characterizations and constructions for pseudo-decomposition majorants, decomposition majorants and the generic decomposition majorant are given.

1. Introduction

While the theory of $*$-representations of $*$-algebras $\mathcal{A}$ on Hilbert spaces by both $*$-algebras of bounded operators and such of unbounded operators is well-developed ([16], [12], [19]), it is wishful to generalize that theory to the more general case of $*$-representations on (possibly indefinite) inner product spaces. For the case that $\mathcal{A}$ is a C^*-algebra, such a generalization is contained in [14]. However a rigorous foundation of gauge field theories within the Wightman framework of general (axiomatic) quantum field theory (GQFT) (see [4, ch. 10], [15], [20]) needs $*$-representations of more general $*$-algebras (Theorems 3.1, 3.3, and Remark 3.2a)).

While every pre-Hilbert (unitary) space $\mathcal{D}$ can be chosen as the common domain of a $*$-algebra of (possibly unbounded) operators, and $\mathcal{D}$ is densely embedded relative to its canonical norm $\|\cdot\|$ in a Hilbert space $\mathcal{H}$, the case at hand is more delicate since a canonical norm is not available in general. Suitable generalizations for $\mathcal{D}$, $\mathcal{H}$ and $\|\cdot\|$ are pseudo-Krein spaces (Def. 4.1), Krein spaces and self-polar Hilbertian norms, respectively. Along these lines a generalization of the well-known

---

AMS Subject Classification (1991): 46C20, 46K10, 81T05.

GNS (Gelfand–Neumark–Segal) representation ([16, §17.3]) is contained in [8]. Furthermore, inner characterizations of pseudo-Krein spaces and the construction of self-polar Hilbertian norms are given in [9] and [10], respectively.

The investigations of the present paper are focused on non-degenerate inner product spaces $E, (.,.)$ with complete Hilbertian majorant τ. For those inner product spaces, a useful characterization of all Hilbertian self-polar norms q being weaker than τ is given in Theorem 5.4. Using that characterization, special self-polar Hilbertian norms such as quasi-decomposition norms (Def. 7.4) and the generic decomposition norm (Def. 6.2) are distinguished. Our interest in those norms is mainly due to the fact that the completed hull $\widetilde{E}$ of E endowed with the continuously extended inner product $(.,.)^{\sim}$ relative to such a norm is a Krein space, and furthermore, $\widetilde{E}$ is the orthogonal direct sum of the intrinsically completed hull of a positive definite and a negative definite linear subspace of E. Hence those norms are suitable from the point of view of inner product spaces, and convenient to handle (Prop. 6.1, 7.5). Finally, characterizations of quasi-decomposition norms and of the generic decomposition norm are given (Prop. 8.2, 8.3, 8.5).

Let us mention that it is supposed that while the (possibly indefinite) inner product $(.,.)$ on the above inner product space $E, (.,.)$ is relevant for physical interpretations, the (positive definite) scalar product $[.,.]$ defined by τ is not intrinsic to the physical problem under consideration. Thus $[.,.]$ is only of auxiliary character (see also [1, p. 122]). This is the motivation for our point of view: the inner product $(.,.)$ is the intrinsic one to start all the further investigations with.

The pattern of the present paper is as follows. After fixing some notations of the theory of locally convex (l.c.) vector spaces and of that of inner product spaces in Section 2, Sections 3 and 4 motivate the special interest in self-polar Hilbertian norms defined on inner product spaces. While Section 3 contains generalizations of the classical GNS Theorem ([16, §17.3, Theorem 2]), which are motivated by their application to GQFT, Section 4 illustrates that the self-polar Hilbertian norms are the suitable generalization of the canonical norm if one steps from the class of pre-Hilbert spaces to that of non-degenerate inner product spaces. Characterizations of self-polar Hilbertian norms are given in Section 4. While Section 5 is devoted to constructions of the generic decomposition norm (see Remark 6.3 a), b)), pseudo-decomposition norms and their pseudo-decomposition cores are investigated in Section 6. Section 7 is finally aimed at characterizations of those norms. Let us mention that some of our considerations, especially in Sections 5, 7, are based on ideas due to Hansen ([7]). Using the theory of induced Krein spaces, self-polar Hilbertian norms are also considered by Constantinescu and Gheondea in [6] (see Remark 8.6).

2. Preliminaries

Letting E be a (complex) vector space, l.c. topologies denoted by τ are introduced on E (see [11]). Furthermore, if p is a norm on E, then the l.c. topology defined by p is denoted by $\tau(p)$. A norm p is called *Hilbertian* (or quadratic) if a scalar product $[.,.]$ exists on E such that $p(\cdot) = \sqrt{[\cdot,\cdot]}$. For a subset $M \subset E$, its closed hull (in E) relative to τ is denoted by $\overline{M}^{\tau}$. In contrast, its completed hull relative to τ is denoted by $\widetilde{M}^{\tau}$. If two l.c. topologies τ_j, $j = 1, 2$, are defined on E, then $\tau_1 \prec \tau_2$ denotes that τ_1 is *weaker* (or coarser) than τ_2.

The notions and concepts from the theory of inner product spaces are used in the sense of [3], [2]. A vector space E is then equipped with a sesquilinear Hermitian form $(.,.)$, and it is assumed that the inner product $(.,.)$ is antilinear in its first variable. Furthermore, let us confine that the generic inner product is always denoted by $(.,.)$. Sometimes, auxiliary scalar products denoted by $[.,.]$ or $[.,.]_B$ are also introduced on E. Furthermore, since all the following considerations are concerned with *non-degenerate* inner product spaces $E, (.,.)$ (i.e., there are no isotropic vectors $x \neq 0$ in E), it is always assumed that an inner product space is non-degenerate without explicitely referring to it.

Recall that an l.c. topology τ on E is called *majorant* (resp. *partial majorant*) for an inner product space $E, (.,.)$ if the inner product $(.,.)$ is continuous (resp. separately continuous) relative to τ at each $x \in E$. If a norm p exists on E such that $\tau(p)$ is a majorant (resp. partial majorant), then p is also called majorant (resp. partial majorant).

If p is a partial majorant on $E, (.,.)$, then

$$x \to p'(x) := \sup_{0 \neq y \in E} \frac{|(x,y)|}{p(y)},$$

$x \in E$, is called the *polar norm* for p, and obviously, $0 < p'(z) < \infty$ for $0 \neq z \in E$. If $p = p'$ (resp. $\tau(p) = \tau(p')$) on E, the norm p (resp. the l.c. topology $\tau(p)$) is called *self-polar*.

An inner product space $E, (.,.)$ is called *decomposable* if a fundamental decomposition

$$E = E^{(+)} \dot{(+)} E^{(-)} \tag{1}$$

of E exists, where $E^{(\pm)}$ are, respectively, some positive and negative definite linear subspaces, and $\dot{(+)}$ refers to the orthogonal (relative to $(.,.)$) direct sum. Remember then that a *fundamental symmetry* $J = P^+ - P^-$ is defined by (1), where $P^\pm : E \to E^{(\pm)}$ are the fundamental projections. Furthermore,

$$x \to \|x\|_J := \sqrt{(x, Jx)} \tag{2}$$

is a Hilbertian self-polar majorant on $E, (.,.)$, and it is denoted as the *decomposition norm* (belonging to (1)). If an auxiliary scalar product $[.,.]$ is also defined on a decomposable inner product space $E, (.,.)$, then

$$E = E^{(+)} \, [\oplus] \, E^{(-)} \tag{3}$$

refers to a special fundamental decomposition, where $E^{(+)}$ and $E^{(-)}$ are orthogonal relative to both the generic inner product $(.,.)$ and $[.,.]$. (3) is called *Phillips-decomposition* of $E, (.,.)$ (relative to $[.,.]$).

If $A: E \to E$ is a (linear) operator, then its domain and range are denoted by $\mathrm{dom}(A)$ and $\mathrm{ran}(A)$, respectively. Further, the adjoint of A relative to an auxiliary scalar product $[.,.]$ is denoted by A^*.

3. On generalizations of the GNS Theorem

Throughout this section let $\mathcal{A}$ be a $*$-algebra with unity $\mathbf{1}$. Recall further that a linear functional T on $\mathcal{A}$ is called *positive* (resp. *hermitian*) if

$$T(f^* f) \geq 0 \tag{4}$$

(resp. $T(f^*) = \overline{T(f)}$, here the bar denotes the complex conjugate) applies to each $f \in \mathcal{A}$ ([16, §10.2]).

In order to apply representation theory of $*$-algebras to the Wightman framework of GQFT (see [4, ch. 8]), the following two generalizations of the famous GNS Theorem were considered by H.-J. Borchers, A. Uhlmann and E. Scheibe about 40 years ago.

THEOREM 3.1. (Borchers [5], Uhlmann [21]).

a) *Let W be a positive linear functional on $\mathcal{A}$ satisfying $W(\mathbf{1}) = 1$. Then a cyclic $*$-representation of $\mathcal{A}$ exists by (possibly unbounded) operators $\pi_W(.)$ defined on a common, dense and invariant domain $\mathcal{D}_W$ in Hilbert space $\mathcal{H}_W$ endowed with scalar product $[.,.]_W$.*

b) *Considering the GNS-data $(\pi_W(.), \mathcal{D}_W, \mathcal{H}_W, [.,.]_W, \psi_0)$, one has*

$$W(a) = [\psi_0, \pi_W(a)\psi_0]_W, \; a \in \mathcal{A}. \tag{5}$$

(Here $\psi_0 \in \mathcal{D}_W$ denotes a cyclic vector for π_W.) By (5), the GNS-data are uniquely given up to unitary equivalence.

REMARK 3.2. a) Introducing locally convex topologies on both the $*$-algebra $\mathcal{A}$ and the set of operators $\{\pi_W(a);\ a \in \mathcal{A}\}$ making them to *topological* $*$-algebras ([16, §8.1]), continuity properties of the $*$-representation $a \to \pi_W(a)$, $a \in \mathcal{A}$, were also considered by Borchers and Uhlmann ([5], [21],[19]).

b) Considering the *Gelfand ideal*

$$I_W = \{f \in \mathcal{A}; W(f^* f) = 0\}, \tag{6}$$

take the quotient space $\mathcal{D}_W = \mathcal{A}/I_W$ and introduce a scalar product

$$[\hat{x}, \hat{y}]_W = W(x^* y),$$

$\hat{x}, \hat{y} \in \mathcal{D}_W$, $x \in \hat{x}, y \in \hat{y}$, where $\hat{x}$ denotes the residue class containing the element $x \in \mathcal{A}$. Noting that I_W is a left ideal in $\mathcal{A}$, it is straightforwardly checked that the above definition of $[.,.]_W$ does not depend on the choice of the representatives x and y of the residue classes $\hat{x}$ and $\hat{y}$, respectively. Since the linear functional W is positive by assumption, (4) yields that

$$\hat{x} \to \|\hat{x}\|_W = \sqrt{W(x^* x)} \tag{7}$$

is a norm on $\mathcal{D}_W$. $\|\cdot\|_W$ is called the *canonical norm.* The Hilbert space $\mathcal{H}_W$ is then given by $\mathcal{H}_W = \widetilde{\mathcal{D}_W}$ (completed hull of $\mathcal{D}_W$ relative to $\|.\|_W$).

c) The operators of the $*$-representation are given by $\pi_W(a)\hat{x} = \widehat{ax}$, $a \in \mathcal{A}$, $\hat{x} \in \mathcal{D}_W$, and it is straightforwardly checked that the above definition is correct. Finally $\psi_0 = \hat{\mathbf{1}}$ is a cyclic vector for the $*$-representation π_W.

THEOREM 3.3. (E. Scheibe [18]). *Let ω be an hermitian linear functional defined on $\mathcal{A}$ and satisfying $\omega(\mathbf{1}) = 1$.*

a) *Then there are*

(i) *a vector space $\mathcal{D}_\omega$ with a non-degenerate inner product $(.,.)_\omega$,*

(ii) *a vector $\psi_0 \in \mathcal{D}_\omega$ satisfying $(\psi_0, \psi_0)_\omega = 1$,*

(iii) *a cyclic $*$-representation $f \to \pi_\omega(f)$ of $\mathcal{A}$ by linear operators on $\mathcal{D}_\omega$ such that*

$$\omega(f) = (\psi_0, \pi_\omega(f)\psi_0)_\omega,$$
$$\mathcal{D}_\omega = \operatorname{span}\{\pi_\omega(f)\psi_0;\ f \in \mathcal{A}\}, \quad \textit{cyclicity of } \psi_0,$$
$$(\phi, \pi_\omega(f)\psi)_\omega = (\pi_\omega(f^*)\phi, \psi)_\omega, \quad f \in \mathcal{A},\ \phi, \psi \in \mathcal{D}_\omega.$$

b) *Furthermore, $\mathcal{D}_\omega$, $(.,.)_\omega$, ψ_0, and $\pi_\omega(.)$ are uniquely defined by* (i),...,(iii) *up to isometric linear isomorphisms.*

Sketch of the proof. Replacing the Gelfand ideal given in (6) by the isotropic part of $\mathcal{A}$:

$$\mathcal{A}^{(0)} = \{f \in \mathcal{A}; \omega(f^*g) = 0, \forall g \in \mathcal{A}\},$$

take the quotient space $\mathcal{D}_\omega = \mathcal{A}/\mathcal{A}^{(0)}$ endowed with the non-degenerate inner product

$$(\hat{x}, \hat{y})_\omega = \omega(x^*y),$$

$\hat{x}, \hat{y} \in \mathcal{D}_\omega$, $x \in \hat{x}, y \in \hat{y}$. After checking that $\mathcal{A}^{(0)}$ is a left ideal in $\mathcal{A}$, the remainder of the proof is the same as that of the classical GNS Theorem (see [16, pp. 275–277], or Remark 3.2 b), c)).

□

REMARK 3.4. a) If ω in Theorem 3.3 is positive (and consequently also hermitian), then the Cauchy–Schwarz inequality applies to ω, and $\mathcal{A}^{(0)} = I_W$ follows. Consequently, Theorem 3.3 is a generalization of Theorem 3.1 to the case of hermitian linear functionals.

b) If one likes to proceed further as described in Remark 3.2 b), a suitable generalization of the canonical norm given in (7) is needed.

4. Generalizations of the canonical norm

Considering a pre-Hilbert (unitary) space $E, (.,.)$ (i.e., the inner product $(.,.)$ is positive definite), a canonical norm

$$p(x) = \sqrt{(x,x)}, \tag{8}$$

$x \in E$, exists on E. Remember that the following properties then follow:

(a) p is the unique self-polar norm on $E, (.,.)$, i.e., $p(x) = \sup_{0\neq y \in E} \frac{|(x,y)|}{p(y)}$, $x \in E$ (cf. [7, pp 891 -892]);

(b) p satisfies the parallelogram identity;

(c) the completed hull $\widetilde{E}^{\tau(p)}$ endowed with the continuously extended inner product $(.,.)^\sim$ is a Hilbert space.

Motivated by the considerations of Section 3, especially of Remark 3.4 b), it is of interest to find the corresponding substitutes when the pre-Hilbert space is replaced by some inner product space $E, (.,.)$.

If the inner product $(.,.)$ is indefinite, then (8) does not define a norm, and thus, no canonical norm is available. The role of the canonical norm (8) is now played by the self-polar Hilbertian norms q on E, i.e., it holds

$$q(x) = \sup_{0\neq y \in E} \frac{|(x,y)|}{q(y)}, \tag{9}$$

and furthermore, there is a positive definite inner product $[.,.]$ on $E \times E$ such that

(10) $$q(x) = \sqrt{[x,x]}\,,$$

$x \in E$. Recall also that a norm is Hilbertian (i.e., (10) applies) if and only if it satisfies the parallelogram identity (cf. [11, p. 241]). The above (a), (b) show that the self-polar Hilbertian norms are the natural generalization of the canonical norm when the class of pre-Hilbert spaces is generalized to non-degenerate inner product spaces.

The significance of self-polar and self-polar Hilbertian norms is due to the following:

(I) If q is a self-polar norm on a non-degenerate inner product space $E, (.,.)$, then
  (i) the inner product $(.,.)$ extends $\tau(q)$-continuously onto $\widetilde{E} \times \widetilde{E}$,
  (ii) the inner product space $\widetilde{E}, (.,.)^\sim$ is non-degenerate.

(II) If q is a self-polar Hilbertian norm on a non-degenerate inner product space $E, (.,.)$, then $\widetilde{E}, (.,.)^\sim$ is a Krein space ([8, Th.1(iii), Prop. 2]), i.e., $\widetilde{E}, [.,.]$ ($q(.) = \sqrt{[.,.]}$) is a Hilbert space and the Gram operator J defined by $(x,y)^\sim = [x, Jy]$, $x, y \in \widetilde{E}$ is a symmetry ($J = J^* = J^{-1}$).

Notice also that (c) is generalized from the case of positive definite inner products $[.,.]$ to (possibly indefinite) inner products in the following:

(c') if q is a Hilbertian self-polar norm on E, then the completed hull $\widetilde{E}^{\tau(q)}$ endowed with the continuously extended inner product $(.,.)^\sim$ is a Krein space ([3, Theorem V.1.3]).

Generalizing the positive definite case, where the canonical norm is uniquely defined, the following facts about self-polar norms have to be taken into account. There are examples of inner product spaces $E, (.,.)$ such that:

(E1) there is no self-polar norm on $E, (.,.)$, (cf. [3, Example III.3.2]);

(E2) there is exactly one self-polar norm on $E, (.,.)$, (cf. [9], and references cited there);

(E3) there is a whole family of self-polar norms defining non-equivalent topologies on E, (such a family of self-polar norms was explicitely constructed by Araki in [1]).

Motivated by the above (II) and (E1), ..., (E3), the following notations are introduced:

Definition 4.1.

a) An inner product space $E, (.,.)$ is called *pseudo-Krein space* if a self-polar Hilbertian norm exists on $E, (.,.)$.

b) A pseudo-Krein space $E, (.,.)$ is called *pre-Krein space* if all Hilbertian self-polar norms on $E, (.,.)$ are equivalent.

Using Hilbertian majorants and Hilbertian partial majorants, one gets the following characterization of pseudo-Krein spaces.

PROPOSITION 4.2. *Letting $E, (.,.)$ be a non-degenerate inner product space, the following are equivalent:*

(i) *$E, (.,.)$ is a pseudo-Krein space,*

(ii) *a Hilbertian majorant exists on $E, (.,.)$,*

(iii) *a Hilbertian partial majorant exists on $E, (.,.)$,*

(iv) *a Hilbertian majorant τ exists such that the completed hull $\widetilde{E}^{\tau}$ endowed with the continuously extended inner product $(.,.)^{\sim}$ is a non-degenerate inner product space.*

Proof. (i)⇒(iv): Assuming that (i) applies, a self-polar Hilbertian norm q exists on $E, (.,.)$. Then $q = q'$ yields

$$|(x,y)| \leq \Big(\sup_{0 \neq y \in E} \frac{|(x,y)|}{q(y)} \Big) q(y) = q(x)q(y)$$

for $x, y \in E$. Hence q is a majorant on $E, (.,.)$. For proving that $\widetilde{E}^{\tau(q)}, (.,.)^{\sim}$ is non-degenerate, assume that there is a $0 \neq z \in \widetilde{E}^{\tau(q)}$ such that

$$(z\ w)^{\sim} = 0 \tag{11}$$

for each $w \in \widetilde{E}^{\tau(q)}$. Letting $\{z_n\}_{n=1}^{\infty}, z_n \in E$, be a Cauchy-sequence with $z_n \to z$ relative to $\tau(q)$ as $n \to \infty$, a contradiction follows from

$$\begin{aligned} 0 < \tilde{q}(z) = \lim_{n\to\infty} q(z_n) &= \lim_{n\to\infty} \sup_{0\neq y\in E} \frac{|(z_n,y)|}{q(y)} = \lim_{n\to\infty} \sup_{0\neq y\in E} \frac{|(z-z_n,y)^{\sim}|}{q(y)} \\ &\leq \lim_{n\to\infty} \tilde{q}(z-z_n) = 0\,, \end{aligned}$$

where the $\tau(q)$-continuous extension of q on $\widetilde{E}^{\tau(q)}$ is denoted by $\tilde{q}$.

(iv)⇒(iii): Assuming that (iv) applies, a Hilbertian norm p exists such that $|(x,y)| \leq p(x)p(y)$ for $x, y \in E$. Setting $c_y = p(y)$, (iii) follows from the estimate

$$|(x,y)| \leq c_y\, p(x) \tag{12}$$

for $x \in E$.

(iii)$\Rightarrow$(ii): Assuming that (12) holds with a Hilbertian norm p and some constant $0 < c_y < \infty$, it follows

$$p'(y) = \sup_{0 \neq x \in E} \frac{|(x,y)|}{p(x)} \leq c_y$$

for $y \in E$. Recalling that the polar norm p' of p is Hilbertian by [3, Lemma III.4.3], it follows that the norm $q(x) = \sqrt{p(x)^2 + p'(x)^2}$ is Hilbertian, too. Now (ii) is a consequence of

$$|(x,y)| \leq p(x)p'(y) \leq q(x)q(y) \tag{13}$$

for $x, y \in E$.

(ii)$\Rightarrow$(i): Assuming that $|(x,y)| \leq q(x)q(y)$, x, y, with some Hilbertian norm q, the Aronszajn–Schatten iteration process gives a self-polar Hilbertian norm q_∞ on $E, (.,.)$ (see [8, Lemma 2] , [10]). Hence (i) holds true.

□

REMARK 4.3. a) For inner characterizations of both pseudo-Krein and pre-Krein spaces, the reader is referred to [9].

b) Returning once more to Theorem 3.3 and Remark 3.4 b), the above (E1), ..., (E3) lead us to the following questions:

(Q1) Under which conditions upon ω do Hilbertian self-polar norms exist on $\mathcal{D}_\omega, (.,.)_\omega$?

(Q2) Under which conditions does the Gram operator J satisfy $J : \mathcal{D}_\omega \to \mathcal{D}_\omega$?

(Q3) Under which conditions does exactly one selfpolar Hilbertian topology exist on $\mathcal{D}_\omega, (.,.)_\omega$?

For answers to (Q1), ... , (Q3), the reader is referred to [8], [22].

5. Characterizations of self-polar Hilbertian norms

In all what follows, let $E, (.,.)$ be a (non-degenerate) inner product space which is complete relative to a Hilbertian majorant τ. Noticing that the above assumptions on $E, (.,.)$ are motivated by Proposition 4.2 (iv), the present section is aimed at characterizing all self-polar Hilbertian norms q on E satisfying

$$\tau(q) \prec \tau\,. \tag{14}$$

Note that a Hilbertian norm p and a scalar product $[.,.]$ exist on E such that

$$p(x) = \sqrt{[x,x]}\,, \quad |(x,y)| \leq p(x)p(y)\,, \quad (x,y) = [x, Gy]$$

for $x, y \in E$, where G is a self-adjoint and bounded operator (Gram operator) on the Hilbert space $E, [.,.]$. Notice that G allows the spectral resolution

$$G = \int_{-1}^{1} \lambda \, dP(\lambda)\,. \tag{15}$$

The following characterization of the degeneratedness of an inner product space with complete Hilbertian majorant in terms of the range of the Gram operator immediately follows from Hilbert space theory.

LEMMA 5.1. *Let $H, (.,.)$ be a possibly degenerate inner product space with complete Hilbertian majorant $\tau(p)$, $p(.) = \sqrt{[.,.]}$, and Gram operator G, $(.,.) = [., G.]$. Then the range* $\operatorname{ran}(G)$ *is dense in $H[\tau(p)]$ if and only if $H, (.,.)$ is non-degenerate.*

The following characterization of Hilbertian self-polar norms is due to F. Hansen (see [7, Theorem 2.5]).

PROPOSITION 5.2. *Let q be a norm on a (non-degenerate) inner product space $E, (.,.)$ with complete Hilbertian majorant $\tau(p)$, $p(.) = \sqrt{[.,.]}$, and Gram operator G, $(x, y) = [x, Gy]$, $x, y \in E$. Then the following are equivalent:*

(i) *q is Hilbertian and self-polar, and $\tau(q) \prec \tau(p)$,*

(ii) *there is a bounded operator A on $E[\tau(p)]$ such that*
   (1) *A is positive relative to the scalar product* $[.,.]$,
   (2) *0 is no eigenvalue of A,*
   (3) $\operatorname{ran}(G) \subset \operatorname{dom}(A^{-1/2})$,
   (4) *$p(A^{1/2}x) = p(A^{-1/2}Gx)$, $x \in E$,*
   (5) *$q(x) = p(A^{1/2}x)$, $x \in E$.*

Some properties of the above operator A are collected in the following

LEMMA 5.3. *Let $E, (.,.)$ be an inner product space with complete Hilbertian majorant $\tau(p)$, $p(.) = \sqrt{[.,.]}$, Gram operator G, $(x, y) = [x, Gy]$, $x, y \in E$, and A a linear operator on E such that* (1) ... (4) *of Proposition* 5.2 (ii) *apply. Then,*

a) *$A^{-1/2}$ is a closed operator,* $\operatorname{dom}(A^{-1/2})$ *is dense in $E[\tau(p)]$, and* $\operatorname{ran}(G)$ *is a core for $A^{-1/2}$.*

b) *$\{A^{-1/2}Gx;\ x \in E\}$ is dense in $E[\tau(p)]$.*

Proof. See [7, Prop. 2.6]. □

Taking the Gram operator G from (15), a further characterization of Hilbertian self-polar norms follows.

THEOREM 5.4. *Let q be a norm on an inner product space $E, (.,.)$ with complete Hilbertian majorant $\tau(p)$, $p(.) = \sqrt{[.,.]}$, and Gram operator G, $(x, y) = [x, Gy]$, $x, y \in E$. Then the following are equivalent:*

(i) *q is Hilbertian and self-polar,*

(ii) *a symmetry S $(= S^{-1} = S^*)$ and a bounded linear operator B exist on the Hilbert space $E, [.,.]$ such that*

  (1) *B is positive on $E, [.,.]$,*

  (2) *$Gx = BSBx$, $x \in E$,*

  (3) *$q(x) = p(Bx)$, $x \in E$.*

Proof. In view of Proposition 5.2, it is enough to show the equivalence of Proposition 5.2 (ii) and Theorem 5.4 (ii). $(\Rightarrow)$: Assuming that Proposition 5.2 (ii) apply, and setting $B = A^{1/2}$, (1) and (3) of the theorem under consideration are obvious. For showing (2), a symmetry S will be defined on the Hilbert space $E, [.,.]$. Using (4) of Proposition 5.2 (ii), an isometric operator $S : \overline{\mathrm{ran}(B^{-1}G)} \to \overline{\mathrm{ran}(B)}$ is defined by

$$S(B^{-1}Gx) = Bx\,, \tag{16}$$

$x \in E$ (here $\bar{\cdot}$ denotes closed hull in the Hilbert space $E, [.,.]$). Using Lemma 5.3 b), $E = \overline{\mathrm{ran}(B^{-1}G)}$ follows, and thus, $\mathrm{dom}(S) = E$. Noticing that $\mathrm{ran}(B) = \mathrm{dom}(A^{-1/2}) \supset \mathrm{ran}(G)$ by (3) of Proposition 5.2 (ii), and $\overline{\mathrm{ranG}} = E$ by Lemma 5.1, it follows that S is a unitary operator on $E, [.,.]$, and consequently, $S^* = S^{-1}$. Hence, $S^*Bx = B^{-1}Gx$, $x \in E$, by (16). For $y = Bx \in \mathrm{dom}(B^{-1})$, it then follows

$$S^*y = B^{-1}Gx = B^{-1}G(B^{-1}y)\,. \tag{17}$$

(17) yields

$$Sy = S^{**}y = (B^{-1}GB^{-1})^*y \overset{(*)}{=} B^{-1}GB^{-1}y = S^*y\,,$$

where $(*)$ holds since $(B^{-1})^* = B^{-1}$ and $G^* = G$. Consequently S is a symmetric unitary operator, and thus a symmetry. Using $S = S^{-1}$, (16) yields $Gx = BSBx$, $x \in E$, and thus (2) of the theorem under consideration applies.

$(\Leftarrow)$: Assume that (ii) of the theorem under consideration applies. Noticing that the non-degeneratedness of $E, (.,.)$ implies that 0 is not an eigenvalue of G,

(2) yields that 0 is not an eigenvalue of B, too. Hence 0 is not an eigenvalue of $A = B^2$ verifying (2) of Proposition 5.2 (ii). Furthermore,

$$\operatorname{ran}(B) = \operatorname{dom}(B^{-1}) = \operatorname{dom}(A^{-1/2})$$

is implied, and thus, (2) of the theorem under consideration yields

$$\operatorname{ran}(G) \subset \operatorname{ran}(B) = \operatorname{dom}(A^{-1/2}),$$

i.e., (3) of Proposition 5.2 (ii) applies. Setting

$$A = B^2, \tag{18}$$

it follows

$$p(A^{-1/2}Gx) = p(B^{-1}(BSB)x) = \sqrt{[SBx, SBx]} = \sqrt{[Bx, Bx]} = p(A^{1/2}x),$$

$x \in E$, verifying (4) of Proposition 5.2 (ii). The other properties listed in Proposition 5.2 (ii) are obvious.

□

REMARK 5.5. a) If a self-polar Hilbertian norm q is given by (3) of Theorem 5.4 (ii) (resp. (5) of Proposition 5.2 (ii)), then its defining operator B (resp. A) is unique.

b) Assuming that two self-polar Hilbertian norms $q_j(\cdot) = p(B_j\cdot)$, $j = 1, 2$, are defined by Theorem 5.4 (ii) (3), it follows that $\tau(q_1) = \tau(q_2)$ if and only if $\operatorname{ran}(B_1) = \operatorname{ran}(B_2)$ ([7, Cor. 2.8]).

c) Considering "($\Leftarrow$)" of the above proof, the following is obvious: If Theorem 5.4 (ii) applies, then the linear mapping $B : E \to \operatorname{ran}(B)$ is bijective, and $\operatorname{ran}(B)$ is $\tau(p)$-dense in E. Since $B: E \to \operatorname{ran}(B)$ is bijective, B^{-1} exists and

$$\operatorname{dom}(B^{-1}) = \operatorname{ran}(B). \tag{19}$$

If a Hilbertian self-polar norm is given by Theorem 5.4 (ii) (3), then it will be written

$$q_B(\cdot) = p(B\cdot). \tag{20}$$

6. On the construction of the generic decomposition norm

A special solution of the operator equation $G = BSB$ of Theorem 5.4 (ii) (2) is given next.

PROPOSITION 6.1. *Letting the inner product space $E, (.,.)$ satisfy the properties given at the beginning of Section 5, put $B = \sqrt{|G|}$, $S = \text{sign}\,(G)$. Then,*

a) *B is a positive bounded operator and S a symmetry on the Hilbert space $E, [.,.]$,*
b) *the operator equation $G = BSB$ considered in Theorem 5.4 (ii) (2) applies,*
c) *$q_{\sqrt{|G|}}$ is the decomposition norm related to the fundamental symmetry $S = \text{sign}\,(G)$,*
d) *all decomposition majorants on $E, (.,.)$ are equivalent to $\tau(q_{\sqrt{|G|}})$.*

Proof. a) Taking the spectral resolution $G = \int_{-1}^{1} \lambda\, dP(\lambda)$ from (15), it follows

$$B = \int_{-1}^{1} \sqrt{|\lambda|}\, dP(\lambda)\,, \quad S = \int_{-1}^{1} \text{sign}\,(\lambda)\, dP(\lambda)\,.$$

It is now obvious that B and S satisfy the properties listed in a).

b) follows readily from

$$BSB = \int_{-1}^{1} \sqrt{|\lambda|}\ \text{sign}\,(\lambda)\, \sqrt{|\lambda|}\, dP(\lambda) = \int_{-1}^{1} \lambda\, dP(\lambda) = G\,.$$

c) Using [3, Theorem IV.52], it follows that $E, (.,.)$ is decomposable and a fundamental decomposition is given by

$$E = E^{(+)}\,(+)\,E^{(-)} \tag{21}$$

with $E^{(+)} = (I - P(0))E$, $E^{(-)} = P(0)E$. Consider then the corresponding decomposition norm $\|.\|_J = \sqrt{(., J.)}$, where

$$J = P^{(+)} - P^{(-)} = I - 2P(0)\,,$$

and $P^{(\pm)} : E \to E^{(\pm)}$ are projections related to (21). For proving c), it is enough to verify that $J = S$ and $\|.\|_J = q_{\sqrt{G}}$. Noticing that

$$\text{sign}\,(G) = I - 2P(0)\,,$$

it follows $J = S$. The proof of c) is completed by

$$\|x\|_J^2 = (x, Sx) = [x, GSx] \overset{(*)}{=} [x, |G|x] = p(\sqrt{|G|}x) = q_{\sqrt{|G|}}(x),$$

$x \in E$, where $(*)$ follows from

$$GS = \int_{-1}^{1} \lambda \operatorname{sign}(\lambda)\, dP(\lambda) = \int_{-1}^{1} |\lambda|\, dP(\lambda) = |G|.$$

d) readily follows from c) and [3, Theorem IV.6.4]. □

The following is motivated by the foregoing proposition.

DEFINITION 6.2. Letting $E, (.,.)$ be an inner product space such that all the properties listed at the beginning of Section 5 apply, the norm $q_{\sqrt{|G|}}$ and topology $\tau(q_{\sqrt{|G|}})$ are called *generic decomposition norm* and *generic decomposition majorant* on $E, (.,.)$, respectively.

REMARK 6.3. a) Recalling the Aronszajn–Schatten iteration, it is known that if one starts with a Hilbertian majorant $\tau(p)$, then the iteration process yields a self-polar Hilbertian norm q satisfying $\tau(q) \prec \tau(p)$ (see [3, Lemma IV.2.5, Section IV.4]), [8, Lemma 3]). It follows from [8, Lemma 4] that q coincides with the generic decomposition norm $q_{\sqrt{|G|}}$.

b) While a first construction of the generic decomposition norm is given in Proposition 6.1 c), a second one not explicitely using the Gram operator G follows from the above a).

7. On quasi-decomposition norms and the construction of the quasi-decomposition core

Assuming that the inner product space $E, (.,.)$ satisfies the assumptions given at the beginning of Section 5, the present section is aimed at the construction of the quasi-decomposition core of a self-polar Hilbertian norm q_B satisfying $\tau(q_B) \prec \tau(p)$.

The following considerations are again based on an analysis of the interplay between the linear operators B, S used in the operator equation

$$G = BSB \tag{22}$$

(see Theorem 5.4 (ii)), and especially, between certain linear subspaces introduced in Notation 7.1, and characterized in Proposition 7.3 a). It will be shown that a Phillips-decomposition $\mathcal{E} = \mathcal{E}^{(+)} [\oplus]_{q_B} \mathcal{E}^{(-)}$ of a linear subspace $\mathcal{E} \subset E$ exists. If $\mathcal{E}$ is $\tau(q_B)$-dense in E, then it will be called the quasi-decomposition core related to q_B (see Definition 7.4). Some properties of the fundamental decomposition J related to the above Phillips-decomposition of $\mathcal{E}$ are finally collected in Lemma 7.6.

Taking the symmetry S from (22), consider the orthogonal projections

$$P^{[\pm]} = \tfrac{1}{2}(I \pm S) \tag{23}$$

in the Hilbert space $E, [.,.]$, and the corresponding orthogonal direct decomposition

$$E = E^{[+]} [\dot{+}] E^{[-]}, \tag{24}$$

where

$$E^{[\pm]} := \mathrm{ran}(P^{[\pm]}) = \{x \in E;\ Sx = \pm x\}. \tag{25}$$

Notice also that $S\restriction_{E^{[\pm]}} = \pm I$ follow from (25).

In order to construct a Phillips-decomposition of a linear subspace $\mathcal{E} \subset E$, the following linear subspaces are introduced.

Notation 7.1. Considering

$$\mathcal{D}^{[\pm]} = \mathrm{ran}(B) \cap E^{[\pm]}, \tag{26}$$

set

$$\mathcal{D} = \mathcal{D}^{[+]} [\dot{+}] \mathcal{D}^{[-]}. \tag{27}$$

Consider furthermore

$$\mathcal{E} = B^{-1}\mathcal{D} := \{B^{-1}d;\ d \in \mathcal{D}\}, \tag{28}$$

$$\mathcal{E}^{(\pm)} = B^{-1}\mathcal{D}^{[\pm]}. \tag{29}$$

Remark 7.2. It is immediate that

$$S\restriction_{\mathcal{D}} : \mathcal{D} \to \mathcal{D}, \tag{30}$$

and also, $S\restriction_{\mathcal{D}^{[\pm]}} : \mathcal{D}^{[\pm]} \to \mathcal{D}^{[\pm]}$.

Some properties of the linear subspaces $\mathcal{E}$, $\mathcal{D}$, $\mathcal{E}^{(\pm)}$, $\mathcal{D}^{[\pm]}$ introduced in Notation 7.1 are collected in the following proposition.

PROPOSITION 7.3. *Using the Hilbertian self-polar norm q_B given in* (20), *the following apply:*

a) *characterizations of $\mathcal{E}$, $\mathcal{D}$, $\mathcal{E}^{(\pm)}$, $\mathcal{D}^{[\pm]}$ are given by*

$$\mathcal{E} = G^{-1}(\mathrm{dom}(G^{-1}) \cap \mathrm{dom}(B^{-2}))\,, \tag{31}$$

$$\mathcal{D} = \{y \in \mathrm{ran}(B); Sy \in \mathrm{ran}(B)\} = \mathrm{ran}(B) \cap S(\mathrm{ran}(B))\,, \tag{32}$$

$$\mathcal{E}^{(\pm)} = \left\{x \in E;\ (x,x) = \pm q_B(x)^2\right\}\,, \tag{33}$$

$$\mathcal{D}^{[\pm]} = \left\{y \in \mathrm{dom}(B^{-1});\ (B^{-1}y, B^{-1}y) = \pm p(y)^2\right\}\,, \tag{34}$$

b) *$\mathcal{E}^{(\pm)}$ are closed in $E[\tau(q_B)]$,*

c) *a Phillips-decomposition of $\mathcal{E}, (.,.)|_{\mathcal{E}\times\mathcal{E}}$ is given by*

$$\mathcal{E} = \mathcal{E}^{(+)}\, [\oplus]_B\, \mathcal{E}^{(-)}\,, \tag{35}$$

where $[\oplus]_B$ refers to orthogonality relative to both the inner product $(.,.)$ and the scalar product $[.,.]_B = q_B(.)^2$,

d) *considering the decomposition norm $\|.\|_J$ related to the fundamental decomposition* (35), *it follows*

$$\|x\|_J = \sqrt{(x, Jx)} = q_B(x) \tag{36}$$

for $x \in \mathcal{E}$.

Proof. a) To prove (32) take any $y \in \mathcal{D}$. Due to (27), $y^\pm \in \mathcal{D}^{[\pm]}$ exist such that $y = y^+ + y^-$. Now, $y \in \mathrm{ran}(B)$ and $Sy = y^+ - y^- \in \mathrm{ran}(B)$ follow from (26) and (25). To prove the converse inclusion, take any $z \in \mathrm{ran}(B)$ such that $Sz \in \mathrm{ran}(B)$. Due to (24) there are $z^\pm \in E^{[\pm]}$ such that $z = z^+ + z^-$. Since $\mathrm{ran}(B)$ is a linear subspace, it follows

$$z^\pm = \tfrac{1}{2}\,(z \pm Sz) \in \mathrm{ran}(B)\,,$$

and thus, $z^\pm \in \mathcal{D}^{[\pm]}$. (27) now yields $z \in \mathcal{D}$ proving the first equality in (32). The second equality in (32) follows readily from $S^{-1} = S$.

To prove (33), take $x^\pm \in \mathcal{E}^{(\pm)}$ from (29). Using the operator equation $G = BSB$, it follows

$$\begin{aligned}(x^\pm, x^\pm) = [x^\pm, Gx^\pm] &= [Bx^\pm, SBx^\pm] \overset{(*)}{=} \pm\,[Bx^\pm, Bx^\pm]\\ &= \pm p(Bx^\pm)^2 = \pm q_B(x^\pm)^2\,,\end{aligned}$$

where $(*)$ is a consequence of $Bx^\pm \in \mathcal{D}^{[\pm]}$, $\mathcal{D}^{[\pm]} \subset E^{[\pm]}$ and (25). To prove the converse, take $y^\pm \in E$ satisfying $(y^\pm, y^\pm) = \pm q_B(y^\pm)^2$, and notice that it remains to verify $z^\pm := By^\pm \in \mathcal{D}^{[\pm]}$. Note that

$$\begin{aligned}[z^\pm, z^\pm] = q_B(y^\pm)^2 &= \pm\,(y^\pm, y^\pm) = \pm\,[y^\pm, Gy^\pm]\\ &= \pm\,[By^\pm, SBy^\pm] = \pm\,[z^\pm, Sz^\pm]\,.\end{aligned} \tag{37}$$

Using the $[.,.]$-orthogonal direct decomposition (27), and noticing $S = P^{[+]} - P^{[-]}$ by (23), (37) implies

(38) $$[P^{[\mp]}z^{\pm}, P^{[\mp]}z^{\pm}] = -[P^{[\mp]}z^{\pm}, P^{[\mp]}z^{\pm}].$$

Since $p(.) = \sqrt{[.,.]}$ is a norm, $P^{[\mp]}z^{\pm} = 0$ follows from (38). Hence, $z^{\pm} \in \mathcal{D}^{[\pm]}$ verifying (33). While (34) readily follows from (33) and $\mathcal{D}^{[\pm]} = B\mathcal{E}^{(\pm)}$, (31) will be shown at the end of the present subsection.

b) Assuming that $\{x_n^{\pm}\}_{n=1}^{\infty}$, $x_n^{\pm} \in \mathcal{E}^{(\pm)}$ are sequences satisfying

(39) $$q_B(z^{\pm} - x_n^{\pm}) \to 0$$

as $n \to \infty$ for some $z^{\pm} \in E$, it remains to verify that $z^{\pm} \in \mathcal{E}^{(\pm)}$. Since q_B is self-polar by Theorem 5.4, (39) yields

$$\begin{aligned} |(z^{\pm}, z^{\pm}) - (x_n^{\pm}, x_n^{\pm})| &\le |(z^{\pm} - x_n^{\pm}, z^{\pm})| + |(x_n^{\pm}, z^{\pm} - x_n^{\pm})| \\ &\le q_B(z^{\pm})\, q_B(z^{\pm} - x_n^{\pm}) + q_B(x_n^{\pm})\, q_B(z^{\pm} - x_n^{\pm}) \\ &\to 0 \end{aligned}$$

as $n \to \infty$. Using that $q_B(x_n^{\pm})^2 = \pm(x_n^{\pm}, x_n^{\pm})$ by (33), it follows

$$|q_B(z^{\pm})^2 \mp (z^{\pm}, z^{\pm})| \le |q_B(z^{\pm})^2 - q_B(x_n^{\pm})^2| + |(z^{\pm}, z^{\pm}) - (x_n^{\pm}, x_n^{\pm})| \to 0$$

as $n \to \infty$, and consequently, $z^{\pm} \in \mathcal{E}^{(\pm)}$ due to (33).

c) Since $B\colon E \to \operatorname{ran}(B)$ is bijective, the algebraically direct decomposition $\mathcal{E} = \mathcal{E}^{(+)} \dotplus \mathcal{E}^{(-)}$ follows from (27). Using that q_B is a norm, (33) yields that $\mathcal{E}^{(\pm)}$ are respectively, positive and negative definite. To show the orthogonality of $\mathcal{E}^{(+)}$ and $\mathcal{E}^{(-)}$ relative to both $(.,.)$ and $[.,.]_B$, take any $x^{\pm} \in \mathcal{E}^{(\pm)}$. Then,

$$[Bx^+, Bx^-] \overset{(*)}{=} [SBx^+, Bx^-] = [Bx^+, SBx^-] \overset{(*)}{=} -[Bx^+, Bx^-],$$

and thus

(40) $$[Bx^+, Bx^-] = [Bx^+, SBx^-] = 0,$$

where $(*)$ was already explained in a). Using (40), the proof of c) is completed by

$$\begin{aligned} (x^+, x^-) &= [x^+, Gx^-] = [Bx^+, SBx^-] = 0, \\ [x^+, x^-]_B &= [Bx^+, Bx^-] = 0. \end{aligned}$$

d) Using the Phillips-decomposition (35), the definition of a decomposition norm yields $\|x\|_J = \sqrt{(x, Jx)}$, $x \in \mathcal{E}$, by (2). Considering any $x = x^+ + x^- \in \mathcal{E}$, $x^\pm \in \mathcal{E}^{(\pm)}$, the remainder of the proof follows from

$$\begin{aligned}(x, Jx) &= (x^+ + x^-, x^+ - x^-) = (x^+, x^+) - (x^-, x^-) \\ &\stackrel{(*)}{=} q_B(x^+)^2 + q_B(x^-)^2 = [x^+, x^+]_B + [x^-, x^-]_B \\ &\stackrel{(**)}{=} [x, x]_B = q_B(x)^2 ,\end{aligned}$$

where $(*)$ follows from (33), and $(**)$ is a consequence of (35) and $\mathcal{E}^{(+)} [\perp]_B \, \mathcal{E}^{(-)}$ (here $[\perp]_B$ refers to orthogonality relative to $[.,.]_B$).

□

DEFINITION 7.4. If the linear subspace $\mathcal{E}$ is $\tau(q_B)$-dense in E, then q_B is called *quasi-decomposition norm.* Furthermore, $\mathcal{E}$ is said to be a *quasi-decomposition core* (relative to the Hilbertian norms q_B and p).

(Let us mention that the above definition of a quasi-decomposition norm is different from the one given in [7, Def. 3.9], where density relative to $\tau(p)$ is postulated. Since $\widetilde{E}^{\tau(q_B)}, (.,.)^\sim$ is a Krein space, the above definition seems to be more suitable for further applications.)

Considering the intrinsic topologies $\tau^\pm$ on $\mathcal{E}^{(\pm)}$ defined by the intrinsic norms

$$x^\pm \to |x^\pm|_{int} := \sqrt{|(x^\pm, x^\pm)|} , \tag{41}$$

$x^\pm \in \mathcal{E}^{(\pm)}$, the significance of a quasi-decomposition norm and its quasi-decomposition core follows from the following

PROPOSITION 7.5. *Assume that the inner product space $E, (.,.)$ satisfies all the properties listed at the beginning of Section 5, and q_B taken from (20) is a quasi-decomposition norm satisfying $\tau(q_B) \prec \tau(p)$. Then,*

a) *the Phillips-decomposition*

$$\widetilde{E}^{\tau(q_B)} = \widetilde{\mathcal{E}^{(+)}}^{\tau^+} [\oplus]_B \, \widetilde{\mathcal{E}^{(-)}}^{\tau^-}$$

follows, where $[\oplus]_B$ refers to orthogonality relative to both the $\tau(q_B)$-continuously extended inner product $(.,.)^\sim$ and the scalar product $[.,.]_B = q_B(.)^2$ on $\widetilde{E}^{\tau(q_B)}$.

b) $\mathcal{E}^{(\pm)} = \widetilde{\mathcal{E}^{(\pm)}}^{\tau^\pm} \cap E$ *is true.*

Proof. a) Using that $\mathcal{E}$ is $\tau(q_B)$-dense in E, and then (35) and (33), it follows

$$\widetilde{E}^{\tau(q_B)} = \widetilde{\mathcal{E}}^{\tau(q_B)} = \widetilde{\mathcal{E}^{(+)}}^{\tau(q_B)} [\oplus]_B \widetilde{\mathcal{E}^{(-)}}^{\tau(q_B)}$$
$$= \widetilde{\mathcal{E}^{(+)}}^{\tau^+} [\oplus]_B \widetilde{\mathcal{E}^{(-)}}^{\tau^-} .$$

b) follows from Proposition 7.3 b) and (33).

□

The fundamental symmetry J related to the Phillips-decomposition (35), the linear operator

$$\Gamma = BSB^{-1}, \quad \mathrm{dom}(\Gamma) = \mathrm{dom}(B^{-1}), \tag{42}$$

and its adjoint Γ^* on the Hilbert space $E, [.,.]$ will be considered next.

LEMMA 7.6. a) *The fundamental symmetry belonging to the Phillips-decomposition* (35) *is given by*

$$J = B^{-1}SB = B^{-2}G, \quad \mathrm{dom}(J) = \mathcal{E}. \tag{43}$$

b) *Taking J from* (43), *it follows that $\Gamma^* = J$, $\mathrm{dom}(\Gamma^*) = \mathcal{E}$.*
c) *The linear operator Γ is closable in $E[\tau(p)]$ if and only if the linear subspace $\mathcal{E}$ is $\tau(p)$-dense in E.*
d) *J is closed in $E[\tau(p)]$.*

Proof. a) Taking any $x \in \mathcal{E}$, consider the Phillips-decomposition

$$x = x^+ + x^- ,$$

$x^\pm \in \mathcal{E}^{(\pm)}$. The fundamental symmetry J is then defined by

$$Jx = x^+ - x^- .$$

Considering now $d^\pm = Bx^\pm \in \mathcal{D}^{[\pm]}$, $Sd^\pm = \pm d^\pm$ follows from $\mathcal{D}^{[\pm]} \subset E^{[\pm]}$ and (25). Noticing that

$$B^{-1}SBx^\pm = B^{-1}Sd^\pm = \pm B^{-1}d^\pm = \pm x^\pm ,$$

the first equality of (43) follows from

$$B^{-1}SBx = B^{-1}SBx^+ + B^{-1}SBx^- = x^+ - x^- = Jx .$$

Using (22), the second equality of (43) is obvious.

b) Since $\mathrm{dom}(\Gamma) = \mathrm{dom}(B^{-1}) = \mathrm{ran}(B)$ is dense in $E[\tau(p)]$ by Remark 5.5 c), the adjoint Γ^* exists. Because B and S are self-adjoint operators on the Hilbert space $E, [.,.]$, it follows $\Gamma^* \supset B^{-1}SB = J$, and it remains to verify that $\mathrm{dom}(\Gamma^*) = \mathrm{dom}(J) = \mathcal{E}$. Considering the linear functionals

$$x \to T_y(x) := [y, \Gamma x]\,,$$

$x \in \mathrm{dom}(\Gamma)$, $y \in E$, remember that

$$\mathrm{dom}(\Gamma^*) = \{y \in E;\ T_y \text{ is } \tau(p)\text{-continuous on } E\}\,.$$

Since B^{-1} is self-adjoint, it follows that the linear functional

$$x \to S_w(x) := [w, B^{-1}x]$$

is $\tau(p)$-continuous on E if and only if $w \in \mathrm{dom}(B^{-1})$. Noticing $T_y(x) = [SBy, B^{-1}x]$ and setting $w = SBy$, it follows now that T_y is $\tau(p)$-continuous on E if and only if

$$SBy \in \mathrm{dom}(B^{-1})\,. \tag{44}$$

Noticing that (44) yields $S(By) \in \mathrm{ran}(B)$ and obviously $By \in \mathrm{ran}(B)$, $By \in \mathcal{D}$ follows from (32). Using (28), it follows $y \in \mathcal{E}$ completing the proof of b). c) readily follows from b) and Hilbert space theory ([13, Ch. III, Theorem 5.28]). Since the adjoint of any linear operator in a Hilbert space is closed, b) implies d).

□

Proof of (31). Using (43), $\mathrm{ran}(G) = \mathrm{dom}(G^{-1})$ and $G = BSB$, the relation under consideration follows from

$$\begin{aligned}\mathcal{E} &= \mathrm{dom}(J) = \mathrm{dom}(B^{-2}G) = \{x \in E;\ Gx \in \mathrm{dom}(B^{-2})\}\\ &= G^{-1}(\mathrm{dom}(G^{-1}) \cap \mathrm{dom}(B^{-2}))\,.\end{aligned}$$

□

8. Characterizations of quasi-decomposition norms

Continuing the investigations of the foregoing sections, characterizations such that q_B is a quasi-decomposition norm, a decomposition norm and the generic decomposition norm will be given in the following Propositions 8.2, 8.3 and 8.5, respectively.

As a prerequisite, some properties of the l.c. vector spaces $\mathcal{D}[\tau(p)]$ and $\mathcal{E}[\tau(q_B)]$ are considered next.

LEMMA 8.1.

a) *The topological isomorphism* $\mathcal{D}[\tau(p)] \cong_{top} \mathcal{E}[\tau(q_B)]$ *holds.*

b) $\mathcal{D}$ *is dense in* E *relative to* $\tau(p)$ *if and only if* $\mathcal{E}$ *is dense in* E *relative to* $\tau(q_B)$.

Proof. a) Using Remark 5.5 c) and noticing that $\mathcal{D} \subset \mathrm{ran}(B) = \mathrm{dom}(B^{-1})$, the bijectivity of the linear mapping

$$B^{-1}|_{\mathcal{D}} : \mathcal{D} \to \mathcal{E} \tag{45}$$

follows. Hence it remains to verify that $B|_{\mathcal{E}} : \mathcal{E}[\tau(q_B)] \to \mathcal{D}[\tau(p)]$ is both continuous and open. Those properties readily follow from $p(Bx) = q_B(x)$, $x \in \mathcal{E}$.

b) ($\Leftarrow$): Take any $x \in E$, $\varepsilon > 0$, and assume that the right-hand side of b) of the lemma under consideration applies. Since $\mathrm{ran}(B)$ is $\tau(p)$-dense in E by Remark 5.5 c), an element $y_\varepsilon \in \mathrm{ran}(B)$ exists such that

$$p(x - y_\varepsilon) < \varepsilon/2 .$$

Using that $B : E \to \mathrm{ran}(B)$ is bijective by Remark 5.5 c), a $z_\varepsilon \in E$ exists such that

$$y_\varepsilon = Bz_\varepsilon .$$

Using now that $\mathcal{E}$ is $\tau(q_B)$-dense in E, there is a $w \in \mathcal{E}$ satisfying

$$p(B(z_\varepsilon - w)) = q_B(z_\varepsilon - w) < \varepsilon/2 .$$

Hence,

$$\begin{aligned} p(x - Bw) &\leq p(x - y_\varepsilon) + p(y_\varepsilon - Bw) < \varepsilon/2 + p(B(z_\varepsilon - w)) \\ &< \varepsilon/2 + \varepsilon/2 = \varepsilon . \end{aligned} \tag{46}$$

Noticing $Bw \in \mathcal{D}$, (46) yields the required implication.

($\Rightarrow$): Taking again any $x \in E$, $\varepsilon > 0$, assume now that the left-hand side applies. Then $x_\varepsilon \in \mathcal{D}$ exists such that

$$p(Bx - x_\varepsilon) < \varepsilon\,.$$

Considering $v_\varepsilon = B^{-1}x_\varepsilon \in \mathcal{E}$, the right-hand side follows from

$$q_B(x - v_\varepsilon) = p(B(x - v_\varepsilon)) = p(Bx - x_\varepsilon) < \varepsilon\,.$$

□

PROPOSITION 8.2. *Let $E, (.,.)$ be an inner product space with complete Hilbertian majorant $\tau(p)$, $p(.) = \sqrt{[.,.]}$, and Gram operator G, $(x,y) = [x, Gy]$, $x, y \in E$. Assuming that $G = BSB$ considered in Theorem 5.4 (ii) (2) is satisfied, take the norm q_B given in (20) and the linear subspaces introduced in Notation 7.1.*

Let us consider the statements

(i) *q_B is a quasi-decomposition norm on $E, (.,.)$ with decomposition core*

$$\mathcal{E} = \mathcal{E}^{(+)}\,[\oplus]_B\,\mathcal{E}^{(-)}\,,$$

(ii) *$\mathcal{D}^{[\pm]}$ are $\tau(p)$-dense in $E^{[\pm]}$,*

(iii) *$\mathcal{D}$ is $\tau(p)$-dense in E,*

(iv) *$E \subset \widetilde{\mathcal{E}^{(+)}}^{\tau^{(+)}}\,[\oplus]_B\,\widetilde{\mathcal{E}^{(-)}}^{\tau^{(-)}}$, where $\tau^{(\pm)} = \tau(|.|_{\rm int})$ are taken from (41),*

(v) *$E \subset \widetilde{\mathcal{E}^{(+)}}^{\tau(q_B)}\,[\oplus]_B\,\widetilde{\mathcal{E}^{(-)}}^{\tau(q_B)}$,*

(vi) *the linear operator Γ introduced in (42) is closable in $E[\tau(p)]$.*

Then the implications (i) $\Leftrightarrow$ (ii) $\Leftrightarrow$ (iii) $\Leftrightarrow$ (iv) $\Leftrightarrow$ (v) $\Leftarrow$ (vi) are true.

Proof. (i)$\Rightarrow$(iv): Assuming (i), remember that $q_B(x) = \|x\|_J$, $x \in \mathcal{E}$, by (36). Proposition 7.5 a) now yields (iv).

(iv)$\Rightarrow$(v): Since $\tau^{(\pm)} = \tau(\|.\|_J)|_{\mathcal{E}^{(\pm)}} = \tau(q_B)|_{\mathcal{E}^{(\pm)}}$ by (36), the implication under consideration is obvious.

(i)$\Leftrightarrow$(iii): follows from Lemma 8.1 b).

(iii)$\Leftrightarrow$(ii): follows from (24), (26) and (27).

(v)$\Rightarrow$(i): Assuming (v), the orthogonality $\mathcal{E}^{(+)}\,[\perp]_B\,\mathcal{E}^{(-)}$ yields $\widetilde{\mathcal{E}}^{\tau(q_B)} = \widetilde{\mathcal{E}^{(+)}}^{\tau(q_B)}\,[\oplus]_B\,\widetilde{\mathcal{E}^{(-)}}^{\tau(q_B)}$. Consequently, $\widetilde{\mathcal{E}}^{\tau(q_B)} \supset E$ verifying (i).

(vi)$\Rightarrow$(i): Assuming (vi), Lemma 7.6 c) yields that $\mathcal{E}$ is $\tau(p)$-dense in E. Since $\tau(p) \succ \tau(q_B)$, (i) follows.

□

The special case when the quasi-decomposition norm q_B, considered in Proposition 8.2, is a decomposition norm is characterized in the following

PROPOSITION 8.3. *Letting all the assumptions of Proposition 8.2 be satisfied, the following are equivalent:*

(i) q_B *is a decomposition norm on* $E, (.,.)$,
(ii) $\mathcal{E} = E$,
(iii) $\mathcal{D} = \operatorname{ran}(B)$,
(iv) $S \operatorname{ran}(B) \subset \operatorname{ran}(B)$,
(v) J *is* $\tau(p)$*-continuous on* E,
(vi) Γ *is* $\tau(p)$*-continuous on* E.

Proof. (i)⇔(ii): follows from Proposition 7.3 c), d).

(ii)⇒(iii): readily follows from (28) and the first half of Remark 5.5 c).

(iii)⇒(iv): is a consequence of (30).

(iv)⇒(v): Since J is $\tau(p)$-closed by Lemma 7.6 d), (iv), $J = B^{-1}SB$ and $\operatorname{dom}(B^{-1}) = \operatorname{ran}(B)$ yield $\operatorname{dom}(J) = E$. Using again that J is $\tau(p)$-closed, (v) follows from Hilbert space theory ([17, Kap.VIII, Nr.117]).

(v)⇒(vi): Assuming (v), Γ is closable by [13, Theorem III.5.28] since $J = \Gamma^*$ by Lemma 7.6 b) and $\operatorname{dom}(J)$ is dense in $E[\tau(p)]$. Noticing then that $\operatorname{dom}(\Gamma) = \operatorname{dom}(B^{-1})$ is dense in $E[\tau(p)]$ by Remark 5.5 c), $\bar{\Gamma} = \Gamma^{**}$ follows from [13, Theorem III.5.29]. The $\tau(p)$-continuity of $J = \Gamma^*$ implies that of $\bar{\Gamma} = \Gamma^{**}$ ([17, Kap.V, Nr.84]). It is now obvious that (vi) applies.

(vi)⇒(v): Using $J = \Gamma^*$, the implication under consideration again follows from [17, Kap.V, Nr.84].

(v)⇒(ii): Since J is $\tau(p)$-closed, (v) implies $\operatorname{dom}(J) = E$. Hence $\mathcal{E} = E$ by (43). □

REMARK 8.4. Due to $S^2 = I$ (see Theorem 5.4 (ii)), the above condition (iv) is equivalent to $S \operatorname{ran}(B) = \operatorname{ran}(B)$.

The special case when the quasi-decomposition norm q_B considered in Proposition 8.2 is the generic decomposition norm is characterized next.

PROPOSITION 8.5. *Letting all the assumptions of Proposition 8.2 be satisfied, the following are equivalent:*

(i) q_B *is the generic decomposition norm on* $E, (.,.)$,
(ii) $\mathcal{E}^{(\pm)} = E^{[\pm]}$,
(iii) $J = S$,
(iv) $BSx = SBx$ *for* $x \in E$,
(v) Γ *is closable and* $\bar{\Gamma} = J$ *applies (here the bar denotes the closure of* Γ).

Proof. (i)⇒(ii): Assuming (i), $B = \sqrt{|G|}$, $S = \text{sign}\,(G)$ follow from Definition 6.2 and Proposition 6.1. Using the spectral resolution $G = \int_{-1}^{1} \lambda\, dP(\lambda)$, one gets $P^{[+]} = \int_0^1 dP(\lambda)$ and $P^{[-]} = \int_{-1}^{0} dP(\lambda) = I - P^{[+]}$. Taking $E^{[\pm]} = \text{ran}(P^{[\pm]})$ from (24) and (25), it follows that the linear mappings

$$(\sqrt{|G|})\big|_{E^{[\pm]}} : E^{[\pm]} \to E^{[\pm]} \cap \text{ran}(\sqrt{|G|})$$

are bijective. Now (ii) follows from

$$E^{[\pm]} = ((\sqrt{|G|})\big|_{E^{[\pm]}})^{-1}(E^{[\pm]} \cap \text{ran}(\sqrt{|G|})) \overset{(*)}{=} \mathcal{E}^{[\pm)},$$

where $(*)$ is a consequence of (26) and (29).

(ii)⇒(iii): Recalling that J is the fundamental symmetry related to the Phillips decomposition $\mathcal{E} = \mathcal{E}^{(+)}\,[\oplus]\,\mathcal{E}^{(-)}$ by Proposition 7.3 c), d), it follows that $J = \mathcal{P}^{(+)} - \mathcal{P}^{(-)}$, where $\mathcal{P}^{(\pm)}\colon \mathcal{E} \to \mathcal{E}^{(\pm)}$ are the corresponding fundamental projections. Assuming that (ii) applies, $\mathcal{P}^{(\pm)} = P^{[\pm]}$ follow. (iii) is now a consequence of

$$S \overset{(**)}{=} P^{[+]} - P^{[-]} = \mathcal{P}^{(+)} - \mathcal{P}^{(-)} = J\,,$$

where $(**)$ follows from (23).

(iii)⇒(ii): Assuming (iii), (ii) immediately follows from

$$P^{[\pm]} = \tfrac{1}{2}(I \pm S) = \tfrac{1}{2}(I \pm J) = \mathcal{P}^{(\pm)}$$

and $E^{[\pm]} = P^{[\pm]}E = \mathcal{P}^{(\pm)}E = \mathcal{E}^{(\pm)}$.

(ii)⇒(iv): Assuming (ii), $BE^{[\pm]} = B\mathcal{E}^{(\pm)} = \mathcal{D}^{[\pm]} \subset E^{[\pm]}$ follow from (29), (26). Remember also that

$$S|_{E^{[\pm]}} = \pm I \tag{47}$$

by (25). Since for each $x \in E$, a unique representation $x = x^+ + x^-$, $x^\pm \in E^{[\pm]}$, is given by (24), it follows that

$$BSx = B(x^+ - x^-) = Bx^+ - Bx^- = SBx^+ + SBx^- = SBx$$

proving (iv).

(iv)⇒(v): Assuming (iv), we infer $J = B^{-1}SB = S$, $\text{dom}(J) = E$, and $\Gamma = BSB^{-1} = S|_{\text{ran}(B)}$, $\text{dom}(\Gamma) = \text{ran}(B)$. Now (v) follows from

$$\bar{\Gamma} = \overline{S|}_{\text{ran}(B)} = S = J\,.$$

(v)⇒(iv): Assuming (v), it follows as well that $\operatorname{ran}(B) = \operatorname{dom}(\Gamma) \subset \operatorname{dom}(J) = \mathcal{E}$ as $BSB^{-1}y = B^{-1}SBy$ for $y \in \operatorname{ran}(B)$. Consequently, $B^2Sz = SB^2z$ for $z \in E$. It then follows from Hilbert space theory (see [17, p. 250]), that $\sqrt{B^2}Sz = S\sqrt{B^2}z$. Since B is a positive operator on Hilbert space $E, [.,.]$, $\sqrt{B^2} = B$ follows. Hence (iv) applies.

(iv)⇒(i): Assuming (iv), consider $B_\pm = B|_{E^{[\pm]}} \oplus 0|_{E^{[\mp]}}$, and notice that $B_\pm E^{[\pm]} \subset E^{[\pm]}$, $B_+B_- = B_-B_+ = 0$, where 0 denotes the 0 operator on E. Setting $G_\pm = G|_{E^{[\pm]}} \oplus 0|_{E^{[\mp]}}$, it follows $G_\pm E^{[\pm]} \subset E^{[\pm]}$, $G_+G_- = G_-G_+ = 0$. Then,

$$G_\pm = (BSB)|_{E^{[\pm]}} = \pm(B_\pm)^2$$

follows from (47). Hence,

$$|G| = G_+ - G_- = (B_+)^2 + (B_-)^2 = (B_+ + B_-)^2 = B^2\,.$$

Since B is positive, it follows $B = \sqrt{B^2} = \sqrt{|G|}$ proving that (i) applies.

□

REMARK 8.6. Using the theory of induced Krein spaces, three possibilities to construct Krein spaces are considered by Constantinescu and Gheondea in [6, Ex. 2.1, 2.3, 2.5]. Let us mention that the norms considered in those examples are equivalent to our generic decomposition norm.

ACKNOWLEDGEMENTS. The author would like to thank Professor H. Langer for helpful comments on the present paper. Stimulating discussions with Dr. P. Jonas and K. Luig are also acknowledged.

References

[1] H. Araki, On a pathology in indefinite inner product spaces, *Commun. Math. Phys.*, **85** (1982), 121–128.

[2] T. Y. Azizov and I. S. Iokhvidov, *Linear operators in spaces with indefinite metric*, John Wiley & Sons Inc., New York, 1989.

[3] J. Bognár, *Indefinite inner product spaces*, Springer-Verlag, Berlin, 1974.

[4] N. N. Bogoljubov, A. A. Logunov, A. I. Oksak and I. T. Todorov, *General principles of quantum field theory*, Kluwer, Dordrecht, 1990.

[5] H. J. Borchers, On the structure of the algebra of field operators, *Nuovo Cimento*, **24** (1962), 214–236.

[6] T. Constantinescu and A. Gheondea, Representations of hermitian kernels by means of Krein spaces, *Publ. RIMS Kyoto Univ.*, **33** (1997), 917–951.

[7] F. HANSEN, Selfpolar norms on an indefinite inner product space, *Publ. RIMS Kyoto Univ.*, **16** (1980), 889–913.

[8] G. HOFMANN, On GNS representations on inner product spaces; I. The structure of the representation space, *Commun. Math. Phys.*, **191** (1998), 299–323.

[9] G. HOFMANN, On inner characterizations of pseudo-Krein and pre-Krein spaces, *Commun. Math. Phys.*, submitted; preprint: Leipzig 1998.

[10] G. HOFMANN and K. LUIG, On the construction of selfpolar and selfpolar Hilbertian norms on inner product spaces, preprint, Leipzig 1999.

[11] H. JARCHOW, *Locally Convex Spaces*, B. G. Teubner, Stuttgart, 1981.

[12] P. E. T. JORGENSEN, *Operators and Representation Theory*, North-Holland (Elsevier), Amsterdam, 1988.

[13] T. KATO, *Pertubation Theory for Linear Operators*, Springer-Verlag, Berlin – Heidelberg, 1995.

[14] E. KISSIN and V. SHULMAN, *Representations on Krein spaces and derivations of C^*-algebras*, Addison Wesley Longman Limited, 1997.

[15] N. NAKANISHI and I. OJIMA, *Covariant operator formalism of gauge theories and quantum gravity*, World Scientific, Singapore, 1990.

[16] M. A. NEUMARK, *Normierte Algebren*, Verlag Harri Deutsch, Thun – Frankfurt am Main, 1990.

[17] F. RIESZ and B. SZ.-NAGY, *Vorlesungen über Funktionalanalysis*, Deutscher Verlag der Wissenschaften, Berlin, 1968.

[18] E. SCHEIBE, *Über Feldtheorien in Zustandsräumen mit indefiniter Metrik*, mimeographed notes of MPI, Max-Planck Institut für Physik und Astrophysik, München, 1960.

[19] K. SCHMÜDGEN, *Unbounded Operator Algebras and Representation Theory*, Akademie-Verlag, Berlin, 1990.

[20] F. STROCCHI, *Selected Topics on the General Properties of QFT*, Lecture Notes in Physics Vol. **51**, World Scientific, Singopore – New Jersey – Hong Kong, 1993.

[21] A. UHLMANN, Über die Definition der Quantenfelder nach Wightman und Haag, *Wiss. Zeitschr. d. Univ. Leipzig*, **11** (1962), 213–217.

[22] J. YNGVASON, Remarks on the reconstruction theorem for field theories with indefinite metric, *Reports on Math. Physics*, **12** (1977), 57–64.

G. HOFMANN, F.-Siemon-Str. 7, D-04317 Leipzig, Germany;
e-mail: hofmann@mathematik.uni-leipzig.de

Operator Theory: Advances and Applications, Vol. 127, 375–386

Schur norms and the multivariate von Neumann inequality

JOHN A. HOLBROOK

In memory of Béla Szőkefalvi-Nagy

Starting from some classical counterexamples to the von Neumann inequality for several variables, we are led to especially simple examples of this phenomenon. We display three commuting 4-dimensional contractions C_k and a polynomial $p(z_1, z_2, z_3)$ such that

$$\|p(C_1, C_2, C_3)\| = \tfrac{6}{5} \max\{|p(z_1, z_2, z_3)| : |z_k| \leq 1\}. \tag{1}$$

We find that this phenomenon depends on the norm of Schur multiplication by certain matrices and on the related Haagerup factorizations. It is easy to perturb the example to a triple of generic commuting contractions and so provide an answer to a question of Lewis and Wermer.

1. Background

The original von Neumann inequality comes from John von Neumann's paper [vN51]. Over the years it has provoked a remarkable variety of related results and stimulating questions. Here we are mainly concerned with the multivariate extensions of the inequality and, indirectly, with the associated unitary dilation theory. The result from [vN51] may be phrased as follows: if C is a contraction operator on a complex Hilbert space and $p(z)$ is any polynomial, then

$$\|p(C)\| \leq \max\{|p(z)| : |z| \leq 1\}. \tag{2}$$

This result is easily reduced to the finite-dimensional case, where we may think of C as a matrix. Many proofs have been devised for (2). For example, both Gilles

AMS Subject Classification (1991): 47A13, 47A20, 47A30.

Pisier (see [Pi95], chapter 1) and Edward Nelson (see [N61]) have pointed out that the maximum modulus principle allows us to reduce (2) to the case where C is unitary, for which one need only invoke the spectral theorem; note that the unitary here is not a unitary dilation in the sense of Sz.-Nagy, but simply a unitary matrix on the same Hilbert space. On the other hand, the approach of Sz.-Nagy (via unitary power dilations) has been crucial for subsequent developments of (2), and particularly for the multivariate versions.

The natural multivariate question suggested by (2) concerns contractions $C_1, \ldots, C_n$ that commute on a common Hilbert space. Is it true that, for any polynomial $p(z_1, \ldots, z_n)$ in n variables,

$$\|p(C_1, \ldots, C_n)\| \leq \max\{|p(z_1, \ldots, z_n)| : |z_k| \leq 1\}? \tag{3}$$

Major steps in answering this question are due to Tsuyoshi Ando and Nicholas Varopoulos. In [A63] Ando showed (by constructing unitary dilations) that (3) holds for $n = 2$. In [V74] Varopoulos found that (3) fails in general. He first established the existence of counterexamples by a probabilistic argument, but he and S. Kaijser (see addendum to [V74]) also provided specific counterexamples which seemed about as simple as one could expect: commuting triples of 5×5 matrices with a quadratic polynomial! About the same time M. J. Crabb and A. Davie (see [C-D75]) found another type of counterexample that involved larger matrices (8×8 triples) but had, nevertheless, certain advantages. In Sections 2 and 3 we shall 'optimize' the Kaijser–Varopoulos construction, and see that it works in dimension 4.

Noting that the classical counterexamples all involved nilpotent matrices, Keith Lewis and John Wermer asked in [L-W92] whether (3) could fail for generic matrices (that is, for matrices with distinct eigenvalues). With Matjaž Omladič we showed that the Kaijser–Varopoulos matrices can be perturbed to (commuting) generics (see [L-W92], p.276). B. Lotto and T. Steger also perturbed classical counterexamples in [L-S94]. We shall see that it is especially easy to make such perturbations starting from the new 4×4 counterexamples; furthermore, such perturbations must exist on general grounds (the effect of dimension on commuting triples is treated in our forthcoming paper [H-O96] with M. Omladič).

By the Schur norm $\|M\|_S$ of a matrix M we mean the norm of the operation of Schur (entry-wise) multiplication $M \circ T$ with respect to the operator norm of T:

$$\|M\|_S = \max\{\|M \circ C\| : \|C\| \leq 1\}. \tag{4}$$

This concept has, of course, a substantial history in the classical work of Schur and others, and it continues to play an important role in matrix analysis. To represent recent trends we may mention various works of Carl Cowen et al. (for example, [A-C-N92] and [C-D-P94]), [B-C-D89] by Rajendra Bhatia, Man-Duen Choi, and Chandler Davis, and [P-P-S89] by Vern Paulsen, Steve Power, and Roger Smith. Of particular interest is an observation of Uffe Haagerup [Haa84] (or see [P86], pp. 110–116):

$$\|M\|_S = \min\{\|X\|_r \|Y\|_c : XY = M\}, \tag{5}$$

where XY is the ordinary matrix product of square matrices X,Y, $\|X\|_r$ is the maximum (Euclidean) length of the rows of X, and $\|Y\|_c$ is the maximum length of the columns of Y. We shall refer to a relation $M = XY$ where $\|M\|_S = \|X\|_r \|Y\|_c$ as a 'Haagerup factorization' of M.

We contend that it is often instructive to find explicit Haagerup factorizations. As an initial example of this idea, let us relate it to the following result of Bhatia, Choi, and Davis.

PROPOSITION 1. (See [B-C-D89]). *If M_n is the $n \times n$ matrix with 0's on the diagonal and 1's elsewhere (so that $T \circ M_n$ is the 'off-diagonal part' of T), then*

$$\|M_n\|_S = 2\Big(1 - \frac{1}{n}\Big). \tag{6}$$

To describe a Haagerup factorization of M_n, we use the structure of n 'maximally spread' unit vectors u_k in $\mathbb{R}^{n-1}$:

$$u_k \cdot u_j = -\frac{1}{n-1} \quad (k \neq j). \tag{7}$$

We are not aware of an established name for this structure (which we may picture as the distribution of n identical, mutually repelling particles confined to the unit sphere in $\mathbb{R}^{n-1}$), but it is occasionally useful (see also the application to statistical sampling in our paper with Ken Gerow [G-H96]). Several possible constuctions for the vectors u_k will no doubt occur to the reader; let us mention only the inductive procedure: the choice $u_1 = 1, u_2 = -1$ is natural for $n = 2$, and given any solution $\{u_k\}$ for n we can define $v_1 = (1, \vec{0})$ and

$$v_k = \left(-\frac{1}{n}, u_{k-1}\sqrt{1 - \frac{1}{n^2}}\right) \quad (k = 2, \ldots, n+1) \tag{8}$$

to obtain a solution for $n+1$. It is easy to verify that the analogue of (7) holds for the v_k.

To obtain a Haagerup factorization of M_n, consider unit vectors u_k as in (7) and construct related unit vectors in $\mathbb{R}^n$ as follows:

$$r_k = \left(\frac{1}{\sqrt{2}}, \frac{1}{\sqrt{2}}u_k\right), \quad c_k = \left(\frac{1}{\sqrt{2}}, -\frac{1}{\sqrt{2}}u_k\right). \tag{9}$$

We compute that

$$r_k \cdot c_j = \frac{1}{2(1-\frac{1}{n})} \quad (k \neq j), \tag{10}$$

whereas $r_k \cdot c_k = 0$, i.e. $r_k \perp c_k$. If X has the vectors $2(1-\frac{1}{n})r_k$ as its rows, and Y has c_k as columns, we see that $M_n = XY$ and it follows that

$$\|M_n\|_S \leq \|X\|_r \|Y\|_c = 2(1-\frac{1}{n}). \tag{11}$$

To show that we have equality in (11) (i.e. that the factorization is really a *Haagerup* factorization), we may use the same extremal matrices as in [B-C-D89].

It may be of interest to note that our argument in (11) involves only real matrices, whereas the quite different approach in [B-C-D89] appears to depend on matrices with complex entries. Since the extremal matrices for (6) can be chosen to be real, the value of $\|M_n\|_S$ does not depend on whether the context is $\mathbb{R}^n$ or $\mathbb{C}^n$, but it does not seem clear that this is the case for every real matrix.

The writer is pleased to acknowledge support from NSERC of Canada, via a research grant. He is also grateful to many colleagues for stimulating discussions; Man-Duen Choi, Ken Davidson, Farid Farid, Douglas Harder, Matjaž Omladič, and Yong Zhong were especially important in this regard.

2. Minimal counterexamples

The Kaijser–Varopoulos counterexamples (see [V74], addendum) made use of the following quadratic polynomial:

$$p(z_1, z_2, z_3) = z_1^2 + z_2^2 + z_3^2 - 2(z_1 z_2 + z_2 z_3 + z_3 z_1). \tag{12}$$

We also work with this specific polynomial, and we need to evaluate

$$\|p\|_\infty = \max\{|p(z_1, z_2, z_3)| : |z_k| \leq 1\}. \tag{13}$$

The treatment of this point in [V74] is somewhat elusive; as an alternative, we offer the argument in the following proposition (which has doubtless occurred to others, as well).

PROPOSITION 2. *If p is the polynomial of* (12), *then* $\|p\|_\infty = 5$.

By the maximum modulus principle, we may assume in (13) that $|z_k| = 1$. Choosing square roots s_k for the z_k, we may write

$$p(z_1, z_2, z_3) = z_1 z_2\Big(\Big(\frac{s_1}{s_2} + \frac{s_2}{s_1} - \frac{z_3}{s_1 s_2}\Big)^2 - 4\Big). \tag{14}$$

Thus,

$$\|p\|_\infty = \max\{|z^2 - 4| : z = t - w, t \in [-2, 2], |w| = 1\}, \tag{15}$$

where, with $s = s_1/s_2$, we interpret t as $s + s^{-1} = s + \overline{s}$. Now we need only examine the maximum in (15) along the boundary of the bacilliform region mapped out by $t - w$. At the rounded ends of this boundary $|z \pm 2| = 1$ so that $|z \mp 2| \leq 5$ and $|z^2 - 4| \leq 5$. The rest of the boundary consists of points of the form $z = t \pm i$, where $|t| \leq 2$; in this case we compute: $|z^2 - 4|^2 = t^2(t^2 - 6) + 25$, so that $|z^2 - 4| \leq 5$ whenever $|t| \leq \sqrt{6}$. The bound 5 is attained at certain points: at $z_1 = z_2 = 1, z_3 = -1$, for example.

Turning to the construction of the matrices, consider a Hilbert space spanned by unit vectors e, f_1, f_2, f_3, h, where e and h are orthogonal and both are orthogonal to the span F of the f_k. Define operators T_k on this space by linear extension of the relations $T_k e = f_k$,

$$T_k f_j = \begin{cases} h & \text{if } k = j \\ -\frac{1}{2}h & \text{if } k \neq j, \end{cases} \tag{16}$$

and $T_k h = 0$. Evidently, the products $T_k T_j$ are null on $e^\perp$ and

$$T_k T_j e = \begin{cases} h & \text{if } k = j \\ -\frac{1}{2}h & \text{if } k \neq j, \end{cases} \tag{17}$$

so that the operators commute and

$$\|p(T_1, T_2, T_3)\| = \|p(T_1, T_2, T_3)e\| = \Big\|3h - 6\Big(-\frac{h}{2}\Big)\Big\| = 6. \tag{18}$$

Finally, the T_k will be contractions if we have an appropriate geometry in F. To this end, let F be two-dimensional and let the f_k be 'maximally spread' unit vectors as in (7) (with $n = 3$). Then (7) and (16) imply that, for any $f \in F$, $T_k f = (f, f_k)h$, so that for any vector v in our (four-dimensional) Hilbert space

$$T_k v = (v, e)f_k + (v, f_k)h. \tag{19}$$

Thus $\|T_k v\| = \|P_k v\|$, where P_k is othogonal projection onto the span of e and f_k. Certainly, then, the T_k are contractions. In view of Proposition 2, our contractions T_k demonstrate:

PROPOSITION 3. *There exist three commuting contractions T_k on a four-dimensional Hilbert space such that*

$$\|p(T_1, T_2, T_3)\| = \frac{6}{5}\|p\|_\infty. \tag{20}$$

In describing our operators T_k, we have followed the presentation of Kaijser–Varopoulos [V74] rather closely, making directly those changes that result in two improvements: a Hilbert space setting of dimension four rather than five, and a 'violation' of von Neumann's inequality (3) by a full 20% (in [V74] the corresponding figure is a little less than 4%, since $3\sqrt{3}/5 \approx 1.039$).

It is also instructive to present the T_k in matrix form. With respect to the orthonormal basis $\{h, f_1, g, e\}$, where $g = (f_2 + \frac{1}{2}f_1)/\|f_2 + \frac{1}{2}f_1\|$, it is easy to check that the T_k have the form

$$\begin{pmatrix} 0 & v_k & - & 0 \\ 0 & 0 & 0 & v_k^* \\ 0 & 0 & 0 & | \\ 0 & 0 & 0 & 0 \end{pmatrix}, \tag{21}$$

where $v_1 = (1, 0)$, $v_2 = (-\frac{1}{2}, \frac{\sqrt{3}}{2})$, and $v_3 = (-\frac{1}{2}, -\frac{\sqrt{3}}{2})$. Note that in (21) and (22) the lines beside vector entries merely indicate that the vectors occupy several matrix positions. In Section 3 we consider a class of such matrix forms that includes those above as well as the Kaijser–Varopoulos examples, and allows an analysis in terms of Schur norms and Haagerup factorizations.

3. Variations on Kaijser–Varopoulos

We contend that an appropriate setting for the examples discussed in Section 2 is the class of matrix triples

$$C_k = \begin{pmatrix} 0 & - & v_k & - & 0 \\ 0 & 0 & 0 & 0 & | \\ 0 & 0 & 0 & 0 & w_k \\ 0 & 0 & 0 & 0 & | \\ 0 & 0 & 0 & 0 & 0 \end{pmatrix}, \tag{22}$$

where the v_k are real unit row vectors of length 3 and the w_k are real unit column vectors of length 3 ($k = 1, 2, 3$). Because these are unit vectors it is clear that

$\|C_k\| = 1$, and we also see that

$$C_kC_j = \begin{pmatrix} 0 & 0 & 0 & 0 & v_k \cdot w_j \\ 0 & 0 & 0 & 0 & 0 \\ 0 & 0 & 0 & 0 & 0 \\ 0 & 0 & 0 & 0 & 0 \\ 0 & 0 & 0 & 0 & 0 \end{pmatrix}, \tag{23}$$

so that the C_k commute and $p(C_1, C_2, C_3)$ (where p is again the polynomial in (12)) depends only on the dot products $v_k \cdot w_j$.

Now, a comparison with [V74] reveals that the Kaijser–Varopoulos examples correspond to a choice of vectors v_k, w_k such that $v_k \cdot w_j = \rho$ when $k = j$ and $v_k \cdot w_j = -\rho$ when $k \neq j$, with $\rho = 1/\sqrt{3}$. Thus

$$\frac{\|p(C_1, C_2, C_3)\|}{\|p\|_\infty} = \frac{3\rho + 6\rho}{5} = \frac{3\sqrt{3}}{5} \approx 1.039, \tag{24}$$

so that (3) is violated by nearly 4%.

If we let V denote the 3×3 matrix whose rows are the v_k and let W denote the 3×3 matrix whose columns are the w_j, the relations of the last paragraph take the form:

$$VW = \rho(I_3 - M_3), \tag{25}$$

where M_3 is the 'off-diagonal' matrix of Proposition 1. Recalling our discussion of Schur norms and Haagerup factorizations in Section 1, we see that a natural way to 'improve' the Kaijser–Varopoulos construction might be to increase ρ so that $\|\rho(I_3 - M_3)\|_S = 1$ and to choose V and W so that VW is a Haagerup factorization of $\rho(I_3 - M_3)$ (with $\|V\|_r = \|W\|_c = 1$). In particular, we need the value of $\|I_3 - M_3\|_S$.

PROPOSITION 4. *If M_n is the $n \times n$ 'off-diagonal' matrix (see Proposition 1), then (for $n \geq 2$)*

$$\|I_n - M_n\|_S = 3 - \frac{4}{n}. \tag{26}$$

One way to establish (26) is to further exploit a technique from [B-C-D89]: for any $n \times n$ matrix T,

$$I_n \circ T = \frac{1}{n} \sum_{k=0}^{n-1} U^k T U^{-k}, \tag{27}$$

where U is the unitary matrix with the n-th roots of unity on the diagonal. Thus

$$\|(I_n - M_n) \circ T\| = \|2I_n \circ T - T\| = \left\|\left(\frac{2}{n} - 1\right)T + \frac{2}{n}\sum_{k=1}^{n-1} U^k T U^{-k}\right\|, \tag{28}$$

and, since $\|U^k T U^{-k}\| = \|T\|$,

$$\|I_n - M_n\|_S \leq \left(1 - \frac{2}{n}\right) + \frac{2}{n}(n-1) = \frac{3n-4}{n}. \tag{29}$$

Let J_n denote the matrix whose entries are all 1's; note that J_n is n times a projection, and so is self-adjoint with eigenvalues n and 0. A routine computation with $T = J_n - \frac{n}{2}I_n$ shows that the estimate (29) is sharp, so that the proposition follows.

For $n = 3$ we obtain in (26) a Schur norm of 5/3, so that we may replace ρ in the Kaijser–Varopoulos construction by 3/5. Thus, invoking the Haagerup theorem, we know that unit vectors v_k, w_j exist such that (22) yields

$$\frac{\|p(C_1, C_2, C_3)\|}{\|p\|_\infty} = \frac{9}{5} \cdot \frac{3}{5} = \frac{27}{25}, \tag{30}$$

a violation of (3) by 8%.

Returning to our contention that one should seek *explicit* Haagerup factorizations when possible, let us consider unit vectors u_k $(k = 1, \ldots, n)$ in $\mathbb{R}^{n-1}$ that are 'maximally spread', that is, satisfy (7). Define, for real a, b with $a^2 + b^2 = 1$, unit vectors v_k, w_j in $\mathbb{R}^n$ by

$$v_k = (au_k, b), \quad w_j = (au_j, -b). \tag{31}$$

With V, W defined as before, we have

$$VW = \rho I_n - \sigma M_n, \tag{32}$$

where

$$\rho = a^2 - b^2 \text{ and } \sigma = \frac{a^2}{n-1} + b^2. \tag{33}$$

By Haagerup's theorem we conclude that

$$\|\rho I_n - \sigma M_n\|_S \leq 1, \tag{34}$$

for each choice of a, b such that $a^2 + b^2 = 1$. A simple computation, again with $T = J_n - \frac{n}{2}I_n$, shows that (34) is sharp. We thus obtain:

PROPOSITION 5. *Whenever $n \geq 2$ and real a, b satisfy $a^2 + b^2 = 1$ we have*

$$\|\rho I_n - \sigma M_n\|_S = 1, \tag{35}$$

where ρ and σ are as in (33).

It is easy to check that we recover Propositions 1 and 4 from the cases where $\rho = 0$ and $\rho = \sigma$, respectively. Other proofs of Proposition 5 are certainly possible (via (27), for example), but it is the factorization VW that we wish to exploit.

In the case $n = 3$, $\rho = \sigma$ occurs when $a^2 = 4/5$ and $b^2 = 1/5$, so that we may find the 8% solution explicitly via (31). Moreover, we may consider the construction where $\rho \neq \sigma$, varying a and b so as to maximize

$$\frac{\|p(C_1, C_2, C_3)\|}{\|p\|_\infty} = \frac{3\rho + 6\sigma}{5} = \frac{3(a^2 - b^2) + 6(a^2/2 + b^2)}{5} = \frac{6a^2 + 3b^2}{5}. \tag{36}$$

Recalling that $a^2 + b^2 = 1$, we see that this ratio is $\frac{6}{5} - \frac{3}{5}b^2$, which is maximized at $6/5$ when $b = 0$.

Thus the most extreme examples of this type violate (3) by 20% and, most surprisingly, occur where $b = 0$, so that we may as well use the two-dimensional vectors $v_k = w_k = u_k$. Thus it is the explicit Haagerup factizations that lead us to the 'minimal' counterexamples of Section 2.

4. The 3×3 case

We cannot quite justify the term 'minimal' as applied to the examples in Section 2, because the 3×3 case is unsettled, as far as we know! What evidence there is suggests that (3) always holds when the commuting contractions are 3×3 matrices. The writer, Farid Farid, Yong Zhong, Man-Duen Choi, and others no doubt, have checked a variety of special cases. In addition, computer experiments, though limited, appear to support (3) in the 3×3 case; in particular, trials devised by Douglas Harder (supported by an NSERC Undergraduate Student Research Award, summer 1994) revealed no counterexamples.

In [H92] we highlighted a question about 2×2 matrices which, if answered, would either lead to 4×4 counterexamples to (3) or verify (3) in the 3×3 case. In view of Section 2, the first alternative is now of less interest; the second is still an interesting possibility.

Choi and Zhong (private communication) have found a class of examples that lead to an important distinction between the 3×3 and 2×2 cases. Any finite family of commuting 2×2 contractions can be expressed in terms of contractive (analytic) functions of a single contraction; it follows that the family is 'dilatable', that is, there are simultaneous commuting unitary dilations for the contractions of such a family (and this is one way to see that (3) holds). In contrast, Choi and Zhong have examples of four commuting 3×3 contractions that are *not* dilatable (although (3) holds for them, and any subset of three *is* dilatable). If a family of contractions can be represented as contractive functions of two fixed commuting contractions, then Ando's dilation theorem (from [A63]) allows us to construct a dilation for the whole family (and (3) follows). The examples of Choi and Zhong show that this representation is not always possible in the 3×3 case. In fact we have:

PROPOSITION 6. *There exist three commuting* 3×3 *contractions* C_k *that cannot be represented in the form*

$$C_k = f_k(T_1, T_2), \tag{37}$$

where T_1, T_2 *are commuting contractions and each* f_k *is a contractive analytic function:*

$$f_k : D \times D \to D, \tag{38}$$

(D *denotes the complex unit disc).*

We sketch one possible argument. If Proposition 6 were not true, then given any number of commuting 3×3 contractions C_k, we could iterate the representation procedure, working first with C_1, C_2, C_3, then T_1, T_2, C_4, then S_1, S_2, C_5, and so on. Commutativity of the successive triples could be ensured by first perturbing the C_k, if necessary, so that they have distinct eigenvalues (compare [H-O96], for example). We would obtain in this way a two-contraction representation of the whole family. Then the Ando theorem combined with the dilation theorem for commuting *normal* contractions would yield a simultaneous dilation for all of the C_k. This, of course, conflicts with the Choi–Zhong examples.

5. Generic examples

As noted in Section 1, the question raised by Lewis and Wermer in [L-W92] as to whether (3) can fail for *generic* contractions (that is, contractions with distinct eigenvalues) may be answered in several ways. It turns out that *all* commuting triples of $n \times n$ matrices can be perturbed to (commuting) generic triples if n is sufficiently small. Since $n = 4$ is certainly small enough, nothing more needs to be said once we have examples such as those of Section 2. It is a striking fact that the general perturbation result *fails* if n is large (if $n = 30$, for example). This phenomenon is examined in [H-O96]; Robert Guralnick's paper [Gu92] is crucial.

Instead of relying on such general results, however, we may exploit some special features of the operators T_1, T_2, T_3 from Section 2 to describe a more explicit generic perturbation. Note that (21) implies that we have

$$T_2 = -\frac{1}{2}T_1 + \frac{\sqrt{3}}{2}Q, \quad T_3 = -\frac{1}{2}T_1 - \frac{\sqrt{3}}{2}Q, \tag{39}$$

where

$$Q = \begin{pmatrix} 0 & 0 & 1 & 0 \\ 0 & 0 & 0 & 0 \\ 0 & 0 & 0 & 1 \\ 0 & 0 & 0 & 0 \end{pmatrix}. \tag{40}$$

Consider the following perturbation of T_1:

$$T_1(\epsilon) = \begin{pmatrix} 0 & 1 & \epsilon & 0 \\ 0 & \epsilon & 0 & 1 \\ 0 & 0 & 0 & \epsilon \\ 0 & 0 & 0 & 0 \end{pmatrix}. \tag{41}$$

For each $\epsilon \neq 0$, Q may be expressed as a polynomial $p_\epsilon(T_1(\epsilon))$ in $T_1(\epsilon)$. In fact, a routine calculation reveals that

$$p_\epsilon(z) = \epsilon^{-1}z + \epsilon^{-4}z^2 - (\epsilon^{-3} + \epsilon^{-5})z^3 \tag{42}$$

is appropriate. Thus we can approximate T_1 arbitrarily well by $T_1' = T_1(\epsilon)$ for some small $\epsilon \neq 0$ and simultaneously approximate T_2, T_3 by $T_2' = q_2(T_1')$, $T_3' = q_3(T_1')$, where

$$q_2(z) = -\frac{1}{2}z + \frac{\sqrt{3}}{2}p_\epsilon(z), \quad q_3(z) = -\frac{1}{2}z - \frac{\sqrt{3}}{2}p_\epsilon(z). \tag{43}$$

Finally, we perturb T_1' to generic T_1'' and set $T_k'' = q_k(T_1'')$ $(k = 2, 3)$.

References

[A63] T. ANDO, On a pair of commutative contractions, *Acta Sci. Math. (Szeged)*, **24** (1963), 88–90.

[A-C-N92] J. R. ANGELOS, C. C. COWEN and S. K. NARAYAN, Triangular truncation and finding the norm of a Hadamard multiplier, *Linear Algebra Appl.*, **170** (1992), 117–136.

[B-C-D89] R. BHATIA, M.-D. CHOI and CH. DAVIS, Comparing a matrix to its off-diagonal part, *Operator Theory: Advances and Applications*, **40** (1989), 151–164.

[C-D-P94] C. C. COWEN, M. A. DRITSCHEL and R. C. PENNEY, Norms of Hadamard multipliers, *SIAM J. Matrix Anal. Appl.*, **15** (1994), 313–320.

[C-D75] M. CRABB and A. DAVIE, Von Neumann's inequality for Hilbert space operators, *Bull. London Math. Soc.*, **7** (1975), 49–50.

[G-H96] K. GEROW and J. HOLBROOK, Statistical sampling and fractal distributions, *The Mathematical Intelligencer*, **18** (1996), 12–22.

[Gu92] R. GURALNICK, A note on commuting pairs of matrices, *Lin. Mult. Alg. (LAMA)*, **31** (1992), 71–75.

[Haa84] U. HAAGERUP, Decompositions of completely bounded maps on operator algebras, *preprint* (1984).

[H92] J. HOLBROOK, Inequalities of von Neumann type for small matrices, *Function Spaces* (ed. K. Jarosz), Marcel Dekker 1992, 189–193.

[H-O96] J. HOLBROOK and M. OMLADIČ, Approximation of commuting operators, *Linear Algebra and Appl.*, to appear.

[L-W92] K. LEWIS and J. WERMER, On the theorems of Pick and von Neumann, *Function Spaces* (ed. K. Jarosz), Marcel Dekker 1992, 273–280.

[L-S94] B. LOTTO and T. STEGER, Von Neumann's inequality for commuting diagonalizable contractions. II, *Proc. Amer. Math. Soc.*, **120** (1995), 897–901.

[N61] E. NELSON, The distinguished boundary of the unit operator ball, *Proc. Amer. Math. Soc.*, **12** (1961), 994–995.

[P86] V. I. PAULSEN, *Completely Bounded Maps and Dilations,* Pitman Research Notes in Mathematics **146**, John Wiley and Sons, New York, 1986.

[P-P-S89] V. I. PAULSEN, S. C. POWER and R. R. SMITH, Schur products and matrix completions, *J. Funct. Anal.*, **85** (1989), 151–178.

[Pi95] G. PISIER, *Similarity problems and completely bounded maps,* Springer Lecture Notes Math., **1618**, Springer-Verlag, 1995.

[V74] N. VAROPOULOS, On an inequality of von Neumann and an application of the metric theory of tensor products to operator theory, *J. Funct. Anal.*, **16** (1974), 83–100.

[vN51] J. VON NEUMANN, Eine Spektraltheorie für allgemeine Operatoren eines unitären Raumes, *Math. Nachr.*, **4** (1951), 258–281.

J. A. HOLBROOK, Mathematics & Statistics Department, University of Guelph, Guelph, ON Canada N1G 2WI; *e-mail:* jholbroo@math.uoguelph.ca

Operator Theory: Advances and Applications, Vol. 127, 387–397

Spectral properties of selfadjoint Jacobi matrices coming from birth and death processes

Jan Janas* and Serguei Naboko⋆

To the memory of Béla Szőkefalvi-Nagy

Let J_a be the selfadjoint Jacobi matrix in $l^2(\mathbb{N})$ with the weights $\lambda_n = n+a$ and the diagonal $q_n = -2(n+a)$, $a \in \mathbb{R}$. If $a \geq 0$ then $J_a \leq -I$ (the identity) and it is shown that its spectrum is purely absolutely continuous in $(-\infty, -1]$. For $a < 0$ the point spectrum in $(-1, +\infty)$ is always nonempty. Moreover, J_a represents simple example of the spectral phase transition of the first order (see Introduction). The point spectrum of J_a (in particular its asymptotis behaviour as a tends to $-\infty$) is also studied. The method of proof is based on finding asymptotic behaviour at $\{\infty\}$ of generalized eigenvectors of the equation $J_a u = \lambda u$, $\lambda \in \mathbb{R}$.

1. Introduction

In the paper of Karlin and McGregor [5] a class of birth and death processes was studied for which the infinitesimal matrix is a Jacobi matrix. This matrix induces an unbounded Jacobi operator J in $l^2(\mathbb{N})$ given by

$$(Ju)_n = \mu_{n-1}u_{n-1} + q_n u_n + \lambda_n u_{n+1}, \qquad n = 1, 2\cdots,$$

where $\mu_n = \mu n + a$, $\lambda_n = \lambda n + b$, $q_n = -(\mu_{n-1} + \lambda_n)$, $\mu > 0$, $\lambda > 0$, $a, b \in \mathbb{R}$ and $u_0 = \lambda_0 = \mu_0 = 0$.

AMS Subject Classification (1991): 47B37, 47A10.

* The research was supported by Grant PB 2 PO3A 002 13 of the Komitet Badań Naukowych, Warszawa.

⋆ Support of the Swedish Royal Academy of Sciences and (partially) the grant RFFI 00-01-00479 are gratefully appreciated.

In particular, if $\lambda = \mu$ and $a = b$ then J is essentially selfadjoint (on the subspace of finite vectors), see [1]. The presence of $\mu = \lambda$ is trivially inessential for us and we may assume that $\lambda = \mu = 1$ and $q_n = -2(n+a)$. Strictly speaking q_n should be equal to $-2(n+a)+1$ but one can subtract $-I$ (the identity) from J and in what follows q_n will be as above. To indicate the dependence on a, which is not trivial, we shall denote below J by J_a.

In our recent papers on unbounded selfadjoint Jacobi operators [3], [4] we studied spectral properties of them. In particular, for a rich class of sequences $\{\lambda_k\}$ and $\{q_k\}$ it was shown that J has purely absolutely continnuous spectrum provided $\lim_n |q_n/\lambda_n| < 2$ and J has discrete spectrum if $\underline{\lim}_n |q_n/\lambda_n| > 2$. The last fact is rather simple and can be proved by a straightforward estimation of the quadratic form of J.

It is clear that J_a does not satisfy any of the above conditions ($q_n/\lambda_n = -2$) and it is of interest to study its spectral character. Introducing the new parameter $\delta \in \mathbb{R}$ one can consider the operator $\tilde{J}_a(\delta)$ with weights $\lambda_n = n+a$ and $q_n = \delta(n+a)$. Then $\tilde{J}_a(-2) = J_a$ and J_a exhibits the spectral phase transition of the first order. Indeed, the spectral picture of $\tilde{J}_a(\delta)$ changes from pure absolutely continuous spectrum for $|\delta| < 2$ to discrete spectrum accumulating at $\{\infty\}$ provided $|\delta| > 2$. At $\delta = -2$ a jump appears as δ crosses this point. The same spectral phase transition occurs at $\delta = 2$ because $\tilde{J}_a(2)$ is unitarily equivalent to $-\tilde{J}_a(-2)$. The operator J_0 was also studied in [2], where in particular it has been proved that its spectrum is purely absolutely continuous and fills the interval $(-\infty, -1)$. The authors of [2] used different technique of orthogonal polynomials.

One of the standard tools used in study of spectral properties of J's is the theory, developed by Khan–Pearson in [6], of the subordinated solution of the infinite system of equations

$$(1.1) \qquad (Ju)_n = \lambda u_n, \ \lambda \in \mathbb{R}, \ n = 2, 3 \cdots .$$

(Note that (1.1) holds only for $n > 1$.) This approach has been already used in [3] and [4].

Let $B_n := \begin{pmatrix} 0 & 1 \\ -\frac{\lambda_{n-1}}{\lambda_n} & \frac{\lambda - q_n}{\lambda_n} \end{pmatrix}$ be the transfer matrix. Then (1.1) can be rewritten as

$$\vec{u}_{n+1} = B_n \vec{u}_n, \quad \text{where} \quad \vec{u}_n = \begin{pmatrix} u_{n-1} \\ u_n \end{pmatrix}.$$

The main idea exploited in [3] and [4] was based on finding asymptotics of $\vec{u}_n$ using the discrete version of the Levinson Theorem [4]. The method worked fine because the eigenvalues of the limit matrix $B_\infty = \lim_n B_n$ happend to be different. But this is not the case for J_a! In fact $B_\infty = \begin{pmatrix} 0 & 1 \\ -1 & -2 \end{pmatrix}$ has the multiple eigenvalue $x = -1$ and so it is similar to a Jordan box. This is the main difficulty in our case. The same situation appears for Discrete Schrödinger Operator ($\lambda_n = 1$) with decreasing potential q_n for the boundary $\{2, -2\}$ of the essential spectrum. For J_a this phenomenon holds for any point of the real line. Therefore a new approach is necessary. This is one of the main reasons why so special class of J_a is studied in the present paper. We associate with (1.1) a suitable linear differential equation (continuous version of (1.1)) for which there exist classical methods of finding the WKB asymptotics of solutions at $\{+\infty\}$. Knowing this asymptotics at $\{+\infty\}$ we can make reasonable ansatz about asymptotics of $\vec{u}_n$, as $n \to \infty$. This point of view is explained briefly in Section 2. Section 3 contains the main result on asymptotic form of the solutions of (1.1) and its application to spectral analysis of J_a. Section 4 presents some elementary spectral properties of J_a which depend on a (in particular as $a \to -\infty$).

2. Preliminaries and heuristic ideas

As we already mentioned in the Introduction the presence of multiple roots of B_∞ does not allow to apply the discrete version of the Levinson Theorem for computing asymptotic behaviour at $\{\infty\}$ of the solutions of (1.1). We overcome this difficulty by considering a suitable linear differential equation corresponding to (1.1). It turns out to be a special case of the classical hypogeometric equation. Let us explain this in detail. Note that (1.1) reads as

$$(n - 1 + a)u_{n-1} - 2(n + a)u_n + (n + a)u_{n+1} = \lambda u_n. \tag{2.1}$$

Divide both sides of (2.1) by $n + a$. Then (2.1) takes the form

$$u_{n+1} + u_{n-1} - 2u_n + \Big(\frac{-1}{n} + \frac{a}{n^2} + \cdots\Big)u_{n-1} - \lambda\Big(\frac{1}{n} - \frac{a}{n^2} + \cdots\Big)u_n = 0. \tag{2.2}$$

Denote by $(\Delta u)(k) := u_{k+1} - u_k$. Adding and subtracting $\Big(-\frac{1}{n} + \frac{a}{n^2} + \cdots\Big)u_n$ in (2.2) we have

$$\Delta^2 u(n-1) - \Big(-\frac{1}{n} + \frac{a}{n^2} + \cdots\Big)\Delta u(n-1) + \Big(-\frac{1}{n} + \frac{a}{n^2} + \cdots\Big)(1+\lambda)u_n = 0 \tag{2.3}$$

Therefore the associated differential equation (for $x = n \gg 1$) is given by

$$\frac{d^2u}{dx^2} - \Big(-\frac{1}{x} + \frac{a}{x^2} + \cdots\Big)\frac{du}{dx} + (1+\lambda)\Big(-\frac{1}{x} + \frac{a}{x^2} + \cdots\Big)u(x) = 0. \tag{2.4}$$

How to find asymptotics at $\{+\infty\}$ of the solutions of (2.4)? Since $\{+\infty\}$ turns out to be the irregular singular point, (2.4) is not of the Fuchs type. However, the answer is provided by the classical theory of such peculiar points for the second order linear differential operators with analytic coefficients, see [7], [8], [9]. The WKB asymptotics of two linearly independent solutions $u^{\pm}(\cdot)$ of (2.4) is given (for $\lambda < -1$) by

$$u^{\pm}(x) = x^{-1/4} \exp\Big(\pm 2\sqrt{(1+\lambda)x}\Big)(1+o(1)), \text{ as } x \to +\infty. \tag{2.5}$$

The above considerations indicate that the spectral phase transition of the first order can be explained by the appearance of the irregular singular point with double roots of the characteristic equation, [7],[9]. Therefore the model presented here displays mathematical "machinery" of this phenomenon. In our opinion this is the most essential part of the paper.

In what follows J_a means the closure of the operator, given by the Jacobi matrix with $\lambda_n = n + a$ and $q_n = -2(n+a)$, defined on the subspace F of finite vectors. A finite vector $f = \{f_i\}_{i=1}^{\infty}$ means that $f_i \equiv 0$ for all i sufficiently large. The numbers λ_n are sometimes called below weights, and $\lambda_0 = 0$ by definition. Recall that a solution u of (1.1) is subordinated if and only if for any linearly independent solution v of (1.1)

$$\lim_{N\to\infty} \left(\sum_{i=1}^{N} |u_i|^2 \Big/ \sum_{i=1}^{N} |v_i|^2 \right) = 0.$$

3. Asymptotics of generalized eigenvectors

This section contains a detailed exposition of asymptotics of generalized eigenvectors of (1.1). It is based on the above mentioned (in Section 2) analogy with the theory of ordinary differential equations having irregular singular points. It turns out that the analogy is not misleading and we have

THEOREM 3.1. *For any $a \in \mathbb{R}$ and $\lambda < -1$ the equation* (1.1) *has two linearly independent solutions $u_n^{\pm}$ with asymptotics given by*

$$u_n^{\pm} = n^{-1/4} \exp\Big[\pm 2\sqrt{(\lambda+1)n}\Big](1+o(1)), \quad \text{as } n \to +\infty. \tag{3.1}$$

Note that the appearance of $\sqrt{n}$ here is a new phenomenon (due to the Jordan box of the asymptotic transfer matrix B_∞) in comparison to n in the asymptotics given by the Levinson Theorem. The branch of $\sqrt{(\lambda+1)n}$ is chosen as: $i\times$ (positive number).

Proof. Fix $\lambda < -1$ and denote for brevity by

$$z_n := n^{-1/4} \exp\left(2\sqrt{(\lambda+1)n}\right), \quad \alpha := 2\sqrt{1+\lambda}, \quad \rho := -\frac{1}{4}.$$

Let

$$S_n := \begin{pmatrix} \overline{z}_{n-1} & z_{n-1} \\ \overline{z}_n & z_n \end{pmatrix}, \ n = 2, 3, \cdots.$$

We claim that $\vec{u}_n = S_n \vec{v}_n$, where $\vec{v}_n$ is a sequence of vectors which tends to a nonzero vector $\vec{c}$. Observe that the above form of S_n appears naturally if we replace the exact solutions $u_n^{\pm}$ by their "tentative" asymptotics z_n from (3.1). Since $\vec{u}_{n+1} = B_n \vec{u}_n$ we have $\vec{v}_{n+1} = S_{n+1}^{-1} B_n S_n \vec{v}_n$. Note that $\det S_n \neq 0$ so S_n^{-1} exists provided $\operatorname{Im}(z_n \overline{z}_{n-1}) \neq 0$. Since λ is fixed it is easy to verify that this condition always holds for sufficiently large n. By the definition of S_n and the form of B_n we find

$$S_{n+1}^{-1} = (\det S_{n+1})^{-1} \begin{pmatrix} z_{n+1}, & -z_n \\ -\overline{z}_{n+1}, & \overline{z}_n \end{pmatrix}$$

and

$$B_n = \begin{pmatrix} 0 & 1 \\ -1 & 2 \end{pmatrix} + \frac{1}{n+a} \begin{pmatrix} 0 & 0 \\ 1 & \lambda \end{pmatrix}.$$

Hence

(3.2) $$S_{n+1}^{-1} \begin{pmatrix} 0 & 1 \\ -1 & 2 \end{pmatrix} S_n = (\det S_{n+1})^{-1} \begin{pmatrix} x_n & y_n \\ -\overline{y}_n & -\overline{x}_n \end{pmatrix};$$

here

$$x_n := z_n \overline{z}_{n-1} + \overline{z}_n z_{n+1} - 2|z_n|^2, \qquad y_n := z_n z_{n-1} + z_n z_{n+1} - 2z_n^2,$$

and

(3.3) $$S_{n+1}^{-1} \begin{pmatrix} 0 & 0 \\ 1 & \lambda \end{pmatrix} S_n = (\det S_{n+1})^{-1} \begin{pmatrix} s_n, & t_n \\ -\overline{t}_n, & -\overline{s}_n \end{pmatrix},$$

where

$$s_n := -z_n(\overline{z}_{n-1} + \lambda \overline{z}_n), \ t_n := -z_n(z_{n-1} + \lambda z_n).$$

Combining (3.2) and (3.3) we have

(3.4) $$S_{n+1}^{-1} B_n S_n = (\det S_{n+1})^{-1} \begin{pmatrix} a_n, & b_n \\ -\overline{b}_n, & -\overline{a}_n \end{pmatrix},$$

where

$$a_n := x_n + (n+a)^{-1} s_n, \qquad b_n := y_n + (n+a)^{-1} t_n.$$

Write

$$b_n(\det S_{n+1})^{-1}$$
$$=[z_n z_{n-1} + z_{n+1} z_n - 2z_n^2 - z_n(z_{n-1} + \lambda z_n) z_n (n+a)^{-1}] \cdot (\overline{z}_n z_{n+1} - z_n \overline{z}_{n+1})^{-1}.$$

Dividing the nominator and the denominator by z_n^2 ($z_n \neq 0!$) we get

$$\begin{aligned} b_n(\det S_{n+1})^{-1} &= [z_{n-1}z_n^{-1} + z_{n+1}z_n^{-1} - 2 - (\lambda + z_{n-1}z_n^{-1})(n+a)^{-1}] \cdot \\ &\cdot (\overline{z}_n z_{n+1} z_n^{-2} - \overline{z}_{n+1} z_n^{-1})^{-1} \end{aligned} \tag{3.5}$$

Formula (3.5) allows to prove (and we will make it below) that $\{b_n(\det S_{n+1})^{-1}\} \in l^1$. Due to the definition of a_n and b_n it is easy to check that

$$|a_n(\det S_{n+1})^{-1} - 1| = |b_n(\det S_{n+1})^{-1}|.$$

Therefore the matrix $S_{n+1}^{-1} B_n S_n$ will have the form $I + T_n$, where $\{\|T_n\|\} \in l^1$, provided we show that $\{b_n(\det S_{n+1})^{-1}\} \in l^1$.

The convergence of the product $\prod_k (I + T_k)$ gives the desired asymptotics of $\vec{u}_n$ by trivial case of the discrete version of the Levinson Theorem [4].

It remains to prove that $\{b_n(\det S_{n+1})^{-1}\} \in l^1$. First note that the modulus of the denominator in (3.5) has asymptotic behaviour at ∞ as $|\alpha| n^{-1/2} + O(\frac{1}{n})$. The estimation of the nominator in (3.5) is more delicate.

Write

$$(n+1)^{1/2} - n^{1/2} = (4n)^{-1/2} + cn^{-3/2} + O(n^{-5/2}), \ c \in \mathbb{R}. \tag{3.6}$$

Using (3.6), the expansion of $\exp x$ up to the third term, where $x = \alpha[(n+1)^{1/2} - n^{1/2}]$, and the formula $(1 + \frac{1}{n})^\rho = 1 + \frac{\rho}{n} + O(\frac{1}{n^2})$ one can easily compute that

$$z_{n+1} z_n^{-1} = 1 + \alpha(4n)^{-1/2} + \Big(\rho + \frac{\alpha^2}{8}\Big) n^{-1} + \alpha\tilde{c} n^{-3/2} + O\Big(\frac{1}{n^2}\Big), \ \tilde{c} \in \mathbb{R}. \tag{3.7}$$

It follows that

$$\begin{aligned} z_{n-1} z_n^{-1} &= (z_n z_{n-1}^{-1})^{-1} \\ &= 1 - \alpha[4(n-1)]^{-1/2} - \Big(\rho + \frac{\alpha^2}{8}\Big) n^{-1} + \alpha^2 (4n)^{-1} - \\ &\quad - \alpha\tilde{c} n^{-3/2} + \alpha\Big(\rho + \frac{\alpha^2}{8}\Big) n^{-3/2} - \frac{\alpha^3}{8} n^{-3/2} + O\Big(\frac{1}{n^2}\Big). \end{aligned} \tag{3.8}$$

Since

$$\alpha[4(n-1)]^{-1/2} = \alpha(4n)^{-1/2} + \frac{\alpha}{4}n^{-3/2} + O\Big(\frac{1}{n^2}\Big),$$

we can rewrite (3.8) as

$$\begin{aligned} z_{n-1}z_n^{-1} =& 1 - \alpha(4n)^{-1/2} - \Big(\rho + \frac{\alpha^2}{8}\Big)n^{-1} + \alpha^2(4n)^{-1} - \alpha\tilde{c}n^{-3/2} + \\ &+ \alpha\rho n^{-3/2} - \frac{\alpha}{4}n^{-3/2} + O\Big(\frac{1}{n^2}\Big). \end{aligned} \tag{3.9}$$

Combining (3.7) and (3.9) we get

$$z_{n+1}z_n^{-1} + z_{n-1}z_n^{-1} = 2 + \alpha^2(4n)^{-1} + \alpha\Big(\rho - \frac{1}{4}\Big)n^{-3/2} + O\Big(\frac{1}{n^2}\Big).$$

Thus the nominator in (3.5) is equal to

$$\frac{1}{4}[\alpha^2 - 4(1+\lambda)](n+a)^{-1} + \Big(\alpha\rho + \frac{1}{4}\alpha\Big)n^{-3/2} + O\Big(\frac{1}{n^2}\Big) = O\Big(\frac{1}{n^2}\Big)$$

(by definitions of ρ and α). This completes the proof.

□

Observe that in the above proof we had to compute expansion of functions up to the third order (the order two is not enough). Seemingly "curious" cancellation of the lower order terms is caused by the validity of the asymptotics given in (3.1).

COROLLARY 3.2. *For any $a \in \mathbb{R}$ the spectrum of J_a in $(-\infty, -1)$ is purely absolutely continuous.*

Proof. Fix any $\lambda < -1$. Since every solution of (1.1) is a linear combination of the solutions $\{u_n^{\pm}\}$ given by (3.1) it is clear that (1.1) has no subordinated solutions and the Khan–Pearson Theorem [6, Theorem 2] completes the proof.

□

REMARK 3.3. Note that the method used in the proof of Theorem 3.1 can also be applied in other "boundary" situations with $\lim_n |q_n/\lambda_n| = 2$, in particular for $\lambda_n = n + a$ and $q_n = 2(n+a)$. We intend to consider in the future other classes of Jacobi matrices which fall into the "boundary" case.

4. Elementary spectral properties of J_a

It turns out that one can easily derive some interesting spectral properties of J_a (in terms of population behaviour for the birth and death processes) which depend on $a \in \mathbb{R}$. As it will be shown below the spectral picture of J_a depends on the sign of a. We start with

PROPOSITION 4.1. *For any $a \in \mathbb{R}$ and $u \in D(J_a)$*
(i) $(J_a u, u) = -a|u_1|^2 - \|u\|^2 - \sum_{n=1}^{\infty}(n+a)|u_{n+1} - u_n|^2$,
(ii) $-1 \notin \sigma_p(J_a)$.

Proof. (i) If $u \in D(J_a)$, then

$$(J_a u, u) = \sum_{n=1}^{\infty}(\lambda_{n-1} + q_n + \lambda_n)|u_n|^2 + \\ + \sum_{n=1}^{\infty}[\lambda_{n-1}u_{n-1}\overline{u}_n + \lambda_n u_{n+1}\overline{u}_n - (\lambda_n + \lambda_{n-1})|u_n|^2] \\ = (\lambda_1 + q_1)|u_1|^2 - \sum_{n=2}^{\infty}|u_n|^2 - \sum_{n=1}^{\infty}\lambda_n|u_{n+1} - u_n|^2,$$

as was to be shown.

(ii) If $J_a u = -u$, then, in particular, for all $n \geq |a| + 1$ we have

$$(n-1+a)u_{n-1} + (n+a)u_{n+1} + [-2(n+a)+1]u_n = 0 \tag{4.1}$$

Observe that $v_n \equiv 1$ and $w_n = \sum_{k=1}^{n-1}\lambda_k^{-1}$ are two linearly independent solutions of (4.1). Since $(u)_{n>|a|} = \alpha v + \beta w$, for some $\alpha, \beta \in \mathbb{R}$ and $u \in l^2$, it follows from the logarythmic divergence of w_n that $\alpha = \beta = 0$, and $u_n \equiv 0$ for $n > |a|$. Therefore u can be written as $u = (u_1, \cdots, u_N, 0, 0, \cdots)$, where $u_N \neq 0$.

Since $(N+1+a-1)u_N = 0$ (put $n = N+1$ in (4.1)) so $a = -N$. It follows that

$$J_a + I = T_N \oplus J_0 + I;$$

here T_N is the finite $N \times N$ Jacobi matrix with diag$(2(N-1)+1, 2(N-2)+1, \ldots, 3, 1)$ and weights $-(N-1), -(N-2), \ldots, -1$. Note that for $f = (u_1, \ldots, u_N)$, $0 = (T_N f, f) \geq N|u_1|^2$ (use the finite version of the above identity (i)).

Hence $u_1 = 0$. Applying induction (and the equality $T_N f = 0$) one can check that $u_2 = \cdots = u_N = 0$ and this completes the proof.

□

COROLLARY 4.2. *For any* $a \geq 0, (J_a u, u) \leq -\|u\|^2$ *and* $\sigma(J_a)$ *is purely absolutely continuous in* $(-\infty, -1]$ *and coincides with* $(-\infty, -1]$.

REMARK 4.3. For a general selfadjoint Jacobi matrix J with weights $\{\lambda_k\}$ and diagonal $\{q_k\}$ such that $\lambda_{k-1} + q_k + \lambda_k = M$, $k \geq 1$, it has been proved in [2] that $M \notin \sigma_p(J)$. In our case $q_1 + \lambda_1 = -1 - a = -1$ only for $a = 0$, and therefore the above mentioned conditions are valid only for $a = 0$. Moreover, if $\lambda_{k-1} + q_k + \lambda_k \leq M$ for all $k \geq 1$, then using the above (i) it is clear that $(Jf, f) \leq M\|f\|$, $f \in D(J)$.

It turns out that for $a < 0$ the spectral picture of J_a looks different. By $[a]$ we mean the smallest integer number greater than or equal to $a \in \mathbb{R}$.

PROPOSITION 4.4.
(i) *If* $a < b$, *then* $J_a > J_b$.
(ii) *For any* $a < 0$, *the point spectrum* $\sigma_p(J_a) \subset (-1, +\infty)$ *is not empty,* $\sigma(J_a)$ *is purely point in* $(-1, +\infty)$ *and* $-[a] \leq \sharp\sigma_p(J_a) < -[a] + 1$.

Proof. Let S be the unilateral shift in the canonical basis of l^2.

(i) Since $J_a = J_0 + a[S + S^* - 2I]$ the implication is obvious because 0 is not in the point spectrum of $S + S^* - 2I$.

(ii) Assume first that $a = -n \in \mathbb{Z}$. Then J_a is unitarily equivalent to $F_n \oplus J_0$, where F_n is the finite $n \times n$ Jacobi matrix with entries coming from J_a. By the definition of F_n it is easy to check (again by using the form of $(F_n f, f)$ and Proposition 4.1 (i)) that $\sigma(F_n) \subset (-1, \|F_n\|]$. Hence there are exactly n simple eigenvalues of F_n (and J_a, $a = -n$) greater than -1.

In the case $a < 0$ is not integer, let $s = -[a]$. Define the sequence $u^N = (u_1, \ldots, u_N, 0, 0, \cdots)$, where $u_r = (\ln N)^{-1} \sum_{k=r}^{N} 1/k$, $r = 1, \ldots, N$, and consider $(J_a u^N, u^N)$. Applying Proposition 4.1.(i) we have

$$((J_a + I)u^N, u^N) = -a(1 + o(1)) + O\Big(\frac{1}{\ln N}\Big), \quad \text{as } N \to \infty.$$

In fact

$$\sum_{k=1}^{N}(k + a)|u_k - u_{k-1}|^2 \leq \frac{1}{\ln^2 N} \sum_{k=1}^{N} \frac{k}{k^2} \sim \frac{1}{\ln N}, \quad \text{as } N \to \infty.$$

Thus $\sigma(J_a) \cap (-1, +\infty) \neq \emptyset$. Since J_a is unitarily equivalent to a finite rank perturbation of J_{s+a}, so $\sigma_{ess}(J_a) = \sigma_{ess}(J_{s+a}) = (-\infty, 1]$. Using (i) we have $J_{[a]} < J_a < J_{[a]-1}$ and so $s \leq \sharp\sigma_p(J_a) \leq s + 1$. The proof is complete.

□

One can also analyse spectral behaviour of the eigenvalues of J_a, as $a \to -\infty$. Denote by $\lambda_k(a)$ the k-th eigenvalue of J_a (enumarated in monotonic way). Note that due to Proposition 4.4.(i), $\lambda_k(a)$ depends monotonically on a. The new eigenvalues appear from the boundary -1 of the essential spectrum of J_a for some critical values of the parameter a: $a_1, a_2, \ldots$, where $n-1 < a_n < n$. It seems that detailed asymptotic behaviour of $\lambda_k(a)$ is not so obvious but we have

PROPOSITION 4.5. *If $a < 0$, then*
(i) $\sum_k \lambda_k(a) \sim a^2$, *as $a \to -\infty$,*
(ii) *and the largest eigenvalue $\lambda_1(a) \sim 4|a|$, as $a \to -\infty$.*

Proof. (i) Again $[-1+a] < a \le [a]$ implies that $(J_{[a-1]}u, u) > (J_a u, u) \ge (J_{[a]}u, u)$ so $\lambda_s(J_{[a-1]}) > \lambda_s(a) \ge \lambda_s(J_{[a]})$ for all suitable s. Hence

$$\sum_s \lambda_s(J_a) \sim \sum_s \lambda_s(J_{[a]}) = \operatorname{Tr} F_{[-a]} = [a]^2 + [a] \sim a^2.$$

(ii) Since $J_a = J_0 + a(S + S^* - 2I)$ it is clear that

$$\lambda_1(a) \le 4|a| - 1. \tag{4.2}$$

On the other hand we define $f := (f_i)_{i=1}^{\infty}$, where

$$f_i := \begin{cases} (-1)^i, & 1 \le i \le [\sqrt{|a|}] \\ 0, & i > [\sqrt{|a|}]. \end{cases}$$

We have $\|f\|^2 = [\sqrt{|a|}]$, $((S + S^* - 2I)f, f) = -4[\sqrt{|a|}] + 2$, and for large $|a|$

$$(J_0 f, f) = \sum_{k=1}^{[\sqrt{|a|}]} (-4k+1) + O(\sqrt{|a|}) \ge -2|a| + O(\sqrt{|a|}).$$

Then the term $O(\sqrt{|a|})$ is of the form $c_1[\sqrt{|a|}]$ or $c_2[\sqrt{|a|}]$, where c_s are constans which depend on the evenness of $[\sqrt{|a|}]$. It follows that

$$\begin{aligned} (J_a f, f)\|f\|^{-2} &= (J_0 f, f)\|f\|^{-2} + a((S + S^* - 2I)f, f)\|f\|^{-2} \\ &\ge -2|a| \cdot |a|^{-1/2} + O(\sqrt{|a|})|a|^{-1/2} + 4|a| = 4|a| + O(\sqrt{|a|}). \end{aligned} \tag{4.3}$$

Combining (4.2) and (4.3) we obtain the desired estimation. The proof is complete. □

REMARK 4.6. Note that all eigenvalues $\lambda_k(a)$ of J_a are estimated from above and from below and these are the eigenvalues of the finite Jacobi matrix $F_{[-a]}$ defined above. Thus the question of asymptotic behaviour of $\lambda_k(a)$ (as $a \to -\infty$) reduces to the same problem for $F_{[-a]}$.

References

[1] YU. M. BEREZIANSKII, *Expansions in Eigenfunctions of Selfadjoint Operators*, Naukova Dumka, Kiev, 1965 (Russian).

[2] J. DOMBROWSKI and S. PEDERSEN, Orthogonal polynomials, spectral measures, and absolute continuity, *J. Comp. App. Math.*, **65** (1995), 115–124.

[3] J. JANAS and S. NABOKO, Multithreshold spectral phase transition examples in a class of unbounded Jacobi matrices, *Res. Reports in Math. Stockholm Univ.*, **7** (1999).

[4] J. JANAS and S. NABOKO, Spectral analysis of selfadjoint Jacobi matrices with periodically modulated entries, Preprint.

[5] S. KARLIN and J. MCGREGOR, Linear growth, birth and death processes, *J. Math. Mechan.*, **7** (1958), 643–662.

[6] S. KHAN and D. B. PEARSON, Subordinacy and Spectral Theory for Infinite Matrices, *Helv. Phys. Acta,* **65** (1992), 505–527.

[7] N. M. MATVEEV, *Analytic Theory of Differential Equations,* Leningrad Pedagogic University, 1989 (Russian).

[8] F. W. OLVER, *Asymptotic and special functions,* Academic Press, New York – London, 1974.

[9] W. WASOW, *Asymptotic Expansions for Ordinary Differential Equations,* A Series of Texts and Monographs Vol. XIV, J. Wiley, 1965.

J. JANAS, Institute of Mathematics, Polish Academy of Sciences, Cracow Branch, Św.Tomasza 30, 31-027 Kraków, Poland; *e-mail:* najanas@cyf-kr.edu.pl

S. NABOKO, Department of Mathematical, Physics, Institute of Physics, St. Petersburg University, Ulianovskaia 1, 198904, St. Petergoff, St. Petersburg, Russia; *e-mail:* naboko@snoopy.phys.spbu.ru

Operator Theory: Advances and Applications, Vol. 127, 399–422

On the hyperinvariant subspace problem for asymptotically nonvanishing contractions

LÁSZLÓ KÉRCHY*

Dedicated to the memory of my master
Professor Béla Szőkefalvi-Nagy

Our aim in the present note is to obtain new information on the structure of contractions of class $C_{1\cdot}$, and to develop new ways for obtaining hyperinvariant subspaces for these operators.

1. Introduction

Let $\mathcal{H}$ be a complex, separable Hilbert space, and let $\mathcal{L}(\mathcal{H})$ stand for the set of all bounded, linear operators acting on $\mathcal{H}$. An operator $T \in \mathcal{L}(\mathcal{H})$ is a contraction if $\|T\| \leq 1$. We say that the contraction T is *asymptotically nonvanishing*, in notation: $T \in C_{*\cdot}$, if there exists a vector $x_0 \in \mathcal{H}$ such that $\lim_{n\to\infty} \|T^n x_0\| > 0$. It is a longstanding open problem *whether every asymptotically nonvanishing contraction T, which is not scalar multiple of the identity, has a nontrivial hyperinvariant subspace $\mathcal{M}$.* We recall that the (closed) subspace $\mathcal{M}$ of $\mathcal{H}$ is called nontrivial, if $\{0\} \neq \mathcal{M} \neq \mathcal{H}$; and $\mathcal{M}$ is hyperinvariant for T, if it is invariant for every operator Q belonging to the commutant $\{T\}' := \{A \in \mathcal{L}(\mathcal{H}) : AT = TA\}$ of T. The hyperinvariant subspace lattice of T is denoted by $\operatorname{Hlat} T$.

It is easy to see that for any contraction $T \in \mathcal{L}(\mathcal{H})$, the subspace $\mathcal{H}_0(T) := \{x \in \mathcal{H} : \lim_{n\to\infty} \|T^n x\| = 0\}$ belongs to $\operatorname{Hlat} T$. We write $T \in C_{0\cdot}$ if $\mathcal{H}_0(T) = \mathcal{H}$, and we write $T \in C_{1\cdot}$ if $\mathcal{H}_0(T) = \{0\}$. For any $j = 0, *, 1$, by definition $T \in C_{\cdot j}$ if $T^* \in C_{j\cdot}(\mathcal{H})$ is true for the adjoint operator. Finally, for any choice of $i, j = 0, *, 1$, we consider the set $C_{ij} := C_{i\cdot} \cap C_{\cdot j}$. This classification of contractions, according

AMS Subject Classification (1991): 47A15, 47A20, 47A45, 47A60.

* Research partially supported by Hungarian NFS Research grant no. T 022920.

to the asymptotic behaviour of the iterates, was introduced by Béla Szőkefalvi-Nagy and Ciprian Foias (see [SzNF1, Section II.4]). They showed that if $T \in C_{**}$ is nonscalar, then $\operatorname{Hlat} T$ is nontrivial, that is $\operatorname{Hlat} T \neq \{\{0\}, \mathcal{H}\}$ (see [SzNF1, Theorem II.5.4]). Therefore, addressing the hyperinvariant subspace problem for asymptotically nonvanishing contractions, we can assume that T is of class C_{10}. In what follows, we shall consider mainly contractions of class $C_{1\cdot}$.

The advantage of the assumption $T \in C_{*\cdot}$ is that a nonzero unitary asymptote $T^{(a)} \in \mathcal{L}(\mathcal{H}_T^{(a)})$ can be associated with T. Namely, if $\langle \cdot, \cdot \rangle$ stands for the original inner product on $\mathcal{H}$, then a new semi-inner product can be introduced on $\mathcal{H}$ by the formula $[x, y] := \lim_{n\to\infty} \langle T^n x, T^n y \rangle$ $(x, y \in \mathcal{H})$. Forming quotient space and taking completion, we arrive at a Hilbert space $\mathcal{H}_{T,+}^{(a)}$, where T can be continuously extended to an isometry $T_+^{(a)}$. The natural embedding $X_T^+ : \mathcal{H} \to \mathcal{H}_{T,+}^{(a)}$, $x \mapsto x + \mathcal{H}_0(T)$ is contractive and intertwines T with $T_+^{(a)}$, in notation: $X_T^+ \in \mathcal{I}(T, T_+^{(a)}) := \{A \in \mathcal{L}(\mathcal{H}, \mathcal{H}_{T,+}^{(a)}) : AT = T_+^{(a)} A\}$. The *unitary asymptote* $T^{(a)} \in \mathcal{L}(\mathcal{H}_T^{(a)})$ is defined as the minimal unitary extension of the isometry $T_+^{(a)}$, and the *canonical intertwining mapping* $X_T \in \mathcal{I}(T, T^{(a)})$ is defined by $X_T x := X_T^+ x$ $(x \in \mathcal{H})$. Clearly, $\ker X_T = \mathcal{H}_0(T)$ and $(\operatorname{ran} X_T)^- = \mathcal{H}_{T,+}^{(a)}$. Thus, if $T \in C_{*\cdot}(\mathcal{H})$ then the unitary operator $T^{(a)}$ is nonzero (i.e. acts on a nonzero space), and if $T \in C_{1\cdot}(\mathcal{H})$ then the mapping X_T is an injection. For more details in connection with these concepts, we refer to [SzNF1, Chapter II], [Bea, Chapter XII], [K4] and [K6].

It is well-known that any contraction $T \in \mathcal{L}(\mathcal{H})$ can be uniquely decomposed into the orthogonal sum $T = T_1 \oplus T_2 \oplus T_3$ of a completely nonunitary (c.n.u.) contraction T_1, an absolutely continuous (a.c.) unitary operator T_2 and a singular unitary operator T_3 (see [SzNF1, Theorem I.3.2] and [Hal]). Applying the Lifting Theorem of Sz.-Nagy and Foias it can be easily verified that the hyperinvariant subspace lattice of T splits into the direct sum $\operatorname{Hlat} T = \operatorname{Hlat}(T_1 \oplus T_2) \oplus \operatorname{Hlat} T_3$ (see [SzNF1, Theorem II.2.3] and [Dou, Corollary 5.1 and Theorem 3]). Thus, we can (and shall) assume in the sequel that the singular unitary component T_3 is zero, that is the contraction T is absolutely continuous. In that case the unitary asymptote $T^{(a)}$ is also a.c. (see [SzNF1, Theorem II.6.4] and [K6, Theorem 4], or [Bea, Proposition XII.2.1]).

Our aim in the present note is to obtain new information on the structure of contractions of class $C_{1\cdot}$, as well as to obtain new criteria for the existence of hyperinvariant subspaces for these operators. In particular, we identify a new spectral invariant $\pi(T)$, which plays a relevant role in this topic and seems to be a useful tool in the study of $C_{1\cdot}$-contractions.

In the paper [K7] we obtained a factorization result of the following type. Let $\mathcal{E}_n$ be an n-dimensional Hilbert space, $n \in \mathbb{N}_\infty := \mathbb{N} \cup \{\infty\}$. Let $L^2(\mathcal{E}_n)$ denote the

Hilbert space of square-integrable, vector-valued functions, defined with respect to the normalized Lebesgue measure m on the unit circle $\mathbb{T}$; and let $H^2(\mathcal{E}_n)$ stand for the Hardy subspace of the analytic functions in $L^2(\mathcal{E}_n)$. Let $\mathcal{B}_1$ denote the system of Borel subsets of $\mathbb{T}$ with positive Lebesgue measure, and for any $\omega \in \mathcal{B}_1$, let χ_ω mean the characteristic function of ω. Finally, the operators of multiplication by the identical function $\chi(z) = z$ on the Hilbert spaces $H^2(\mathcal{E}_n)$ and $\chi_\omega L^2(\mathcal{E}_n)$ are denoted by S_n and $M_{n,\omega}$, respectively. Therefore, S_n is the unilateral shift of multiplicity n, and $M_{n,\omega}$ is an a.c. unitary operator with spectral multiplicity function $n\chi_\omega$. Now, the factorization theorem claims that if $T \in \mathcal{L}(\mathcal{H})$ is an a.c. contraction such that the spectral multiplicity function of the unitary asymptote $T^{(a)}$ dominates the function $n\chi_\omega$ $(n \in \mathbb{N}_\infty$ and $\omega \in \mathcal{B}_1)$, then the natural embedding $J_{n,\omega}\colon H^2(\mathcal{E}_n) \to \chi_\omega L^2(\mathcal{E}_n)$, $f \mapsto \chi_\omega f$ can be factorized into the product $J_{n,\omega} = ZY$, where $Y \in \mathcal{I}(S_n, T)$ and $Z \in \mathcal{I}(T, M_{n,\omega})$, and we have a control on the norms of Y and Z.

The hyperinvariant subspace lattices of the operators S_n and $M_{n,\omega}$ are dramatically different. Namely, $\operatorname{Hlat} S_n$ is isomorphic to the lattice of (equivalence classes of) inner functions, while $\operatorname{Hlat} M_{n,\omega}$ is isomorphic to the Boolean lattice of (equivalence classes of) Borel subsets of ω. (See e.g. [K9].) Now, the question is how the hyperinvariant subspace lattice of the intermediate operator T behaves. We are going to examine under what conditions the properties of T show similarities with those of S_n and when the properties of T are closer to those of $M_{n,\omega}$.

Our paper is organized in the following way. The concept of the quasianalytical spectral set $\pi(T)$ of a $C_{1\cdot}$-contraction T is introduced in Section 2. This is a Borel subset of the unit circle, which plays central role in our investigations. The connection of $\pi(T)$ with the support $\rho(T)$ of the spectral measure of $T^{(a)}$ (called the residual set of T) is examined. One of the main results in this section is that $\operatorname{Hlat} T$ is nontrivial, if $\pi(T) \neq \rho(T)$. In Section 3 we study the transformation laws concerning these sets in the Sz.-Nagy–Foias functional calculus. As a result, we obtain an abundance of examples for possible pairs of $\pi(T)$ and $\rho(T)$. It is shown in Section 4 that the contraction T exhibits a 'quasianalytic property' on the quasianalytical spectral set $\pi(T)$. Section 5 is devoted to different intertwining relations. Finally, in Section 6 the operators in the commutant of T are studied. It is shown that every nonzero operator in $\{T\}'$ is injective in the quasianalytic, cyclic case. Furthermore, a sufficient condition is given for the existence of an operator $0 \neq Q \in \{T\}'$ with nondense range.

2. Asymptotically strongly nonvanishing contractions

Let H^∞ denote the Hardy space of bounded, analytic functions, defined on the open unit disc $\mathbb{D}$. We recall that, for any $u \in H^\infty$, the radial limit $\lim_{r\to 1-} u(rz)$ exists for almost every (a.e.) $z \in \mathbb{T}$; the limit function will be also denoted by u. In connection with the basic properties of H^∞, we refer to [Hof] and [SzNF1, Chapter III]. Given any $u, v \in H^\infty$, we say that *the function u is smaller than v in absolute value*, in notation: $u \overset{a}{\prec} v$, if $|u(z)| \le |v(z)|$ holds, for every $z \in \mathbb{D}$. This relation can be characterized in the following way.

LEMMA 1. *For any functions $u, v \in H^\infty$, the following conditions are equivalent:*
(a) $u \overset{a}{\prec} v$;
(b) *there exists $w \in H^\infty$ such that $u = vw$ and $\|w\|_\infty \le 1$;*
(c) *$|u(z)| \le |v(z)|$ is true for a.e. $z \in \mathbb{T}$, and the inner component of v divides the inner component of u.*

In a similar fashion, given any $A, B \in \mathcal{L}(\mathcal{H})$, we say that *the operator A is smaller than B in absolute value*, in notation: $A \overset{a}{\prec} B$, if $\|Ax\| \le \|Bx\|$ is true, for every $x \in \mathcal{H}$. This relation can be also easily characterized.

LEMMA 2. *For any operators $A, B \in \mathcal{L}(\mathcal{H})$, the following conditions are equivalent:*
(a) $A \overset{a}{\prec} B$;
(b) *there exists $C \in \mathcal{L}(\mathcal{H})$ such that $A = CB$, $\|C\| \le 1$;*
(c) $A^*A \le B^*B$.

Clearly, these relations on H^∞ and $\mathcal{L}(\mathcal{H})$ are reflexive and transitive. We shall consider the following sets of decreasing sequences:

$$\mathcal{D}(H^\infty) := \{F = \{f_n\}_{n=1}^\infty : \ f_n \in H^\infty, \ f_{n+1} \overset{a}{\prec} f_n \ \text{ for every } n \in \mathbb{N}\}$$

and

$$\mathcal{D}(\mathcal{H}) := \{\mathcal{A} = \{A_n\}_{n=1}^\infty : \ A_n \in \mathcal{L}(\mathcal{H}), \ A_{n+1} \overset{a}{\prec} A_n \ \text{ for every } n \in \mathbb{N}\}.$$

To any sequence $F = \{f_n\}_{n=1}^\infty \in \mathcal{D}(H^\infty)$ we can associate the *limit function*

$$\varphi_F(z) := \lim_{n\to\infty} |f_n(z)|,$$

defined almost everywhere on the unit circle $\mathbb{T}$. Similarly, to any sequence $\mathcal{A} = \{A_n\}_{n=1}^{\infty} \in \mathcal{D}(\mathcal{H})$ there corresponds the *limit operator*

$$\Phi_{\mathcal{A}} := \left(\lim_{n\to\infty} A_n^* A_n\right)^{1/2}.$$

Here the convergence is meant in the strong operator topology; furthermore, $\|\Phi_{\mathcal{A}} x\| = \lim_{n\to\infty} \|A_n x\|$ is true, for every $x \in \mathcal{H}$.

Let $T \in \mathcal{L}(\mathcal{H})$ be an a.c. contraction, and let us consider the Sz.-Nagy–Foias functional calculus for T. We remind the reader that this calculus is a uniquely determined unital algebra-homomorphism $\Psi_T: H^\infty \to \mathcal{L}(\mathcal{H})$, $f \mapsto f(T)$, which assigns T to the identical function χ, and which is continuous in the weak-$*$ topologies (see [SzNF1, Theorem III.2.1] or [BFP, Theorem 4.1]). It is easy to verify that Ψ_T is monotone with respect to the relations introduced before. Indeed, if $u \overset{a}{\prec} v$ $(u, v \in H^\infty)$, then $u = vw$ is true with some $w \in H^\infty$, $\|w\|_\infty \le 1$, and since Ψ_T is contractive, we infer that $\|u(T)x\| = \|w(T)v(T)x\| \le \|v(T)x\|$ $(x \in \mathcal{H})$, that is $u(T) \overset{a}{\prec} v(T)$. Therefore, given any $F = \{f_n\}_{n=1}^{\infty} \in \mathcal{D}(H^\infty)$, we can form the sequence $F(T) := \{f_n(T)\}_{n=1}^{\infty} \in \mathcal{D}(\mathcal{H})$. The subspace

$$\mathcal{H}_0(T, F) := \ker \Phi_{F(T)} = \{x \in \mathcal{H} : \lim_{n\to\infty} \|f_n(T)x\| = 0\}$$

is clearly hyperinvariant for T. The following propositions show that under specific conditions this subspace is trivial. We note that for a function $\varphi \in L^\infty = L^\infty(m)$, the notation $\varphi = 0$ or $\varphi > 0$ means that $\varphi(z) = 0$ or $\varphi(z) \in (0, \infty)$, respectively, for a.e. $z \in \mathbb{T}$.

PROPOSITION 3. *Let $T \in \mathcal{L}(\mathcal{H})$ be an a.c. contraction, and $F = \{f_n\}_{n=1}^{\infty} \in \mathcal{D}(H^\infty)$. If $\varphi_F = 0$, then $\mathcal{H}_0(T, F) = \mathcal{H}$.*

Proof. By Sz.-Nagy's celebrated dilation theorem, there exists a unitary operator U on a larger Hilbert space $\mathcal{K}$, such that $T^n = P_{\mathcal{H}} U^n | \mathcal{H}$ is true for every $n \in \mathbb{N}$, and $\mathcal{K} = \vee_{n=-\infty}^{\infty} U^n \mathcal{H}$. Furthermore, this minimal unitary dilation of T is uniquely determined and absolutely continuous (see [SzN1] and [SzNF1, Theorems I.4.2, II.6.4]). Let E denote the spectral measure of U, and let E_x be the localization of E at the vector $x \in \mathcal{H}$. Then applying Lebesgue's dominated convergence theorem, we obtain that

$$\|\Phi_{F(T)} x\| = \lim_{n\to\infty} \|f_n(T)x\| = \lim_{n\to\infty} \|P_{\mathcal{H}} f_n(U)x\|$$
$$\le \lim_{n\to\infty} \|f_n(U)x\| = \lim_{n\to\infty} \left(\int_{\mathbb{T}} |f_n|^2 \, dE_x\right)^{1/2} = 0,$$

and so $\Phi_{F(T)} = 0$. □

PROPOSITION 4. *Given any a.c. contraction $T \in \mathcal{L}(\mathcal{H})$, the following conditions are equivalent:*

(a) $T \in C_{1\cdot}$;

(b) *for any $F \in \mathcal{D}(H^\infty)$, the relation $\varphi_F > 0$ implies that $\mathcal{H}_0(T, F) = \{0\}$.*

Proof. (b)$\Longrightarrow$(a): This implication is trivial since $\varphi_{F_0} = 1 > 0$ holds for the sequence $F_0 = \{\chi^n\}_{n=1}^\infty \in \mathcal{D}(H^\infty)$, and since $\mathcal{H}_0(T, F_0) = \mathcal{H}_0(T)$.

(a)$\Longrightarrow$(b): Let us suppose that $T \in C_{1\cdot}$, and let us consider the unitary asymptote $T^{(a)} \in \mathcal{L}(\mathcal{H}_T^{(a)})$ of T and the canonical intertwining mapping $X_T \in \mathcal{I}(T, T^{(a)})$. Let $F = \{f_n\}_{n=1}^\infty \in \mathcal{D}(H^\infty)$ be a sequence such that $\varphi_F > 0$. Then, for any vector $x \in \mathcal{H}$ and for any $n \in \mathbb{N}$, we have

$$\begin{aligned}\|f_n(T)x\| \geq \|X_T\|^{-1}\|X_T f_n(T)x\| &= \|X_T\|^{-1}\|f_n(T^{(a)})X_T x\| \\ &= \|X_T\|^{-1}\left(\int_{\mathbb{T}} |f_n|^2\, dE_{X_T x}\right)^{1/2},\end{aligned}$$

where E is the spectral measure of the a.c. unitary operator $T^{(a)}$. Lebesgue's theorem yields that

$$\|\Phi_{F(T)}x\| \geq \|X_T\|^{-1}\left(\int_{\mathbb{T}} \varphi_F^2\, dE_{X_T x}\right)^{1/2} \quad (x \in \mathcal{H}).$$

Taking into account that X_T is injective and that the measure $E_{X_T x}$ is absolutely continuous, it readily follows that $\ker \Phi_{F(T)} = \{0\}$. □

Let $T \in \mathcal{L}(\mathcal{H})$ be an a.c. contraction. We say that T is *asymptotically strongly nonvanishing with respect to the Borel set* $\alpha \in \mathcal{B}_1$, in notation: $T \in C_{1\cdot}(\alpha)$, if $\mathcal{H}_0(T, F) = \{0\}$ is fulfilled for every sequence $F \in \mathcal{D}(H^\infty)$ satisfying the condition $\chi_\alpha \varphi_F \neq 0$. Since $\chi_\alpha \varphi_{F_0} = \chi_\alpha \neq 0$ is true with $F_0 = \{\chi^n\}_{n=1}^\infty$, we can see that $C_{1\cdot}(\alpha)$ $(\alpha \in \mathcal{B}_1)$ is a subclass of $C_{1\cdot}$. It also follows immediately from the definition that $C_{1\cdot}(\alpha) \supset C_{1\cdot}(\beta)$ holds, whenever $\alpha \subset \beta$ $(\alpha, \beta \in \mathcal{B}_1)$. Furthermore, given any sequence $\{\alpha_n\}_{n=1}^\infty$ of Borel sets in $\mathcal{B}_1$, we have

$$\cap_{n=1}^\infty C_{1\cdot}(\alpha_n) = C_{1\cdot}(\cup_{n=1}^\infty \alpha_n).$$

Let $\tilde{C}_{1\cdot} := \cup_{\alpha \in \mathcal{B}_1} C_{1\cdot}(\alpha)$ be the set of those a.c. contractions, which are asymptotically strongly nonvanishing with respect to some $\alpha \in \mathcal{B}_1$. We make the convention that two Borel subsets α, β of $\mathbb{T}$ are considered equal, if $\chi_\alpha = \chi_\beta$, that is if the symmetric difference $\alpha \triangle \beta$ is of measure zero.

PROPOSITION 5. *For any contraction $T \in \tilde{C}_{1\cdot}$, there exists a (unique) largest set $\alpha_T \in \mathcal{B}_1$, such that $T \in C_{1\cdot}(\alpha_T)$.*

Proof. Setting $\delta := \sup\{m(\alpha) : \alpha \in \mathcal{B}_1, T \in C_{1\cdot}(\alpha)\}$, let us consider a sequence $\{\alpha_n\}_{n=1}^{\infty} \subset \mathcal{B}_1$ such that $\lim_{n\to\infty} m(\alpha_n) = \delta$, and $T \in C_{1\cdot}(\alpha_n)$, for every $n \in \mathbb{N}$. Since $T \in \cap_{n=1}^{\infty} C_{1\cdot}(\alpha_n) = C_{1\cdot}(\cup_{n=1}^{\infty}\alpha_n)$, we can easily see that $\alpha_T = \cup_{n=1}^{\infty}\alpha_n$ possesses the required properties.

□

The Borel set $\alpha_T \in \mathcal{B}_1$, appearing in the previous proposition, will be called the *quasianalytical spectral set* of the contraction $T \in \tilde{C}_{1\cdot}$, and it will be denoted by $\pi(T)$. If the a.c. contraction T is of class $C_{1\cdot}\backslash\tilde{C}_{1\cdot}$, then by definition $\pi(T) := \emptyset$.

PROPOSITION 6.
(i) *If $S_n \in \mathcal{L}(H^2(\mathcal{E}_n))$ is the unilateral shift of multiplicity $n \in \mathbb{N}_\infty$, then $\pi(S_n) = \mathbb{T}$.*
(ii) *If $U \in \mathcal{L}(\mathcal{H})$ is an a.c. unitary operator, then $\pi(U) = \emptyset$.*

Proof. (i): Let us consider a sequence $F = \{f_k\}_{k=1}^{\infty} \in \mathcal{D}(H^\infty)$ such that $\varphi_F \neq 0$, and a vector $0 \neq x \in H^2(\mathcal{E}_n)$. Taking into account that $x(z) \neq 0$ for a.e. $z \in \mathbb{T}$, we obtain that

$$\begin{aligned}\|\Phi_{F(S_n)}x\|^2 &= \lim_{k\to\infty} \|f_k(S_n)x\|^2 = \lim_{k\to\infty} \int_{\mathbb{T}} |f_k(z)|^2 \|x(z)\|^2 \, dm(z) \\ &= \int_{\mathbb{T}} \varphi_F(z)^2 \|x(z)\|^2 \, dm(z) > 0.\end{aligned}$$

(ii): Let $\chi_\omega dm$ ($\omega \in \mathcal{B}_1$) be a scalar spectral measure of the a.c. unitary operator U. Given any $\alpha \in \mathcal{B}_1$, we can find $F \in \mathcal{D}(H^\infty)$ such that $\chi_\alpha \varphi_F \neq 0$ and the set $\omega_0 := \{z \in \omega : \varphi_F(z) = 0\}$ is of positive measure. Since

$$\ker \Phi_{F(U)} = \ker \varphi_F(U) = \operatorname{ran} \chi_{\omega_0}(U) \neq \{0\},$$

it follows that $U \notin C_{1\cdot}(\alpha)$.

□

Given any a.c. contraction $T \in \mathcal{L}(\mathcal{H})$ of class $C_{*\cdot}$, there exists a (unique) Borel set $\beta_T \in \mathcal{B}_1$ such that $\chi_{\beta_T} dm$ is a scalar spectral measure of the unitary asymptote $T^{(a)}$. This set is called the *residual set* of the contraction T, and is denoted by $\rho(T)$. We note here that $T^{(a)}$ is unitarily equivalent to the $*$-residual part of the minimal unitary dilation of T (see [K6]), and that, for c.n.u. C_{11}-contractions, $\rho(T)$ is the smallest Borel set on $\mathbb{T}$ which is residual for T in the sense used in

[SzNF1, Section VII.5]. We note also that the residual set $\rho(T)$ is included in the spectrum $\sigma(T)$ of T; see e.g. [K4, Theorem 4]. The description of possible spectra of contractions of class $C_1.$ was given in [BK2] and [K2].

We proceed with exploring the connection of the sets $\pi(T)$ and $\rho(T)$.

THEOREM 7. *For any a.c. contraction $T \in C_1.(\mathcal{H})$, we have $\pi(T) \subset \rho(T)$.*

Proof. Let us assume that the set $\alpha := \pi(T)\backslash\rho(T)$ is of positive measure. Let us choose a strictly decreasing sequence $\{c_n\}_{n=1}^{\infty}$ of real numbers, converging to zero; and, for any $n \in \mathbb{N}$, let $f_n \in H^\infty$ be an outer function such that $|f_n| = \chi_\alpha + c_n\chi_{\mathbb{T}\backslash\alpha}$. Now, we consider the sequence $F = \{f_n\}_{n=1}^{\infty} \in \mathcal{D}(H^\infty)$, with limit function $\varphi_F = \chi_\alpha$.

In view of the equality $\chi_{\rho(T)}|f_n| = c_n\chi_{\rho(T)}$, we know that $c_n^{-1}f_n(T^{(a)})$ is a unitary operator ($n \in \mathbb{N}$). Thus, given any $0 \neq x_0 \in \mathcal{H}$ and $n \in \mathbb{N}$, we have

$$\begin{aligned}\lim_{j\to\infty} \|T^j f_n(T)x_0\| = \|X_T f_n(T)x_0\| = \|f_n(T^{(a)})X_T x_0\| = c_n\|X_T x_0\| \\ \leq c_n\|x_0\|,\end{aligned}$$

and so there exists $j(n) \in \mathbb{N}$ such that $\|T^j f_n(T)x_0\| \leq 2c_n\|x_0\|$ holds, whenever $j \geq j(n)$. Let $k: \mathbb{N} \to \mathbb{N}$ be an increasing mapping such that $k(n) \geq j(n)$ is true, for every $n \in \mathbb{N}$; and let us consider the sequence $G = \{g_n\}_{n=1}^{\infty} \in \mathcal{D}(H^\infty)$, where $g_n := \chi^{k(n)}f_n$ ($n \in \mathbb{N}$). The relations $\|g_n(T)x_0\| \leq 2c_n\|x_0\|$ ($n \in \mathbb{N}$) imply that $0 \neq x_0 \in \mathcal{H}_0(T,G)$. On the other hand, the limit function $\varphi_G = \varphi_F = \chi_\alpha$. Therefore, $\chi_{\pi(T)}\varphi_G = \chi_\alpha \neq 0$ is fulfilled, which yields that $\mathcal{H}_0(T,G) = \{0\}$ must hold, what is a contradiction.

□

It turns out that if the quasianalytical spectral set and the residual set do not coincide, then we have affirmative answer for the hyperinvariant subspace problem.

THEOREM 8. *Let $T \in \mathcal{L}(\mathcal{H})$ be an a.c. contraction of class $C_1.$. If $\pi(T) \neq \rho(T)$, then T has a nontrivial hyperinvariant subspace.*

Proof. Let us suppose that the set $\beta := \rho(T)\backslash\pi(T)$ is of positive measure. We can find a sequence $F = \{f_n\}_{n=1}^{\infty} \in \mathcal{D}(H^\infty)$ such that $\chi_\beta\varphi_F \neq 0$ and $\mathcal{H}_0(T,F) \neq \{0\}$. Let us consider the Borel set $\omega_0 := \{z \in \rho(T) : \varphi_F(z) = 0\}$. Since $X_T f_n(T) = f_n(T^{(a)})X_T$ holds for every $n \in \mathbb{N}$, it follows that

$$X_T\mathcal{H}_0(T,F) \subset \mathcal{H}_0^{(a)}(T^{(a)},F) = \operatorname{ran}\chi_{\omega_0}(T^{(a)}) \neq \mathcal{H}_T^{(a)}.$$

Taking into account that the subspace $\mathcal{H}_0^{(a)}(T^{(a)},F)$ is reducing for $T^{(a)}$, and applying the relation $\vee_{n\in\mathbb{N}}(T^{(a)})^{-n}X_T\mathcal{H}=\mathcal{H}_T^{(a)}$, we infer that $\mathcal{H}_0(T,F)\neq\mathcal{H}$, and so $\mathcal{H}_0(T,F)$ is a proper hyperinvariant subspace of T.

□

Now, we examine how the corresponding quasianalytical spectral and residual sets relate if there is some weak similarity connection between two contractions.

Let $T_1\in\mathcal{L}(\mathcal{H}_1)$ and $T_2\in\mathcal{L}(\mathcal{H}_2)$ be arbitrary operators. We say that T_1 can be injected into T_2, in notation: $T_1\overset{i}{\prec}T_2$, if $\mathcal{I}(T_1,T_2)$ contains an injection. The operators T_1 and T_2 are called injection-similar, in notation: $T_1\overset{i}{\sim}T_2$, if both $T_1\overset{i}{\prec}T_2$ and $T_2\overset{i}{\prec}T_1$ hold. We say that T_1 can be densely mapped into T_2, in notation: $T_1\overset{d}{\prec}T_2$, if $\mathcal{I}(T_1,T_2)$ contains a transformation with dense range. The operators T_1 and T_2 are densely-similar, in notation: $T_1\overset{d}{\sim}T_2$, if $T_1\overset{d}{\prec}T_2$ and $T_2\overset{d}{\prec}T_1$ hold simultaneously. The operator T_1 is called a quasiaffine transform of T_2, in notation: $T_1\prec T_2$, if $\mathcal{I}(T_1,T_2)$ contains a quasiaffinity, that is an injection with dense range. Finally, T_1 and T_2 are quasisimilar, in notation: $T_1\sim T_2$, if both $T_1\prec T_2$ and $T_2\prec T_1$ are fulfilled. These relations played important role in establishing canonical models for special classes of operators; see e.g. [SzNF1, Section II.3], [SzNF3], [K1], [Ber] or [DH].

PROPOSITION 9. *Let $T_1\in\mathcal{L}(\mathcal{H}_1)$ and $T_2\in\mathcal{L}(\mathcal{H}_2)$ be a.c. contractions of class $C_{1\cdot}$. Then the following statements are true:*

(i) *$T_1\overset{i}{\prec}T_2$ implies $\pi(T_1)\supset\pi(T_2)$,*
(ii) *$T_1\overset{i}{\sim}T_2$ implies $\pi(T_1)=\pi(T_2)$,*
(iii) *$T_1\overset{d}{\prec}T_2$ implies $\rho(T_1)\supset\rho(T_2)$,*
(iv) *$T_1\overset{d}{\sim}T_2$ implies $\rho(T_1)=\rho(T_2)$,*
(v) *$T_1\prec T_2$ implies $\pi(T_1)\supset\pi(T_2)$ and $\rho(T_1)\supset\rho(T_2)$,*
(vi) *$T_1\sim T_2$ implies $\pi(T_1)=\pi(T_2)$ and $\rho(T_1)=\rho(T_2)$.*

Proof. (i): Let $Y\in\mathcal{I}(T_1,T_2)$ be an injection. Given any sequence $F=\{f_n\}_{n=1}^\infty\in\mathcal{D}(H^\infty)$, the relations $Yf_n(T_1)=f_n(T_2)Y$ ($n\in\mathbb{N}$) yield that $Y((\mathcal{H}_1)_0(T_1,F))\subset(\mathcal{H}_2)_0(T_2,F)$. Thus, if $(\mathcal{H}_2)_0(T_2,F)=\{0\}$ then $(\mathcal{H}_1)_0(T_1,F)=\{0\}$ must be also true; whence the inclusion $\pi(T_2)\subset\pi(T_1)$ readily follows.

(iii): Let $Z\in\mathcal{I}(T_1,T_2)$ be a transformation with dense range. Since $X_{T_2}Z\in\mathcal{I}(T_1,T_2^{(a)})$, it follows by the universality property of the pair $(X_{T_1},T_1^{(a)})$ (see [K4, Theorem 2]) that there exists a transformation $W\in\mathcal{I}(T_1^{(a)},T_2^{(a)})$ such that $X_{T_2}Z=WX_{T_1}$. Taking into account that Z has dense range, we infer that

$$(W\mathcal{H}_{T_1}^{(a)})^-=\vee_{n\in\mathbb{N}}(T_2^{(a)})^{-n}W\mathcal{H}_{T_1}^{(a)}=\mathcal{H}_{T_2}^{(a)}.$$

We know that the subspace $\ker W$ is reducing for $T_1^{(a)}$, and that the restriction $T_1^{(a)}|(\ker W)^\perp$ is unitarily equivalent to $T_2^{(a)}$; see [Dou, Lemma 4.1]. Hence $\rho(T_1^{(a)}) \supset \rho(T_2^{(a)})$ must be true; see e.g. [C1, Section IX.10].

The remaining statements are immediate consequences of (i) and (iii). □

We note that if $\omega, \omega' \subset \mathbb{T}$ are Borel sets such that $\emptyset \neq \omega \neq \omega'$, $\omega \subset \omega'$, then $S_1 \overset{i}{\prec} M_{1,\omega} \overset{i}{\prec} M_{1,\omega'}$ and $\rho(S_1) = \mathbb{T}$, $\rho(M_{1,\omega}) = \omega$, $\rho(M_{1,\omega'}) = \omega'$. Therefore, the relation $T_1 \overset{i}{\prec} T_2$ does not yield $\rho(T_1) \subset \rho(T_2)$ nor $\rho(T_2) \subset \rho(T_1)$.

Now, we are able to extend the validity of Proposition 6. We recall that the defect operators of the contraction $T \in \mathcal{L}(\mathcal{H})$ are defined by $D_T := (I - T^*T)^{1/2}$ and $D_{T^*} := (I - TT^*)^{1/2}$. The defect spaces of T are $\mathcal{D}_T := (\operatorname{ran} D_T)^-$ and $\mathcal{D}_{T^*} := (\operatorname{ran} D_{T^*})^-$; and the defect indices of T are given by $d_T := \dim \mathcal{D}_T$ and $d_{T^*} := \dim \mathcal{D}_{T^*}$. The characteristic function Θ_T of T, introduced by Sz.-Nagy and Foias in [SzNF1], is an $\mathcal{L}(\mathcal{D}_T, \mathcal{D}_{T^*})$-valued, bounded, analytic function, defined on the open unit disc $\mathbb{D}$ by the formula $\Theta_T(z) := (-T + zD_{T^*}(I - zT^*)^{-1}D_T)|\mathcal{D}_T$.

COROLLARY 10. *Let $T \in \mathcal{L}(\mathcal{H})$ be an a.c. contraction of class $C_{1\cdot}$.*

(i) *Let us assume that $T \in C_{10}$. Then $\pi(T) = \mathbb{T}$ is true if $d_T < \infty$, or if $\dim \ker T^* < \infty$ and there exists a nonzero $\delta \in H^\infty$ such that $\Psi\Theta_T = \delta I$ is fulfilled with some $\mathcal{L}(\mathcal{D}_{T^*}, \mathcal{D}_T)$-valued, bounded, analytic function Ψ.*

(ii) *If $T \in C_{11}$, then $\pi(T) = \emptyset$.*

Proof. (i): Under both conditions T is injection-similar to a unilateral shift; the case $d_T < \infty$ was discussed in [SzN2], while the other assumption was considered in [T2]. Thus, Propositions 6 and 9 yield the statement.

(ii): If $T \in C_{11}$, then T is quasisimilar to its unitary asymptote $T^{(a)}$; see [SzNF1, Proposition II.3.5] and [K6]. Hence, Propositions 6 and 9 can be applied again. □

Let $T \in \mathcal{L}(\mathcal{H})$ be an a.c. contraction of class $C_{1\cdot}$. Let us assume that $\mathcal{M}$ is a proper invariant subspace of T, and let $T = \begin{bmatrix} T_1 & * \\ 0 & T_2 \end{bmatrix}$ be the matrix of T in the decomposition $\mathcal{H} = \mathcal{M} \oplus \mathcal{M}^\perp$. We know that the unitary asymptote $T^{(a)}$ is unitarily equivalent to the orthogonal sum of the corresponding unitary asymptotes of T_1 and T_2: $T^{(a)} \cong T_1^{(a)} \oplus T_2^{(a)}$; see [BK1] and [K6]. Hence $\rho(T) = \rho(T_1) \cup \rho(T_2)$ readily follows.

On the other hand, it is clear that the a.c. contraction $T_1 \in \mathcal{L}(\mathcal{H})$ is of class $C_{1\cdot}$, and that $\pi(T_1) \supset \pi(T)$ is fulfilled. Furthermore, it can be easily verified that $\pi(T) = \pi(T_1) \cap \pi(T_2)$ holds, if $T = T_1 \oplus T_2$. As a consequence, we obtain that if the set $\{\pi(T_1), \pi(T_2), \rho(T_1), \rho(T_2)\}$ is not a singleton, and if $T = T_1 \oplus T_2$, then $\pi(T) \neq \rho(T)$, and so T has a nontrivial hyperinvariant subspace by Theorem 8.

We note that the contraction T_2, occurring in the previous triangulation, need not be of class $C_{1\cdot}$. Of course, the definition of quasianalytical spectral set can be easily extended for arbitrary a.c. contractions. Namely, given any a.c. contraction $T \in \mathcal{L}(\mathcal{H})$, let us consider the matrix $T = \begin{bmatrix} T_0 & * \\ 0 & T_1 \end{bmatrix}$ of T in the decomposition $\mathcal{H} = \mathcal{H}_0(T) \oplus \mathcal{H}_0(T)^{\perp}$. Then $T_0 \in C_{0\cdot}$, $T_1 \in C_{1\cdot}$, and the unitary asymptote $T^{(a)}$ can be identified with $T_1^{(a)}$. Hence $\rho(T) = \rho(T_1)$, and it is natural to work with the definition $\pi(T) := \pi(T_1)$. In particular, if $T \in C_{0\cdot}$, then T_1 acts on the zero space, and then $\rho(T) := \emptyset$ and $\pi(T) := \emptyset$. In the special case, when $T = S_1 \in \mathcal{L}(H^2)$ is the simple unilateral shift and $\mathcal{M} = \vee_{n=1}^{\infty} \chi^n$, we can infer that $\pi(T) = \pi(T_1) = \mathbb{T}$ and $\pi(T_2) = \emptyset$, thus $\pi(T) \neq \pi(T_1) \cap \pi(T_2)$.

3. Quasianalytical spectral and residual mapping theorems

Let $T \in \mathcal{L}(\mathcal{H})$ be an a.c. contraction, and let us consider a nonconstant function $u \in H^{\subset}$ with $\|u\|_{\subset} \leq 1$. Taking into account that any a.c. unitary operator is similar to a c.n.u. contraction (see [K5, Lemma 7]), we infer by [SzNF1, Theorem III.2.1] that the operator $u(T)$ is also an a.c. contraction. The spectrum of $u(T)$ was characterized in terms of $\sigma(T)$ and u in the paper [FM]. Now, we are going to describe the residual and quasianalytical spectral sets of $u(T)$. Let us introduce the measurable set $\Omega(u) := \{z \in \mathbb{T} : |u(z)| = 1\}$. First of all, we note that if T is of class $C_{*\cdot}$ and $\Omega(u) \cap \rho(T) \neq \emptyset$ (that is $\Omega(u) \cap \rho(T)$ is of positive measure by our convention), then $u(T)$ is of class $C_{*\cdot}$, as well. Indeed, the transformation $Q \in \mathcal{L}(\mathcal{H}, \operatorname{ran} \chi_\omega(T^{(a)}))$, defined by $Qx = \chi_\omega(T^{(a)}) X_T x$ with $\omega := \Omega(u) \cap \rho(T)$, is nonzero and intertwines T with the unitary operator $T^{(a)} | \operatorname{ran} \chi_\omega(T^{(a)})$.

First, we consider the special case, when u is a Möbius transformation.

PROPOSITION 11. *If $T \in \mathcal{L}(\mathcal{H})$ is an a.c. contraction and if we are given $u(z) = k(z-a)(1-\bar{a}z)^{-1}$ $(k \in \mathbb{T}, a \in \mathbb{D})$, then $\pi(u(T)) = u(\pi(T))$ and $\rho(u(T)) = u(\rho(T))$; furthermore, $u(T)^{(a)}$ is unitarily equivalent to $u(T^{(a)})$.*

Proof. In view of [SzNF1, Sections I.4, VI.1 and Proposition VI.3.5] we know that T is of class $C_{0\cdot}$ ($C_{1\cdot}$) if and only if $u(T)$ is of class $C_{0\cdot}$ ($C_{1\cdot}$, respectively). Considering the canonical triangulation of T, the proof can be easily reduced to the case when T and $A := u(T)$ are of class $C_{1\cdot}$.

Let us assume that $\pi(T) \neq \emptyset$ and that $\varphi_F \chi_{u(\pi(T))} \neq 0$ is true, for the sequence $F = \{f_n\}_{n=1}^{\infty} \in \mathcal{D}(H^{\infty})$. Clearly, $\varphi_F \circ u = \varphi_{F \circ u}$ is valid, where $F \circ u := \{f_n \circ$

$u\}_{n=1}^{\infty} \in \mathcal{D}(H^\infty)$. Since $\varphi_{F\circ u}\chi_{\pi(T)} \neq 0$ is fulfilled, it follows that $\mathcal{H}_0(T, F\circ u) = \{0\}$. In virtue of the equations $(f_n \circ u)(T) = f_n(A)$ $(n \in \mathbb{N})$, we infer that $\mathcal{H}_0(A, F) = \mathcal{H}_0(T, F\circ u) = \{0\}$. We obtain that $\pi(A) \supset u(\pi(T))$. Taking into account that $T = v(A)$ is true, with the Möbius transformation $v(z) = \bar{k}(z + ka)(1 + \overline{ka}z)^{-1}$, the equality $\pi(u(T)) = u(\pi(T))$ follows.

As for the residual sets, the equation $X_T T = T^{(a)} X_T$ implies that $X_T u(T) = u(T^{(a)}) X_T$. Applying the universality property of the pair $(X_A, A^{(a)})$, we infer that there exists a transformation $Y \in \mathcal{I}(A^{(a)}, u(T^{(a)}))$ such that $X_T = Y X_A$. As in the proof of Proposition 9.(iii), we can deduce that $u(T^{(a)}) \overset{i}{\prec} A^{(a)}$. Considering the previous inverse function v of u, we obtain in a similar way that $v(A^{(a)}) \overset{i}{\prec} v(A)^{(a)} = T^{(a)}$, whence $A^{(a)} = u(v(A^{(a)})) \overset{i}{\prec} u(T^{(a)})$ follows. Therefore, the unitary operators $u(T^{(a)})$ and $A^{(a)}$ are unitarily equivalent (see [KS]), and so $\rho(u(T)) = u(\rho(T))$. □

We are able to extend the transformation law of the quasianalytical spectral sets for a larger class of functions. We shall say that the nonconstant function $u \in H^\infty$ is *regular*, if (i) $\|u\|_\infty = 1$, (ii) $m(\Omega(u)) > 0$, (iii) $u(\omega)$ is measurable, whenever $\omega \subset \Omega(u)$ is measurable, and (iv) $m(u^{-1}(\omega')) > 0$, provided $\omega' \subset u(\Omega(u))$ is measurable and $m(\omega') > 0$. (If $\omega \subset \Omega(u)$ is of positive measure, then $u(M_{1,\omega})$ is a nonzero a.c. unitary operator, and so $m(u(\omega)) > 0$.) We note that a function $u \in H^\infty$, with $\|u\|_\infty = 1$, $|u(0)| < 1$ and $m(\Omega(u)) > 0$, is regular, if u is of bounded variation on $\mathbb{T}$; and that this is the case when u is a Riemann map onto a simple Jordan region with rectifiable boundary. (See [C2, Theorem 20.4.8] and [C2, Theorems 14.5.6, 14.5.8].)

THEOREM 12. *Let $T \in \mathcal{L}(\mathcal{H})$ be an a.c. contraction such that $\pi(T) \neq \emptyset$. If $u \in H^\infty$ is a regular function and $\Omega(u) \cap \pi(T) \neq \emptyset$, then $\pi(u(T)) \supset u(\Omega(u) \cap \pi(T))$.*

Proof. In view of Proposition 11, we can assume that $u(0) = 0$, that is $u = \chi v$, with $v \in H^\infty$, $\|v\|_\infty = 1$. Since the quasianalytical spectral set $\pi(T)$ is nonempty, we know that T is of class $C_{*\cdot}$. Let us consider the triangulation $T = \begin{bmatrix} T_0 & * \\ 0 & T_1 \end{bmatrix}$, where T_0 is of class $C_{0\cdot}$ and T_1 belongs to $C_{1\cdot}$. Then $u(T)$ is of the form $u(T) = \begin{bmatrix} u(T_0) & * \\ 0 & u(T_1) \end{bmatrix}$. Since $u(T_0) = v(T_0)T_0$, we can see that $u(T_0) \in C_{0\cdot}$. On the other hand, $\Omega(u) \cap \pi(T_1) = \Omega(u) \cap \pi(T) \neq \emptyset$ implies that $\varphi_{F_u}\chi_{\pi(T_1)} \neq 0$ is true for the sequence $F_u = \{u^n\}_{n=1}^{\infty} \in \mathcal{D}(H^\infty)$, and so $u(T_1)$ is of class $C_{1\cdot}$. Therefore, we may assume that T and $u(T)$ are a.c. contractions belonging to $C_{1\cdot}$. Now, the inclusion $u(\Omega(u) \cap \pi(T)) \subset \pi(u(T))$ can be verified as in the proof of Proposition 11. □

We proceed with the description of $\rho(u(T))$, for an arbitrary $u \in H^\infty$. We shall need the following lemma.

LEMMA 13. *Let $T \in \mathcal{L}(\mathcal{H})$ be an a.c. contraction of class $C_{*\cdot}$, and let us assume that the functions $f, g, h \in H^\infty$ satisfy the conditions $\|f\|_\infty = \|g\|_\infty = 1$, $|f(0)| < 1$, $h = fg$ and $\Omega(h) \cap \rho(T) \neq \emptyset$. Let us consider the a.c. contractions $B = f(T)$ and $C = h(T)$ of class $C_{*\cdot}$. Then, there exists a unique operator $B_C \in \{C^{(a)}\}'$, such that $X_C B = B_C X_C$; furthermore, B_C is an a.c. isometry.*

Proof. The existence of a unique $B_C \in \{C^{(a)}\}'$, satisfying the condition $X_C B = B_C X_C$, follows from the universality property of $(X_C, C^{(a)})$; see [K4, Section II]. For any $n \in \mathbb{N}$, the relations $h^{n+1} \overset{a}{\prec} h^n f \overset{a}{\prec} h^n$ imply $h^{n+1}(T) \overset{a}{\prec} h^n(T) f(T) \overset{a}{\prec} h^n(T)$. Tending n to infinity, we obtain that $\|X_C x\| \leq \|X_C B x\| \leq \|X_C x\|$ is true, for every $x \in \mathcal{H}$. Thus, given any vector $y = (C^{(a)})^{-k} X_C x$ $(k \in \mathbb{N}, x \in \mathcal{H})$, we infer that

$$\|B_C y\| = \|(C^{(a)})^k B_C y\| = \|B_C X_C x\| = \|X_C B x\| = \|X_C x\| = \|y\|.$$

Since the set $\cup_{k\in\mathbb{N}} (C^{(a)})^{-k} X_C \mathcal{H}$ is dense in $\mathcal{H}_C^{(a)}$, it follows that B_C is an isometry.

Let us consider the decomposition $\mathcal{H}_C^{(a)} = \mathcal{M}_a \oplus \mathcal{M}_s$, reducing for B_C, such that $B_C|\mathcal{M}_a$ is an a.c. isometry and $B_C|\mathcal{M}_s$ is a singular unitary operator. Let $P_s \in \mathcal{L}(\mathcal{H}_C^{(a)})$ denote the orthogonal projection onto $\mathcal{M}_s$. Since B is an a.c. contraction and $(P_s X_C)B = P_s B_C X_C = B_C(P_s X_C)$, we infer that $P_s X_C = 0$, that is $\mathcal{M}_a$ contains the subspace $\mathcal{H}_{C,+}^{(a)}$. Taking into account that $\mathcal{M}_a$ is hyperinvariant for B_C, and that the operators $C^{(a)}, (C^{(a)})^{-1}$ commute with B_C, we obtain that $\mathcal{M}_a = \mathcal{H}_C^{(a)}$, and so the isometry B_C is a.c..

□

To formulate the transformation law for the residual sets we introduce some notation. Given a set $\omega \subset \mathbb{T}$ of positive measure and a unimodular measurable function $h: \omega \to \mathbb{T}$, the *properly essential range* of h is defined by

$$\text{pe-ran}h := \left\{ z \in \mathbb{T} : \lim_{r\to 0+} m(h^{-1}(D(z,r)))r^{-1} > 0 \right\},$$

where $D(z,r) := \{\zeta \in \mathbb{T} : |\zeta - z| < r\}$. We note that if the Borel measure μ on $\mathbb{T}$, given by $\mu(\omega') := m(h^{-1}(\omega'))$ $(\omega' \subset \mathbb{T})$, is absolutely continuous, then μ is equivalent to the measure $\chi_\omega dm$, where $\omega = \text{pe-ran}h$; see [R, Chapter 8]. If $u \in H^\infty$ is a regular function, then $\text{pe-ran}(u|\omega) = u(\omega)$ is true, for any $\omega \in \mathcal{B}_1$, included in $\Omega(u)$. (Note that in the latter case both $\chi_{u(\omega)} dm$ and $m \circ (u|\omega)^{-1}$ are scalar spectral measures of the a.c. unitary operator $u(M_{1,\omega})$.)

THEOREM 14. *Let $T \in \mathcal{L}(\mathcal{H})$ be an a.c. contraction of class $C_{*\cdot}$, and let $u \in H^\infty$ be a function satisfying the conditions $\|u\|_\infty = 1$, $|u(0)| < 1$ and $\Omega(u) \cap \rho(T) \neq \emptyset$. Then, the unitary asymptote $A^{(a)}$ of the a.c. contraction $A = u(T)$ is unitarily equivalent to the restriction of the normal operator $u(T^{(a)})$ to its hyperinvariant subspace $\operatorname{ran} \chi_\omega(T^{(a)})$, where $\omega = \Omega(u) \cap \rho(T)$. Therefore, $\rho(u(T)) =$ pe-ran$(u|\omega)$; in particular, if u is regular, then $\rho(u(T)) = u(\Omega(u) \cap \rho(T))$.*

Proof. In view of Proposition 11, we may assume that $u(0) = 0$. Hence u is of the form $u = \chi v$, where $v \in H^\infty$ and $\|v\|_\infty = 1$. Applying Lemma 13 with χ, v, u in place of f, g, h, respectively, we obtain that there exists an a.c. isometry $T_A \in \{A^{(a)}\}'$ such that $X_A T = T_A X_A$. In virtue of the equations $A^{(a)} X_A = X_A A = X_A u(T) = u(T_A) X_A$, we infer that the subspace $\ker(A^{(a)} - u(T_A))$ — which is reducing for $A^{(a)}$ — contains the subspace $\mathcal{H}_{A,+}^{(a)}$. Thus $\ker(A^{(a)} - u(T_A)) = \mathcal{H}_A^{(a)}$, and so $A^{(a)} = u(T_A)$.

Assuming that the isometry T_A is not unitary, the Wold-decomposition of T_A yields that $T_A \cong U \oplus S_n$, where U is an a.c. unitary operator and $n \in \mathbb{N}_\infty$. Since $A^{(a)} = u(T_A) \cong u(U) \oplus u(S_n)$ and since $u(S_n)$ is a c.n.u. contraction, we arrive at a contradiction. Therefore, T_A must be an a.c. unitary operator.

In view of the universality of $(X_T, T^{(a)})$, there exists a mapping $Y \in \mathcal{I}(T^{(a)}, T_A)$ such that $Y X_T = X_A$. We can easily verify that Y has dense range; see the proof of Proposition 9.(iii). Furthermore, the equations $Y u(T^{(a)}) = u(T_A) Y = A^{(a)} Y$ imply that $A^{(a)} \cong u(T^{(a)})|(\ker Y)^\perp$.

Since the restriction $u(T^{(a)})|(\ker Y)^\perp$ is unitary, we can see that the subspace $(\ker Y)^\perp$ is contained in the spectral subspace $\mathcal{M}_\omega := \operatorname{ran} \chi_\omega(T^{(a)})$, where $\omega := \Omega(u) \cap \rho(T)$. Let $Q \in \mathcal{L}(\mathcal{H}_T^{(a)})$ denote the orthogonal projection onto $\mathcal{M}_\omega$, and let $R \in \mathcal{I}(T, T^{(a)}|\mathcal{M}_\omega)$ be defined by $Rx := Q X_T x$ ($x \in \mathcal{H}$). Taking into account that $u(T^{(a)})|\mathcal{M}_\omega$ is unitary and that $R \in \mathcal{I}(A, u(T^{(a)})|\mathcal{M}_\omega)$, we infer by the universality of $(X_A, A^{(a)})$ that there exists a transformation $Z \in \mathcal{I}(A^{(a)}, u(T^{(a)})|\mathcal{M}_\omega)$ such that $R = Z X_A$. Clearly, $\operatorname{ran} Z \supset \operatorname{ran} R$ and

$$\vee_{n=1}^\infty (T^{(a)})^{-n} R\mathcal{H} = \vee_{n=1}^\infty Q(T^{(a)})^{-n} X_T \mathcal{H} = (Q\mathcal{H}_T^{(a)})^- = \mathcal{M}_\omega.$$

Since $(\operatorname{ran} Z)^-$ is reducing for $u(T^{(a)}|\mathcal{M}_\omega) = u(T^{(a)})|\mathcal{M}_\omega$, so it is for $T^{(a)}|\mathcal{M}_\omega$ as well. Thus, the transformation Z must have dense range, and so we conclude that $u(T^{(a)})|\mathcal{M}_\omega \cong A^{(a)}|(\ker Z)^\perp$. Now, an application of [KS] results in that the operators $A^{(a)}$ and $u(T^{(a)})|\mathcal{M}_\omega$ are unitarily equivalent.

It is known that the scalar spectral measure of the a.c. unitary operator $u(T^{(a)})|\mathcal{M}_\omega$ is $\mu \circ (u|\omega)^{-1}$, where $\mu = \chi_\omega dm$; see e.g. [C1, Proposition IX.8.12]. Thus, we conclude that $\rho(u(T)) = \rho(u(T^{(a)})|\mathcal{M}_\omega) =$ pe-ran$(u|\omega)$.

□

As an immediate consequence of Theorems 7, 12, 14 and Proposition 6, we obtain the following

COROLLARY 15. *If $T \in \mathcal{L}(\mathcal{H})$ is an a.c. contraction satisfying the condition $\pi(T) = \rho(T) \neq \emptyset$, then $\pi(u(T)) = \rho(u(T)) = u(\Omega(u) \cap \rho(T))$ is true, for any regular function $u \in H^\infty$ such that $\Omega(u) \cap \rho(T) \neq \emptyset$. In particular, if $T_u = u(S_1) \in \mathcal{L}(H^2)$ is the analytic Toeplitz operator with symbol u, and $u \in H^\infty$ is regular, then $\pi(T_u) = \rho(T_u) = u(\Omega(u))$.*

We note here that, in contrast with Proposition 3, the condition $\varphi_F \chi_{\pi(T)} = 0$ does not imply $\mathcal{H}_0(T, F) \neq \{0\}$. Indeed, in view of Corollary 15, we can find a.c. contractions A, B of class $C_1.$, such that the sets $\pi(A) \cap \pi(B)$, $\pi(A) \backslash \pi(B)$ and $\pi(B) \backslash \pi(A)$ are of positive measure. Let us consider the orthogonal sum $T = A \oplus B$; we know that $\pi(T) = \pi(A) \cap \pi(B)$ and $\rho(T) = \rho(A) \cup \rho(B)$. Let $f \in H^\infty$ be an outer function such that $|f| = (1/2)\chi_{\pi(T)} + \chi_{\mathbb{T} \backslash \pi(T)}$, and let us form the sequence $F = \{f^n\}_{n=1}^\infty \in \mathcal{D}(H^\infty)$. Since $\varphi_F = \chi_{\mathbb{T} \backslash \pi(T)}$, we obtain that $\varphi_F \chi_{\pi(T)} = 0$ and $\mathcal{H}_0(T, F) = \mathcal{H}_0(A, F) \oplus \mathcal{H}_0(B, F) = \{0\}$.

4. Quasianalytic contractions

We are going to show that the property of an a.c. contraction T being asymptotically strongly nonvanishing on a set $\alpha \in \mathcal{B}_1$ is equivalent to a quasianalytic behaviour of T on α. To be more precise, let us introduce some notation. Given an a.c. unitary operator $U \in \mathcal{L}(\mathcal{K})$, we know that $\chi_{\rho(U)} dm$ is a scalar spectral measure of U. Let E denote the spectral measure of U. For any vector $x \in \mathcal{K}$, there exists a unique Borel set $\omega(U, x) \subset \mathbb{T}$ such that the measure $\chi_{\omega(U,x)} dm$ is equivalent to the localization E_x of E at x. Clearly, $\omega(U, x) \neq \emptyset$ exactly when $x \neq 0$. Furthermore, the vector x is cyclic for the commutant of the restriction of U to the spectral subspace $\operatorname{ran} \chi_{\omega(U,x)}(U)$, that is $\vee\{Cx : C \in \{U\}'\} = \operatorname{ran} \chi_{\omega(U,x)}(U)$.

Let $T \in \mathcal{L}(\mathcal{H})$ be an a.c. contraction of class $C_1.$, and let $\alpha \in \mathcal{B}_1$. We say that *the contraction T is quasianalytic on the set α*, if $\omega(T^{(a)}, X_T x) \supset \alpha$ is fulfilled, for any nonzero vector $x \in \mathcal{H}$.

THEOREM 16. *Given an a.c. contraction $T \in \mathcal{L}(\mathcal{H})$ of class $C_1.$ and a Borel set $\alpha \in \mathcal{B}_1$, the following conditions are equivalent:*
(a) *$T \in C_1.(\alpha)$,*
(b) *T is quasianalytic on α.*

Proof. (a)$\Longrightarrow$(b): Let us assume that the contraction T is not quasianalytic on α. Then, there exists a nonzero vector $x_0 \in \mathcal{H}$ such that the set $\omega := \alpha \backslash \omega(T^{(a)}, X_T x_0)$ is of positive measure. Let us consider a sequence $F = \{f_n\}_{n=1}^{\infty} \in \mathcal{D}(H^\infty)$, with limit function $\varphi_F = \chi_\omega$. Since

$$\lim_{n\to\infty} \|X_T f_n(T) x_0\| = \lim_{n\to\infty} \|f_n(T^{(a)}) X_T x_0\| = \|\chi_\omega(T^{(a)}) X_T x_0\| = 0,$$

we can choose an increasing mapping $k\colon \mathbb{N} \to \mathbb{N}$ such that $\lim_{n\to\infty} \|T^{k(n)} f_n(T) x_0\| = 0$; see the proof of Theorem 7. Taking into account that $\varphi_G \chi_\alpha = \chi_\omega \neq 0$ is true for the sequence $G = \{g_n = \chi^{k(n)} f_n\}_{n=1}^{\infty} \in \mathcal{D}(H^\infty)$, we conclude that the contraction T is not of class $C_{1\cdot}(\alpha)$.

(b)$\Longrightarrow$(a): Assuming that T is quasianalytic on α, let us consider a sequence $F = \{f_n\}_{n=1}^{\infty} \in \mathcal{D}(H^\infty)$ such that $\varphi_F \chi_\alpha \neq 0$. Let E denote the spectral measure of $T^{(a)}$. Given any nonzero vector $x \in \mathcal{H}$, we know that the measure $E_{X_T x}$ is of the form $E_{X_T x} = g_x dm$, where $g_x(z) > 0$ holds for a.e. $z \in \alpha$. Thus, we have

$$\|\Phi_{F(T)} x\|^2 \geq \|X_T\|^{-2} \int_{\mathbb{T}} |\varphi_F|^2 g_x \, dm > 0,$$

and so T is asymptotically strongly nonvanishing on α. □

In view of this theorem, the a.c. $C_{1\cdot}$-contraction T will be called *quasianalytic* if $\rho(T) = \pi(T)$.

We note that Theorems 7 and 8 can be also derived from Theorem 16. For example, if $\pi(T) \neq \rho(T)$, then by Theorem 16 there exists a nonzero vector $x_0 \in \mathcal{H}$ such that the set $\omega := \rho(T) \backslash \omega(T^{(a)}, X_T x_0)$ is of positive measure. Since the nonzero hyperinvariant subspace $\mathcal{M} = \{C x_0 : C \in \{T\}'\}^-$ is transformed into the subspace $\operatorname{ran} \chi_{\rho(T)\backslash\omega}(T^{(a)})$ by X_T, it follows that $\mathcal{M}$ is a proper hyperinvariant subspace of T. We mention that if T is of class C_{11}, then to every spectral subspace of $T^{(a)}$ there corresponds a hyperinvariant subspace of T; see [SzNF1, Sections II.5, VII.5] and [K4, Theorem 5]. Existence of infinitely many disjoint nontrivial hyperinvariant subspaces of nonquasianalytic type was proved also in [K8].

We close this section by posing the following problem.

QUESTION 17. Is it true that the unitary asymptote $T^{(a)}$ has uniform spectral multiplicity on the quasianalytical spectral set $\pi(T)$, for any a.c. contraction T of class $C_{1\cdot}$?

Since T is quasianalytic on $\pi(T)$, we have some evidence to guess that the answer is positive. An affirmative answer to Question 17 would imply that every a.c. contraction T of class $C_{1\cdot}$ has a nontrivial hyperinvariant subspace, provided the spectral multiplicity function of the unitary asymptote $T^{(a)}$ is not constant on $\rho(T)$.

5. Intertwining relations

Applying Theorem 16, we are able to prove the following statement.

THEOREM 18. *Let $A \in \mathcal{L}(\mathcal{H})$ and $B \in \mathcal{L}(\mathcal{K})$ be a.c. contractions of class $C_{1\cdot}$.*
(i) *If $\mathcal{I}(A,B) \neq \{0\}$, then $\rho(A) \supset \pi(B)$.*
(ii) *If $\pi(A) = \rho(A)$ and $\pi(B)\backslash\pi(A) \neq \emptyset$, then $\mathcal{I}(A,B) = \{0\}$.*

Proof. (i): Let $Y \in \mathcal{I}(A,B)$ be a nonzero transformation. By the universality of $(X_A, A^{(a)})$, there exists a transformation $Z \in \mathcal{I}(A^{(a)}, B^{(a)})$ such that $X_B Y = Z X_A$. Let $x_0 \in \mathcal{H}$ be a vector such that $Y x_0 \neq 0$. Since the contraction B is quasianalytic on the set $\pi(B)$, we know that $\omega(B^{(a)}, X_B Y x_0) \supset \pi(B)$. Taking into account that $X_B Y x_0 = Z X_A x_0$, we infer by [Dou, Lemma 4.1] that $\rho(A) \supset \omega(B^{(a)}, X_B Y x_0)$.
Statement (ii) is an immediate consequence of (i).
□

As a first application, we prove the following proposition, establishing connection between the commutant of a contraction and its n-th power.

PROPOSITION 19. *Let $T \in \mathcal{L}(\mathcal{H})$ be an a.c. contraction of class $C_{1\cdot}$, and let us assume that $\pi(T) = \rho(T) = \alpha_n$, where $\alpha_n := \{e^{it} : 0 \leq t \leq 2\pi/n\}$, $n \in \mathbb{N}$. Then, for the a.c. $C_{1\cdot}$-contraction T^n, we have $\pi(T^n) = \rho(T^n) = \mathbb{T}$ and $\{T\}' = \{T^n\}'$.*

Proof. The equation $\pi(T^n) = \rho(T^n) = \mathbb{T}$ follows from Corollary 15. Setting $\lambda_n := e^{i2\pi/n}$, we know that $\pi(\lambda_n^j T) = \lambda_n^j \pi(T) \neq \pi(T)$ is true, for every $0 < j < n$, $j \in \mathbb{N}$. We infer by Theorem 18 that $\mathcal{I}(T, \lambda_n^j T) = \mathcal{I}(\lambda_n^j T, T) = \{0\}$ holds, whenever $0 < j < n$. Thus, a result of Cowen yields that $\{T^n\}' = \{T\}'$; see [Cow].
□

In view of Proposition 19 and Theorem 8, the hyperinvariant subspace problem for a.c. $C_{1\cdot}$-contractions, with an arc on $\mathbb{T}$ as a residual set, has an affirmative answer if it has positive answer for a.c. $C_{1\cdot}$-contractions satisfying the condition $\pi(T) = \rho(T) = \mathbb{T}$. This latter situation seems to be more tractable, since the assumption $\rho(T) = \mathbb{T}$ implies existence in abundance of invariant subspaces, where T is similar to the simple unilateral shift S_1; see [K7, Theorem 3]. Analogous statements of reductive type were proved in [FPSzN].

Let us assume now that the a.c. $C_{1\cdot}$-contraction $T \in \mathcal{L}(\mathcal{H})$ is cyclic. We know from [SzNF2] that the commutant of T is abelian, that is $\{T\}' = \{T\}''$, and that the adjoint T^* is cyclic, as well. Alternative proofs of these facts can be given in the following way. Since T is cyclic and the transformation $X_T^+ \in \mathcal{I}(T, T_+^{(a)})$ has dense range, it follows that the isometry $T_+^{(a)}$ is cyclic; hence $\{T_+^{(a)}\}'$ is abelian. By the universality of $(X_T^+, T_+^{(a)})$ (see [K4, Theorem 1]), there exists an injective algebra-homomorphism $\gamma_+ \colon \{T\}' \to \{T_+^{(a)}\}'$ such that $X_T^+ C = \gamma_+(C) X_T^+$ is true, for every $C \in \{T\}'$. Thus, the commutativity of $\{T\}'$ is implied by that of $\{T_+^{(a)}\}'$. On the other hand, if the isometry $T_+^{(a)}$ is cyclic, then so is its adjoint; and since $(T_+^{(a)})^* \prec T^*$, we obtain that T^* is cyclic. For the characterization of cyclic C_{11}-contractions, see [K3, Theorem 15].

It is known that commutativity of $\{T\}'$ does not imply cyclicity, in general; see [Ded]. In view of Theorem 18 we can easily provide a large class of noncyclic $C_{1\cdot}$-contractions with abelian commutant.

EXAMPLE 20. Let $A_j \in \mathcal{L}(\mathcal{H}_j)$, $j \in \mathbb{N}$, be a sequence of cyclic a.c. contractions of class $C_{1\cdot}$ such that $\pi(A_j) = \rho(A_j)$ is true for every $j \in \mathbb{N}$, $\pi(A_j) \setminus \pi(A_k) \neq \emptyset$ whenever $j \neq k$, and $\cap_{j \in \mathbb{N}} \rho(A_j) \neq \emptyset$. (Corollary 15 ensures the existence of such a sequence.) Let us form the orthogonal sum $A = \sum_{j=1}^{\infty} \oplus A_j$. In virtue of Theorem 18, the commutant of A splits into the direct sum of the commutants of the operators A_j : $\{A\}' = \sum_{j=1}^{\infty} \oplus \{A_j\}'$; thus $\{A\}'$ is abelian. On the other hand, the condition $\cap_{j \in \mathbb{N}} \rho(A_j) \neq \emptyset$ readily implies that the multiplicity of A is infinite.

We say that the a.c. $C_{1\cdot}$-contraction T is a *quasiunitary operator*, if the canonical intertwining mapping $X_T \in \mathcal{I}(T, T^{(a)})$ has dense range. These operators are characterized in the following proposition. We recall that Θ_T stands for the characteristic function of T, and that $\tilde{\Theta}_T(z) := \Theta_T(\bar{z})^*$ $(z \in \mathbb{D})$.

PROPOSITION 21. *Let $T \in \mathcal{L}(\mathcal{H})$ be an a.c. contraction of class $C_{1\cdot}$. Then the following conditions are equivalent:*

(i) *T is quasiunitary,*

(ii) $\ker \tilde{\Theta}_T \cap H^2(\mathcal{D}_{T^*}) = \{0\}$,

(iii) *$\Phi\Theta_T = 0$ implies $\Phi = 0$, for any bounded, analytic, $\mathcal{L}(\mathcal{D}_{T^*}, \mathcal{E})$-valued function Φ,*

(iv) *$\mathcal{I}(T, S_n) = \{0\}$ is true, for every $n \in \mathbb{N}_\infty$,*

(v) $\mathcal{I}(T, S_1) = \{0\}$.

Proof. The implications (ii)$\Longrightarrow$(iii) and (iv)$\Longrightarrow$(v) are trivial. For the equivalence of (i) and (ii), see [K6, Section 2]. It was shown in [T1] that condition (iii) implies (iv). Finally, if T is not quasiunitary then $T_+^{(a)}$ is a nonunitary isometry containing S_1 on a reducing subspace, and so (v) implies (i).

□

We note that the assumption $T \in C_{1\cdot}$ yields that the characteristic function Θ_T is $*$-outer, that is $(\tilde{\Theta}_T H^2(\mathcal{D}_{T^*}))^- = H^2(\mathcal{D}_T)$. The conditions (ii) and (iii) express injectivity properties of $\tilde{\Theta}_T$.

If the a.c. $C_{1\cdot}$-contraction T is not quasiunitary, then it follows by Proposition 21 that $\mathcal{I}(T, S_1) \neq \{0\}$. Taking into account that the restriction of S_1 to any of its nonzero invariant subspaces is unitarily equivalent to S_1, we obtain that $T \overset{d}{\prec} S_1$. Thus $S_1^* \overset{i}{\prec} T^*$, and so the point spectrum $\sigma_p(T^*)$ of T^* covers the open unit disc $\mathbb{D}$. We conclude that $(\ker(T^* - \lambda I))^\perp$ is a nontrivial hyperinvariant subspace of T, for any $\lambda \in \mathbb{D}$. Therefore, the hyperinvariant subspace problem for $C_{*\cdot}$-contractions can be reduced to the case, when T is a quasiunitary operator.

It can be easily seen that the a.c. $C_{1\cdot}$-contraction T is quasiunitary, if the residual set $\rho(T)$ does not cover the unit circle $\mathbb{T}$. Indeed, the condition $\rho(T) \neq \mathbb{T}$ implies that the unitary asymptote $T^{(a)}$ is reductive, that is $\mathrm{Lat}T^{(a)} = \mathrm{Lat}(T^{(a)})^*$, and so the transformation X_T has dense range. It may even happen that such T is of class C_{10}; see [K2]. Now, we exhibit an example for a quasiunitary C_{10}-contraction T, with the property $\pi(T) = \rho(T) = \sigma(T) = \mathbb{T}$.

Example 22. Let $w: \mathbb{Z} \to [1, \infty)$ be a dissymetric weight, considered by Esterle in [E]; that is w is a decreasing sequence such that $\limsup_{n\to-\infty} w(n-1)/w(n) < \infty$, $\lim_{n\to-\infty} w(n)^{1/|n|} = 1$ and $w(n) = 1$, for every $n \geq 0$. Let us assume that w is submultiplicative and quasianalytic, the latter meaning that $\sum_{n=1}^\infty (\log w(-n))n^{-2} = \infty$. (For concrete examples, see [E, Remark 4.3].)

Let us consider the Hilbert space

$$L^2(w) := \left\{ f \in L^2(\mathbb{T}) : \|f\|_w^2 := \sum_{n=-\infty}^{\infty} |\hat{f}(n)|^2 w_n^2 < \infty \right\},$$

and the operator $T_w \in \mathcal{L}(L^2(w))$, defined by $T_w f := \chi f$. The equations $(T_w f)\hat{}(n) = \hat{f}(n-1)$ $(n \in \mathbb{Z})$ are valid for the Fourier coefficients, so T_w is a weighted bilateral shift. It is easy to see that T_w is a contraction of class C_{10}. Furthermore, [E, Propositions 2.3, 2.4] and [E, Theorem 4.2] imply that $\sigma(T_w) = \mathbb{T}$ and that $f(z) \neq 0$ is true a.e. on $\mathbb{T}$, for any function $0 \neq f \in L^2(w)$.

The pair $(X_{T_w}, T_w^{(a)})$ is equivalent to the pair (X_0, T_{w_0}), where $w_0 \equiv 1$ and $X_0: L^2(w) \to L^2, f \mapsto f$ is the natural embedding; that is there exists a unitary transformation $Z \in \mathcal{I}(T_{w_0}, T_w^{(a)})$ such that $X_{T_w} = ZX_0$. Thus, we infer that T_w is quasianalytic on $\mathbb{T}$, and so $\pi(T_w) = \mathbb{T}$ holds by Theorem 16. Taking into account that the trigonometric polynomials are contained in $L^2(w)$, we obtain that T_w is a quasiunitary operator.

The invariant subspace problem is open for quasiunitary operators, in general. On the other hand, it is known from [T3] that if the residual set $\rho(T)$ covers the unit circle, then the quasiunitary operator T has disjoint nontrivial invariant subspaces.

6. Operators in the commutant

Let us assume that the a.c. $C_{1\cdot}$-contraction $T \in \mathcal{L}(\mathcal{H})$ is cyclic. Since the commutant $\{T\}'$ is abelian, we know that the subspaces $(\operatorname{ran} Q)^-$ and $\ker Q$ belong to Hlat T, for every operator $Q \in \{T\}'$. The next proposition claims that the nullspaces are all trivial, if the contraction T is quasianalytic. Furthermore, it is sufficient to assume only the cyclicity of $T^{(a)}$, which is a slighter condition than the cyclicity of T; see [SzNF3].

PROPOSITION 23. *Let $T \in \mathcal{L}(\mathcal{H})$ be an a.c. contraction of class $C_{1\cdot}$, and let us assume that $\pi(T) = \rho(T)$ and that the unitary assymptote $T^{(a)}$ is cyclic. Then, every nonzero operator $Q \in \{T\}'$ is injective.*

Proof. Let $Q \in \{T\}'$ be an operator with $\ker Q \neq \{0\}$. Let $R \in \{T^{(a)}\}'$ be the uniquely determined operator, satisfying the condition $X_T Q = RX_T$. Since the unitary operator $T^{(a)}$ is cyclic, there exists a function $\psi \in \chi_{\rho(T)} L^\infty$ such that $R = \psi(T^{(a)})$; see [C1, Chapter IX]. Taking into account that X_T is injective and that $X_T(\ker Q) \subset \ker R$, we infer that the set $\omega := \{z \in \rho(T) : \psi(z) = 0\}$ is of positive measure. Thus, $\omega(T^{(a)}, X_T Qx) = \omega(T^{(a)}, \psi(T^{(a)})X_T x) \subset \rho(T) \backslash \omega \neq \rho(T)$ is fulfilled, for every vector $x \in \mathcal{H}$. Since the contraction T is quasianalytic, we conclude that $Q = 0$.

□

A. Atzmon posed us his conjecture that *there exists a nonzero operator $Q \in \{T\}'$ with nondense range, for every cyclic a.c. contraction $T \in \mathcal{L}(\mathcal{H})$ of class C_{10}.* Verification of that statement would solve the hyperinvariant subspace problem for cyclic a.c. contractions of class $C_{*\cdot}$. It is natural to start the quest for an appropriate Q in the set of functions of T. It is known that if $v \in H^\infty$ is an outer function, then the operator $v(T)$ is a quasiaffinity, see [SzNF1, Proposition III.3.1]. Hence, we should concentrate on the class of operators $u(T)$, where $u \in H^\infty$ is an inner function.

For any operator $Q \in \mathcal{L}(\mathcal{H})$, let $\gamma(Q) := \inf\{\|Qx\| : x \in \mathcal{H}, \|x\| = 1\}$ denote the lower norm of Q; in the case $\mathcal{H} = \{0\}$, let $\gamma(Q) := 1$. It is known that $\gamma(Q^*) = 0$ is valid exactly when $\operatorname{ran} Q \neq \mathcal{H}$. We are going to give a sufficient condition for $\gamma(u(T)^*) = 0$ in terms of u and the characteristic function Θ_T of T. To that end we introduce the quantity $\eta_*(u,T) := \inf\{|u(\lambda)| + \gamma(\Theta_T(\lambda)^*) : \lambda \in \mathbb{D}\}$.

LEMMA 24. *Let $T \in \mathcal{L}(\mathcal{H})$ be an a.c. contraction of class $C_{*\cdot}$, and let $u \in H^\infty$ be an inner function. If $\eta_*(u,T) = 0$, then $\gamma(u(T)^*) = 0$.*

Proof. We shall apply the technique, introduced in [F]. First of all, we may assume that the contraction T is c.n.u.. We know from the Sz.-Nagy–Foias model theory of contractions that T is unitarily equivalent to the operator $S(\Theta_T)$, defined on the space $\mathcal{H}(\Theta_T) := \mathcal{K}_+ \ominus \{\Theta_T w \oplus \Delta_T w : w \in H^2(\mathcal{D}_T)\}$ as the compression $S(\Theta_T) := P_{\mathcal{H}(\Theta_T)} U_+ | \mathcal{H}(\Theta_T)$, where $\mathcal{K}_+ := H^2(\mathcal{D}_{T^*}) \oplus (\Delta_T L^2(\mathcal{D}_T))^-$, $\Delta_T := (I - \Theta_T^* \Theta_T)^{1/2}$ and $U_+ \in \mathcal{L}(\mathcal{K}_+)$ is the operator of multiplication by χ. Hence, we can work with $S(\Theta_T)$ instead of T. For short, let us write $\hat{S} = S(\Theta_T)$, $\hat{\mathcal{H}} = \mathcal{H}(\Theta_T)$, $\Theta = \Theta_T$, $\Delta = \Delta_T$, $\mathcal{D} = \mathcal{D}_T$ and $\mathcal{D}_* = \mathcal{D}_{T^*}$. We note that $T \in C_{*\cdot}$ implies that $\mathcal{D}_* \neq \{0\}$.

Given any $\lambda \in \mathbb{D}$ and $0 \neq x \in \mathcal{D}_*$, let us consider the vector-valued function $E_{\lambda,x} \in H^2(\mathcal{D}_*) \subset \mathcal{K}_+$, defined by $E_{\lambda,x}(z) := (1 - |\lambda|^2)^{1/2}(1 - \bar{\lambda} z)^{-1} x$, $z \in \mathbb{D}^-$. Forming power series expansion, we can see that $\|E_{\lambda,x}\|^2 = \|x\|^2$ and $U_+^* E_{\lambda,x} = \bar{\lambda} E_{\lambda,x}$. Let us introduce the projection $F_{\lambda,x} := P_{\hat{\mathcal{H}}} E_{\lambda,x} \in \hat{\mathcal{H}}$. A straightforward computation yields that $F_{\lambda,x} = E_{\lambda,x} - G_{\lambda,x}$, where $G_{\lambda,x} = \Theta\Theta(\lambda)^* E_{\lambda,x} \oplus \Delta\Theta(\lambda)^* E_{\lambda,x}$. Thus, we have

$$\begin{aligned} u(\hat{S})^* F_{\lambda,x} &= P_{\hat{\mathcal{H}}} u(U_+)^* F_{\lambda,x} = P_{\hat{\mathcal{H}}} \tilde{u}(U_+^*) E_{\lambda,x} - P_{\hat{\mathcal{H}}} \tilde{u}(U_+^*) G_{\lambda,x} \\ &= P_{\hat{\mathcal{H}}} \overline{u(\lambda)} E_{\lambda,x} - P_{\hat{\mathcal{H}}} \tilde{u}(U_+^*) G_{\lambda,x} = \overline{u(\lambda)} F_{\lambda,x} - P_{\hat{\mathcal{H}}} \tilde{u}(U_+^*) G_{\lambda,x}, \end{aligned}$$

whence

$$\|u(\hat{S})^* F_{\lambda,x}\| \leq |u(\lambda)| \|F_{\lambda,x}\| + \|G_{\lambda,x}\| = |u(\lambda)| \|F_{\lambda,x}\| + \|\Theta(\lambda)^* x\|$$

follows. Taking into account that $\|F_{\lambda,x}\|^2 = \|x\|^2 - \|\Theta(\lambda)^* x\|^2 > 0$, we obtain that

$$\gamma(u(\hat{S})^*) \leq |u(\lambda)| + \frac{\|\Theta(\lambda)^* x\|}{(\|x\|^2 - \|\Theta(\lambda)^* x\|^2)^{1/2}}.$$

In view of this inequality, we can easily verify that $\eta_*(u,T) = 0$ implies $\gamma(u(T)^*) = \gamma(u(\hat{S})^*) = 0$.

□

The following theorem claims that the range of $u(T)$ is not dense for some inner function u, if the characteristic function Θ_T satisfies some boundary conditions.

Theorem 25. *Let $T \in \mathcal{L}(\mathcal{H})$ be a contraction of class C_{10}, and let us assume that there exist $z_0 \in \rho(T)$ and $\{\lambda_n\}_{n=1}^\infty \subset \mathbb{D}$ such that $\Theta_T(z_0)$ is a nonunitary isometry, $\lim_{n\to\infty}\lambda_n = z_0$, the sequence $\{\Theta_T(\lambda_n)^*\}_{n=1}^\infty$ converges to $\Theta(z_0)^*$ in the strong operator topology, and $\lim_{n\to\infty}\gamma(\Theta_T(\lambda_n)) = 1$. Then, there exits an inner function $u \in H^\infty$ such that the nonzero operator $u(T)$ has nondense range.*

We note that the assumption $T \in C_{10}$ yields for a.e. $z \in \rho(T)$ that $\Theta_T(z)$ is a nonunitary isometry, and that $\{\Theta_T(\lambda_n)^*\}_{n=1}^\infty$ converges strongly to $\Theta(z)^*$, whenever λ_n tends to z nontangentially; see [SzNF1, Sections V.2, VI.3]. Furthermore, if $\Theta_T(z)$ is a nonunitary isometry, then $\gamma(\Theta_T(z)) = 1$ and $\gamma(\Theta_T(z)^*) = 0$.

Proof. Turning to a suitable subsequence, if necessary, we can assume that $\sum_{n=1}^\infty(1-|\lambda_n|) < \infty$ and that $\lambda_n \neq 0$, $\gamma(\Theta_T(\lambda_n)) > 1 - 2^{-n}$ are true, for every $n \in \mathbb{N}$. Setting

$$u_n(z) := \frac{\overline{\lambda_n}}{|\lambda_n|}\frac{\lambda_n - z}{1-\overline{\lambda_n}z} \quad (n \in \mathbb{N}),$$

let us form the Blaschke product $u = \prod_{n=1}^\infty u_n$. Let x_0 be a unit vector in $\ker\Theta_T(z_0)^*$. Since $\lim_{n\to\infty}\|\Theta_T(\lambda_n)^*x_0\| = \|\Theta_T(z_0)^*x_0\| = 0$ and $u(\lambda_n) = 0$ $(n \in \mathbb{N})$, we can see that $\eta_*(u,T) = 0$, and so $\gamma(u(T)^*) = 0$ by Lemma 24.

It is clear by [SzNF1, Section VI.1] that, for any $n \in \mathbb{N}$, we have

$$\gamma(u_n(T)) = \gamma\left((T-\lambda_n I)(I-\overline{\lambda_n}T)^{-1}\right) = \gamma(\Theta_T(\lambda_n)) \geq 1 - 2^{-n}.$$

Let us form the partial products $v_N := \prod_{n=1}^N u_n$, $N \in \mathbb{N}$. Then

$$\gamma(v_N(T)) \geq \prod_{n=1}^N \gamma(u_n(T)) \geq \prod_{n=1}^\infty (1-2^{-n}) =: c > 0$$

holds, for every $N \in \mathbb{N}$. We can select a subsequence $\{v_{N_k}\}_{k=1}^\infty$ such that $\lim_{k\to\infty} v_{N_k}(z) = u(z)$ is true, for a.e. $z \in \mathbb{T}$ (see [C2, Corollary 20.2.7]). Now, we infer by [SzNF1, Theorem III.2.1] that the operators $\{v_{N_k}(T)\}_{k=1}^\infty$ converge to $u(T)$ in the strong operator topology, and so $\gamma(u(T)) \geq c > 0$ must be also true. The relations $\gamma(u(T)) > 0$ and $\gamma(u(T)^*) = 0$ imply that the nonzero operator $u(T)$ has closed range, which is a nontrivial subspace of $\mathcal{H}$.

□

References

[Bea] B. Beauzamy, *Introduction to Operator Theory and Invariant Subspaces*, North Holland, Amsterdam, 1988.

[Ber] H. Bercovici, *Operator Theory and Arithmetic in H^∞*, Math. Surveys Monographs **26**, Amer. Math. Soc., Providence, R. I., 1988.

[BFP] H. Bercovici, C. Foias and C. Pearcy, *Dual Algebras with Applications to Invariant Subspaces and Dilation Theory,* CBMS Regional Conference Series in Math. No. **56**, Amer. Math. Soc., Providence, R. I., 1985.

[BK1] H. Bercovici and L. Kérchy, Quasisimilarity and properties of the commutant of C_{11} contractions, *Acta Sci. Math. (Szeged)*, **45** (1983), 67–74.

[BK2] H. Bercovici and L. Kérchy, On the spectra of C_{11}-contractions, *Proc. Amer. Math. Soc.*, **95** (1985), 412–418.

[C1] J. B. Conway, *A Course in Functional Analysis*, Springer Verlag, New York, 1990.

[C2] J. B. Conway, *Functions of One Complex Variable. II*, Springer Verlag, New York, 1995.

[Cow] C. C. Cowen, Commutants and the operator equation $AX = \lambda XA$, *Pacific J. Math.*, **80** (1979), 337–340.

[DH] K. R. Davidson and D. A. Herrero, The Jordan form of a bitriangular operator, *J. Funct. Anal.*, **94** (1990), 27–73.

[Ded] J. A. Deddens, Intertwining analytic Toeplitz operators, *Michigan Math. J.*, **18** (1971), 243–246.

[Dou] R. G. Douglas, On the operator equation $S^*XT = X$ and related topics, *Acta Sci. Math. (Szeged)*, **30** (1960), 19–32.

[E] J. Esterle, Singular inner functions and biinvariant subspaces for dissymetric weighted shifts, *J. Funct. Anal.*, **144** (1997), 64–104.

[F] P. A. Fuhrmann, On the Corona Theorem and its application to spectral problems in Hilbert space, *Trans. Amer. Math. Soc.*, **132** (1968), 55–66.

[FM] C. Foias and W. Mlak, The extended spectrum of completely nonunitary contractions and the spectral mapping theorem, *Studia Math.*, **26** (1966), 239–245.

[FPSzN] C. Foias, C. M. Pearcy and B. Sz.-Nagy, Contractions with spectral radius one and invariant subspaces, *Acta Sci. Math. (Szeged)*, **43** (1981), 273–280.

[Hal] P. R. Halmos, *Introduction to Hilbert Space and the Theory of Spectral Multiplicity*, New York, 1951.

[Hof] K. Hoffman, *Banach Spaces of Analytic Functions*, Dover Publications, Inc., New York, 1988.

[KS] R. V. Kadison and I. M. Singer, Three test problems in operator theory, *Pacific J. Math.*, **7** (1957), 1101–1106.

[K1] L. Kérchy, Injection-similar isometries, *Acta Sci. Math. (Szeged)*, **44** (1982), 157–163.

[K2] L. KÉRCHY, On the spectra of contractions belonging to special classes, *J. Funct. Anal.*, **67** (1986), 153–166.

[K3] L. KÉRCHY, Contractions being weakly similar to unitaries, *Advances in Invariant Subspaces and Other Results of Operator Theory,* OT **17,** Birkhäuser Verlag, Basel – Boston – Stuttgart, 1986, 187–200.

[K4] L. KÉRCHY, Isometric asymptotes of power bounded operators, *Indiana Univ. Math. J.*, **38** (1989), 173–188.

[K5] L. KÉRCHY, On the functional calculus of contractions with nonvanishing unitary asymptotes, *Michigan Math. J.*, **37** (1990), 323–338.

[K6] L. KÉRCHY, Unitary asymptotes of Hilbert space operators, *Banach Center Publications* **30,** Polish Academy of Sciences, Warszawa, 1994, 191–201.

[K7] L. KÉRCHY, Injection of unilateral shifts into contractions with nonvanishing unitary asymptotes, *Acta Sci. Math. (Szeged)*, **61** (1995), 443–476.

[K8] L. KÉRCHY, Hyperinvariant subspaces of operators with nonvanishing orbits, *Proc. Amer. Math. Soc.*, **127** (1999), 1363–1370.

[K9] L. KÉRCHY, Isometries with isomorphic invariant subspace lattices, *J. Funct. Anal.*, **170** (2000), 475–511.

[R] W. RUDIN, *Real and Complex Analysis,* McGraw-Hill Book Company, New York, 1966.

[SzN1] B. SZ.-NAGY, Sur les contraction de l'espace de Hilbert, *Acta Sci. Math. (Szeged)*, **15** (1953), 87–92.

[SzN2] B. SZ.-NAGY, Diagonalization of matrices over H^∞, *Acta Sci. Math. (Szeged)*, **38** (1976), 223–238.

[SzNF1] B. SZ.-NAGY and C. FOIAS, *Harmonic Analysis of Operators on Hilbert Space*, North Holland and Akadémiai Kiadó, Amsterdam – Budapest, 1970.

[SzNF2] B. SZ.-NAGY and C. FOIAS, Vecteurs cyclique et commutativité des commutants, *Acta Sci. Math. (Szeged)*, **32** (1971), 177–183.

[SzNF3] B. SZ.-NAGY and C. FOIAS, Jordan model for contractions of class $C_{\cdot 0}$, *Acta Sci. Math. (Szeged)*, **36** (1974), 305–322.

[T1] K. TAKAHASHI, Contractions with the bicommutant property, *Proc. Amer. Math. Soc.*, **93** (1985), 91–95.

[T2] K. TAKAHASHI, Quasiaffine transforms of unilateral shifts, *Proc. Amer. Math. Soc.*, **100** (1987), 683–687.

[T3] K. TAKAHASHI, On contractions without disjoint invariant subspaces, *Proc. Amer. Math. Soc.*, **110** (1990), 935–937.

L. KÉRCHY, Bolyai Institute, University of Szeged, Aradi vértanúk tere 1, H-6720 Szeged, Hungary; *e-mail*: kerchy@math.u-szeged.hu

Operator Theory: Advances and Applications, Vol. 127, 423–435

Unstable dynamics on a Markov background and stability in average

ALEXANDER M. KRÄGELOH and BORIS S. PAVLOV

Dynamical systems which switch between several different modes of unitary evolution via a Markov process are simple mathematical models of irreversible systems, when considering the averaged evolution. If some of the modes are unstable evolutions, still the averaged evolution might be stable.

From the theory of random evolution the generator $\mathcal{A}$ of the averaged evolution is obtained, and a definition for stability in average is suggested. In this context the generator A is investigated, a result for a special case is proved, and a conjecture is made about sufficient and necessary conditions for stability in average. A simple example supports this conjecture.

Introduction

In the paper "Quantum Dynamics on a Markov Background and Irreversibility" ([10]) the second named author considered a certain class of dynamical systems which possess a natural property of irreversibility. It was assumed that the generator $A(t)$ of the corresponding evolution depends on time via a continuous-time n-state Markov process generated by a certain stochastic matrix Q. The generator 'jumps' between n different self-adjoint operators $A_1, \dots, A_n$ defined on a Hilbert space H. If the trajectory of the generator A on some time interval $[0, t]$ is fixed as

$$A(t) = A_{j_k} \qquad \text{for} \quad t_k \le t < t_{k+1},$$

with $0 = t_0 < \cdots < t_N = t$, $k = 0, \dots, N-1$, and $j_k \in \{1, \dots, n\}$, then the evolution

$$\prod_{k=0}^{N-1} e^{iA_{j_k}(t_{k+1}-t_k)} =: U_A(t)$$

AMS Subject Classification (1991): 47D06, 60K37, 34D99, 47B44 .

along this trajectory is represented by a family of unitary operators, however $(U_A(t))_{t\geq 0}$ does not form a group. Following Cheremshantsev ([1]) we consider the 'averaged evolution' over all trajectories with fixed ends $A(0) = A_{j_0}, A(t) = A_{j_N}$. This averaging yields matrix elements $Z(t)_{j_0,j_N}$, $t > 0$, from some matrix operators which form a semi-group of operators generated by the dissipative operator

$$\mathcal{A} = \operatorname{diag}(A_1, \dots, A_n) - iQ^t \times \mathrm{id}_H$$

defined on the direct product $\mathcal{H}$ of the initial Hilbert space H and the stochastic space P. The same result is obtained when we fix only one end of the trajectory and consider component-wise the expected outcome as the result of the action of the matrix operator $(Z(t)_{jk})_{j,k=1}^{n}$, this was done by Griego and Hersh in [2]. The operator $\mathcal{A}$ may possess a dense complex spectrum. The paper [10] was dedicated to the spectral analysis of this dissipative operator by methods of dilation theory and functional model. The minimal selfadjoint dilation of the original operator $\mathcal{A}$ is a selfadjoint generator $\mathcal{L}$ of the unitary group $e^{i\mathcal{L}t}$ which, being compressed onto $\mathcal{H}$, gives $e^{i\mathcal{A}t}$. The functional model of $\mathcal{A}$ is obtained from the selfadjoint dilation $\mathcal{L}$ as a generator of the compression of the group $e^{i\mathcal{L}t}$ onto the coinvariant subspace, i.e. the initial space $\mathcal{H}$, represented via the so called 'incoming' spectral representation of the dilation, see Sz.-Nagy–Foias [11].

In the actual paper we assume that some of the operators $A_1, \dots, A_n$ are selfadjoint or dissipative, but others are not: they may have some complex eigenvalues with negative imaginary part. Now the evolution jumps between some *stable modes* (generated by the selfadjoint or dissipative operators) and *non-stable modes* (generated by the operators with eigenvalues with negative imaginary parts). One may expect that under certain conditions the averaged evolution (in the probabilistic sense) is a stable evolution. Then we say that *stability in average* takes place.

We suggest a natural definition and analyse the problem of stability in average based on the notion of random evolution, which was introduced by Griego and Hersh in 1969 ([2]). We formulate a theorem that relates strong dissipativity of the operator $\mathcal{A}$ on $\mathcal{H}$ to the strong dissipativity of a certain weighted sum of the operators $A_1, \dots, A_n$ for a special situation. This theorem is then extended to the case when one requires only similarity to strongly dissipative operators. Simple examples show, however, that we need to consider much more general situations, thus we suggest a conjecture regarding conditions for possible stability in average for the case of bounded operators $A_1, \dots, A_n$. We study one example for which we obtain the condition of stability in a simple and natural form, in accordance with the conjecture.

1. Random evolutions

For each $j = 1, \dots, n$ let $(T_j(t))_{t\geq 0}$ be a strongly continuous semi-group of bounded linear operators on the Hilbert space $(H_j, \langle .,. \rangle_j)$. We assume that these spaces coincide with each other as sets of vectors, i.e. $H_1 = \cdots = H_n =: H$, but unlike in [2] and [10] we allow different inner products $\langle .,. \rangle_j$ on each space, which, however, are assumed to be equivalent. For each j let A_j be the generator of $(T_j(t))_{t\geq 0}$, defined on some dense domain $\mathcal{D}(A_j) \subset H$. Further let $(\mathcal{H}, [.,.])$ be the cartesian product $(H, \langle .,. \rangle_1) \times \cdots \times (H, \langle .,. \rangle_n)$ with elements $\tilde{f} = (f_1, \dots, f_n)^t$.

For a stationary continuous-time Markov process on the state space $S = \{1, \dots, n\}$ with infinitesimal generator[1)] $Q = (q_{jk})_{j,k\in S}$ we denote a sample path with $\omega\colon [0, \infty) \to S$, the jump time of the k-th jump of ω with $\tau_k(\omega)$, and the number of jumps on the time interval $[0, t]$ with $N(t)$. For the following calculations we need the measure defined on all trajectories of the Markov process and the corresponding measure restricted onto the class of all trajectories with fixed initial and/or final states. We reserve the notation $E[X(t); \omega(t_0) = k]$ for the expectation value of the random variable $X(t)$ taken over all paths ω with the property $\omega(t_0) = k$.

The random evolutions along these trajectories are constructed as solutions of the Cauchy problem with the time-dependent generator $A_{\omega(t)}$:

$$\begin{cases} \dfrac{1}{i}\dfrac{d}{dt}y(t) = A_{\omega(t)}y(t) \\ \quad y(0) = f. \end{cases} \tag{1}$$

For a given path ω with the sequence $(\tau_k)_{k\in\mathbb{N}}$ of jump times the *backward random evolution* is defined as (Griego–Hersh [3])

$$M_b(t,\omega) := T_{\omega(0)}(\tau_1)T_{\omega(\tau_1)}(\tau_2 - \tau_1)\cdots T_{\omega(\tau_{N(t)})}(t - \tau_{N(t)}), \quad t \geq 0,$$

and the *forward random evolution* is (Keepler [5] and [6])

$$M_f(t,\omega) := T_{\omega(\tau_{N(t)})}(t - \tau_{N(t)})\cdots T_{\omega(\tau_1)}(\tau_2 - \tau_1)T_{\omega(0)}(\tau_1), \quad t \geq 0.$$

Let $\tilde{f} = (f_1, \dots, f_n)^t \in \mathcal{H}$, then the *backward expectation operators* are defined component-wise for $k = 1, \dots, n$ as ([3])

$$\left(\mathcal{E}_b(t)\tilde{f}\right)_k = E\Big[M_b(t,\omega)f_{\omega(t)} \; ; \; \omega(0) = k\Big], \quad t \geq 0,$$

---

1) We adopt the standard notation for the transition probabilities, i.e. $p_{jk}(t) = Pr(X(t) = k | X(0) = j)$, with $q_{jk} = \frac{d}{dt}p_{jk}(t)|_{t=0}$.

and, similarly, the *forward expectation operators* are defined as ([6])

$$\left(\mathcal{E}_f(t)\tilde{f}\right)_k = E\left[M_f(t,\omega)f_{\omega(0)}\ ;\ \omega(t)=k\right], \quad t \geq 0.$$

Griego and Hersh showed in [3] that the family $(\mathcal{E}_b(t))_{t\geq 0}$ is a strongly continuous semi-group of bounded operators, the same holds for $(\mathcal{E}_f(t))_{t\geq 0}$ (cf. [6]). They also showed that the vector $\mathcal{E}_b(t)\tilde{f}$ solves the system of Cauchy problems

$$\begin{cases} \dfrac{1}{i}\dfrac{d}{dt}y_j(t) = A_j y_j - i\displaystyle\sum_{k=1}^{n} q_{jk}\, y_k(t) \\ y_j(0) = f_j \end{cases} \qquad j = 1,\ldots,n$$

or, in other words, the infinitesimal generator of $\mathcal{E}_b(t)$ is

$$\mathcal{A}_b = \operatorname{diag}(A_1,\ldots,A_n) - iQ \times \mathrm{id}_H,$$

where with $Q \times \mathrm{id}_H$ we denote the matrix operator

$$(2) \qquad Q \times \mathrm{id}_H = \begin{pmatrix} q_{11}\,\mathrm{id}_H & \cdots & q_{1n}\,\mathrm{id}_H \\ \vdots & \ddots & \vdots \\ q_{n1}\,\mathrm{id}_H & \cdots & q_{nn}\,\mathrm{id}_H \end{pmatrix}.$$

The forward expectation semi-group is generated by, (cf. [6]),

$$\mathcal{A}_f = \operatorname{diag}(A_1,\ldots,A_n) - iQ^t \times \mathrm{id}_H,$$

thus it follows immediately that $(\mathcal{E}_b(t))_{t\geq 0} \equiv (\mathcal{E}_f(t))_{t\geq 0}$ if and only if Q is symmetric.

The original proofs were done under the assumption that all norms are equal, but they are true also in the case of equivalent norms.

REMARK 1. If $\tilde{f} = (p_1 f,\ldots,p_n f)^t$ with some initial probability vector $p = (p_1,\ldots,p_n)^t$ and $f \in H$, then it is easily seen that the vector

$$\Psi[f,p](t) := \sum_{k=1}^{n}\left(\mathcal{E}_f(t)\tilde{f}\right)_k \in H$$

is the expected value at time $t \geq 0$ of the solution to the Cauchy problem (1).

REMARK 2. In the theory of tensor products (e.g. Treves [12]), the above vector $(p_1 f, \dots, p_n f)^t$ is denoted by $p \otimes f$. However, since we do not deal with tensor product spaces here, we shall use the notations $p \odot f$ for $p \in \mathbb{C}^n$ and $f \in H$, and $Y \odot F := \{y \odot f : y \in Y, f \in F\} \subset \mathcal{H}$ for two subsets $Y \subset \mathbb{C}^n$ and $F \subset H$. With view to Remark 3 we do not adopt this notation for $Q \times \mathrm{id}_H$.

2. Some properties of the operator $Q \times \mathrm{id}_H$

In this section we let, unless stated otherwise, Q be a general complex $n \times n$-matrix.

REMARK 3. (i) The operators id_H in (2) represent the identity operator in H. Should H be of finite dimension, and the operators A_j be represented as some matrices with respect to different bases, then the operators id_H are not represented as identity matrices, but rather as some change of basis matrices, according to their position in the matrix operator $Q \times \mathrm{id}_H$.

(ii) In the case when the inner products $\langle .,. \rangle_1, \dots, \langle .,. \rangle_n$ are different, the operator id_H in the j-th row and k-th column of $Q \times \mathrm{id}_H$ is the embedding

$$I_{jk}: (H, \langle .,. \rangle_k) \to (H, \langle .,. \rangle_j) \ , \quad u \mapsto u.$$

These embeddings are bounded if and only if the inner products are mutually equivalent, and then the operator norms are not necessarily equal to 1.

LEMMA 4.

(i) *If all inner products are mutually equivalent then $Q \times \mathrm{id}_H$ is a bounded operator.*

(ii) *Conversely, if Q generates an irreducible Markov process, and if $Q \times \mathrm{id}_H$ is bounded, then all inner products are mutually equivalent.*

The first part of Lemma 4 can be proved by direct calculation. Part (ii) is proved by contradiction with help of the following fact about irreducible Markov process with generator $(q_{jk})_{j,k=1}^n$: For any two disjoint subsets J and K of the state space S there exist $j_0 \in J$ and $k_0 \in K$ such that $q_{j_0 k_0} \neq 0$ (e.g. Grimmett–Stirzaker [4]).

Lemme 5 is proved similarly.

LEMMA 5. *If all inner products $\langle .,.\rangle_1, \ldots, \langle .,.\rangle_n$ are equal, then the adjoint of $Q \times \mathrm{id}_H$ in $(\mathcal{H}, [.,.])$ is $\overline{Q}^t \times \mathrm{id}_H$.*

Conversely, if Q generates an irreducible n-state Markov process and $(Q \times \mathrm{id}_H)^* = (\overline{Q}^t \times \mathrm{id}_H)$, then all inner products are equal.

COROLLARY 6. *If all inner products are equal, then $\Re(Q \times \mathrm{id}_H) = (\Re Q) \times \mathrm{id}_H$ and $\Im(Q \times \mathrm{id}_H) = (\Im Q) \times \mathrm{id}_H$.*

LEMMA 7. *Assume all inner products are equal.*

(i) *If $x, y \in \mathbb{C}^n$ with $Qx \perp_{\mathbb{C}^n} y$ then $Q \times \mathrm{id}_H(x \odot f) \perp_{\mathcal{H}} (y \odot f)$ for all $f \in H$.*

(ii) *Let Y be a subspace of $\mathbb{C}^n$ and $m \in \mathbb{R}$ such that $\langle Qx, x\rangle_{\mathbb{C}^n} \leq m\|x\|^2_{\mathbb{C}^n}$ for all $x \in \mathbb{C}^n$ with $x \perp_{\mathbb{C}^n} Y$, then we have*

$$[Q \times \mathrm{id}_H\, u, u] \leq m[u, u]$$

for all $u \in \mathcal{H}$ such that $u \perp_{\mathcal{H}} (Y \odot H)$.

Lemma 7.(i) is proved directly, and the proofs of the other results can be found in Section 2.3 of [8], which is based on the thesis [7] of the first named author.

3. Stability in average

Our concept of stability is motivated from the physical context where the (suitably chosen) norm of a vector function $u(t)$ may represent the physical energy of a certain object, e.g. a particle or a string. We call an evolution given by a semigroup $(T(t))_{t\geq 0}$ of bounded operators *stable* if for each $u \in H$ the norms $\|T(t)u\|_H$ are uniformly bounded for all times $t > 0$.

In general the matrix Q generating the Markov process in Section 1 will not be symmetric, but then the forward and backward expectation semi-groups are not equal. Considering their definitions and Remark 9.(ii) below it seems more natural to connect stability in average with the forward evolution, thus we suggest the

DEFINITION 8. The dynamical system given as the solution of (1) is said to be *stable in average* if the operators $\mathcal{E}_f(t)$, $t \geq 0$, of the forward expectation semi-group are uniformly bounded, i.e. if there exists a constant $K > 0$ such that $\|\mathcal{E}_f(t)\| \leq K$ for all $t \geq 0$. We call the system *conditionally stable in average*, if there exists $\kappa > 0$ such that the system affected by the scaled Markov process generated by κQ is stable in average.

REMARK 9. (i) It would be interesting to find an example of a system (with non-symmetric Q) where the forward expectation semi-group is uniformly bounded, and the backward is not.

(ii) If the operator norms of the family $(\mathcal{E}_f(t))_{t\geq 0}$ are uniformly bounded, then the norms of the expected values $\Psi[f,p](t)$ are uniformly bounded for every initial vector $f \in H$ and initial distribution p, which corresponds to the described concept of stability.

It follows from the Theorem of Phillips and Lumer (e.g. Yosida [13, Ch.IX]), that the system is stable in average if the generator of the forward expectation semigroup is dissipative. The following theorem gives sufficient and necessary conditions for strong dissipativity of the operator $\mathcal{A}_f(\kappa) := \Big(\mathrm{diag}(A_1, \dots, A_n) - i\kappa Q^t \times \mathrm{id}_H\Big)$, $\kappa > 0$, hence conditional stability in average, for a special case.

In the following two theorems we will assume that the generator Q of the Markov process has the property (P), i.e.:

(P) There exists a unique vector of equilibrium distribution $\pi \in \mathbb{R}^n$ of the Markov process, i.e. $Q^t\pi = 0$, such that additionally $Q^t x \perp_{\mathbb{C}^n} \pi$ for all $x \in R := \mathbb{C}^n \ominus \{\pi\}$. Further, there exists $m_0 > 0$ so that $\Re(Q^t|_R) \leq -m_0 \,\mathrm{id}_R$.

REMARK 10. If Q is symmetric with eigenvalues $0 = \lambda_1 > \lambda_2 \geq \dots \geq \lambda_n$ then Q has property (P) with $m_0 = |\lambda_2|$. If the Markov process is also irreducible then $\pi = \frac{1}{n}(1, \dots, 1)^t$.

THEOREM 11. *For each $j = 1, \dots, n$ let A_j be a bounded generator of a strongly continuous semigroup $(T_j(t))_{t\geq 0}$ of bounded operators on the Hilbert space $(H, \langle .,. \rangle)$, and assume that not all operators are self-adjoint. Further assume that the generator A_ω of the dynamical system* (1) *jumps between the operators $A_1, \dots, A_n$ via an irreducible Markov process which is generated by the $n \times n$-matrix Q with property* (P) *and the vector of equilibrium distribution $\pi = (\pi_1, \dots, \pi_n)$.*

Now, if for some $\alpha > 0$

$$\Im\Big(\sum_{j=1}^{n} \pi_j^2 A_j\Big) \geq \alpha\Big(\sum_{j=1}^{n} \pi_j^2\Big)\,\mathrm{id}_H, \tag{3}$$

then for each $0 < \lambda < \alpha$ there exists $\kappa_0 > 0$ such that for all $\kappa \geq \kappa_0$

$$\Im\Big(\mathcal{A}_f(\kappa)\Big) \geq \lambda \,\mathrm{id}_{\mathcal{H}}\,.$$

Conversely, if $\Im \mathcal{A}_f(\kappa) \geq \lambda \, \mathrm{id}_{\mathcal{H}}$ *for some* $\lambda > 0$ *and* $\kappa \in \mathbb{R}$, *then* $\Im\Big(\sum_{j=1}^n \pi_j^2 A_j\Big) \geq \lambda\Big(\sum_{j=1}^n \pi_j^2\Big) \, \mathrm{id}_H$.

Proof. Since the operators $A_1, \dots, A_n$ generate strongly continuous semigroups we can define $\mu_j := \sup\{\beta \in \mathbb{R} : \Im A_j \geq \beta \, \mathrm{id}_H\}$ for each $1 \leq j \leq n$ and $\mu_0 := \min_{1 \leq j \leq n} \mu_j$, and it follows $\Im\Big(\mathrm{diag}(A_1, \dots, A_n)\Big) \geq \mu_0 \, \mathrm{id}_{\mathcal{H}}$. We further define $\tilde{A} := \mathrm{diag}(A_1, \dots, A_n)$ and $\sigma := \|\Im \tilde{A}\| = \max_{1 \leq j \leq n} \|\Im(A_j)\|$, note that σ is positive since not all operators A_j are selfadjoint.

From property (P) it follows that the null space of $M := -iQ^t \times \mathrm{id}_H$ is

$$\mathcal{N} := \{\pi\} \odot H = \Big\{(\pi_1 f, \dots, \pi_n f)^t : f \in H\Big\};$$

and if we let $\mathcal{R} := \mathcal{H} \ominus \mathcal{N}$, Lemma 7 then implies that $M\mathcal{R} \perp_{\mathcal{H}} \mathcal{N}$ and

$$\mathrm{Im}(M|_{\mathcal{R}}) = -\Re(Q^t \times \mathrm{id}_H |_{\mathcal{R}}) \geq m_0 \, \mathrm{id}_{\mathcal{R}} \, .$$

Now let $u \in \mathcal{H}$, and find $v \in \mathcal{R}$, $w = \pi \odot f \in \mathcal{N}$ with $f \in H$ such that $u = v + w$. We get for $\kappa > 0$

$$\begin{aligned}
\Im\Big[(\tilde{A} + \kappa M)u, u\Big] &= \Im\Big[(\tilde{A} + \kappa M)v, v\Big] + \Im\Big[\tilde{A}v, w\Big] + \Im\Big[\tilde{A}w, v\Big] + \Im\Big[\tilde{A}w, w\Big] \\
&= \Im\Big[(\tilde{A} + \kappa M)v, v\Big] + \Im\Big[(\tilde{A} - \tilde{A}^*)v, w\Big] + \Im\Big(\sum_{j=1}^n \langle A_j \pi_j f, \pi_j f\rangle\Big) \\
&\geq (\mu_0 + \kappa m_0)\|v\|^2 - \|\Im \tilde{A}\| \, \|v\| \, \|w\| + \alpha\Big(\sum_{j=1}^n \pi_j^2\Big) \|f\|_H^2 .
\end{aligned}$$

With $\sigma = \|\Im \tilde{A}\|$ and

$$\|v\| \, \|w\| \leq \varepsilon \, \|v\|^2 + \frac{1}{\varepsilon} \, \|w\|^2$$

for any $\varepsilon > 0$ it follows

$$\Im[(\tilde{A} + \kappa M)u, u] \geq (\mu_0 + \kappa m_0 - \varepsilon\sigma)\|v\|^2 + \Big(\alpha - \frac{\sigma}{\varepsilon}\Big)\|w\|^2 =: C.$$

We require that

$$C \geq \lambda\|u\|^2 = \lambda\Big(\|v\|^2 + \|w\|^2\Big),$$

which is satisfied if

$$\kappa m_0 + \mu_0 - \varepsilon\sigma \geq \lambda \qquad \text{and} \tag{4}$$

$$\alpha - \frac{\sigma}{\varepsilon} \geq \lambda. \tag{5}$$

Equation (5) is fulfilled for $\varepsilon = \frac{\sigma}{\alpha-\lambda}$ and any $0 < \lambda < \alpha$, and then (4) is satisfied for

$$\kappa \geq \kappa_0 := \frac{1}{m_0}\left(\lambda - \mu_0 + \frac{\sigma^2}{\alpha - \lambda}\right). \tag{6}$$

For the converse, choose any $u = \pi \odot f \in \{\pi\} \odot H$ and observe that

$$\Im\Big\langle \sum_{j=1}^{n} \pi_j^2 A_j f, f\Big\rangle_H = \Im\Big[(\tilde{A} + \kappa M)u, u\Big] \geq \lambda \|u\|_{\mathcal{H}}^2 = \lambda\Big(\sum_{j=1}^{n} \pi_j^2\Big)\|f\|_H^2.$$

This proves the theorem.

□

The expectation semigroup is also uniformly bounded if $\mathcal{A}_f$ is only similar to a strongly dissipative operator, which is considered in the next theorem.

THEOREM 12. *In the situation of Theorem* 11 *let P_0 and P_1 be the orthogonal projections of $\mathcal{H}$ onto $\mathcal{N}$ and $\mathcal{R}$ respectively, then:*

The operator $\mathcal{A}_f(\kappa)$ is similar to a strongly dissipative operator for some $\kappa > 0$ if the component $P_0\tilde{A}P_0$ is similar to a strongly dissipative operator B_0 on $\mathcal{N} = \{\pi\} \odot H$, i.e. if there exists a bounded and invertible operator $X{:}\,\mathcal{N} \to \mathcal{N}$ with a bounded inverse, and $\beta > 0$ such that $B_0 = XP_0\tilde{A}P_0X^{-1}$ and $\Im\Big(B_0\Big) \geq \beta\, \mathrm{id}_{\mathcal{N}}$.

Proof. First we note that the operator $X{:}\,\mathcal{N} \to \mathcal{N}$ can be identified with an operator on H. Since a vector $(\pi_1 f, \ldots, \pi_n f)^t \in \mathcal{N}$ is mapped to a vector $(\pi_1 g, \ldots, \pi_n g)^t \in \mathcal{N}$, the essential part of X is the mapping from $f \in H$ to $g \in H$. We also denote this operator with X.

We transform the operator $\mathcal{A}_f(\kappa)$, $\kappa > 0$, on $\mathcal{R} \oplus \mathcal{N}$ to

$$\mathcal{B}_f(\kappa) := \begin{pmatrix} \mathrm{id}_{\mathcal{R}} & 0 \\ 0 & X \end{pmatrix} \begin{pmatrix} P_1(\tilde{A} + \kappa M)P_1 & P_1\tilde{A}P_0 \\ P_0\tilde{A}P_1 & P_0\tilde{A}P_0 \end{pmatrix} \begin{pmatrix} \mathrm{id}_{\mathcal{R}} & 0 \\ 0 & X^{-1} \end{pmatrix},$$

and by similar calculations as in the proof of Theorem 11 we can find

$$\kappa \geq \frac{1}{m_0}\left(\lambda - \mu_0 + \frac{\sigma^2}{\beta - \lambda}\right)$$

with $0 < \lambda < \beta$ and $\sigma = \frac{1}{2}\|\tilde{A}\|\Big(\|X\| + \|X^{-1}\|\Big)$ so that

$$\Im(\mathcal{B}_f(\kappa)) \geq \lambda\ \mathrm{id}_{\mathcal{H}}.$$

□

Theorems 11 and 12 deal with the situation of equal norms. However, in many interesting situations, like the example in Section 4 and the other examples to be found in [8], the canonical choice of the norms $\|.\|_j = \sqrt{\langle .,. \rangle_j}$ on each copy of H results in mere equivalent norms, and the Markov process generally is not symmetric. Additionally, it is sufficient for stability in average that the generator $\mathcal{A}_f$ of the expectation semi-group be similar to a simply dissipative operator. Thus with view to Theorem 12 we pose the following conjecture.

CONJECTURE 13. *Let $A_1, \dots, A_n$ be bounded operators on some Hilbert spaces $(H, \langle .,. \rangle_1), \dots (H, \langle .,. \rangle_n)$ with mutually equivalent norms $\|.\|_1, \dots, \|.\|_n$, and let Q be the generator of an irreducible Markov process with equilibrium distribution $\pi = (\pi_1, \dots, \pi_n)$. Then the following statements are equivalent:*

(i) *The dynamical system (1) is conditionally stable in average.*

(ii) *The operator $\hat{A} := \pi_1^2 A_1 + \cdots + \pi_n^2 A_n$ generates a semi-group $\left(e^{i\hat{A}t}\right)_{t \geq 0}$ of uniformly bounded operators.*

4. An example

Imagine the one-dimensional movement of a point mass m in a potential $V(x) = cx^2$, $c \in \mathbb{R}$, which is described by the differential equation

$$m\ddot{x} + 2cx = 0. \tag{7}$$

Depending on the initial conditions $x(0)$ and $\dot{x}(0)$ the point mass will always be in a fixed interval around the origin for $c > 0$; we call this evolution stable. For $c < 0$ the point mass will in general approach to infinity, i.e. we have an unstable dynamic (see Figure 1).

With help of the Cauchy data $y(t) = (y_0, y_1)^t = (x(t), \dot{x}(t))^t$ on $\mathbb{C}^2$ equipped with the standard basis and some constants $q_1, q_2 > 0$ we can write (7) as the Cauchy problem:

$$\text{for } c = c_1 > 0,\ q_1^2 = \tfrac{2c_1}{m} \quad : \ \tfrac{1}{i}\tfrac{d}{dt}y(t) = \begin{pmatrix} 0 & -i \\ iq_1^2 & 0 \end{pmatrix} y(t) =: A_1 y(t),\ y(0) = f,$$

$$\text{for } c = -c_2 < 0,\ q_2^2 = -\tfrac{2c_2}{m} \ : \ \tfrac{1}{i}\tfrac{d}{dt}y(t) = \begin{pmatrix} 0 & -i \\ -iq_2^2 & 0 \end{pmatrix} y(t) =: A_2 y(t),\ y(0) = f.$$

The canonical choices for the (*different*) inner products are

$$\langle y, z \rangle_1 = \frac{1}{2}(q_1^2 y_0 \overline{z_0} + y_1 \overline{z_1}) \qquad \text{and} \qquad \langle y, z \rangle_2 = \frac{1}{2}(q_2^2 y_0 \overline{z_0} + y_1 \overline{z_1}),$$

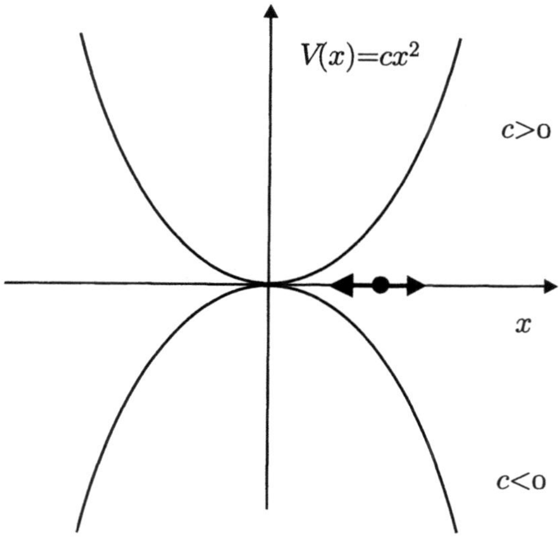

Figure 1. One-dimensional movement of a point mass in different potentials.

since then we have $A_1^* = A_1$ in $(\mathbb{C}^2, \langle .,. \rangle_1)$ and $A_2^* = -A_2$ in $(\mathbb{C}^2, \langle .,. \rangle_2)$. Hence $(T_1(t))_{t\geq 0} = (e^{iA_1 t})_{t\geq 0}$ and $(T_2(t))_{t\geq 0} = (e^{iA_2 t})_{t\geq 0}$ are stable and unstable evolution semi-groups resp. (the eigenvalues of A_2 are $\pm i q_2$).

We assume that the potential jumps between $V_1(x) = c_1 x^2, c_1 > 0$, and $V_2(x) = -c_2 x^2, c_2 > 0$ via the Markov process generated by

$$Q = \begin{pmatrix} -1 & 1 \\ 1 & -1 \end{pmatrix}$$

with vector of equilibrium distribution $\pi = (\frac{1}{2}, \frac{1}{2})^t$. With reference to Remark 3(i) we note that both A_1 and A_2 are represented with respect to the standard basis, so that the expectation semi-group $\mathcal{E}_f(t)$ is generated by, for $\kappa > 0$,

$$\mathcal{A}_f(\kappa) = \operatorname{diag}(A_1, A_2) - i\kappa Q^t \times \mathrm{id}_{\mathbb{C}^2} = \begin{pmatrix} i\kappa & -i & -i\kappa & 0 \\ iq_1^2 & i\kappa & 0 & -i\kappa \\ -i\kappa & 0 & i\kappa & -i \\ 0 & -i\kappa & -iq_2^2 & i\kappa \end{pmatrix}.$$

In order to check the Conjecture 13 in this case, we need to find conditions for which a $\kappa > 0$ exists such that the family $(\mathcal{E}_f(t))_{t\geq 0}$ is uniformly bounded. This is the case when all four (complex) eigenvalues $\lambda_1, \ldots, \lambda_4$ of $\mathcal{A}_f(\kappa)$ have positive imaginary part.

With $r := \frac{|c_1|}{|c_2|} = \frac{q_1^2}{q_2^2}$, $\alpha := -\kappa^2 - \frac{1}{2}q_2^2(1-r)$ and $\beta := q_2^2(1+r)^2 + 8\kappa^2(1-r)$ we obtain

$$\lambda_1 = i\kappa + \sqrt{\alpha + \tfrac{q_2}{2}\sqrt{\beta}} \quad , \quad \lambda_2 = i\kappa - \sqrt{\alpha + \tfrac{q_2}{2}\sqrt{\beta}} \, ,$$
$$\lambda_3 = i\kappa + \sqrt{\alpha - \tfrac{q_2}{2}\sqrt{\beta}} \quad , \quad \lambda_4 = i\kappa - \sqrt{\alpha - \tfrac{q_2}{2}\sqrt{\beta}} \, .$$

The calculations done in [8] show that it is possible to find $\kappa > 0$ as in Conjecture 13.(i) if and only if $r > 1$ ($\Leftrightarrow c_1 > c_2 \Leftrightarrow q_1^2 > q_2^2$). Geometrically this means that the system is conditionally stable in average if and only if the "stable potential" V_1 is represented by a steeper parabola than the "unstable potential" V_2, see Figure 1.

Now we investigate the two eigenvalues of

$$\hat{A} = \frac{1}{4}A_1 + \frac{1}{4}A_2 = \frac{1}{2}\begin{pmatrix} 0 & -i \\ \frac{i}{2}(q_1^2 - q_2^2) & 0 \end{pmatrix},$$

which are $\mu_1 = \frac{1}{4}\sqrt{2(q_1^2 - q_2^2)}$ and $\mu_2 = -\frac{1}{4}\sqrt{2(q_1^2 - q_2^2)}$. For $q_1^2 < q_2^2$ it is $\Im(\mu_2) < 0$, so that the operator norms of $e^{i\hat{A}t}$ increase to infinity as $t \to \infty$. This is also the case for $q_1^2 = q_2^2$, since then $\hat{A}$ is a two-dimensional Jordan block with zero eigenvalues. However, for $q_1^2 > q_2^2$ the eigenvalues μ_1 and μ_2 are real, different and non-zero, so that $(e^{i\hat{A}t})_{t\geq 0}$ is a uniformly bounded semi-group. Thus $(e^{i\hat{A}t})_{t\geq 0}$ is uniformly bounded if and only if $q_1^2 > q_2^2$, which is exactly the condition for conditional stability in average obtained with the help of the expectation semi-group.

Conjecture 13 is supported by the example, further investigations will be done.

ACKNOWLEDGEMENT. We wish to thank Professor V. Adamjan for his valuable remarks and suggestions, especially towards the definition of stability in average.

The authors recognise support in parts from the Marsden Fund grant 3368152 awarded by the Royal Society of New Zealand, and the grant RFFI 97-01-01149 from the Russian Academy.

The authors are grateful to Professor I. Antoniou and the Solvay Institutes, Brussels, for the hospitality and the inspiring discussions during the conference on the EC-project ESPRIT in June 1998.

References

[1] S. E. CHEREMSHANTSEV, *Theory of Scattering by a Brownian Particle,* Proceedings of the Steklov Institute of Mathematics, **2**, 1991.

[2] R. GRIEGO and R. HERSH, Random evolutions, Markov chains and systems of partial differential equations, *Proc. Nat. Acad. Sci. USA*, **62** (1969), 305–308.

[3] R. Griego and R. Hersh, Theory of random evolutions with applications to partial differential equations, *Trans. Amer. Math. Soc.*, **156** (1971), 405–418.

[4] G. R. Grimmett and D. R. Stirzaker, *Probability and Random Processes*, Clarendon Press, Oxford, 1982.

[5] M. Keepler, Backward and forward random evolutions, *Indiana Univ. Math. J.*, **24** (1975), 937–947.

[6] M. Keepler, On the sample path approach to random evolutions, *Bull. Malaysian Math. Soc.*, **5** (1982), 79–86.

[7] A. M. Krägeloh, *Unstable Dynamics on a Markov Background and Stability in Average,* Master Thesis in Mathematics, The University of Auckland, New Zealand, 1998.

[8] A. M. Krägeloh, *Unstable Dynamics on a Markov Background and Stability in Average*, CDMTCS Research Report Series 097, http://www.cs.auckland.ac.nz/staff-cgi-bin/mjd/secondcgi.pl, March 1999.

[9] P. D. Lax and R. S. Phillips, *Scattering Theory for Automorphic Functions,* Annals of Mathematical Studies, Princeton University Press, 1976.

[10] B. S. Pavlov, *Quantum Dynamics on a Markov Background and Irreversibility,* Nonlinear Dynamics Chaotic and Complex Systems, Proceedings of an International Conference held in Zakopane, Poland, 1995, Cambridge University Press, Cambridge, 1997.

[11] B. Sz.-Nagy and C. Foias, *Harmonic analysis of operators in Hilbert space*, North-Holland, Amsterdam, 1970.

[12] F. Treves, *Topological Vector Spaces, Distributions and Kernels*, Academic Press, 1967.

[13] K. Yosida, *Functional Analysis*, Springer, 1965.

A. M. Krägeloh, Friedrich-Alexander-Universität Erlangen/Nürnberg, Mathematisches Institut, Bismarckstr. 1 1/2, 91054 Erlangen, Germany; *e-mail*: kraegeloh@mi.uni-erlangen.de

B. S. Pavlov, University of Auckland, Department of Mathematics, The University of Auckland, Private Bag 92019, Auckland, New Zealand;
e-mail: pavlov@math.auckland.ac.nz

Operator Theory: Advances and Applications, Vol. 127, 437–445

A relation for the spectral shift function of two self-adjoint extensions

H. LANGER, H. S. V. DE SNOO and V. A. YAVRIAN

Dedicated to the memory of Béla Szőkefalvi-Nagy

The spectral shift function for two self-adjoint extensions of a symmetric operator with defect index $(1,1)$ is expressed by means of the normalized spectral functions of all 'intermediate' self-adjoint extensions.

1. Introduction and main result

Let S be a closed symmetric operator in a Hilbert space $\mathfrak{H}$ with defect index $(1,1)$. The set of all self-adjoint extensions of S in the space $\mathfrak{H}$ can be described by M. G. Kreĭn's formula which gives a one-to-one correspondence between all $\tau \in \mathbb{R} \cup \{\infty\}$ and all self-adjoint extensions $A(\tau)$ of S via

$$(A(\tau)-z)^{-1} = (A-z)^{-1} - \chi(z)\,\frac{1}{Q(z)+\tau}\,[\,\cdot\,,\chi(\bar{z})], \quad z \in \mathbb{C}\setminus\mathbb{R}. \tag{1.1}$$

For $\tau = \infty$ the second term in the righthand side is assumed to vanish. Furthermore, $A = A(\infty)$ is a fixed self-adjoint extension of S in $\mathfrak{H}$, $\chi(z_0)$ is a nonzero vector of the defect space $\ker(S^*-z_0)$ where $z_0 \in \rho(A)$ is chosen, and the family of defect vectors $\chi(z) \in \ker(S^*-z)$, $z \in \rho(A)$, is determined by the relation

$$\chi(z) = (I + (z-z_0)(A-z)^{-1})\chi(z_0).$$

Furthermore, $Q(z)$ is the Q-function of S and A: it is a Nevanlinna function, uniquely defined, up to a real constant, by the relation

$$\frac{Q(z)-\overline{Q(w)}}{z-\overline{w}} = [\chi(z),\chi(w)], \quad z, w \in \rho(A). \tag{1.2}$$

AMS Subject Classification (1991): 47A55, 47B25, 34B20, 47A10, 47B10.

Here, Nevanlinna function means that $Q(z)$ is holomorphic on $\mathbb{C} \setminus \mathbb{R}$ and has there the properties

$$Q(\overline{z}) = \overline{Q(z)}, \quad \frac{\operatorname{Im} Q(z)}{\operatorname{Im} z} \geq 0.$$

For a densely defined symmetric operator S all its self-adjoint extensions in $\mathfrak{H}$ are operators, while if the symmetric operator S is nondensely defined one of these self-adjoint extensions is a linear relation (that is a multivalued operator with a nontrivial multivalued part).

With the Q-function $Q(z)$ define for $\tau \in \mathbb{R} \cup \{\infty\}$ the fractional linear transformation

$$Q_\tau(z) = \frac{\tau Q(z) - 1}{\tau + Q(z)}, \quad Q_\infty(z) = Q(z). \tag{1.3}$$

For $z \in \mathbb{C} \setminus \mathbb{R}$ it follows that

$$\operatorname{Im} Q_\tau(z) = \frac{(1+\tau^2) \operatorname{Im} Q(z)}{|Q(z) + \tau|^2}.$$

Hence $Q_\tau(z)$, $\tau \in \mathbb{R} \cup \{\infty\}$, is again a Nevanlinna function, and thus there exist $\alpha_\tau \in \mathbb{R}$, $\beta_\tau \geq 0$, and a nondecreasing function $\sigma_\tau(t)$ with $\int_{\mathbb{R}} d\sigma_\tau(t)/(t^2+1) < \infty$, such that

$$Q_\tau(z) = \alpha_\tau + \beta_\tau z + \int_{\mathbb{R}} \left(\frac{1}{t-z} - \frac{t}{t^2+1} \right) d\sigma_\tau(t), \quad z \in \mathbb{C} \setminus \mathbb{R}. \tag{1.4}$$

Here $\sigma_\tau(t)$ is normalized such that it is left-continuous and $\sigma_\tau(0) = 0$; it is called the *spectral function* of $Q_\tau(z)$. In applications to differential operators the function $Q_\tau(z)$ is often called the *Titchmarsh–Weyl coefficient* of the extension $A(\tau)$, and then $\sigma_\tau(t)$ is the corresponding *spectral function*.

Let $A(\tau)$ and $A(\rho)$, $\rho, \tau \in \mathbb{R} \cup \{\infty\}$, be two self-adjoint extensions of S. Since the difference of their resolvents is one-dimensional, there exists the spectral shift function $\xi(\,\cdot\,; A(\tau), A(\rho))$ with the property $\xi(\,\cdot\,; A(\tau), A(\rho)) = -\xi(\,\cdot\,; A(\rho), A(\tau))$. The aim of this note is to prove the following connection between the spectral shift function $\xi(\,\cdot\,; A(\tau), A(\rho))$ and the spectral functions $\sigma_s(t)$, $\rho < s < \tau$.

MAIN THEOREM. *Let $\rho < \tau \leq \infty$. For each $t \in \mathbb{R}$ the function $s \mapsto \sigma_s(t)/(s^2+1)$ is Riemann-integrable over the interval (ρ, τ), and*

$$\int_0^t \xi(u; A(\tau), A(\rho))\, du = -\int_\rho^\tau \sigma_s(t) \frac{ds}{s^2+1}. \tag{1.5}$$

The possibly improper integral on the righthand side of (1.5) can be understood as the Riemann integral

$$\int_{\operatorname{Arctan}\rho}^{\operatorname{Arctan}\tau} \sigma_{\tan u}(t)\, du,$$

where Arctan takes its values in the interval $(-\pi/2, \pi/2]$. For the case where the self-adjoint extensions correspond to different boundary conditions at 0 of a 2×2 canonical system on the halfline $[0, \infty)$, formula (1.5) was proved by one of the authors [Y]. However, in [Y] the denominator $s^2 + 1$ does not appear since there the self-adjoint extensions are parametrized through $(-\pi/2, \pi/2]$ instead of through $\mathbb{R} \cup \{\infty\}$.

Corollary. *Let τ, $\rho \in \mathbb{R}$, $\rho < \tau \leq \infty$. If for $s \in [\rho, \tau]$ the functions $\sigma_s(\cdot)$ are absolutely continuous in a neighbourhood of $u \in \mathbb{R}$, then the relation*

$$\xi(u; A(\tau), A(\rho)) = -\int_{\rho}^{\tau} \sigma_s'(u)\, \frac{ds}{s^2+1}\,. \tag{1.6}$$

holds.

This paper is dedicated to the memory of Béla Szőkefalvi-Nagy, whose work and personality have been of great influence to us. The first author will never forget the support and friendship shown to him.

2. Proof of the main theorem and its corollary

First recall the definition of the spectral shift function, see [BP], [BY], [GMN], [K]. Let A, B be self-adjoint operators in $\mathfrak{H}$ such that for some (and hence for all) $z \in \mathbb{C} \setminus \mathbb{R}$ the difference $(B-z)^{-1} - (A-z)^{-1}$ is a trace class operator. Then there exists a measurable function $\xi(\,\cdot\,; B, A)$ on $\mathbb{R}$ such that

$$\operatorname{tr}\left((B-z)^{-1} - (A-z)^{-1}\right) = -\int_{\mathbb{R}} \frac{\xi(t; B, A)}{(t-z)^2}\, dt, \quad z \in \mathbb{C} \setminus \mathbb{R}. \tag{2.1}$$

The function $\xi(\,\cdot\,; B, A)$ is uniquely determined by this relation up to an additive constant. It is called the *spectral shift function* of the ordered pair A, B. In the following theorem a unique choice for the spectral shift function $\xi(\,\cdot\,; B, A)$ will be given in terms of a Q-function, when A and B are selfadjoint extensions of a symmetric operator with defect index $(1, 1)$.

In the sequel the notation $\operatorname{Log} z$, $z \in \mathbb{C}$, stands for $\operatorname{Log} z := \ln|z| + i \arg z$, with $-\pi < \arg z \leq \pi$.

THEOREM 1. *Let S be a symmetric operator in the Hilbert space $\mathfrak{H}$ with defect index $(1,1)$, and denote by $Q(z)$ a Q-function of S corresponding to some self-adjoint extension A of S in $\mathfrak{H}$. For $\rho < \tau \leq \infty$ the spectral shift function $\xi(t; A(\tau), A(\rho))$ of the pair $A(\rho)$, $A(\tau)$ of self-adjoint extensions of S in $\mathfrak{H}$ is given by*

$$\xi(t; A(\tau), A(\rho)) = -\frac{1}{\pi} \lim_{\varepsilon \to 0} \arg\left(\frac{Q(t+i\varepsilon)+\rho}{Q(t+i\varepsilon)+\tau}\right), \quad \tau < \infty, \tag{2.2}$$

and by

$$\xi(t; A(\infty), A(\rho)) = -\frac{1}{\pi} \lim_{\varepsilon \to 0} \arg\left(Q(t+i\varepsilon)+\rho\right), \quad \tau = \infty. \tag{2.3}$$

Moreover, for $\rho < \tau \leq \infty$,

$$-1 \leq \xi(t; A(\tau), A(\rho)) \leq 0. \tag{2.4}$$

Proof. Step 1. It follows from (1.1) that for $\rho \in \mathbb{R}$

$$\operatorname{tr}\left((A(\rho)-z)^{-1} - (A-z)^{-1}\right) = -\frac{[\chi(z), \chi(\bar{z})]}{Q(z)+\rho}. \tag{2.5}$$

The identity (1.2) shows that $[\chi(z), \chi(\bar{z})] = Q'(z)$, and hence (2.5) is equivalent to

$$\operatorname{tr}\left((A(\rho)-z)^{-1} - (A-z)^{-1}\right) = -\frac{d}{dz}\operatorname{Log}\left(Q(z)+\rho\right). \tag{2.6}$$

Since $0 \leq \arg(Q(z)+\rho) \leq \pi$, the function $\operatorname{Log}(Q(z)+\rho)$ in (2.6) is a Nevanlinna function. Hence, there exists a measurable function $\xi(t, A(\rho))$ with

$$0 \leq \xi(t; A(\rho)) \leq 1, \tag{2.7}$$

and $\gamma_\rho \in \mathbb{R}$ such that

$$\operatorname{Log}\left(Q(z)+\rho\right) = \gamma_\rho + \int_{\mathbb{R}} \left(\frac{1}{t-z} - \frac{t}{t^2+1}\right) \xi(t; A(\rho))\, dt, \tag{2.8}$$

and

$$\xi(t, A(\rho)) = \frac{1}{\pi} \lim_{\varepsilon \to 0} \arg\left(Q(t+i\varepsilon)+\rho\right), \tag{2.9}$$

cf. [D]. It follows from (2.6) and (2.8) that

$$\text{tr}\left((A(\rho)-z)^{-1}-(A-z)^{-1}\right)=-\int_{\mathbb{R}}\frac{\xi(t;A(\rho))}{(t-z)^2}\,dt. \tag{2.10}$$

Step 2. For $\tau,\rho\in\mathbb{R}$ define

$$\xi(t;A(\tau),A(\rho))=\xi(t;A(\tau))-\xi(t;A(\rho)),$$

so that it follows from (2.10) that

$$\text{tr}\left((A(\tau)-z)^{-1}-(A(\rho)-z)^{-1}\right)=-\int_{\mathbb{R}}\frac{\xi(t;A(\tau),A(\rho))}{(t-z)^2}\,dt, \tag{2.11}$$

as an analogue of (2.1). For $\rho<\tau<\infty$ the function $\text{Log}\,\frac{Q(z)+\rho}{Q(z)+\tau}$ is a Nevanlinna function, and it follows from (2.8) that

$$\text{Log}\,\frac{Q(z)+\rho}{Q(z)+\tau}=\gamma_\rho-\gamma_\tau-\int_{\mathbb{R}}\left(\frac{1}{t-z}-\frac{t}{t^2+1}\right)\xi(t;A(\tau),A(\rho))\,dt. \tag{2.12}$$

Therefore (2.2) follows (cf. (2.9)) and the estimate (2.4) evidently holds (cf. (2.7)).

Step 3. For $\tau=\infty$ define

$$\xi(t;A(\infty),A(\rho))=-\xi(t;A(\rho)),$$

so that (2.10) (recall $A=A(\infty)$) can be rewritten as

$$\text{tr}\left((A(\infty)-z)^{-1}-(A(\rho)-z)^{-1}\right)=-\int_{\mathbb{R}}\frac{\xi(t;A(\infty),A(\rho))}{(t-z)^2}\,dt, \tag{2.13}$$

as an analogue of (2.1). The identity (2.8) is equivalent to

$$\text{Log}\,(Q(z)+\rho)=\gamma_\rho-\int_{\mathbb{R}}\left(\frac{1}{t-z}-\frac{t}{t^2+1}\right)\xi(t;A(\infty),A(\rho))\,dt, \tag{2.14}$$

Hence, (2.3) follows (cf. (2.9)) and (2.4) is satisfied (cf. (2.7)). □

Proof of the Main Theorem. Step 1. Observe that the function $Q_\tau(z)$ in (1.3) satisfies the following identity:

$$Q_\tau(z)\,\frac{1}{\tau^2+1}=\frac{\tau Q(z)-1}{(\tau+Q(z))(\tau^2+1)}=-\frac{d}{d\tau}\mathrm{Log}\left(\frac{\tau+Q(z)}{\sqrt{\tau^2+1}}\right).$$

Hence, using (2.12) for $\rho<\tau<\infty$, and (2.14) for $\rho<\tau=\infty$, one obtains

$$\begin{aligned}\int_\rho^\tau Q_s(z)\frac{ds}{s^2+1}&=\mathrm{Log}\frac{Q(z)+\rho}{\sqrt{\rho^2+1}}-\mathrm{Log}\frac{Q(z)+\tau}{\sqrt{\tau^2+1}}\\ &=-\mathrm{Log}\frac{1}{\sqrt{\rho^2+1}}+\mathrm{Log}\frac{1}{\sqrt{\tau^2+1}}+\gamma_\rho-\gamma_\tau-\\ &\quad-\int_{\mathbb{R}}\left(\frac{1}{t-z}-\frac{t}{t^2+1}\right)\xi(t;A(\tau),A(\rho))\,dt,\end{aligned}\tag{2.15}$$

where the terms $\mathrm{Log}\,(1/\sqrt{\tau^2+1})-\gamma_\tau$ are absent in case $\tau=\infty$. Clearly, the lefthand side of (2.15) is a Nevanlinna function. Since $Q_s(z)$ for fixed z is a rational function in s, it can be written as a Riemann integral

$$\int_{\mathrm{Arctan}\,\rho}^{\mathrm{Arctan}\,\tau} Q_{\tan u}(z)\,du.\tag{2.16}$$

Step 2. The Riemann integral (2.16) can be approximated, pointwise in $z\in\mathbb{C}\setminus\mathbb{R}$, by Nevanlinna functions which are obtained as Riemann sums of the form

$$\sum_{i=1}^{n}(u_i-u_{i-1})\,Q_{\tan u_{i-1}}(z),\tag{2.17}$$

where $u_0<u_1<\ldots<u_n$ is a partition of the interval $[\mathrm{Arctan}\,\rho,\mathrm{Arctan}\,\tau]$. In fact, each summand in (2.17) is a Nevanlinna function with corresponding normalized spectral function

$$(u_i-u_{i-1})\,\sigma_{\tan u_{i-1}}(t),$$

according to the integral representation in (1.4). Hence, the Nevanlinna functions in (2.17) have normalized spectral functions

$$\sum_{i=1}^{n}(u_i-u_{i-1})\,\sigma_{\tan u_{i-1}}(t).\tag{2.18}$$

Step 3. Pointwise convergence of a sequence of Nevanlinna functions implies that their normalized spectral functions converge to the normalized spectral function of the limit function in the continuity points of the latter, see [Y] and for a formal proof [CG]. Therefore the Riemann sums in (2.18) converge to

$$\int_{\mathrm{Arctan}\,\rho}^{\mathrm{Arctan}\,\tau} \sigma_{\tan u}(t)\,du, \tag{2.19}$$

and for each $t \in \mathbb{R}$ this limit is equal to the normalized spectral function of the function in the last integral in (2.15), which is

$$-\int_0^t \xi(u; A(\tau), A(\rho))\,du.$$

This completes the proof of the Main Theorem.

□

Proof of the Corollary. In order to prove the corollary of the Main Theorem, observe that $Q_s(z)$ is measurable in (s, z). Hence, by the Stieltjes inversion formula, also $\sigma_s(t)$ is measurable in (s, t). Due to the assumption on $\sigma_s(\cdot)$, also the nonnegative function $\sigma_s'(t)$ is measurable in (s, t) for $s \in [\rho, \tau]$ and t in a neighborhood of u. For t_0 sufficiently close to t, $t_0 < t$, it follows from (1.5) and Fubini's theorem that

$$\begin{aligned}\int_{t_0}^t \xi(u; A(\tau), A(\rho))\,du &= -\int_\rho^\tau (\sigma_s(t) - \sigma_s(t_0))\,\frac{ds}{s^2+1} \\ &= -\int_\rho^\tau \left(\int_{t_0}^t \sigma_s'(u)\,du\right)\frac{ds}{s^2+1} \\ &= -\int_{t_0}^t \left(\int_\rho^\tau \sigma_s'(u)\,\frac{ds}{s^2+1}\right) du,\end{aligned}$$

and the relation (1.6) follows.

□

3. An elementary example

Let $\mathfrak{H} = \mathbb{C}$ and let S be the zero operator in $\mathfrak{H}$. Then S is symmetric and has defect index $(1, 1)$. Since S is nondensely defined, there is precisely one self-adjoint extension which is not an operator. In fact, all self-adjoint extensions $A(\tau)$, $\tau \in \mathbb{R} \cup \{\infty\}$, of S are given by

$$A(\tau) = \{\,\{u, -\tau u\} : u \in \mathbb{C}\,\}, \quad \tau \in \mathbb{R}, \tag{3.1}$$

and

$$A(\tau) = \{\,\{0, u\} : u \in \mathbb{C}\,\}, \quad \tau = \infty. \tag{3.2}$$

Clearly, for $A = A(\infty)$ the Q-function $Q(z)$ is given by $Q(z) = z$ and in (1.4) with $\tau = \infty$ one has

$$\alpha_\infty = 0, \quad \beta_\infty = 1, \quad \sigma_\infty(t) = 0.$$

For $\tau \in \mathbb{R}$ the Q-function $Q_\tau(z)$ of $A(\tau)$ in (1.3) has the representation

$$Q_\tau(z) = \tau - \frac{1+\tau^2}{z+\tau} = \tau + \int_{\mathbb{R}} \frac{1}{t-z}\, d\sigma_\tau(t),$$

cf. (1.4), where the *normalized* spectral function $\sigma_\tau(t)$ is a bounded function, given for $\tau \geq 0$ by

$$\text{(3.3)} \qquad \begin{cases} \sigma_\tau(t) = -(\tau^2+1) & t \leq -\tau, \\ \sigma_\tau(t) = 0 & t > -\tau, \end{cases}$$

and for $\tau < 0$ by

$$\text{(3.4)} \qquad \begin{cases} \sigma_\tau(t) = 0 & t \leq -\tau, \\ \sigma_\tau(t) = \tau^2 + 1, & t > -\tau. \end{cases}$$

It follows from (2.2) that for $\rho < \tau < \infty$

$$\text{(3.5)} \qquad \begin{cases} \xi(t; A(\tau), A(\rho)) = -1, & -\tau < t < -\rho, \\ \xi(t; A(\tau), A(\rho)) = 0, & t < -\tau \text{ or } t > -\rho, \end{cases}$$

and from (2.3) that

$$\text{(3.6)} \qquad \begin{cases} \xi(t; A(\infty), A(\rho)) = -1, & t < -\rho, \\ \xi(t; A(\infty), A(\rho)) = 0, & t > -\rho. \end{cases}$$

The resolvent operators of the self-adjoint extensions in (3.1) and (3.2) satisfy the resolvent identities:

$$\text{(3.7)} \qquad (A(\tau)-z)^{-1} - (A(\rho)-z)^{-1} = \frac{\tau-\rho}{(\tau+z)(\rho+z)}, \quad \rho < \tau < \infty,$$

and

$$\text{(3.8)} \qquad (A(\infty)-z)^{-1} - (A(\rho)-z)^{-1} = \frac{1}{\rho+z}, \quad \rho < \infty,$$

(recall that $(A(\infty)-z)^{-1} = 0$). The relations (3.7) and (3.8) are equivalent to the spectral shift property, expressed by (2.11) and (2.13) with (3.5) and (3.6), respectively. In the present case the Main Theorem can be verified directly by means of (3.3), (3.4), (3.5), and (3.6).

References

[BP] M. S. Birman and A. B. Pushnitski, Spectral shift function, amazing and multifaceted, *Integral Equations Operator Theory,* **30** (1998), 191–199.

[BY] M. S. Birman and D. R. Yafaev, The spectral shift function. The work of M. G. Kreĭn and its further development, *Algebra i Analiz,* **4** (1992), 1–44 (Russian); English translation: *St. Petersburg Math. J.*, **4** (1993), 833–870.

[CG] E. A. Coddington and R. C. Gilbert, Generalized resolvents of ordinary differential operators, *Trans. Amer. Math. Soc.*, **93** (1959), 216–241.

[D] W. F. Donoghue, Jr., *Monotone matrix functions and analytic continuation,* Springer-Verlag, Berlin – Heidelberg – New York, 1974.

[GMN] F. Gesztesy, K. Makarov and S. Naboko, The spectral shift operator, *Operator Theory: Adv. Appl.*, **108** (1999), 59–90.

[K] M. G. Kreĭn, On certain new studies in the perturbation theory for selfadjoint operators, *Operator Theory: Adv. Appl.*, **7** (1983), 107–172; First mathematical summer school (Kanev, 1963), Part I, Naukova Dumka, Kiev, 1964 (Russian).

[Y] V. A. Yavrian, On the Kreĭn spectral shift function for canonical systems of differential operators, *Operator Theory: Adv. Appl.*, **117** (2000), 393–417.

H. Langer, Institut für Analysis und Technische Mathematik, Technische Universität Wien, Wiedner Hauptstrasse 8–10/114, A–1040 Wien, Austria; *e-mail:* hlanger@mail.zserv.tuwien.ac.at

H. S. V. de Snoo, Department of Mathematics, University of Groningen, Postbus 800, 9700 AV Groningen, Nederland; *e-mail:* desnoo@math.rug.nl

V. A. Yavrian, Department of Mathematics, Yerevan State University, Alek Manukian Street 1, Yerevan 375049, Armenia

Operator Theory: Advances and Applications, Vol. 127, 447–464

Beppo Levi and Lebesgue type theorems for bundle convergence in noncommutative L_2-spaces

Barthélemy Le Gac and Ferenc Móricz*

Dedicated to the memory of Professor Béla Szőkefalvi-Nagy

Our subject is noncommutative measure theory in the context of von Neumann algebras. We consider a von Neumann algebra $\mathfrak{A}$ equipped with a faithful normal state. The notion of bundle convergence for sequences of operators in $\mathfrak{A}$ as well as for sequences of vectors in its L_2-completion were introduced in 1996 by Hensz, Jajte and Paszkiewicz [2], and this notion turned out to be an appropriate substitute for almost everywhere convergence in the noncommutative setting. In terms of bundle convergence, we prove the noncommutative versions of the standard Beppo Levi Theorem (for an increasing sequence of operators), the Lebesgue Dominated Convergence Theorem, in particular, the Arzela–Osgood Theorem (for a bounded sequence of operators). Only classical results of von Neumann algebras are required in our proofs. As an auxiliary result, we present a proof for the existence of a bundle which is contained in each of a countable number of given bundles in $\mathfrak{A}$.

0. Introduction and notations

Let $\mathfrak{H}$ be a Hilbert space over the field $\mathbb{C}$ of complex numbers, $\mathcal{L}(\mathfrak{H})$ the family of all bounded linear operators acting on $\mathfrak{H}$, $\mathfrak{A}$ a von Neumann algebra in $\mathcal{L}(\mathfrak{H})$, and

$$\|A\|_\infty := \sup\{\|Ah\|_{\mathfrak{H}} : \|h\|_{\mathfrak{H}} \leq 1\}$$

AMS Subject Classification (2000): 46L51, 46L10, 28A20.

* This research was partially supported by the Hungarian NFSR under Grant T 029094; and it was completed while the author visited the Department of Mathematics at the University of Tennessee, Knoxville, in the academic year 1999/2000.

the operator norm in $\mathfrak{A}$, where $\|.\|_{\mathfrak{H}}$ means the norm in $\mathfrak{H}$.

If ϕ is a faithful and normal state acting on $\mathfrak{A}$, then the Cauchy–Schwarz inequality holds true:

$$(0.1)\qquad |\phi(B^*A)|^2 \le \phi(B^*B)\phi(A^*A), \quad A, B \in \mathfrak{A};$$

furthermore, for each $A \in \mathfrak{A}$ we have

$$(0.2)\qquad |\phi(A)| \le \|A\|_\infty,$$

and

$$(0.3)\qquad \phi(A^*A) = 0 \quad \text{implies} \quad A = 0.$$

Thus,

$$\langle A, B\rangle := \phi(B^*A), \quad A, B \in \mathfrak{A},$$

defines an inner product on $\mathfrak{A}$, and $(\mathfrak{A}, \langle\cdot,\cdot\rangle)$ is a prehilbert space over $\mathbb{C}$. We denote by $L_2 = L_2(\mathfrak{A}, \phi)$ its completion, by $(\cdot,\cdot)$ the inner product, and by $\|\cdot\|$ the norm in L_2.

By the Gelfand–Naimark–Segal Representation Theorem, there exist a $*$-homeomorphism π of $\mathfrak{A}$ into the algebra $\mathcal{L}(L_2)$ and a cyclic vector ω in L_2 such that

$$(0.4)\qquad \phi(A) = (\pi(A)\omega, \omega), \quad A \in \mathfrak{A}.$$

By (0.3), this ω is also a separating vector for $\pi(\mathfrak{A})$. Consequently, this $*$-homeomorphism π is necessarily one-to-one (but not "onto" in general).

By (0.1), (0.2) and (0.4), it follows that

$$(0.5)\qquad \|\pi(A)\omega\| \le \{\phi(A^*A)\}^{1/2} \le \|A^*A\|_\infty^{1/2} = \|A\|_\infty, \quad A \in \mathfrak{A}.$$

Therefore, $\mathfrak{A}$ can be identified with the subset $\pi(\mathfrak{A})\omega$ of L_2, while the closure of this subset with respect to the norm $\|\cdot\|$ coincides with L_2 itself. This means that for each vector ξ in L_2 there exists a sequence $(A_n : n \ge 1)$ of operators in $\mathfrak{A}$ such that

$$\sum_{n=1}^\infty \pi(A_n)\omega = \xi,$$

where the series is strongly convergent in L_2.

The reader interested in details may consult the books by Dixmier [1], Jajte [3] and Takesaki [6, pp. 39-41].

In 1996 Hensz, Jajte and Paszkiewicz [2] introduced a new notion of convergence, so-called bundle convergence, in $\mathfrak{A}$ and $L_2(\mathfrak{A},\phi)$. Bundle convergence seems to be an adequate substitute for almost sure convergence in the attempt to extend convergence results from the traditional commutative setting to the noncommutative one.

We briefly recall their definitions. To this end, denote by Proj$\mathfrak{A}$ the class of all (self-adjoint) projections P in $\mathfrak{A}$, and by $\mathfrak{A}_+$ the cone of the positive operators in $\mathfrak{A}$.

DEFINITION 1. Let $(D_n : n \geq 1)$ be a sequence of operators in $\mathfrak{A}_+$ such that

$$\sum_{n=1}^{\infty} \phi(D_n) < \infty. \tag{0.6}$$

The *bundle* $\mathcal{P} = \mathcal{P}(D_n)$ associated with this sequence is defined by

$$\begin{aligned} \mathcal{P} := \Big\{P \in \operatorname{Proj}\mathfrak{A} : \sup_{n\geq 1} \Big\|P\Big(\sum_{k=1}^{n} D_k\Big)P\Big\|_\infty < \infty \\ \text{and} \quad \|PD_nP\|_\infty \to 0 \ \text{ as } \ n \to \infty\Big\}. \end{aligned} \tag{0.7}$$

The crucial fact (see [2, Corollary 2.2]) is that given any sequence (D_n) of operators in $\mathfrak{A}_+$ satisfying (0.6), the bundle $\mathcal{P}$ defined by (0.7) is "abundant" enough in the sense that $\mathcal{P}$ contains projections P arbitrarily close to the identity operator $1_{\mathfrak{H}}$ in $\mathfrak{A}$.

DEFINITION 2. Let (A_n) be a sequence of operators in $\mathfrak{A}$. We say that (A_n) is *bundle convergent* to some $A \in \mathfrak{A}$, symbolically:

$$A_n \overset{\mathrm{b},\mathfrak{A}}{\to} A \quad \text{as} \quad n \to \infty, \tag{0.8}$$

if there exists a bundle $\mathcal{P}$ such that for every P in $\mathcal{P}$ we have

$$\|(A_n - A)P\|_\infty \to 0 \quad \text{as} \quad n \to \infty. \tag{0.9}$$

DEFINITION 3. Let (ξ_n) be a sequence of vectors in $L_2 = L_2(\mathfrak{A},\phi)$. We say that (ξ_n) is *bundle convergent* to some ξ in L_2, symbolically:

$$\xi_n \overset{\mathrm{b}}{\to} \xi \quad \text{as} \quad n \to \infty,$$

if there exists a sequence (A_n) of operators in $\mathfrak{A}$ such that we have (0.8) with $A = 0$ and

$$\sum_{n=1}^{\infty} \|\xi_n - \xi - \pi(A_n)\omega\|^2 < \infty.$$

In this paper, $\mathfrak{A}$ will be a fixed von Neumann algebra with a faithful and normal state ϕ, $L_2 = L_2(\mathfrak{A},\phi)$ the completion of $\mathfrak{A}$ under the norm $A \to \phi(A^*A)^{1/2}$, π a $*$-homeomorphism of $\mathfrak{A}$ into $\mathcal{L}(L_2)$, and ω a cyclic and separating vector in L_2.

1. Auxiliary results

In the sequel, we shall use the notation

$$|A| := (A^*A)^{1/2}, \quad A \in \mathfrak{A}. \tag{1.1}$$

The square root makes sense because A^*A is in $\mathfrak{A}_+$.

We note that the usual triangle inequality does not hold in general (see, e.g. [4]). Fortunately, the following weaker substitute for the triangular inequality (a noncommutative version of the Cauchy inequality) is available in any von Neumann algebra $\mathfrak{A}$ (see, e.g. [3, p. 4]).

LEMMA 1. *If $c_n \in \mathbb{C}$ and $A_n \in \mathfrak{A}$ for $1 \le n \le N$, then*

$$\Big|\sum_{n=1}^{N} c_n A_n\Big|^2 \le \sum_{n=1}^{N} |c_n|^2 \sum_{n=1}^{N} |A_n|^2.$$

The following lemma is a noncommutative version of the well-known Arzela–Osgood Theorem, which is valid for uniformly bounded sequences of integrable (in the sense of Riemann) functions on a finite interval.

LEMMA 2. *If $(A_n : n \ge 1)$ is a bounded sequence of operators in $\mathfrak{A}$ such that*

$$A_n \overset{\mathrm{b},\mathfrak{A}}{\to} A \quad as \quad n \to \infty \tag{1.2}$$

for some A in $\mathfrak{A}$, then

$$\phi(|A_n - A|^2) \to 0 \quad as \quad n \to \infty. \tag{1.3}$$

Proof. The limit relation (1.3) is a simple consequence of known results. In fact, setting $B_n := A_n - A$, by [2, Proposition 3.7] and [3, Theorem 1.1.3 on p. 5], the sequence (B_n) converges strongly to the zero operator 0. Since ϕ is faithful, it follows from [1, Proposition 4 on p. 58] that

$$\phi(B_n^* B_n) \to 0 \quad \text{as} \quad n \to \infty,$$

which is (1.3) (cf. (1.1)).

□

If we weaken the boundedness condition imposed on (A_n), we can prove a conclusion weaker than (1.3), even in the special case when the operators A_n are in $\mathfrak{A}_+$.

PROPOSITION. *Let $(A_n : n \geq 1)$ be a sequence of operators in $\mathfrak{A}_+$ such that*

$$c := \sup_{n \geq 1} \phi(A_n^2) < \infty. \tag{1.4}$$

If

$$A_n \overset{\mathrm{b},\mathfrak{A}}{\to} 0 \quad as \quad n \to \infty, \tag{1.5}$$

then

$$\phi(A_n) \to 0 \quad as \quad n \to \infty. \tag{1.6}$$

Proof. By defintion, (1.5) means that there exists a boundle $\mathcal{P}$ such that for every P in $\mathcal{P}$ we have

$$\|A_n P\|_\infty \to 0 \quad \text{as} \quad n \to \infty. \tag{1.7}$$

By (0.1) and (0.2), we conclude that

$$0 \leq \phi(A_n) \leq |\phi(A_n P)| + |\phi(A_n P^\perp)| \leq \|A_n P\|_\infty + \phi(A_n^2)^{1/2}\phi(P^\perp)^{1/2}. \tag{1.8}$$

As a consequence of [2, Corollary 2.2], given any $\varepsilon > 0$, there exists a projection P in $\mathcal{P}$ such that

$$\phi(P^\perp) < c^{-1/2}\varepsilon/2, \tag{1.9}$$

where c is from (1.4). Combining (1.7)–(1.9) gives

$$0 \leq \phi(A_n) < \varepsilon$$

for n large enough. This proves (1.6).

□

Remark. We cannot expect the stronger conclusion

$$\phi(A_n^2) \to 0 \quad \text{as} \quad n \to \infty$$

under the conditions of the above Proposition. In fact, such a conclusion fails even in the commutative case. To see this, let $\mathfrak{A} := L_\infty(0,1)$ and define

$$\phi(A) := \int_0^1 A(t)\,dt, \quad A \in L_\infty(0,1).$$

Setting

$$A_n(t) := \frac{1}{t}\chi_{(1/(n+1),1/n)}(t),$$

where $\chi_{(a,b)}$ represents the characteristic function of the interval (a,b), we have

$$A_n(t) \to 0 \ \text{a.e.} \quad \text{and} \quad \phi(A_n^2) = 1, \quad n \geq 1.$$

However, if (A_n) is an increasing and bounded sequence of operators in $\mathfrak{A}_+$, then the situation is more satisfactory. Indeed, in this case (A_n) is strongly convergent and its limit A is also in $\mathfrak{A}$, due to the fact that $\mathfrak{A}$ is strongly closed.

Lemma 3. *If $(A_n : n \geq 1)$ is an increasing and bounded sequence of operators in $\mathfrak{A}_+$, then both* (1.2) *and* (1.3) *are satisfied, where A is the strong limit of (A_n).*

Proof. The same known results used in the proof of Lemma 2 apply here, as well, establishing (1.3). It remains to prove (1.2).

To this end, by (0.4) and (1.3), we have

$$\|\pi(A_n - A)\omega\|^2 = \phi((A_n - A)^2) \to 0 \quad \text{as} \quad n \to \infty.$$

We select a subsequence $(A_{n_k} : k \geq 1)$ of (A_n) for which

$$\sum_{k=1}^{\infty} \|\pi(A_{n_k} - A)\omega\|^2 < \infty.$$

We may assume that $n_1 = 1$. By [2, Propositions 3.5 and 3.6], it follows that

$$A_{n_k} \overset{\text{b},\mathfrak{A}}{\to} A \quad \text{as} \quad k \to \infty.$$

This means that there exists a bundle $\mathcal{P}$ such that for every P in $\mathcal{P}$ we have

$$\|(A_{n_k} - A)P\|_\infty \to 0 \quad \text{as} \quad k \to \infty. \tag{1.10}$$

Given an integer $n > 1$, there is a unique integer k such that $n_k < n \le n_{k+1}$. Relying on the additive property of bundle convergence, it is enough to prove that

$$(A_n - A_{n_k}) \stackrel{\mathrm{b},\mathfrak{A}}{\to} 0 \quad \text{as} \quad n \to \infty. \tag{1.11}$$

To achieve this goal, we fix a projection P in $\mathcal{P}$ and a vector h_0 in $\mathfrak{H}$ with $\|h_0\|_{\mathfrak{H}} \le 1$. Let $h := Ph_0$. By the Cauchy–Schwarz inequality (applied in the case of the inner product associated with the positive operator $A_n - A_{n_k}$, $n_k < n \le n_{k+1}$, on $\mathfrak{H}$), in the same manner as in [5, p. 261] we obtain the following:

$$\begin{aligned}\|(A_n - A_{n_k})h\|^4 &= ((A_n - A_{n_k})h,\, (A_n - A_{n_k})h)^2 \\ &\le ((A_n - A_{n_k})h, h)((A_n - A_{n_k})^2 h,\, (A_n - A_{n_k})h) \\ &\le ((A_n - A_{n_k})h, h)\|A_n - A_{n_k}\|_\infty^3 \le 8\|A\|_\infty^3((A_{n_{k+1}} - A_{n_k})h,\, h),\end{aligned}$$

whenever $n_k < n \le n_{k+1}$. Since $h = Ph_0$ and h_0 is an arbitrary vector in $\mathfrak{H}$ with $\|h_0\|_{\mathfrak{H}} \le 1$, it follows that

$$\|(A_n - A_{n_k})P\|_\infty^4 \le 8\|A\|_\infty^3 \|(A_{n_{k+1}} - A_{n_k})P\|_\infty. \tag{1.12}$$

Combining (1.10) and (1.12) yields

$$\|(A_n - A_{n_k})P\|_\infty \to 0 \quad \text{as} \quad n \to \infty.$$

Since P is arbitrary in $\mathcal{P}$, this proves (1.11).

□

Our last auxiliary result is stated in [2] without any proof or hint.

LEMMA 4. *Given a sequence of bundles $(\mathcal{P}_n : n \ge 1)$ in $\mathfrak{A}$, there exists a bundle $\mathcal{P}$ in $\mathfrak{A}$ such that*

$$\mathcal{P} \subseteq \mathcal{P}_n \quad \textit{for each} \quad n \ge 1. \tag{1.13}$$

Since we shall rely heavily on this lemma in Section 2, we present a proof here.

Proof. By definition (see (0.6) and (0.7)), for each $n \geq 1$ there exists a sequence $(D_{nk} : k \geq 1)$ of operators in $\mathfrak{A}_+$ such that

$$\sum_{k=1}^{\infty} \phi(D_{nk}) < \infty$$

and

$$\mathcal{P}_n = \Big\{P \in \operatorname{Proj}\mathfrak{A} : \sup_{m\geq 1} \Big\| P\Big(\sum_{k=1}^{m} D_{nk}\Big)P\Big\|_\infty < \infty \quad \text{and} \quad \|PD_{nk}P\|_\infty \to 0 \quad \text{as} \quad k \to \infty\Big\}.$$

For each $n \geq 1$, we define

$$D'_{nk} := \frac{2^{-n}}{a_n} D_{nk}, \quad \text{where} \quad k \geq 1 \quad \text{and} \quad a_n := \sum_{k=1}^{\infty} \phi(D_{nk}).$$

It is evident that the bundle $\mathcal{P}'_n$ associated with $(D'_{nk} : k \geq 1)$ is identical with $\mathcal{P}_n$, since the factor $2^{-n}/a_n$ does not affect either the boundedness of the sums $\Big\|P\Big(\sum_{k=1}^{m} D'_{nk}\Big)P\Big\|_\infty$ or the limit relation $\|PD'_{nk}P\|_\infty \to 0$ as $k \to \infty$. Consequently, we have $\mathcal{P}'_n = \mathcal{P}_n$ for each $n \geq 1$.

Now, we rearrange the double sequence $(D'_{nk} : n, k \geq 1)$ into a single one $(D''_j : j \geq 1)$ by making use of the diagonalization process going back to Cantor. Since

$$\sum_{j=1}^{\infty} \phi(D''_j) = \sum_{n=1}^{\infty} \sum_{k=1}^{\infty} \phi(D'_{nk}) = \sum_{n=1}^{\infty} 2^{-n} = 1,$$

the sequence (D''_j) also determines a bundle, say $\mathcal{P}$. We claim that $\mathcal{P}$ is contained in each $\mathcal{P}_n$.

To this end, we fix a projection P in $\mathcal{P}$ and an integer $n \geq 1$. For each $m \geq 1$, we define

$$j(m) := \inf\Big\{j : \{D'_{n1}, D'_{n2}, \ldots, D'_{nm}\} \subseteq \{D''_1, D''_2, \ldots, D''_j\}\Big\}.$$

Clearly, $j(m)$ depends on n, as well; but since n is fixed, we do not indicate this dependence in our notation. Taking into account that if A is a self-adjoint operator, then

$$\|A\|_\infty = \sup\{(Ah, h)_{\mathfrak{H}} : \|h\|_{\mathfrak{H}} \leq 1\},$$

it follows immediately that

$$\Big\|P\Big(\sum_{k=1}^{m} D'_{nk}\Big)P\Big\|_\infty \leq \Big\|P\Big(\sum_{j=1}^{j(m)} D''_j\Big)P\Big\|_\infty.$$

Since P is in $\mathcal{P}$, the right-hand side in this inequality is bounded by some finite constant c, not depending on m (c may depend on n). Therefore, the same is true for the left-hand side.

Furthermore, by definition, we have

$$\|PD_j''P\|_\infty \to 0 \quad \text{as} \quad j \to \infty.$$

Since $(\|PD_{nk}'P\|_\infty : k \geq 1)$ is a subsequence of $(\|PD_j''P\|_\infty : j \geq 1)$, we also have

$$\|PD_{nk}'P\|_\infty \to 0 \quad \text{as} \quad k \to \infty.$$

This means that $P \in \mathcal{P}_n$. Since P is an arbitrary projection in $\mathcal{P}$, we also have $\mathcal{P} \subseteq \mathcal{P}_n$. Since $n \geq 1$ is an arbitrary integer, we have proved (1.13).

□

2. A Beppo Levi type theorem

Our first main result is a noncommutative version of the classical Beppo Levi Theorem (relating to the integral in the sense of Lebesgue) in the framework of bundle convergence.

THEOREM 1. *Let* $(A_n : n \geq 1)$ *be an increasing sequence of operators in* $\mathfrak{A}$. *If there exist a sequence* $(P_k : k \geq 1)$ *of projections in* Proj$\mathfrak{A}$ *and sequence* $(B_k : k \geq 1)$ *of operators in* $\mathfrak{A}_+$ *such that*

(2.1) $P_k P_j = 0$ *for all* k *and* j *with* $k \neq j$,

(2.2) $\sum_{k=1}^{\infty} P_k h = h$ *in* $\|\cdot\|_{\mathfrak{H}}$ *norm, for all* $h \in \mathfrak{H}$,

(2.3) $P_k A = A P_k$ *for all* A *in* $\mathfrak{A}$ *and* $k \geq 1$,

(2.4) $P_k A_n^2 P_k \leq B_k$ *for all* k *and* n,

(2.5) $\sum_{k=1}^{\infty} \phi(B_k) < \infty$,

then there exists a vector ξ *in* $L_2 = L_2(\mathfrak{A},\phi)$ *such that*

(2.6) $$\|\pi(A_n)\omega - \xi\| \to 0 \quad \text{and} \quad \pi(A_n)\omega \xrightarrow{\text{b}} \xi \quad \text{as} \quad n \to \infty.$$

A few comments are appropriate here.

(i) The conditions imposed in Theorem 1 are fairly restrictive, especially the assumption that the projections involved are central. We guess that Theorem 1 can be proved under essentially weaker conditions, but so far we have been unable to find such a proof.

(ii) From conditions (2.1) and (2.2) it follows that the underlying Hilbert space $\mathfrak{H}$ (respectively L_2) can be represented in the form of a direct sum of the subspaces $\mathrm{Im}(P_k)$ (respectively $\mathrm{Im}(\pi(P_k))$), where $k = 1, 2, \ldots$. Condition (2.3) ensures that the von Neumann algebra $\mathfrak{A}$ (respectively $\pi(\mathfrak{A})$) can also be represented in the form of the corresponding direct sum.

(iii) In the classical commutative case, where $\mathfrak{A} := L_\infty(\Omega, \mu)$ over a probability space (Ω, μ), $\mathfrak{A}$ is acting on $\mathfrak{H} := L_2(\Omega, \mu)$ by pointwise (almost sure, in abbreviation: a.s.) multiplication, and the faithful, normal state ϕ is defined on $\mathfrak{A}$ by

$$\phi(A) := \int_\Omega A(t)p(t)\, d\mu(t), \quad A \in L_\infty(\Omega, \mu),$$

where p is a measurable function on Ω such that $p > 0$ a.s. and

$$\int_\Omega p(t)\, d\mu(t) = 1.$$

In this case, we can identify $L_2(\mathfrak{A},\phi)$ with $L_2(\Omega, \mu)$ with the choice $\omega := \sqrt{p}$ as a cyclic vector. Furthermore, we set $d\lambda := p d\mu$.

Now, the familiar Beppo Levi Theorem is valid: If $(f_n : n \geq 1)$ is an increasing sequence of (a.s.) positive, measurable functions on Ω satisfying the condition

$$\sup_{n\geq 1} \int_\Omega f_n(t)\, d\lambda(t) < \infty,$$

then there exists a λ-integrable function f on Ω such that

$$\int_\Omega f_n(t)\, d\lambda(t) \uparrow \int_\Omega f(t)\, d\lambda(t) \quad \text{as} \quad n \to \infty.$$

Let $g_n := \sqrt{f_n}$ and $g := \sqrt{f}$. Since $(b-a)^2 \leq b^2 - a^2$ whenever $0 \leq a \leq b$ are real numbers, it follows that

$$\int_\Omega (g_n(t) - g(t))^2\, d\lambda(t) \to 0 \quad \text{as} \quad n \to \infty.$$

The truncated functions defined by $A_n := \min(g_n, g)$ are in $L_\infty(\Omega, \mu)$ and we also have

$$\int_\Omega (A_n(t) - g(t))^2\, d\lambda(t) \to 0 \quad \text{as} \quad n \to \infty.$$

Finally, setting

$$\Omega_k := \{k-1 \le f < k\}, \quad P_k := 1_{\Omega_k}, \quad B_k := f 1_{\Omega_k}, \quad \text{for} \quad k \ge 1,$$

conditions (2.1)–(2.5) are automatically satisfied, and Theorem 1 turns out to be a particular case of the classical Beppo Levi Theorem when applied to an increasing sequence of bounded, measurable functions.

Proof of Theorem 1. From (2.3) it follows that

$$P_k A_n P_k = A_n P_k,$$

whence

$$(P_k A_n P_k)^2 = A_n^2 P_k \quad \text{for all} \quad k, n \ge 1.$$

Since the operators occurring in these equalities are all positive, by (2.4) we infer (see, e.g. [6, p. 25]) that

$$A_n P_k \le B_k^{1/2} \quad \text{for all} \quad k, n \ge 1.$$

Keeping the value of $k \ge 1$ fixed, the sequence $(A_n P_k : n \ge 1)$ is also increasing and has an upper bound. Consequently, it converges strongly to some operator C_k in $\mathfrak{A}_+$, say:

$$\|(A_n P_k - C_k)h\|_{\mathfrak{H}} \to 0 \quad \text{as} \quad n \to \infty \quad \text{for all} \quad h \in \mathfrak{H}, \tag{2.7}$$

and $C_k \le B_k^{1/2}$. It is also obvious that $C_k P_k = C_k$, for all $k \ge 1$.

By virtue of [1, Corollary 1 on p. 54], the sequence $(\pi(A_n P_k) : n \ge 1)$ is also strongly convergent to $\pi(C_k)$ in L_2. Setting

$$\omega_k := \pi(P_k)\omega, \quad k \ge 1, \tag{2.8}$$

we clearly have

$$\pi(A_n)\omega_k = \pi(A_n P_k)\omega \to \pi(C_k)\omega = \pi(C_k)\omega_k \quad \text{as} \quad n \to \infty \tag{2.9}$$

(in the strong topology of L_2). By (0.4) and (2.4), we also have

$$\|\pi(A_n)\omega_k\|^2 = \|\pi(A_n P_k)\omega\|^2 = \pi(P_k A_n^2 P_k) \le \phi(B_k), \quad k \ge 1. \tag{2.10}$$

Let

$$\xi := \sum_{k=1}^{\infty} \pi(C_k)\omega_k. \tag{2.11}$$

By (2.2), (2.3) and (2.8), we may write

$$\pi(A_n)\omega = \sum_{k=1}^{\infty} \pi(P_k)\pi(A_n)\omega = \sum_{k=1}^{\infty} \pi(A_n)\omega_k.$$

The series here converges in the strong (norm) topology of L_2. By this, (2.1), (2.4), (2.10) and (2.11), we estimate as follows:

$$\begin{aligned}
&\|\pi(A_n)\omega - \xi\|^2 \\
&\quad = \sum_{k=1}^{\infty} \|\pi(A_n)\omega_k - \pi(C_k)\omega_k\|^2 \\
&\quad \le \sum_{k=1}^{K} \|\pi(A_n)\omega_k - \pi(C_k)\omega_k\|^2 + 2 \sum_{k=K+1}^{\infty} (\|\pi(A_n)\omega_k\|^2 + \|\pi(C_k)\omega_k\|^2) \\
&\quad \le \sum_{k=1}^{K} \|\pi(A_n)\omega_k - \pi(C_k)\omega_k\|^2 + 4 \sum_{k=K+1}^{\infty} \phi(B_k),
\end{aligned}$$

where $K \ge 1$ is an arbitrary integer. Taking into account (2.5) and (2.9), we conclude

$$\|\pi(A_n)\omega - \xi\| \to 0 \quad \text{as} \quad n \to \infty.$$

This proves the first limit relation in (2.6).

In order to prove the second limit relation in (2.6), we have to construct a sequence $(X_n : n \ge 1)$ of operators in $\mathfrak{A}$ such that the following two conditions are satisfied:

$$\sum_{n=1}^{\infty} \|\pi(A_n)\omega - \xi - \pi(X_n)\omega\|^2 < \infty \tag{2.12}$$

and

$$X_n \overset{\text{b},\mathfrak{A}}{\to} C \quad \text{as} \quad n \to \infty. \tag{2.13}$$

To begin with, for each $n \ge 1$ we choose an integer $N(n)$ such that

$$\sum_{k>N(n)} \phi(B_k) < 2^{-n}. \tag{2.14}$$

By (2.5), this choice is possible. We may assume that $N(1) < N(2) < \ldots$. Then we set

$$X_n := A_n - \sum_{j=1}^{N(n)} C_j, \quad n \geq 1. \tag{2.15}$$

By (2.1), (2.2) and (2.8) we have

$$\phi(P_k) = \|\omega_k\|^2 \quad \text{and} \quad \sum_{k=1}^{\infty} \|\omega_k\|^2 < \infty.$$

Thus, for each $m \geq 1$ we can select an integer $k(m)$ such that

$$\sum_{k>k(m)} \phi(P_k) < 2^{-m},$$

while $k(1) < k(2) < \ldots$. Setting

$$R_m := \sum_{k>k(m)} P_k = 1_{\mathfrak{H}} - \sum_{k=1}^{k(m)} P_k, \quad m \geq 1 \tag{2.16}$$

(cf. (2.2)), it is clear that the operators R_m are in $\mathfrak{A}_+$ and we have

$$\sum_{m=1}^{\infty} \phi(R_m) < \infty. \tag{2.17}$$

Now, let $\mathcal{P}$ be the bundle associated with the sequence $(R_m : m \geq 1)$. We consider an arbitrary projection P in $\mathcal{P}$. By (2.16), we may write

$$\begin{aligned}\|PR_mP\|_\infty &= \sup\{(PR_mPh, h)_{\mathfrak{H}} : \|h\|_{\mathfrak{H}} \leq 1\} \\ &= \sup\Big\{ \sum_{k>k(m)} (P_k h', h')_{\mathfrak{H}} : \quad h' := Ph, \quad \|h\|_{\mathfrak{H}} \leq 1\Big\}.\end{aligned}$$

From this, (2.1) and (2.2) it follows that

$$\|PR_mP\|_\infty \to 0 \quad \text{as} \quad m \to \infty. \tag{2.18}$$

Next, we claim that there exists an integer $K(P)$ depending on the projection P such that

$$\operatorname{Im}(P) \subseteq \bigoplus_{k=1}^{K(P)} E_k, \quad \text{where} \quad E_k := \operatorname{Im}(P_k). \tag{2.19}$$

In order to justify (2.19), first we observe that, by (2.3),

$$Q_k := PP_k$$

is a projection onto the closed subspace

$$F_k := \operatorname{Im}(P) \cap E_k, \quad k \geq 1.$$

By (2.2), we have

$$Ph = \sum_{k=1}^{\infty} Q_k h,$$

and by (2.16),

$$PR_m Ph = R_m Ph = \sum_{k>k(m)} Q_k h \quad \text{for all} \quad h \in \mathfrak{H} \quad \text{and} \quad m \geq 1.$$

Hence it follows that

$$\|PR_m Ph\|_{\mathfrak{H}}^2 = \sum_{k>k(m)} \|Q_k h\|_{\mathfrak{H}}^2,$$

and that (2.19) is equivalent to the following:

$$\operatorname{Im}(P) = \bigoplus_{k=1}^{K(P)} F_k,$$

for some integer $K(P)$ depending on P.

Suppose to the contrary of (2.19) that there exists an infinite subset J of the positive integers such that $F_j \neq \{0\}$ for $j \in J$. Then, by using the Axiom of Choice, there exists a family $(h_j : j \in J)$ of vectors in $\mathfrak{H}$ such that $h_j \in F_j$ and $\|h_j\|_{\mathfrak{H}} = 1$. Now, if $j > k(m)$, then

$$\|PR_m Ph_j\|_{\mathfrak{H}}^2 = \sum_{k>k(m)} \|Q_k h_j\|_{\mathfrak{H}}^2 = \|h_j\|_{\mathfrak{H}}^2 = 1,$$

whence it follows that $\|PR_m P\|_\infty = 1$, which contradicts (2.18). This contradiction proves (2.19).

As a consequence of (2.7) and Lemma 3, for each $k \geq 1$ we have

$$A_n P_k \overset{\mathrm{b},\mathfrak{A}}{\to} C_k \quad \text{as} \quad n \to \infty.$$

Thus, for each $k \geq 1$ there exists a bundle, say $\mathcal{P}_k$, such that for every projection P in $\mathcal{P}_k$ we have

$$\|(AP_k - C_k)P\|_\infty \to 0 \quad \text{as} \quad n \to \infty. \tag{2.20}$$

By Lemma 4, there exists a bundle $\mathcal{Q}$ such that

$$\mathcal{Q} \subseteq \mathcal{P} \quad \text{and} \quad \mathcal{Q} \subseteq \mathcal{P}_k \quad \text{for each} \quad k \geq 1.$$

So, it remains to check the fulfillment of (2.12) and (2.13).

First, we note that, by (2.3) and by what we have said just after (2.7), we have

$$C_j P_j = P_j C_j = C_j, \quad j \geq 1. \tag{2.21}$$

By (2.1), it follows that

$$C_j P_k = P_k C_j = 0, \quad j \neq k. \tag{2.22}$$

Since

$$\pi(C_j)\omega = \sum_{k=1}^{\infty} \pi(C_j)\pi(P_k)\omega = \pi(C_j P_j)\omega = \pi(C_j)\omega_j, \quad j \geq 1,$$

by making use of (2.11), (2.14) and (2.15), we can estimate as follows:

$$\begin{aligned}\|\pi(A_n)\omega - \xi - \pi(X_n)\omega\|^2 &= \Big\| \sum_{j=1}^{N(n)} \pi(C_j)\omega - \sum_{k=1}^{\infty} \pi(C_k)\omega_k \Big\|^2 \\ &= \Big\| \sum_{k>N(n)} \pi(C_k)\omega_k \Big\|^2 \leq \sum_{k>N(n)} \phi(B_k) < 2^{-n}.\end{aligned}$$

This proves (2.12).

Second, let P be an arbitrary projection in $\mathcal{Q}$. By (2.19),

$$P = \sum_{k=1}^{K(P)} P_k P,$$

whence, taking into account also (2.21) and (2.22), it follows that

$$\begin{aligned}X_n P = \sum_{k=1}^{K(P)} X_n P_k P &= \sum_{k=1}^{K(P)} A_n P_k P - \sum_{k=1}^{K(P)} \sum_{j=1}^{N(n)} C_j P_k P \\ &= \sum_{k=1}^{K(P)} A_n P_k P - \sum_{k=1}^{K(P)} C_k P,\end{aligned}$$

provided that $N(n) > K(P)$. Combining this and (2.7) yields

$$\|X_n P\|_\infty = \Big\| \sum_{k=1}^{K(P)} (A_n P_k - C_k) P \Big\|_\infty \to 0 \quad \text{as} \quad n \to \infty.$$

This proves (2.13) (cf. (0.9)), and completes the proof of Theorem 1. □

REMARK. Conditions (2.1)–(2.5) seem to be rather restrictive. It may happen that some of them are superfluous to conclude (2.6). Likewise, we are unable to prove or disprove the following

PROBLEM. Let $(A_n : n \geq 1)$ be an increasing sequence of operators in $\mathfrak{A}_+$ such that

$$\sup_{n\geq 1} \phi(A_n^2) < \infty. \tag{2.23}$$

Does there exist a vector ξ in $L_2(\mathfrak{A},\phi)$ such that both limit relations in (2.6) are satisfied?

It is evident that, under conditions (2.1)–(2.5), we necessarily have (2.23):

$$\phi(A_n^2) = \sum_{k=1}^{\infty} \phi(A_n^2 P_k) \leq \sum_{k=1}^{\infty} \phi(B_k) < \infty.$$

3. A dominated convergence type theorem

In Theorem 2 below, we do not require the sequence $(A_n : n \geq 1)$ of operators to be increasing or the operators A_n to belong to $\mathfrak{A}_+$. On the other hand, instead of (2.4) we assume now that

$$P_k|A_n|^2 P_k \leq B_k \quad \text{for all } k \text{ and } n. \tag{2.4'}$$

Our second main result is a noncommutative version of the classical dominated convergence theorem in the framework of bundle convergence.

THEOREM 2. *Let $(A_n : n \geq 1)$ be a sequence of operators in $\mathfrak{A}$ such that*

$$A_n \overset{\mathrm{b},\mathfrak{A}}{\longrightarrow} A \quad \textit{as} \quad n \to \infty, \tag{3.1}$$

for some A in $\mathfrak{A}$. If there exist a sequence $(P_k : k \geq 1)$ of projections in Proj$\mathfrak{A}$ *and a sequence $(B_k : k \geq 1)$ of operators in $\mathfrak{A}_+$ such that conditions* (2.1) – (2.3), (2.4′) *and* (2.5) *are satisfied, then*

$$\phi(|A_n - A|^2) \to 0 \quad \textit{as} \quad n \to \infty. \tag{3.2}$$

Proof. Without loss of generality, we may assume that $A = 0$ in (3.1). Indeed, set $A'_n := A_n - A$. By Lemma 1, (2.4) and (0.1), we have

$$\begin{aligned}\phi(|A'_n|^2 P_k) &\leq 2\phi(|A_n|^2 P_k) + 2\phi(|A|^2 P_k)\\ &\leq 2\phi(B_k) + 2\|A\|_\infty^2 \phi(P_k) = \phi(B'_k),\end{aligned}$$

where $B'_k := 2B_k + 2\|A\|_\infty^2 P_k$, and, by (2.2) and (2.5),

$$\sum_{k=1}^{\infty} \phi(B'_k) < \infty.$$

This justifies our claim above that it is enough to prove Theorem 2 in the case when $A = 0$ in (3.1).

To obtain (3.2) we start with the equality

$$|A_n|^2 = \sum_{k=1}^{\infty} |A_n|^2 P_k,$$

the series being strongly convergent. By (2.4′), we may estimate as follows:

$$\phi(|A_n|^2) \leq \sum_{k=1}^{N} \phi(|A_n|^2 P_k) + \sum_{k>N} \phi(B_k), \tag{3.3}$$

where $N \geq 1$ is an arbitrary integer at our disposal.

Now, let k be a fixed integer. By (3.1) with $A = 0$, there exists a bundle $\mathcal{P}$ such that for any projection P in $\mathcal{P}$ we have

$$\|A_n P\|_\infty \to 0 \quad \text{as} \quad n \to \infty.$$

By (2.3) (with $A := P$), we conclude that

$$\|A_n P_k P\|_\infty = \|A_n P P_k\|_\infty \leq \|A_n P\|_\infty \|P_k\|_\infty \leq \|A_n P\|_\infty \to 0 \quad \text{as} \quad n \to \infty.$$

This proves that for each $k \geq 1$ we have

$$A_n P_k \overset{\mathrm{b},\mathfrak{A}}{\to} 0 \quad \text{as} \quad n \to \infty.$$

Moreover, for each $k \geq 1$, the sequence $(A_n P_k : n \geq 1)$ is bounded. In fact, by (2.3) and (2.4′) we can estimate as follows:

$$\begin{aligned}\|A_n P_k\|^2 &= \|(A_n P_k)^* A_n P_k\|_\infty = \|P_k A_n^* A_n P_k\|_\infty\\ &= \|P_k |A_n|^2 P_k\|_\infty \leq \|B_k\|_\infty.\end{aligned}$$

Thus, we have verified that the conditions of Lemma 2 are satisfied. As a consequence, for each $k \geq 1$ we have

$$\phi(|A_n|^2 P_k) \to 0 \quad \text{as} \quad n \to \infty. \tag{3.4}$$

Combining (2.5), (3.3) and (3.4) yields (3.2) immediately. □

References

[1] J. DIXMIER, *Les algèbres d'opérateurs dans l'espace Hilbertien (Algèbres de von Neumann)*, Gauthier-Villars, Paris, 1969.

[2] E. HENSZ, R. JAJTE and A. PASZKIEWICZ, The bundle convergence in von Neumann algebras and their L_2-spaces, *Studia Math.*, **120** (1996), 23–46.

[3] R. JAJTE, *Strong limit theorems in non-commutative L_2-spaces,* Lecture Notes in Math. **1477**, Springer-Verlag, Berlin – Heidelberg – New York, 1991.

[4] F. MÓRICZ and B. LE GAC, On the bundle convergence of double orthogonal series in noncommutative L_2-spaces, *Studia Math.*, **140** (2000), 177–190.

[5] F. RIESZ and B. SZŐKEFALVI-NAGY, *Lecons d'analyse fonctionelle*, Gauthier-Villars et Akadémiai Kiadó, Paris et Budapest, 1968.

[6] M. TAKESAKI, *Theory of operator algebras.* Vol. 1, Springer-Verlag, New York – Heidelberg – Berlin, 1979.

B. LE GAC, Centre de Mathématiques et d'Informatique, Université de Provence, 39 rue Joliot-Curie, 13 453 Marseille, Cedex 13, France

F. MÓRICZ, Bolyai Institute, University of Szeged, Aradi vértanúk tere 1, 6720 Szeged, Hungary

Operator Theory: Advances and Applications, Vol. 127, 465–472

$*$-semigroup endomorphisms of $B(H)$

LAJOS MOLNÁR*

Dedicated to the memory of Professor Béla Szőkefalvi-Nagy

Let H be a complex separable infinite dimensional Hilbert space. We describe the form of all $*$-semigroup endomorphisms ϕ of $B(H)$ which are uniformly continuous on every commutative C^*-subalgebra. In particular, we obtain that if ϕ satisfies $\phi(0) = 0$, then ϕ is additive.

1. Introduction

In the paper [2] the author studied the question of additivity of $*$-semigroup isomorphisms between operator algebras.

The problem that when multiplicativity implies additivity was previously investigated also from the purely algebraic point of view [7]. Using that result, the semigroup isomorphisms between standard operator algebras were completely described in [10]. The common characteristic of the mentioned investigations is that they considered multiplicative bijections, or, as in [7], at least the surjectivity was assumed.

In our present paper we get rid of this very restrictive algebraic condition and, instead, we impose a certain continuity assumption.

In contrast to the proofs of the previously mentioned results which were based mainly on algebraic manipulations, our argument needs much more functional analysis and this is the reason why we obtain our result "only" for the full operator

---

AMS Subject Classification (1991): 47B49.

* This research was supported from the following sources: 1) Hungarian National Foundation for Scientific Research (OTKA), Grant No. T–030082 F–019322, 2) A grant from the Ministry of Education, Hungary, Reg. No. FKFP 0304/1997.

algebra over a separable infinite dimensional Hilbert space (as for the finite dimensional case, we should refer to the paper [5] where all so-called non-degenerate multiplicative maps of a matrix algebra were determined).

2. Results

The main result of the paper reads as follows.

THEOREM. *Let H be a complex separable infinite dimensional Hilbert space. Let $\phi: B(H) \to B(H)$ be a $*$-semigroup endomorphism which is uniformly continuous on the commutative C^*-subalgebras of $B(H)$. Then ϕ can be written in the form*

$$\phi(A) = \begin{bmatrix} 0 & 0 & \dots & \dots & \dots & \dots & \dots & \dots & \dots \\ 0 & I & 0 & \dots & \dots & \dots & \dots & \dots & \dots \\ 0 & 0 & A & 0 & \dots & \dots & \dots & \dots & \dots \\ 0 & 0 & 0 & A & 0 & \dots & \dots & \dots & \dots \\ \vdots & \vdots & \vdots & \vdots & \ddots & \vdots & \vdots & \vdots & \vdots \\ 0 & \dots & \dots & \dots & 0 & A^{*tr} & 0 & \dots & \dots \\ 0 & \dots & \dots & \dots & \dots & 0 & A^{*tr} & 0 & \dots \\ \vdots & \vdots & \vdots & \vdots & \vdots & \vdots & \vdots & \ddots & \vdots \end{bmatrix} \qquad (A \in B(H)).$$

Proof. Clearly, ϕ sends projections to projections. So, $\phi(I)$ and $\phi(0)$ are projections. Since

$$\phi(I)\phi(A) = \phi(A)\phi(I) = \phi(A)$$

and

$$\phi(0)\phi(A) = \phi(A)\phi(0) = \phi(0),$$

it is easy to see that ϕ can be written in the form

$$\phi(A) = \begin{bmatrix} 0 & 0 & 0 \\ 0 & I & 0 \\ 0 & 0 & \phi'(A) \end{bmatrix} \qquad (A \in B(H))$$

where ϕ' is a $*$-semigroup endomorphism of $B(H)$ having the same continuity property as ϕ which sends 0 to 0 and maps I into I. Therefore, we can assume that our original map ϕ satisfies $\phi(0) = 0$ and $\phi(I) = I$. The main step of the proof which follows is to prove that ϕ is orthoadditive on the set of all projections. This means that $\phi(P+Q) = \phi(P) + \phi(Q)$ for any mutually orthogonal projections

$P, Q \in B(H)$. To see this, let $P \in B(H)$ be an arbitrary projection and set $Q = I - P$. Consider the map

$$\lambda \longmapsto \phi(e^{\lambda P}) = \phi(Q + e^{\lambda} P)$$

from $\mathbb{R}$ into the group of all invertible operators in $B(H)$. This is a continuous one-parameter group and hence there is an operator $T \in B(H)$ such that

$$\phi(Q + e^{\lambda} P) = e^{\lambda T} \qquad (\lambda \in \mathbb{R})$$

(see, for example, [9, 6.4.6 Proposition]). Since ϕ is $*$-preserving, we obtain that $e^{\lambda T}$ is self-adjoint for all $\lambda \in \mathbb{R}$. This yields that T is also self-adjoint. The norm and the spectral radius of any self-adjoint operator coincide. So, from the continuity property of ϕ we deduce that for every $\epsilon > 0$ there is a $\delta > 0$ such that $\sup_{t \in \sigma(T)} |e^{\lambda t} - e^{\mu t}| < \epsilon$ if $|e^{\lambda} - e^{\mu}| < \delta$. Therefore, the function $x \mapsto x^t$ is uniformly continuous on the positive half-line for every $t \in \sigma(T)$. This gives us that $\sigma(T) \subset \{0, 1\}$. Consequently, T is a projection. Then we have $e^{\lambda T} = (I - T) + e^{\lambda} T$ and thus

$$\phi(Q + e^{\lambda} P) = (I - T) + e^{\lambda} T \qquad (\lambda \in \mathbb{R})$$

or, equivalently,

$$\phi(Q + \epsilon P) = (I - T) + \epsilon T \tag{1}$$

for every positive ϵ. By the continuity property of ϕ we obtain $\phi(Q) = I - T$. As a particular case, we have

$$\phi(\epsilon I) = \epsilon I \qquad (\epsilon > 0). \tag{2}$$

Indeed, this follows from $\phi(0) = 0$ and (1). Therefore, ϕ is positive homogeneous, and referring to (1) again, if we divide by ϵ and use the continuity property of ϕ, then we arrive at $\phi(P) = T$. Thus, we obtain

$$\phi(P) + \phi(I - P) = I \tag{3}$$

for every projection P in $B(H)$. If P, Q are arbitrary projections with $PQ = QP = 0$, then we infer from the multiplicativity of ϕ and (3) that

$$\begin{aligned} \phi(P) + \phi(Q) &= \phi(P)\phi(P + Q) + \phi(I - P)\phi(P + Q) \\ &= (\phi(P) + \phi(I - P))\phi(P + Q) = \phi(P + Q). \end{aligned}$$

Consequently, ϕ is orthoadditive on the set of all projections.

Since $\phi(I) = I$, it follows that ϕ sends unitaries to unitaries. Consider the map $t \mapsto \phi(e^{it}I)$. Clearly, this is a continuous one-parameter unitary group. By Stone's theorem there is a self-adjoint operator $S \in B(H)$ such that

$$\phi(e^{it}I) = e^{itS} \qquad (t \in \mathbb{R}).$$

Since $\phi(I) = I$, we have $e^{2\pi i S} = I$. By spectral mapping theorem this yields that the spectrum of S consists of integers. So, S can be written in the form $S = \sum_{k=-n}^{n} kP_k$, where the P_k's are pairwise orthogonal projections with $\sum_{k=-n}^{n} P_k = I$ and n is a suitable positive integer. We compute

$$\begin{aligned}\phi(e^{it}I) &= e^{it\sum_{k=-n}^{n} kP_k} = \prod_{k=-n}^{n} e^{itkP_k} = \prod_{k=-n}^{n} (I + (e^{itk} - 1)P_k) \\ &= I + \sum_{k=-n}^{n} (e^{itk} - 1)P_k = \sum_{k=-n}^{n} e^{itk} P_k \qquad (t \in \mathbb{R}).\end{aligned}$$

Consequently, we have

$$\phi(\lambda I) = \sum_{k=-n}^{k=n} \lambda^k P_k \qquad (\lambda \in \mathbb{C}, |\lambda| = 1). \tag{4}$$

From (2) and (4) we infer that

$$\phi(\lambda I) = \sum_{k=-n}^{n} |\lambda| \left(\frac{\lambda}{|\lambda|} \right)^k P_k \qquad (\lambda \in \mathbb{C}, \lambda \neq 0). \tag{5}$$

We know that $\phi(\lambda I)$ commutes with $\phi(A)$ for every $\lambda \in \mathbb{C}$. Thus, for any $A \in B(H)$ we have

$$\begin{aligned}\sum_{k,l=-n}^{n} \lambda^k P_k \phi(A) P_l &= \sum_{k,l=-n}^{n} \lambda^k P_k P_k \phi(A) P_l = \phi(\lambda I)\phi(A) = \phi(A)\phi(\lambda I) \\ &= \sum_{k,l=-n}^{n} P_k \phi(A) P_l P_l \lambda^l P_l = \sum_{k,l=-n}^{n} \lambda^l P_k \phi(A) P_l\end{aligned}$$

for every $\lambda \in \mathbb{C}$ of modulus 1. This implies that $P_k \phi(A) P_l = 0$ if $k \neq l$. Consequently, ϕ can be written in the form

$$\phi(A) = \sum_{k=-n}^{k=n} P_k \phi(A) P_k \qquad (A \in B(H))$$

or, in another way,

$$\phi(A) = \begin{bmatrix} \phi_{-n}(A) & 0 & \dots & 0 \\ 0 & \phi_{-n+1}(A) & \dots & 0 \\ \vdots & \vdots & \ddots & \vdots \\ 0 & \dots & \dots & \phi_n(A) \end{bmatrix} \qquad (A \in B(H)).$$

Here, every ϕ_k $(k = -n, \dots, n)$ is a $*$-semigroup endomorphism of $B(H)$ which is uniformly continuous on the commutative C^*-subalgebras.

Every orthoadditive projection valued measure on the set of all projections in $B(H)$ can be extended to a linear map on $B(H)$. This is a particular case of the solution of the Mackey–Gleason problem obtained in [1]. Since this extension is linear and sends projections to projections, it is a standard argument to verify that this is in fact a Jordan $*$-homomorphism (see, for example, the proof of [8, Theorem 2]). A linear map J between $*$-algebras $\mathcal{A}$ and $\mathcal{B}$ is called a Jordan $*$-homomorphism if it satisfies

$$J(x)^2 = J(x^2), \quad J(x)^* = J(x^*) \quad (x \in \mathcal{A}).$$

So, we have Jordan $*$-homomorphisms $\psi_{-n}, \dots, \psi_n$ of $B(H)$ such that

$$\phi(P) = \begin{bmatrix} \psi_{-n}(P) & 0 & \dots & 0 \\ 0 & \psi_{-n+1}(P) & \dots & 0 \\ \vdots & \vdots & \ddots & \vdots \\ 0 & \dots & \dots & \psi_n(P) \end{bmatrix}$$

for every projection $P \in B(H)$.

Let $R_1, \dots, R_m$ be pairwise orthogonal projections whose sum is I and pick nonzero scalars $\lambda_1, \dots, \lambda_m \in \mathbb{C}$. Using the orthoadditivity of ϕ, for any $k = -n, \dots, n$ we compute

$$\begin{aligned} &\phi_k(\lambda_1 R_1 + \dots + \lambda_m R_m) \\ &\quad = \phi_k(\lambda_1 R_1 + \dots + \lambda_m R_m)\phi_k(R_1 + \ldots + R_m) \\ &\quad = \phi_k(\lambda_1 R_1 + \dots + \lambda_m R_m)(\phi_k(R_1) + \ldots + \phi_k(R_m)) \\ &\quad = \phi_k(\lambda_1 R_1) + \dots + \phi_k(\lambda_m R_m) = \phi_k(\lambda_1 I)\phi_k(R_1) + \dots + \phi_k(\lambda_m I)\phi_k(R_m) \\ &\quad = |\lambda_1|\left(\frac{\lambda_1}{|\lambda_1|}\right)^k \phi_k(R_1) + \dots + |\lambda_m|\left(\frac{\lambda_m}{|\lambda_m|}\right)^k \phi_k(R_m) \\ &\quad = |\lambda_1|\left(\frac{\lambda_1}{|\lambda_1|}\right)^k \psi_k(R_1) + \dots + |\lambda_m|\left(\frac{\lambda_m}{|\lambda_m|}\right)^k \psi_k(R_m) \\ &\quad = \psi_k\left(|\lambda_1|\left(\frac{\lambda_1}{|\lambda_1|}\right)^k R_1 + \dots + |\lambda_m|\left(\frac{\lambda_m}{|\lambda_m|}\right)^k R_m\right). \end{aligned}$$

Using the continuity property of ϕ_k, the automatic continuity of Jordan $*$-homomorphisms between C^*-algebras and the spectral theorem of normal operators, we deduce that

$$\phi_k(N) = \psi_k(|N|(N|N|^{-1})^k) \tag{6}$$

holds for every invertible normal operator $N \in B(H)$ (note that N and the range of its spectral measure generate a commutative C^*-subalgebra).

Every Jordan $*$-homomorphism of $B(H)$ is the direct sum of a $*$-homomorphism and a $*$-antihomomorphism (see [11, Theorem 3.3]). Let ψ_k^h denote the $*$-homomorphic and let ψ_k^a denote the $*$-antihomomorphic part of ψ_k. Let $N, M \in B(H)$ be invertible normal operators whose product is also normal. By the multiplicativity of ϕ_k and (6) we have

$$\begin{aligned}\psi_k(|NM|(NM|NM|^{-1})^k) &= \phi_k(NM) = \phi_k(N)\phi_k(M)\\ &= \psi_k(|N|(N|N|^{-1})^k)\psi_k(|M|(M|M|^{-1})^k).\end{aligned}$$

This implies that

$$\begin{aligned}\psi_k^h(|NM|(NM|NM|^{-1})^k) &= \psi_k^h(|N|(N|N|^{-1})^k)\psi_k^h(|M|(M|M|^{-1})^k)\\ &= \psi_k^h(|N|(N|N|^{-1})^k|M|(M|M|^{-1})^k)\end{aligned} \tag{7}$$

and

$$\begin{aligned}\psi_k^a(|NM|(NM|NM|^{-1})^k) &= \psi_k^a(|N|(N|N|^{-1})^k)\psi_k^a(|M|(M|M|^{-1})^k)\\ &= \psi_k^a(|M|(M|M|^{-1})^k|N|(N|N|^{-1})^k).\end{aligned} \tag{8}$$

Any $*$-homomorphism or $*$-antihomomorphism of $B(H)$ is either injective or identically 0 which follows from the form of representations of $B(H)$ on separable Hilbert spaces (see [6, 10.4.14. Corollary]). Now, taking (7) and (8) into account, one can verify that the only values of k for which ψ_k can be nonzero are -1 and 1. Moreover, because of the same reasons, for $k = 1$ we have $\psi_1^a = 0$ and for $k = -1$ we have $\psi_{-1}^h = 0$. Observe that $|N|(N|N|^{-1})^{-1} = N^*$. Therefore, ϕ can be written in the form

$$\phi(N) = \begin{bmatrix} \psi'(N) & 0 \\ 0 & \psi''(N^*) \end{bmatrix} \tag{9}$$

for every invertible normal operator N, where ψ' is a $*$-endomorphism and ψ'' is a $*$-antiendomorphism of $B(H)$. By continuity and spectral theorem we clearly have (9) for every normal operator in $B(H)$. Define

$$\psi(A) = \begin{bmatrix} \psi'(A) & 0 \\ 0 & \psi''(A^*) \end{bmatrix}$$

for every $A \in B(H)$. Clearly, ψ is an additive $*$-semigroup endomorphism of $B(H)$ (it is not linear unless ψ'' is missing).

It is easy to see that every rank-one operator is the product of at most three (rank-one) normal operators. This gives us that $\phi(A) = \psi(A)$ for every rank-one operator $A \in B(H)$. Now, let $A \in B(H)$ be arbitrary. Pick rank-one projections $P, Q \in B(H)$. Since PAQ is of rank at most 1, we compute

$$\begin{aligned} \phi(P)\phi(A)\phi(Q) &= \phi(PAQ) = \psi(PAQ) = \psi(P)\psi(A)\psi(Q) \\ &= \phi(P)\psi(A)\phi(Q). \end{aligned} \tag{10}$$

Since every $*$-endomorphism of $B(H)$ is normal, that is, weakly continuous on the bounded subsets of $B(H)$ (see [6, 10.4.14. Corollary]), it follows from (9) that for every maximal family $(P_n)_n$ of pairwise orthogonal rank-one projections we have $\sum_n \phi(P_n) = I$. Therefore, we infer from (10) that $\phi(A) = \psi(A)$ for every $A \in B(H)$. Finally, in order to get the explicite form of ϕ we refer once again to the form of linear $*$-endomorphisms of $B(H)$ appearing in [6, 10.4.14. Corollary] and note that one could get the form of linear $*$-antiendomorphisms of $B(H)$ in a similar way.

□

As for the additivity of $*$-semigroup endomorphisms of $B(H)$ we obtain the following result.

COROLLARY. *Let H be a complex separable infinite dimensional Hilbert space. Let $\phi: B(H) \to B(H)$ be a $*$-semigroup endomorphism which is uniformly continuous on the commutative C^*-subalgebras of $B(H)$. If $\phi(0) = 0$, then ϕ is additive.*

REMARK. In the paper [3] the author studied the problem of additivity of Jordan $*$-isomorphisms between operator algebras under the same continuity assumption that we have used in our paper. It was conjectured by S. Sakai and proved in [4] that this condition is in fact redundant. Therefore, it would be interesting to study our problem without assuming any kind of continuity. However, taking into account the results in [5], one can get evidence to see that our problem in that case is much more complicated.

Finally, we note that it would also be interesting to consider more general algebras like von Neumann algebras in place of $B(H)$ and thus generalize our result for that case.

References

[1] L. J. BUNCE and D. M. WRIGHT, The Mackey–Gleason problem, *Bull. Amer. Math. Soc.*, **26** (1992), 288–293.

[2] J. HAKEDA, Additivity of $*$-semigroup isomorphisms among $*$-algebras, *Bull. London Math. Soc.*, **18** (1986), 51–56.

[3] J. HAKEDA, Additivity of Jordan $*$-maps on AW^*-algebras, *Proc. Amer. Math. Soc.*, **96** (1986), 413–420.

[4] J. HAKEDA and K. SAITÔ, Additivity of Jordan $*$-maps between operator algebras, *J. Math. Soc. Japan*, **38** (1986), 403–408.

[5] M. JODEIT JR. and T. Y. LAM, Multiplicative maps of matrix semi-groups, *Arch. Math.*, **20** (1969), 10–16.

[6] R. V. KADISON and J. R. RINGROSE, *Fundamentals of the Theory of Operator Algebras,* Vol II., Academic Press, 1986.

[7] W. S. MARTINDALE III, When are multiplicative mappings additive?, *Proc. Amer. Math. Soc.*, **21** (1969), 695–698.

[8] L. MOLNÁR, The set of automorphisms of $B(H)$ is topologically reflexive in $B(B(H))$, *Studia Math.*, **122** (1997), 183–193.

[9] T. W. PALMER, *Banach Algebras and The General Theory of $*$-Algebras,* Vol. I., Encyclopedia Math. Appl. **49**, Cambridge University Press, 1994.

[10] P. ŠEMRL, Isomorphisms of standard operator algebras, *Proc. Amer. Math. Soc.*, **123** (1995), 1851–1855.

[11] E. STØRMER, On the Jordan structure of C^*-algebras, *Trans. Amer. Math. Soc.*, **120** (1965), 438–447.

L. MOLNÁR, Institute of Mathematics and Informatics, University of Debrecen, 4010 Debrecen, P.O. Box 12, Hungary; *e-mail:* molnarl@math.klte.hu

Operator Theory: Advances and Applications, Vol. 127, 473–490

Spectral singularities, Szőkefalvi-Nagy–Foias functional model and the spectral analysis of the Boltzmann operator

S. Naboko and R. Romanov

Dedicated to the memory of Prof. Szőkefalvi-Nagy

The problem of separation of an invariant subspace of a dissipative operator corresponding to spectral singularities is studied by means of the functional model. The results obtained are applied to Boltzmann operator. The asymptotic behaviour of the semigroup generated by dissipative operator with spectral singularities is discussed.

1. Introduction

One of the important questions in the theory of nonselfadjoint operators with real spectrum is the estimation of the deviation of a given operator from a selfadjoint one. In particular, one may check if a dissipative operator is similar to a selfadjoint one using the following Sz.-Nagy–Foias criterion [1]: *A maximal dissipative operator L is similar to a selfadjoint operator if and only if $\sigma(L) \subset \mathbb{R}$ and*

$$\left\|(L-z)^{-1}\right\| \leq \frac{C}{\operatorname{Im} z}, \quad \operatorname{Im} z > 0. \tag{1.1}$$

A dissipative operator with real spectrum admits spectral decomposition in the original Hilbert space if and only if the condition (1.1) is satisfied. According to this criterion, a distinguished role is played by those points on the real axis at

---

AMS Subject Classification (1991): 47A45, 82D75.

The work of R. V. R. was supported in part by EPSRC Grant No. GR/ M00549.

vicinities of which the estimate (1.1) fails to hold. For an operator with absolutely continuous spectrum [2] these points are called spectral singularities. The role of spectral singularities was first discovered by Naimark [3] in course of the study of the Schrödinger operator on the halfaxis with exponentially decreasing complex potential. In this case the spectral singularities are finite in number and the bounded spectral projection is defined for any interval whose ends are not the spectral singularities. Then it turns out that the norm of the spectral projection with respect to the ε-vicinity of any of the spectral singularities grows to infinity as $\varepsilon \to 0$. The system of generalized eigenfunctions of the absolutely continuous spectrum is, however, known to be complete in the absolutely continuous subspace [4] if the operator is dissipative. Note that this property fails in general, if the operator is not dissipative [5]. Two circumstances should be remarked here:

1. Spectral singularities for the Schrödinger operator appear only ad hoc, for specially constructed potentials [4], [6]

2. Attempts [6] to construct a spectral decomposition for a dissipative operator with spectral singularities lead to expansions which converge only for a dense set of vectors whose "spectral representatives" have sufficiently regular behaviour at the spectral singularities. In the paper [7] Pavlov suggested a method of summation of the expansion in eigenfunctions of the absolutely continuous spectrum of an abstract dissipative operator which converges for all vectors from the Hilbert space. This summation method is essentially reduced to the convergence in the topology of a Hilbert space with an operator weight. It is however not easy to figure out the analytical content of the method in concrete problems.

Our work is motivated by these. First, we solve an abstract problem of separation of an invariant subspace of the operator on which it is similar to a selfadjoint one under a nonzero angle. The second complementary invariant subspace possesses all the information about spectral singularities. In the case of relatively compact perturbations, under some mild assumption about smoothness of the perturbation which is often fulfilled in applications, the spectral multiplicity of the complementary component turns out to be finite. This statement is the main abstract result of the paper. It is formulated in Lemma 3.5 and Corollary 3.6. The proof is based on an explicit construction of the splitting in terms of the functional model of Sz.-Nagy and Foias [1].

As an example, we then consider an operator with spectral singularities arising in physics, the one-speed Boltzmann operator. Compared to the Schrödinger operator, the presence of a spectral singularity for Boltzmann operator is quite natural in a sense described in the main body of the text.

The above described difficulties in constructing spectral decompositions for operators with spectral singularities lead to a natural assumption that the impact of spectral singularities can be observed through the properties of functions of the operator. In particular, we believe that spectral singularities are revealed in the asymptotic behaviour of the semigroup generated by the operator, which is the main object of interest from the physical viewpoint. There are good reasons to expect, for instance, that in the "simplest" case of finitely many spectral singularities, each of finite power order, the spectral singularities can be localised in terms of the asymptotics, that is, their orders and locations can be calculated from the asymptotics, similar to what takes place in the case of discrete spectrum.

The structure of the paper is the following. In section 2 some preliminaries are given. Then in section 3 we describe the functional model and prove the abstract result on splitting of the absolutely continuous component. The description of the model and of the construction of invariant subspaces closely follows our paper [8]. The proof of the result about splitting taken from the manuscript [9], was not published before. In section 4 we present results about the spectral singularities of the Boltzmann operator. A detailed exposition of them with complete proofs can be found in [8], [9]. Here we only reproduce the proof of absolute continuity of the essential spectrum of the operator in order to illustrate the use of the functional model in the analysis of concrete operators. Finally, in section 5 we discuss links between spectral singularities and asymptotics of the semigroup. We prove a simple result in this direction which can informally be stated as follows. The uniform power estimate of the resolvent along the directions nontangential to the real axis of the order p implies the power estimate of the semigroup generated by $-L$ of the order $p-1$. The proof of this result is patterned after the elementary proof of the Sz.-Nagy–Foias criterion obtained in the papers [10], [11].

2. Preliminaries

Throughout the paper $U_\delta(z) = \{z' \in \mathbb{C} : \ |z - z'| < \delta\}$, $\omega_\delta(z) = U_\delta(z) \bigcap \mathbb{C}_+$, $\mathbb{C}_+$ being the open upper half plane. The term *measurable* refers to the Lebesgue measure on $\mathbb{R}$. Parentheses $(\cdot, \cdot)$ stand for the angle between subspaces of a Hilbert space. We adopt the definition of the essential spectrum of a closed operator used in [12], vol. 4. Given a bounded measurable function Ξ: $\mathbb{R} \to \mathbf{B}(E)$, E being a Hilbert space, we often use Ξ without arguments to denote the operator of multiplication by this function in $L^2(\mathbb{R}, E)$.

Let $S: \mathbb{C}_+ \to \mathbf{B}(E)$, E being a Hilbert space, be a bounded analytic operator-function. A scalar function $m(z) \not\equiv 0$ in $\mathbb{C}_+$ is called a *scalar multiple for* S if there exists a bounded analytic operator-function $\Omega(z)$ such that $m(z)I = S(z)\Omega(z) = \Omega(z)S(z)$ for all $z \in \mathbb{C}_+$ [13].

The determinant $\det(I+A)$ is defined for any $A \in \mathbf{S}^1$ [14]. If $A(z)$ is an $\mathbf{S}^1$-valued function on a domain $\mathcal{D} \subset \mathbb{C}$ analytic in the trace norm, then $a(z) = \det(I + A(z))$ is a scalar analytic function on $\mathcal{D}$.

3. Functional model

We use the symmetric form of the functional model (see [4]). The following exposition of the model is taken, with minor changes, from our papers [8], [13].

Let L be a closed dissipative operator with bounded imaginary part $V = \operatorname{Im} L$ such that $\sigma_{ess}(L) \subset \mathbb{R}$ and let $E = \overline{\operatorname{Ran} V}$. By the *characteristic function of* L we call a contractive analytic function $S(z): E \to E$ in $\mathbb{C}_+$ defined by the formula

$$S(z) = I + 2i\sqrt{V}\,(L^* - z)^{-1}\sqrt{V}, \quad z \in \mathbb{C}_+.$$

This function has boundary values on the real axis, $S(k) \equiv S(k+i0)$, in the strong sense for a.e. $k \in \mathbb{R}$. For $z \in \mathbb{C}_+ \bigcap \rho(L)$ the operator $S^{-1}(z)$ exists on E and is bounded, and, moreover [14],

$$\left\|S^{-1}(z)\right\| \asymp \operatorname{Im} z \left\|(L-z)^{-1}\right\|. \tag{3.1}$$

The matrix $\begin{pmatrix} I & S^*(k) \\ S(k) & I \end{pmatrix}$ defines a nonnegative operator in $E \oplus E$ for a.e. $k \in \mathbb{R}$. Denote by $\mathcal{X} = L^2\begin{pmatrix} I & S^* \\ S & I \end{pmatrix}$ the Hilbert space obtained by the closure of the linear set $L^2(\mathbb{R}, E \oplus E)$ in the metric given by the weight $\begin{pmatrix} I & S^* \\ S & I \end{pmatrix}$, after factoring modulo the set of elements of zero norm. Let $H^2_\pm(E)$ be the Hardy classes of E-valued functions f analytic in $\mathbb{C}_\pm$, respectively, and satisfying $\sup_{\varepsilon>0} \int_{\mathbb{R}} \|f(k \pm i\varepsilon)\|^2_E \, dk < \infty$. Define for a representative $(\tilde{g}, g) \in L^2(\mathbb{R}, E \oplus E)$ from the equivalence class of an element of $\mathcal{X}$

$$P_{\mathcal{K}} \begin{pmatrix} \tilde{g} \\ g \end{pmatrix} = \begin{pmatrix} \tilde{g} - P_+ (\tilde{g} + S^* g) \\ g - P_- (S\tilde{g} + g) \end{pmatrix}, \tag{3.2}$$

where $P_\pm$ are the Riesz projections on $H^2_\pm(E)$ in $L^2(E)$. Then [4], $P_{\mathcal{K}}$ is a correctly defined operator in $\mathcal{X}$ and its closure in $\mathcal{X}$, which will be denoted with the same letter, is an orthogonal projection. Let $\mathcal{K}$ be the range of $P_{\mathcal{K}}$ in $\mathcal{X}$. Define a unitary group $\mathcal{U}_t$ of operators in $\mathcal{X}$ by the formula $(\mathcal{U}_t f)(k) = e^{ikt} f(k)$. Let L_{pure} be the completely nonselfadjoint part of L. Then L_{pure} is unitarily equivalent to the generator of the contraction semigroup $Z_t = P_{\mathcal{K}}\mathcal{U}_t|_{\mathcal{K}}$, $t \geq 0$. This generator is called the functional model of the operator L_{pure}.

Let $R_0(z) = (\operatorname{Re} L - z)^{-1}$. For $z \in \mathbb{C}_+$ we define

$$Q(z) = i\sqrt{V}R_0(z)\sqrt{V}\Big|_E. \tag{3.3}$$

For all $z \in \mathbb{C}_+$ the operator $I - Q(z)$ has bounded inverse, $\left\|(I - Q(z))^{-1}\right\| \leq 1$, since $\Re Q(z) \leq 0$. It can be shown by simple manipulations with the resolvent identity that $S(z)$ is represented as [15]

$$S(z) = \frac{I + Q(z)}{I - Q(z)}, \quad z \in \mathbb{C}_+. \tag{3.4}$$

Let $\Delta = I - S^*S$. Since $\Delta(k) \geq 0$ for a.e. $k \in \mathbb{R}$ one can define the space $L^2(\mathbb{R}; \Delta)$ as the closure of $L^2(\mathbb{R}; E)/\{f\colon\ \Delta f = 0\}$ in the metric given by the weight Δ. We do not distinguish in our notation an element $f \in L^2(\mathbb{R}, E)$ and the corresponding equivalence class in $L^2(\mathbb{R}; \Delta)$. Note that $\sqrt{\Delta}u$ for $u \in L^2(\mathbb{R}; \Delta)$ is a well-defined element of $L^2(\mathbb{R}, E)$. Throughout the rest of the section $\|\cdot\|$ stands for the norm in $L^2(\mathbb{R}, E)$ for E-valued functions on the real axis.

We use theorems about dissipative operators corresponding to theorems stated for contractions in the book [1] without special explanations. Also the definitions of the functional-theoretic objects related to the upper half plane like inner and outer functions, regular factorizations etc. are adopted corresponding naturally to those introduced in [1] for the unit circle. For example, a bounded analytic function $\Theta\colon \mathbb{C}_+ \to \mathbf{B}(E)$ is called *outer* if $\overline{\Theta H^2_+(E)} = H^2_+(E)$.

Throughout the paper, by an invariant subspace of an operator we mean a regular invariant subspace; that is, a subspace $\mathcal{H}$ is called an *invariant subspace* of L if $\overline{(L-\lambda)^{-1}\mathcal{H}} = \mathcal{H}$ for all $\lambda \in \rho(L)$. Using the functional calculus for L one can easily show [9] that $\mathcal{D}(L) \bigcap \mathcal{H} = (L-\lambda)^{-1}\mathcal{H}$ for all $\lambda \in \rho(L)$, and, therefore, $L\,(\mathcal{D}(L) \bigcap \mathcal{H}) \subset \mathcal{H}$, for any regular invariant subspace $\mathcal{H}$. It follows that the restriction $L_{\mathcal{H}} = L|_{\mathcal{H}}$ with the domain $\mathcal{D}(L) \bigcap \mathcal{H}$ is closed and densely defined.

In the following, we define some invariant subspaces of the completely nonselfadjoint part of L in terms of its functional model. In doing so, to keep notation at minimum, we omit in formulas the operator that accomplishes the unitary equivalence between L_{pure} and its functional model.

Note that for any $\tilde{g} \in L^2(\mathbb{R}; \Delta)$ the vector $\begin{pmatrix} \tilde{g} \\ -S\tilde{g} \end{pmatrix}$ defines an element of $\mathcal{X}$ of the norm $\|\tilde{g}\|_{L^2(\mathbb{R};\Delta)}$. *The absolutely continuous subspace* $\mathcal{N}_e \subset \mathcal{K}$ *of the operator* L [16] is defined as the closure of the set $\widetilde{\mathcal{N}_e}$ of smooth vectors:

$$\mathcal{N}_e = \overline{\widetilde{\mathcal{N}_e}}, \qquad \widetilde{\mathcal{N}_e} \equiv \left\{ P_{\mathcal{K}} \begin{pmatrix} \tilde{g} \\ -S\tilde{g} \end{pmatrix}, \ \tilde{g} \in L^2(\mathbb{R}; \Delta) \right\}.$$

Then $\mathcal{N}_e$ coincides with the invariant subspace of L corresponding to the canonical factorization [1] $S = S_i S_e$ of the characteristic function in the sense that the characteristic function of $L|_{\mathcal{N}_e}$ coincides with the pure part of S_e (see [16] and the discussion below). We call the spectrum of L *absolutely continuous* if $L = L|_{H_0} \oplus L|_{\mathcal{N}_e}$, where H_0 is an invariant subspace of L such that $L|_{H_0}$ is a selfadjoint operator with absolutely continuous spectrum in the sense of the spectral theory of selfadjoint operators [12]. A motivation of the above definition of the absolutely continuous subspace from the viewpoint of the theory of smooth perturbations can be found in [13], [17] along with a model-free description of the set of smooth vectors.

Let us define the bounded operator $W: L^2(\mathbb{R};\Delta) \to \mathcal{K}$ by setting

$$W: \tilde{g} \mapsto P_{\mathcal{K}} \begin{pmatrix} \tilde{g} \\ -S\tilde{g} \end{pmatrix}. \tag{3.5}$$

The norm of a smooth vector $W\tilde{g}$ is given by [7]

$$\|W\tilde{g}\|^2 = \left\|S\sqrt{\Delta}\tilde{g}\right\|^2 + \|P_-\Delta\tilde{g}\|^2. \tag{3.6}$$

Then [13, Theorem 4, Corollary 1],

$$(L-z)^{-1} W = W(A_0 - z)^{-1}, \quad z \in \rho(L), \tag{3.7}$$

where A_0 is the operator of multiplication by the independent variable in $L^2(\mathbb{R};\Delta)$. The following lemma describes a class of invariant subspaces of the operator L in $\mathcal{N}_e$. Let $\{X(k)\}$ be a measurable family of subspaces of E defined for a.e. $k \in \mathbb{R}$, and let $X = \{f \in L^2(\mathbb{R}, E) : f(k) \in X(k) \text{ for a.e. } k\}$. Define H to be the closure of X in $L^2(\mathbb{R};\Delta)$. For an operator A in a Hilbert space $\mathcal{A}$ a subspace $\mathcal{J} \subset \mathcal{A}$ is called generating if $\mathcal{A} = \bigvee_{\lambda\in\rho(A)} (A-\lambda)^{-1}\mathcal{J}$. We call the *spectral multiplicity* of an operator A the number $m(A) = \inf \dim \mathcal{N}$, where $\mathcal{N}$ ranges over the generating subspaces of A.

LEMMA 3.1. *$\mathcal{H} = \overline{WH}$ is an invariant subspace of the operator L. If $S(k)$ is invertible a.e. on the real axis in the wide sense, that is,* $\ker S(k) = \{0\}$ *for a.e. $k \in \mathbb{R}$, then*

$$m(L_{\mathcal{H}}) = \operatorname*{ess\,sup}_{k\in\mathbb{R}} \dim \Delta(k)X(k). \tag{3.8}$$

If $\operatorname{ess\,sup}_{k\in\mathbb{R}} \left\| \left(S(k)|_{\Delta(k)X(k)}\right)^{-1} \right\| < \infty$, *then $W|_H$ has a bounded inverse, and thus $L_{\mathcal{H}}$ is similar to a selfadjoint operator.*

We omit here the quite elementary proof of this lemma which is based on the intertwining relation (3.7) (see [9] for details). Note that a partial case of the formula (3.8) was obtained in the paper [2] when studying the triangle model of a dissipative operator. In more general context a study of the multiplicity of spectrum was undertaken by Tikhonov [18].

Applying Lemma 3.1 to $H = L^2(\mathbb{R}, \Delta)$ one gets the "if" part of the Sz.-Nagy–Foias criterion in the following form.

COROLLARY 3.2. *$L|_{\mathcal{N}_e}$ is similar to a selfadjoint operator if and only if* $\sup_{k\in\mathbb{R}} \|S^{-1}(k)\| < \infty$.[1)]

In the usual form of the criterion [1] the supremum is taken over the upper half plane and the statement refers, correspondingly, to the operator L itself, which is equivalent but less convenient for our proposes. The equivalence can easily be established directly by reduction to an obvious statement for the unit circle through the Caley transform.

A point $k \in \mathbb{R}$ is called a *proper point of the operator L* if $\sup_{z\in\omega_\delta(k)} \|S_e^{-1}(z)\| < \infty$ for some $\delta > 0$. Following [16] we call a point $k \in \mathbb{R}$ a *spectral singularity* if it is not proper. For operators with absolutely continuous spectrum $S = S_e$, and thus this definition is equivalent to the one described in the Introduction in view of the estimate (3.1). The following proposition is given here to clarify the definition of spectral singularity. It is not used in the rest of the paper. For a Borel set $\omega \subset \mathbb{R}$ put $X_\omega = \{f \in L^2(\mathbb{R}, E) : f(k) = 0 \text{ for a.e. } k \in \mathbb{R} \setminus \omega\}$. Define H_ω to be the closure of X_ω in $L^2(\mathbb{R}; \Delta)$. By Lemma 3.1, $\mathcal{H}_\omega = \overline{WH_\omega}$ is an invariant subspace of the operator L.

PROPOSITION 3.3. [1], [7]. *If S admits scalar multiple then*

(i) $\sigma(L|_{\mathcal{H}_\omega}) \subset \overline{\omega}$,

(ii) *any invariant subspace $\mathcal{H} \subset \mathcal{N}_e$ of L such that $\sigma(L|_{\mathcal{H}}) \subset \overline{\omega}$ is contained in $\mathcal{H}_\omega$.*

A source of the existence of a scalar multiple important in applications is provided by the following lemma essentially proved in [4]. Let $\|\cdot\|_p$, $p \geq 1$, be the norm in the class $\mathbf{S}^p$ of compact operators.

1) The invertibility on the image is assumed.

LEMMA 3.4. *If $A(\cdot)$ is an $\mathbf{S}^p$-valued ($p \in \mathbb{N}$) function on a domain $\mathcal{D} \subset \mathbb{C}$, analytic in the operator norm and satisfying* $\sup_{z\in\mathcal{D}} \|A(z)\|_p < \infty$, *then* $s(z) = \det(I - A^p(z))$ *is a scalar multiple for* $\Xi(z) = I - A(z)$ *provided* $s(z) \not\equiv 0$.

One should mention that in general the absolutely continuous spectrum is not local, that is, Proposition 3.3 requires some assumptions about S. We refer to the paper [19] for a study of conditions ensuring locality of spectrum on the real axis beyond the assumption of its absolute continuity.

Given X and defining the corresponding invariant susbspace $\mathcal{H}$ as in Lemma 3.1, one can always construct another invariant subspace $\mathcal{H}^\sim$ of L such that $\mathcal{N}_e = \overline{\mathcal{H} \dot{+} \mathcal{H}^\sim}$ by the formula $\mathcal{H}^\sim = \overline{WH^\sim}$, $H^\sim$ being the closure of $X^\perp = \{f \in L^2(\mathbb{R}, E) : f(k) \in X(k)^\perp \text{ for a.e. } k\}$ in $L^2(\mathbb{R}; \Delta)$. We now choose $\{X(k)\}$ in a special way so that the angle $(\mathcal{H}, \mathcal{H}^\sim)$ admits an estimate in terms of S.

Define the function $D = S^*S$. Given a measurable function $k \mapsto \gamma_k$, $\gamma_k \in [0,1]$, $k \in \mathbb{R}$, let $P_1(k)$, $P_2(k)$ be the spectral projections of $D(k)$ for the intervals $[0, \gamma_k)$, $[\gamma_k, 1]$, respectively, and let $X_{1,2}(k) = \operatorname{Ran} P_{1,2}(k)$, hence $E = X_1(k) \oplus X_2(k)$ since $0 \le D \le I$. Put[2)] $\widetilde{S}_1 = \sqrt{D}\,|_{X_1} \oplus I\,|_{X_2}$, $\widetilde{S}_2 = I\,|_{X_1} \oplus \sqrt{D}\,|_{X_2}$.

Define $X_{1,2} = \operatorname{Ran} P_{1,2}$. By construction, the pair of orthogonal subspaces $X_{1,2} \subset L^2(\mathbb{R}, E)$ reduces Δ. Let $H_{1,2}$ be the closures of $X_{1,2}$ in $L^2(\mathbb{R}; \Delta)$, hence $L^2(\mathbb{R}; \Delta) = H_1 \oplus H_2$. By Lemma 3.1, $\mathcal{H}_{1,2} = \overline{WH_{1,2}}$ are invariant subspaces of the operator L.

LEMMA 3.5.

$$\sin(\mathcal{H}_1, \mathcal{H}_2) \ge \operatorname*{ess\,inf}_{k\in\mathbb{R}} \left\| \left(S(k)\,|_{X_2(k)} \right)^{-1} \right\|^{-1}. \tag{3.9}$$

Proof. The assertion is nontrivial if $\left\| \widetilde{S}_2^{-1} \right\| = \operatorname{ess\,sup}_{k\in\mathbb{R}} \left\| \left(S(k)\,|_{X_2(k)} \right)^{-1} \right\|$ is finite. For $\tilde{u} \in X_1$, $\tilde{v} \in X_2$ we have

$$\|W\tilde{v}\| \le \left\| \sqrt{\Delta}\tilde{v} \right\| \le \left\| \widetilde{S}_2^{-1} \right\| \left\| \widetilde{S}_2 \sqrt{\Delta}\tilde{v} \right\|$$

and

$$\begin{aligned} \left\| \widetilde{S}_2 \sqrt{\Delta}\tilde{v} \right\|^2 &= \left\| \sqrt{D\Delta}\tilde{v} \right\|^2 \\ &\le \left\| \sqrt{D\Delta}\tilde{u} \right\|^2 + \left\| \sqrt{D\Delta}\tilde{v} \right\|^2 = \left\| \sqrt{D\Delta}(\tilde{u} - \tilde{v}) \right\|^2 = \left\| S\sqrt{\Delta}(\tilde{u} - \tilde{v}) \right\|^2 \\ &\le \left\| W(\tilde{u} - \tilde{v}) \right\|^2. \end{aligned}$$

---

2) We omit the argument k after all the spaces and operators in the following two equalities which are assumed to hold for a.e. $k \in \mathbb{R}$.

Here we used (3.6) and the fact that $X_{1,2}$ reduce Δ. Thus

$$\left\|\widetilde{S_2}^{-1}\right\|^{-1} \|W\tilde{v}\| \leq \|W(\tilde{u}-\tilde{v})\|.$$

Now, using the definition of the angle between subspaces X, Y of a Hilbert space [14], $\sin(X,Y) = \inf_{u\in X, v\in Y:\|v\|=1} \|u-v\|$, we obtain

$$\sin(\mathcal{H}_1, \mathcal{H}_2) = \inf_{\tilde{u}\in X_1, \tilde{v}\in X_2:\|W\tilde{v}\|=1} \|W\tilde{u} - W\tilde{v}\| \geq \left\|\widetilde{S_2}^{-1}\right\|^{-1},$$

since $X_{1,2}$ are dense in $H_{1,2}$.

□

In the standard setup of the Sz.-Nagy–Foias theory (see [1, Theorem VII.1.1]) invariant subspaces are related to regular factorizations of the characteristic function. Namely, if $S = S_1 S_2$ is a regular factorization of the characteristic function with an outer S_2, then an invariant subspace $\mathcal{H}$ corresponds to this factorization if and only if the characteristic function of the restriction $L|_{\mathcal{H}}$ coincides with the pure part of S_2 (see [1, VII, §4.2]). Recall [1, Prop. VII.4.4.] that since S_2 is outer, such a subspace is unique. Let us explain how to establish correspondence between the constructed invariant subspaces $\mathcal{H}_{1,2}$ and regular factorizations. Basically, this is reduced to an operator analogy of the theorem about representation of a non-negative function on the boundary as the square modulus of the boundary values of an outer function. We henceforth assume that $S(k)$ is invertible a.e. on the real axis in the wide sense.

Applying Proposition V.4.2 in [1] to the nonnegative function $\widetilde{S_2}$ we get that there exists a contractive outer function S_2 such that $\widetilde{S_2}^2 \geq S_2^* S_2$ (bearing in mind the inequality for operators in $L^2(\mathbb{R}, E)$) and the equality is achieved if $\bigcap_{n\geq 0} e^{int}\overline{\widetilde{S_2} H_+^2(E)} = \{0\}$. Let us verify this condition. Let $S = U\sqrt{D}$ be the polar representation of S and $\widehat{S_1} = U\widetilde{S_1}$, so $S = \widehat{S_1}\widetilde{S_2}$ since $\widetilde{S_1}\widetilde{S_2} = \widetilde{S_2}\widetilde{S_1} = \sqrt{D}$. We then have $\widehat{S_1}\left(\bigcap_{n\geq 0} e^{int}\overline{\widetilde{S_2} H_+^2(E)}\right) \subset \bigcap_{n\geq 0} e^{int}\overline{S H_+^2(E)} = \{0\}$. Since $\ker \widehat{S_1} \subset \ker D = \{0\}$ by the above assumption on S, the result follows immediately. For the case when γ_k is the indicator of a Borel set this argument is contained in the proof of Proposition V.4.3 in [1]. Thus $\widetilde{S_2}^2 = S_2^* S_2$, and obviously $S_2^* S_2 \geq S^* S$. By Proposition V.4.1 in [1] one concludes that there exists a contractive analytic function S_1 such that $S = S_1 S_2$. The regularity of the factorization constructed can easily be verified from the definition. Let us show that the characteristic function of the restriction $L|_{\mathcal{H}_2}$ coincides with the pure part of S_2. Define $\Delta_2 = 1 - S_2^* S_2$. Since $\Delta_2 = 0|_{X_1} \oplus \Delta|_{X_2}$, the identification operator $J\colon H_2 \to L^2(\mathbb{R}, \Delta_2)$, $J\tilde{g} = \tilde{g}$

is unitary. Let $\mathcal{N}_e^{(2)}$ be the space of the model representation of the dissipative operator[3] L_2 whose characteristic function is the pure part of S_2. Since S_2 is outer, the space $\mathcal{N}_e^{(2)}$ coincides with the absolutely continuous subspace of L_2. Define the operator $W_2 : L^2(\mathbb{R}, \Delta_2) \to \mathcal{N}_e^{(2)}$ by (3.5) for $L := L_2$. Then the closure of $W_2 J W^{-1}|_{W\mathcal{H}_2}$ carries out a unitary equivalence between $L|_{\mathcal{H}_2}$ and L_2. Thus $\mathcal{H}_2$ is the invariant subspace corresponding to the factorization $S = S_1 S_2$. The space $\mathcal{H}_1$ is treated in the same way starting from the function $\widetilde{S_1}$ instead of $\widetilde{S_2}$.

For any $\beta \in (0,1)$ one can construct a decomposition of $L_e = L|_{\mathcal{N}_e}$ into a linear sum $L_e = L_1^\beta \dot{+} L_2^\beta$ of operators $L_{1,2}^\beta = L|_{\mathcal{H}_{1,2}^\beta}$ acting in the invariant subspaces $\mathcal{H}_{1,2}^\beta$ such that (i) L_2^β is similar to a selfadjoint operator, and (ii) $\sin(\mathcal{H}_1^\beta, \mathcal{H}_2^\beta) \ge \beta \ne 0$, by setting $\gamma_k \equiv \beta$ in the construction above. If the perturbation V satisfies some assumptions of relative compactness and smoothness with respect to L_0 formulated in the following corollary in terms of the characteristic function, then the multiplicity $m(L_1^\beta) < \infty$, for any $\beta \in (0,1)$. According to this corollary, in this case one can choose β appropriately to give $m(L_1^\beta)$ its smallest value and thus to obtain a decomposition of L_e with the least possible spectral multiplicity of the component L_1^β such that L_2^β is still similar to a selfadjoint operator, at the penalty of loss of the quantitative estimate of the angle $(\mathcal{H}_1, \mathcal{H}_2)$.

COROLLARY 3.6. *Suppose that S is norm continuous in the closed upper half-plane, $S(k) - I \in \mathbf{S}^\infty$ for all $k \in \mathbb{R}$, and $S(k) \to I$ as $|k| \to \infty$ in the operator norm. Then $\mathcal{N}_e$ can be represented as a linear sum $\mathcal{N}_e = \mathcal{H}_1 \dot{+} \mathcal{H}_2$ of invariant subspaces $\mathcal{H}_{1,2}$ of the operator L such that*

1. $L_1 = L|_{\mathcal{H}_1}$ *has spectral multiplicity* $\max_{k\in\mathbb{R}} \dim\ker S(k) < \infty$,
2. $L_2 = L|_{\mathcal{H}_2}$ *is similar to a selfadjoint operator,*
3. $(\mathcal{H}_1, \mathcal{H}_2) > 0$.

Proof. Under the formulated assumptions the eigenvalues $\xi_j(k)$, $0 \le \xi_j(k) \le \xi_{j+1}(k)$, $j \ge 1$ of $D(k)$ are continuous in k. Put $n(k) = \dim\ker S(k)$. Then the set $\{k : n(k) \le m\}$ is open for any $m \ge 0$. From this fact and the asymptotic behaviour of S at infinity one gets that $n = \max_{k\in\mathbb{R}} n(k) < \infty$ and $\xi_{n+1}(k) \ge 2\beta > 0$. Define the subspaces $\mathcal{H}_{1,2}$ according to the construction above with $\gamma_k \equiv \beta$. Then, since $\operatorname{rank}\left(P_{[0,\beta)}(k)\right) = n$ for $k \in \omega$ with an interval $\omega \subset \mathbb{R}$ and $\ker S(k) = \{0\}$ for a.e. $k \in \mathbb{R}$ (see e. g. [17]), one gets the result from (3.8) and Lemma 3.5. □

---

[3] Since the factorization $S = S_1 S_2$ is regular, the operator L_2 is unitarily equivalent to a part of the operator L and therefore has bounded imaginary part. Thus L_2 admits functional model of the form described in the paper.

4. Transport operator

A representative example of an operator with spectral singularities is provided by the one-speed transport operator [20]. This operator arises in the transport theory of neutrons [21] and governs the evolution of particles density. A density is a function of variables $x \in \mathbb{R}$ (position) and $\mu \in \Omega$, $\Omega \equiv [-1,1]$ (cosine of the angle between the momentum and the x axis), belonging to the Hilbert space $H = L^2(\mathbb{R} \times \Omega)$. Physical parameters in the problem are the total cross-section $\sigma \geq 0$ and the mean number of secondaries per collision $c\colon \mathbb{R} \to \mathbb{R}_+$. It is assumed that $c \in L^\infty(\mathbb{R})$.

Define the bounded operator K in H by the formula $Kf = \frac{1}{2}\int_\Omega f\, d\mu$, where $d\mu$ is the Lebesgue measure on Ω. We then use the identification $L^2(\mathbb{R} \times \Omega) = L^2(\mathbb{R}) \otimes L^2(\Omega)$ and do not distinguish an operator A in $L^2(\mathbb{R})$ and the operator $A \otimes I$ in $L^2(\mathbb{R}\times\Omega)$ in our notation. We denote by L_0^+ the set of compactly supported nonnegative functions from $L^\infty(\mathbb{R})$.

The evolution of densities is given by a semigroup $Z_t = e^{iLt}$, $t \geq 0$, of operators in H. The generator L, which is called the one-speed transport operator, is defined in H by the following expression

$$L = i\mu\partial_x + i\sigma(1 - c(x)K)$$

on the domain $\mathcal{D} = \{f \in H : (1) f(\cdot,\mu)$ is absolutely continuous for a.e. $\mu \in \Omega$; (2) $\mu\partial_x f \in H\}$. The imaginary part of L is bounded, while its real part $L_0 = i\mu\partial_x$ is a selfadjoint operator on the domain $\mathcal{D}$.

Instead of L, it is convenient to deal with the dissipative operator $T = L^* + i\sigma = i\mu\partial_x + i\sigma c(x)K$. The spectral analysis of L is reduced to that of T since L and $-T+i$ are unitarily equivalent, $L = J(-T+i)J^*$ with $(Jf)(x,\mu) = f(x,-\mu)$. Without loss of generality, one can set $\sigma = 1$; then $V = \Im T = c(x)K$. We henceforth assume that $c \in L_0^+$ which corresponds to the physically important case of the problem. The spectrum of the operator T is described by the following theorem essentially established in [20] (see [8] for a contemporary exposition of the proof). Put $\sigma_+(T) = \sigma(T) \bigcap \mathbb{C}_+$.

THEOREM 4.1. *Assume $c \in L_0^+$. Then the nonreal spectrum $\sigma_+(T)$ of the operator T consists of finitely many eigenvalues lying on the imaginary axis. The dimension of the invariant subspace H_d corresponding to $\sigma_+(T)$ is finite. The essential spectrum of T coincides with the real axis: $\sigma_{ess}(T) = \mathbb{R}$.*

We refer to [8] for some further results concerning the point spectrum of T. Define $\mathcal{P}_d$ to be the Riesz projection corresponding to $\sigma_+(T)$, so $H_d = \mathcal{P}_d H$.

Then $H = H_d \dot{+} H_{ess}$, where $H_{ess} = (I - \mathcal{P}_d)H$ is the invariant subspace of T corresponding to the essential spectrum. To apply the functional model to the study of the component $T_{ess} = T|_{H_{ess}}$ we shall need some information about boundary behaviour of the resolvent summarized below.

One can naturally identify the range of K with the space $L^2(\mathbb{R})$ of functions of the x variable, $KH = L^2(\mathbb{R}) \otimes \mathbf{1} \simeq L^2(\mathbb{R})$, where $\mathbf{1}$ stands for the indicator of the set Ω, and hence $E \equiv \overline{\mathrm{Ran} V} \subset L^2(\mathbb{R})$. Set $R_0(z) = (L_0 - z)^{-1}$. Define the operator $Q(z)$, $z \in \mathbb{C}_+$, by the formula (3.3). Its integral kernel has the form[4] $-\frac{1}{2}\sqrt{c(x)}E(-iz|x-y|)\sqrt{c(y)}$, where $E(s) = \int_1^\infty e^{-st}\frac{dt}{t}$ for $\Re s > 0$. Since $E(s) = O(\ln|s|)$ as $s \to 0$ and c has compact support, it follows that $Q(z)$ is of the Hilbert–Schmidt class $\mathbf{S}^2$ for all $z \in \mathbb{C}_+$. The following properties of the function Q are important when analysing the essential spectrum of the operator T [21], [22].

PROPOSITION 4.2. (i) *The function $Q(z)$ has an analytical continuation from $\mathbb{C}_+$ to $\mathcal{O} = \mathbb{C} \setminus \{-it,\ t \geq 0\}$ given by the following formula*

$$Q(z) = \widetilde{Q}(z) + \frac{1}{2}\Theta(z), \tag{4.1}$$

where $\widetilde{Q}(z)$ is the operator with the kernel $\frac{1}{2}\sqrt{c(x)}(\ln(-iz|x-y|) + \gamma)\sqrt{c(y)}$, γ is the Euler constant, and $\Theta(z)$ is an entire $\mathbf{S}^2$-valued function such that $\Theta(0) = 0$. Here the cut of the logarithm is chosen to be the negative real axis, hence $\widetilde{Q}(z)$ is analytic in $\mathcal{O}$.

(ii) $\ker(I + Q(k)) = \{0\}$ *for real* $k \neq 0$.

(iii) $Q(it)$ *is a negative selfadjoint operator for $t > 0$ and a monotone increasing function of t.*

(iv) $\|Q(k)\| \to 0$ *as* $k \to \infty$, $k \in \mathbb{R}$.

It follows from (ii) and the formula (3.4) that all points on the real axis except for at most the point 0 are proper points of T. Let $\{\eta_n(z)\}_{n=1}^\infty$, $|\eta_n| \geq |\eta_{n+1}| > 0$, be the eigenvalues of $Q(z)$ and $\mathcal{B}_c = \{k_n = \lim_{\varepsilon \downarrow 0} \eta_n(i\varepsilon) : |k_n| < \infty\}$. Define the singular set $\mathcal{E} \subset L_0^+$ as follows, $\mathcal{E} = \{c : -1 \in \mathcal{B}_c\}$. Note that for any $c \in L_0^+$, the function $\kappa c \in \mathcal{E}$ for the infinite discrete set $\{-k_n^{-1},\ k_n \in \mathcal{B}_c\}$ of values of the constant $\kappa > 0$. It is easy to see from (4.1) that 0 is a spectral singularity of T if and only if $c \in \mathcal{E}$. Taking into account (iv) one concludes from the Sz.-Nagy–Foias criterion that if $c \notin \mathcal{E}$ then $T|_{\mathcal{N}_e}$ is similar to a selfadjoint operator.

In the following theorem we show that T_{ess} has absolutely continuous spectrum. The proof of this fact is based on the functional model. Let S be the characteristic

[4] This kernel has essentially been calculated in [20, §2].

function of the operator T. Note that S is continuous at 0 which can easily be shown to follow from (i).

THEOREM 4.3. *The operator T_{ess} has absolutely continuous spectrum, that is, $H_{ess} = \mathcal{N}_e \oplus H_0$ where H_0 is the maximal reducing subspace of T on which it coincides with L_0. If $c \in \mathcal{E}$ then T_{ess} can be represented as a linear sum $T_{ess} = T_1 \dot{+} T_2$ of operators $T_{1,2}$ acting in invariant subspaces $\mathcal{H}_{1,2}$ such that*

1. *T_1 has the spectral multiplicity $\mathcal{M} = \dim\ker S(0)$,*
2. *T_2 is similar to a selfadjoint operator,*
3. *the angle of $(\mathcal{H}_1, \mathcal{H}_2) > 0$.*

Proof. It follows from the relation (3.7) that $\mathcal{N}_e \subset H_{ess}$. Indeed, integrating (3.7) over a closed contour $\Gamma \subset \mathbb{C}_+$, $\sigma_+(T) \subset \operatorname{Int}\Gamma$, we get that $\mathcal{P}_d W = 0$, and the closure gives $\mathcal{P}_d \mathcal{N}_e = 0$.

It is obvious from (3.4) that $T(z) \equiv S(z) - I \in \mathbf{S}^2$ for $z \in \mathbb{C}_+$, since so does $Q(z)$. It then turns out [8] that the function $m(z) \equiv \det\left(I - T^2(z)\right)$ is a scalar multiple for S. This conclusion is drawn from Lemma 3.4 through the estimates of the $\mathbf{S}^2$-norm of $T(z)$ in the upper half plane. It is obvious from the representation (4.1) that the analytical continuation of the function $Q(z)$ constructed in Proposition 4.2 is also analytic in the $\mathbf{S}^2$-norm. By Lemma 3.4 this implies that the scalar multiple $m(z)$ is analytic on the real axis outside 0. This means that the singular component m_s in the canonical factorization of m has the form

$$m_s(z) = e^{i\mu_1 z - \mu_2 \frac{i}{z}}$$

with some $\mu_1, \mu_2 \geq 0$.

It follows from (i) and (iii) in Proposition 4.2 that $(I - \widetilde{Q}(z))^{-1}$ is a bounded analytic function in $\widetilde{U}_\delta(0) \equiv U_\delta(0) \setminus [0, -i\delta)$ for sufficiently small δ. Taking into account that $\|\Theta(z)\| = o(1)$ as $z \to 0$, we find that $(I - Q(z))^{-1}$, and therefore $S(z)$, extends to a bounded analytic function on $\widetilde{U}_\delta(0)$ for sufficiently small δ. Then by (4.1),

$$T(z) = (A + \Theta(z))\,(I - Q(z))^{-1} + M(z)$$

where rank $M(z) = 1$ and $A \in \mathbf{S}^2$. Since $\|T(z)\| \leq 1 + \|S(z)\| \leq C$, we conclude that the Hilbert–Schmidt norm $\|M(z)\|_2 = \|M(z)\|$ is bounded in $\widetilde{U}_\delta(0)$. Hence the analytic continuation of $T(z)$ into $\widetilde{U}_\delta(0)$ is bounded in the $\mathbf{S}^2$-norm, and therefore m admits analytic continuation into $\widetilde{U}_\delta(0)$ which is bounded. It follows that $\mu_2 = 0$. Indeed, otherwise an application of the Carlson Theorem [23] to the function $g\left(z^{-1}\right)$, where $g(z) = e^{\frac{\mu_2}{2}\frac{i}{z}} m(z)$, implies $g \equiv 0$. Thus, the inner factor of m is $e^{i\mu_1 z} b(z)$, b being a Blaschke product.

Let $\mathcal{N}_i$ be the inner subspace of T (see [16]). It can be easily seen that if $\mu_1 \neq 0$ then $m(z)e^{-i\mu_1 z}$ is also a scalar multiple for S, otherwise there would exist an invariant subspace of T such that the restriction of T to this subspace has empty spectrum which is impossible because T has bounded imaginary part. Thus the inner factor of a scalar multiple for S is a pure Blaschke product, and hence $\mathcal{N}_i$ is generated by the root vectors of T. It follows that $\mathcal{N}_i = H_d$. Since [16] $H \ominus H_0 = \mathcal{N}_e \dot{+} \mathcal{N}_i$, we therefore have $H_{ess} = \mathcal{N}_e \oplus H_0$. The details can be found in [9].

The second assertion of the theorem follows immediately from Corollary 3.6. □

Similar results were obtained [9] for the transport operator in the three-dimensional space. In this case the characteristic function is analytic on the real axis which permits us to define order of the spectral singularity. In fact, it turns out that the point 0 is a simple zero of the characteristic function for $c \in \mathcal{E}$.

5. Estimate of the induced semigroup

As it is known [6], when attempting to separate out the terms corresponding to spectral singularities in the spectral decomposition of an operator we get an expansion which converges only for dense linear sets of test functions, for example, for smooth vectors. This suggests to study spectral singularities in terms different from spectral decompositions. The idea to consider a spectral singularity as a kind of resonance embedded into the continuum gives a hint that the properties of spectral singularities are revealed in the asymptotics of the semigroup generated by the operator. Here we prove a rather simple result in this direction. In the following theorem L is a maximal dissipative operator with real spectrum satisfying $\|(L-z)^{-1}\| \leq (\operatorname{Im} z - \omega)^{-1}$ for some $\omega > 0$ and all z with $\operatorname{Im} z > \omega$. The latter condition guarantees that the operator $Z_t = e^{iLt}$, $t \geq 0$, has bounded inverse. In particular, it is obviously satisfied if L has bounded imaginary part.

THEOREM 5.1. *Assume*

$$\|(L-z)^{-1}\| \leq C_0(\operatorname{Im} z)^{-p-1} \tag{5.1}$$

for all $z \in \Pi$ *where* $\Pi = \{z : \ 0 < \operatorname{Im} z < b\}$ *with some* $b > \omega$. *Then for all* $t \geq 0$

$$\|Z_t^{-1}\| \leq C + 2eC_0t^p \tag{5.2}$$

with a constant $C \geq 1$.

Proof. Proof of the theorem is based on the following estimate which is valid for any dissipative operator L [13]:

$$\int_{\mathbb{R}} \|R_L(k - i\varepsilon)u\|^2 \, dk \leq \frac{\pi}{\varepsilon} \|u\|^2 , \; \varepsilon > 0, \tag{5.3}$$

where $R_L(z) = (L - z)^{-1}$. This estimate is deduced in [13] from the existence of a selfadjoint dilation of L. An elementary derivation of it is outlined in [10].

By the Hille–Yosida theorem one has for all u from the domain of L and $t > 0$

$$Z_t^{-1}u = -\frac{1}{2\pi i} \lim_{N\to\infty} \int_{-N+i\varepsilon}^{N+i\varepsilon} e^{-izt} R_L(z)u \, dz, \tag{5.4}$$

where $\varepsilon > 0$ is large enough. We first show that (5.4) holds for any $\varepsilon > 0$. Remark that the inequality (5.3) implies that $R_L(\cdot - i\varepsilon)u \in H^2_-$ for any $\varepsilon > 0$. Then, it follows from the resolvent identity

$$R_L(z) = [I + 2i\varepsilon R_L(z)] R_L(\overline{z}), \; \varepsilon = \operatorname{Im} z,$$

by the estimate (5.1) that

$$\|R_L(z)u\| \leq \left(1 + 2C_0\varepsilon^{-p}\right) \|R_L(\overline{z})u\| \tag{5.5}$$

for $\operatorname{Im} z \in (0, b)$. Hence, $R_L(\cdot + i\nu)u \in H^2_+$ for any $\nu > 0$, which allows to deform the contour of integration in (5.4) in the upper half plane.

Notice that

$$\int_{-N+i\varepsilon}^{N+i\varepsilon} e^{-izt} \langle R_L(\overline{z})u, v\rangle \, dz = e^{\varepsilon t} \int_{-N}^{N} e^{-ikt} \langle R_L(k - i\varepsilon)u, v\rangle \, dk \underset{N\to\infty}{\longrightarrow} 0$$

for a.e. $t > 0$ by the Paley–Wiener theorem. We then have for a.e. $t > 0$

$$\begin{aligned} \left\langle Z_t^{-1}u, v\right\rangle &= -\frac{1}{2\pi i} \lim_{N\to\infty} \int_{-N+i\varepsilon}^{N+i\varepsilon} e^{-izt} \langle R_L(z)u, v\rangle \, dz \\ &= -\frac{1}{2\pi i} \lim_{N\to\infty} \int_{-N+i\varepsilon}^{N+i\varepsilon} e^{-izt} \langle (R_L(z) - R_L(\overline{z}))u, v\rangle \, dz \\ &= -\frac{\varepsilon}{\pi} \int_{\mathbb{R}+i\varepsilon} e^{-izt} \langle R_L(z)u, R_L^*(\overline{z})v\rangle \, dz. \end{aligned} \tag{5.6}$$

The latter integral exists in the usual Lebesgue sense and admits the following estimate

$$\begin{aligned} \left| \int_{\mathbb{R}+i\varepsilon} \cdots \right| &\leq e^{\varepsilon t} \left(\int_{\mathbb{R}+i\varepsilon} \|R_L(z)u\|^2 \, dz \right)^{1/2} \left(\int_{\mathbb{R}+i\varepsilon} \|R_L^*(\overline{z})v\|^2 \, dz \right)^{1/2} \\ &\leq e^{\varepsilon t} \left(\int_{\mathbb{R}+i\varepsilon} \|R_L(z)u\|^2 \, dz \right)^{1/2} \sqrt{\frac{\pi}{\varepsilon}} \, \|v\| . \end{aligned}$$

In the last step we have used the inequality (5.3) for the dissipative operator $-L^*$. By virtue of (5.5) one can estimate for $\varepsilon < b$

$$\int_{\mathbb{R}+i\varepsilon} \|R_L(z)u\|^2\, dz \le \left(1+2C_0\varepsilon^{-p}\right)^2 \int_{\mathbb{R}+i\varepsilon} \|R_L(\overline{z})u\|^2\, dz$$
$$\le \frac{\pi}{\varepsilon}\left(1+2C_0\varepsilon^{-p}\right)^2 \|u\|^2 .$$

Thus,

$$\left|\int_{\mathbb{R}+i\varepsilon} \cdots\right| \le \frac{\pi}{\varepsilon} e^{\varepsilon t}\left(1+2C_0\varepsilon^{-p}\right)\|u\|\,\|v\| .$$

Inserting this into (5.6) we obtain

$$\left|\left\langle Z_t^{-1}u, v\right\rangle\right| \le \frac{\varepsilon}{\pi}\frac{\pi}{\varepsilon} e^{\varepsilon t}\left(1+2C_0\varepsilon^{-p}\right)\|u\|\,\|v\| ,$$

and this is valid by the continuity of the left hand side for all $t > 0$. Now setting $\varepsilon = t^{-1}$ for t large enough one gets the result.

□

In the "simplest" case, when the characteristic function is analytic on the real axis, and, for example, Im L is bounded and relatively compact with respect to Re L, the estimate (5.2) can be shown to be sharp in the power scale. This means that if p is the maximum over the (algebraic) orders of zeroes of S on the real axis, then for any sufficiently small $\varepsilon > 0$ there exists a vector $u \in H$ such that $\left\|Z_t^{-1}u\right\| = t^{p-\varepsilon}(1+o(1))$. Proof of this result, which is basically an explicit construction of the u, will be published elsewhere. Note that problems related to the above studies are treated in a recent paper by M. Malejki [24].

References

[1] B. SZ.-NAGY and C. FOIAS, *Analyse harmonique des operateurs de l'espase de Hilbert,* Masson et Cie – Akadémiai Kiadó, 1967.

[2] L. A. SAHNOVIČ, Dissipative operators with absolutely continuous spectrum, *Trans. Moscow Math. Soc.*, **19** (1968), 2?3–298.

[3] M. A. NAĬMARK, Investigation of the spectrum and the expansion in eigenfunctions of a nonselfadjoint operator of the second order on a semi-axis, *Trudy Mosk. Mat. Obsch.*, **3** (1954), 181–270 (Russian); English transl.: *Amer. Math. Soc. Transl.*, **16** (1960), 103–193.

[4] B. S. PAVLOV, Selfadjoint dilation of the dissipative Schrödinger operator and its resolution in terms of eigenfunctions, *Math. USSR Sbornik*, **31** (1977), 457.

[5] V. F. Veselov and S. N. Naboko, The determinant of the characteristic function and the singular spectrum of a nonselfadjoint operator, *Math. USSR-Sb.*, **57** (1987), 21–41.

[6] V. E. Ljance, On a differential operator with spectral singularities. I, *Mat. Sb.* 64 (106), **4** (1964), 521–561; II, *Mat. Sb.* 65 (107), **1** (1964), 47–103 (Russian); English transl.: *Amer. Math. Soc. Transl.*, **60** (1967).

[7] B. S. Pavlov, On expansions in eigenfunctions of the absolutely continuous spectrum of a dissipative operator, *Vestnik Leningrad Univ.* (Mat. Mech. Astronom), **1** (1975), 130–137 (Russian); English transl.: *Vestnik Leningrad Univ. Math.*, **8** (1980), 135–143.

[8] Yu. A. Kuperin, S. N. Naboko and R. V. Romanov, Spectral analysis of the one-speed transport operator and the functional model, *Funct. Anal. Appl.*, **33** (1999), 199–207.

[9] Yu. A. Kuperin, S. N. Naboko and R. V. Romanov, Spectral analysis of the one-speed transport operator, Submitted for publication.

[10] S. N. Naboko, Conditions for similarity to unitary and selfadjoint operators, *Funct. Anal. Appl.*, **18** (1984), 13–21.

[11] J. A. van Casteren, Operators similar to unitary or selfadjoint ones, *Pacific J. Math.*, **104** (1983), 241–255.

[12] M. Reed and B. Simon, *Methods of Modern Mathematical Physics* Vol. 1-4, Academic Press, New York, 1972–1978.

[13] S. N. Naboko, A functional model of perturbation theory and its application to scattering theory, *Trudy MIAN*, **147** (1980), 86–114 (Russian); English transl.: *Proc. Steklov Inst. Math.*, **2** (1981), 85–116.

[14] I. Gohberg and M. Krein, *Introduction to the theory of linear non-selfadjoint operators in Hilbert space*, AMS, Providence, 1969.

[15] M. S. Brodskiĭ, *Triangular and Jordan Representations of Linear Operators*, AMS Transl. Math. Monographs Vol. **32**, AMS, Providence, 1971.

[16] B. S. Pavlov, On separation conditions for the spectral components of a dissipative operator, *Math. USSR Izvestija*, **9** (1975), 113–137.

[17] S. N. Naboko, On the conditions for existence of wave operators in the nonselfadjoint case, *Probl. Mat. Fiz.*, **12** (1987), 132–155 (Russian); English transl.: *Amer. Math. Soc. Transl.*, **157** (1993), 127–149.

[18] A. S. Tikhonov, *Absolutely continuous spectrum of a linear operator and problems of scattering theory and factorization of operator-functions*, Ph. D. Thesis, Simferopol State University, Simferopol, 1989 (Russian).

[19] Yu. I. Lyubič and V. I. Macaev, On the spectral theory of linear operators in Banach space, *Soviet Math. Dokl.*, **1** (1960), 184–186.

[20] J. Lehner, The spectrum of the neutron transport operator for the infinite slab, *J. Math. Mech.*, **11** (1962), 173–181.

[21] J. Lehner and G. Wing, On the spectrum of an unsymmetric operator arising in the transport theory of neutrons, *Comm. Pure Appl. Math.*, **8** (1955), 217–234.

[22] J. LEHNER and G. WING, Solution of the linearized Boltzmann equation for the slab geometry, *Duke Math. J.*, **23** (1956), 125–142.
[23] E. C. TITCHMARSH, *The theory of functions*, Oxford Univ. Press, Oxford, 1939.
[24] M. MALEJKI, C_0-semigroups with polinomial growth, submitted.

S. NABOKO, Department of Mathematical Physics, Faculty of Physics, St.Petersburg State University, 198904, St.Petersburg, Russia; *e-mail*: naboko@snoopy.phys.spbu.ru

R. ROMANOV, School of Mathematical Sciences, University of Sussex, Falmer, Brighton BN1 9QH, UK; Laboratory of Complex Systems Theory, Institute for Physics, Saint Petersburg State University, 198904, Saint Petersburg, Russia

Operator Theory: Advances and Applications, Vol. 127, 491–517

Uniqueness of invariant measures for the stochastic Cauchy problem in Banach spaces

J. M. A. M. VAN NEERVEN*

We study uniqueness for invariant measures of the stochastic abstract Cauchy problem

$$\begin{aligned} du(t) &= Au(t)\,dt + B\,dW_H(t), \quad t \ge 0,\\ u(0) &= x, \end{aligned}$$

where A is the generator of a C_0-semigroup $\{S(t)\}_{t\ge0}$ on a separable real Banach space, $\{W_H(t)\}_{t\ge0}$ is a cylindrical Wiener process with Cameron–Martin space H, and $B \in \mathcal{L}(H,E)$ is a bounded linear operator. Under a nondegeneracy assumption, it is shown that a Gaussian invariant measure, if one exists, is unique if the adjoint orbits $t \mapsto S^*(t)x^*$ are bounded for all $x^* \in \bigcap_{n\ge1}\mathcal{D}(A^{*n})$.

1. Introduction

In this paper we study uniqueness of invariant measures for the stochastic abstract Cauchy problem

$$\begin{aligned} du(t) &= Au(t)\,dt + B\,dW_H(t), \quad t \ge 0,\\ u(0) &= x, \end{aligned} \tag{1.1}$$

where A is the generator of a C_0-semigroup $\mathbf{S} = \{S(t)\}_{t\ge0}$ on a separable real Banach space E, $\{W_H(t)\}_{t\ge0}$ is a cylindrical Wiener process with Cameron–Martin

---

AMS Subject Classification (1991): 60H15, 47D03, 35R15, 60J35, 93C25.

* This work was started while attending the Memorial Conference for Professor B. Sz.-Nagy, to the memory of whom I would like to dedicate this paper.

space H, and $B \in \mathcal{L}(H, E)$ is a bounded linear operator. For Hilbert spaces E, a variety of conditions is known that imply uniqueness of the invariant measure if one exists:

- $\mathbf{S}$ is strongly stable, i.e. $\lim_{t\to\infty} \|S(t)x\| = 0$ for all $x \in E$;
- $\mathbf{S}$ is compact;
- $\mathbf{S}$ is null controllable with respect to the control pair (B, H).

In the last two cases an additional nondegeneracy condition needs to be imposed; cf. Example 2.8 below. We refer to [7], [8], and the references cited there for the proofs and more detailed information.

We will give alternative, purely functional analytic proofs of these results which work for arbitrary separable real Banach spaces E. Among other things we prove uniqueness of the invariant measure under a nondegeneracy assumption and the following rather mild boundedness condition:

- For all $x^* \in \bigcap_{n\geq 1} \mathcal{D}(A^{*n})$, the orbit $t \mapsto S^*(t)x^*$ is bounded.

This result seems to be new also in the Hilbert space framework.

Finally we mention [4] and the papers cited there, where uniqueness of invariant measures in the finite-dimensional setting is studied from a more general perspective.

2. The Lyapunov equation $AX + XA^* = -Q$

Let E be a real Banach space. An operator $Q \in \mathcal{L}(E^*, E)$ is called *positive* if $\langle Qx^*, x^*\rangle \geq 0$ for all $x^* \in E^*$, and *symmetric* if $\langle Qx^*, y^*\rangle = \langle Qy^*, x^*\rangle$ for all $x^*, y^* \in E^*$.

In our study of (1.1), we will be mostly interested the following example:

EXAMPLE 2.1. If H is a real Hilbert space and $B \in \mathcal{L}(H, E)$ is a bounded linear operator, then $Q := B \circ B^* \in \mathcal{L}(E^*, E)$ is positive and symmetric.

It is easily checked that on the range of Q, the formula

$$[Qx^*, Qy^*]_{H_Q} := \langle Qx^*, y^*\rangle \tag{2.1}$$

defines an inner product $[\cdot,\cdot]_{H_Q}$. Note that if $Qx^* = 0$, then $[Qx^*, Qy^*]_{H_Q} = \langle Qx^*, y^*\rangle = 0$, and if $Qy^* = 0$, then $[Qx^*, Qy^*]_{H_Q} = \langle Qx^*, y^*\rangle = \langle Qx^*, y^*\rangle = 0$ by the symmetry of Q. This shows that $[\cdot,\cdot]_{H_Q}$ is well defined.

We denote by H_Q the real Hilbert space obtained by completing the range of Q with respect to $[\cdot,\cdot]_{H_Q}$. This space is usually called the *reproducing kernel Hilbert space* associated with Q. From

$$\|Qx^*\|_{H_Q}^2 = \langle Qx^*, x^*\rangle = |\langle Qx^*, x^*\rangle| \le \|Q\|_{\mathcal{L}(E^*,E)}\|x^*\|^2$$

we see that Q, as an operator from E^* into H_Q, is bounded with norm $\le \|Q\|_{\mathcal{L}(E^*,E)}^{1/2}$. From

$$|\langle Qx^*, y^*\rangle| \le \|Qx^*\|_{H_Q}\|Qy^*\|_{H_Q} \le \|Qx^*\|_{H_Q}\|Q\|_{\mathcal{L}(E^*,H_Q)}\|y^*\|$$

it then follows that

$$\|Qx^*\| \le \|Q\|_{\mathcal{L}(E^*,H_Q)}\|Qx^*\|_{H_Q}.$$

Thus, the inclusion mapping from the range of Q into E is continuous with respect to the inner product $[\cdot,\cdot]_{H_Q}$ and extends to a bounded linear operator i_Q from H_Q into E of norm $\le \|Q\|_{\mathcal{L}(E^*,H_Q)}$. We will show next that i_Q is an injection from H_Q into E and that

$$Q = i_Q \circ i_Q^*. \tag{2.2}$$

Given an element $x^* \in E^*$ we denote by h_{x^*} the element in H represented by Qx^*. With this notation we have

$$i_Q(h_{x^*}) = Qx^*$$

and (2.1) becomes

$$[h_{x^*}, h_{y^*}]_{H_Q} = \langle Qx^*, y^*\rangle.$$

For all $y^* \in E^*$ we then have

$$[h_{x^*}, h_{y^*}]_{H_Q} = \langle Qx^*, y^*\rangle = \langle i_Q(h_{x^*}), y^*\rangle = [h_{x^*}, i_Q^* y^*]_{H_Q}.$$

Since the elements h_{x^*} span a dense subspace of H_Q it follows that

$$h_{y^*} = i_Q^* y^*.$$

Therefore,

$$Qy^* = i_Q(h_{y^*}) = i_Q(i_Q^* y^*)$$

for all $y^* \in E^*$, and (2.2) follows. Finally if $i_Q g = 0$ for some $g \in H_Q$, then for all $y^* \in E^*$ we have

$$[g, h_{y^*}]_{H_Q} = [g, i_Q^* y^*]_{H_Q} = \langle i_Q g, x^*\rangle = 0,$$

and therefore $g = 0$. This proves that i_Q is injective.

Being an adjoint operator, i_Q^* is weak*-continuous, hence weak*-to-weakly continuous since H is reflexive. Since i_Q is continuous, hence weakly continuous, it follows that $Q = i_Q \circ i_Q^*$ is weak*-to-weakly continuous. A similar argument shows that H_Q is separable whenever E is separable. These observations will be used repeatedly below.

If E is a Hilbert space, then upon identifying E^* with E we may regard Q as a positive selfadjoint operator on E and as such we have $i_Q(H_Q) = Q^{1/2}(E)$.

We refer to [17] and the references cited there for more details.

Throughout the rest of this paper, E is a *separable* real Banach space, $\mathbf{S} = \{S(t)\}_{t\geq 0}$ is a C_0-semigroup of bounded linear operators on E with infinitesimal generator A, and $Q \in \mathcal{L}(E^*, E)$ is a positive symmetric operator. For each $t > 0$ we define a positive symmetric operator $Q_t \in \mathcal{L}(E^*, E)$ by

$$Q_t x^* := \int_0^t S(s)QS^*(s)x^*\,ds, \qquad x^* \in E^*.$$

This integral is easily seen to exist as a Bochner integral; cf. [15]. For $t > 0$ and $h \geq 0$ we have the following algebraic identity:

$$Q_{t+h} = Q_t + S(t)Q_h S^*(t).$$

From [10] we recall the following result.

Proposition 2.2. *The following assertions are equivalent:*

1. *There exists an operator $Q_\infty \in \mathcal{L}(E^*, E)$ such that*

$$\langle Q_\infty x^*, y^* \rangle = \lim_{t\to\infty} \langle Q_t x^*, y^* \rangle, \qquad x^*, y^* \in E^*;$$

2. *There exists a positive symmetric solution $X \in \mathcal{L}(E^*, E)$ to the operator equation*

$$AX + XA^* = -Q.$$

In this situation Q_∞ is a positive symmetric solution of the equation $AX + XA^* = -Q$.

Here, we say that an operator $X \in \mathcal{L}(E^*, E)$ is a *solution* of $AX + XA^* = -Q$ if for all $x^* \in \mathcal{D}(A^*)$ we have $Xx^* \in \mathcal{D}(A)$ and $AXx^* + XA^*x^* = -Qx^*$.

For a systematic study of the so-called *Lyapunov equation*

$$AX - XB = C$$

we refer to [1], [18], and the references cited there. We will need the following observation; cf. [18], [10]:

PROPOSITION 2.3. *If $X \in \mathcal{L}(E^*, E)$ is a positive symmetric solution of the equation $AX + XA^* = -Q$, then for all $t > 0$ we have*

$$X - S(t)XS^*(t) = Q_t.$$

In what follows, we shall say that *Hypothesis* $(\mathbf{H}Q_\infty)$ *holds* if the equivalent conditions of Proposition 2.2 are satisfied. Whenever we assume $(\mathbf{H}Q_\infty)$, this will be indicated by the notation '$(\mathbf{H}Q_\infty)$' in the statement of the results.

Hypothesis $(\mathbf{H}Q_\infty)$ trivially holds if $\mathbf{S}$ is *uniformly exponentially stable*, i.e. if there are constants $M \geq 1$ and $a > 0$ such that $\|S(t)\| \leq Me^{-at}$ for all $t \geq 0$. In this case Q_∞ is the *unique* positive symmetric solution of $AX + XA^* = -Q$. Indeed, by Proposition 2.3 for all $x^*, y^* \in E^*$ we have

$$\langle Q_\infty x^*, y^* \rangle = \lim_{t\to\infty} \langle Q_t x^*, y^* \rangle = \lim_{t\to\infty} \langle Xx^* - S(t)XS^*(t)x^*, y^* \rangle = \langle Xx^*, y^* \rangle.$$

The following result shows that conversely, Hypothesis $(\mathbf{H}Q_\infty)$ implies a stability property of the adjoint semigroup $\mathbf{S}^*$:

LEMMA 2.4. $(\mathbf{H}Q_\infty)$. *For all $x^* \in \mathcal{D}(A^*)$ we have*

$$\lim_{t\to\infty} \|i_Q^* S^*(t)x^*\|_{H_Q} = 0.$$

Proof. Let $x^* \in \mathcal{D}(A^*)$ be fixed. Put $y^* := (\lambda - A^*)x^*$, where λ is a sufficiently large fixed positive real number. Then by the Cauchy–Schwarz inequality,

$$\|i_Q^* S^*(t)x^*\|_{H_Q}^2 = \left\| \int_0^\infty e^{-\lambda s} i_Q^* S^*(t+s)y^* \, ds \right\|_{H_Q}^2 \leq \frac{1}{2\lambda} \int_0^\infty \|i_Q^* S^*(t+s)y^*\|_{H_Q}^2 \, ds.$$

As $t \to \infty$, the right hand side tends to 0. □

We shall use the notation $\sigma_p(B)$ for the point spectrum of the complexification $B_{\mathbb{C}}$ of a linear operator B.

PROPOSITION 2.5. $(\mathbf{H}Q_\infty)$. *For all $\lambda \in \mathbb{C}$ with $\operatorname{Re}\lambda \geq 0$ we have*

$$\ker(\lambda - A_{\mathbb{C}}^*) \subseteq \ker((Q_\infty)_{\mathbb{C}}).$$

In particular, if Q_∞ is injective then

$$\sigma_p(A^*) \cap \{\operatorname{Re} z \geq 0\} = \emptyset.$$

Proof. First note that since by assumption Hypothesis ($\mathbf{H}Q_\infty$) holds for $\mathbf{S}$, for all $\alpha \geq 0$ it also holds for the rescaled semigroup $\mathbf{S}_\alpha = \{e^{-\alpha t}S(t)\}_{t\geq 0}$. Indeed, this follows from

$$\int_0^\infty e^{-2\alpha t}\langle QS^*(t)x^*, S^*(t)x^*\rangle\, dt \leq \int_0^\infty \langle QS^*(t)x^*, S^*(t)x^*\rangle\, dt, \qquad x^* \in E^*,$$

and polarization. Therefore it is enough to check that $\ker(i\omega - A^*_{\mathbb{C}}) \subseteq \ker((Q_\infty)_{\mathbb{C}})$ for all $\omega \in \mathbb{R}$.

Let $x^*_{\mathbb{C}} \in \ker(i\omega - A^*_{\mathbb{C}})$ be fixed. By the spectral mapping theorem for the point spectrum [12, Theorem A.III.6.3] we have $S^*_{\mathbb{C}}(t)x^*_{\mathbb{C}} = e^{i\omega t}x^*_{\mathbb{C}}$ for all $t \geq 0$. Write $x^*_{\mathbb{C}} = y^* + iz^*$ with $y^*, z^* \in E^*$ and note that $y^*, z^* \in \mathcal{D}(A^*)$. By Lemma 2.4, applied to the elements $S^*(t)y^*, S^*(t)z^* \in \mathcal{D}(A^*)$,

$$\|(i_Q)^*_{\mathbb{C}} S^*_{\mathbb{C}}(t)x^*_{\mathbb{C}}\|_{(H_Q)_{\mathbb{C}}} = \lim_{s\to\infty} \|(i_Q)^*_{\mathbb{C}} S^*_{\mathbb{C}}(t+s)x^*_{\mathbb{C}}\|_{(H_Q)_{\mathbb{C}}} = 0.$$

It follows that $i^*_Q S^*(t)y^* = i^*_Q S^*(t)z^* = 0$ for all $t \geq 0$. By the identity

$$\int_0^\infty \|i^*_Q S^*(t)x^*\|^2_{H_Q}\, dt = \langle Q_\infty x^*, x^*\rangle, \qquad x^* \in E^*,$$

this implies

$$\langle Q_\infty y^*, y^*\rangle = \langle Q_\infty z^*, z^*\rangle = 0.$$

Since Q_∞ is positive and symmetric it follows that $Q_\infty y^* = Q_\infty z^* = 0$ and therefore $(Q_\infty)_{\mathbb{C}}\, x^*_{\mathbb{C}} = 0$.

□

THEOREM 2.6. ($\mathbf{H}Q_\infty$). *If Q_∞ is injective and $S(t)$ is compact for all $t > 0$, then $\mathbf{S}$ is uniformly exponentially stable.*

Proof. Let

$$E^\odot := \Big\{x^* \in E^* : \lim_{t\downarrow 0}\|S^*(t)x^* - x^*\| = 0\Big\}.$$

This is a closed, $\mathbf{S}^*$-invariant linear subspace of E^*. The restriction $\mathbf{S}^\odot$ of $\mathbf{S}^*$ to $E^\odot$ is strongly continuous, and for the spectrum of its generator $A^\odot$ we have $\sigma(A) = \sigma(A^\odot)$; cf. [13, Chapter 1].

The operators $S^\odot(t)$ being compact for all $t > 0$, by [12, Theorem A.II.1.25 and Proposition A.III.2.5] and the Riesz–Schauder theory of compact operators we have $\sigma(A^\odot) = \sigma_p(A^\odot)$; cf. [11, Theorem 5.14.2]. Hence,

$$\sigma(A) = \sigma(A^\odot) = \sigma_p(A^\odot) = \sigma_p(A^*) \subset \{\mathrm{Re}\, z < 0\}.$$

Moreover, $\mathbf{S}$ is uniformly continuous for $t > 0$ by [12, A.II.1.25]. Hence by [12, Theorem A.II.1.20], for all $\varepsilon > 0$ the set $\{\lambda \in \sigma(A) : \operatorname{Re}\lambda \geq -\varepsilon\}$ is compact. It follows that the spectral bound $s(A) = \sup\{\operatorname{Re}\lambda : \lambda \in \sigma(A)\}$ is strictly negative. The spectral mapping theorem for eventually uniformly continuous semigroups [12, Theorem A.III.6.6] now implies the desired result.

□

COROLLARY 2.7. ($\mathbf{H}Q_\infty$). *If Q_∞ is injective and $S(t)$ is compact for all $t > 0$, then Q_∞ is the unique positive symmetric solution of the equation $AX - XA^* = -Q$.*

The following example shows that the injectivity of Q_∞ cannot be omitted:

EXAMPLE 2.8. Let $E = \mathbb{R}^2$ and define $\mathbf{S}$ and Q by

$$S(t)(x,y) = (e^{-t}x, y), \quad t \geq 0,$$
$$Q(x,y) = (x,0).$$

Note that Q is positive and symmetric. Hypothesis ($\mathbf{H}Q_\infty$) holds and we have

$$Q_\infty(x,y) = (-\tfrac{1}{2}x, 0).$$

For $c \geq 0$ define positive symmetric operators X_c by

$$X_c(x,y) := (\tfrac{1}{2}x, cy).$$

Since $A(x,y) = A^*(x,y) = (-x,0)$ we have

$$AX_c(x,y) + X_cA^*(x,y) = (-\tfrac{1}{2}x, 0) + (-\tfrac{1}{2}x, 0) = -(x,0) = -Q(x,y),$$

and therefore each X_c is a solution of $AX + XA^* = -Q$.

By [14, Proposition 5.1.15], for *bounded* semigroups we have $\sigma_p(A) \cap i\mathbb{R} \subseteq \sigma_p(A^*) \cap i\mathbb{R}$. For later use we need a slightly sharpened version of this result which we state next.

PROPOSITION 2.9. *If there exists a weak*-dense linear subspace Y of E^* such that $t \mapsto S^*(t)x^*$ is bounded for all $x^* \in Y$, then $\sigma_p(A) \cap i\mathbb{R} \subseteq \sigma_p(A^*) \cap i\mathbb{R}$.*

Proof. Let $i\omega \in \sigma_p(A) \cap i\mathbb{R}$. By rescaling we may assume that $\omega = 0$. Let $x_0 \in D(A)$ be an eigenvector with eigenvalue 0. Then $S(t)x_0 = x_0$ for all $t \geq 0$. Here and in the rest of the paper, for the sake of notational simplicity we will suppress subscripts indicating that we are using complexifications.

Since Y is weak*-dense in E^* we may choose $y_0^* \in Y$ such that $\langle x_0, y_0^* \rangle = 1$. Then $t \mapsto \langle S(t)x, y_0^* \rangle$ defines an element of $C_b[0,\infty)$, the space of all bounded continuous functions on $[0,\infty)$. Let $\phi \in (C_b[0,\infty))^*$ be a left invariant mean and define $x_0^* \in X^*$ by

$$\langle x, x_0^* \rangle := \phi(\langle S(\cdot)x, y_0^* \rangle), \quad x \in X.$$

For all $x \in X$ we have

$$\langle x, S^*(t)x_0^* \rangle = \phi(\langle S(t+\cdot)x, y_0^* \rangle) = \phi(\langle S(\cdot)x, y_0^* \rangle) = \langle x, x_0^* \rangle.$$

Hence, $S^*(t)x_0^* = x_0^*$ for all $t \geq 0$, so $x_0^* \in D(A^*)$ and $A^*x_0^* = 0$. Also,

$$\langle x_0, x_0^* \rangle = \phi(\langle S(\cdot)x_0, y_0^* \rangle) = \phi(\langle x_0, y_0^* \rangle \mathbf{1}) = \phi(\mathbf{1}) = 1,$$

which shows that $x_0^* \neq 0$. This proves that x_0^* is an eigenvector of A^* with eigenvalue 0.

□

Combining Propositions 2.5 and 2.9 we have proved:

THEOREM 2.10. ($\mathbf{H}Q_\infty$). *If Q_∞ is injective and the linear subspace of all $x^* \in E^*$ for which $t \mapsto S^*(t)x^*$ is bounded is weak*-dense in E^*, then $\sigma_p(A) \cap i\mathbb{R} = \emptyset$.*

Recall that if $S(t)$ is compact for all $t > 0$, then the resolvent operator $(\lambda - A)^{-1}$ is compact for all $\lambda \in \varrho(A)$. In the following result we prove uniqueness of positive symmetric solutions under a weaker compactness assumption. The price to pay is a mild boundedness condition.

Let

$$\mathcal{E} := \bigcap_{n\geq 1} \mathcal{D}\big(A^{*n}\big)$$

denote the set of *entire vectors* for A^*. As is well known, $\mathcal{E}$ is weak*-dense in E^*.

THEOREM 2.11. *Assume that Q_∞ is injective and that for all $x^* \in \mathcal{E}$ the orbit $t \mapsto S^*(t)x^*$ is bounded. Let R_1 and R_2 be positive symmetric solutions of the equation*

$$AX - XA^* = -Q.$$

If the operator $(\lambda - A)^{-1}(R_1 - R_2)$ is compact for some $\lambda \in \varrho(A)$, then $R_1 = R_2$.

Proof. Step 1. Fix $x^* \in \mathcal{E}$. For all $y^* \in E^*$ we have

$$\begin{aligned}|\langle Q_\infty y^*, S^*(t)x^*\rangle| &= \left|\int_0^\infty [i_Q^* S^*(s)y^*, i_Q^* S^*(t+s)x^*]_{H_Q}\, ds\right| \\ &\le \left(\int_0^\infty \|i_Q^* S^*(s)y^*\|_{H_Q}^2\, ds\right)^{1/2} \left(\int_t^\infty \|i_Q^* S^*(s)x^*\|_{H_Q}^2\, ds\right)^{1/2}\end{aligned}$$

It follows that

$$\lim_{t\to\infty} \langle Q_\infty y^*, S^*(t)x^*\rangle = 0, \qquad y^* \in E^*.$$

Since Q_∞ is injective and positive symmetric, its range is dense in E. The boundedness of $t \mapsto S^*(t)x^*$ therefore implies that

$$\lim_{t\to\infty} \langle x, S^*(t)x^*\rangle = 0, \qquad x \in E.$$

Step 2. It follows from Proposition 2.3 that

$$R_1 - R_2 = S(t)(R_1 - R_2)S^*(t), \qquad t \ge 0.$$

For $x^* \in \mathcal{E}$ define

$$V_{x^*} := \Big\{(R_1 - R_2)S^*(t)x^* : \ t \ge 0\Big\}.$$

We will show that V_{x^*} is relatively compact in E.

Let $y^* := (\lambda - A^*)x^*$. Then $y^* \in \mathcal{E}$, so $t \mapsto S^*(t)y^*$ is bounded. It follows that the set

$$W_{y^*} := \Big\{(R_1^* - R_2^*)(\lambda - A^*)^{-1}S^*(t)y^* : t \ge 0\Big\}$$

is relatively compact in E. But R_1^* and R_2^* map E^* into E, and as operators from E^* into E we have $R_1^* = R_1$ and $R_2^* = R_2$. Hence,

$$(R_1 - R_2)S^*(t)x^* = (R_1^* - R_2^*)(\lambda - A^*)^{-1}S^*(t)y^*,$$

and $V_{x^*} = W_{y^*}$ is relatively compact in E.

Step 3. For all $x^* \in \mathcal{E}$ and $y^* \in \mathcal{E}$ we have

$$\langle (R_1 - R_2)x^*, y^*\rangle = \langle (R_1 - R_2)S^*(t)x^*, S^*(t)y^*\rangle.$$

By Step 1, $\lim_{t\to\infty} S^*(t)y^* = 0$ weak*. By the boundedness of $t \mapsto S^*(t)y^*$, the weak*-convergence is uniform on compact subsets of E. In particular, it is uniform on the relatively compact set V_{x^*}. Hence by Step 2,

$$\lim_{t\to\infty} \langle (R_1 - R_2)S^*(t)x^*, S^*(t)y^*\rangle = 0.$$

This shows that $\langle (R_1 - R_2)x^*, y^* \rangle = 0$ for all $x^* \in \mathcal{E}$ and $y^* \in \mathcal{E}$.

Since $\mathcal{E}$ is weak*-dense in E^*, it follows that $(R_1 - R_2)x^* = 0$ for all $x^* \in \mathcal{E}$. But the operator $R_1 - R_2$ is weak*-to-weakly continuous, being the difference of two positive symmetric operators, and therefore $R_1 - R_2 = 0$. □

The following example shows that without any compactness assumption, non-uniqueness may occur. At the same time it shows that Hypothesis $(\mathbf{H}Q_\infty)$ need not imply any strong stability of the orbits of $\mathbf{S}^*$ itself; compare with Lemma 2.4.

EXAMPLE 2.12. Let $E = L^2[0,\infty)$ and let $\mathbf{S}$ be the left shift semigroup on E:

$$S(t)g(s) := g(s+t), \qquad s,t \geq 0.$$

This is a strongly stable C_0-semigroup on E with generator A given by $Af = f'$ for $f \in \mathcal{D}(A) = W^{1,2}[0,\infty)$. Its adjoint is the right shift semigroup,

$$S^*(t)f(s) = \begin{cases} 0, & 0 \leq s < t; \\ f(s-t), & s \geq t. \end{cases}$$

This is a semigroup of partial isometries and for $f \in \mathcal{D}(A^*) = W_0^{1,2}[0,\infty)$ we have $A^*f = -f'$.

Let ψ be a nonnegative measurable function on $[0,\infty)$. Let $Q \in \mathcal{L}(E)$ be the operator of multiplication with ψ. Then Q is positive and selfadjoint, hence positive and symmetric in the sense defined at the beginning of this section. Defining

$$\psi_t(s) := \int_0^t \psi(\tau + s)\, d\tau,$$

an easy calculation shows that

$$Q_t f = \int_0^t S(\tau)QS^*(\tau)f\, d\tau = \psi_t f.$$

Hence $(\mathbf{H}Q_\infty)$ holds if and only if the function $\psi_\infty \colon [0,\infty) \to [0,\infty]$ defined by

$$\psi_\infty(s) := \int_0^\infty \psi(\tau + s)\, d\tau$$

belongs to $L^\infty[0,\infty)$. Since ψ is nonnegative, ψ_∞ is nonincreasing, and therefore $\psi_\infty \in L^\infty[0,\infty)$ if and only if

$$\int_0^\infty \psi(\tau)\, d\tau < \infty,$$

i.e., if and only if $\psi \in L^1[0,\infty)$.

Let us from now on assume that $\psi \in L^1[0,\infty)$, and let ψ_∞ be defined as above. Then $Q_\infty f = \psi_\infty f$. Since ψ_∞ is nonincreasing, Q_∞ is injective if and only if ψ_∞ is strictly positive, which happens if and only if ψ is not compactly supported.

For $c \geq 0$ let us define

$$X_c f := (Q_\infty + c)f = (\psi_\infty + c)f, \qquad f \in L^2[0,\infty).$$

This is a positive selfadjoint operator on $L^2[0,\infty)$. For all $f \in \mathcal{D}(A^*)$ we have $Q_\infty f \in \mathcal{D}(A)$ (because Q_∞ is a solution of $AX + XA^* = -Q$) and $cf \in \mathcal{D}(A)$ (because $\mathcal{D}(A^*) \subseteq \mathcal{D}(A)$). Therefore $X_c f \in \mathcal{D}(A)$ for all $f \in \mathcal{D}(A^*)$ and

$$\begin{aligned} AX_c f + X_c A^* f &= \Big((\psi_\infty + c)f\Big)' - (\psi_\infty + c)f' = \Big(\psi_\infty f\Big)' - \psi_\infty f' \\ &= AQ_\infty f + Q_\infty A^* f = -Qf. \end{aligned}$$

Thus for all $c \geq 0$ the operator X_c is a positive selfadjoint solution of $AX + XA^* = -Q$.

Theorem 2.11 applies in particular to semigroups with compact resolvent. Compact semigroups always have a compact resolvent, but for such semigroups we can apply Corollary 2.7. On the other hand, *eventually* compact semigroups need not have a compact resolvent; cf. [12, Remark A.II.1.26]. The following result gives a sufficient condition implying uniqueness for eventually compact semigroups.

Corollary 2.13. $(\mathbf{H}Q_\infty)$. *Assume that Q_∞ is injective and that for all $x^* \in \mathcal{E}$ the orbit $t \mapsto S^*(t)x^*$ is bounded. Let R_1 and R_2 be positive symmetric solutions of the equation*

$$AX - XA^* = -Q.$$

If $\mathbf{S}$ is eventually compact, then $R_1 = R_2$.

Proof. For all $t > 0$ we have $(R_1 - R_2) - S(t)(R_1 - R_2)S^*(t) = 0$. Taking t sufficiently large, it follows that $R_1 - R_2$ is compact. We may now apply Theorem 2.11.

□

We conclude this section with an example where $(\mathbf{H}Q_\infty)$ is satisfied with Q_∞ injective, even though $\mathbf{S}$ is unbounded.

EXAMPLE 2.14. First let $\mathbf{S}$ be an arbitrary C_0-semigroup on E. Choose a fixed $x_0 \in E$ and let

$$Q := x_0 \otimes x_0;$$

cf. [7, Example 11.9]. The operator Q is positive and symmetric, and

$$Q_t x^* = \int_0^t \langle S(s)x_0, x^*\rangle S(s)x_0 \, ds, \qquad t > 0, \ x^* \in E^*.$$

In particular,

$$\langle Q_t x^*, x^*\rangle = \int_0^t \langle S(s)x_0, x^*\rangle^2 \, ds,$$

and therefore $(\mathbf{H}Q_\infty)$ holds if and only if

$$\int_0^\infty \langle S(s)x_0, x^*\rangle^2 \, ds < \infty, \qquad x^* \in E^*.$$

Let E_0 denote the closed linear span in E of the orbit of x_0. Clearly, $Q_\infty x^* \in E_0$ for all $x^* \in E^*$. Hence $E_0 = E$ if Q_∞ has dense range. If Q_∞ fails to have dense range, then by the Hahn–Banach theorem there exists a non-zero $x^* \in E^*$ such that

$$\langle Q_\infty x^*, x^*\rangle = \int_0^\infty \langle S(s)x_0, x^*\rangle^2 \, ds = 0.$$

Hence $\langle S(s)x_0, x^*\rangle = 0$ for almost all $s \geq 0$, and then by continuity $\langle S(s)x_0, x^*\rangle = 0$ for all $s \geq 0$. It follows that x^* annihilates E_0, showing that $E_0 \neq E$. On the other hand, since Q_∞ is positive and symmetric, Q_∞ is injective if and only if Q_∞ has dense range. Summarizing, Q_∞ is injective if and only if the orbit of x_0 spans a dense linear subspace of E.

We will now give an example of an unbounded semigroup $\mathbf{S}$ with a square integrable orbit whose span is dense. It is a minor modification of [2, Example 3.4].

Take $E = C_0[0,\infty)$, the space of continuous functions on $[0,\infty)$ vanishing at infinity. Define

$$S(t)f(s) := e^t f(t+s), \qquad t, s \geq 0, \quad f \in C_0[0,\infty).$$

Let $\{f^{(k)} : k = 1, 2, \ldots\}$ be a dense subset of the unit ball of

$$C_{00}[0,\infty) := \{g \in C_0[0,\infty) : \ g(0) = 0\}$$

such that each $f^{(k)}$ has support in $[0, 2^k]$, and let

$$f(t) = \exp\left(-(2^{n+1} + 2^{k+1})^2\right) f^{(k)}(t - 2^{n+1} - 2^k),$$
$$2^{n+1} + 2^k \leq t < 2^{n+1} + 2^{k+1}; \quad k = 1, \ldots, n; \ n = 1, 2, \ldots,$$

with $f(t) = 0$ for all remaining $t \geq 0$. The construction of f implies that closed linear span of the translates of f contains each $f^{(k)}$. It follows that $C_{00}[0,\infty)$ is contained in the closed linear span of the translates of f, and the same is true for the the closed linear span of the orbit of f. Noting that $C_{00}[0,\infty)$ has codimension one in $C_0[0,\infty)$ and that the orbit of f also contains elements that do not vanish at 0, it follows that the closed linear span of the orbit of f is all of $C_0[0,\infty)$.

Finally, from

$$|f(t)| \leq \exp\left(-(2^{n+1}+2^{k+1})^2\right), \qquad 2^{n+1}+2^k \leq t < 2^{n+1}+2^{k+1},$$

it follows that $|f(t)| \leq e^{-t^2}$ for all $t \geq 0$. Therefore,

$$\int_0^\infty \|S(t)f\|^2\,dt = \int_0^\infty (e^t f(t))^2\,dt \leq \int_0^\infty e^{2t-2t^2}\,dt < \infty.$$

3. Cylindrical invariant measures

In what follows, we will assume the reader to be familiar with elements of the theory of cylindrical Gaussian measures. For more information and an explanation of the terminology we are using we refer to [17]. We just recall the following elementary facts.

An operator $R \in \mathcal{L}(E^*, E^{**})$ is called *positive* if $\langle Rx^*, x^*\rangle \geq 0$ for all $x^* \in E^*$ and *symmetric* if $\langle Rx^*, y^*\rangle = \langle Ry^*, x^*\rangle$ for all $x^*, y^* \in E^*$. The Fourier transform establishes a one-to-one correspondence between centred cylindrical Gaussian measures μ on E and positive symmetric operators R from $\mathcal{L}(E^*, E^{**})$ via the formula

$$\widehat{\mu}(x^*) := \int_E \exp(i\langle x, x^*\rangle)\,d\mu(x) = \exp\left(-\frac{1}{2}\langle Rx^*, x^*\rangle\right), \qquad x^* \in E^*.$$

Let $m \in E$. We say that μ is a *cylindrical Gaussian measure with mean* m if there exists a positive symmetric $R \in \mathcal{L}(E^*, E^{**})$ such that the Fourier transform of μ is given by

$$\widehat{\mu}(x^*) = \exp\left(i\langle m, x^*\rangle - \frac{1}{2}\langle Rx^*, x^*\rangle\right), \qquad x^* \in E^*.$$

Now let us return to the setting we have been considering so far. For each $t > 0$, the operator $Q_t \in \mathcal{L}(E^*, E)$ is positive and symmetric, and as such it is the covariance of a unique centred cylindical Gaussian measure on E, which we denote by μ_t. Similarly, the operator $Q_\infty \in \mathcal{L}(E^*, E)$, if it exists, is the covariance of a unique centred cylindical Gaussian measure μ_∞ on E. The following result describes the relationship between the μ_t and μ_∞:

PROPOSITION 3.1. *The following assertions are equivalent:*

1. *There exists a centred cylindrical Gaussian measure μ_∞ on E such that for all bounded continuous cylindrical functions $f: E \to \mathbb{R}$ we have*

$$\lim_{t\to\infty} \int_E f(x)\, d\mu_t(x) = \int_E f(x)\, d\mu_\infty(x);$$

2. *Hypothesis* $(\mathbf{H}Q_\infty)$ *holds.*

In this situation, Q_∞ is the covariance of μ_∞.

Proof. 1.$\Rightarrow$2.: Let $Q_\infty \in \mathcal{L}(E^*, E^{**})$ denote the covariance operator of μ_∞. By considering the real- and imaginary parts of the function $x \mapsto \exp(i\langle x, x^*\rangle)$ separately, it follows from 1. that

$$\begin{aligned}\lim_{t\to\infty} \exp\left(-\frac{1}{2}\langle Q_t x^*, x^*\rangle\right) &= \lim_{t\to\infty} \int_E \exp(i\langle x, x^*\rangle)\, d\mu_t(x) \\ &= \int_E \exp(i\langle x, x^*\rangle)\, d\mu_\infty(x) = \exp\left(-\frac{1}{2}\langle Q_\infty x^*, x^*\rangle\right).\end{aligned}$$

Hence,

$$\lim_{t\to\infty} \langle Q_t x^*, x^*\rangle = \langle Q_\infty x^*, x^*\rangle, \qquad x^* \in E^*.$$

By polarization, it follows that

$$\lim_{t\to\infty} \langle Q_t x^*, y^*\rangle = \langle Q_\infty x^*, y^*\rangle, \qquad x^*, y^* \in E^*.$$

By [5, Proposition 2.2], this implies that Q_∞ actually takes values in E.

2.$\Rightarrow$1.: Let μ_∞ be the unique centred cylindrical Gaussian measure on E whose covariance operator is Q_∞.

Fix $x^* \in E^*$. Since $\lim_{t\to\infty}\langle Q_t x^*, x^*\rangle = \langle Q_\infty x^*, x^*\rangle$, for the Fourier transforms of μ_t and μ_∞ we have

$$\lim_{t\to\infty} \widehat{\mu_t}(x^*) = \widehat{\mu_\infty}(x^*).$$

By Levi's theorem [16, Corollary 2.8], this implies 1. $\square$

If $f: E \to \mathbb{R}$ is a bounded continuous cylindrical function, then the function $P(t)f: E \to \mathbb{R}$ defined by

$$P(t)f(x) := \int_E f(S(t)x + y)\, d\mu_t(y)$$

is a bounded continuous cylindrical function as well. Hence it makes sense to call a cylindrical measure ν on E $P(t)$*-invariant* if for all bounded continuous cylindrical $f: E \to \mathbb{R}$ we have

$$\int_E P(t)f(x)\,d\nu(x) = \int_E f(x)\,d\nu(x). \tag{3.1}$$

We say that ν is $\mathbf{P}$-invariant, or briefly *invariant*, if ν is $P(t)$-invariant for all $t > 0$. We have the following simple observation:

PROPOSITION 3.2. ($\mathbf{H}Q_\infty$). *The centred cylindrical Gaussian measure μ_∞, whose covariance operator is Q_∞, is invariant.*

Proof. The algebraic identity $Q_\infty = Q_t + S(t)Q_\infty S^*(t)$ implies $\mu_\infty = \mu_t * S(t)\mu_\infty$; here $S(t)\mu_\infty$ denotes the image cylindrical measure of μ_∞ under the operator $S(t)$.

Now let $f: E \to \mathbb{R}$ be a bounded continuous cylindrical function. Then,

$$\begin{aligned}
\int_E P(t)f(x)\,d\mu_\infty(x) &= \int_E \int_E f(S(t)x + y)\,d\mu_t(y)\,d\mu_\infty(x) \\
&= \int_E \int_E f(\xi + y)\,d\mu_t(y)\,d(S(t)\mu_\infty)(\xi) \\
&= \int_E f(\eta)\,d(\mu_t * S(t)\mu_\infty)(\eta) \\
&= \int_E f(\eta)\,d\mu(\eta).
\end{aligned}$$

□

In the converse direction we have the following result:

THEOREM 3.3. *Let ν be a cylindrical Gaussian measure on E with covariance operator $Q_\nu \in \mathcal{L}(E^*, E)$ and mean $m \in E$. If ν is invariant, then Q_ν is a solution of the equation $AX - XA^* = -Q$ and its mean satisfies $S(t)m = m$ for all $t \geq 0$.*

Proof. Applying (3.1) to the real- and imaginary parts of the functions $\exp(i\langle x, x^*\rangle)$, it follows that the Fourier transform of ν satisfies

$$\widehat{\nu}(x^*) = \widehat{\nu}(S^*(t)x^*) \exp\left(-\frac{1}{2}\langle Q_t x^*, x^*\rangle\right), \qquad t > 0,\ x^* \in E^*;$$

cf. [7, p. 307]. Because of $\widehat{\nu}(x^*) = \exp(i\langle m, x^*\rangle) - \frac{1}{2}\langle Q_\nu x^*, x^*\rangle)$, for $t \ge 0$ this implies

$$\langle m, x^*\rangle = \langle m, S^*(t)x^*\rangle, \qquad x^* \in E^*,$$

and

$$\langle Q_\nu x^*, x^*\rangle = \langle Q_\nu S^*(t)x^*, S^*(t)x^*\rangle + \langle Q_t x^*, x^*\rangle, \qquad x^* \in E^*.$$

The first of these identities implies that $S(t)m = m$ for all $t \ge 0$. By polarization, the second gives

$$\langle Q_\nu x^*, y^*\rangle = \langle Q_\nu S^*(t)x^*, S^*(t)y^*\rangle + \langle Q_t x^*, y^*\rangle, \qquad x^*, y^* \in E^*.$$

Now take $x^*, y^* \in \mathcal{D}(A^*)$, differentiate both sides with respect to t and evaluate at $t = 0$. It follows that

$$\langle Q_\nu x^*, A^*y^*\rangle + \langle Q_\nu A^*x^*, y^*\rangle = -\langle Qx^*, y^*\rangle.$$

This shows that $Q_\nu x^* \in \mathcal{D}(A)$ for all $x^* \in \mathcal{D}(A^*)$ and that Q_ν is a solution of $AX + XA^* = -Q$.

□

From Theorem 2.10 and the fact that the subspace $\mathcal{E}$ of entire vectors for A^* is weak*-dense in E^* we infer:

COROLLARY 3.4. ($\mathbf{H}Q_\infty$). *Assume that Q_∞ is injective and $t \mapsto S^*(t)x^*$ is bounded for all $x^* \in \mathcal{E}$. If ν is an invariant cylindrical Gaussian measure with covariance $Q_\nu \in \mathcal{L}(E^*, E)$ and mean $m \in E$, then $m = 0$, i.e. ν is centred.*

We say that a cylindrical Gaussian measure γ is $\mathbf{S}$-*invariant* if $S(t)\gamma = \gamma$ for all $t \ge 0$; here $S(t)\gamma$ denotes the image cylindrical measure of γ under the linear transformation $S(t)$. We now have the following representation theorem extending [7, Theorem 11.7], [8, Proposition 7.2.5].

PROPOSITION 3.5. *Let ν be an invariant cylindrical Gaussian measure on E with covariance operator $Q_\nu \in \mathcal{L}(E^*, E)$ and mean m. Then ($\mathbf{H}Q_\infty$) holds and there exists a unique cylindrical Gaussian measure γ such that*

$$\nu = \mu_\infty * \gamma,$$

where μ_∞ is the cylindrical Gaussian measure on E whose covariance operator is Q_∞. Moreover, γ is $\mathbf{S}$-invariant and has mean m.

Proof. Let $R := Q_\nu - Q_\infty$. By Theorem 3.3, the positive symmetric operator Q_ν solves the equation $AX - XA^* = -Q$. Hence,

$$\langle Q_\nu x^*, x^*\rangle \geq \langle Q_\nu x^*, x^*\rangle - \langle Q_\nu S^*(t)x^*, S^*(t)x^*\rangle = \langle Q_t x^*, x^*\rangle.$$

It follows that

$$\langle Rx^*, x^*\rangle = \lim_{t\to\infty}\Big(\langle Q_\nu x^*, x^*\rangle - \langle Q_t x^*, x^*\rangle\Big) \geq 0, \qquad x^* \in E^*.$$

Since R is also symmetric, there exists a unique cylindrical Gaussian measure γ whose covariance operator is R and whose mean is m. From

$$\begin{aligned}(\widehat{\mu_\infty * \gamma})(x^*) &= \exp(-\tfrac{1}{2}\langle Q_\infty x^*, x^*\rangle)\exp(i\langle m, x^*\rangle - \tfrac{1}{2}\langle Rx^*, x^*\rangle)\\ &= \exp(i\langle m, x^*\rangle - \tfrac{1}{2}\langle Q_\nu x^*, x^*\rangle) = \widehat{\nu}(x^*), \qquad x^* \in E^*,\end{aligned}$$

it follows that $\nu = \mu_\infty * \gamma$. For all $t \geq 0$ we have

$$S^*(t)RS(t) = S^*(t)(Q_\nu - Q_\infty)S(t) = Q_\nu - Q_\infty = R,$$

because both Q_ν and Q_∞ solve $AX + XA^* = -Q$. Since $S^*(t)RS(t)$ is the covariance of $S(t)\gamma$, and $S(t)m = m$ is the mean of $S(t)\gamma$, this proves the **S**-invariance of γ.

Finally, if γ' is a cylindrical Gaussian measure for which $\nu = \mu_\infty * \gamma'$, then by taking Fourier transforms we see that its covariance operator R' is equal to $Q_\nu - Q_\infty = R$ and that it has a mean equal to m. It follows that $\gamma' = \gamma$. □

We conclude with the following converse of Proposition 3.5; the proof is similar and is left to the reader.

PROPOSITION 3.6. ($\mathbf{H}Q_\infty$). *If γ is an* **S***-invariant cylindrical Gaussian measure with covariance operator $R \in \mathcal{L}(E^*, E)$ and mean $m \in E$, then $\mu_\infty * \gamma$ is invariant and has mean m.*

4. Uniqueness of invariant measures

As before let A be the generator of a C_0-semigroup **S** on a separable real Banach space E. Let H be a separable real Hilbert space and let $B \in \mathcal{L}(H, E)$ be bounded.

In this section we will apply our results to obtain uniqueness results for invariant measures of the stochastic abstract Cauchy problem (1.1),

$$\begin{aligned} du(t) &= Au(t)\,dt + B\,dW_H(t), \quad t \geq 0, \\ u(0) &= x, \end{aligned}$$

where $\{W_H(t)\}_{t\geq 0}$ is a cylindrical Wiener process with Cameron–Martin space H. For a precise meaning of these concepts we refer to [5].

The operator $Q := B \circ B^* \in \mathcal{L}(E^*, E)$ is positive and symmetric. As a subset of E, the reproducing kernel Hilbert space H_Q associated with Q can be identified with range B; the inner product in H_Q is given by

$$[Bg, Bh]_{H_Q} = [Pg, Ph]_H, \qquad g, h \in H,$$

where P is the orthogonal projection in H onto $(\ker B)^\perp$. We define the operators $Q_t \in \mathcal{L}(E^*, E)$ as before:

$$Q_t x^* := \int_0^t S(s) Q S^*(s) x^*\, ds, \qquad x^* \in E^*.$$

We have the following existence and uniqueness result [5]:

PROPOSITION 4.1. *The problem (1.1) has a weak solution $\{u(t,x)\}_{t\geq 0}$ if and only if for all $t > 0$ the operator Q_t is the covariance operator of a centred Gaussian Borel measure μ_t on E. In this case the solution is unique and μ_t is the distribution of the random variable $u(t,0)$. Moreover, $\{u(t,x)\}_{t\geq 0}$ is a Markov process and the semigroup $\mathbf{P}$ is its transition semigroup:*

$$P(t)f(x) = \mathbb{E}\,(f(u(t,x))), \qquad t \geq 0, \quad x \in E,$$

for all bounded Borel functions $f: E \to \mathbb{R}$. If $(\mathbf{H}Q_\infty)$ holds and Q_∞ is the covariance operator of a centred Gaussian Borel measure μ_∞ on E, then μ_∞ is an invariant measure.

In what follows we will consider the following Hypotheses:

- $(\mathbf{H}\mu_t)$: For all $t > 0$, the operator Q_t is the covariance of a centred Gaussian Borel measure μ_t on E.
- $(\mathbf{H}\mu_\infty)$: Hypothesis $(\mathbf{H}Q_\infty)$ holds, and the operator Q_∞ is the covariance of a centred Gaussian Borel measure μ_∞ on E.

Trivially, $(\mathbf{H}\mu_\infty)$ implies $(\mathbf{H}Q_\infty)$, and by standard tightness arguments $(\mathbf{H}\mu_\infty)$ implies $(\mathbf{H}\mu_t)$. Conversely, if $(\mathbf{H}\mu_t)$ and $(\mathbf{H}Q_\infty)$ hold, it easily follows from Levi's continuity theorem that the centred cylindrical Gaussian measure μ_∞ whose covariance operator is Q_∞ extends uniquely to a Gaussian Borel measure on E if and only if the family of measures $\{\mu_t : t \geq 1\}$ is tight. In this situation, $\lim_{t\to\infty} \mu_t = \mu_\infty$ weakly.

The measure μ_∞, if it exists, is invariant.

THEOREM 4.2. ($\mathbf{H}\mu_\infty$). *If* $\mathbf{S}$ *is strongly stable, then* μ_∞ *is the unique invariant Gaussian Borel measure on* E.

Proof. First we recall that Gaussian measures have a mean and that their covariance operator is compact; this follows from [3, Theorem 3.2.3 and Corollary 3.2.4] and (2.2).

Let ν be in invariant Gaussian measure. By Theorem 3.3 and the strong stability of $\mathbf{S}$, ν is centred. Denote by Q_ν the covariance operator of ν. From

$$Q_\nu - S(t)Q_\nu S^*(t) = Q_t, \qquad t > 0,$$

and the compactness of Q_ν we obtain

$$\langle Q_\nu x^*, y^*\rangle = \lim_{t\to\infty} \langle S(t)Q_\nu S^*(t)x^* + Q_t x^*, y^*\rangle = \lim_{t\to\infty} \langle Q_t x^*, y^*\rangle = \langle Q_\infty x^*, y^*\rangle;$$

we used here that the strong stability of $\mathbf{S}$ is uniform on compacts.

□

We call μ_∞ *nondegenerate* if $\mu_\infty(O) \neq 0$ for every open subset $O \subseteq E$. As is well known [3, Theorem 3.6.1], this happens if and only if the covariance operator Q_∞ of μ_∞ is injective.

The next result is an immediate consequence of Corollary 2.7, Theorem 3.3 and Corollary 3.4.

THEOREM 4.3. ($\mathbf{H}\mu_\infty$). *If* μ_∞ *is nondegenerate and* $\mathbf{S}$ *is compact, then* $\mathbf{S}$ *is uniformly exponentially stable and* μ_∞ *is the unique invariant Gaussian Borel measure on* E.

Example 2.8 shows that the injectivity of Q_∞ cannot be omitted.

From Theorem 2.11, Corollary 3.4 and the fact that covariance operators of Gaussian measures are compact, we obtain the following uniqueness result:

Theorem 4.4. ($\mathbf{H}\mu_\infty$). *If μ_∞ is non-degenerate and $t \mapsto S^*(t)x^*$ is bounded for all entire vectors $x^* \in \mathcal{E}$, then μ_∞ is the unique invariant Gaussian Borel measure on E.*

This result is applicable in the following situation:

Corollary 4.5. ($\mathbf{H}\mu_\infty$). *If $\mathcal{D}(A^n) \subseteq \operatorname{range} B$ for some $n \geq 1$, then μ_∞ is the unique invariant Gaussian Borel measure on E.*

Proof. By the strong continuity of $\mathbf{S}$ and density of $\mathcal{D}(A^n)$ in E, range $B = H_Q$ is dense in E. It follows that Q is injective, and this easily implies that Q_∞ is injective; cf. [9]. Hence, μ_∞ is nondegenerate.

Fix $\lambda \in \varrho(A)$ and define $T \in \mathcal{L}(E, H_Q)$ by

$$Tx := (\lambda - A)^{-n}x \in \mathcal{D}(A^n) \subseteq \operatorname{range} B = H_Q.$$

Fix $x^* \in \mathcal{E}$ and put $y^* := (\lambda - A^*)^n x^*$. Noting that $y^* \in \mathcal{D}(A^*)$, it follows from Lemma 2.4 that

$$\lim_{t\to\infty} \langle S^*(t)x^*, x\rangle = \lim_{t\to\infty} \langle S^*(t)y^*, (\lambda - A)^{-n}x\rangle = \lim_{t\to\infty} [i_Q^* S^*(t)y^*, Tx]_{H_Q} = 0$$

for all $x \in E$. Hence by the uniform boundedness theorem, the orbit $t \mapsto S^*(t)x^*$ is bounded. Now apply Theorem 4.4. □

For reasons of completeness we state the following σ-additive analogue of Proposition 3.5 and its converse:

Proposition 4.6. *Let ν be an invariant Gaussian Borel measure on E. Then ($\mathbf{H}\mu_\infty$) holds and there exists a unique Gaussian Borel measure γ such that*

$$\nu = \mu_\infty * \gamma.$$

Moreover, γ is $\mathbf{S}$-invariant and has the same mean as ν. Conversely, if ($\mathbf{H}\mu_\infty$) holds and γ is an $\mathbf{S}$-invariant Gaussian Borel measure, then $\mu_\infty * \gamma$ is invariant.

Proof. Let γ be the unique invariant cylindrical Gaussian measure from Proposition 3.5, and let R be its covariance operator. Let Q_ν be the covariance operator of ν. From

$$0 \le \langle Rx^*, x^* \rangle = \langle (Q_\nu - Q_\infty)x^*, x^* \rangle \le \langle Q_\nu x^*, x^* \rangle, \qquad x^* \in E^*$$

and the fact that Q_ν is the covariance of a Gaussian Borel measure, it follows that R is the covariance of a Gaussian Borel measure; cf. [17, Proposition VI.3.4]. This measure is the unique extension of γ to a Gaussian Borel measure on E. The converse follows immediately from Proposition 3.5.

□

5. Null controllability

Let A be the generator of a C_0-semigroup $\mathbf{S}$ on a separable real Banach space E, let H be a separable real Hilbert space and let $B \in \mathcal{L}(H, E)$ be a bounded linear operator. We consider the inhomogeneous Cauchy problem

$$(5.1) \qquad \begin{aligned} u'(t) &= Au(t) + Bf(t), \qquad 0 \le t \le T, \\ u(0) &= x, \end{aligned}$$

where $x \in E$ and $f \in L^2([0,T]; H)$. The unique mild solution of this problem is given by

$$u(t, x; f) := S(t)x + \int_0^t S(t-s)Bf(s)\, ds, \qquad 0 \le t \le T.$$

We say that $\mathbf{S}$ is *null controllable* with respect to the *control pair* (B, H) if for all $x \in E$ and $T > 0$ there exists a function $f \in L^2([0,T]; H)$ such that

$$u(T, x; f) = 0.$$

As before we let $Q = B \circ B^*$. For $0 < t \le \infty$ let H_t denote the reproducing kernel Hilbert space associated with the positive symmetric operator Q_t; thus, H_t is Hilbert space completion of the range of Q_t with respect to the inner product $[Q_t x^*, Q_t y^*]_{H_t} := \langle Q_t x^*, y^* \rangle$. The inclusion mapping from H_t into E will be denoted by i_t. Whenever it is convenient, we will identify H_t with its image $i_t(H_t)$ in E.

We have the following characterization of null controllability in terms of the spaces H_t:

PROPOSITION 5.1. **S** *is null controllable with respect to the control pair* (B, H) *if and only if*

$$S(t)E \subseteq H_t, \qquad t > 0.$$

This is an immediate consequence of the following simple lemma; cf. [21], [7, Appendix B]. For the convenience of the reader we include the proof.

LEMMA 5.2. *For all* $t > 0$ *we have*

$$H_t = \left\{ \int_0^t S(t-s)Bf(s)\,ds : \ f \in L^2([0,t];H) \right\}.$$

Proof. Define $A_t\colon L^2([0,t];H) \to E$ by

$$A_t f := \int_0^t S(t-s)Bf(s)\,ds, \qquad f \in L^2([0,t];H).$$

It is trivially checked that

$$A_t^* x^* = B^* S(t-\cdot)x^*, \qquad x^* \in E^*.$$

Hence,

$$A_t(A_t^* x^*) = \int_0^t S(t-s)QS^*(t-s)x^*\,ds = Q_t x^* = i_t(i_t^* x^*)$$

from which it follows that $A_t(A_t^* x^*) \in i_t(H_t)$ for all $x^* \in E^*$. We have

$$\begin{aligned} \|A_t(A_t^* x^*)\|_{H_t}^2 &= \|i_t^* x^*\|_{H_t}^2 = \langle Q_t x^*, x^* \rangle \\ &= \|B^* S(t-\cdot)x^*\|_{L^2([0,t];H)}^2 = \|A_t^* x^*\|_{L^2([0,t];H)}^2. \end{aligned}$$

Hence A_t maps $\overline{\text{range}\, A_t^*}$ isometrically into H_t, and since the range of i_t^* is dense in H_t, this mapping is onto. Finally, for $f \perp \text{range}\, A_t^*$ it is clear that $A_t f = 0$. □

We refer to [6], [7], and [21] for a further discussion of the concept of null controllability.

Next we show that for differentiable semigroups **S**, the assumption of Corollary 4.5 implies null controllability.

PROPOSITION 5.3. *If* $\mathcal{D}(A^n) \subseteq \text{range}\, B$ *for some* $n \geq 1$ *and* **S** *is a differentiable semigroup, then* **S** *is null controllable with respect to the control pair* (B, H).

Proof. We will prove that $D(A^n) \subseteq \operatorname{range} B$ implies $D(A^{n+1}) \subseteq H_t$ for all $t > 0$; this does not require any regularity of the semigroup. If $\mathbf{S}$ is differentiable, it follows that $S(t)E \subseteq \mathcal{D}(A^{n+1}) \subseteq H_t$ for all $t > 0$.

Fix $t > 0$. First note that for all $\omega \in \mathbb{R}$, and $h \in H$ we have

$$\int_0^t e^{-\omega s} S(s)Bh\, ds = \int_0^t S(t-s)Bf_{\omega,h}(s)\, ds,$$

where $s \mapsto f_{\omega,h}(s) := e^{-\omega(t-s)} \otimes h \in L^2([0,t];H)$. Hence by Lemma 5.2,

$$\int_0^t e^{-\omega s} S(s)Bh\, ds \in H_t.$$

Choose $\omega \in \varrho(A) \cap \mathbb{R}$ so large that $\|e^{-\omega t}S(t)|_{D(A^n)}\|_{\mathcal{L}(D(A^n))} < 1$. Then the restriction to $D(A^n)$ of $I - e^{-\omega t}S(t)$ is invertible in $D(A^n)$, and for $x \in D(A^n)$ we have

$$\begin{aligned}(\omega - A)^{-1}x &= (I - e^{-\omega t}S(t))(\omega - A)^{-1}(I|_{D(A^n)} - e^{-\omega t}S(t)|_{D(A^n)})^{-1}x \\ &= -\int_0^t e^{-\omega s}S(s)(I|_{D(A^n)} - e^{-\omega t}S(t)|_{D(A^n)})^{-1}x\, ds \in H_t,\end{aligned}$$

noting that

$$(I|_{D(A^n)} - e^{-\omega t}S(t)|_{D(A^n)})^{-1}x = \sum_{n=0}^{\infty} e^{-n\omega t}S(nt)x \in D(A^n) \subseteq \operatorname{range} B,$$

the sum being absolutely convergent in $D(A^n)$. □

In particular, if in this situation $(\mathbf{H}\mu_\infty)$ holds, then μ_∞ is the unique invariant Gaussian Borel measure on E by Corollary 4.5. More generally we have:

THEOREM 5.4. $(\mathbf{H}\mu_\infty)$. *If* $\mathbf{S}$ *is null controllable with respect to the control pair* (B,H), *then* $\mathbf{S}$ *is uniformly exponentially stable and* μ_∞ *is the unique invariant Gaussian Borel measure on* E.

Proof. Let $t > 0$ be fixed. Denoting by $T(t)$ the operator $S(t)$, regarded as a map from E into H_t, we have

$$S(t) = i_t \circ T(t).$$

Since the triple (i_t, H_t, E) is an abstract Wiener space, the inclusion operator i_t is compact [3, Corollary 3.2.4]. Therefore $S(t)$ is compact as well.

It remains to check that μ_∞ is nondegenerate. But from $S(t)E \subseteq H_t \subseteq H_\infty$ (the second inclusion follows from [7, Appendix B]) and the strong continuity of $\mathbf{S}$ it follows that H_∞ is dense in E, and therefore Q_∞ is injective. This implies that μ_∞ is nondegenerate.

□

For Hilbert spaces E, uniqueness of the invariant measure in the null controllable case is well known; cf. [20], [7, Theorem 11.3]. Note that more generally, by a similar argument the uniqueness of μ_∞ follows if we only assume that $S(t)E \subseteq H_\infty$ for all $t > 0$.

6. Stabilizability

We call the problem (5.1) *stabilizable* if there exists a bounded linear operator $K: E \to H$ such that the operator

$$A_K := A + B \circ K$$

generates a uniformly exponentially stable semigroup.

If E is a *Hilbert space*, null controllability of $\mathbf{S}$ implies stabilizability of (5.1), and stabilizability of (5.1) together with $(\mathbf{H}Q_\infty)$ implies uniform exponential stability of $\mathbf{S}$; see [6] and [19]. In particular, if (5.1) is stabilizable and $(\mathbf{H}\mu_\infty)$ holds, then μ_∞ is the unique invariant Gaussian Borel measure on E.

The proofs of the results just cited are Hilbert space theoretic and as far as we know they do not extend to Banach spaces. Nevertheless, as an application of Theorem 4.4 we can prove directly:

THEOREM 6.1. $(\mathbf{H}\mu_\infty)$. *If μ_∞ is nondegenerate and the problem* (5.1) *is stabilizable, then μ_∞ is the unique invariant Gaussian Borel measure on E.*

Proof. We will prove that for all $x^* \in \mathcal{D}(A^*)$ the orbit $t \mapsto S^*(t)x^*$ is bounded. By Theorem 4.4, this will give the result.

Let $\mathbf{S}_K$ denote the semigroup generated by A_K and choose constants $M \geq 0$ and $\omega > 0$ such that $\|S_K(t)\| \leq Me^{\omega t}$ for all $t \geq 0$. Then by dualizing the variation of constants formula,

$$S_K^*(t)x^* = S^*(t)x^* + \int_0^t S_K^*(t-s)K^*B^*S^*(s)x^*\,ds.$$

Now let $x^* \in \mathcal{D}(A^*)$. Then by Lemma 2.4,

$$\lim_{t\to\infty} \|B^*S^*(t)x^*\|_H = \lim_{t\to\infty} \|i_Q^*S^*(t)x^*\|_{H_Q} = 0.$$

If $\varepsilon > 0$ is given, then we may choose $t_0 \geq 0$ so large that $\|B^*S^*(t)x^*\|_H \leq \varepsilon$ for all $t \geq t_0$. Then for $t \geq t_0$ we have

$$\begin{aligned}
&\left\|\int_0^t S_K^*(t-s)K^*B^*S^*(s)x^*\,ds\right\| \\
&\quad\leq \int_0^{t_0} \|S_K^*(t-s)K^*B^*S^*(s)x^*\|\,ds + \int_{t_0}^t \|S_K^*(t-s)K^*B^*S^*(s)x^*\|\,ds \\
&\quad\leq \left(\sup_{t\geq 0}\|B^*S^*(t)x^*\|_H\right)\int_0^{t_0} \|S_K^*(t-s)K^*\|\,ds + \varepsilon\int_{t_0}^t \|S_K^*(t-s)K^*\|\,ds \\
&\quad\leq \left(\sup_{t\geq 0}\|B^*S^*(t)x^*\|_H\right)\cdot\frac{M\|K\|}{\omega}\left(e^{-\omega(t-t_0)} - e^{-\omega t}\right) + \varepsilon\,\frac{M\|K\|}{\omega}.
\end{aligned}$$

Since $\varepsilon > 0$ was arbitrary and $\mathbf{S}_K$ is uniformly exponentially stable, it follows that for all $x^* \in \mathcal{D}(A^*)$,

$$\lim_{t\to\infty} \|S^*(t)x^*\| = \lim_{t\to\infty} \left\|S_K^*(t)x^* - \int_0^t S_K^*(t-s)K^*B^*S^*(s)x^*\,ds\right\| = 0. \qquad \square$$

Acknowledgement. The author thanks Professor G. Da Prato for pointing out an improvement to Proposition 2.5, and Professor Vũ Quôc Phóng for an interesting discussion about the topic of this paper.

References

[1] W. Arendt, F. Räbiger and A. Sourour, Spectral properties of the operator equation $AX + XB = Y$, *Quart. J. Math. Oxford*, **45** (1994), 133–149.

[2] C. J. K. Batty, J. M. A. M. van Neerven and F. Räbiger, Tauberian theorems and stability of solutions of the Cauchy problem, *Trans. Amer. Math. Soc.*, **350** (1998), 2087–2103.

[3] V. I. Bogachev, *Gaussian measures*, Math. Surveys and Monographs, **62**, Amer. Math. Soc., Providence, R.I., 1998.

[4] V. I. Bogachev, M. Röckner and W. Stannat, Uniqueness of invariant measures and essential m-dissipativity of diffusion operators on L^1, *Infinite dimensional stochastic analysis* (Ph. Clément, F. den Hollander, J. M. A. M. van Neerven and B. de Pagter, Eds.), Proceedings of the Colloquium (Amsterdam, 1998), Royal Netherlands Academy of Sciences and Arts, Amsterdam, 2000, 39–54.

[5] Z. Brzeźniak and J. M. A. M. van Neerven, Stochastic convolution in separable Banach spaces and the stochastic linear Cauchy problem, *Studia Math.*, **143** (2000), 43–74.

[6] R. F. Curtain and A. J. Pritchard, *Infinite-dimensional linear systems theory,* Lecture Notes in Control and Information Sciences **8**, Springer–Verlag, Berlin – New York, 1978.

[7] G. Da Prato and J. Zabczyk, *Stochastic equations in infinite dimensions,* Encyclopedia of Mathematics and its Applications, Cambridge University Press, Cambridge, 1992.

[8] G. Da Prato and J. Zabczyk, *Ergodicity for infinite-dimensional systems,* London Math. Soc. Lect. Note Series **229**, Cambridge University Press, Cambridge, 1996.

[9] B. Goldys, F. Gozzi and J. M. A. M. van Neerven, On closability of directional gradients, *Potential Anal.*, submitted.

[10] B. Goldys and J. M. A. M. van Neerven, Symmetric and nonsymmetric Ornstein–Uhlenbeck semigroups in Banach spaces and their reproducing kernel Hilbert spaces, in preparation.

[11] E. Hille and R. S. Phillips, *Functional analysis and semi-groups,* Amer. Math. Soc. Colloq. Publ. **31** (revised edition), Amer. Math. Soc., Providence, R.I., 1957.

[12] R. Nagel (Ed.), *One-parameter semigroups of positive operators,* Springer Lect. Notes Math. **1184**, Springer Verlag, 1986.

[13] J. M. A. M. van Neerven, *The adjoint of a semigroup of linear operators,* Springer Lect. Notes Math. **1529**, Springer Verlag, 1992.

[14] J. M. A. M. van Neerven, *The asymptotic behaviour of semigroups of linear operators,* Operator Theory: Adv. and Appl. **88**, Birkhäuser Verlag, Basel – Boston – Berlin, 1996.

[15] J. M. A. M. van Neerven, Nonsymmetric Ornstein-Uhlenbeck semigroups in Banach spaces, *J. Funct. Anal.*, **155** (1998), 495–535.

[16] K. R. Stromberg, *Probability for analysts*, Chapman and Hall, 1994.

[17] N. N. Vakhania, V. I. Tarieladze and S. A. Chobanyan, *Probability distributions in banach spaces*, D. Reidel Publishing Company, Dordrecht – Boston – Lancaster – Tokyo, 1987.

[18] Vũ Quôc Phóng, The operator equation $AX - XB = C$ with unbounded operators A and B related to abstract Cauchy problems, *Math. Z.*, **208** (1991), 567–588.
[19] J. Zabczyk, Remarks on the algebraic Ricatti equation in Hilbert spaces, *J. Appl. Math. Optimiz.*, **2** (1976), 251–258.
[20] J. Zabczyk, Linear stochastic systems in Hilbert spaces, *Spectral properties and limit behaviour,* Banach Center Publications **41**, 1985, 591–609.
[21] J. Zabczyk, *Mathematical control theory: an introduction,* Systems & Control: Foundations & Applications, Birkhäuser, Boston, 1992.

J. M. A. M. van Neerven, Department of Applied Mathematical Analysis, Technical University of Delft, P.O. Box 5031, 2600 GA Delft, The Netherlands; *e-mail:* J.vanNeerven@its.tudelft.nl

Operator Theory: Advances and Applications, Vol. 127, 519–551

The Friedrichs operator of a planar domain. II

MIHAI PUTINAR and HAROLD S. SHAPIRO

Paper dedicated to the memory of Béla Szőkefalvi-Nagy

We consider relations between the Friedrichs operator and constructive aspects of the Dirichlet problems for the Laplace and $\overline{\partial}^2$-operator. Then we investigate the Fourier expansions in the eigenfunctions of the Friedrichs operator. A link between a generalized Friedrichs operator and minimal nodes quadratures for complex polynomials of a fixed degree is explained. We initiate a discussion of the boundary Friedrichs operator on the Hardy space of a domain. The transformation law of the Friedrichs operator under conformal mappings leads to a modified version of it, based on a symbol function; this object will turn out to be closely related to Hankel operators. We obtain some results concerning which symbols correspond to compact operators.

1. Introduction

Let Ω be a planar domain and let $AL^2(\Omega), CAL^2(\Omega)$ be the Bergman space and its complex conjugated space in $L^2(\Omega, dA)$, where dA stands for the Lebesgue measure in $\mathbf{C}$. Motivated by some planar elasticity problems, Friedrichs [7], has studied the "gap operator" (in the sense of Kato [12] Chapter IV) between these subspaces, and a natural anti-linear square root F of it, called below the *Friedrichs operator* of the domain Ω. To be more specific, the operator F acts as follows:

$$\langle f, Fg \rangle = \langle g, Ff \rangle = \int_\Omega fg \, dA, \qquad f, g \in AL^2(\Omega).$$

AMS Subject Classification (1991): 30D55, 47B35, 35J25.
Partially supported by the National Science Foundation Grant DMS-9800666.

A first paper [25] was devoted to some basic spectral theory questions for the operator F, and their relevance to the geometry and analysis on Ω. The present note is a direct continuation of [25], touching other aspects of this rich field.

The contents is as follows. In Section 2, for a Jordan domain with rectifiable boundary, one starts with a function f in the Hardy space $H^2(\Omega)$ and obtains the decomposition $\overline{z}f(z) = \overline{\int_{z_0}^z Ff(u)\,du} + h(z)$, with $h \in H^2(\Omega)$ and $z_0 \in \Omega$. This simple fact has remarkable consequences: for instance if Fz is a linear polynomial, then Ω must be an ellipse. Or a solution of the Dirichlet problem with boundary data of the form $\overline{z}^n f(z)$ can recurrently be described in terms of the action of the Friedrichs operator on such special functions and quadratures. Similarly one obtains necessary conditions for the existence of bi-analytic functions with prescribed boundary data, see also [2] and [5] for related topics.

Section 3 deals with some simple aspects of the Fourier series expansions in the eigenfunctions of the operator $\sqrt{S}$, where $S = F^2$. Whenever the *Friedrichs inequality* holds, see Theorem 4.1 for the precise statement, one can prove that any square summable harmonic function in the simply connected domain Ω is the sum of a square summable analytic function and the conjugate of such a function. This opens a natural way of computing the reproducing kernel for the space of square summable harmonic functions. This is sketched in Section 4.

Section 5 is based on the remark that, given a degree n, to find minimal quadrature formulas on a bounded domain Ω, for all complex polynomials of degree less than or equal to n, involves only the bilinear form $E(p,q) = \langle p, Fq\rangle = \int_\Omega pq\,dA$. A canonical space which parametrizes all these minimal quadratures is described in terms of n and the action of a generalized Friedrichs operator on polynomials (of degree less than n).

An extension to Friedrichs operators with L^∞-symbol is considered in Section 6. This is motivated, for instance, by the pull back of the Friedrichs operator via a conformal map. The compactness of such an operator is discussed with the methods of Chapter 8 of [30]. In particular we obtain another proof of Friedrichs' result on the essential spectrum of F on a domain with corners.

The analogous Friedrichs operators with symbols, acting on the Hardy space of a domain with rectifiable boundary are defined in Section 7. On the unit disk, treated in Section 8, these operators are rank-one perturbations of some conjugated Hankel operators. For completeness we include the well known Nehari's compactness criterion, adapted to the setting of Friedrichs operators.

We are indebted to the two anonymous referees, whose accurate observations and recommendations led to an improved version of this paper.

Notation and terminology. (Consistent with those in [25].)

Throughout this paper Ω will be a planar domain; the regularity assumptions on its boundary $\partial\Omega$ will be made clear in each section separately. The area measure in $\mathbf{C}$ will be denoted by dA and the associated Lebesgue spaces by $L^p(\Omega) = L^p(\Omega, dA)$. We put $W^{s,p}(\Omega)$, $\overset{\circ}{W}{}^{s,p}(\Omega)$ for the Sobolev spaces of functions, or distributions, in Ω having the derivative up to order s in $L^p(\Omega, dA)$, respectively the closure of the Schwartz space $D(\Omega)$ in $W^{s,p}(\Omega)$. The Bergman space associated to Ω will be denoted by $AL^2(\Omega)$ and the Bergman projection by P, that is the orthogonal projection of $L^2(\Omega) = L^2(\Omega, dA)$ onto $AL^2(\Omega)$. The Hardy spaces of a domain Ω with sufficiently regular boundary will be denoted by $H^p(\Omega)$, $1 \le p \le \infty$. The arc length element on $\partial\Omega$ will be ds. To be more precise, if Ω is a simply-connected bounded domain with smooth C^2 boundary, then the Hardy space $H^2(\Omega)$ will be by definition the closure of complex polynomials in $L^2(\partial\Omega, ds)$.

Let $C: L^2(\Omega) \longrightarrow L^2(\Omega)$ be the complex conjugation operator: $Cf = \overline{f}, f \in L^2(\Omega)$. The *Friedrichs operator* of the domain Ω is then the anti-linear operator $F = PC: AL^2(\Omega) \longrightarrow AL^2(\Omega)$; we put as before $S = F^2$, so that S is a complex linear, non-negative and contractive operator. Throughout this paper we refer to $\sqrt{S}$ as the *modulus* of the Friedrichs operator F.

2. The Friedrichs operator and certain boundary value problems

We show that the Friedrichs operator of a planar domain is intimately connected to the Dirichlet problems for the Laplacian and for the square of the Cauchy–Riemann operator. The main observation we develop below is that, knowing the action of the Friedrichs operator on a space of functions f, such as the complex polynomials, it is a matter of finitely many quadratures to solve the Dirichlet problem for Δ with boundary data $\overline{z}^n f(z)$, for arbitrary $n \ge 1$. Then a solution to the $\overline{\partial}^2$ Dirichlet problem is also at hand.

Throughout this section we assume that Ω is a Jordan domain with rectifiable boundary $\Gamma = \partial\Omega$, which moreover is sufficiently regular that the space $W^{1,2}(\Omega)$ has traces in $L^2(\Gamma, ds)$. For some purposes even more regularity will be needed. In this initial study we shall not strive for maximal generality. Therefore we shall sometimes invoke a rather vague hypothesis: "sufficiently smooth", leaving for future study the sharpest possible results. Certainly, C^2 regularity is more than enough to justify all our assertions.

The next proposition is a simple application of the Friedrichs–Havin Lemma, see [7], [11].

PROPOSITION 2.1. *Let $f \in H^2(\Omega)$, where $\partial\Omega$ is sufficiently smooth, and let G be a primitive function of $g = Ff$. Then*

$$\overline{\zeta} f(\zeta) = \overline{G(\zeta)} + h(\zeta), \qquad \zeta \in \Gamma \quad (ds - a.e.), \tag{1}$$

where $h \in H^2(\Omega)$.

Proof. By definition, the function $g - \overline{f}$ is orthogonal to $AL^2(\Omega)$. So, by Friedrichs–Havin lemma, there exists a distribution $v \in \overset{\circ}{W}{}^{1,2}(\Omega)$ such that $\overline{\partial} v = \overline{g} - f$, in Ω, that is:

$$\overline{\partial}[\overline{z} f(z) - \overline{G(z)} - v(z)] = 0, \qquad z \in \Omega.$$

Thus there exists a holomorphic function $h(z)$, $z \in \Omega$, with $h(z) = \overline{z} f(z) - \overline{G(z)} + v(z), z \in \Omega$. Our regularity assumptions imply that G, h are in $H^2(\Omega)$, hence, by taking traces on Γ and using the fact that v vanishes there a.e., we obtain relation (1).

□

Note that the decomposition (1) can be reversed and used in computing the value Ff of the Friedrichs operator from the solution u of the Dirichlet problem with boundary data $\overline{\zeta} f(\zeta)$. Indeed, the harmonic function u can always be decomposed, uniquely up to an additive constant, as $\overline{G} + h$, with G, h in the Hardy space $H^2(\Omega)$, and then $Ff = G'$.

The decomposition (1) has several interesting consequences. First of all we note the following regularity result.

THEOREM 2.2. *Let Ω be a Jordan domain with sufficiently smooth boundary $\Gamma = \partial\Omega$, so that the space $W^{1,2}(\Omega)$ has traces in $L^2(\Gamma, ds)$.*

If Fz is the derivative of a rational function, then Γ is a subset of an algebraic curve.

Proof. Write $\phi = Fz$ and let Φ be a primitive of ϕ, assumed to be rational. Thus relation (1) yields

$$\overline{\zeta}\zeta = \overline{\Phi(\zeta)} + h(\zeta), \qquad \zeta \in \Gamma, \quad \text{a.e.}, \tag{2}$$

hence the function $h - \Phi \in H^2(\Omega)$ has boundary values $\overline{\zeta}\zeta - \overline{\Phi(\zeta)} - \Phi(\zeta)$, $\zeta \in \Gamma$, which are real. Therefore $h - \Phi$ is constant, and we can write:

$$|\zeta|^2 - 2\Re\Phi(\zeta) = c, \qquad \zeta \in \Gamma,$$

where c is a real constant. It remains to remark that the function $R(z,\overline{z}) = |z|^2 - 2\Re\Phi(z) - c$ is not constant since its Laplacian equals 4.

□

Inspection of the above proof shows that the simple assumption that Fz extends analytically across the boundary of Ω implies that the latter is real analytic, locally or globally.

As an application of the reasoning used to prove Theorem 2.2 we state the following characterization of ellipses, including among them the disks.

COROLLARY 2.3. *In the conditions of Theorem* 2.2, *assume that Fz is a linear polynomial. Then Ω is an ellipse.*

Proof. In this situation $\Phi(\zeta)$ is a quadratic polynomial, and by invoking the proof of the theorem, the boundary Γ of Ω is contained in the set of zeros of a real polynomial in $z, \overline{z}$, of total degree 2. Hence Ω must be an ellipse.

□

By combining this corollary with the analysis of the eigenvalues of the Friedrichs operator of an ellipse, cf. [25] Proposition 6.1, we obtain the following result. Below we denote by P_n the space of complex polynomials of degree less than or equal to n.

COROLLARY 2.4. *Let Ω be Jordan domain as in Theorem* 2.2. *The following conditions are equivalent:*

a) *The space P_1 is invariant under F;*
b) *For all $n \geq 1$, the space P_n is invariant under F;*
c) *Ω is an ellipse.*

Next we turn to the Dirichlet problem for the Laplace operator. Let f, G, h be as in Proposition 2.1, and let $z \in \Omega$. Then, since $h \in H^2(\Omega)$:

$$h(z) = \frac{1}{2\pi i}\int_\Gamma \frac{h(\zeta)}{\zeta - z}\,d\zeta = \frac{1}{2\pi i}\int_\Gamma \frac{\overline{\zeta}f(\zeta) - \overline{G(\zeta)}}{\zeta - z}\,d\zeta. \tag{3}$$

Consequently we can state the following observation.

COROLLARY 2.5. *The solution to the Dirichlet problem:*

$$\Delta u = 0 \quad in \quad \Omega; \quad u(\zeta) = \overline{\zeta} f(\zeta), \ \zeta \in \Gamma,$$

is $u(z) = \overline{G(z)} + h(z)$, *where the function* h *is defined by* (3).

Observe that, once $g = Ff$ is known, both G and h are determined by integrations. This opens the way to recursively solving the Dirichlet problem with data of the form $\overline{z}^m f(z)$, with $m \geq 0$ and $f \in H^2(\Omega)$, in terms of the Friedrichs operator and finitely many integration operations. Indeed, starting from (1) we obtain:

$$\overline{\zeta}^2 f(\zeta) = \overline{\zeta G(\zeta)} + \overline{\zeta} h(\zeta), \qquad \zeta \in \Gamma.$$

Since the function $zG(z)$ is analytic, solving the Dirichlet problem with boundary data $\overline{\zeta}^2 f(\zeta)$ is reduced to that for $\overline{\zeta} h(\zeta)$, and hence to a second application of Corollary 2.5. In particular, choosing $f(z) = z^k, k \geq 1$, we see that the solution of the Dirichlet problem with polynomial data on the boundary is reduced to the knowledge of the Friedrichs operator and a finite number of integrations. This is of special interest in quadrature domains, when F has finite rank and in principle can be calculated once its values are known on a finite set of functions, see [25].

To make the above remark more precise, let us define the operators A, B, acting on the Hardy space $H^2(\Omega)$ as:

$$(Af)(z) = \int_{z_0}^{z} (Ff)(w)\, dw, \qquad Bf)(z) = \frac{1}{2\pi i} \int_\Gamma \frac{\overline{\zeta} f(\zeta) - \overline{Af(\zeta)}}{\zeta - z} d\zeta,$$

where z_0 is a fixed point in Ω and the integration in the expression of A is taken on a path fully contained in Ω. The operators

$$CA: H^2(\Omega) \longrightarrow CH^2(\Omega), \qquad B: H^2(\Omega) \longrightarrow H^2(\Omega)$$

are complex linear and bounded on the Hardy space and they satisfy, in the sense of boundary values, the identity:

$$\overline{\zeta} f(\zeta) = CAf(\zeta) + Bf(\zeta), \qquad \zeta \in \Gamma. \tag{4}$$

By iterating this relation we obtain, for an arbitrary polynomial $P(z)$, the next formula, in which the standard notation $P^*(z) = \overline{P(\overline{z})}$ appears:

$$\overline{P(\zeta)} f(\zeta) = [P^*(B)f](\zeta) + C\Big[A\frac{P(z) - P(B)}{z - B} f\Big](\zeta), \qquad \zeta \in \Gamma.$$

The symbolic notation $\frac{P(z)-P(B)}{z-B}$ means the value of the polynomial function $\frac{P(z)-P(w)}{z-w}$ at $w = B$. The first term in the decomposition above extends analytically inside Ω, while the second term, understood as a polynomial in z, A and B, with the multiplication operator by z acting to the *left* of A and B, admits an anti-analytic extension in Ω. Therefore, we are led to the following "explicit" formula for solving the Dirichlet problem.

COROLLARY 2.6. *Let $P(z)$ be a polynomial and let $f \in H^2(\Omega)$. The solution to the Dirichlet problem*

$$\Delta u(z) = 0, \quad z \in \Omega; \quad u(\zeta) = \overline{P(\zeta)} f(\zeta), \quad \zeta \in \Gamma,$$

is

$$u(z) = [P^*(B)f](z) + C\Big[A\frac{P(z)-P(B)}{z-B}f\Big](z). \tag{5}$$

Finally, let us remark that formula (5) can be extended to more general functions instead of polynomials, as soon as the functional calculus $P(B)$ and the evaluations $P(z), z \in \Omega \cup \Gamma$, are well defined (as for instance for entire functions).

Another application of decomposition (1) is to the $\overline{\partial}^2$ operator. Suppose we wish to solve the boundary value problem:

$$\overline{\partial}^2 u = 0 \quad \text{in} \quad \Omega; \quad u(\zeta) = p(\zeta), \quad \zeta \in \Gamma, \tag{6}$$

for some function p on Γ.

The most general solution to $\overline{\partial}^2 u = 0$ is of the form $u(z) = f_1(z) + \overline{z} f_2(z)$, with f_1, f_2 analytic functions in Ω. If we suppose the Dirichlet problem for the Laplacian with boundary data p solved, then we can write $p = a + \overline{b}$, with a, b analytic in Ω. Then clearly (6) reduces to $p = \overline{b}$, and the decomposition (1) just solves this problem. To be more specific we have proved, modulo the regularity details which are left to the reader, the following proposition.

PROPOSITION 2.7. *Assume that $p \in L^2(\Gamma, ds)$. Then problem (6) is solvable if and only if $p = \overline{G}$, where G is the primitive of an element $g \in \mathrm{Ran}(F)$, $g = Ff$, and in this case the solution is $u(z) = [(\overline{z} - B)f](z)$.*

Thus, the $\overline{\partial}^2$ Dirichlet problem can be reduced to the Laplace Dirichlet problem and knowledge of the Friedrichs operator. Again, in quadrature domains this reduces to finitely many quadratures and algebraic computations, modulo a fixed finite "table" of values of F.

3. Fourier expansions in the Friedrichs eigenfunctions

The aim of this section is to continue the study, started in [25], of the elements in the kernel and the range of the Friedrichs operator.

First we derive from Proposition 2.1 a characterization of quadrature domains in terms of the location of zeros of functions in the kernel of F.

PROPOSITION 3.1. *Let Ω be a bounded domain with rectifiable boundary Γ. Then Ω is a quadrature domain if and only if there exists a function $f \in H^2(\Omega)$ satisfying $Ff = 0$ and such that there exists $\epsilon > 0$ with $|f(z)| > \epsilon$ for* $\operatorname{dist}(z, \Gamma) < \epsilon$.

Proof. If Ω is a quadrature domain of order d, then the Friedrichs operator F has rank d, [25]. We also know that any element $f \in \ker F$ vanishes, counting multiplicities, at the quadrature nodes. Since a non-trivial linear combination of the monomials $1, z, z^2, \ldots, z^d$ lies in $\ker F$, there exists a monic polynomial p of degree d, which satisfies: $Fp = 0$ and p is zero on d points belonging to Ω. Consequently $p(z)$ is bounded away from zero on $\partial\Omega$.

Conversely, assume that the function f as in the statement exists. Then, according to decomposition (1) we obtain:

$$\overline{\zeta} f(\zeta) = h(\zeta), \quad \zeta \in \Gamma,$$

with $h \in H^2(\Omega)$. Consequently, the function $\overline{\zeta}$ admits the meromorphic extension $h(z)/f(z)$ throughout Ω, and this is equivalent to Ω being a quadrature domain, see [30].

□

A consequence of the previous proof is that functions $f \in \ker(F)$ which are bounded away from zero in a neighbourhood of the boundary Γ must have zeros in Ω.

Also, the *apriori* weaker hypothesis that to each point $\zeta \in \partial\Omega$ there exists $f \in \ker F$ and $\epsilon > 0$ so that $|f(z)| > \epsilon$ for all $z \in \Omega \cap D(\zeta, \epsilon)$ suffices to show that Ω is a quadrature domain.

Next we consider a bounded domain Ω with compact and injective Friedrichs operator F, so that the same is true for the square $S = F^2$. (For examples and necessary conditions for these assumptions, see [25].) Let $1 = \lambda_0 \geq \lambda_1 \geq \ldots$ be the eigenvalues of $\sqrt{S}$, with associated eigenfunctions $\phi_n \in AL^2(\Omega)$, $n \geq 1$, which form an orthonormal basis of the Bergman space.

The following general criterion is well known for any compact operator and we omit its proof.

LEMMA 3.2. *The necessary and sufficient condition for a function $g \in AL^2(\Omega)$ to belong to the range of F is:*

$$\sum_{n=0}^{\infty} \lambda_n{}^{-2} |\langle g, \phi_n \rangle|^2 < \infty. \tag{7}$$

In the specific situation of the Friedrichs operator this leads to the next proposition.

PROPOSITION 3.3. *With assumptions as above,*

$$\sum_{n=0}^{\infty} \frac{|\phi_n(z)|^2}{\lambda_n{}^2} = \infty, \tag{8}$$

for each $z \in \Omega$.

Proof. In view of Lemma 3.2, the finiteness of the left hand side of (8) would imply that the Bergman evaluation function k_z belongs to the range of F, and we know from [25] Corollary 4.2 that this cannot happen.

□

The same idea leads to more general results, as follows.

PROPOSITION 3.4. *Assume F compact and let μ be a measure supported on the zero set of a nontrivial function $f \in H^\infty(\Omega)$, such that μ is bounded as a functional on $AL^2(\Omega)$. Then:*

$$\sum_{n=0}^{\infty} \frac{|\int_\Omega \phi_n \, d\mu|^2}{\lambda_n{}^2} = \infty. \tag{9}$$

Proof. Recall from [25] that the Toeplitz operator T_f corresponding to f satisfies $FT_f = T_f^* F$. Let a be the representing vector for μ:

$$\int_\Omega g \, d\mu = \langle g, a \rangle, \quad g \in AL^2(\Omega).$$

Then, by our assumption, $T_f^* a = 0$.

Assume by contradiction that $a = Fh$ for some $h \in AL^2(\Omega)$. Then $FT_f h = T_f^* F h = T_f^* a = 0$, which contradicts the injectivity of F, because $T_f h = fh \neq 0$.

□

4. The harmonic kernel

This section is mainly devoted to the computation of the orthogonal projection of the Lebesgue space $L^2(\Omega, dA)$ onto the closed subspace $HL^2(\Omega)$ of square-summable harmonic functions in Ω. We prove that, if Ω is simply connected, then whenever the Friedrichs inequality holds, the space $AL^2(\Omega)+CAL^2(\Omega)$ is closed and equals $HL^2(\Omega)$. In addition, a general formula due to Lenard, [16], gives a rather explicit formula for the orthogonal projection onto the sum of two non-orthogonal subspaces.

First we recall some basic facts and conventions concerning Friedrichs' inequality. For more details we refer to [29], [30] and [25].

THEOREM 4.1. (H. S. Shapiro [29]). *Let Ω be a bounded domain satisfying an interior cone condition and let $z_0 \in \Omega$. Let $f \in AL^2(\Omega)$ be a function subject to one of the following normalizations:*

$$\text{a)} \quad \int_\Omega f\, dA = 0, \quad \textit{or} \quad \text{b)} \quad f(z_0) = 0.$$

Then the Friedrichs inequality holds in either of the forms:

$$\left| \int_\Omega f^2\, dA \right| \le c \int_\Omega |f|^2\, dA, \tag{10}$$

for some constant $c < 1$, or:

$$\int_\Omega |\tilde{u}|^2\, dA \le C \int_\Omega |u|^2\, dA, \tag{11}$$

for some finite constant C, where $u = \Re f$ and $\tilde{u} = \Im f$.

The constants c, C above depend only on Ω if the normalization a) is used, and respectively on Ω and z_0 under the normalization b). Inequality (10) holds also if f^2 is replaced by an arbitrary function in $AL^1(\Omega)$ normalized to have mean zero, or to vanish at z_0. Moreover, (11) holds also for exponents $p > 1$, and then C depends also on p. Finally, in (11) it is sufficient to normalize $\tilde{u}$ by one of the conditions a) or b).

For the sake of completeness we state below some consequences of the Friedrichs inequality, part of them previously known in one form or another.

COROLLARY 4.2. *If Ω is a simply connected, bounded domain satisfying an interior cone condition, then for every real valued $u \in HL^2(\Omega)$, its harmonic conjugate satisfies $\tilde{u} \in HL^2(\Omega)$.*

Proof. We only sketch the idea of the proof. One can show that Ω is the union of an increasing sequence Ω_j, $j \geq 1$, of simply connected domains which uniformly satisfy an interior cone condition. Let z_0 be a point of Ω_1. We can assume that $\tilde{u}(z_0) = 0$. Then, by the Friedrichs inequality (11) the norms $\|\tilde{u}\|_{2,\Omega_j}$ are uniformly bounded in j, hence $\tilde{u} \in L^2(\Omega)$.

□

COROLLARY 4.3. *If Ω is simply connected, then every $u \in HL^2(\Omega)$ can be written, uniquely up to an additive constant, as $f + \overline{g}$, with $f, g \in AL^2(\Omega)$.*

Proof. It suffices to deal with a real valued u. By the preceding corollary the harmonic conjugate function $\tilde{u}$ is square integrable in Ω, hence $2u = f + \overline{f}$, where $f = u + i\tilde{u} \in AL^2(\Omega)$.

□

COROLLARY 4.4. *In any domain Ω for which the Friedrichs inequality holds (in particular, for Ω satisfying an interior cone condition) we have, whenever $f, g \in AL^2(\Omega)$ and g satisfies one of the normalizations in Theorem* 4.1 :

$$\|f\| + \|\overline{g}\| \leq C\|f + \overline{g}\|, \tag{12}$$

where the norm is in $L^2(\Omega)$ and C is a constant depending only on Ω (or Ω and z_0, as the normalization demands).

Proof. In virtue of Friedrichs' inequality and Schwarz' inequality we obtain:

$$\begin{aligned}\|f + \overline{g}\|^2 = \|f\|^2 + \|g\|^2 + 2\Re \int_\Omega fg\, dA &\geq \|f\|^2 + \|g\|^2 - 2c\int_\Omega |fg|\, dA \\ &\geq (1-c)[\|f\|^2 + \|g\|^2] + c(\|f\| - \|g\|)^2.\end{aligned}$$

Since the last term is non-negative, relation (12) follows with $C = \sqrt{\frac{1}{1-c}}$.

□

COROLLARY 4.5. *With the same hypotheses on Ω, analytic polynomials are dense in $AL^2(\Omega)$ if and only if harmonic polynomials are dense in $HL^2(\Omega)$.*

Proof. If analytic polynomials are dense in the Bergman space, then any $u \in HL^2(\Omega)$ can be decomposed as before $u = f + \overline{g}$ and therefore it is approximable by harmonic polynomials.

Conversely, assume that the harmonic polynomials are dense in $HL^2(\Omega)$ and let $f \in AL^2(\Omega)$ be given. Let $\epsilon > 0$ be arbitrary. Then there exists a harmonic polynomial h with the property $\|f - h\| < \epsilon$. Now, $h = f_1 + \overline{g_1}$, where f_1, g_1 are analytic polynomials and, without loss of generality, $g_1(z_0) = 0$ for some fixed point z_0. Thus, $\|(f - f_1) - \overline{g_1}\| \leq \epsilon$, and, using (12),

$$\|f - f_1\| \leq C\epsilon,$$

which completes the proof. □

The observation that P is the orthogonal projection onto $AL^2(\Omega)$, $Q = CF = CPC$ is the orthogonal projection onto $CAL^2(\Omega)$, and $PQ = S = F^2$, leads to the following simple but useful partial conclusion of the above results. For details about the geometric relations of two subspaces of a Hilbert space we refer to [12], Section IV.2.

PROPOSITION 4.6. *Let Ω be a bounded planar domain. Then the following assertions are equivalent:*

1) *The Friedrichs inequality holds;*
2) *Restricted to the subspace of $AL^2(\Omega)$ consisting of functions with mean value zero, the operator S has norm less than* 1;
3) *The space $AL^2(\Omega) + CAL^2(\Omega)$ is closed.*

In addition, if Ω is simply connected, then the above assertions are equivalent to:

4) *The vector sum $AL^2(\Omega) + CAL^2(\Omega)$ equals $HL^2(\Omega)$.*

Our next aim is to compute the reproducing kernel $H(z,w)$ of the space $HL^2(\Omega)$, under the assumption that the domain Ω is simply connected and satisfies Friedrichs inequality.

In general, if P and Q are two orthogonal projections of a Hilbert space H, such that, for instance $\|PQ\| < 1$, then $\mathrm{Ran}(P) \cap \mathrm{Ran}(Q) = 0$, the subspace $V = \mathrm{Ran}(P) + \mathrm{Ran}(Q)$ is closed and the operators $I - PQ, I - QP$ are invertible. By a formula due to Lenard, [16], the orthogonal projection R onto V is:

$$R = (I - Q)(I - PQ)^{-1}P + (I - P)(I - QP)^{-1}Q.$$

We could simply use this formula for the subspace H of functions $f \in L^2(\Omega)$ having mean 0 (so that $\mathrm{Ran}(P) \cap \mathrm{Ran}(Q) = 0$) and deduce the form of the harmonic reproducing kernel $H(z,w)$. Below we denote $AL^2_0(\Omega) = \{f \in AL^2(\Omega);\ \int_\Omega f\, dA = 0\}$.

As a matter of fact, assuming the spectral decomposition of the Friedrichs operator known, our situation is simpler.

Assume first that F is compact and injective, and let $1 = \lambda_0 > \lambda_1 \geq \lambda_2 \geq \dots$ be the eigenvalues of $\sqrt{S}$ with corresponding eigenfunctions ϕ_n, $n \geq 0$. The system of functions:

$$\psi_n = \frac{\overline{\phi_n} - \lambda_n \phi_n}{\sqrt{1-\lambda_n^2}}, \quad n \geq 1, \tag{13}$$

is orthonormal and, by the very definition of F, $\langle \phi_m, \psi_n \rangle = 0$ for all $m \geq 1$ and $n \geq 0$. Thus ϕ_0, ϕ_n, ψ_n, $n \geq 1$ is an orthonormal basis of $HL^2(\Omega)$. As a first conclusion of these simple remarks we note the following lemma.

LEMMA 4.7. *Assume that Ω is a simply connected domain which satisfies Friedrichs' inequality. Let $\phi_n, n \geq 0$, be the eigenfunctions of the modulus of the Friedrichs operator and let $\psi_n, n \geq 1$, be defined as in* (13).

Then the reproducing kernel of $HL^2(\Omega)$ is:

$$H(z,w) = \phi_0(z)\overline{\phi_0(w)} + \sum_{n=1}^{\infty}(\phi_n(z)\overline{\phi_n(w)} + \psi_n(z)\overline{\psi_n(w)}). \tag{14}$$

Second, we assume that Ω is a quadrature domain, so that, $1 = \lambda_0 > \lambda_1 \geq \lambda_2 \geq \dots \lambda_d > \lambda_{d+1} = 0$. Then the finite dimensional space $\mathrm{Ran}(F)$ is spanned by the vectors ϕ_n, $n \leq d$, and the Bergman kernel $K(z,w)$ of Ω can be written as:

$$K(z,w) = N(z,w) + \sum_{n=0}^{d} \phi_n(z)\overline{\phi_n(w)}, \tag{15}$$

where $N(z,w)$ is the reproducing kernel of the space $\mathrm{Ker}(F)$.

The harmonic space can be decomposed into an orthogonal sum as:

$$HL^2(\Omega) = [\mathrm{Ker}(F) + C\,\mathrm{Ker}(F)] \oplus [\mathrm{Ran}(F) + C\,\mathrm{Ran}(F)].$$

Moreover, the reproducing kernel of the first summand is precisely $N(z,w) + N(w,z)$, because $Ff = 0$ implies

$$\int_\Omega f(z)(N(z,w) + N(w,z))\,dA = 0 + f(w),$$

and by symmetry,

$$\int_\Omega \overline{f(z)}(N(z,w) + N(w,z))\,dA = \overline{f(z)} + 0.$$

In conclusion we can state the following result.

LEMMA 4.8. *If Ω is a quadrature domain, then the reproducing kernel of $HL^2(\Omega)$ is:*

$$(16)\quad H(z,w) = K(z,w)+K(w,z)-\phi_0(w)\overline{\phi_0(z)}+\sum_{n=1}^{d}(\psi_n(z)\overline{\psi_n(w)}-\phi_n(w)\overline{\phi_n(z)}).$$

Further on, recall ([25]) that ϕ_n is a linear combination of the evaluation vectors $K(z,a_i), 1 \le i \le d$, where $a_i, 1 \le i \le d$, are the quadrature nodes. Thus the real kernel H has the form:

$$H(z,w) = 2\Re K(z,w) + \Re \sum_{i,j=1}^{d} K(z,a_i)C_{ij}K(a_j,w) + \Re \sum_{i,j=1}^{d} K(z,a_i)D_{ij}K(w,a_j),$$

with the unknown matrices C_{ij}, D_{ij}, to be determined by a linear system of reproducing formulas for the functions $f = K(z,a_i)$ and their complex conjugates. We do not expand here these details.

5. Interpolatory quadrature formulas

Since the Friedrichs operator encodes the geometrical relationship of two subspaces, it is a flexible concept, and can usefully be introduced in other contexts than that of Friedrichs. From this point on we explore such generalizations. We emphasize that in the following sections the term "Friedrichs operator" and symbol F thus get a special meaning. We hope this will cause no confusion.

A generalized Friedrichs operator is related below to Gaussian cubatures for complex polynomials. This is a direct extension of the relation between orthogonal polynomials and Gaussian cubatures on the line.

To be more specific, let μ be a positive measure on $\mathbf{C}$, rapidly decreasing at infinity, and with infinite support. These conditions assure that any complex analytic polynomial p is integrable with respect to μ, and $\int_{\mathbf{C}} |p|\, d\mu = 0$, if and only if $p = 0$. Let P denote the orthogonal projection of $L^2(\mu)$ onto the closure $P^2(\mu)$ of all polynomials. The Friedrichs operator F is then well defined:

$$Ff = PCf = P(\overline{f}), \qquad f \in P^2(\mu).$$

Note that $\mathbf{C}[z] \subset L^2(\mu)$. For a fixed integer $d \ge 0$ we adopt the following notation:

$$P_d = \{p \in \mathbf{C}[z];\ \ \deg(p) \le d\}, \qquad FP_d = \{Fp;\ \ p \in P_d\}.$$

We will show that the angles between the elements of the increasing chains of subspaces $(P_d)_{d=0}^{\infty}$ and $(FP_d)_{d=0}^{\infty}$ are responsible for the existence of quadrature formulas of prescribed degree, for the measure μ.

Fix a degree $n \geq 1$. A *quadrature formula* of degree n, for the measure μ, has the form:

$$\int_{\mathbf{C}} p\, d\mu = \sum_{k=1}^{m} \sum_{j=1}^{n_k-1} c_k^j p^{(j)}(a_k), \qquad p \in P_n.$$

The number $N = \sum_{k=1}^{m} n_k$ is called the *order* of the quadrature formula.

A simple linear dependence argument shows that always, for fixed n, such formulas exist. We will be interested in minimal quadrature formulas, that is those with minimal order N. In this case we can assume that the formula is also *interpolatory*, that is the map:

$$P_n \longrightarrow \mathbf{C}^N, \qquad p \mapsto (p^{(j)}(a_k))_{j,k},$$

is onto. Details about quadrature, or after some authors cubature, formulas can be found in [19]. Some aspects of the related theory of complex orthogonal polynomials is developed in [31].

We start with a series of simple remarks. For every d, $\dim P_d = d+1$, while $\dim FP_{d+1} \leq \dim FP_d + 1$. Indeed, denoting by M the multiplication operator on $\mathbf{C}[z]$ by the complex variable z and by $\mathbf{1}$ the function equal to 1, we have:

$$FP_d = \{p(M^*)\mathbf{1};\ p \in P_d\}. \tag{17}$$

We have to be careful with the possible unboundedness of M on the whole space $\mathbf{C}[z]$; however, for arbitrary polynomials $p, q \in \mathbf{C}[z]$, the definition of M^* is unambigous:

$$\langle Mp, q\rangle = \langle p, M^*q\rangle = \langle p, P(\overline{z}q)\rangle.$$

Assume that for a given d, $\dim FP_{d+1} = \dim FP_d$. Then, according to formula (17), the subspace FP_d is invariant under the linear operator M^*. In virtue of the same formula we find that $FP_{d+k} = FP_d$ for all $k \geq 0$. Thus the second tower $(FP_d)_d$ is stationary. Let $T = M^*|_{FP_d}$, regarded as an operator from FP_d to itself. Let

$$\sigma(T) = \{\overline{a_1}, \overline{a_2}, \ldots, \overline{a_m}\}$$

be the spectrum of T. Let v_k be the corresponding generalized eigenvectors:

$$(T - \overline{a_k})^{n_k} v_k = 0.$$

In virtue of the fact that T admits $\mathbf{1}$ as a cyclic vector, for each a_k there exists exactly one such v_k. A basis of FP_d is then given by the vectors $(T-\overline{a_k})^j v_k$, $1 \leq k \leq m, 0 \leq j \leq n_k - 1$. By decomposing $\mathbf{1}$ in this basis:

$$\mathbf{1} = \sum_{j,k} \overline{d_k^j} (T-\overline{a_k})^j v_k,$$

we obtain a universal quadrature formula:

$$\int_{\mathbf{C}} p\, d\mu = \langle p, \mathbf{1}\rangle = \sum_{j,k} d_k^j \langle p, (T-\overline{a_k})^j v_k\rangle.$$

After rearranging the terms above, we get:

$$\int_{\mathbf{C}} p\, d\mu = \sum_{j,k} c_{jk} p^{(j)}(a_k), \quad p \in \mathbf{C}[z]. \tag{18}$$

Thus we can state the following proposition.

PROPOSITION 5.1. *Let μ be a positive, rapidly decreasing measure on $\mathbf{C}$. The following assertions are equivalent:*

a) *The measure μ admits a universal quadrature formula* (18);

b) *There exists a degree d_0 with the property that $FP_{d_0} = FP_{d_0+1}$;*

c) *The Friedrichs operator F annihilates a non-trivial polynomial.*

Let dA be the planar Lebesgue measure, let Ω be a bounded quadrature domain and let χ_Ω be the characteristic function of Ω. Then the measure $\chi_\Omega dA$ satisfies the conditions in the above proposition, see [25]. More generally, assume that $\Omega_k, 1 \leq k \leq m$, are quadrature domains and that $\alpha_k > 0, 1 \leq k \leq m$. Then the measure $\mu = \sum_{k=1}^m \alpha_k \chi_{\Omega_k}$ still satisfies the conditions in Proposition 5.1.

From now on we assume that $\dim FP_d = d+1$ for all $d \geq 0$. Assume that the quadrature formula of degree n is given. We denote by I_d the intersection of the polynomial ideal I vanishing at the quadrature nodes with P_d:

$$I_d = \{p \in P_d;\ p^{(j)}(a_k) = 0,\ 1 \leq k \leq m, 0 \leq j \leq n_k - 1\}.$$

Note that for all $d \geq 0$, $I_{d+1} \cap P_d = I_d$, hence the maps induced by inclusion:

$$P_d/I_d \longrightarrow P_{d+1}/I_{d+1}$$

are all one to one. Since we assume the quadrature formula to be interpolatory, the evaluation map:

$$P_n/I_n \longrightarrow \mathbf{C}^N, \quad p \mapsto (p^{(j)}(a_k))_{j,k},$$

is an isomorphism. Thus

$$\dim(P_d/I_d) \leq N, \quad 0 \leq d \leq n,$$

with equality for $d = n$.

The following simple remark will be essential for the characterization of minimal quadrature formulas:

$$I_d \subset P_d \ominus FP_{n-d}, \quad 0 \leq d \leq n. \tag{19}$$

Indeed, if $p \in I_d$ and $q \in P_{n-d}$, then:

$$\langle p, Fq \rangle = \langle pq, \mathbf{1} \rangle = 0.$$

Thus we obtain a lower bound for the number N of nodes in the quadrature formula (18):

$$N \geq \max_{0 \leq d \leq n} \dim[P_d/(P_d \ominus FP_{n-d})]. \tag{20}$$

Our next aim is to study the function:

$$\kappa_d = \dim[P_d/(P_d \ominus FP_{n-d})].$$

Since $FP_0 = P_0 = \mathbf{C1}$ we find $\kappa_0 = \kappa_n = 1$.

LEMMA 5.2. *Let d_0 be the largest integer $d \in [0, n]$ satisfying $P_d \ominus FP_{n-d} = 0$. Then $\kappa_{d_0+l} \leq \kappa_{d_0} = d_0 + 1$, for all $l \in [0, n - d_0]$.*

Proof. Assume that $P_d \ominus FP_{n-d} \neq 0$ and $d < n$. Let a be an arbitrary complex number. Then the multiplication map by $z - a$ is well defined and injective:

$$P_d \ominus FP_{n-d} \xrightarrow{z-a} P_{d+1} \ominus FP_{n-d-1}.$$

Indeed, if $p \in P_d$ and $q \in P_{n-d-1}$, then

$$\langle (z-a)p, Fq \rangle = \langle p, (M^* - \overline{a})Fq \rangle = \langle p, F((z-a)q) \rangle = 0.$$

By choosing a to be distinct from the roots of a non-zero element of $P_{d+1} \ominus FP_{n-d-1}$, the map $z - a$ above will not be surjective.

Thus we have proved that

$$\dim(P_{d+1} \ominus FP_{n-d-1}) > \dim(P_d \ominus FP_{n-d}).$$

But $\dim P_{d+1} = \dim P_d$, so that $\kappa_{d+1} \leq \kappa_d$.

Let d_0 be as in the statement. Then $\kappa_{d_0} = d_0 + 1$ and $\kappa_{d_0+1} \leq d_0 + 2 - 1 = d_0 + 1$. □

The ideal I is principal, generated by a monic polynomial p of degree N:

$$p(z) = \prod_{k=1}^{m} (z - a_k)^{n_k}.$$

We know already that $d_0 < N \leq n$, and that $p \in P_d \ominus FP_d$ for all $d \geq N$. If we seek minimal interpolatory quadrature formulas, then $N = d_0 + 1$ and p is subject to the only restriction $p \in P_{d_0+1} \ominus FP_{n-d_0-1}$.

Indeed, with such a choice, the ideal $I = (p)$ satisfies $\dim(P_n/I_n) = N$ and the integration functional on P_n factors through P_n/I_n: if $q \in P_{n-N}$, then

$$\langle pq, \mathbf{1} \rangle = \langle p, Fq \rangle = 0.$$

Therefore the existence of a quadrature formula with nodes, counting multiplicities, equal to the zeroes of p, is established.

In conclusion, we have proved the following result.

THEOREM 5.3. *Let μ be a rapidly decreasing, positive measure on $\mathbf{C}$ and let F be the associated generalized Friedrichs operator on $P^2(\mu)$. Let n be a positive integer and let d_0 be the largest $d \in [0, n)$ with the property that $P_d \ominus FP_{n-d} = 0$.*

The set of all minimal interpolatory formulas of degree n of μ is in bijective correspondence with the monic polynomials $p \in P_{d_0+1} \ominus FP_{n-d_0-1}$.

For such a polynomial p, the corresponding quadrature formula has nodes equal to the roots of p.

Thus, the theorem above proves in particular the existence of minimal quadratures. For a fixed polynomial p, the corresponding weights in the quadrature formula can be obtained from the moments of degree n of μ, as in the classical case, by linear algebra. A simple dimension count shows that $\dim[P_{d_0+1} \ominus FP_{n-d_0-1}] \leq 2$.

The case when the support of μ is real corresponds to classical Gaussian cubatures. Indeed, since $\text{supp}(\mu) \subset \mathbf{R}$ implies $Fp = p$ for all real polynomials p, the above method of finding minimal cubatures yields the next well known result.

COROLLARY 5.4. *Let μ be a rapidly decreasing measure supported by a subset of the real line and let n be a positive integer. Then:*

a) *If $n = 2k+1$, then there exists a unique minimal cubature formula of degree n, whose nodes coincide with the (simple) zeros of the orthogonal polynomial $p \in P_{k+1} \ominus P_k$;*

b) *If $n = 2k$, then the set of minimal cubature formulas of degree n is parametrized by all monic polynomials $p \in P_{k+1} \ominus P_{k-1}$.*

In parallel with the theory of "mother bodies" of planar domains with real analytic boundary, see [25], we analyze below the following situation. Suppose, besides the measure μ above, a second positive measure $\hat{\mu}$, rapidly decaying at infinity is given, so that:

$$\int p d\mu = \int p d\hat{\mu}, \qquad p \in \mathbf{C}[z]. \tag{21}$$

As in the case of μ, we assume that the support of $\hat{\mu}$ is infinite.

The typical example for such a pair of measures is $\mu = \chi_E dA$ and $\hat{\mu} = \text{const} \cdot \sqrt{1-x^2}dx$, where E is an ellipse with foci at ± 1, see [25].

Let P_d, $\hat{P}_d$ be the spaces of polynomials of degree less than or equal to d, regarded as subspaces of $P^2(\mu)$, respectively $P^2(\hat{\mu})$. Let $F, \hat{F}$ be the corresponding Friedrichs operators. The following simple remark is all we need.

The identity map id: $P_d \longrightarrow \hat{P}_d$ is an isomorphism, and for every pair of non-negative integers k, l, we have:

$$\mathrm{id}[P_k \ominus FP_l] = \hat{P}_k \ominus \hat{F}\hat{P}_l.$$

Indeed, if $p \in P_k$ and $q \in P_l$, then

$$\langle p, \hat{F}q\rangle_{P^2(\hat{\mu})} = \int pqd\hat{\mu} = \int pqd\mu = \langle p, Fq\rangle_{P^2(\mu)}.$$

Thus we have in particular proved the following result.

THEOREM 5.5. *Let $\mu, \hat{\mu}$ be a pair of measures with the same complex moments, both of infinite support.*

For every positive integer n the minimal interpolatory quadrature formulas of degree n, of μ and $\hat{\mu}$, coincide.

Moreover, the spaces $P_d \ominus FP_{n-d}$ and $\hat{P}_d \ominus \hat{F}\hat{P}_{n-d}$ which parametrize these quadratures, are identical.

For instance, in the case of the ellipse E, the minimal quadrature formulas are supported by the zeros of the Chebyshev polynomials, i.e. orthogonal polynomials with respect to the measure $\sqrt{1-x^2}dx$, on the interval $[-1, +1]$. A remarkable feature of this example is that the polynomials in $P_d \ominus FP_{n-d}$ are the same for all confocal ellipses E.

The same phenomenon holds for a continuous family of equipotential domains satisfying a generalized quadrature identity given by a positive measure ν compactly supported on the real axis. To be more specific, let $\Omega_t, 0 < t < t_0$, be the domains constructed by inverse balayge as in [27] and satisfying for all $t \in (0, t_0)$:

$$\int_{\Omega_t} f dA = e^t \int_{\mathbf{R}} f d\nu, \quad f \in \mathbf{C}[z].$$

Then the orthogonal complement $P_d \ominus FP_{n-d}$, considered in the metric of the Hilbert space $L^2(\Omega_t, dA)$ does not depend on t. Consequently the nodes of the minimal quadrature formula are located on the real axis, and can be identified, in terms of the measure ν only, as in Corollary 5.4.

Finally we mention that the location of the zeros of the polynomials $p \in P_{d_0+1} \ominus FP_{n-d_0-1}$, for a fixed measure μ as in Theorem 5.3, seems to be the most difficult part of this theory.

6. Friedrichs operators with symbol

In analogy to the theories of Hankel and Toeplitz operators, we introduce below a class of "Friedrichs operators with symbol". The simplest motivation for this generalization is the pull back of the standard Friedrichs operator by a conformal map.

If not otherwise stated, throughout this section Ω is an arbitrary planar domain. For a function $a \in L^\infty(\Omega)$ we denote by $M_a \in L(L^2(\Omega, dA))$ the multiplication operator $(M_a f)(z) = a(z)f(z)$, $z \in \Omega$, $f \in L^2(\Omega)$. As before, P stands for the Bergman projection.

The *Friedrichs operator* F_a *with symbol* a is then the anti-linear operator :

$$F_a\colon AL^2(\Omega) \longrightarrow AL^2(\Omega), \qquad F_a = PM_aC.$$

It is clear that $\|F_a\| \leq \|a\|_\infty$.

The next property appears implicitly in [17].

LEMMA 6.1. *Let $\phi\colon \Omega^\natural \longrightarrow \Omega$ be a conformal map and let $U\colon L^2(\Omega) \longrightarrow L^2(\Omega^\natural)$ be the associated unitary map $Uf = (f \circ \phi)\phi'$.*

Then for every function $a \in L^\infty(\Omega)$, the corresponding Friedrichs operators satisfy:

$$F_a = U^* F_b^\natural U, \tag{22}$$

where $b = (a \circ \phi)\phi'/\overline{\phi'}$.

The proof is straightforward and it is left to the reader.

In analogy to the compactness criteria known for the original Friedrichs operator, see [7], [28], [30], [17], [25], we investigate below such a property for the operators F_a. The spirit of the proofs below is that of [30].

THEOREM 6.2. *Let Ω be a bounded planar domain with boundary Γ and suppose that $a \in C(\overline{\Omega})$. If for each point $\zeta \in \Gamma$ (at least) one of the following assertions is true:*

(i) *for some $\epsilon > 0$, the set $\Gamma \cap \{z;\ |\zeta - z| < \epsilon\}$ is a $C^{1+\alpha}$ Jordan arc, for some $\alpha > 0$;*

or,

(ii) $a(\zeta) = 0$,

then the operator F_a is compact.

Proof. As in [30], we go back to the definition of a compact operator and prove that under the stated conditions, if (f_j) is a sequence in $AL^2(\Omega)$ which converges weakly to zero, then the sequence $\|Ff_j\|$ tends to zero.

For $f, g \in L^2(\Omega)$ we have:

$$\langle F_a f, g\rangle = \langle M_a \overline{f}, g\rangle = \langle a, fg\rangle,$$

therefore

$$\|F_a f\| = \sup_{\|g\|=1} \left| \int \overline{a} f g \, dA \right|.$$

We have to prove that

$$\lim_{j\to\infty} \sup_{\|g\|=1} \left| \int \overline{a} f_j g \, dA \right| = 0.$$

Assume by contradiction that this is false. Hence there exists a constant $c > 0$ and a sequence $g_j \in AL^2(\Omega)$, $\|g_j\| = 1$, such that:

$$\left| \int \overline{a} f_j g_j \, dA \right| \geq c, \quad j \geq 1. \tag{23}$$

Write $h_j = f_j g_j$ so that h_j is a bounded sequence in $AL^1(\Omega)$, and $h_j(z) \to 0$ for every point $z \in \Omega$. By theorems contained in Chapter 8 of [30], there exists a subsequence, still denoted by h_j, such that the measures $h_j dA$ converge weak* to a signed measure ν carried on Γ_{irr}, that is on the set of those points $\zeta \in \Gamma$ where condition (i) does not hold.

Thus we find that $|\int \overline{a} \, d\nu| \geq c$, but $\overline{a}$ vanishes on supp(ν), a contradiction. □

Thus, in studying the compactness behaviour of the operator F_a, one can localize the symbol a at points of Γ. As a direct application of the above proof we state the following result.

COROLLARY 6.3. *Let $a, b \in C(\overline{\Omega})$. If the operator F_a is compact, so is F_{ab}.*

As Friedrichs showed, if $\partial\Omega$ is smooth except for one "corner" where there is an angle $\alpha \neq \pi, 2\pi$, then the operator $F = F_1$ cannot be compact. He did this by showing that the self-adjoint operator $S = F^2$ has in its essential spectrum the number $(\frac{\sin\alpha}{\alpha})^2$. Here we outline another, computationally perhaps simpler way to establish non-compactness in this situation, that is also applicable with slight modifications to the boundary Friedrichs operator, to be discussed in the next section.

PROPOSITION 6.4. *Assume that the boundary of the bounded domain Ω contains a corner of angle $\alpha \neq \pi, 2\pi$. Then the operator F is not compact.*

Proof. We will assume for simplification, although not necessary, that the boundary has only one corner z_0. Let $\rho > 0$ be a sufficiently small radius so that the set $\Omega \cap B(z_0, \rho)$ has a simply connected component Ω_0 having z_0 on its boundary.

Let ψ be a non-negative C^∞ function on $\mathbf{C}$ supported by a compact subset of the ball $B(z_0, \rho)$ and satisfying $\psi = 1$ on $B(z_0, \rho/2)$. Denote $a = \psi\chi_\Omega$. Then $F = F_a + F_{1-a}$; by the above theorem the operator F_{1-a} is compact since its symbol vanishes at the irregular point of $\partial\Omega$. Thus it suffices to show that F_a is not compact.

Next, map the wedge $W = \{w : |\arg w| < \alpha/2\}$ by a conformal map $z = \psi(w)$ on Ω_0, so that $\psi(0) = z_0$. According to Lemma 6.1, the operator F_a is unitarily equivalent to the Friedrichs operator on W with symbol $b = (a \circ \phi)(\phi'/\overline{\phi'})$. It is easy to see that ϕ' is continuous on $\overline{W}$ and non-vanishing near $w = 0$.

Thus our assertion is reduced to proving that if the function $b \in C(\overline{W})$ is bounded, with $b(0) \neq 0$, then the operator F_b on W is not compact.

Let $f \in AL^2(W)$ be a function of order $O(|w|^{-2})$ near ∞ and satisfying $\int_\Omega f^2\, dA \neq 0$. This is possible because $\alpha \neq \pi$ and hence W is not a "null quadrature domain". For instance $f(w) = (w+1)^{-3}$ is such a function.

Then the functions $f_n(w) = nf(nw)$, $n \geq 1$, have equal L^2 norms on W and tend weakly to zero in $AL^2(W)$.

But $F_b f_n$ do not tend in norm to zero. Indeed,

$$\begin{aligned}\langle f_n, F_b f_n\rangle &= \int_W \bar{b} f_n^2\, dA = \int_W \overline{b(w)} n^2 f(nw)^2\, dA \\ &= \int_W \overline{b(w/n)} f(w)^2\, dA \to \overline{b(0)} \int_W f^2\, dA \neq 0,\end{aligned}$$

by Lebesgue's dominated convergence theorem.

□

In view of Theorem 6.2 we obtain the next application.

COROLLARY 6.5. *In the conditions of Proposition 6.4, let $a \in C(\overline{\Omega})$. If $a(z_0) \neq 0$, then the operator F_a is not compact.*

Exactly as in the case of the Friedrichs operator, we note the following useful intertwining property. We denote as usual by $T_h = PM_h \colon AL^2(\Omega) \longrightarrow AL^2(\Omega)$ the *Toeplitz operator* with symbol h.

LEMMA 6.6. *Let $a \in L^\infty(\Omega)$ and let $h \in H^\infty(\Omega)$. Then:*

$$T_h^* F_a = F_a T_h. \tag{24}$$

Proof. The proof is a simple chain of operator identitites:

$$T_h^* F_a = PM_{\overline{h}} PM_a CP = PM_{\overline{h}a} CP = PM_a CM_h P = F_a T_h.$$

□

By using this identity and following the proof in [25] one can show that for any symbol $a \in L^\infty(\Omega)$ the corresponding operator F_a cannot be Fredholm.

7. The boundary Friedrichs operator

Let Ω be a bounded domain with smooth boundary and let $H^2(\Omega)$ be the Hardy space, regarded as a closed subspace of $L^2(\Omega, ds)$; let P denote the corresponding orthogonal projection. To simplify the coming computations we asume that the measure ds is normalized and has mass one. By considering the complex conjugation operator $C: L^2(\Omega, ds) \longrightarrow L^2(\Omega, ds)$, a boundary Friedrichs operator $B = PC: H^2(\Omega) \longrightarrow H^2(\Omega)$ can therefore be defined in analogy to the Bergman space situation considered at the beginning of this note. The purpose of the present section is to outline a few relevant properties of this generalized Friedrichs operator.

First we consider the ideal situation of a domain bounded by a real analytic Jordan curve. Our aim is to draw a parallel to the classical Riesz–Herglotz representation in the unit disk, cf. for instance [6], and show that the boundary Friedrichs operator naturally appears in at least one such formula.

Let $S(z, \overline{w})$ be the Szegő kernel, that is the reproducing kernel of the Hardy space $H^2(\Omega)$, cf. [3]. Then any analytic function $f \in H^2(\Omega)$, assimilated with its non-tangential boundary values, satisfies:

$$f(z) = \int_{\partial\Omega} f(\zeta) S(z, \overline{\zeta})\, ds(\zeta), \qquad z \in \Omega.$$

Similarly we have:

$$Bf(z) = \int_{\partial\Omega} \overline{f(\zeta)} S(z, \overline{\zeta})\, ds(\zeta), \qquad z \in \Omega.$$

By denoting $u = \Re f$ and adding the above equations we obtain:

$$\frac{1}{2}(f + Bf)(z) = \int_{\partial\Omega} u(\zeta)S(z,\overline{\zeta})\, ds(\zeta), \qquad z \in \Omega. \tag{25}$$

An equally simple argument leads to the following companion formula:

$$\frac{1}{2}(f(z) + \overline{f(w)})S(z,\overline{w}) = \int_{\partial\Omega} u(\zeta)S(\zeta,\overline{w})S(z,\overline{\zeta})\, ds(\zeta), \qquad z, w \in \Omega. \tag{26}$$

Both the above representations can be interpreted as generalizations of the classical Riesz–Herglotz formula in the disk. The first one is of interest for instance when the operator B has finite rank, in which case a solution of the Dirichlet problem for the Laplace operator can be deduced from it by linear algebra in a finite dimensional space.

Also, under the generous conditions imposed on Ω we will see below that 1 is an isolated point in the spectrum of B^2, so that the operator $(I + B)$ can be inverted on the subspace $H_0^2(\Omega)$ of functions f having zero mean on $\partial\Omega$: $\langle f, \mathbf{1}\rangle = 0$. This yields an arbitrary function $f(z)$ as an integral operator applied to $u(\zeta)$. To be more precise, let $a = \langle f, \mathbf{1}\rangle$, so that the function $f - a\mathbf{1}$ has zero mean on the boundary and $\langle u, \mathbf{1}\rangle = \Re a$. By subtracting formula (25) written for the constant function a from that of f we obtain a Riesz–Herglotz type formula.

LEMMA 7.1. *Let Ω be a simply connected domain with real analytic boundary and let $f \in H^2(\Omega)$. Then*

$$f = \langle f, \mathbf{1}\rangle\mathbf{1} + \left(\frac{I + B}{2}\right)^{-1} \int_{\partial\Omega} (S(*,\overline{\zeta}) - \mathbf{1})\Re f(\zeta)\, ds(\zeta). \tag{27}$$

Note that the inverse of $(I + B)$ acts on an element of $H_0^2(\Omega)$ and that $\Re a = \Re\langle f, \mathbf{1}\rangle$ is determined by the boundary values $u(\zeta) = \Re f(\zeta)$, $\zeta \in \partial\Omega$, while $\Im a = \Im\langle f, \mathbf{1}\rangle$ remains, as expected, free. We leave it to the interested reader to develop the ramifications of this formula.

The second formula, (26), contains a more symmetric, positive definite Poisson type kernel $P(z,\zeta)$ of the domain Ω: $P(z,\zeta) = \frac{|S(z,\overline{\zeta})|^2}{S(z,\overline{z})}$; see [3] for more details of this nature. In the (only) well studied case of a disk (or polydisk), formula (26) is the link between the Riesz–Herglotz representation and classical interpolation problems, see [15].

Returning to formula (25), a conformal mapping argument, or simply the inspection of the regularity of the integral kernel of the Friedrichs operator, shows that B extends uniquely (by continuity) to any function $f \in H^1(\Omega)$, where now $Bf \in H^p(\Omega), p < 1$, and therefore it has almost everywhere boundary values. We will denote the extension by the same symbol Bf. This immediately leads to the following representation formula, also of classical flavour.

PROPOSITION 7.2. *Let Ω be an analytic, smooth, Jordan domain and let f be an analytic function in Ω satisfying $\Re f > 0$ in Ω. Let μ be the positive measure supported by $\partial\Omega$ obtained as distributional boundary values of $\Re f(\zeta)$. Then*

$$\frac{1}{2}(f+Bf)(z) = \int_{\partial\Omega} S(z,\overline{\zeta})\, d\mu(\zeta), \qquad z \in \Omega. \tag{28}$$

Conversely, any positive measure μ on $\partial\Omega$ produces by the above formula an analytic function f in Ω, unique up to an imaginary constant, and satisfying $\Re f > 0$ in Ω.

Proof. It is known that any positive measure μ on $\partial\Omega$ is the distributional boundary limit of a positive harmonic function $\Re f > 0$. To be more specific, this means that μ is the weak* limit of measures $\Re f ds$ carried on suitable curves bounding domains exhausting Ω, see [6].

We have only to prove the fact that, if $f \in H^1(\Omega)$ and $(I+B)f = 0$, then f is a purely imaginary constant. Let $h \in H^\infty(\Omega)$. By continuity we infer:

$$\int_{\partial\Omega} fh\, ds = -\int_{\partial\Omega} \overline{f} h\, ds.$$

But this shows that the harmonic function $\Re f$ is orthogonal to $H^\infty(\Omega)+CH^\infty(\Omega)$. The regularity assumption on the boundary, and the simply connectedness of Ω imply then that $\Re f = 0$, hence $f = iC$ with $C \in \mathbf{R}$. □

From this point on we assume that Ω is a domain with rectifiable boundary Γ. The Szegő projector will be denoted by $P\colon L^2(\Gamma, ds) \longrightarrow H^2(\Gamma, ds)$, and for a function $a \in L^\infty(\Gamma)$ we define the *boundary Friedrichs operator with symbol* a as: $B_a = PM_aC$. In parallel to the case of the Bergman space we will be interested in the compactness of this operator.

Again, if $\phi\colon \Omega^\natural \longrightarrow \Omega$ is a conformal map, then the operator $B_a \in L(H^2(\Omega))$ is unitarily equivalent to the operator $B_b^\natural \in L(H^2(\Omega^\natural))$ with symbol $b = (a\circ\phi)(\phi'/\overline{\phi'})$. We omit the nearly identical details.

THEOREM 7.3. *Let Ω be a bounded domain with C^1 smooth boundary Γ and let $a \in C(\Gamma)$. Then the operator B_a is compact.*

Proof. The proof follows the same scheme as in the Bergman space case: assume that (f_j) is a weakly convergent to zero sequence in $H^2(\Omega)$. We have to show that

$$\lim_{j\to\infty} \sup_{\|g\|_{2,\Gamma}=1} \left| \int_\Gamma \overline{a(z)} f_j(z) g(z)\, ds \right| = 0.$$

By our regularity assumption, on the boundary Γ one can write $dz = \tau ds$, where τ denotes the unit tangent vector at $z \in \Gamma$, and the function $\tau(z)$ is continuous. Thus we must prove that

$$\lim_{j\to\infty} \sup_{\|g\|_{2,\Gamma}=1} \left| \int_\Gamma \overline{a(z)\tau(z)} f_j(z) g(z)\, dz \right| = 0.$$

Now $h_j = f_j g$ are in $H^1(\Omega)$ with bounded norms and it is easy to check that $h_j(z) \to 0$ for each $z \in \Omega$. Since $a\tau \in C(\Gamma)$, the proof will be completed by the next proposition.

□

Proposition 7.4. *Let (h_j) be a sequence in $H^1(\Omega)$ which is bounded and tends pointwise to zero in Ω. Then the measures $(h_j dz)$ on Γ tend to zero in the weak* topology.*

Proof. We are given that

$$h_j(\zeta) = \frac{1}{2\pi i} \int_\Gamma \frac{h_j(z)dz}{z-\zeta} \to 0, \qquad \zeta \in \Omega.$$

Moreover,

$$0 = \int_\Gamma \frac{h_j(z)dz}{z-\zeta}, \qquad \zeta \in \mathbf{C} \setminus \overline{\Omega}.$$

Hence

$$\lim_{j\to\infty} \int_\Gamma h_j(z) R(z) dz = 0,$$

for any rational function R with poles off Γ.

Since the area of Γ is zero by our regularity assumption, and the $L^1(\Gamma, ds)$ norms of h_j are bounded, the Hartogs–Rosenthal Approximation Theorem ([8]) allows us to replace R in the last equation by an arbitrary continuous function on Γ, which concludes the proof.

□

Exactly as in the preceding section we can treat the situation of domains with finitely many corners and obtain the next result.

PROPOSITION 7.5. *Let Ω be a domain with smooth boundary, except a corner of angle $\alpha \neq \pi, 2\pi$. Then the boundary Friedrichs operator B_1 is not compact.*

Finally, let us remark that Lemma 6.6 has a counterpart for the boundary Friedrichs operator: if $a \in L^\infty(\Gamma)$ and $h \in H^\infty(\Omega)$, then

$$T_h^* B_a = B_a T_h,$$

where $T_h = PM_hP$ is the corresponding Toeplitz operator. Thus, as before, the operator B_a cannot be Fredholm.

8. Friedrichs operators on the unit disk

Of special interest is the case of the unit disk $\Omega = \mathbf{D}$. We denote by $z = e^{i\theta}$ the coordinate on the unit circle $\mathbf{T}$, and work with the normalized arc length measure $ds = \frac{1}{2\pi} d\theta$. The standard orthonormal basis of $L^2 = L^2(\mathbf{T}, ds)$ is $(z^n)_{n\in\mathbf{Z}}$ and the vector z^0 will also be denoted $\mathbf{1}$. We recall that for a function $a \in L^\infty(\mathbf{T})$ the associated *Hankel operator* $H_a: H^2 \longrightarrow L^2 \ominus H^2$ is $H_a = (I-P)M_aP$. Henceforth we regard H_a, by natural coextension, as an operator from H^2 to L^2.

LEMMA 8.1. *Let $a \in L^\infty(\mathbf{T})$. The boundary Friedrichs operator B_a and the Hankel operator $H_{\overline{a}}$ are related by the following equation:*

$$B_a = CH_{\overline{a}} + \mathbf{1} \otimes a^*, \tag{29}$$

where $a^*(z) = \overline{a(\overline{z})}$.

Proof. It is enough to check the identity on a basis vector z^n, $n \in \mathbf{Z}$. Let $a(z) = \sum_{k=-\infty}^{\infty} a_k z^k$ be the Fourier decomposition of the function a.
Then

$$B_a z^n = P\Big(\sum_{k=-\infty}^{\infty} a_k z^{k-n} \Big) = \sum_{k\geq n} a_k z^{k-n},$$

and on the other hand,

$$CH_{\overline{a}} z^n = C(I-P) \sum_{k=-\infty}^{\infty} \overline{a_k} z^{n-k} = \sum_{k>n} a_k z^{k-n}.$$

Thus the operator $B_a - CH_{\overline{a}}$ has rank one and can be identified with $\langle *, a^* \rangle \mathbf{1}$, that is the operator which maps z^n to $a_n \mathbf{1}$ for every $n \geq 0$.

$\square$

Thus the twisted Hankel operator $CH_{\overline{a}}$ is a rank-one perturbation of B_a, and therefore if one is compact the other is, too. Now the compactness or Schatten–von Neumann norm estimates of Hankel operators on the unit disk is a well studied area, with important applications to function theory, operator theory, approximation theory and stochastic processes, see for references [4], [20], [22], [23], [24].

We present below as a continuation of the previous proofs a slightly different approach to the compactness of the Friedrichs operators B_a and F_a on the unit disk.

THEOREM 8.2. *Let $a \in L^\infty(\mathbf{T})$. Consider the following conditions:*
(i) *the boundary Friedrichs operator B_a is compact;*
(ii) *every maximizing sequence for the extremal problem* (iii) *is norm convergent;*
(iii) *the linear functional $\Lambda_a : H^1(\mathbf{T}) \longrightarrow \mathbf{C}$,*

$$\Lambda_a g = \frac{1}{2\pi i}\int_{\mathbf{T}} \overline{a} g\, dz$$

attains a maximum on the unit sphere of H^1;
(iv) *$\overline{a}$ has a unique closest element in H^∞;*
(v) *$\overline{a} \in H^\infty + C(\mathbf{T})$.*
Then (i)$\Rightarrow$(ii)$\Rightarrow$(iii)$\Rightarrow$(iv) *and* (v)$\Leftrightarrow$(i).

The equivalence between (i) and (v) is due to Nehari [20]; the implication from (iii) to (iv) is standard in the theory of dual problems, cf. [6] pp. 133; the implication from (iv) to (v) is not true as it was for instance remarked in [1].

We are grateful to one of the referees for the following couple of counterexamples. Quite specifically, let $f \in L^\infty = L^\infty(\mathbf{T})$ satisfy $\mathrm{dist}(f, H^\infty + C) < \mathrm{dist}(f, H^\infty)$. Then it is well known and it is easy to prove that f has a unique closest element in H^∞. However, under this condition f does not have to belong to $H^\infty + C$. To see that the implication (iii)$\Rightarrow$(v) is false, consider a function $h \in H^2$ such that $f = h/\overline{h} \notin H^\infty + C$. It is easy to see then that the functional $L_{\overline{f}}$ attains a maximum at $g = h^2$.

In what follows we merely indicate the principal ideas in the proof of Theorem 8.2. As in the proofs contained in the last two sections, the main ingredient is the following lemma. The notations are unchanged.

LEMMA 8.3. *A necessary and sufficient condition for the operator B_a to be compact is: for every sequence (h_n) in $H^1(\mathbf{T})$ with bounded norms such that $h_n(z) \to 0$ for all $z \in \mathbf{D}$ we have $\Lambda_a h_n \to 0$.*

Proof. Suppose that B_a is compact, and let $h_n \in H^1$ be a sequence with $\|h_n\|_1 = 1$ and $h_n(z) \to 0$ pointwise in $\mathbf{D}$. One can factor $h_n = f_n g_n$ with $f_n, g_n \in H^2$ and $\|f_n\|_2 = \|g_n\|_2 = 1,\ n \geq 1$. By passing to subsequences we can assume that $f_n \to f$ and $g_n \to g$, in the weak topology of H^2. Then $f_n g_n \to fg$ pointwise, hence $fg = 0$, that is $f = 0$ or $g = 0$. Without loss of generality we can assume $f = 0$. Since $\|B_a f_n\|_2 \to 0$, we conclude:

$$\Lambda_a h_n = \langle g_n, B_a f_n\rangle \to 0.$$

In the opposite direction, assuming $\Lambda_a h_n \to 0$ for every sequence $(h_n) \subset H^1$ converging weak* to zero, let f_n be a sequence of H^2 which converges weakly to zero. Then for every bounded sequence $(g_n) \subset H^2$ we have

$$\langle g_n, B_a f_n\rangle = \Lambda_a(f_n g_n) \to 0,$$

verifying the compactness of B_a. □

In the proof of (i)⇒(ii) we need the following known result [21].

Lemma 8.4. *Let $h_n, h \in H^1$ satisfy $h_n \to h$ in the weak* topology and $\|h_n\| \to \|h\|$. Then $\|h_n - h\|_1 \to 0$.*

Proof. We can assume $\|h_n\|_1 = \|h\|_1,\ n \geq 1$. Factor then $h_n = f_n g_n$ with $f_n, g_n \in H^2$, $\|f_n\|_2 = \|g_n\|_2 = 1$. By passing to subsequences we can assume that $f_n \to f$ and $g_n \to g$, weakly in H^2. Thus $1 \leq \|f\|_2\|g\|_2 \leq 1$, therefore $\|f\|_2 = \|g\|_2 = 1$.

Then

$$\begin{aligned}\int_{\mathbf{T}} |h_n - h|\, ds = \int_{\mathbf{T}} |f_n g_n - f_n g + f_n g - fg|\, ds &\leq \int_{\mathbf{T}} (|f_n||g_n - g| + |f_n - f||g|)\, ds \\ &\leq \|f_n - f\|_2 + \|g_n - g\|_2.\end{aligned}$$

But

$$\|f_n - f\|_2^2 = \|f_n\|_2^2 + \|f\|_2^2 - 2\Re \int_{\mathbf{T}} f\overline{f_n}\, ds \to 0,$$

and similarly $\|g_n - g\|_2 \to 0$.

In conclusion $\|h_n - h\|_1 \to 0$. □

Next we discuss the proof of (i)⇒(ii) in Theorem 8.2. Suppose B_a is compact and let

$$M = \|\Lambda_a\| = \sup_{h \in (H^1)_1} \left|\int_{\mathbf{T}} \overline{a} h\, ds\right|.$$

For a Banach space X, we have denoted above by X_1 its closed unit ball. Let (h_n) be a sequence in H^1, satisfying $\|h_n\|_1 = 1$ and $\Lambda_a h_n \to M$. We can assume that $M \neq 0$, otherwise the assertion is trivial. By passing to a subsequence we can assume that $h_n \to h$ pointwise, for some element $h \in H^1$. If we can show that $\|h - h_n\|_1 \to 0$ we are done since then $\Lambda_a h = M$.

Assume by contradiction, in view of the preceding lemma, that $\|h\|_1 < 1$. Then

$$\Lambda_a h = \Lambda_a h_n + \Lambda_a(h - h_n),$$

and the last integral tends to zero by Lemma 8.3. Thus $\Lambda_a h = M$ which contradicts $\|h\| < 1$.

The implication (v)⇒(i) follows from Theorem 7.3, and the rest of the proof can be derived from the cited references: [1] or [20] for the implication (i)⇒(v) and [6] for the implication (iii)⇒(iv).

Theorem 8.2 and its proof show that the compactness of the boundary Friedrichs operator B_a is easier to characterize than its Bergman space counterpart. For the operator F_a a Nehari type condition is not yet known.

We can sum up what we know about the compactness of F_a for the unit disk as follows. We denote by

$$N(\mathbf{D}) = \{u \in L^\infty(\mathbf{D}) : \int_\Omega fu\,dA = 0,\ f \in AL^1(\mathbf{D})\},$$

the annihilator of the space $AL^1(\mathbf{D})$.

Proposition 8.5. *Let $a \in L^\infty(\mathbf{D})$, and consider the assertions:*

(i) *F_a is compact;*

(ii) *any maximizing sequence for the extremal problem* (iii) *is norm convergent;*

(iii) *the linear functional*

$$L_a\colon AL^1(\Omega) \longrightarrow \mathbf{C}, \qquad L_a(f) = \int_\Omega \overline{a} f\,dA$$

attains a maximum on the unit sphere of $AL^1(\Omega)$;

(iv) *$\overline{a}$ has a unique closest element in $N(\mathbf{D})$, in sup norm;*

(v) *$\overline{a}$ belongs to $C(\overline{\mathbf{D}}) + N(\mathbf{D})$.*

Then (i)⇒(ii)⇒(iii)⇒(iv) *and* (v)⇒(i).

REMARKS. Observe that, since the annihilator space $N(\mathbf{D})$ is weak* closed in $L^\infty(\mathbf{D})$, every function in $L^\infty(\mathbf{D})$ does have at least one closest element in $N(\mathbf{D})$. For aspects of this see [14] and further references there.

Functions a in $L^\infty(\mathbf{D})$, for which the maximum of L_a is attained on the unit sphere of $AL^1(\mathbf{D})$, have a special interest in connection with an extremal problem in quasiconformal mapping, which translates into the question: for which a in $L^\infty(\mathbf{D})$ of norm one the functional L_a has norm one? In those cases where L_a attains a maximum this can only happen if a has the special form $|g|/g$ for some $g \in AL^1(\mathbf{D})$, and then a corresponds to the complex dilation of a so-called Teichmüller mapping, see e.g. the references in [28].

Proof of Proposition 8.5. (outline) The implications (i)⇒(ii)⇒(iii) are proved in a similar way to the corresponding steps in Theorem 8.2, based on the (known) analog of Lemma 8.4. The implication (iii)⇒(iv) is standard approximation theory, see [14]. And (v)⇒(i) has been done. □

References

[1] S. AXLER, I. D. BERG, N. P. JEWELL and A. SHIELDS, Approximation by compact operators and the space $H^\infty + C$, *Ann. Math.*, **109** (1979), 601–612.
[2] M. B. BALK, *Polyanalytic Functions*, Akademie Verlag, Berlin, 1991.
[3] S. BERGMAN, *The Kernel Function and Conformal Mapping*, Amer. Math. Soc., Providence, R. I., 1970.
[4] J. BOURGAIN, On the similarity problem for polynomially bounded operators on Hilbert space, *Israel J. Math.*, **53** (1986), 315–332.
[5] J. J. CARMONA, K. YU. FEDOROVSKI and P. V. PARAMONOV, *On uniform approximation by polyanalytic polynomials and the Dirichlet problem for bianalytic functions*, preprint, 1999.
[6] P. DUREN, *Theory of H^p Spaces*, Academic Press, New York, 1970.
[7] K. FRIEDRICHS, On certain inequalities and characteristic value problems for analytic functions and for functions of two variables, *Trans. Amer. Math. Soc.*, **41** (1937), 321–364.
[8] T. GAMELIN, *Uniform Algebras*, Prentice Hall, Englewood Cliffs New Jersey, 1969.
[9] B. GUSTAFSSON, Quadrature identities and the Schottky double, *Acta Appl. Math.*, **1** (1983), 209–240.
[10] B. GUSTAFSSON, On mother bodies of convex polyhedra, *SIAM J. Math. Analysis*, **29** (1998), 1106–1117.
[11] V. HAVIN, Approximation in the mean by analytic functions, *Dokl. Akad. Nauk. SSSR*, **178** (1968), 1025–1028.
[12] T. KATO, *Perturbation Theory for Linear Operators*, Springer Verlag, Berlin, 1995.

[13] D. Khavinson, J. McCarthy and H. S. Shapiro, Best approximation in the mean by analytic and harmonic functions, *Indiana Univ. Math. J.*, to appear.

[14] D. Khavinson and H. S. Shapiro, Best approximation in the supremum norm by analytic and harmonic functions, *Ark. för Mat.*, to appear.

[15] A. Korányi and L. Pukánsky, Holomorphic functions with positive real part on polycylinders, *Trans. Amer. Math. Soc.*, **108** (1963), 449–456.

[16] A. Lenard, The numerical range of a pair of projections, *J. Funct. Analysis*, **10** (1972), 410–413.

[17] P. Lin and R. Rochberg, On the Friedrichs operator, *Proc. Amer. Math. Soc.*, **123** (1995), 3335–3342.

[18] D. Luecking, The compact Hankel operators form an M-ideal in the space of Hankel operators, *Proc. Amer. Math. Soc.*, **79** (1980), 222–224.

[19] I. P. Mysovskikh, *Interpolatory Cubature Formulas*, Nauka, Moscow, 1981 (in Russian).

[20] Z. Nehari, On bounded bilinear forms, *Ann. Math.*, **65** (1957), 153–162.

[21] D. J. Newman, Pseudo-uniform convexity in H^1, *Proc. Amer. Math. Soc.*, **14** (1963), 676–679.

[22] N. K. Nikolskii, *Treatise on the Shift Operator*, Springer, Berlin, 1986.

[23] V. V. Peller, Estimates of functions of power bounded operators on Hilbert spaces, *J. Operator Theory*, **7** (1982), 341–472.

[24] V. V. Peller and S. V. Hruscev, Hankel operators, best approximations and stationary Gaussian processes, *Uspehi Math. Nauk*, **37** (1982), 53–124.

[25] M. Putinar and H. S. Shapiro, The Friedrichs operator of a planar domain, *S. A. Vinogradov Memorial Volume,* Birkhäuser Verlag, Basel, 2000, 303–330.

[26] F. Riesz and B. Sz.-Nagy, *Functional Analysis*, Dover, New York, 1990.

[27] M. Sakai, *Quadrature Domains,* Lect. Notes Math. Vol. **934**, Springer, Berlin, 1982.

[28] H. S. Shapiro, Some inequalities for analytic functions integrable over a plane domain, *Approximation and Function Spaces,* Proc. Conf. Gdansk 1979, North Holland, 1981, 645–666.

[29] H. S. Shapiro, On some Fourier and distribution-theoretic methods in approximation theory, *Approximation Theory.* III, Proc. Conf. Austin, Texas, 1980, Academic Press, San Diego, 1980, 87–124.

[30] H. S. Shapiro, *The Schwarz Function and its Generalization to Higher Dimensions*, Wiley-Interscience, New York, 1992.

[31] P. K. Suetin, *Polynomials Orthogonal over a Region and Bieberbach Polynomials,* Proc. Steklov Inst. Vol. **100** (1971), Amer. Math. Soc., Providence, R.I., 1974.

M. Putinar, Department of Mathematics, University of California, Santa Barbara, CA 93106; *e-mail*: mputinar@math.ucsb.edu

H. S. Shapiro, Department of Mathematics, The Royal Institute of Technology, 10044 Stockholm, Sweden; *e-mail*: shapiro@math.kth.se

Operator Theory: Advances and Applications, Vol. 127, 553–560

Localization of the Wielandt–Wintner Theorem

MIHAI ŞABAC

The Wielandt–Wintner Theorem states impossibility to write nonzero scalar multiple of the identity operator on some complex Banach space $\mathcal{X}$ as the commutator of two bounded operators on $\mathcal{X}$. In what follows we give some extensions of this result in which we consider the representations of the canonical commutation relations in some fixed non zero point of $\mathcal{X}$.

Let $\mathcal{X}$ be a complex Banach space, $\mathcal{B}(\mathcal{X})$ the algebra of all bounded operators on $\mathcal{X}$, and $P, Q \in \mathcal{B}(\mathcal{X})$ two bounded operators on $\mathcal{X}$. $\mathbb{C}$ will be the field of complex numbers. We will denote by $\sigma(A)$ the spectrum of $A \in \mathcal{B}(\mathcal{X})$ and $[P,Q] = PQ-QP$.

A mathematical formulation of the Heisenberg uncertainty principle can be phrased:

"Is the scalar multiple of the identity a commutator?"

i.e.

Do there exist operators P, Q such that $PQ - QP = \alpha I$?

If P, Q are bounded operators on a complex Banach space, the answer to the above question is negative and that is exactly the Wielandt–Wintner theorem. For this case, two different proofs are known:

The first, Wintner's proof (1947) used the compactness of the spectrum of bounded operators, the equality between $\sigma(PQ)$ and $\sigma(QP)$ when P is invertible and the triviality of the translations that can leave a non-empty compact subset of the complex plane invariant.

The second, Wielandt's proof (1949) used a norm evaluation of the equality $[P^n, Q] = n\alpha P^{n-1}$ derived from $[P,Q] = \alpha I$.

Our aim is to give a slight extension of the Wielandt–Wintner Theorem, i.e. to prove the following implication:

$$\text{I.} \qquad PQ \overset{sp}{\sim} QP + \alpha I, \quad \alpha \in \mathbb{C}, \quad P, Q \in \mathcal{B}(\mathcal{X}) \Longrightarrow \alpha = 0,$$

AMS Subject Classification (1991): 47B47, 47D45.

and mainly to study the consequences of the following local version of Heisenberg's commutation relations:

$$\text{II.} \qquad [P,Q]x = \alpha x \text{ for some } \alpha \in \mathbb{C} \text{ and } 0 \neq x \in \mathcal{X}.$$

The assertions I and II are obtained by replacing the equality in the classical formulation of Heisenberg principle by spectral equivalence or by equality in some fixed nonzero point of $\mathcal{X}$.

1. Spectral form of the Wielandt–Wintner Theorem

First we recall that $\sigma(PQ) \setminus \{0\} = \sigma(QP) \setminus \{0\}$ for every $P, Q \in \mathcal{B}(\mathcal{X})$. Indeed, if $0 \neq \lambda \notin \sigma(PQ)$ and $U = (\lambda I - PQ)^{-1}$, it is easy to verify that $\lambda^{-1}(QUP + I)$ is $(\lambda I - QP)^{-1}$.

We recall also the implication

$$A \overset{sp}{\sim} B \Longrightarrow \sigma(A) = \sigma(B),$$

for every $A, B \in \mathcal{B}(\mathcal{X})$, where $A \overset{sp}{\sim} B$ means

$$\lim_{n\to\infty} \|C^n(A,B)(I)\|^{1/n} = \lim_{n\to\infty} \|C^n(B,A)(I)\|^{1/n} = 0$$

and $C(A,B)\colon \mathcal{B}(\mathcal{X}) \longrightarrow \mathcal{B}(\mathcal{X})$ is $C(A,B)(X) = AX - XB$ (see [1]).

THEOREM 1. *If $P, Q \in \mathcal{B}(\mathcal{X})$, then the following implication holds:*

$$\sigma(PQ) = \sigma(QP) + \alpha, \quad \alpha \in \mathbb{C} \Longrightarrow \alpha = 0.$$

The following corollaries of this theorem are obvious.

COROLLARY 1. *If $P, Q \in \mathcal{B}(\mathcal{X})$, then the following implication holds:*

$$PQ \overset{sp}{\sim} QP + \alpha I, \quad \alpha \in \mathbb{C} \Longrightarrow \alpha = 0.$$

WIELANDT–WINTNER THEOREM. *If $P, Q \in \mathcal{B}(\mathcal{X})$ then the following implication holds:*

$$[P,Q] = \alpha I, \quad \alpha \in \mathbb{C} \Longrightarrow \alpha = 0.$$

Proof of Theorem 1. We will adapt the proof of Wintner ([5]). We will prove the following lemma.

LEMMA. *Let* $M_1, M_2 \subseteq \mathbb{C}$ *be two bounded subsets of the complex plane such that* $M_1 \setminus \{0\} = M_2 \setminus \{0\}$. *The following implication holds*:

$$\alpha \in \mathbb{C}, \quad M_2 = M_1 + \alpha \Longrightarrow \alpha = 0.$$

Indeed, if $\mathbb{C} \ni \alpha \neq 0$ and $M_2 = M_1 + \alpha$ let $\lambda \in M_1$. There are two possibilities: (a) $\lambda = 0$ or, (b) $\lambda \neq 0$.

In case (a) we have by hypothesis

$$0 \neq \alpha \in M_2 \setminus \{0\} = M_1 \setminus \{0\} \Rightarrow 0 \neq \alpha + \alpha = 2\alpha \in M_2 \Rightarrow \cdots \Rightarrow n\alpha \in M_2$$

for every $n \in \mathbb{N}$, which is a contradiction because M_2 is bounded.

In the case (b) there are two possibilities: (b_1) $\lambda + n\alpha \neq 0$ for every $n \in \mathbb{N}$, which implies as above, $\lambda + \alpha, \lambda + 2\alpha, \ldots, \lambda + n\alpha \in M_2$ for every $n \in \mathbb{N}$, which is a contradiction; (b_2) $\lambda + n\alpha = 0$ for some natural number $n \neq 0$.

But $M_2 = M_1 + \alpha$ and we can write from (b_2)

$$0 \neq \lambda = -n\alpha \in M_1 \setminus \{0\} = M_2 \setminus \{0\} \Rightarrow -n\alpha = \mu + \alpha, \quad \mu \in M_1.$$

Hence

$$\mu = -(n+1)\alpha \neq 0,\ \mu \in M_1 \setminus \{0\} = M_2 \setminus \{0\} \Rightarrow -(n+1)\alpha = \theta + \alpha, \quad \theta \in M_1$$

and

$$\theta = -(n+2)\alpha \neq 0,\ \theta \in M_1 \Rightarrow \theta \in M_2, \ldots \text{ etc.}$$

Therefore $-(n+p)\alpha \in M_2$ for every $p \in \mathbb{N}$, which is a contradiction because M_2 is bounded. So our lemma is proved.

For proving Theorem 1 we put $M_2 = \sigma(PQ)$, $M_1 = \sigma(QP)$. □

2. Punctual form of the Wielandt–Wintner Theorem

We will describe some situations when II also implies $\alpha = 0$. For one case we use the ideas of Wielandt's proof. For another case a new proof idea is presented; particularly, a new proof of the Wielandt–Wintner Theorem is given.

Now we suppose that $\|P\| = 1$ and $\sigma(P) \subset \mathbb{T}_1 = \{z \mid |z| = 1\}$. For instance, every unitary operator on a Hilbert space has these properties.

THEOREM 2. *Let us consider* $P, Q \in \mathcal{B}(\mathcal{X})$, $\|P\| = 1$, $\sigma(P) \subset \mathbb{T}_1, 0 \neq x \in \mathcal{X}$ *such that* $Px = \lambda x$ *and* $[P, Q]x = \alpha x$ *where* λ *and* α *are scalars in* $\mathbb{C}$. *Under these conditions we have*:

1. $\alpha = 0$ *i.e.* $[P, Q]x = 0$;
2. $PQx = \lambda Qx$ *and* $Q\mathcal{X}_0 \subset Q\mathcal{X}_0$, *for* $\mathcal{X}_0 = \{y \mid Py = \lambda y\}$.

Proof. In this case Wielandt's proof ([6]) will be adapted. We can prove by induction the following formula:

$$[P^n, Q]x = n\alpha\lambda^{n-1}x. \tag{1}$$

Indeed, by the hypothesis we have

$$[P, Q]x = \alpha x.$$

It is obvious that $P^n x = \lambda^n x$ and the above equality gives

$$P^n[P, Q]P^k x = \alpha P^{k+n}x. \tag{2}$$

If (1) holds for n, we can compute

$$[P^n, Q]P^{k+1}x = n\alpha P^{k+n}x. \tag{3}$$

Then (2) and (3) give by addition $[P^{n+1}, Q]P^k x = (n+1)\alpha P^{k+n}x$ and

$$[P^{n+1}, Q]x = (P^{n+1}Q - QP^{n+1})x = (n+1)\alpha\lambda^n x.$$

Now we observe that

$$|(n+1)\alpha| \cdot |\lambda|^n \|x\| \leq 2\|P\|^{n+1}\|Q\|\|x\|$$

and by the hypothesis we deduce

$$|(n+1)\alpha| \leq 2\|Q\|, \qquad \text{for every } n \in \mathbb{N}.$$

Hence we deduce that $\alpha = 0$. Assertion 2 of the conclusion of the theorem is obvious and the proof is finished. □

For arbitrary $P, Q \in \mathcal{B}(\mathcal{X})$ we can describe a similar result. We will consider

$$\operatorname{ad} Q\colon \mathcal{B}(\mathcal{X}) \longrightarrow \mathcal{B}(\mathcal{X}),\ \operatorname{ad} Q(T) = QT - TQ.$$

THEOREM 3. *If* $P, Q \in \mathcal{B}(\mathcal{X})$ *and* $x_0 \neq 0$ *is a common eigenvector for the operators* $(\operatorname{ad} Q)^n(P), n = 0, 1, 2, \ldots$, *then* $(\operatorname{ad} Q)^n(P)x_0 = 0$ *for any* $n \geq 1$. *This means also*

$$\operatorname{ad} Q(E)x_0 = 0 \quad \text{for any} \quad E \in \overline{sp}\{(\operatorname{ad} Q)^n(P) \mid n = 0, 1, 2, \ldots\},$$

and the closed subspace of all common eigenvectors for $\{(\operatorname{ad} Q)^n(P) \mid n = 0, 1, 2, \ldots\}$ *is invariant for* Q.

Proof. We denote

$$\mathcal{E} = \operatorname{sp}\{(\operatorname{ad} Q)^n(P) \mid n \geq 0\} \subset \mathcal{B}(\mathcal{X}).$$

By the hypothesis there exists $x_0 \neq 0$ and $\varphi\colon \mathcal{E} \longrightarrow \mathbb{C}$ such that

$$Ex_0 = \varphi(E)x_0 \qquad \text{for any } E \in \mathcal{E}.$$

We shall denote

$$\mathcal{X}_0 = \{x \in \mathcal{X} \mid Ex = \varphi(E)x \text{ for every } E \in \mathcal{E}\}.$$

By the hypothesis $0 \neq \mathcal{X}_0$. Obviously $\mathcal{X}_0$ is a closed subspace, $\varphi\colon \mathcal{E} \longrightarrow \mathbb{C}$ is a linear function which can be extended to a linear continuous function $\tilde{\varphi}\colon \overline{\mathcal{E}} \longrightarrow \mathbb{C}$, where $\overline{\mathcal{E}}$ is the uniform closure of $\mathcal{E}$ in $\mathcal{B}(\mathcal{X})$ and

$$\tilde{E}x = \tilde{\varphi}(\tilde{E}) \quad \text{for every } x \in \mathcal{X}_0 \text{ and } \tilde{E} \in \overline{\mathcal{E}}.$$

Now we recall the well known Rosenblum formula

$$(e^{z \operatorname{ad} Q})(E) = e^{-zQ}Ee^{zQ}, \qquad z \in \mathbb{C} \text{ and } E, Q \in \mathcal{B}(\mathcal{X}).$$

For $E \in \overline{\mathcal{E}}$ we have $(e^{z \operatorname{ad} Q})(E) = \tilde{E} \in \overline{\mathcal{E}}$ and the above equality gives

$$\tilde{\varphi}(e^{z \operatorname{ad} Q}(E))x = e^{-zQ}Ee^{zQ}x \quad \text{for every } x \in \mathcal{X}_0.$$

Hence we can write for a fixed $E \in \overline{\mathcal{E}}$

$$E(e^{zQ}x) = \tilde{\varphi}(e^{z \operatorname{ad} Q}(E))e^{zQ}x, \tag{4}$$

which means that $\tilde{\varphi}(e^{z \operatorname{ad} Q}(E)) \in \sigma(E)$ because $e^{zQ}x \neq 0$ for any $z \in \mathbb{C}$.

Therefore for every $E \in \overline{\mathcal{E}}$ we have

$$|\tilde{\varphi}(e^{z \operatorname{ad} Q}(E))| \leq \|E\|, \text{ any } z \in \mathbb{C}.$$

On the other hand, for every $z \in \mathbb{C}$ we can write

$$\tilde{\varphi}(e^{z \operatorname{ad} Q}(E)) = \sum_{n=0}^{\infty} \frac{z^n}{n!} \varphi((\operatorname{ad} Q)^n(E))$$

because $\tilde{\varphi}$ is a linear continuous function.

By Liouville's theorem we deduce

$$\tilde{\varphi}(e^{z \operatorname{ad} Q}(E)) = \tilde{\varphi}(E), \quad \text{i.e.} \quad e^{z \operatorname{ad} Q}(E)x = \tilde{\varphi}(E)x$$

for every $E \in \overline{\mathcal{E}}$, $x \in \mathcal{X}_0$ and $z_0 \in \mathbb{C}$.

The last equality gives by derivation in $z = 0$

$$\operatorname{ad} Q(E)x = 0 \text{ for every } x \in \mathcal{X}_0 \text{ and } E \in \overline{\mathcal{E}}.$$

Therefore we have $(\operatorname{ad} Q)^n(P)x = 0$ for every $n \geq 1$ and $x \in \mathcal{X}_0$, in particular for $x = x_0$. We can also explicitly write

$$QEx = EQx$$

or

$$EQx = \varphi(E)Qx \text{ for } E \in \overline{\mathcal{E}} \text{ and } x \in \mathcal{X}_0$$

which means that $\mathcal{X}_0$ is invariant for Q. The proof is now finished. □

REMARK. The idea of the proof may be also found in [4] and [3].

The following corollary of the last theorem is obvious.

COROLLARY 2. *If $P, Q \in \mathcal{B}(\mathcal{X})$ and $0 \neq x_0 \in \mathcal{X}$ is a common eigenvector of the operators $(\operatorname{ad} Q)^n(P)$ for any $n \geq k$, $k \in \mathbb{N}$ fixed, then we have*

$$(\operatorname{ad} Q)^n(P)x_0 = 0 \text{ for any } n \geq k+1,$$

i.e. $(\operatorname{ad} Q)(E)x = 0$ *for any common eigenvector x of the operators $E \in \overline{sp}\{(\operatorname{ad} Q)^n(P) | n \geq k\}$, and the closed span of these common eigenvectors is invariant to Q.*

As in [2] we can write the following asymptotic formula:

$$[f(Q), X] = \sum_{n=1}^{\infty} \frac{(-1)^{n-1}}{n!} f^{(n)}(Q)(\operatorname{ad} Q)^n(X), X \in \mathcal{B}(\mathcal{X})$$

for any $f \in \mathcal{O}(\overline{B(\sigma(Q), \epsilon)})$, where $\epsilon > 0$ is such that $\sigma(\operatorname{ad} Q) \subset B(0, \epsilon)$.

Using this formula we can deduce by Theorem 3 and Corollary 2 the following results.

THEOREM 4. *If $P, Q \in \mathcal{B}(\mathcal{X})$ and $0 \neq x_o \in \mathcal{X}$ is a common eigenvector of the set $\{(\operatorname{ad} Q)^n(P) \mid n \geq k\}$ for some fixed $k \in \mathbb{N}$, then for f as above and $E \in \overline{sp}\{(\operatorname{ad} Q)^n(P) | n \geq k\}$ we have*

$$[f(Q), E]x_0 = 0,$$

and the closed subspace of all common eigenvectors for $\{(\operatorname{ad} Q)^n(P) \mid n \geq k\}$ is invariant to $f(Q)$.

ACKNOWLEDGMENT. I thank Professor C. Foiaş for his suggestions which have improved the first version of this paper.

References

[1] I. COLOJOARĂ and C. FOIAŞ, *Theory of generalized spectral operators*, Gordon and Breach, New York, 1968.

[2] D. BELTIŢĂ and M. ŞABAC, An asymptotic formula for the commutators, *J. Funct. Anal.*, **153** (1998), 262–275.

[3] ŞT. FRUNZĂ, Generalized weights for operator Lie algebras, *Spectral theory*, Banach Center Publications, Warsaw (1982), 281–287.

[4] M. HAUSNER and J. T. SCHWARTZ, *Lie groups, Lie algebras,* Gordon and Breach, Sc. Publishers, New York – London – Paris, 1968.

[5] H. WIELANDT, Über die Unbeschränktheit der operatoren im Hilbertschen Raum, *Götingen Nachr.* (1949), 5–6.

[6] A. WINTNER, The unboundedness of quantum-mechanical matrices, *Phys. Rev.*, **71** (1947), 738–739.

M. ŞABAC, University of Bucharest, Department of Mathematics, Str. Academiei 14, Bucharest, Romania; *e-mail:* msabac@math.math.unibuc.ro

Operator Theory: Advances and Applications, Vol. 127, 561–584

Order and square roots in hermitian Banach $*$-algebras

P. G. Spain

Dedicated to the memory of Professor Béla Szőkefalvi-Nagy

Classical results of Löwner, Sherman, Topping and Ogasawara are extended to unital hermitian $*$-normed Banach $*$-algebras. A new norm condition is presented for such an algebra to be C^*-equivalent.

1. Introduction

The theme of this paper is the interplay between order and lattice structure, commutativity, existence of square roots, and monotonicity of powers, in the *real part* (the hermitian elements) of a unital hermitian Banach $*$-algebra.

These questions first appeared in the context of operators on Hilbert space. Löwner [Lö] established the monotonicity of the pth power for $0 \leq p \leq 1$ for positive matrices, that is, for positive operators on a finite dimensional Hilbert space. Heinz [H] extended this to infinite dimensions. At this stage the order structure was determined spatially: only after Gel'fand–Naimark was the order relation on the hermitians characterised spectrally.

Ogasawara [Og] showed that this is best possible in a noncommutative C^*-algebra: the pth power can be monotone for $p = 2$ (or for any other $p > 1$) only when the algebra is commutative.

Sherman [Sh] showed that the real part of a C^*-algebra can be a lattice only when the algebra is commutative. His proof features an analysis of the extremal positive functionals on the algebra and makes appeal to the Krein–Mil'man Theorem. It would be interesting to find a mode of proof internal to the algebra, not involving

AMS Subject Classification (1991): 46H99, 46L05.

its dual. It should be noticed that Sherman makes crucial use of a theorem stated in [KK] for which no proof is offered. I have supplied this deficiency.

Sherman's work was extended by Curtis [Cur]: the recent [LY] also deals with these themes.

Topping [To] investigated the order and lattice structure in great detail, for (real) linear spaces of hermitians as well as for real parts of C^*-algebras.

An alternative treatment of some of these topics, in [CDM], is based on a result attributed to Kaplansky (see [D, p. 58]): if A is a C^*-algebra in which 0 is the only nilpotent, then A is commutative. The proof of this depends on representation theory, and the use made of it is explicitly spatial: it seems unlikely that this mode of proof can be generalised. There, however, may be found what is perhaps the simplest example of a pair of positive hermitians that have no infimum: when $h = \begin{pmatrix} 2 & \sqrt{2} \\ \sqrt{2} & 1 \end{pmatrix} \geq 0$ and $k = \begin{pmatrix} 1 & 0 \\ 0 & 0 \end{pmatrix} \geq 0$ we have $(h-k)^2 = \begin{pmatrix} 1 & \sqrt{2} \\ \sqrt{2} & 1 \end{pmatrix}^2 = \begin{pmatrix} \sqrt{2} & 1 \\ 1 & \sqrt{2} \end{pmatrix}^2$; so $2(h \wedge k) = h + k - |h-k| = \begin{pmatrix} 3-\sqrt{2} & \sqrt{2}-1 \\ \sqrt{2}-1 & 1-\sqrt{2} \end{pmatrix} \not\geq 0$: but if h and k had an infimum it would have to be $h \wedge k$ (see Theorem 9.3).

Ptak [Pt] demonstrated the existence of the C^*-seminorm $|\cdot|_\Sigma$ on a hermitian Banach $*$-algebra, and made it possible to extend some of the results for C^*-algebras to hermitian Banach $*$-algebras: Okayasu [Ok] has recently extended Löwner's result.

A crucial tool in proving some of the results just mentioned is that in a C^*-algebra any positive element has a unique positive square root. One cannot expect this in any more general Banach algebra: Katznelson [Ka] showed that if all positive elements in a semisimple commutative Banach $*$-algebra admit square roots then the algebra can be renormed to become a C^*-algebra; Cuntz [Cun] removed the hypothesis of commutativity. Nevertheless the holomorphic functional calculus provides positive square roots for *strictly positive* elements: and, used carefully, this tool suffices in hermitian Banach $*$-algebras.

In this article I show how the classical results mentioned above can be generalised to hermitian Banach $*$-algebras. Since there is no unified exposition of these themes, even for C^*-algebras, I have included enough (some proofs are simplifications or improvements of the original proofs for C^*-algebras, some are quite new) to make the paper relatively self-contained.

I also exhibit a new characterisation, **NC**, of C^*-equivalent algebras: that $\|1-h\| \leq 1$ whenever $h \geq 0$ and $\|h\| \leq 1$. This allows me to improve the known result [Pt, Theorem 8.4] that an algebra is C^*-equivalent if, for some β, we have $\|k\| \leq \beta\,|k|_\sigma$ for every hermitian k: it is sufficient to know that $\|k\| \leq \beta\,|k|_\sigma$ for every *positive* k. Proofs previously published do not seem to adapt to prove this, although the equivalent condition, that $\sup_{t\in\mathbb{R}} \|e^{itk}\| < \infty$ for every hermitian k, can be relaxed trivially to the requirement that $\sup_{t\in\mathbb{R}} \|e^{itk}\| < \infty$ for every positive k.

2. Preliminaries

I suggest [BD] as a reference for all results stated without proof, here and below.

DEFINITION. A unital Banach $*$-algebra A is

hermitian		$\sigma(h) \subset \mathbb{R}$	$(\forall h \in A^h)$
symmetric	if	$\sigma(a^*a) \subset \mathbb{R}^+$	$(\forall a \in A)$
$*$-*normed*		$\|a^*\| = \|a\|$	$(\forall a \in A)$.

REMARK. If A is a Banach $*$-algebra with continuous involution then putting $\|a\|' = \max\{\|a\|, \|a^*\|\}$ defines an equivalent algebra norm on A for which the algebra is $*$-normed. So there is no topological loss of generality in assuming that a continuous involution is isometric.

REMARK. A is symmetric if and only if A is hermitian. This is the *Shirali–Ford Theorem* [BD, §41].

From now on A will be a unital $*$-normed Banach $*$-algebra.

NOTATION. The following notation will be employed routinely:

$$\begin{aligned} a \star b &= (ab+ba)/2 \\ |a|_\sigma &= \sup|\sigma(a)| \\ |a|_\Sigma &= |a^*a|_\sigma^{1/2} \quad \left(= |aa^*|_\sigma^{1/2} = |a^*|_\Sigma\right) \\ A^h &= \{h \in A : h = h^*\} \\ A^+ &= \{h \geq 0\} = \{h \in A^h : \sigma(h) \subset \mathbb{R}^+\} \\ h \gg 0 &\iff h \geq \varepsilon \quad (\exists \varepsilon > 0). \end{aligned}$$

3. Various axioms

The following axioms are listed together for ease of reference.

C A is commutative.

AC A is almost commutative; that is, $A/\mathrm{rad}(A)$ is commutative.

ss A is semisimple.

MS $0 \leq h \leq k \Longrightarrow h^2 \leq k^2$.

SQ $-k \leq h \leq k \Longrightarrow h^2 \leq k^2$.

M$_p$ $0 \ll h \leq k \Longrightarrow 0 \ll h^p \leq k^p$.

SRS $\forall h \in A^h \quad \exists!\ k \geq 0$ such that $k^2 = h^2$. Given **SRS** we can define $|h| = \left(h^2\right)^{1/2} \quad (\forall h \in A^h)$.

SR $\forall h \geq 0 \quad \exists!\ k \geq 0$ such that $k^2 = h$.

BSR $\forall h \geq 0 \quad \exists!\ k \geq 0$ such that $k^2 = h$ and $\|k\| \leq \|h\|^{1/2}$.

BSR$_\beta$ $\forall h \geq 0 \quad \exists!\ k \geq 0$ such that $k^2 = h$ and $\|k\| \leq \beta \|h\|^{1/2}$.

J A^+ is closed under Jordan multiplication, that is $h \geq 0,\ k \geq 0 \Longrightarrow h \star k \geq 0$.

J$^+$ $0 \leq h \leq k \Longrightarrow h \star k \geq 0$.

C*e A is C^*-equivalent.

NC $\left\{\begin{array}{c} 0 \leq h \\ \|h\| \leq 1 \end{array}\right\} \Longrightarrow \|1-h\| \leq 1$.

NR $\|h\| = |h|_\sigma \quad (h \in A^h)$.

NRP $\|h\| = |h|_\sigma \quad (h \geq 0)$.

NRP$_\beta$ $\|h\| \leq \beta\, |h|_\sigma \quad (h \geq 0)$ that is, $\sup\{\|h\| : 0 \leq h \leq 1\} \leq \beta$.

N$_\beta$ A^h is β-normal (see [AE, Chapter 2]), that is, $h \leq k \leq l \Longrightarrow \|k\| \leq \beta \max\{\|h\|, \|l\|\}$.

PT Positive functionals are traces.

L A^h is a vector lattice *with respect to* $\geq$. If so, then, for any $h, k, l \in A^h$,

$$\sup\{h,\ k\} = -\inf\{-h,\ -k\}$$
$$\sup\{h,\ k\} + l = \sup\{h+l,\ k+l\}$$
$$h + k = \sup\{h,\ k\} + \inf\{h,\ k\}.$$

Further, for $h \in A^h$ we may define $h^{(+)} = \sup\{h, 0\}$ and $h^{(-)} = \sup\{-h, 0\}$. Then $h^{(+)} \wedge h^{(-)} = 0$ and the expression $h = h^{(+)} - h^{(-)}$ is the unique representation of h as a difference of disjoint positive elements (see [La, §1]).

SL Following Topping [To], who, studying real linear spaces of hermitians, not necessarily the real parts of algebras, found it necessary to distinguish *special vector lattices*, I say that A satisfies **SL** if it satisfies **L**, **SRS**, and if $|h| = \sup\{h, -h\} \quad (h \in A^h)$.

It turns out that **L** $\Longrightarrow$ **SL** for unital semisimple hermitian Banach ∗-algebras: see Theorem 10.14. Indeed, **SRS** $\Longleftrightarrow$ **L** given **C** +ss.

The commutative semisimple Banach algebra $AC[0, 1]$, all *absolutely continuous* functions on the interval [0, 1], with the bounded variation norm, satisfies **L** and **SRS** but not **SR**.

Simple implications. The following results are immediate.

LEMMA 3.1. $\mathbf{C} \Longrightarrow \mathbf{SQ} \Longrightarrow \mathbf{J} \Longrightarrow \mathbf{MS} \Longleftrightarrow \mathbf{J}^+$.

Proof. Use the identities $(k+h)^2-(k-h)^2 = 4h\star k$ and $k^2-h^2 = (k+h)\star(k-h)$. □

REMARK. One might ask whether it can ever happen that $hk+kh \leq 0$ for some $h \geq 0$, $k \geq 0$. This cannot be unless $hk = 0$. For, in a concrete C^*-algebra, if $hk+kh \leq 0$ then the numerical range $V(hk)$ of hk lies in the left half plane, whence $\sigma(k^{1/2}hk^{1/2}) \subseteq \big(\sigma(hk) \cup \{0\}\big) \cap \mathbb{R}^+ = \{0\}$. (I am indebted to Prof. A. R. Sourour for this proof.)

LEMMA 3.2. $\mathbf{N}_\beta \Longrightarrow \mathbf{NRP}_\beta$.

Proof. $0 \leq h \Longrightarrow 0 \leq h/\,|h|_\sigma \leq 1 \Longrightarrow \|h/\,|h|_\sigma\| \leq \beta$. □

REMARK. $\mathbf{NRP}_\beta$ forces $\mathbf{C^*e}$ (see §11) but does not force **C**, although the condition $\|a\| \leq |a|_\sigma$ $(\forall a \in A)$ does force **AC** (see [Pa, 2.4.11]).

THEOREM 3.3. $\mathbf{BSR}_\beta \Longrightarrow \mathbf{NRP}_{\beta^2}$ *(for any β)*.

Proof. Suppose $h \geq 0$: then $\|h\| \leq \beta \left\|h^2\right\|^{\frac{1}{2}} \leq \beta^{1+\frac{1}{2}} \left\|h^4\right\|^{\frac{1}{4}} \ldots$ □

Modulus and ± decomposition. Assume **SRS**, so that $|h| := (h^2)^{1/2}$ is defined $(\forall\, h \in A^h)$; suppose further that $|h|\, h = h\, |h|$ (**ss** guarantees this, see Corollary 5.16): put

$$h^{\pm} := (|h| \pm h)/2.$$

By elementary spectral theory $\phi(|h|) = |\phi(h)|$ for every character defined on the (commutative) unital closed algebra generated by h. It follows that $h^{\pm} \geq 0$.

Then

$$h = h^+ - h^-, \qquad |h| = h^+ + h^-, \qquad h^+h^- = h^-h^+ = 0.$$

If $h,\ k \in A^h$ and if $|h-k|$ is defined put

$$h \vee k = (h + k + |h-k|)/2 \qquad \text{and} \qquad h \wedge k = (h + k - |h-k|)/2.$$

Then

$$h \vee k \geq \begin{Bmatrix} h \\ k \end{Bmatrix} \qquad \text{and} \qquad h \wedge k \leq \begin{Bmatrix} h \\ k \end{Bmatrix}.$$

Note that if $|h|$ is defined, then

$$h^+ \vee h^- = |h| \qquad \text{and} \qquad h^+ \wedge h^- = 0.$$

Remark. $\sup\{h,k\}$ may or may not be defined, even if $h \vee k$ is defined; see also Theorem 9.3.

Theorem 3.4. *Let A be a commutative unital semisimple ($*$-normed) Banach $*$-algebra satisfying* **SRS**. *Then A is hermitian and satisfies* **SL**.

Proof. Let Φ be the character space of A.

Suppose that A satisfies **SRS**. Consider $h \in A^h$. Then there is a *positive* k such that $h^2 = k^2$. It follows that, for every $\phi \in \Phi$, $\phi(h)^2 = \phi(h^2) = \phi(k^2) = \phi(k)^2 \geq 0$: hence $\phi(h) \in \mathbb{R}$. Thus $\sigma(h) \subset \mathbb{R}$, which shows that A is hermitian.

Consider $h,\ k \in A^h$ and any $m \geq h,\ k$. Then, for any $\phi \in \Phi$, we have $\phi(m) \geq \phi(h),\ \phi(k)$ so that $\phi(m) \geq \max\{\phi(h), \phi(k)\} = \phi(h) \vee \phi(k) = \phi(h \vee k)$. Thus $m \geq h \vee k$. This shows that $\sup\{h,k\}$ is defined and equals $h \vee k$.

Further, as remarked above, $\phi(|h|) = |\phi(h)| = \max\{\phi(h), \phi(-h)\} = \phi(h \vee -h)$: so $|h| = h \vee -h$.

□

4. Uniform approximation of the square root by polynomials

The real polynomial functions π_n, defined recursively by

$$\pi_0(t) = 0$$
$$2\pi_{n+1}(t) = t - 2\pi_n(t) - \pi_n(t)^2,$$

are a standard tool in proving the Stone–Weierstrass Theorem. They have the useful property that each is a monotonically increasing polynomial and that on $[0,\ 1]$ they provide a monotonically increasing uniform approximation to the square root function.

LEMMA 4.1. *$\pi_n(t) \nearrow_{\text{unif}} t^{1/2}$ and $\pi_n(t^2) \nearrow_{\text{unif}} t$ (uniform convergence on $[0, 1]$).*

COROLLARY 4.2. *If $h \gg 0$ then $\pi_n(h) \nearrow_{\text{norm}} h^{1/2}$, $\pi_n(h^2) \nearrow_{\text{norm}} h$.*

Proof. Holomorphic functional calculus. □

COROLLARY 4.3. *$\forall \varepsilon > 0$ there is a polynomial p_ε with real coefficients such that*

$$\|t - p_\varepsilon(t^2)\|_{[0,1]} = \|t^{1/2} - p_\varepsilon(t)\|_{[0,1]} < \varepsilon.$$

COROLLARY 4.4. $0 \le h \le 1 \Longrightarrow \left|h - p_\varepsilon(h^2)\right|_\sigma < \varepsilon$.

THEOREM 4.5. *For any β,* $\quad \mathbf{NRP}_\beta \Longrightarrow \mathbf{BSR}_\beta$.

Proof. Suppose $h \ge 0$: without loss of generality we may assume $h \le 1$. Then $\left(\pi_n - \pi_m\right)(h) \ge 0$ for $n \ge m$. So

$$\|\pi_n(h) - \pi_m(h)\| \le \beta\,|\pi_n(h) - \pi_m(h)|_\sigma \le \beta\,\|\pi_n - \pi_m\|_{[0,1]} \to 0.$$

Hence $h^{1/2} = \lim \pi_n(h)$ is defined and $\|h^{1/2}\| \le \lim \beta\,|\pi_n(h)|_\sigma \le \beta\,|h|_\sigma \le \beta\,\|h\|$. If $k \ge 0$ and $k^2 = h$ then

$$\|h^{1/2} - k\| \le \|h^{1/2} - \pi_n(h)\| + \|\pi_n(k^2) - k\| \to 0.$$

□

COROLLARY 4.6. $\mathbf{NRP} \iff \mathbf{BSR}$.

Proof. Theorems 4.5 and 3.3. □

5. Ptak's results and consequences

From now on *A will be a unital hermitian $*$-normed Banach $*$-algebra.*

One may note that all semisimple such algebras are A^*-algebras, but not conversely [BD, §41], and that the group algebra $L^1(G)$ of a locally compact group G is always A^*. While $L^1(G)$ is hermitian when G is compact, abelian or nilpotent, it is not so in general: for instance, $L^1(G)$ is not hermitian for any noncompact real semisimple Lie group (see [Pa, Vol. II, Chapter 12]).

The results now to be listed are due to [Pt]: an exposition of this cycle of ideas is presented in [BD].

THEOREM 5.1. (Ptak). *Suppose that A is a unital hermitian Banach $*$-algebra. Then for all $a \in A$ and all h, $k \in A^h$:*

$$|hk|_\Sigma = |kh|_\Sigma, \quad |a|_\sigma \le |a|_\Sigma, \quad |hk|_\sigma \le |h|_\sigma |k|_\sigma$$

$$h \ge 0,\, k \ge 0 \Longrightarrow h+k \ge 0, \quad |h+k|_\sigma \le |h|_\sigma + |k|_\sigma, \quad |\mathcal{R}a|_\sigma \le |a|_\Sigma \quad (a \in A);$$

$|\cdot|_\Sigma$ *is a C^*-seminorm on A, is the maximal C^*-seminorm on A, and is $\|\cdot\|$ continuous.*

COROLLARY 5.2. *If $h_n \ge 0$ and $h_n \to h$, then $h \ge 0$.*

Proof. $\forall n$ $\exists n_0$ such that $\|h_n - h\| \le \varepsilon$ $(n \ge n_0)$. Then $h - h_n \ge -\varepsilon$, so $h \ge h_n - \varepsilon \ge -\varepsilon$ $(n \ge n_0)$, so $h \ge 0$. □

COROLLARY 5.3. *A^+ is a closed wedge.*

LEMMA 5.4. **MS** $\Longrightarrow$ **J**, *that is,* $0 \le h,\, k \Longrightarrow h \star k \ge 0$.

Proof. Given $\varepsilon > 0$ $\exists$ $\lambda > 0$ such that $\lambda[k+\varepsilon] \ge \|h\|$. Then $0 \le \lambda[k+\varepsilon] - h \le \lambda[k+\varepsilon] + h$, so $(\lambda[k+\varepsilon] - h)^2 \le (\lambda[k+\varepsilon]+h)^2$, from which $h \star (k+\varepsilon) \ge 0$; so, by Corollary (5.2), $h \star k \ge 0$ (and then $(h-k)^2 \le (h+k)^2$). □

COROLLARY 5.5. (See also Lemma 3.1.) **J** $\Longleftrightarrow$ $\mathbf{J}^+$ [$\Longleftrightarrow$ **MS**].

THEOREM 5.6.

$$\begin{aligned}\operatorname{rad}(A) &= \{q : |aq|_\sigma = 0 \quad (\forall a \in A)\}\\ &= \bigcap\{\textit{kernels of all irreducible (star) representations}\}\\ &= \operatorname{srad}(A) = \{q \in A : |q|_\Sigma = 0\}.\end{aligned}$$

Proof. $|aq|_\sigma \le |aq|_\Sigma \le |a|_\Sigma\, |q|_\Sigma$ [BD, §41, Theorem 9]. □

COROLLARY 5.7. **ss** $\Longrightarrow$ $|\cdot|_\Sigma$ *is a norm.*

THEOREM 5.8. **ss** $\Longrightarrow$ $\ge$ *is a partial order (that is, A^+ is a cone).*

Proof. If $h \geq 0$ and $h \leq 0$ then $\sigma(h) = \{0\}$, so $|h|_\Sigma = 0$, so $h \in \mathrm{rad}(A)$, so $h = 0$. □

COROLLARY 5.9. $h \in \mathrm{rad}(A) \iff |h|_\sigma = 0 \quad (h \in A^h)$.

COROLLARY 5.10. $\mathbf{NRP}_\beta \Longrightarrow \mathbf{ss} \quad$ *(any β).*

Proof. $h+ik \in \mathrm{rad}(A) \Longrightarrow h,\ k \in \mathrm{rad}(A)$ and $h \geq 0,\ k \geq 0$. So $\|h\| \leq \beta\,|h|_\sigma = 0$ that is, $h = 0$; similarly $k = 0$.

□

COROLLARY 5.11. $\mathbf{BSR}_\beta \Longrightarrow \mathbf{ss} \quad$ *(any β).*

COROLLARY 5.12. $\mathbf{L} \Longrightarrow \mathbf{ss}$.

Proof. $a \in \mathrm{rad}(A) \Longrightarrow a^* \in \mathrm{rad}(A) \Longrightarrow \mathcal{R}a,\ \mathcal{I}a \in \mathrm{rad}(A) \Longrightarrow \mathcal{R}a \leq 0,\ \mathcal{R}a \geq 0,$ $\mathcal{I}a \leq 0,\ \mathcal{I}a \geq 0 \Longrightarrow a = 0$.

□

THEOREM 5.13. *Assume* ss: *then* $h = (h^2)^{1/2} \quad (\forall h \geq 0)$.

Proof. Suppose that $h^2 = k^2$ and $h \geq 0,\ k \geq 0$. Without loss of generality $\|h\| \leq 1$ and $\|k\| \leq 1$, and then, $\forall \varepsilon > 0$,

$$|h-k|_\sigma \leq \left|h - p_\varepsilon(h^2)\right|_\sigma + \left|p_\varepsilon(k^2) - k\right|_\sigma < 2\varepsilon.$$

□

COROLLARY 5.14. *Assume* ss. *Then positive square roots are unique, when they exist.*

THEOREM 5.15. *Assume* ss. *If $h^2k = kh^2$ and $h \in A^+,\ k \in A^h$, then $hk = kh$.*

Proof. Without loss of generality $\|h\| \leq 1$ and $\|k\| \leq 1$: and then $|i(hk-kh)|_\sigma \leq \left|[h - p_\varepsilon(h^2)]k\right|_\sigma + \left|k[h - p_\varepsilon(h^2)]\right|_\sigma < 2\varepsilon \ (\forall \varepsilon > 0)$.

□

As a consequence

COROLLARY 5.16. $\mathbf{SRS} + \mathbf{ss} \Longrightarrow |h|\,h = h\,|h| \quad (\forall h \in A^h)$.

REMARK. **SRS** + ss guarantees ± decomposition — but without control of norms of h^+ and h^-.

COROLLARY 5.17. **BSR** $\Longrightarrow |h|\, h = h\, |h|$, *in which case we have a* ± *decomposition with norm control:*

$$\begin{aligned}|||h||| &= \||h|\|_\sigma = |h|_\sigma \\ &= \max\left\{ \left|h^+\right|_\sigma, \left|h^-\right|_\sigma \right\} = \max\{\left\|h^+\right\|, \left\|h^-\right\|\} \le \|h\| \qquad (h \in A^h).\end{aligned}$$

Proof. By **NRP** $\left[\iff\right.$ **BSR**$\left.\right]$ we have $|h^\pm|_\sigma = \|h^\pm\|$, so, using elementary spectral theory,

$$|||h||| = \||h|\|_\sigma = |h|_\sigma = \max\left\{ \left|h^+\right|_\sigma, \left|h^-\right|_\sigma \right\} = \max\{\left\|h^+\right\|, \left\|h^-\right\|\}.$$

Also $|||h||| \le \|\, |h|^2 \,\|^{1/2} = \|h^2\|^{1/2} \le \|h\|$.

□

6. Fundamental order lemma

The later sections depend crucially on the next lemma and three theorems. The ideas are parallel to [Pe2, §1.3]. There, however, it is assumed that $0 \le h \le k \Longrightarrow \|h\| \le \|k\|$; that is, that (A^h, A^+) is 1-normal; that is, $\mathbf{N}_1$. This is of course true in a C^*-algebra, but this condition forces **NRP**; and even the weaker condition of β-normality, $\mathbf{N}_\beta$, for any β (≥ 1), forces $\mathbf{NRP}_\beta$ and so **C*e** (see Theorem 11.6 below).

LEMMA 6.1. $h \ge 0,\ k \ge 0 \Longrightarrow khk \ge 0.$

Proof. Let $\varepsilon > 0$. Then $(h+\varepsilon)^{1/2}$ is defined, so

$$k(h+\varepsilon)k = [(h+\varepsilon)^{1/2}\, k]^*[(h+\varepsilon)^{1/2}\, k] \ge 0,$$

and A^+ is closed.

□

THEOREM 6.2. $0 \le h \le k \Longrightarrow 0 \le a^*ha \le a^*ka \quad (\forall a).$

Proof. $a^*(k-h+\varepsilon)^{1/2}(k-h+\varepsilon)^{1/2}a \geq 0$ $(\forall \varepsilon > 0)$. □

THEOREM 6.3. *If* $0 \leq h$, $0 \ll k$ *and* $|hk^{-1}|_\sigma < 1$, *then* $h < k$.

Proof. $|k^{-1/2}hk^{-1/2}|_\sigma = |hk^{-1}|_\sigma < 1$, so $1 - k^{-1/2}hk^{-1/2} > 0$, and therefore $k - h = k^{1/2}(1 - k^{-1/2}hk^{-1/2})k^{1/2} > 0$. □

REMARK. If $0 \leq h$ and $\exists\, h^{-1}$ then $0 \ll h$.

THEOREM 6.4. $0 \ll h \leq k \Longrightarrow 0 \ll k^{-1} \leq h^{-1}$.

Proof. $(h^{1/2}k^{-1/2})^*(h^{1/2}k^{-1/2}) = k^{-1/2}hk^{-1/2} \leq k^{-1/2}kk^{-1/2} = 1$; so $h^{1/2}k^{-1}h^{1/2} = (h^{1/2}k^{-1/2})(h^{1/2}k^{-1/2})^* \leq 1$; so $k^{-1} \leq h^{-1/2}\,1\,h^{-1/2} = h^{-1}$. □

7. The Löwner–Heinz inequality (after Pedersen and Okayasu)

Okayasu has generalised the Löwner–Heinz Inequality, following the plan of [Pe1]. I prefer to model my proof, which is less intricate, on the treatment in [Pe2, §1.3].

DEFINITION. For $\alpha > 0$ define

$$f_\alpha \colon \left(-\frac{1}{\alpha}, \infty\right) \to \mathbb{R}\,,\ t \mapsto \frac{t}{1+t\alpha} = \frac{1}{\alpha}\left\{1 - \frac{1}{1+t\alpha}\right\}.$$

LEMMA 7.1. (Pedersen). *f_α is monotone increasing, that is,* $0 \leq h \leq k \Longrightarrow 0 \leq f_\alpha(h) \leq f_\alpha(k)$.

Proof. This is immediate from Theorem 6.4. □

THEOREM 7.2. (Löwner–Heinz–Okayasu).

$$0 \ll h \leq k \Longrightarrow 0 \ll h^p \leq k^p \qquad (0 \leq p \leq 1).$$

Proof. First

$$\int_0^\infty f_\alpha(t)\alpha^{-p}\,d\alpha = t^p\int_0^\infty (1+\alpha)^{-1}\alpha^{-p}\,d\alpha = \gamma\,t^p \qquad (\exists\ \gamma=\gamma_p>0).$$

Next, given $\varepsilon > 0$, we can choose a large integer n and equispaced $0 = \alpha_0 < \alpha_1 < \cdots < \alpha_m = n$ such that

$$\left| t^p - \frac{n}{\gamma m}\sum_{k=1}^m f_{\alpha_k}(t)\alpha_k^{-p} \right| < \varepsilon, \qquad 0 \le t \le \|k\|\,.$$

It follows that $k^p - h^p \ge -2\varepsilon$.

□

The next two results are routine to prove.

COROLLARY 7.3. $\mathbf{M}_p$ *and* $\mathbf{M}_q$ *are equivalent* $(p,\ q>1)$.

COROLLARY 7.4. $\mathbf{M}_p$ $(\exists\ p>1)$ *is equivalent to* **MS**.

COROLLARY 7.5. *Assume* **MS**. *Then* $h \ge 0,\ k \ge 0 \Longrightarrow h \wedge k \ge 0$.

Proof. For if $h \ge 0$, $k \ge 0$ then, assuming **J**, which is equivalent to **MS**, we have $(h-k)^2 \le (h+k)^2$ and so $|h-k| \le h+k$.

□

See also the remark after the definition of axiom **SL**, and the example in the introduction.

8. Ogasawara generalised

Ogasawara [Og] showed that a C^*-algebra must be commutative if squaring is monotone on its positive elements. I generalise this result to the case of a unital hermitian Banach $*$-algebra: the proof follows the same lines. If one does not assume the algebra to be semisimple the conclusion is that it is *almost commutative,* commutative modulo its radical.

Suppose then that A satisfies **MS**, which is equivalent to **J**, by Corollary 5.5. Consider any $h,\ k \ge 0$ in A, and let $l = \mathcal{R}(hk),\ m = \mathcal{I}(hk)$.

LEMMA 8.1. $l \geq 0$.

Proof. $2l = hk + kh \geq 0$ (by $\mathbf{J}$). □

LEMMA 8.2. $l^2 \geq m^2$.

Proof. $h,\ khk\ \geq 0$ and $l^2 - m^2 = \mathcal{R}(h(khk))$ [and $lm + ml = \mathcal{I}(h(khk))$]. □

LEMMA 8.3. $lmlm + mlml \geq 0$.

Proof. $l \geq 0,\ mlm \geq 0$ (by $\mathbf{J}$). □

LEMMA 8.4. $l^2m^2 + m^2l^2 \geq 2m^4$.

Proof. $(l^2 - m^2)m^2 + m^2(l^2 - m^2) \geq 0$ (by $\mathbf{J}$). □

DEFINITION. $E = \{\lambda \geq 1 : \mathcal{R}(hk)^2 \geq \lambda^2 \mathcal{I}(hk)^2 \quad (\forall\ h \geq 0,\ k \geq 0)\}$.

LEMMA 8.5. *E is closed and not empty* [$1 \in E$ *by Lemma* 8.2].

So *either* $E = [1, \infty)$ *or* $E = [1, \sup E]$.

THEOREM 8.6. *E is not bounded.*

Proof. If E were bounded and $\mu = \sup E$, then $l^2 \geq \mu^2 m^2$ so $0 \leq m^2(l^2 - \mu^2 m^2) + (l^2 - \mu^2 m^2)m^2 = m^2l^2 + l^2m^2 - 2\mu^4 m^4$.

Now, as above, $h,\ khk \geq 0$ and $h(khk) = l^2 - m^2 + i(lm + ml)$, so $(l^2 - m^2)^2 \geq \mu^2(lm + ml)^2$.

Hence, expanding, $l^4 + m^4 \geq [l^2m^2 + m^2l^2] + \mu^2[lmlm + mlml + lm^2l + ml^2m] \geq [2m^4] + \mu^2[ml^2m] \geq 2m^4 + \mu^4 m^4$, from which $l^4 \geq (1 + \mu^4)m^4$, whence, by $\mathbf{M}_{1/2}$, $l^2 \geq (1 + \mu^4)^{1/2} m^2$, which contradicts the definition of μ. □

COROLLARY 8.7. $m \in \mathrm{rad}(A)$.

Proof. The theorem forces $m^2 \leq 0$: so $\sigma(m^2) = \{0\}$. □

THEOREM 8.8. (Generalised Ogasawara). *Let A be a unital hermitian Banach $*$-algebra, and suppose that squaring is monotone on the positive elements of A. Then A is almost commutative, that is,* **MS** $\Longrightarrow$ **AC**.

Proof. $hk - kh = 2im \in \mathrm{rad}(A)$. Extend from A^+ to A^h and then to A. □

COROLLARY 8.9. **MS** + **ss** $\Longrightarrow$ **C**.

9. After Topping

Topping's results for real linear spaces of hermitians satisfying **MS** can be simplified for algebras of hermitians when one has Ogasawara's theorem to hand. Nevertheless it is instructive to follow his line of argument.

LEMMA 9.1. *Assume* ss. *If* $0 \leq m \leq h$, k *and* $hk = 0$ *then* $m = 0$.

Proof. Without loss of generality $\|h\| < 1$, $\|k\| < 1$, $\|m\| < 1$. First, $0 \leq hmh \leq hkh = 0$ by Theorem 6.2, so $hmh = 0$. Next, for small positive ε,

$$\begin{aligned}
\left|(m+\varepsilon)^{1/2}\, h\, (m+\varepsilon)^{1/2}\right|_\sigma &= \left|[(m+\varepsilon)^{1/2}\, h\, (m+\varepsilon)^{1/2}]^2\right|_\sigma^{1/2} \\
&= \left|(m+\varepsilon)^{1/2}\, h\, (m+\varepsilon)\, h\, (m+\varepsilon)^{1/2}\right|_\sigma^{1/2} \\
&= \varepsilon^{1/2}\left|(m+\varepsilon)^{1/2}\, h^2\, (m+\varepsilon)^{1/2}\right|_\sigma^{1/2} \\
&= \varepsilon^{1/2}\left|h^2(m+\varepsilon)\right|_\sigma^{1/2} \\
&\leq \varepsilon^{1/2}.
\end{aligned}$$

Now $0 \leq (m+\varepsilon)^{1/2}\, m\, (m+\varepsilon)^{1/2} \leq (m+\varepsilon)^{1/2}\, h\, (m+\varepsilon)^{1/2} \leq \varepsilon^{1/2}$. So $|m(m+\varepsilon)|_\sigma \leq \varepsilon^{1/2}$, $m^2 \leq m^2 + \varepsilon m \leq \varepsilon^{1/2}$, $m^2 = 0$, which implies $m = 0$. □

REMARK. Compare with [To, Lemma 3]: if $0 \leq m \leq h$, $0 \leq n \leq k$, and $hk = 0$, then $mn = 0$.

THEOREM 9.2. *Assume* **ss**. *If* $0 \leq h,\ k$ *and* $hk = 0$ *then*

$$\inf\{h,k\} = h \wedge k = 0.$$

Proof. Immediate from Lemma 9.1.

□

THEOREM 9.3. *Assume* **SRS** + **ss**, *and that* h *and* k *have an infimum. Then*

$$\inf\{h,k\} = h \wedge k.$$

Proof. Write $m = \inf\{h,k\}$ and $n = h \wedge k$. Then $n \leq h, k$, so $n \leq m \leq h, k$, hence $0 \leq m - n \leq h - n \leq \frac{1}{2}[h - k + |h-k|]$ and $0 \leq m - n \leq k - n \leq \frac{1}{2}[k - h + |h-k|]$; by Lemma 9.1 we have $m - n = 0$.

□

COROLLARY 9.4. *Assume* **SRS** + **ss**. *Then*

$$\inf\{h^+,\ h^-\} = h^+ \wedge h^- = 0 \qquad (\forall h \in A^h).$$

LEMMA 9.5. *Assume* **MS** + **ss**. *Suppose that* $0 \leq h \leq l$ *and* $0 \leq k \leq l$. *Then* $h + k \leq l$.

Proof. For $m = 1, 2, \ldots$ note that $(h+k)^{2^m} = h^{2^m} + k^{2^m} \leq l^{2^m} + l^{2^m} = 2\, l^{2^m}$. So $h + k \leq 2^{2^{-m}}\, l$, by $\mathbf{M}_{2^{-m}}$.

□

THEOREM 9.6. *Assume* **MS** + **ss**. *Suppose that* $0 \leq h,\ 0 \leq k$, *and that* $hk = kh = 0$. *Then*

$$\sup\{h,k\} = h \vee k = h + k, \qquad \inf\{h,k\} = h \wedge k = 0.$$

Proof. Immediate from Lemmas 9.1 and 9.5.

□

LEMMA 9.7. *Assume* **SRS** + **MS** + **ss**. *If* $0 \leq k$ *and* $h \leq k$ *then* $h^+ \leq k$.

Proof. $h^+ \leq h^- + k$ and $h^- \leq h^- + k$; so $h^+ + h^- \leq h^- + k$, by Lemma 9.5; so $h^+ \leq k$.

□

COROLLARY 9.8. *Assume* **SRS** + **MS** + **ss**. *If* $0 \le h$, $0 \le k$, $m \le h$, $m \le k$ *and* $hk = 0$, *then* $m \le 0$.

Proof. $0 \le m^+ \le h,\ k$, by Lemma 9.7; so $m^+ = 0$ by Lemma 9.1. □

LEMMA 9.9. *Assume* **SRS** + **MS** + **ss**. *Suppose that* $\pm h \le k$. *Then* $|h| \le k$.

Proof. First, $0 = h - h \le k + k$. Second, $h^+,\ h^- \le k$, by Lemma 9.7. So $|h| = h^+ + h^- \le k$, by Lemma 9.5. □

COROLLARY 9.10. *Assume* **SRS** + **MS** + **ss**. *Then*

$$|h| = h^+ \vee h^- = \sup\{h^+,\ h^-\} = h \vee -h = \sup\{h,\ -h\}.$$

THEOREM 9.11. (Topping). **SRS** + **MS** + **ss** $\Longrightarrow$ **SL**.

Proof. Suppose $l \ge h,\ k$. Let $m = h + k + |h - k| - 2l$. Then

$$m \le \begin{cases} |h - k| + h - k \\ |h - k| + k - h \end{cases}$$

and the two elements on the right are positive with product 0; so $m \le 0$, by Corollary 9.8; hence

$$h + k + |h - k| - 2l \le 0,$$

which shows that $h \vee k = \sup\{h, k\}$. The rest is routine. (*cf.* [To, Proposition 5]).

Alternatively since **MS** + **ss** $\Longrightarrow$ **C** we can invoke Theorem 3.4. □

The requirement that $|h| = \sup\{h, -h\}$ in the definition of **SL** is redundant, as remarked in [To]. The next corollary shows this.

COROLLARY 9.12. (Topping). **L** + **SRS** $\Longrightarrow$ **SL**.

Proof. First, $\inf\{h^+, h^-\} = 0$ by Corollary (9.4), so $\sup\{h^+, h^-\} = h^+ + h^- - \inf\{h^+, h^-\} = |h|$.

Next, $\sup\{h, -h\} = \sup\{h^+ - h^-,\ h^- - h^+\} = \sup\{2h^+,\ 2h^-\} - h^+ - h^- = 2\,|h| - |h| = |h|$.

□

In fact, **L** $\iff$ **MS** + **SRS** + **ss** $\iff$ **C** + **SRS** + **ss**; see Theorem 10.14.

10. Positive functionals and states

The plan of my proof that **L** $\Longrightarrow$ **C** follows that of [Sh]: characterise the pure states as those which respect the lattice structure, then invoke the Krein–Milman Theorem. I have modelled my treatment of positive functionals on that in [La].

Throughout, ω will denote a bounded linear functional on A, an element of A'.

DEFINITION. $\omega \geq 0$ means $\omega(h) \geq 0 \quad (\forall\, h \geq 0)$.

THEOREM 10.1. *If* $\omega \geq 0$ (*even if only* $\omega(a^*a) \geq 0$ $(\forall a \in A)$) *then*

$$\begin{aligned}
&\|\omega\| = \omega(1),\\
&\omega(b^*a) = \overline{\omega(a^*b)}, \qquad |\omega(b^*a)|^2 \leq \omega(a^*a)\omega(b^*b) \qquad (a,\, b \in A),\\
&|\omega(h)|^2 \leq \omega(1)\,\omega(h^2) \qquad (h \in A^h).
\end{aligned}$$

Proof. See [BD, §37].

□

DEFINITION. ω (≥ 0) is a *state* if $\|\omega\| = \omega(1) = 1$.

THEOREM 10.2. *Positive functionals separate* (*indeed,* A'^+ *determines* A^+).

Proof. ([La, §2]). If $h \in A^h \setminus A^+$ then, by the Separation Theorem, $\exists\ \eta \in A'$ such that $\eta(h) < \inf_{k \in A^+} \eta(k)$. Thus $\inf_{k \in A^+} \eta(k) = 0$, from which $\eta(h) < 0$ and $\eta \geq 0$.
□

DEFINITION. A functional ω on A is a *trace* if $\omega(ab) = \omega(ba) \quad (a, b \in A)$.

THEOREM 10.3. **PT** $\Longrightarrow$ **C**.

Proof. Suppose $h \geq 0$, $k \geq 0$ and $\omega \geq 0$. Then $\omega[(h+\varepsilon)k] = \omega[(h+\varepsilon)^{1/2}k(h+\varepsilon)^{1/2}] \geq 0$. So $\omega(hk)$ is positive, hence real. Hence $0 \leq \omega(kh) = \overline{\omega(hk)} = \omega(hk)$ and so $\omega(i[hk - kh]) = 0$. Thus $hk = kh$, by Theorem 10.2. (See also [LY].) □

Pure (i.e. extremal) states, given L. Now assume that A satisfies **L**: then A is semisimple, by Corollary 5.13.

Suppose $h \geq 0$, ω is a state, and $\omega(h) > 0$. Define

$$\eta_h(k) := \sup\{\omega(k \wedge nh) \ : \ n = 1, 2, \dots\}.$$

LEMMA 10.4. *$\eta_h \neq 0$, $\quad 0 \leq \eta_h \leq \omega$, $\quad \eta_h(h) = \omega(h)$, and η_h is 'linear$|_{+ve}$'.*

Proof. We know that $\eta_h(\lambda k) = \lambda\, \eta_h(k)$ $(\lambda,\ k \geq 0)$ and

$$\begin{aligned}\eta_h(k_1 + k_2) &= \sup\{\omega[(k_1 + k_2) \wedge nh] \ : \ n = 1, 2, \dots\} \\ &\leq \sup\{\omega[k_1 \wedge nh] \ : \ n = 1, 2, \dots\} + \sup\{\omega[k_2 \wedge nh] \ : \ n = 1, 2, \dots\} \\ &= \eta_h(k_1) + \eta_h(k_2),\end{aligned}$$

while

$$\omega(k_1 \wedge nh) + \omega(k_2 \wedge mh) \leq \omega[(k_1 + k_2) \wedge (n+m)h] \leq \eta_h(k_1 + k_2).$$ □

THEOREM 10.5. *η_h extends to be a bounded positive functional on A, and $\|\eta_h\| \leq 1$.*

Proof. If $k \in A^h$ and $\|k\| \leq 1$ then $1 \pm k \geq 0$ so $|\eta_h(k)| \leq \eta_h(1) \leq \omega(1) = 1$. But, by Theorem 10.1, $\eta_h \geq 0$; so $\|\eta_h\| = \eta_h(1) \leq 1$. □

COROLLARY 10.6. *If ω is pure then $\eta_h = \omega$.*

Proof. We must have $\eta_h = \lambda\omega$ for some $\lambda \geq 0$, but $\eta_h(h) = \omega(h)$ so $\lambda = 1$. □

REMARK. $k \wedge h = 0 \Longrightarrow k \wedge nh = 0$, so $h \wedge k = 0 \Longrightarrow \eta_h(k) = 0$.

THEOREM 10.7. *If ω is pure and if $h,\ k \geq 0$, $h \wedge k = 0$ then $\omega(h \vee k) = \omega(h) \vee \omega(k)$.*

Proof. If $\omega(h) > 0$ then $\omega(k) = \eta_h(k) = 0$; so

$$\omega(h \vee k) = \omega(h + k) = \omega(h) + \omega(k) = \omega(h) = \omega(h) \vee \omega(k).$$

If $\omega(k) > 0$, reverse the roles of h and k.

□

REMARK. $(h - h \wedge k) \wedge (k - h \wedge k) = 0$, $h \vee k = (h - h \wedge k) \vee (k - h \wedge k)$.

THEOREM 10.8. (Krein and Krein). *If ω is pure then ω is lattice-preserving.*

Proof.

$$\begin{aligned}\omega(h \vee k) &= \omega(h + k - h \wedge k) \\ &= \omega[(h - h \wedge k) + (k - h \wedge k)] + \omega(h \wedge k) \\ &= \sup\{\omega(h - h \wedge k),\ \omega(k - h \wedge k)\} + \omega(h \wedge k) \\ &= \sup\{\omega(h),\ \omega(k)\} - \omega(h \wedge k) + \omega(h \wedge k) \\ &= \sup\{\omega(h),\ \omega(k)\}.\end{aligned}$$

□

REMARK. This result is stated as Theorem 1 in [KK], no proof was offered.

REMARK. ω is pure *if* ω is lattice-preserving: $[\ 0 \le \eta \le \omega$ and $\omega(h) = 0$ and $[\omega(h) = 0 \Longrightarrow \omega(|h|) = 0] \Longrightarrow \eta(h) = 0 \Longrightarrow \eta = \lambda\omega \quad (\exists\ \lambda)]$.

For the rest of this section A^h *is a lattice and* ω *is a state on* A.

DEFINITION. $N_\omega = \{a \in A : \omega(a) = 0\}$.

REMARK. $N_\omega = N_\omega^h + i\, N_\omega^h$: for $\omega(h + ik) = 0 \Longrightarrow \omega(h) = \omega(k) = 0$.

LEMMA 10.9. *If $h \ge 0$ and $\omega(h) = 0$ then $\omega(ah) = 0$* $(\forall a \in A)$.

Proof. For any $\varepsilon > 0$ we have

$$\begin{aligned}|\omega[a(h + \varepsilon)]| &= \left|\omega[a(h + \varepsilon)^{1/2}(h + \varepsilon)^{1/2}]\right| \\ &\le [\omega(h + \varepsilon)]^{1/2}\ \left[\omega\left[a(h + \varepsilon)^{1/2}(h + \varepsilon)^{1/2}a^*\right]\right]^{1/2} \\ &\le \varepsilon^{1/2}\ \|a\|\ \|h + \varepsilon\|^{1/2}.\end{aligned}$$

Thus $|\omega(ah)| \le \varepsilon\ \|a\| + \varepsilon^{1/2}\ \|a\|\ \|h + \varepsilon\|^{1/2}$ from which $\omega(ah) = 0$.

□

THEOREM 10.10. *If ω is pure then N_ω is an ideal in A.*

Proof. If $h \in N_\omega^h$ then $\omega(h^{(+)}) = \omega(h \vee 0) = \omega(h) \vee 0 = 0$. So $\omega(h^{(-)}) = 0$. Thus $\omega(ah) = \omega(ah^{(+)}) - \omega(ah^{(-)}) = 0$. □

COROLLARY 10.11. *Pure states are multiplicative.*

THEOREM 10.12. **L** $\Longrightarrow$ **C**.

Proof. Pure states separate (Krein–Milman Theorem). □

COROLLARY 10.13. **L** $\Longrightarrow$ **SRS** *so* **L** $\Longrightarrow$ **SL**.

Proof. Since A is commutative and semisimple we can apply the Gel'fand Theory. Thus, given $h \in A^h$ we have $h^2 = |h|^2$, that is, **SRS**. □

THEOREM 10.14. **L** $\Longleftrightarrow$ **MS** + **SRS** + **ss** $\Longleftrightarrow$ **C** + **SRS** + **ss**.

Proof. Combine Theorems 3.4, 8.8, 10.12. □

It seems natural to seek a direct proof (not involving the structure of A') that **L** $\Longrightarrow$ **MS**; this, however, eludes me.

11. C^*-equivalence

I present a new condition, **NC** [that is, $h \geq 0$ and $\|h\| \leq 1 \Longrightarrow \|1-h\| \leq 1$] for C^*-equivalence, as well as improvements of known results.

Recall that **BSR** $\Longleftrightarrow$ **NRP**, **BSR** $\Longrightarrow$ **ss**.

I show here that **NC** $\Longrightarrow$ **BSR** and **NRP** $\Longrightarrow$ **C*e**.

THEOREM 11.1. *Let A satisfy* **NC** *and consider $h \in A^h$ such that $\|h\| \leq 1$. Put $h_n = \pi_n(h)$ where π_n are the real polynomials described in §4. Then*

$$0 = h_0 \leq h_1 \leq \ldots \leq 1, \qquad h_0^2 \leq h_1^2 \leq \ldots \leq h, \qquad \|h_n\| \leq 1 \quad (n \geq 1).$$

If also $k \in A^h$, $\|k\| \leq 1$ and $hk = kh$ then, putting $k_n = \pi_n(k)$, we have $h_n k_m = k_m h_n$ $(\forall m, n)$ and

$$\|h_n - k_n\| \leq \|h - k\|/2 \qquad (n \geq 1).$$

Proof. Note that

$$
\begin{aligned}
2(1-h_{n+1}) &= 1-h+(1-h_n)^2 \\
4(h-h_{n+1}^2) &= (h-h_n^2)[(2-h_n)^2-h] \\
2(k_{n+1}-h_{n+1}) &= k-h+(k_n-h_n)\,[1-h_n+1-k_n]
\end{aligned}
$$

and use induction. **NC** comes into play to show that if $\|h\| \le 1$ then $\|1-h_n\| \le 1$, so that $\|1-h_{n+1}\| \le 1$, from which $\|h_{n+1}\| \le 1$.

□

THEOREM 11.2. **NC** $\Longrightarrow$ **BSR**, *that is, given $h \ge 0$ such that $\|h\| \le 1$ there is a unique $k \ge 0$ such that $k^2 = h$ and $\|k\| \le 1$. Moreover, $k \in \{h\}''$ (the double commutant of h in A).*

Proof. If $h \gg 0$, then, as mentioned above [§4], $h_n = \pi_n(h) \nearrow_{\text{norm}} h^{1/2}$ (by the holomorphic functional calculus).

Suppose that $h \ge 0$ but not necessarily $h \gg 0$. For any ε such that $0 < \varepsilon < 1$ define $h_\varepsilon = \frac{\varepsilon+h}{\varepsilon+1}$. Then $0 \ll h_\varepsilon \le 1$ and $\|h_\varepsilon\| \le 1$, also $h_\varepsilon - h = \varepsilon\frac{1-h}{\varepsilon+1}$ and $1-h_\varepsilon = \frac{1-h}{\varepsilon+1}$. Since $h_\varepsilon \gg 0$ we have $h_{\varepsilon,n} \nearrow_{\text{norm}} h_\varepsilon^{1/2}$. Also $\|h_{\varepsilon,n}-h_n\| \le \|h_\varepsilon - h\|/2 \le \varepsilon/2$ $(n \ge 1)$ (Theorem 11.1) and hence $\|h_m-h_n\| \le \|h_m-h_{\varepsilon,m}\| + \|h_{\varepsilon,m}-h_{\varepsilon,n}\| + \|h_{\varepsilon,n}-h_n\| \le 2\varepsilon$ eventually. Thus $h_{1/2} = \lim h_n$ exists, $h_{1/2} \ge 0$, $\|h_{1/2}\| \le 1$ and $2h_{1/2} = h + 2h_{1/2} - {h_{1/2}}^2$, so $(h_{1/2})^2 = h$. Further, $h_{1/2}$, being the limit of polynomials in h, belongs to $\{h\}''$.

Suppose that k is another positive square root of h, that is, $k \ge 0$ and $k^2 = h$. We may assume, by scaling, that $\|k\| \le 1$. Now $k_\varepsilon \gg 0$, so ${k_\varepsilon}^2 \gg 0$. Also $\|{k_\varepsilon}^2 - k^2\| \le 2\varepsilon(2+\varepsilon) \le 6\varepsilon$. By the above

$$k_\varepsilon = \lim \pi_n({k_\varepsilon}^2).$$

Hence there is an n_ε such that

$$\|k_\varepsilon - \pi_n({k_\varepsilon}^2)\| < \varepsilon, \qquad \|h_{1/2} - \pi_n(h)\| < \varepsilon \qquad (\forall n > n_\varepsilon).$$

Now

$$
\begin{aligned}
\|k-h_{1/2}\| &\le \|k-k_\varepsilon\| + \|k_\varepsilon - \pi_n({k_\varepsilon}^2)\| + \|\pi_n({k_\varepsilon}^2) - \pi_n(k^2)\| + \|\pi_n(k^2) - h_{1/2}\| \\
&\le \varepsilon + \|k_\varepsilon - \pi_n({k_\varepsilon}^2)\| + \|{k_\varepsilon}^2 - k^2\|/2 + \|\pi_n(h) - h_{1/2}\| \\
&\le 5\varepsilon \qquad (\forall n > n_\varepsilon).
\end{aligned}
$$

So $k = h_{1/2}$.

□

COROLLARY 11.3. **NRP** $\Longrightarrow$ $\|\cdot\|$ *is* C^* *on* A^+.

Proof. For **NRP** $\Longrightarrow \|h\| = |h|_\Sigma$ $(h \geq 0)$ and $|\cdot|_\Sigma$ is a C^*-seminorm. □

EXAMPLE. Let $A = \mathbb{C}[0,1]$ and let $B = \mathcal{R}A$. Construct norm $\|\cdot\|$ on A as complexification of B. Then $\|h+ik\| \geq \sqrt{2}\,\|h+ik\|_\infty$ if $h,\ k \in B,\ hk = 0,\ \|h\|_\infty = \|k\|_\infty$.

So this norm is a C^*-norm on A^+ but is not a C^*-norm on A (Burckel, MR 39 #4685, reviewing [Be]).

THEOREM 11.4. **NRP** $\Longrightarrow$ **C*e**.

Proof. Given $x \in A$ write $x = h + ik = h^+ - h^- + i(k^+ - k^-)$ to get

$$\begin{aligned}\|x\| &\leq \|h^+\| + \|h^-\| + \|k^+\| + \|k^-\| \\ &\leq 2\,\{|h|_\sigma + |k|_\sigma\} = 2\,\{|\mathcal{R}x|_\sigma + |\mathcal{R}ix|_\sigma\} \\ &\leq 4\,|x|_\Sigma \leq 4\,\|x\|.\end{aligned}$$

□

COROLLARY 11.5. **NC** + **MS** $\Longrightarrow$ **C** + **SL**.

The next result is a strengthening of [Pt, Theorem 8.4] where the hypothesis is that $\|k\| \leq \beta\,|k|_\sigma$ $(\forall k \in A^h)$ (cf. also [DB, 34.3]).

THEOREM 11.6. *If A is a unital hermitian Banach $*$-algebra with* $\mathbf{BSR}_\beta$, *or, equivalently,* $\mathbf{NRP}_\beta$, *for some β (≥ 1) that is, $\|k\| \leq \beta\,|k|_\sigma$ $(\forall k \geq 0)$, then A is C^*-equivalent.*

Proof. Split an arbitrary x as in the proof of Theorem 11.4 to obtain $\|x\| \leq 4\beta^2\,|x|_\Sigma \leq 4\beta^2\,\|x\|$. □

COROLLARY 11.7. *If A is a unital hermitian Banach $*$-algebra with* $\mathbf{N}_\beta$, *that is, its real part is β-normal ($h \leq k \leq l \Longrightarrow \|k\| \leq \beta\max\{\|h\|,\|l\|\}$), then A is C^*-equivalent.*

Proof. $\mathbf{N}_\beta \Longrightarrow \mathbf{NRP}_\beta$ by Lemma (3.2). □

12. Concluding remarks

A deep result of Katznelson [Ka] states that if in a commutative semisimple Banach $*$-algebra all elements with real non-negative spectrum have a square root, then the algebra has an equivalent C^*-norm. Cuntz [Cun] extended this result to noncommutative algebras.

Both these papers are intricate. It is natural to wonder whether it might be possible to find a simple proof of a converse to Corollary 5.10 or Corollary 5.11 given **SR**: that is, to show directly that $\mathbf{SR} + \mathbf{ss} \Longrightarrow \mathbf{BSR}_\beta$, or, equivalently, that $\mathbf{SR} + \mathbf{ss} \Longrightarrow \mathbf{NRP}_\beta$ (for some β), and so obtain Cuntz's result.

References

[AE] L. Asimow and A. J. Ellis, *Convexity Theory and its Applications in Functional Analysis*, Academic Press, 1980.

[Be] H. Behncke, A remark on C^*-algebras, *Comm. Math. Phys.*, **12** (1969), 142–144.

[BD] F. F. Bonsall and J. Duncan, *Complete Normed Algebras*, Springer, 1973.

[CDM] M. J. Crabb, J. Duncan and C. M. McGregor, Characterizations of Commutativity for C^*-algebras, *Glasgow Math. J.*, **15** (1974), 172–175.

[Cur] P. C. Curtis, Order and Commutativity in Banach Algebras, *Proc. Amer. Math. Soc.*, **9** (1958), 643–646.

[Cun] J. Cuntz, Locally C^*-Equivalent Algebras, *J. Funct. Anal.*, **23** (1976), 95–106.

[D] J. Dixmier, *Les C^*-algèbres et leurs représentations*, Gauthier-Villars, 1969.

[DB] R. S. Doran and V. A. Belfi, *Characterizations of C^*-Algebras*, Dekker, 1986.

[H] E. Heinz, Beiträge zur Störungstheorie der Spektralzerlegung, *Math. Ann.*, **123** (1951), 415–438.

[Ka] Y. Katznelson, Sur les algèbres dont les éléments non négatifs admettent des racines carrées, *Ann. Sci. École Norm. Sup.*, **77** (1960), 167–174.

[KK] M. Krein and S. Krein, On an inner characteristic of the set of all continuous functions defined on a bicompact Hausdorff space, *Comptes Rendus de l'Acad. Sci. URSS*, **XXVII** (1940), 427–430.

[La] H. E. Lacey, *The Isometric Theory of Classical Banach Spaces*, Springer, 1974.

[LY] L. C. Leinbach and B. Yood, Commutativity Criteria for Banach $*$-Algebras, *Proc. Amer. Math. Soc.*, **125** (1997), 2307–2312.

[Lö] K. Löwner, Über monotone Matrixfunktionen, *Math. Z.*, **38** (1934), 177–216.

[Og] T. Ogasawara, A Theorem on Operator Algebras, *J. Sci. Hiroshima Univ. Ser. A*, **18** (1955), 307–309.

[Ok] T. OKAYASU, The Löwner–Heinz Inequality in Banach $*$-Algebras, *Glasgow Math. J.*, **42** (2000), to appear.

[Pa] T. W. PALMER, *Algebras and Banach Algebras* Vol. 1 and 2, Cambridge University Press, 1994, 2001.

[Pe1] G. K. PEDERSEN, Some operator monotone functions, *Proc. Amer. Math. Soc.*, **36** (1972), 309–310.

[Pe2] G. K. PEDERSEN, *C^*-Algebras and their Automorphism Groups*, Academic Press, 1978.

[Pt] V. PTAK, Banach algebras with involution, *Manuscripta Math.*, **6** (1972), 245–290.

[Sh] S. SHERMAN, Order in Operator Algebras, *Amer. J. Math.*, **73** (1951), 227–232.

[To] D. M. TOPPING, Vector Lattices of Self-Adjoint Operators, *Trans. Amer. Math. Soc.*, **115** (1965), 14–30.

P. G. SPAIN, Mathematics Department, University of Glasgow, Glasgow G12 8QW, Scotland; *e-mail*: pgs@maths.gla.ac.uk

Operator Theory: Advances and Applications, Vol. 127, 585–598

Unitary dilation of several contractions

Jan Stochel and Franciszek Hugon Szafraniec

Béla Szőkefalvi-Nagy in memoriam

A characterization for a commutative family of operators to have a unitary power dilation is given here. The problem is converted into one on extension of positive definite operator valued functions defined on a subset of a $*$-semigroup. As a byproduct, a solution to Krein's extension problem for positive definite functions on subsets of $\mathbb{R}^N$ is proposed.

A question the celebrated Sz.-Nagy dilation theorem [25] has given rise to is a possibility of extending it to several contractions. This has become intriguing since Ando's theorem [1] and Parrott's counterexample [19] appeared. In a number of papers (see [3], [4], [8], [16], [15], [9], for instance) an attempt has been made to provide a solution in particular instances. In the present paper we propose a *test* for a commutative family of operators to have a unitary power dilation (written in terms of the operators in question).

One of the ways to prove the Sz.-Nagy dilation theorem which says that every Hilbert space contraction T has a unitary power dilation, is to formulate it [26], [27] in an equivalent way as follows: the sequence $\{T^n\}_{n=0}^{\infty}$ extends to a positive definite one enumerated by $\mathbb{Z}$ (= the additive group of all integers). The same statement remains true for two commuting contractions S, T: the bisequence $\{S^m T^n\}_{m,n=0}^{\infty}$ extends to a positive definite one enumerated by $\mathbb{Z}^2$. This is due to Ando's unitary dilation theorem which says that every pair of commuting contractions has a unitary power dilation [1]. According to an example of Parrott [19] this assertion is no

AMS Subject Classification (1991): 43A35, 47A20.
The research resulting in this paper was supported by the KBN grant #2 P03A 004 17.

longer true for more than two contractions: three commuting contractions S, T, R may not possess a unitary power dilation (though some triplets may, see [2]). In other words, the 3-sequence $\{S^m T^n R^k\}_{m,n,k=0}^{\infty}$ may not extend to a positive definite one enumerated by $\mathbb{Z}^3$. It was an idea of Brehmer to introduce the concept of the regular unitary dilation for a commutative family of contractions in order to exhibit certain particular cases in which the extension problem is solvable (cf. [5] and [28, Theorem I.9.1]). This is why much of our attention is called here to extending positive definite functions.

The above causes that another classical question of analysis fits into the scope of our work, this is the M. G. Krein problem [12] on extending positive definite functions from a cube to the additive group $\mathbb{R}^N$. Our solution of it (see Corollary 4) completely characterizes such functions, admitting more general (than cubes) subsets of $\mathbb{R}^N$.

Replacing the groups involved in the aforesaid extension problems by a more general algebraic structure, like a $*$-semigroup, an object introduced by Sz.-Nagy in [27], opens new possibilities. In particular, in [23] the authors worked out (on the occasion of solving the complex moment problem) tests for a function defined on a $*$-subsemigroup $\mathfrak{T}$ of a $*$-semigroup $\mathfrak{S}$ to be extendible to a positive definite one on the whole $\mathfrak{S}$. Here we weaken the assumptions imposed on $\mathfrak{T}$ so that the tests remain valid and, moreover, are applicable to the afore-mentioned extension problems.

1. Given a nonempty set $\mathfrak{X}$ and a complex linear space $\mathcal{D}$, we say that a mapping $\omega: \mathfrak{X} \times \mathcal{D} \times \mathcal{D} \to \mathbb{C}$ is a *form* over $(\mathfrak{X}, \mathcal{D})$, if $\omega(s; \cdot, -)$ is sesquilinear for every $s \in \mathfrak{X}$ (cf. [24]). Denote by $\mathrm{F}(\mathfrak{X}, \mathcal{D})$ the linear space (with pointwise defined algebraic operations) of all forms over $(\mathfrak{X}, \mathcal{D})$. In the sequel $\mathbf{B}(\mathcal{H})$ stands for the C^*-algebra of all bounded linear operators on a complex Hilbert space $\mathcal{H}$ and $I = I_{\mathcal{H}}$ stands for the identity operator on $\mathcal{H}$. In case $\omega(s; f, g) = \langle \varphi(s) f, g\rangle$ for $s \in \mathfrak{X}$ and $f, g \in \mathcal{H}$, where $\varphi: \mathfrak{X} \to \mathbf{B}(\mathcal{H})$, we say that the form $\omega \in \mathrm{F}(\mathfrak{X}, \mathcal{H})$ *comes from an operator function.*

Suppose $(\mathfrak{S}, \cdot, *)$ is a $*$-semigroup (cf. [27]), i.e. $(\mathfrak{S}, \cdot)$ is a multiplicative semigroup with a unit ε and the mapping $\mathfrak{S} \ni s \longmapsto s^* \in \mathfrak{S}$, called an *involution* on $\mathfrak{S}$, satisfies the following conditions for all $s, t \in \mathfrak{S}$:

$$(s^*)^* = s, \quad (st)^* = t^* s^*.$$

Since ε^* is a unit of $\mathfrak{S}$, it must be $\varepsilon^* = \varepsilon$. Given a subset $\mathfrak{X}$ of $\mathfrak{S}$, we denote by $\mathfrak{X}_{\mathfrak{h}}$ the set of all hermitian elements of $\mathfrak{X}$, i.e. $\mathfrak{X}_{\mathfrak{h}} \stackrel{\mathrm{df}}{=} \{s \in \mathfrak{X};\ s = s^*\}$, and by $\mathfrak{X}^*$ the set $\{s^*;\ s \in \mathfrak{X}\}$. A form $\omega \in \mathrm{F}(\mathfrak{S}, \mathcal{D})$ is said to be *positive definite* if

$$\sum_{s,t \in \mathfrak{S}} \omega(t^* s; f(s), f(t)) \geq 0 \quad \text{for any } f \in \mathcal{D}[\mathfrak{S}],$$

where $\mathcal{D}[\mathfrak{S}]$ stands for the linear space (with pointwise defined linear operations) of all functions $f\colon \mathfrak{S} \to \mathcal{D}$ of finite support $\{s \in \mathfrak{S};\ f(s) \neq 0\}$. It is worth noting that the form ω is positive definite if and only if

$$\sum_{i,j=1}^{n} \omega(s_j^* s_i; f_i, f_j) \geq 0 \quad \text{for all } n \geq 1,\ s_1, \dots, s_n \in \mathfrak{S},\ f_1, \dots, f_n \in \mathcal{D}.$$

One can easily adopt the notion of positive definiteness to the context of operator valued functions (via the identification from the first paragraph of this section). We say that a form $\omega \in \mathrm{F}(\mathfrak{X}, \mathcal{D})$, $\mathcal{D}$ being a complex inner product space, is *isometric* if $\omega(\varepsilon; f, f) = \|f\|^2$ for all $f \in \mathcal{D}$.

The question of when a form over $(\mathfrak{X}, \mathcal{D})$ extends to a positive definite one over $(\mathfrak{S}, \mathcal{D})$ has been studied in [23]. In case $\mathfrak{X}$ is a $*$-subsemigroup of $\mathfrak{S}$, Theorems 27 and 29 in [23] offer the solutions to this problem written in terms of the initial form. One of our goals is to remove the requirement that $\mathfrak{X}$ be closed under the operation of multiplication, still preserving the conclusions of Theorems 27 and 29 in [23].

2. We always use $\langle \cdot, - \rangle$ for denoting inner product regardless the space is; in particular, the n-dimensional complex Euclidean space $\mathbb{C}^n$ is equipped with the inner product $\langle z, w \rangle = \sum_{j=1}^n z_j \bar{w}_j$ for $z = (z_1, \dots, z_n), w = (w_1, \dots, w_n) \in \mathbb{C}^n$. Moreover, if $\mathcal{D}$ is a complex inner product space and $f, g \in \mathcal{D}$, then $f \boxtimes g$ stands[1)] for the linear mapping from $\mathcal{D}$ into $\mathcal{D}$ defined by

$$(f \boxtimes g)(h) = \langle h, g \rangle f, \quad h \in \mathcal{D}.$$

THEOREM 1. *Let $\mathfrak{X}$ be a subset of a $*$-semigroup $(\mathfrak{S}, \cdot, *)$ with a unit ε such that $\mathfrak{S}_\mathfrak{h} \subset \mathfrak{X}$ and $\mathfrak{X}^* \subset \mathfrak{X}$. Assume $\mathcal{D}$ is a complex linear space and $\omega \in \mathrm{F}(\mathfrak{X}, \mathcal{D})$. Then the following conditions are equivalent*

(i) *ω extends to a positive definite form over $(\mathfrak{S}, \mathcal{D})$;*

(ii) $\sum_{k,l=1}^m \sum_{\substack{s,t \in \mathfrak{S} \\ t^* s \in \mathfrak{X}}} \omega(t^* s; e_k, e_l) \langle \lambda_k(s), \lambda_l(t) \rangle \geq 0$ *for any sequence $\{e_j\}_{j=1}^m$ of linearly independent vectors in $\mathcal{D}$ and for all functions $\lambda_1, \dots, \lambda_m \in \mathbb{C}^n[\mathfrak{S}]$ with arbitrary $n \geq m \geq 1$ such that $\sum_{\substack{s,t \in \mathfrak{S} \\ t^* s = u}} \langle \lambda_k(s), \lambda_l(t) \rangle = 0$ for $u \in \mathfrak{S} \setminus \mathfrak{X}$ and $k, l = 1, \dots, m$.*

Moreover, if $\mathcal{D}$ is an inner product space, then (i) *is equivalent to*

(iii) $\sum_{j=1}^n \sum_{\substack{s,t \in \mathfrak{S} \\ t^* s \in \mathfrak{X}}} \omega(t^* s; f_j(s), f_j(t)) \geq 0$ *for all functions $f_1, \dots, f_n \in \mathcal{D}[\mathfrak{S}]$ with arbitrary $n \geq 1$ such that $\sum_{j=1}^n \sum_{\substack{s,t \in \mathfrak{S} \\ t^* s = u}} f_j(s) \boxtimes f_j(t) = 0$ for $u \in \mathfrak{S} \setminus \mathfrak{X}$.*

[1)] This is to distinguish it from tensor product.

In case the form ω comes from an operator function, so does the extended one.

The condition (ii) *simplifies in the following cases*

(a) *if $\mathcal{D}$ is not finite dimensional, then it suffices to take $m = n$;*

(b) *if $\mathcal{D}$ is an inner product space, it is enough to consider the sequence $\{e_j\}_{j=1}^m$ to be orthonormal.*

We would like to point out that a positive definite extension of a given form ω may not be uniquely determined (cf. Remark 3 in [23]).

Proof of Theorem 1. Utilizing the discussion in the proof of [23, Theorem 29] it is enough to show that (iii)$\Leftrightarrow$(i). We follow the main idea of the proof of Theorem 27 in [23]. Notice first that $\varepsilon \in \mathfrak{S}_\mathfrak{h} \subset \mathfrak{X}$. Denote by $\mathbf{L}^\#(\mathcal{D})$ the $*$-algebra of all linear operators $A: \mathcal{D} \to \mathcal{D}$ for which there exists a linear operator $A^\#: \mathcal{D} \to \mathcal{D}$ such that $\langle Af, g\rangle = \langle f, A^\# g\rangle$ for all $f, g \in \mathcal{D}$; the operator $A^\#$ is uniquely determined and the mapping $A \mapsto A^\#$ is an involution on $\mathbf{L}^\#(\mathcal{D})$. Set

$$\mathbf{F}^\#(\mathcal{D}) \overset{\mathrm{df}}{=} \{A \in \mathbf{L}^\#(\mathcal{D});\ \dim A(\mathcal{D}) < \infty\}.$$

Then $\mathbf{F}^\#(\mathcal{D})$ is a $*$-algebra which is linearly spanned by the set $\{f \boxtimes g : f, g \in \mathcal{D}\}$ (cf. [23, Proposition 25]).

The linear space $\mathbf{F}^\#(\mathcal{D})[\mathfrak{S}]$ becomes a complex $*$-algebra with multiplication of convolution type and involution defined as follows

$$f \star g(u) = \sum_{\substack{s,t \in \mathfrak{S} \\ st=u}} f(s)g(t), \quad f, g \in \mathbf{F}^\#(\mathcal{D})[\mathfrak{S}],\ u \in \mathfrak{S},$$

$$f^*(s) = f(s^*)^\#, \quad f \in \mathbf{F}^\#(\mathcal{D})[\mathfrak{S}],\ s \in \mathfrak{S}.$$

Notice that the $*$-algebra $\mathbf{F}^\#(\mathcal{D})[\mathfrak{S}]$ is $*$-algebra isomorphic to the tensor product $\mathbf{F}^\#(\mathcal{D}) \otimes \mathbb{C}[\mathfrak{S}]$ of $*$-algebras. Set $(\mathbf{F}^\#(\mathcal{D})[\mathfrak{S}])_\mathfrak{h} = \{h \in \mathbf{F}^\#(\mathcal{D})[\mathfrak{S}];\ h = h^*\}$. In the sequel, we identify $\mathbf{F}^\#(\mathcal{D})[\mathfrak{X}]$ with the set $\{f \in \mathbf{F}^\#(\mathcal{D})[\mathfrak{S}];\ f(s) = 0,\ \forall s \in \mathfrak{S} \setminus \mathfrak{X}\}$. The main *difference* between this and what is in [23] is in the fact that $\mathbf{F}^\#(\mathcal{D})[\mathfrak{X}]$ is no longer a subalgebra of $\mathbf{F}^\#(\mathcal{D})[\mathfrak{S}]$ though it is a linear subspace still closed under involution. So what we are going to do now is to recall the essential steps in the proof of [23, Theorem 27] paying attention to how this difference can be dealt with.

Denote by $\mathfrak{P}$ the convex cone generated by the set $\{f^* \star f;\ f \in \mathbf{F}^\#(\mathcal{D})[\mathfrak{S}]\}$. According to Lemma 33 of [23] we have

$$(\mathbf{F}^\#(\mathcal{D})[\mathfrak{S}])_\mathfrak{h} = \mathfrak{P} + (\mathbf{F}^\#(\mathcal{D})[\mathfrak{X}])_\mathfrak{h}.$$

Observe that the proof of that Lemma remains unchanged because the only thing which we use in it is that $\mathfrak{S}_{\mathfrak{h}} \subset \mathfrak{X}$.

The next step is Lemma 34 of [23]. It remains true if one replaces $\mathfrak{T}$ by $\mathfrak{X}$. The only argument in the proof which has to be modified is that

$$\omega(s^*; f, g) = \overline{\omega(s; g, f)}, \quad s \in \mathfrak{X},\ f, g \in \mathcal{D} \tag{1}$$

(it is used in the proof of the part (iii) of the Lemma). Take $h \in \mathcal{D}[\mathfrak{X}]$ such that the support of h is contained in $\{\varepsilon, s\}$, $s \neq \varepsilon$. Then, by $s, s^*, s^*s \in \mathfrak{X}$ (this is due to $\mathfrak{X}^* \subset \mathfrak{X}$ and $\mathfrak{S}_{\mathfrak{h}} \subset \mathfrak{X}$), we have $\sum_{\substack{u,v \in \mathfrak{S} \\ v^*u = w}} h(u) \boxtimes h(v) = 0$ for $w \in \mathfrak{S} \setminus \mathfrak{X}$. Setting $f = h(\varepsilon)$ and $g = h(s)$ we conclude from condition (iii) (with $n = 1$) that

$$\omega(\varepsilon; f, f) + \omega(s^*; f, g) + \omega(s; g, f) + \omega(s^*s; g, g) = \sum_{\substack{u,v \in \mathfrak{S} \\ v^*u \in \mathfrak{X}}} \omega(v^*u; h(u), h(v)) \geq 0.$$

Since f and g are arbitrary, the matrix

$$\begin{pmatrix} \omega(\varepsilon; f, f) & \omega(s; g, f) \\ \omega(s^*; f, g) & \omega(s^*s; g, g) \end{pmatrix} \tag{2}$$

is positive definite which leads to (1) (the case $s = \varepsilon$ is to be treated likewise).

(iii)⇒(i). According to Lemma 34 in [23], there is a unique $\mathbb{C}$-linear functional $\Phi\colon \mathbf{F}^{\#}(\mathcal{D})[\mathfrak{X}] \to \mathbb{C}$ such that $\Phi((f \boxtimes g) \otimes \delta_s) = \omega(s^*; f, g)$ for all $s \in \mathfrak{X}$ and $f, g \in \mathcal{D}$. By parts (ii) and (iii) of Lemma 34 in [23], the functional Φ is non-negative on $\mathfrak{P} \cap (\mathbf{F}^{\#}(\mathcal{D})[\mathfrak{X}])_{\mathfrak{h}}$ and $\Phi(g^*) = \overline{\Phi(g)}$ for $g \in \mathbf{F}^{\#}(\mathcal{D})[\mathfrak{X}]$. This in turn implies that the real-valued functional $\Phi_{\mathfrak{h}} \stackrel{\text{df}}{=} \Phi|_{(\mathbf{F}^{\#}(\mathcal{D})[\mathfrak{X}])_{\mathfrak{h}}}$ is $\mathbb{R}$-linear. Since, moreover, $\mathfrak{P}$ is a convex cone in the $\mathbb{R}$-linear space $(\mathbf{F}^{\#}(\mathcal{D})[\mathfrak{S}])_{\mathfrak{h}}$ such that $(\mathbf{F}^{\#}(\mathcal{D})[\mathfrak{S}])_{\mathfrak{h}} = \mathfrak{P} + (\mathbf{F}^{\#}(\mathcal{D})[\mathfrak{X}])_{\mathfrak{h}}$, we conclude from a theorem of M. G. Krein (cf. [7]) that there exists an $\mathbb{R}$-linear extension $\Psi_{\mathfrak{h}}\colon (\mathbf{F}^{\#}(\mathcal{D})[\mathfrak{S}])_{\mathfrak{h}} \to \mathbb{R}$ of $\Phi_{\mathfrak{h}}$ which is non-negative on $\mathfrak{P}$. It is clear that $\mathbf{F}^{\#}(\mathcal{D})[\mathfrak{S}] = (\mathbf{F}^{\#}(\mathcal{D})[\mathfrak{S}])_{\mathfrak{h}} + \mathrm{i}(\mathbf{F}^{\#}(\mathcal{D})[\mathfrak{S}])_{\mathfrak{h}}$, so there exists a unique $\mathbb{C}$-linear extension $\Psi\colon \mathbf{F}^{\#}(\mathcal{D})[\mathfrak{S}] \to \mathbb{C}$ of $\Psi_{\mathfrak{h}}$. As $\Psi(f) \geq 0$ for every $f \in \mathfrak{P}$, the form $\tilde{\omega} \in \mathrm{F}(\mathfrak{S}, \mathcal{D})$ defined by $\tilde{\omega}(s; f, g) = \Psi((f \boxtimes g) \otimes \delta_{s^*})$ for $s \in \mathfrak{S}$ and $f, g \in \mathcal{D}$ is positive definite (use part (i) of [23, Lemma 34]). It turns out that $\tilde{\omega}$ extends ω. Indeed, since $\mathbf{F}^{\#}(\mathcal{D})[\mathfrak{X}] = (\mathbf{F}^{\#}(\mathcal{D})[\mathfrak{X}])_{\mathfrak{h}} + \mathrm{i}(\mathbf{F}^{\#}(\mathcal{D})[\mathfrak{X}])_{\mathfrak{h}}$, we obtain for all $s \in \mathfrak{X}$ and $f, g \in \mathcal{D}$ with $\alpha \stackrel{\text{df}}{=} (f \boxtimes g) \otimes \delta_{s^*} \in \mathbf{F}^{\#}(\mathcal{D})[\mathfrak{X}]$

$$\begin{aligned}\tilde{\omega}(s; f, g) &= \Psi(\alpha) \\ &= \Psi_{\mathfrak{h}}\Big(\frac{\alpha + \alpha^*}{2}\Big) + \mathrm{i}\Psi_{\mathfrak{h}}\Big(\frac{\alpha - \alpha^*}{2\mathrm{i}}\Big) = \Phi_{\mathfrak{h}}\Big(\frac{\alpha + \alpha^*}{2}\Big) + \mathrm{i}\Phi_{\mathfrak{h}}\Big(\frac{\alpha - \alpha^*}{2\mathrm{i}}\Big) \\ &= \Phi(\alpha) = \omega(s; f, g).\end{aligned}$$

This completes the proof of (iii)⇒(i). The reverse implication can be shown via parts (i) and (ii) of Lemma 34 in [23]. In case ω comes from an operator function, so does any positive definite extension of ω due to Lemma 18 in [23]. □

We have noticed in the proof of Theorem 1 that, under the assumptions of it, if the form $\omega \in \mathrm{F}(\mathfrak{X}, \mathcal{D})$ extends to a positive definite one over $(\mathfrak{S}, \mathcal{D})$, then the matrix (2) is positive definite for all $f, g \in \mathcal{D}$ and $s \in \mathfrak{X}$ and, consequently, the form ω has to satisfy an inequality of Schwarz type (cf. [14, p. 19] and [24, p. 252]):

$$|\omega(s; g, f)| \leq \sqrt{\omega(\varepsilon; f, f)\,\omega(s^*s; g, g)}, \quad s \in \mathfrak{X},\ f, g \in \mathcal{D}. \tag{3}$$

This implies in particular that if $\omega(\varepsilon; \cdot, \text{-})$ is jointly continuous, then every $\omega(s; \cdot, \text{-})$, $s \in \mathfrak{X}$, is separately continuous. In the case when $\mathfrak{S}$ is a group (cf. Section 3) we conclude that $\omega(s; \cdot, \text{-})$ is jointly continuous for every $s \in \mathfrak{X}$ (see also [23, Lemma 18] for a related result).

As in [23, Remark 30] one can consider another condition which completely characterizes forms extendible to positive definite ones and which is given in terms of square matrix-valued positive definite kernels.

Theorem 1 enables us to show the rather pleasant fact that the set of all forms over $(\mathfrak{X}, \mathcal{D})$, which are extendible to positive definite ones over $(\mathfrak{S}, \mathcal{D})$, is closed with respect to the topology of pointwise convergence.

COROLLARY 2. *Let $\mathfrak{X}$ and $\mathfrak{S}$ satisfy the assumptions of Theorem 1 and let $\mathcal{D}$ be a complex linear space. Assume $\omega \in \mathrm{F}(\mathfrak{X}, \mathcal{D})$ and $\{\omega_\sigma\}_{\sigma\in\Sigma} \subset \mathrm{F}(\mathfrak{X}, \mathcal{D})$ is a net such that $\omega(s; f, f) = \lim_\sigma \omega_\sigma(s; f, f)$ for all $s \in \mathfrak{X}$ and $f \in \mathcal{D}$. If every form ω_σ extends to a positive definite one over $(\mathfrak{S}, \mathcal{D})$, then so does the limit form ω.*

Proof. Apply part (ii) of Theorem 1 and the polarization formula. □

In case a net $\varphi_\sigma \colon \mathfrak{X} \to \mathbf{B}(\mathcal{H})$, $\sigma \in \Sigma$, is pointwise convergent in the weak operator topology to $\varphi \colon \mathfrak{X} \to \mathbf{B}(\mathcal{H})$ and each φ_σ extends to a positive definite function on $\mathfrak{S}$, φ extends to a positive definite function on $\mathfrak{S}$ as well.

3. Let $\mathfrak{G}$ be a (multiplicative) group with unit denoted by ε. A natural involution in $\mathfrak{G}$ which can immediately be taken into account is defined as $x^* = x^{-1}$ for $x \in \mathfrak{G}$; in the sequel we consider this involution exclusively. It is clear that an element x of $\mathfrak{G}$ belongs to $\mathfrak{G}_\mathfrak{h}$ if and only if either $x = \varepsilon$ or x is of order 2 (i.e. $x^2 = \varepsilon$ and $x \neq \varepsilon$). In these circumstances Theorem 1 gives us the following extension of a theorem of Naimark (cf. [17] a commutative case, and [26] the general case).

THEOREM 3. *Let $\mathfrak{G}$ be a group with no element of order 2 and let $\mathfrak{X}$ be a subset of $\mathfrak{G}$ such that $\varepsilon \in \mathfrak{X}$ and $\mathfrak{X}^{-1} \subset \mathfrak{X}$. Assume $\mathcal{D}$ is a complex inner product space and ω is an isometric form over $(\mathfrak{X}, \mathcal{D})$. Then any of the conditions* (i), (ii) *and* (iii) *of Theorem* 1 *is equivalent to*[2)]

(iv) *there exists a complex Hilbert space $\mathcal{K}$ and a unitary representation Φ of $\mathfrak{G}$ on $\mathcal{K}$ such that $\mathcal{D} \subset \mathcal{K}$ and $\omega(x; f, g) = \langle \Phi(x) f, g\rangle$ for $x \in \mathfrak{X}$ and $f, g \in \mathcal{D}$.*

Proof. According to Theorem 1 it is sufficient to show that (i)$\Leftrightarrow$(iv). If (iv) holds, then the form $\tilde{\omega} \in \mathrm{F}(\mathfrak{G}, \mathcal{D})$ defined by

$$\tilde{\omega}(x; f, g) = \langle \Phi(x) f, g\rangle, \quad x \in \mathfrak{G},\ f, g \in \mathcal{D}, \tag{4}$$

is evidently positive definite and it extends ω.

(i)$\Rightarrow$(iv) Let $\tilde{\omega} \in \mathrm{F}(\mathfrak{G}, \mathcal{D})$ be a positive definite extension of ω. By [24, Section 4], there exist a complex Hilbert space $\mathcal{K}$, a dense linear subspace $\mathcal{F}$ of $\mathcal{K}$, a linear mapping $V: \mathcal{D} \to \mathcal{F}$ and a $*$-semigroup homomorphism[3)] $\tilde{\Phi}: \mathfrak{G} \to \mathbf{L}^{\#}(\mathcal{F})$ such that $\mathcal{F}$ is equal to the linear span of the set $\{\tilde{\Phi}(x) V h;\ x \in \mathfrak{S},\ h \in \mathcal{D}\}$ and

$$\tilde{\omega}(x; f, g) = \langle \tilde{\Phi}(x) V f, V g\rangle, \quad x \in \mathfrak{G},\ f, g \in \mathcal{D}. \tag{5}$$

Since $x^* x = \varepsilon$ for $x \in \mathfrak{G}$, we deduce from (5) that

$$\Big\|\tilde{\Phi}(x)\Big(\sum_{i=1}^{n} \tilde{\Phi}(y_i) V h_i\Big)\Big\|^2 = \Big\|\sum_{i=1}^{n} \tilde{\Phi}(y_i) V h_i\Big\|^2, \quad x, y_i \in \mathfrak{G},\ h_i \in \mathcal{D}.$$

As $\tilde{\Phi}(x^{-1}) = \tilde{\Phi}(x)^{-1}$, $\tilde{\Phi}(x)$ is a unitary operator on $\mathcal{D}$ $(x \in \mathfrak{G})$. Let $\Phi(x) \in \mathbf{B}(\mathcal{K})$ be a unique unitary extension of $\tilde{\Phi}(x)$ $(x \in \mathfrak{G})$. Then Φ is a unitary representation of $\mathfrak{G}$ on $\mathcal{K}$. Since $\langle V f, V g\rangle = \omega(\varepsilon; f, g) = \langle f, g\rangle$ for $f, g \in \mathcal{D}$, we can identify $\mathcal{D}$, up to a unitary equivalence, with a linear subspace of $\mathcal{K}$ so that (4) be satisfied. □

If $\mathcal{H}$ is a Hilbert space and φ is a function from $\mathfrak{X}$ into $\mathbf{B}(\mathcal{H})$ such that $\varphi(\varepsilon) = I$, then the condition (iv) of Theorem 3 takes the form

(iv′) there exists a complex Hilbert space $\mathcal{K}$ and a unitary representation Φ of $\mathfrak{G}$ on $\mathcal{K}$ such that $\mathcal{H} \subset \mathcal{K}$ and $\varphi(x) = P\Phi(x)|_{\mathcal{H}}$ for $x \in \mathfrak{X}$, where P is the orthogonal projection of $\mathcal{K}$ onto $\mathcal{H}$.

[2)] One can formulate a "nontopological" version of Theorem 3 for forms over $(\mathfrak{X}, \mathcal{D})$ which are not assumed to be isometric, $\mathcal{D}$ being only a linear space (see the proof of Theorem 3).

[3)] For the definition of $\mathbf{L}^{\#}(\mathcal{F})$ see the first paragraph of the proof of Theorem 1.

We now state, as an application of Theorem 3, a result which goes back to an extension problem originated in Krein's theorem [12] on extending positive definite functions from an interval to $\mathbb{R}$; a several dimensional version of this is not true according to examples of Calderón and Pepinsky [6] (the group of lattice points in $\mathbb{R}^N$) and Rudin [20] (the group $\mathbb{R}^N$). On the other hand, there are results (cf. [10], [11], [21], [18]) which provide a positive answer to the question under some more restrictive assumptions on functions in question. Our result (notice that an "operator" version of it is available) offers a complete characterization of those functions defined on an open symmetric neighborhood of the origin in $\mathbb{R}^N$ which are extendible to positive definite ones on the whole group $\mathbb{R}^N$.

Corollary 4. *Let $\mathfrak{X}$ be a subset of $\mathbb{R}^N$ such that 0 is an interior point of $\mathfrak{X}$ and $-\mathfrak{X} \subset \mathfrak{X}$. A function $\varphi\colon \mathfrak{X} \to \mathbb{C}$ extends to a positive definite one on the additive group $\mathbb{R}^N$ if and only if*[4] $\sum_{x-y\in\mathfrak{X}} \varphi(x-y)\langle f(x), f(y)\rangle \geq 0$ *for every* $f \in \mathbb{C}^n[\mathbb{R}^N]$, *with arbitrary* $n \geq 1$, *such that* $\sum_{x-y=u}\langle f(x), f(y)\rangle = 0$ *for* $u \in \mathbb{R}^N \setminus \mathfrak{X}$. *If φ is continuous at 0, then the extended function is continuous on $\mathbb{R}^N$.*

The continuity part of the conclusion of Corollary 4 follows from the well known fact that a positive definite function on $\mathbb{R}^N$ which is continuous at the origin is continuous in its domain.

It follows from Corollary 4 that the set of all scalar valued functions on $\mathfrak{X}$ which extend to positive definite ones on the additive group $\mathbb{R}^N$ is closed in the topology of pointwise convergence.

4. In this section we diminish the set $\mathfrak{X}$ from which an operator valued function is supposed to be extended.

Theorem 5. *Let $(\mathfrak{G}, +)$ be an abelian group with no element of order 2 and let $\mathfrak{P}$ be a subset of $\mathfrak{G}$ such that $\mathfrak{P} \cap (-\mathfrak{P}) = \{0\}$. Assume $\mathcal{H}$ is a complex Hilbert space and φ is a function from $\mathfrak{P}$ to $\mathbf{B}(\mathcal{H})$ such that $\varphi(0) = I$. Then the following conditions are equivalent*

(v) *φ extends to a positive definite operator valued function on $\mathfrak{G}$;*

(vi) *if $\Lambda(\cdot) = \sum_{j=1}^m e_j \otimes \lambda_j(\cdot)$, where $\{e_j\}_{j=1}^m$ is an orthonormal sequence of vectors in $\mathcal{H}$ and $\lambda_1, \ldots, \lambda_m \in \mathbb{C}^n[\mathfrak{G}]$, $n \geq m \geq 1$, are functions such that*

---

[4] By $\sum_{x-y\in\mathfrak{X}}$ (resp. $\sum_{x-y=u}$) we understand the summation over the set of all admissible pairs (x, y) (in this particular case $x, y \in \mathbb{R}^N$) such that $x - y \in \mathfrak{X}$ (resp. $x - y = u$); the same convention is in force in Theorem 5 and Corollary 6.

$\sum_{x-y=u}\langle\lambda_k(x),\lambda_l(y)\rangle = 0$ *for* $u \in \mathfrak{G}\setminus(\mathfrak{P}\cup(-\mathfrak{P}))$ *and* $k,l = 1,\dots,m$, *then*

$$\sum_{x\in\mathfrak{G}}\|\Lambda(x)\|^2 + 2\operatorname{Re}\sum_{x-y\in\mathfrak{P}\setminus\{0\}}\langle(\varphi(x-y)\otimes I_{\mathbb{C}^n})\Lambda(x),\Lambda(y)\rangle \geq 0;$$

(vii) *if* $f_1,\dots,f_n \in \mathcal{H}[\mathfrak{G}]$, $n \geq 1$, *are such that* $\sum_{j=1}^n\sum_{x-y=u} f_j(x)\boxtimes f_j(y) = 0$ *for* $u \in \mathfrak{G}\setminus(\mathfrak{P}\cup(-\mathfrak{P}))$, *then*

$$\sum_{j=1}^n\Big(\sum_{x\in\mathfrak{G}}\|f_j(x)\|^2 + 2\operatorname{Re}\sum_{x-y\in\mathfrak{P}\setminus\{0\}}\langle\varphi(x-y)f_j(x),f_j(y)\rangle\Big) \geq 0;$$

(viii) *there exists a complex Hilbert space* $\mathcal{K}$ *and a unitary representation* Φ *of* $\mathfrak{G}$ *on* $\mathcal{K}$ *such that* $\mathcal{H}\subset\mathcal{K}$ *and* $\varphi(x) = P\Phi(x)|_{\mathcal{H}}$ *for* $x \in \mathfrak{P}$, *where* P *is the orthogonal projection of* $\mathcal{K}$ *onto* $\mathcal{H}$.

If, moreover, $\dim\mathcal{H} \geq \aleph_0$, *then in condition* (vi) *it is sufficient to consider* $m = n$.

Condition (viii) is considered in [13] in the case $\mathfrak{P}\cup(-\mathfrak{P}) = \mathfrak{G}$ and $\mathfrak{P}+\mathfrak{P}\subset\mathfrak{P}$.

Proof of Theorem 5. Set $\mathfrak{X} = \mathfrak{P}\cup(-\mathfrak{P})$. It is obvious that $0 \in \mathfrak{X}$ and $-\mathfrak{X}\subset\mathfrak{X}$. Define a function $\psi\colon\mathfrak{X}\to\mathbf{B}(\mathcal{H})$ by

$$\psi(x) = \begin{cases}\varphi(x) & \text{if } x \in \mathfrak{P}\\ \varphi(-x)^* & \text{if } x \in -\mathfrak{P}.\end{cases}$$

Suppose that φ extends to a positive definite function $\tilde{\varphi}\colon\mathfrak{G}\to\mathbf{B}(\mathcal{H})$. Notice that $\tilde{\varphi}$ is an extension of ψ. Indeed, if $x \in \mathfrak{P}$, then by [14, p. 18] $\tilde{\varphi}(-x) = \tilde{\varphi}(x)^* = \varphi(x)^* = \psi(-x)$. In other words, φ extends to a positive definite function on $\mathfrak{G}$ if and only if so does ψ.

Now we can apply Theorem 3 to $\mathfrak{G}$, $\mathfrak{X}$ and ψ to get the conclusion. The only thing which requires some comment is how we pass from the sums of Theorem 3 to these of Theorem 5. This can be done because, due to the fact that $\mathfrak{P}\cap(-\mathfrak{P}) = \{0\}$, we have

$$\begin{aligned}\{(x,y)\in\mathfrak{G}^2;\ x-y\in\mathfrak{X}\} ={}&\{(x,y)\in\mathfrak{G}^2;\ x=y\}\cup\\ &\cup\{(x,y)\in\mathfrak{G}^2;\ x-y\in\mathfrak{P}\setminus\{0\}\}\cup\\ &\cup\{(x,y)\in\mathfrak{G}^2;\ y-x\in\mathfrak{P}\setminus\{0\}\}\end{aligned}$$

with the ingredients being disjoint. □

5. Suppose now that A is a nonempty set. Then by $\mathbb{Z}[A]$ we understand the group which is the direct sum of card(A) copies of the additive group $\mathbb{Z}$, that is, $\mathbb{Z}[A]$ is the set of all functions from A to $\mathbb{Z}$ of finite support equipped with pointwise group operation. The zero function 0 is a neutral element of $\mathbb{Z}[A]$. Because there is no element in $\mathbb{Z}[A]$ of order 2 Theorem 5 can be further simplified.

Set $\mathbb{Z}[A]_+ = \{x \in \mathbb{Z}[A];\ x(a) \geq 0 \text{ for } a \in A\}$ and $\mathbb{Z}[A]_- = -\mathbb{Z}[A]_+$. We write $x < y$ in case $x, y \in \mathbb{Z}[A]$, $x \neq y$ and $y - x \in \mathbb{Z}[A]_+$. Given a family $\mathbf{T} = \{T_a\}_{a\in A} \subset \mathbf{B}(\mathcal{H})$ of commuting operators, we put

$$\mathbf{T}^x = \prod_{a\in A} T_a^{x(a)}, \quad x \in \mathbb{Z}[A]_+.$$

Following [28, page 32] we say that $\mathbf{T}$ has a *unitary (power) dilation* if there exists a complex Hilbert space $\mathcal{K} \supset \mathcal{H}$ and a family $\mathbf{U} = \{U_a\}_{a\in A}$ of commuting unitary operators on $\mathcal{K}$ such that

$$\mathbf{T}^x = P\mathbf{U}^x\Big|_{\mathcal{H}}, \quad x \in \mathbb{Z}[A]_+,$$

where P stands for the orthogonal projection of $\mathcal{K}$ onto $\mathcal{H}$.

We are now in a position to derive necessary and sufficient conditions for a family of commuting operators to have a unitary dilation.

COROLLARY 6. *A family $\mathbf{T} = \{T_a\}_{a\in A} \subset \mathbf{B}(\mathcal{H})$ of commuting operators has a unitary dilation if and only if at least one of the following two conditions is satisfied*

(α) *if $\Lambda(\cdot) = \sum_{j=1}^m e_j \otimes \lambda_j(\cdot)$, where $\{e_j\}_{j=1}^m$ is an orthonormal sequence of vectors in $\mathcal{H}$ and $\lambda_1, \ldots, \lambda_m \in \mathbb{C}^n[\mathbb{Z}[A]]$, $n \geq m \geq 1$, are functions such that $\sum_{x-y=u}\langle\lambda_k(x), \lambda_l(y)\rangle = 0$ for $u \in \mathbb{Z}[A] \setminus (\mathbb{Z}[A]_+ \cup \mathbb{Z}[A]_-)$ and $k, l = 1, \ldots, m$, then*

$$\sum_{x\in\mathbb{Z}[A]} \|\Lambda(x)\|^2 + 2\operatorname{Re}\sum_{x>y}\langle(\mathbf{T}^{x-y} \otimes I_{\mathbb{C}^n})\Lambda(x), \Lambda(y)\rangle \geq 0;$$

(β) *if $f_1, \ldots, f_n \in \mathcal{H}[\mathbb{Z}[A]]$, $n \geq 1$, are such that $\sum_{j=1}^n \sum_{x-y=u} f_j(x) \boxtimes f_j(y) = 0$ for $u \in \mathbb{Z}[A] \setminus (\mathbb{Z}[A]_+ \cup \mathbb{Z}[A]_-)$, then*

$$\sum_{j=1}^n \Big(\sum_{x\in\mathbb{Z}[A]} \|f_j(x)\|^2 + 2\operatorname{Re}\sum_{x>y}\langle\mathbf{T}^{x-y} f_j(x), f_j(y)\rangle\Big) \geq 0.$$

If, moreover, $\dim\mathcal{H} \geq \aleph_0$, then in condition ($\alpha$) it is sufficient to consider $m = n$.

Proof. Apply Theorem 5 to $\mathfrak{G} = \mathbb{Z}[A]$, $\mathfrak{P} = \mathbb{Z}[A]_+$ and $\varphi(x) = \mathbf{T}^x$. □

Notice that the set $\mathbb{Z}[A] \setminus (\mathbb{Z}[A]_+ \cup \mathbb{Z}[A]_-)$, which appears in both conditions (α) and (β) of Corollary 6, can be interpreted as the set of all $x \in \mathbb{Z}[A]$ which are incomparable with 0 with respect to the partial order relation $\leq$ in $\mathbb{Z}[A]$ given by:

$$x \leq y \iff y - x \in \mathbb{Z}[A]_+.$$

The two results which follow are direct consequences of Corollary 6.

COROLLARY 7. *Suppose $\mathbf{T}_\sigma = \{T_{\sigma,a}\}_{a\in A} \subset \mathbf{B}(\mathcal{H})$, $\sigma \in \Sigma$, is a net of families of commuting operators*[5] *and $\mathbf{T} = \{T_a\}_{a\in A} \subset \mathbf{B}(\mathcal{H})$ is a family such that $T_a = \lim_{\sigma\in\Sigma} T_{\sigma,a}$ in the strong operator topology for every $a \in A$. If each family $\mathbf{T}_\sigma$, $\sigma \in \Sigma$, has a unitary dilation, then so does the limit family $\mathbf{T}$.*

Proof. Since multiplication of operators on the closed unit ball in $\mathbf{B}(\mathcal{H})$ is jointly continuous with respect to the strong operator topology and members of the families $\mathbf{T}_\sigma$ are contractions, we conclude that $\mathbf{T}^x = \lim_{\sigma\in\Sigma} \mathbf{T}_\sigma^x$ in the strong operator topology for every $x \in \mathbb{Z}[A]_+$ (the same argument shows that the family $\mathbf{T}$ is commutative). Applying part (β) of Corollary 6, we get the conclusion. □

COROLLARY 8. *A family $\mathbf{T} = \{T_a\}_{a\in A} \subset \mathbf{B}(\mathcal{H})$ of commuting operators has a unitary dilation if and only if, for every finite nonempty subset A' of A, the subfamily $\{T_a\}_{a\in A'}$ has a unitary dilation*[6].

Proof. Notice that $\mathbb{Z}[A] = \bigcup\{\mathbb{Z}[A'];\ A' \text{ finite subset of } A\}$ and consequently

$$\mathbb{Z}[A] \setminus (\mathbb{Z}[A]_+ \cup \mathbb{Z}[A]_-) = \bigcup\{\mathbb{Z}[A'] \setminus (\mathbb{Z}[A']_+ \cup \mathbb{Z}[A']_-);\ A' \text{ finite subset of } A\}.$$

The conclusion follows from part (β) of Corollary 6. □

We complete this paper with a result which connects the problem of unitary dilatability of families of commuting operators with the problem of extendibility of matrix valued functions to positive definite ones.

---

[5] i.e. $T_{\sigma,a}T_{\sigma,a'} = T_{\sigma,a'}T_{\sigma,a}$ for all $a, a' \in A$ and $\sigma \in \Sigma$.

[6] Notice that the Hilbert spaces in which unitary dilations of $\{T_a\}_{a\in A'}$ act may depend on the set A'.

COROLLARY 9. *A family* $\mathbf{T} = \{T_a\}_{a\in A} \subset \mathbf{B}(\mathcal{H})$ *of commuting operators has a unitary dilation if and only if for every finite dimensional subspace* $\mathcal{H}'$ *of* $\mathcal{H}$, *the function* $\mathbb{Z}[A]_+ \ni x \mapsto P_{\mathcal{H}'}\mathbf{T}^x|_{\mathcal{H}'} \in \mathbf{B}(\mathcal{H}')$ *extends to a positive definite* $\mathbf{B}(\mathcal{H}')$-*valued function on* $\mathbb{Z}[A]$.

Proof. Corollary 9 follows from part (vii) of Theorem 5 and part (β) of Corollary 6 as well as from the fact that

$$\mathcal{H}[\mathbb{Z}[A]] = \bigcup\{\mathcal{H}'[\mathbb{Z}[A]];\ \mathcal{H}' \text{ a finite dimensional subspace of } \mathcal{H}\}. \qquad \square$$

It is worthwhile to point out that any of the conditions (ii) and (iii) of Theorem 1 can serve as a test for a finite sequence $\mathbf{T} = (T_1, \dots, T_N) \subset \mathbf{B}(\mathcal{H})$ of a priori noncommuting operators to have a unitary dilation; one has to set: $\mathfrak{G}$ = the additive group $\mathbb{Z}^N$, $\mathfrak{X} = \mathbb{Z}_+^N \cup -\mathbb{Z}_+^N$ ($\mathbb{Z}_+$ is the set of all non-negative integers), $\mathcal{D} = \mathcal{H}$ and $\omega(x; f, g) = \langle \mathbf{T}^{[x]} f, g\rangle$ for $x \in \mathfrak{X}$ and $f, g \in \mathcal{H}$, where

$$\mathbf{T}^{[x]} = \begin{cases} T_1^{x_1} \cdots T_N^{x_N} & \text{for } x \in \mathbb{Z}_+^N \\ T_1^{*|x_1|} \cdots T_N^{*|x_N|} & \text{for } x \in -\mathbb{Z}_+^N. \end{cases}$$

Indeed, by Theorem 1 and the Naimark Dilation Theorem (cf. [14, Theorem 7.1]) there exists a commutative sequence $\mathbf{U} = (U_1, \dots, U_N)$ of unitary operators on a Hilbert space $\mathcal{K} \supset \mathcal{H}$ such that $\mathbf{T}^{[x]} = P\mathbf{U}^{[x]}|_{\mathcal{H}}$ for $x \in \mathfrak{X}$, where P is the orthogonal projection of $\mathcal{K}$ onto $\mathcal{H}$. It is then a simple matter to verify that the sequence $\mathbf{T}$ must be commutative (cf. [22] for a related question).

Notice that if a family $\mathbf{T} = \{T_a\}_{a\in A}$ satisfies any of the conditions (α) and (β) of Corollary 6, then (by this Corollary) every T_a, $a \in A$, must necessarily be a contraction. In the case of a single contraction T these conditions reduce to positive definiteness of the sequence $\{T^{[n]}\}_{n\in\mathbb{Z}}$ (which has been proved by Sz.-Nagy in [26]). It is an *open* question whether for two commuting contractions one can get the conditions (α) and (β) of Corollary 6 in a direct way as in [27]; this would lead to an alternative proof of the Ando unitary dilation theorem [1].

References

[1] T. ANDO, On a pair of commutative contractions, *Acta Sci. Math. (Szeged)*, **24** (1963), 88–90.

[2] T. ANDO, Unitary dilation for a triple of commuting contractions, *Bull. Acad. Polon. Sci. Ser. Math. Sci. Math. Astronom. Phys.*, **24** (1976), 851–853.

[3] A. Athavale, Holomorphic kernels and commuting operators, *Trans. Amer. Math. Soc.*, **304** (1987), 101–110.

[4] A. Athavale, Unitary and spherical dilations: a hypercontractive perspective, *Rev. Roumaine Math. Pures Appl.*, **38** (1993), 387–400.

[5] S. Brehmer, Über vertauschbare Kontraktionen des Hilbertschen Raumes, *Acta Sci. Math. (Szeged)*, **22** (1961), 106–111.

[6] A. P. Calderón and R. Pepinsky, On the phase of Fourier coefficients for positive real periodic functions, *Computing Methods and the Phase Problem in X-ray Crystal Analysis,* Department of Physics, Pennsylvania State College, State College, Pa. (1952), 339–348.

[7] G. Choquet, *Le problème des moments,* Séminaire d'initiation à l'analyse, 1ère année, Inst. Henri Poincaré, Paris, 1962.

[8] R. E. Curto and F.-H. Vasilescu, Standard operator models in the polydisc, *Indiana Univ. Math. J.*, **42** (1993), 791–810.

[9] R. E. Curto and F.-H. Vasilescu, Standard operator models in the polydisc. II, *Indiana Univ. Math. J.*, **44** (1995), 727–746.

[10] A. Devinatz, On the extensions of positive definite functions, *Acta Math.*, **102** (1959), 109–134.

[11] G. I. Èskin, A sufficient condition for the solvability of a multi-dimensional problem of moments, *Dokl. Akad. Nauk SSSR,* **133** (1960), 540–543.

[12] M. G. Krein, Sur le problème du prolongement des fonctions hermitiennes positives et continues, *Dokl. Akad. Nauk SSSR,* **26** (1940), 17–22 (Russian).

[13] W. Mlak, Unitary dilations in case of ordered groups, *Ann. Polon. Math.*, **17** (1966), 321–328.

[14] W. Mlak, Dilations of Hilbert space operators (General theory), *Dissertationes Math.*, **153** (1978), 65.

[15] V. Müller, Commutant lifting theorem for n-tuples of contractions, *Acta Sci. Math. (Szeged)*, **59** (1994), 465–474.

[16] V. Müller and F.-H. Vasilescu, Standard models for some commuting multioperators, *Proc. Amer. Math. Soc.*, **117** (1993), 979–989.

[17] M. A. Naimark, Positive definite operator functions on a commutative group, *Bulletin (Izvestiya) Acad. Sci. URSS (Sér. Math.)*, **7** (1943), 237–244 (Russian, with English summary).

[18] A. E. Nussbaum, Integral representation of functions and distributions positive definite relative to the orthogonal group, *Trans. Amer. Math. Soc.*, **175** (1973), 355–387.

[19] S. Parrott, Unitary dilations for commuting contractions, *Pacific J. Math.*, **34** (1970), 481–490.

[20] W. Rudin, The extension problem for positive-definite functions, *Illinois J. Math.*, **7** (1963), 532–539.

[21] W. Rudin, An extension theorem for positive-definite functions, *Duke Math. J.*, **37** (1970), 49–53.

[22] J. STOCHEL, Positive definiteness and commutativity of operators, *Proc. Amer. Math. Soc.*, **126** (1998), 431–440.

[23] J. STOCHEL and F. H. SZAFRANIEC, The complex moment problem and subnormality: a polar decomposition approach, *J. Funct. Anal.*, **159** (1998), 432–491.

[24] F. H. SZAFRANIEC, Boundedness of the shift operator related to positive definite forms: an application to moment problems, *Ark. Mat.*, **19** (1981), 251–259.

[25] B. SZ.-NAGY, Sur les contractions de l'espace de Hilbert, *Acta Sci. Math. (Szeged)*, **15** (1953), 87–92.

[26] B. SZ.-NAGY, Transformations de l'espace de Hilbert. Fonctions de type positif sur un groupe, *Acta Sci. Math. (Szeged)*, **15** (1954), 104–114.

[27] B. SZ.-NAGY, Extensions of linear transformations in Hilbert space which extend beyond this space, F. Riesz and B. Sz.-Nagy, *Functional Analysis,* Ungar, New York, 1960, Appendix.

[28] B. SZ.-NAGY and C. FOIAŞ, *Harmonic analysis of operators on Hilbert space,* Akadémiai Kiadó and North-Holland, Budapest and Amsterdam – London, 1970.

J. STOCHEL, Instytut Matematyki, Uniwersytet Jagielloński, ul. Reymonta 4, PL-30059 Kraków; *e-mail*: stochel@im.uj.edu.pl

F. H. SZAFRANIEC, Instytut Matematyki, Uniwersytet Jagielloński, ul. Reymonta 4, PL-30059 Kraków; *e-mail*: fhszafra@im.uj.edu.pl;

Operator Theory: Advances and Applications, Vol. 127, 599–611

Inequalities for semibounded operators and their applications to log-hyponormal operators

MITSURU UCHIYAMA

Dedicated to the memory of B. Sz.-Nagy

Let H, K be selfadjoint operators bounded above. We will show that if $H \geq K$, then for $0 < r$, $0 \leq s \leq t$

$$(e^{\frac{r}{2}K}e^{tH}e^{\frac{r}{2}K})^{\frac{s+r}{t+r}} \geq e^{\frac{r}{2}K}e^{sH}e^{\frac{r}{2}K}, \quad e^{\frac{r}{2}H}e^{sK}e^{\frac{r}{2}H} \geq (e^{\frac{r}{2}H}e^{tK}e^{\frac{r}{2}H})^{\frac{s+r}{t+r}}.$$

As an application, we will prove that if T is log-hyponormal (as defined later), then $(T^{*n}T^n)^{1/n} \leq (T^{*n+1}T^{n+1})^{1/(n+1)}$. This includes a result of Ando.

1. Introduction

Let X be a linear operator on a Hilbert space $\mathfrak{H}$. Then $\mathcal{R}(X)$ and $\mathcal{N}(X)$ stand respectively for the range and the null space of X. Throughout this paper, both A and B represent bounded selfadjoint operators and also H and K do semi-bounded selfadjoint operators.

The notation $A \geq B$ means $(Ax, x) \geq (Bx, x)$ for every x, by definition. It is well-known that $A \geq B \geq 0$ implies

$$A^\alpha \geq B^\alpha \quad (0 < \alpha < 1), \quad \log(A + \beta) \geq \log(B + \beta) \quad (0 < \beta);$$

the first inequality is called *Löwner–Heinz inequality.*

The following inequality was found by Hansen [8]: *if P is a projection, and if $A \geq 0$, then*

$$(PAP)^\alpha \geq PA^\alpha P \quad (0 < \alpha < 1). \tag{1}$$

AMS Subject Classification (1991): 47A63, 47B20, 47B25.

Let H and K be both bounded below with spectral families $\{\Lambda_t\}$ and $\{\Gamma_t\}$ respectively. Then we write $H \geq K$ if

$$\int_{-\infty}^{\infty} t\, d(\Lambda_t x, x) \geq \int_{-\infty}^{\infty} t\, d(\Gamma_t x, x) \quad \text{for every } x \in \mathfrak{H}.$$

It is known that for any real number λ, $H + \lambda \geq K + \lambda$ follows from $H \geq K$ and that if $H \geq K \geq 0$ and if $\mathcal{N}(K) = \{0\}$ then $K^{-1} \geq H^{-1} \geq 0$. If H and K are both bounded above, we denote $H \geq K$ if $-K \geq -H$.

Let A be a bounded self-adjoint operator with the spectral family $\{E_t\}$. If $A \geq 0$ and if $\mathcal{N}(A) = \{0\}$, then $\{0\}$ is a null set with respect to $d(E_t x, x)$ for every $x \neq 0$. Hence $\log A$ is well-defined by the functional calculus and bounded above. The following fact is obvious but important in this paper, so we give a proof for the completeness:

$$A \geq B \geq 0, \quad \mathcal{N}(B) = \{0\} \Rightarrow \log A \geq \log B. \tag{2}$$

To see this, multiplying both A and B by a constant, we may assume that $1/2 \geq A \geq B \geq 0$. Hence we have $1 \geq A + \epsilon \geq B + \epsilon \geq \epsilon$ for $1/2 \geq \epsilon > 0$. Let $\{E_t\}$ and $\{F_t\}$ be the spectral families of A and B respectively. Since $\log(A+\epsilon) \geq \log(B+\epsilon)$, $-\log(B+\epsilon) \geq -\log(A+\epsilon)$. This implies

$$\int_0^{1/2} -\log(t+\epsilon)\, d(F_t x, x) \geq \int_0^{1/2} -\log(t+\epsilon)\, d(E_t x, x) \quad (x \in \mathfrak{H}).$$

Letting ϵ tend to 0, by Lebesgue's theorem, we get

$$\int_0^{1/2} -\log t\, d(F_t x, x) \geq \int_0^{1/2} -\log t\, d(E_t x, x) \quad (x \in \mathfrak{H}).$$

This implies $-\log B \geq -\log A$, and hence $\log A \geq \log B$.

Furuta ([5] and [6]) proved that if $A \geq B \geq 0$, then for $0 < r, 1 \leq t$

$$(B^{\frac{r}{2}} A^t B^{\frac{r}{2}})^{\frac{1+r}{t+r}} \geq B^{1+r}, \quad A^{1+r} \geq (A^{\frac{r}{2}} B^t A^{\frac{r}{2}})^{\frac{1+r}{t+r}},$$

and moreover, for $0 < r$, $0 \leq s \leq t$

$$(B^{\frac{r}{2}} A^t B^{\frac{r}{2}})^{\frac{s+r}{t+r}} \geq B^{\frac{r}{2}} A^s B^{\frac{r}{2}}, \quad A^{\frac{r}{2}} B^s A^{\frac{r}{2}} \geq (A^{\frac{r}{2}} B^t A^{\frac{r}{2}})^{\frac{s+r}{t+r}}. \tag{3}$$

As inequalities related to these inequalities the following were shown (see [3], [4] and [12]): *if $A \geq B$, then for $r, t > 0$*

$$(e^{rB/2} e^{tA} e^{rB/2})^{r/(t+r)} \geq e^{rB}, \quad e^{rA} \geq (e^{rA/2} e^{tB} e^{rA/2})^{r/(t+r)}. \tag{4}$$

We will establish these inequalities for semibounded operators H and K in Section 2. In Section 3, we will modify the definition of log-hyponormality and show that if T is log-hyponormal then

$$(T^{*n}T^n)^{1/n} \leq (T^{*n+1}T^{n+1})^{1/(n+1)} \quad (n = 1, 2, \ldots).$$

In Section 4, we will prove that if T is log-hyponormal and if $|T|^n = |T^n|$ for some $n \geq 3$, then the polar decomposition of T is commutative, that is, T is a quasi-normal operator.

2. Semi-bounded operator inequalities

In this section we establish exponential inequalities for semi-bounded operators in the same way that we have shown (4) in [12]. For convenience, we first consider selfadjoint operators bounded below.

THEOREM 2.1. *Let H and K be bounded below and suppose $H \geq K$. Then for $0 < r$, $0 \leq s \leq t$*

$$\begin{aligned} (e^{-\frac{r}{2}H}e^{-tK}e^{-\frac{r}{2}H})^{\frac{s+r}{t+r}} &\geq e^{-\frac{r}{2}H}e^{-sK}e^{-\frac{r}{2}H}, \\ e^{-\frac{r}{2}K}e^{-sH}e^{-\frac{r}{2}K} &\geq (e^{-\frac{r}{2}K}e^{-tH}e^{-\frac{r}{2}K})^{\frac{s+r}{t+r}}. \end{aligned} \tag{5}$$

Proof. It is clear that $(1 + K/n)^{-1}$ and $(1 + H/n)^{-1}$ are both bounded for sufficiently large n and that $(1 + K/n)^{-1} \geq (1 + H/n)^{-1} \geq 0$. Therefore, from (3) it follows that

$$\begin{aligned} &\left\{\left(1+\frac{H}{n}\right)^{-\frac{nr}{2}}\left(1+\frac{K}{n}\right)^{-nt}\left(1+\frac{H}{n}\right)^{-\frac{n}{2}r}\right\}^{\frac{ns+nr}{nt+nr}} \\ &\qquad \geq \left(1+\frac{H}{n}\right)^{-\frac{nr}{2}}\left(1+\frac{K}{n}\right)^{-ns}\left(1+\frac{H}{n}\right)^{-\frac{n}{2}r}. \end{aligned}$$

Since the sequence of functions $(1 + \lambda/n)^{-n}$ of λ converges uniformly to $e^{-\lambda}$ on $\gamma \leq \lambda < \infty$ as $n \to \infty$, where γ is a lower bound of K, $(1 + H/n)^{-nr/2}$ and $(1 + K/n)^{-nt}$ converge to $e^{-rH/2}$ and e^{-tK} in the norm sense, respectively. Thus the above inequality yields

$$(e^{-\frac{r}{2}H}e^{-tK}e^{-\frac{r}{2}H})^{\frac{s+r}{t+r}} \geq e^{-\frac{r}{2}H}e^{-sK}e^{-\frac{r}{2}H}.$$

One can see the second inequality of the theorem as well. □

THEOREM 2.2. *Let H and K be bounded above and suppose $H \geq K$. Then for $0 < r$, $0 \leq s \leq t$*

$$(6) \qquad (e^{\frac{r}{2}K}e^{tH}e^{\frac{r}{2}K})^{\frac{s+r}{t+r}} \geq e^{\frac{r}{2}K}e^{sH}e^{\frac{r}{2}K}, \qquad e^{\frac{r}{2}H}e^{sK}e^{\frac{r}{2}H} \geq (e^{\frac{r}{2}H}e^{tK}e^{\frac{r}{2}H})^{\frac{s+r}{t+r}}.$$

In particular,

$$(7) \qquad (e^{\frac{r}{2}K}e^{tH}e^{\frac{r}{2}K})^{\frac{r}{t+r}} \geq e^{rK}, \qquad e^{rH} \geq (e^{\frac{r}{2}H}e^{tK}e^{\frac{r}{2}H})^{\frac{r}{t+r}}.$$

Proof. Since both $-H$ and $-K$ are bounded below, and since $-K \geq -H$, (5) yields (6). Put $s = 0$ in (6) to get (7). □

THEOREM 2.3. *Let A and B be bounded non-negative operators such that $\mathcal{N}(A) = \mathcal{N}(B) = \{0\}$, and suppose $\log A \geq \log B$. Then for $0 < r$, $0 \leq s \leq t$*

$$(8) \qquad (B^{\frac{r}{2}}A^tB^{\frac{r}{2}})^{\frac{s+r}{t+r}} \geq B^{\frac{r}{2}}A^sB^{\frac{r}{2}}, \qquad A^{\frac{r}{2}}B^sA^{\frac{r}{2}} \geq (A^{\frac{r}{2}}B^tA^{\frac{r}{2}})^{\frac{s+r}{t+r}}.$$

In particular,

$$(9) \qquad (B^{\frac{r}{2}}A^tB^{\frac{r}{2}})^{\frac{r}{t+r}} \geq B^r, \qquad A^r \geq (A^{\frac{r}{2}}B^tA^{\frac{r}{2}})^{\frac{r}{t+r}}.$$

Proof. We know that $\log A$ is a selfadjoint operator bounded above, and

$$e^{t\log A} = A^t$$

(see Section 128 of [9]). Thus (8) follows clearly from (6), and (9) is obvious. □

3. Log-hyponormal operators

From now on, T represents a bounded operator. T is said to be *subnormal* if T has a normal extension, *quasi-normal* if $T(T^*T) = (T^*T)T$, *hyponormal* if $T^*T \geq TT^*$ and *paranormal* if $\|T^2x\|\,\|x\| \geq \|Tx\|^2$ $(x \in \mathfrak{H})$ (see [7], [2]). The relations among these classes of operators are as follows: normal ⇒ quasi-normal ⇒ subnormal ⇒ hyponormal ⇒ paranormal.

For a subspace $\mathfrak{L} \subseteq \mathfrak{H}$ and the projection P onto $\mathfrak{L}$, we call $PT|_{\mathfrak{L}}$ the *compression* of T to $\mathfrak{L}$. Ando [3] showed that *if*

$$(10) \qquad \mathcal{N}(T) \subseteq \mathcal{N}(T^*) \quad \text{and} \quad \log A \geq \log B,$$

where A and B are compressions of T^*T and TT^* to $\overline{\mathcal{R}(T)}$, then T is a paranormal.

Recently in many papers ([1], [4], [10] and so on) T is called log-hyponormal if T is *invertible* and $\log T^*T \geq \log TT^*$; the invertibility of T is necessary for $\log T^*T$ to be bounded. According to this definition, the class of log-hyponormal operators does not contain all hyponormal operators; in fact, there are many hyponormal operators which are not invertible. Therefore, we remove the condition of the invertibility from the definition of log-hyponormality.

DEFINITION. T is said to be *log-hyponormal* if it satisfies (10).

According to this definition, we have the following relation: hyponormal $\Rightarrow$ log-hyponormal $\Rightarrow$ paranormal.

In [11] it was shown that a subnormal operator T satisfies

$$|T| \leq |T^2|^{1/2} \leq \ldots \leq |T^n|^{1/n} \leq \ldots .$$

Yamazaki [13] showed that if $(T^*T)^s \geq (TT^*)^s$ for some $s > 0$ or if T is an *invertible* log-hyponormal operator, then $T^{*n}T^n \leq (T^{*n+1}T^{n+1})^{\frac{n}{n+1}}$, and $T^nT^{*n} \geq (T^{n+1}T^{*n+1})^{\frac{n}{n+1}}$ $(n = 1, 2, \ldots)$, so

$$T^*T \leq (T^{*2}T^2)^{1/2} \leq \ldots \leq (T^{*n}T^n)^{1/n} \leq \ldots,$$
$$TT^* \geq (T^2T^{*2})^{1/2} \geq \ldots \geq (T^nT^{*n})^{1/n} \geq \ldots .$$

An operator T satisfying $T^*T \leq (T^{*2}T^2)^{1/2}$ is paranormal: indeed, in this case, by Jensen's inequality, for $x \neq 0$

$$\|Tx\|^2 = (T^*Tx, x) \leq ((T^{*2}T^2)^{1/2}x, x) \leq (T^{*2}T^2x, x)^{1/2}\|x\| = \|T^2x\|\|x\|.$$

This says that the result of Yamazaki is a partial extension of that of Ando. We will show the inequalities of Yamazaki without assuming the invertibility of T, which induce a complete extension of the result of Ando. We first state a simple lemma needed later.

LEMMA. *Let $T = U|T|$ be the polar decomposition of a bounded operator T and $f(t)$ a continuous function on $[0, \infty)$ with $f(0) = 0$. Then for any $A \geq 0$*

$$f(U|T|A|T|U^*) = f(TAT^*) = Uf(|T|A|T|)U^*,$$
$$f(U^*|T^*|A|T^*|U) = f(T^*AT) = U^*f(|T^*|A|T^*|)U.$$

Proof. Since $T = U|T| = |T^*|U$ and $T^* = |T|U^* = U^*|T^*|$, calculation shows that the equalities hold in case f is a polynomial which vanishes at $t = 0$. Then for general f, we need only to take a sequence $\{p_n\}$ of polynomials with $p_n(0) = 0$ which converges uniformly to f on $[0, \|T\|^2\|A\|\,]$.

□

THEOREM 3.1. *If T is log-hyponormal, then*

$$T^{*n}T^n \leq (T^{*n+1}T^{n+1})^{\frac{n}{n+1}}. \tag{11}$$

Proof. Let $T = U|T|$ be the polar decomposition of T and P the orthogonal projection onto $\overline{\mathcal{R}(T)}$. We denote the compression of an operator X to $\overline{\mathcal{R}(T)}$ by $[X]$. Then the log-hyponormality of T means $\log[T^*T] \geq \log[TT^*]$. Thus, by the first inequality of (9) we get

$$([TT^*]^{1/2}[T^*T][TT^*]^{1/2})^{1/2} \geq [TT^*],$$

which is equivalent to

$$\{(PTT^*P)^{1/2}(PT^*TP)(PTT^*P)^{1/2}\}^{1/2} \geq PTT^*P.$$

Since $TT^*(1-P) = 0$, this gives

$$\{(TT^*)^{1/2}PT^*TP(TT^*)^{1/2}\}^{1/2} \geq TT^*,$$

and hence

$$U^*(|T^*|PT^*TP|T^*|)^{1/2}U \geq U^*TT^*U.$$

In view of the lemma, the left hand side equals $(T^*PT^*TPT)^{1/2}$. Since $PT = T$ and $U^*TT^*U = T^*T$, from the above inequality we get

$$(T^{*2}T^2)^{1/2} \geq T^*T.$$

This means that (11) holds for $n = 1$.

Assume that (11) holds for $n \leq m-1$. Therefore,

$$T^{*m-1}T^{m-1} \leq (T^{*m}T^m)^{\frac{m-1}{m}}, \tag{12}$$

$$T^*T \leq (T^{*m}T^m)^{\frac{1}{m}}. \tag{13}$$

The inequality (13) implies

$$PT^*TP \leq P(T^{*m}T^m)^{\frac{1}{m}}P \leq (PT^{*m}T^mP)^{\frac{1}{m}},$$

here the last inequality is due to (1). Thus $[T^*T] \leq [T^{*m}T^m]^{\frac{1}{m}}$. Since $\mathcal{N}([T^*T]) = 0$ because $\mathcal{R}(T)$ is orthogonal to $\mathcal{N}(T)$ by definition of log-hyponormality of T, by (2) we have

$$\log[T^*T] \leq \log[T^{*m}T^m]^{\frac{1}{m}}.$$

Since T is log-hyponormal, this gives

$$\log[TT^*] \leq \log[T^{*m}T^m]^{\frac{1}{m}}.$$

By the first inequality of (8), we obtain

$$[TT^*]^{1/2}[T^{*m}T^m]^{\frac{m-1}{m}}[TT^*]^{1/2} \le ([TT^*]^{1/2}[T^{*m}T^m]^{\frac{m}{m}}[TT^*]^{1/2})^{\frac{m-1+1}{m+1}},$$

which is equivalent to

$$\begin{aligned}&(PTT^*P)^{1/2}(PT^{*m}T^mP)^{\frac{m-1}{m}}(PTT^*P)^{1/2}\\ &\qquad\le \{(PTT^*P)^{1/2}(PT^{*m}T^mP)(PTT^*P)^{1/2}\}^{\frac{m}{m+1}}.\end{aligned}$$

Since $PT = T$ and $T^*P = T^*$, this gives

$$|T^*|(PT^{*m}T^mP)^{\frac{m-1}{m}}|T^*| \le \{|T^*|(PT^{*m}T^mP)|T^*|\}^{\frac{m}{m+1}}. \tag{14}$$

From (12) it follows that

$$\begin{aligned}T^{*m}T^m &= T^*(T^{*m-1}T^{m-1})T\\ &\le T^*(T^{*m}T^m)^{\frac{m-1}{m}}T = U^*|T^*|P(T^{*m}T^m)^{\frac{m-1}{m}}P|T^*|U\\ &\le U^*|T^*|(PT^{*m}T^mP)^{\frac{m-1}{m}}|T^*|U \quad \text{(by (1))}\\ &\le U^*\{|T^*|(PT^{*m}T^mP)|T^*|\}^{\frac{m}{m+1}}U \quad \text{(by (14))}\\ &= (U^*|T^*|PT^{*m}T^mP|T^*|U)^{\frac{m}{m+1}} \quad \text{(by Lemma 3.1)}\\ &= (T^{*m+1}T^{m+1})^{\frac{m}{m+1}}.\end{aligned}$$

This completes the proof. □

Note that (11) implies that $\mathcal{N}(T) = \mathcal{N}(T^2) = \ldots = \mathcal{N}(T^n) = \ldots$.

THEOREM 3.2. *If for a natural number m*

$$\mathcal{N}(T) \subseteq \mathcal{N}(T^*) \quad \textit{and} \quad \log[T^*T]_n \ge \log[TT^*]_n \quad (1 \le n \le m), \tag{15}$$

where $[X]_n$ means the compression of X to $\overline{\mathcal{R}(T^n)}$, then

$$T^nT^{*n} \ge (T^{n+1}T^{*n+1})^{\frac{n}{n+1}} \quad (n = 1, 2, \ldots, m). \tag{16}$$

Proof. We prove this by the mathematical induction of m. Suppose (15) holds for $m = 1$. Since then T is log-hyponormal, from the second inequality of (9) it follows that

$$[T^*T]_1 \geq \{[T^*T]_1^{1/2}[TT^*]_1[T^*T]_1^{1/2}\}^{1/2}.$$

Denoting the projection to $\overline{\mathcal{R}(T)}$ by P, the above shows

$$PT^*TP \geq \{(PT^*TP)^{1/2}(PTT^*P)(PT^*TP)^{1/2}\}^{1/2}.$$

Let $TP = U|TP|$ be the polar decomposition of TP. Then the above leads to

$$|TP|^2 \geq \{|TP|(PTT^*P)|TP|\}^{1/2},$$

and hence

$$U|TP|^2U^* \geq U\{|TP|(PTT^*P)|TP|\}^{1/2}U^*.$$

By the Lemma, we obtain

$$|PT^*|^2 \geq \{TP(PTT^*P)(TP)^*\}^{1/2}.$$

Since $PT = T$, the righthand side equals $(T^2T^{*2})^{1/2}$. Thus we get

$$TT^* \geq TPT^* \geq (T^2T^{*2})^{1/2}.$$

Consequently (16) holds for $m = 1$.

Assume that the theorem is valid for $m - 1$. To show that it is valid for m, suppose (15) holds for $n = 1, 2, \ldots, m$. From the inductive assumption, (16) holds for $n = 1, 2, \ldots, m - 1$; therefore, we have

$$T^{m-1}T^{*m-1} \geq (T^mT^{*m})^{\frac{m-1}{m}} \quad \text{and} \quad TT^* \geq (T^mT^{*m})^{1/m}. \tag{17}$$

The second inequality yields

$$[TT^*]_m \geq [T^mT^{*m}]_m^{1/m},$$

and hence

$$\log[TT^*]_m \geq \log([T^mT^{*m}]_m^{1/m}).$$

Therefore, by (15) with $n = m$ we obtain

$$\log[T^*T]_m \geq \log([T^mT^{*m}]_m^{1/m}).$$

Applying this to the second inequality of (8) we get

$$[T^*T]_m^{1/2}([T^mT^{*m}]_m^{1/m})^{m-1}[T^*T]_m^{1/2} \geq \{[T^*T]_m^{1/2}([T^mT^{*m}]_m^{1/m})^m[T^*T]_m^{1/2}\}^{\frac{m-1+1}{m+1}}.$$

Denote the projection to $\overline{\mathcal{R}(T^m)}$ by Q, and rewrite the above as

$$(QT^*TQ)^{1/2}(T^mT^{*m})^{\frac{m-1}{m}}(QT^*TQ)^{1/2} \\ \geq \{(QT^*TQ)^{1/2}(T^mT^{*m})(QT^*TQ)^{1/2}\}^{\frac{m-1+1}{m+1}}.$$

Now let $TQ = V|TQ|$ be the polar decomposition of TQ. Then we have

$$|TQ|(T^mT^{*m})^{\frac{m-1}{m}}|TQ| \geq \{|TQ|(T^mT^{*m})|TQ|\}^{\frac{m-1+1}{m+1}}.$$

Multiplying this inequality by V and V^*, in virtue of the Lemma, yields

$$TQ(T^mT^{*m})^{\frac{m-1}{m}}QT^* \geq \{TQ(T^mT^{*m})QT^*\}^{\frac{m-1+1}{m+1}}.$$

From this it follows that

$$T(T^mT^{*m})^{\frac{m-1}{m}}T^* \geq \{T(T^mT^{*m})T^*\}^{\frac{m-1+1}{m+1}} = (T^{m+1}T^{*m+1})^{\frac{m}{m+1}},$$

for $Q(T^mT^{*m}) = T^mT^{*m} = (T^mT^{*m})Q$. This in conjunction with the first inequality of (17) gives

$$T^mT^{*m} = T(T^{m-1}T^{*m-1})T^* \geq T(T^mT^{*m})^{\frac{m-1}{m}}T^* \geq (T^{m+1}T^{*m+1})^{\frac{m}{m+1}}.$$

This shows that (16) holds for $n = m$. Thus, we conclude the proof. □

As mentioned in the proof, (15) with $m = 1$ is satisfied for a log-hyponormal operator. Therefore,

$$TT^* \geq (T^2T^{*2})^{1/2}$$

is valid for a log-hyponormal operator T. The condition (15) is technical. We do not know if (15) follows from the log-hyponormality of T. However, one can easily see that (15) holds for every m if T is hyponormal. Moreover, $\mathcal{N}(T^*) = \mathcal{N}(T^{*m})$ for every m if $\mathcal{N}(T^*) = 0$ or $\mathcal{N}(T^*) = \mathcal{N}(T^{*2})$, so we get:

COROLLARY 3.3. *If T is a log-hyponormal operator and if $\mathcal{N}(T^*) = 0$ or $\mathcal{N}(T^*) = \mathcal{N}(T^{*2})$, then for every n*

$$T^nT^{*n} \geq (T^{n+1}T^{*n+1})^{\frac{n}{n+1}}.$$

4. Normality

Recall the definition of the quasi-normality of T stated in the previous section. Then we can see that T is quasi-normal if and only if the polar decomposition of T is commutative, so that a quasi-normal operator T is normal if $\mathcal{N}(T) = \mathcal{N}(T^*)$.

In this section, we give a few conditions which guarantee quasi-normality or normality of a log-hyponormal operator.

THEOREM 4.1. *If a log-hyponormal operator T satisfies $|T|^n = |T^n|$ for some $n \geq 3$, then T is quasi-normal.*

Proof. By Theorem 3.1, we get

$$|T| \leq |T^2|^{1/2} \leq \ldots \leq |T^n|^{1/n} \leq \ldots.$$

Thus the assumption implies that

$$(T^*T)^2 = T^{*2}T^2 \quad \text{and} \quad (T^*T)^3 = T^{*3}T^3. \tag{18}$$

From the first equality of (18) it follows that $T^*(TT^* - T^*T)T = 0$, and hence, for the projection P onto $\overline{\mathcal{R}(T)}$

$$PT^*TP = PTT^*P = TT^*. \tag{19}$$

By (18) and (19)

$$T^*(TT^*)^2T = (T^*T)^3 = T^{*3}T^3 = T^*(T^{*2}T^2)T = T^*(T^*T)^2T,$$

from which it follows that

$$P(TT^*)^2P = P(T^*T)^2P.$$

Since the left hand side equals $(TT^*)^2$, this and (19) yield

$$(PT^*TP)^2 = (TT^*)^2 = P(TT^*)^2P = P(T^*T)^2P.$$

Consequently,

$$PT^*T(1-P)T^*TP = 0 \quad \text{and hence} \quad T^*TP = PT^*TP.$$

From this and (19) we derive

$$(T^*T)T = (T^*TP)T = (PT^*TP)T = (TT^*)T = T(T^*T).$$

This means T is quasi-normal.

□

The above theorem does not hold for $n = 2$: for a counterexample see p. 199 of [11].

THEOREM 4.2. *If a log-hyponormal operator T satisfies*

$$|T^*|^n = |T^{*n}| \quad \textit{for some} \quad n \geq 2,$$

then T is normal.

Proof. The assumption implies $T^nT^{*n} = (TT^*)^n$ and hence

$$\mathcal{N}(T^{*n}) = \mathcal{N}(T^nT^{*n}) = \mathcal{N}((TT^*)^n) = \mathcal{N}(TT^*) = \mathcal{N}(T^*),$$

from which $\mathcal{N}(T^{*2}) = \mathcal{N}(T^*)$ follows. By Corollary 3.3

$$(T^mT^{*m})^{1/m} \geq (T^{m+1}T^{*(m+1)})^{1/(m+1)} \quad \text{for every } m.$$

Therefore, by assumption, we have

$$TT^* = (T^2T^{*2})^{1/2} = \ldots = (T^nT^{*n})^{1/n},$$

which implies $T^2T^{*2} = (TT^*)^2$. Denoting the projection onto $\overline{\mathcal{R}(T^*)}$ by Q, this implies that

$$QTT^*Q = QT^*TQ = T^*T.$$

On the other hand, since $\mathcal{N}(T) \subseteq \mathcal{N}(T^*)$ by assumption and since $1-Q$ is the projection to $\mathcal{N}(T)$, we have $T^*(1-Q) = 0$, which implies $TT^* = QTT^*Q$. Consequently, we obtain $TT^* = QTT^*Q = T^*T$.

□

THEOREM 4.3. *A log-hyponormal operator T is normal if T^{*n} is log-hyponormal for some natural number n.*

Proof. As we mentioned after the proof of Theorem 3.1,

$$\mathcal{N}(T^*) \supseteq \mathcal{N}(T) = \mathcal{N}(T^m) \quad \text{for every } m.$$

On the other hand,

$$\mathcal{N}(T^*) \subseteq \mathcal{N}(T^{*2}) \subseteq \cdots \subseteq \mathcal{N}(T^{*n}) \subseteq \mathcal{N}(T^n)$$

since T^{*n} is log-hyponormal. Thus we have $\mathcal{N}(T) = \mathcal{N}(T^{*m})$ for $m = 1, 2, \ldots, n$. Denote the compression of X to $\overline{\mathcal{R}(T)}$ by $[X]$. By Theorems 3.1 and 3.2 we get

$$[T^*T] \le [T^{*n}T^n]^{1/n}, \quad [T^nT^{*n}]^{1/n} \le [TT^*].$$

Since $\log(A^{1/n}) = \frac{1}{n}\log A$ for $A \ge 0$ with $\mathcal{N}(A) = 0$, and since T^{*n} is log-hyponormal,

$$\log[T^*T] \le \frac{1}{n}\log[T^{*n}T^n] \le \frac{1}{n}\log[T^nT^{*n}] \le \log[TT^*].$$

Therefore the log-hyponormality of T yields $\log[T^*T] = \log[TT^*]$. This implies $T^*T = TT^*$.

□

References

[1] A. ALUTHGE and D. WANG, An operator inequalty which implies paranormality, *Math. Inequal. Appl.*, **2** (1999), 113–119.

[2] T. ANDO, Opertors with a norm condition, *Acta Sci. Math. (Szeged)*, **33** (1972), 169–178.

[3] T. ANDO, On some operator inequalities, *Math. Ann.*, **279** (1987), 157–159.

[4] M. FUJII, T. FURUTA and E. KAMEI, Furuta's inequality and its application to Ando's theorem, *Linear Alg. Appl.*, **179** (1883), 161–169.

[5] T. FURUTA, $A \ge B \ge 0$ assures $(B^rA^pB^r)^{1/q}$ for $r \ge 0$, $p \ge 0$, $q \ge 1$ with $(1+2r)q \ge p+2r$, *Proc. Amer. Math. Soc.*, **101** (1987), 85–88.

[6] T. FURUTA, Two operator functions with monotone property, *Proc. Amer. Math. Soc.*, **111** (1991), 511–516.

[7] P. R. HALMOS, *Hilbert space problem book*, Springer Verlag, New York, 1982.

[8] F. HANSEN, An operator inequality, *Math. Ann.*, **246** (1980), 249–250.

[9] F. RIESZ and B. SZ.-NAGY, *Functional Analysis*, New York, 1955.

[10] K. Tanahashi, On log-hyponormal operators, *Integral Equation Operator Theory*, **34** (1999), 364–372.

[11] M. Uchiyama, Operators which have commutative polar decompositions, *Operator Theory: Advances and Appl.*, **62** (1993), 197–208.

[12] M. Uchiyama, Some exponential operator inequalities, *Math. Inequal. Appl.*, **2** (1999), 469–471.

[13] T. Yamazaki, Extension of the results on p-hyponormal and log-hyponormal operators by Althuge and Wang, *SUT J. Math.*, **35** (1999), 139–148.

M. Uchiyama, Department of Mathematics, Fukuoka University of Education, Munakata, Fukuoka, 811-4192, Japan; *e-mail*: uchiyama@fukuoka-edu.ac.jp

Operator Theory: Advances and Applications, Vol. 127, 613–638

Operator moment problems in unbounded sets

F.-H. VASILESCU

Dedicated to the memory of Béla Szőkefalvi-Nagy

Results concerning some moment problems in unbounded sets from the scalar context to the case of linear operator data are extended. The present methods also lead to a theoretic characterization of a class of subnormal multioperators.

0. Introduction

An important contribution of Béla Szőkefalvi-Nagy in the operator moment problems is recorded as early as 1952, and it concerns a solution to the Hausdorff moment problem with operator data (see [SzN1]). Since then, there have occurred many other contributions in this area. See [AtPe], [Fri], [Mac], [Nar], [Sch], [StSz2], [Vas2], [Vas4] etc. for further development.

Recently, some solutions to multivariable moment problems on unbounded semi-algebraic sets, in the scalar case, were given in [Vas3] and [PuVa]. The main aim of this paper is to present operator versions of the results from [Vas3] and [PuVa]. In particular, we extend the corresponding results concerning the solutions of the moment problems of Hamburger and Stieltjes type, in several variables, to the case of operator data.

When trying to extend the results from [Vas3] and [PuVa], one is tempted to replace the numerical sequences by sequences of bounded linear operators. Relevant examples related to the characterization of those unbounded operators which have unbounded normal extensions (see [StSz1]) show that the numerical sequences must

---

AMS Subject Classification (1991): 47A57, 44A60, 42A82.

be replaced by sequences of unbounded operators. As a matter of fact, for technical reasons, it is even better to work in the context of hermitian (or even sesquilinear) forms (see also [Sch], [StSz1] etc. for similar discussions). We shall therefore consider sequences of hermitian (or sesquilinear) forms as moment data, trying to associate them with of what we call a *moment form* (see Definition 1.3), which is a hermitian form on a space of vector-valued functions (regarded as an algebraic tensor product), having an integral representation via an operator-valued positive measure.

In the first section we introduce the notation and terminology, and discuss the basic construction, in the spirit of [Vas4] but in a more general setting. As the solution of a given moment problem is, in general, a positive measure, it seems natural to state a large class of such problems in connection with a space of (universally measurable) functions. This explains the systematic use, in this section, of what is called a *function space* (see [Vas4]) with the additional hypothesis requiring that the elements of such a space be Borel measurable functions.

In the second section, which is the most consistent part of this work, we extend the results from [Vas3] and [PuVa]. Integral representations for positive semi-definite hermitian forms defined on tensor products $\mathcal{R} \otimes \mathcal{D}$ are obtained, where $\mathcal{D}$ is a dense subspace in a given Hilbert space and $\mathcal{R}$ is an algebra of rational functions. As a matter of fact, we are interested in some easily described algebras of rational functions, corresponding to the cases investigated in [Vas3] and [PuVa], respectively. The integral representations mentioned above are then applied to obtain solutions to various moment problems with hermitian form (in particular, with self-adjoint operator) data.

The approaches from [Vas3] and [PuVa] have some common ideas, and we shall try to emphasize the differences. In fact, the more refined method from [PuVa] allows, in addition, the characterization of the representing measures whose support lies in a (not necessarily bounded) domain defined by polynomial inequalities.

The third section contains an application of these results to the study of the normal extensions of subnormal (unbounded) multioperators. Roughly speaking, a subnormal multioperator is the restriction of a normal multioperator (i.e., a multioperator consisting of commuting normal operators) to an invariant subspace, with some precautions in the unbounded case (see [StSz1]). We shall show that our methods lead to a theoretic characterization of a class of subnormal multioperators (see also [AtPe], [Bra], [Hal], [Ito], [StSz1], [StSz2] etc., for various characterizations of subnormal operators and multioperators).

The author is grateful to the referee for a careful reading of the manuscript and for several useful suggestions and some supplementary references. Particularly, his remarks helped the author to state Theorem 3.4 in the actual form.

1. Moment forms on function spaces

We start by fixing the basic terminology and notation. If $\mathcal{L}$ is complex vector space, by *sesquilinear form* on $\mathcal{L}$ we mean a function $\Lambda: \mathcal{L} \times \mathcal{L} \to \mathbb{C}$ such that both maps $\Lambda(*, y), \overline{\Lambda(x, *)}$ are linear for all $x, y \in \mathcal{L}$. If, in addition, $\Lambda(y, x) = \overline{\Lambda(x, y)}$ for all $x, y \in \mathcal{L}$, then Λ is said to be a *hermitian form*. A sesquilinear form Λ on $\mathcal{L}$ is said to be *positive semi-definite* if $\Lambda(x, x) \geq 0$ for all $x \in \mathcal{L}$. A polarization argument shows that every positive semi-definite form is hermitian.

Let $n \geq 1$ be a fixed integer, and let $\mathcal{S}$ be a vector space consisting of complex-valued functions, defined on the Euclidean space $\mathbb{R}^n$, endowed with the usual algebraic operations. As in [Vas4], we shall call $\mathcal{S}$ a *function space* (*on* $\mathbb{R}^n$) if the constant function $1 \in \mathcal{S}$, and if $f \in \mathcal{S}$ then its complex-conjugate $\bar{f} \in \mathcal{S}$.

Let $\mathcal{H}$ be a Hilbert space, whose norm $\| * \|$ is defined by the inner product $\langle *, * \rangle$. We denote by $\mathcal{L}(\mathcal{H})$ the space of all linear bounded operators, acting in $\mathcal{H}$.

Let also $\mathcal{D}$ be a fixed dense linear subspace of $\mathcal{H}$. We shall occasionally use the notation $\mathcal{E}(\mathcal{D}, \mathcal{H})$ to designate the space of all linear maps from $\mathcal{D}$ into $\mathcal{H}$.

Let $\mathcal{S}$ be a fixed function space on $\mathbb{R}^n$ and let Λ be a positive semi-definite form on the (algebraic) tensor product $\mathcal{S} \otimes \mathcal{D}$. In particular, the form Λ satisfies the *Cauchy–Schwarz inequality*

$$(1.1) \qquad | \Lambda(\varphi, \psi) |^2 \leq \Lambda(\varphi, \varphi) \Lambda(\psi, \psi)$$

for all φ, ψ from $\mathcal{S} \otimes \mathcal{D}$.

Using a standard classical procedure (essentially due to Gelfand–Naimark; see also [DuSc], [Fug], [Hal], [Sti] etc.) one can associate the form Λ with a certain Hilbert space $\mathcal{H}_\Lambda$, defined in the following way.

Let

$$(1.2) \qquad \mathcal{N}_\Lambda = \{\varphi \in \mathcal{S} \otimes \mathcal{D}; \Lambda(\varphi, \varphi) = 0\}.$$

Then $\mathcal{N}_\Lambda$ is a linear subspace of $\mathcal{S} \otimes \mathcal{D}$ (via (1.1)), and the quotient $(\mathcal{S} \otimes \mathcal{D}) / \mathcal{N}_\Lambda$ is an inner product space, whose inner product is given by

$$(1.3) \qquad \langle \varphi + \mathcal{N}_\Lambda, \psi + \mathcal{N}_\Lambda \rangle = \Lambda(\varphi, \psi).$$

The completion of the space $(\mathcal{S} \otimes \mathcal{D}) / \mathcal{N}_\Lambda$ with respect to the norm induced by (1.3) is a Hilbert space, denoted in what follows by $\mathcal{H}_\Lambda$.

REMARK 1.1. Let Λ be a sesquilinear form on $\mathcal{S} \otimes \mathcal{D}$. We say that Λ is *unital* if

$$\Lambda(1 \otimes x, 1 \otimes x) = \|x\|^2, \qquad x \in \mathcal{D}.$$

Note that if Λ is a unital positive semi-definite form, then the linear map

$$\mathcal{D} \ni x \to 1 \otimes x + \mathcal{N}_\Lambda \in \mathcal{H}_\Lambda, \tag{1.4}$$

is an isometry, which allows the identification of the space $\mathcal{D}$ with a subspace of $\mathcal{H}_\Lambda$. Moreover, the map (1.4) extends to an isometry from $\mathcal{H}$ into $\mathcal{H}_\Lambda$.

Let $\mathcal{A} \subset \mathcal{S}$ be an algebra which is a function space, and such that $\mathcal{S}$ be an $\mathcal{A}$-module (with respect to the usual operations). Clearly, $\mathcal{S} \otimes \mathcal{D}$ is an $\mathcal{A}$-module. Let Λ be a sesquilinear form on $\mathcal{S} \otimes \mathcal{D}$. We say that Λ is *$\mathcal{A}$-symmetric* if

$$\Lambda(p \cdot \varphi, \psi) = \Lambda(\varphi, \bar{p} \cdot \psi), \qquad \varphi, \psi \in \mathcal{S} \otimes \mathcal{D}, \quad p \in \mathcal{A} \tag{1.5}$$

(see also [Mas2] for a similar property in the context of operator valued kernels).

If Λ is a positive semi-definite form, which is also $\mathcal{A}$-symmetric, then $\mathcal{N}_\Lambda$ is an $\mathcal{A}$-module. Indeed, in this case for all $\varphi, \psi \in \mathcal{S} \otimes \mathcal{D}$ and $p \in \mathcal{A}$ we have

$$| \Lambda(p \cdot \varphi, p \cdot \varphi) |^2 \leq \Lambda(| p |^2 \varphi, | p |^2 \varphi)\Lambda(\varphi, \varphi),$$

by the Cauchy–Schwarz inequality, which implies that $\mathcal{N}_\Lambda$ is an $\mathcal{A}$-module. This shows that the quotient $(\mathcal{S} \otimes \mathcal{D})/\mathcal{N}_\Lambda$ is an $\mathcal{A}$-module too.

REMARK 1.2. Let $\mathrm{Bor}(\mathbb{R}^n)$ be the family of all Borel subset of $\mathbb{R}^n$, and let $F: \mathrm{Bor}(\mathbb{R}^n) \to \mathcal{L}(\mathcal{H})$ be a positive (operator-valued) measure (i.e., a strongly countably additive set function whose values are positive operators). We always assume that $F(\mathbb{R}^n) = 1_\mathcal{H}$ (where $1_\mathcal{H}$ is the identity on $\mathcal{H}$). For every pair $x, y \in \mathcal{H}$ we define the complex measure $F_{x,y}$ by the relation $F_{x,y}(*) = \langle F(*)x, y \rangle$.

Let $\mathcal{I}_F^p$ denote the space of all Borel measurable functions f on $\mathbb{R}^n$ such that $f \in L^p(F_{x,x})$ for all $x \in \mathcal{H}$, $p \in \{1, 2\}$. Obviously, both $\mathcal{I}_F^1$, $\mathcal{I}_F^2$ are function spaces.

As noticed in [Vas4] (see also [Mas1], [Sto1] etc.), there exists a linear map $L_F: \mathcal{I}_F^1 \to \mathcal{L}(\mathcal{H})$ such that

$$\langle L_F(f)x, y \rangle = \int f \, dF_{x,y}, \qquad f \in \mathcal{I}_F^1, \ x, y \in \mathcal{H}. \tag{1.6}$$

To see this, note first that if $f: \mathbb{R}^n \to \mathbb{C}$ is a bounded Borel measurable function, then $\int f \, dF$ exists in the uniform topology of $\mathcal{L}(\mathcal{H})$ (see [Ber], Section 5). Moreover, if $\mathcal{B}_n$ is the algebra of all bounded Borel functions on $\mathbb{R}^n$, which is a C^*-algebra when endowed with the sup norm, then the map $\mathcal{B}_n \ni f \to \int f\,dF \in \mathcal{L}(\mathcal{H})$ is a continuous unital linear map, which preserves the involution, but not necessarily multiplicative (see [Ber], Section 5, Theorem 10).

Let $f\colon \mathbb{R}^n \to \mathbb{C}$ be an arbitrary Borel measurable function, and let $(f_m)_{m\geq 1}$ be a sequence of bounded Borel measurable functions, which is pointwise convergent to f, and such that $\mid f_m \mid\leq\mid f \mid$ for all $m \geq 1$.

If $f \in \mathcal{I}_F^1$, then the sequence $(\int f_m\,dF)_{m\geq 1}$ is at least weakly convergent (by the Lebesgue theorem of dominated convergence), and we denote by $L_F(f)$ its weak limit. Then (1.6) is clearly fulfilled.

Obviously, the operator $L_F(f)$ does not depend on the particular choice of a sequence of bounded Borel measurable functions $(f_m)_{m\geq 1}$, which is pointwise convergent to f, and such that $\mid f_m \mid\leq\mid f \mid$ for all $m \geq 1$.

It is easily seen that L_F is a unital linear map, which preserves the involution.

We can also associate the measure F with a hermitian form Λ_F on $\mathcal{I}_F^2 \otimes \mathcal{H}$, which is unital and $\mathcal{B}_n$-symmetric, by setting

$$\Lambda_F(\varphi, \psi) = \sum_{j,k} \int f_j \bar{g}_k\,dF_{x_j,y_k} = \sum_{j,k} \langle L_F(f_j\bar{g}_k)x_j, y_k\rangle, \tag{1.7}$$

with $\varphi = \sum_j f_j \otimes x_j$, $\psi = \sum_k g_k \otimes y_k$ arbitrary elements from $\mathcal{I}_F^2 \otimes \mathcal{H}$.

DEFINITION 1.3. Let $\mathcal{S}$ be a function space on $\mathbb{R}^n$ consisting of Borel measurable functions, and let $F\colon \mathrm{Bor}(\mathbb{R}^n) \to \mathcal{L}(\mathcal{H})$ be a positive measure.

Let also Λ be a sesquilinear form on $\mathcal{S} \otimes \mathcal{D}$. We say that F is a *representing measure* for Λ if $f \in L^2(F_{x,x})$ for all $f \in \mathcal{S}$ and $x \in \mathcal{D}$, and

$$\Lambda(\varphi, \psi) = \sum_{j,k} \int f_j \bar{g}_k\,dF_{x_j,y_k}, \tag{1.8}$$

where $\varphi = \sum_j f_j \otimes x_j$, $\psi = \sum_k g_k \otimes y_k$ are arbitrary elements from $\mathcal{S} \otimes \mathcal{D}$.

If the sesquilinear form Λ possesses a representing measure, then Λ is said to be a *moment form*.

Let $\mathcal{A} \subset \mathcal{S}$ be an algebra which is a function space, and such that $\mathcal{S}$ is an $\mathcal{A}$-module. If Λ satisfies (1.8), then Λ is unital and $\mathcal{A}$-symmetric.

LEMMA 1.4. *Let Λ be a moment form on $\mathcal{S}\otimes\mathcal{D}$. Then Λ is positive semi-definite.*

Proof. Let F be a representing measure for Λ, and let L_F be given by (1.6). Let also $\varphi = \sum_j f_j \otimes x_j \in \mathcal{I}_F^2 \otimes \mathcal{D}$, with f_j simple functions for all j. With no loss of generality we may assume that $f_j = \chi_{\sigma_j}$, where $\sigma_j \in \mathrm{Bor}(\mathbb{R}^n), \sigma_j \cap \sigma_k = \emptyset$ if $j \neq k$, and χ_{σ_j} is the characteristic function of σ_j for all j. Then, with Λ_F given by (1.7), we have

$$\Lambda_F(\varphi, \varphi) = \sum_{j,k} \langle L_F(\chi_{\sigma_j}\chi_{\sigma_k})x_j, x_k\rangle = \sum_j \langle F(\sigma_j)x_j, x_j\rangle \geq 0. \tag{1.9}$$

If $\varphi = \sum_j f_j \otimes x_j \in \mathcal{S} \otimes \mathcal{D}$ is arbitrary, for every index j we choose a sequence of bounded Borel functions $(f_{j,m})_{m\geq 1}$ such that $|f_{j,m}| \leq |f_j|$ for all m, and which is pointwise convergent to f_j. If $\varphi_m = \sum_j f_{j,m} \otimes x_j$, then we have

$$\begin{aligned}\Lambda(\varphi, \varphi) = \sum_{j,k} \int f_j \bar{f}_k \, dF_{x_j,x_k} &= \lim_{m\to\infty} \sum_{j,k} \int f_{j,m} \bar{f}_{k,m} \, dF_{x_j,x_k} \\ &= \lim_{m\to\infty} \Lambda_F(\varphi_m, \varphi_m) \geq 0\end{aligned}$$

by (1.9), showing that Λ is positive semi-definite. □

REMARK. In the present context, the *operator moment problem* means to characterize those positive semi-definite forms that have representing measures. An approach at this level of generality (i.e. for an arbitrary function space), although fairly natural, seems to be difficult. In the next section, we shall describe the positive semi-definite forms in some particular cases, when the basic function space is a certain algebra of rational functions. Then we shall derive solutions to *power moment problems* with operator data, i.e., when the basic function space is the algebra of polynomial functions in several variables.

EXAMPLES 1.5. 1) Let $\mathcal{S}$ be a function space and assume that $\mathcal{S}$ has an algebraic basis $\mathcal{G} = (g_\alpha)_{\alpha\in\mathbb{Z}_+^n}, g_0 = 1$. Let also $\Theta = (\Theta_{\alpha,\beta})_{\alpha,\beta\in\mathbb{Z}_+^n}$ be a $2n$-sequence of sesquilinear forms on $\mathcal{D}$ such that $\Theta_{0,0}$ is the restriction to $\mathcal{D}$ of the scalar product of $\mathcal{H}$ and $\Theta_{\alpha,\beta}(x,y) = \overline{\Theta_{\beta,\alpha}(y,x)}$ for all $\alpha, \beta \in \mathbb{Z}_+^n$ and $x, y \in \mathcal{D}$.

We can associate the sequence Θ with a hermitian form $\Lambda_{\Theta,\mathcal{G}}$, defined by the equation

$$\Lambda_{\Theta,\mathcal{G}}(\varphi, \psi) = \sum_{\alpha,\beta} \Theta_{\alpha,\beta}(x_\alpha, y_\beta), \tag{1.10}$$

where $\varphi = \sum_\alpha g_\alpha \otimes x_\alpha, \psi = \sum_\beta g_\beta \otimes y_\beta$ are arbitrary elements from $\mathcal{S} \otimes \mathcal{D}$.

Note also that the form $\Lambda_{\Theta,\mathcal{G}}$ is unital.

Using a concept wich goes back to [SzN2] (see also [BCR] etc.), we say that the sequence Θ is of *positive type with respect to* $\mathcal{S}$ if the associated hermitian form $\Lambda_{\Theta,\mathcal{G}}$ is positive semi-definite.

It is easily seen that this property of Θ does not depend on $\mathcal{G}$. Indeed, let $\mathcal{G}' = (g'_\alpha)_{\alpha\in\mathbb{Z}^n_+}, g'_0 = 1$, be another algebraic basis of $\mathcal{S}$. If $\varphi' = \sum_\alpha g'_\alpha \otimes x_\alpha, \psi' = \sum_\beta g'_\beta \otimes y_\beta$ are arbitrary elements from $\mathcal{S}\otimes\mathcal{D}$, it is clear that $\Lambda_{\Theta,\mathcal{G}'}(\varphi',\psi') = \Lambda_{\Theta,\mathcal{G}}(\varphi,\psi)$, with $\varphi = \sum_\alpha g_\alpha \otimes x_\alpha, \psi = \sum_\beta g_\beta \otimes y_\beta$, showing, in particular, that the property of positive semi-definiteness does not depend on a fixed basis of $\mathcal{S}$.

We say that the sequence Θ is a *moment sequence with respect to* $\mathcal{G}$ if the associated hermitian form $\Lambda_{\Theta,\mathcal{G}}$ is a moment form. In this case, a representing measure for $\Lambda_{\Theta,\mathcal{G}}$ will be also called a *representing measure* for Θ. Note that if F is a representing measure for Θ, then we have $\Theta_{\alpha,\beta}(x,y) = \int g_\alpha\bar{g}_\beta dF_{x,y}$ for all $\alpha,\beta\in\mathbb{Z}^n_+, x,y\in\mathcal{D}$. In fact, Θ is a moment sequence with respect to $\mathcal{G}$ if and only if there exists an operator-valued positive measure F such that $g_\alpha\in L^2(F_{x,x})$ and $\Theta_{\alpha,\beta}(x,y) = \int g_\alpha\bar{g}_\beta dF_{x,y}$ for all $\alpha,\beta\in\mathbb{Z}^n_+, x,y\in\mathcal{D}$.

If Θ is a moment sequence (with respect to $\mathcal{G}$), then Θ is of positive type, by Lemma 1.4.

To apply the previous construction to n-sequences of sesquilinear forms, we need some supplementary conditions.

First of all, let us fix an involution $\mathbb{Z}^n_+ \ni \alpha \to \alpha^* \in \mathbb{Z}^n_+$ (i.e., $(\alpha+\beta)^* = \alpha^*+\beta^*$, $\alpha^{**} = \alpha$ for all $\alpha,\beta\in\mathbb{Z}^n_+$).

Secondly, suppose that the algebraic basis $\mathcal{G} = (g_\alpha)_{\alpha\in\mathbb{Z}^n_+}$ has the properties $g_{\alpha^*} = \bar{g}_\alpha, g_{\alpha+\beta} = g_\alpha g_\beta$ for all $\alpha,\beta\in\mathbb{Z}^n_+$. This implies, in particular, that $\mathcal{S}$ is an algebra.

Let $\Theta = (\Theta_\alpha)_{\alpha\in\mathbb{Z}^n_+}$ be a n-sequence of sesquilinear forms on $\mathcal{D}$ such that Θ_0 is the restriction to $\mathcal{D}$ of the scalar product of $\mathcal{H}$ and $\Theta_\alpha(x,y) = \overline{\Theta_{\alpha^*}(y,x)}$ for all $\alpha\in\mathbb{Z}^n_+$ and $x,y\in\mathcal{D}$

We apply the preceding discussion to the sesquilinear forms $\Theta_{\alpha,\beta}(x,y) = \Theta_{\alpha+\beta^*}(x,y)$ for all $\alpha,\beta\in\mathbb{Z}^n_+$ and $x,y\in\mathcal{D}$. Consequently, we associate the sequence Θ with the hermitian form $\Lambda_{\Theta,\mathcal{G}}$ (corresponding to (1.10)), defined by the equation

$$\Lambda_{\Theta,\mathcal{G}}(\varphi,\psi) = \sum_{\alpha,\beta}\Theta_{\alpha+\beta^*}(x_\alpha,y_\beta), \tag{1.10'}$$

where $\varphi = \sum_\alpha g_\alpha\otimes x_\alpha, \psi = \sum_\beta g_\beta\otimes y_\beta$ are arbitrary elements from $\mathcal{S}\otimes\mathcal{D}$. Moreover,

$$\Lambda_{\Theta,\mathcal{G}}(g_\xi(g_\alpha\otimes x), g_\beta\otimes y) = \Theta_{\alpha+(\xi^*+\beta)^*}(x,y) = \Lambda_{\Theta,\mathcal{G}}(g_\alpha\otimes x, \bar{g}_\xi(g_\beta\otimes y))$$

for all $\alpha, \beta, \xi \in \mathbb{Z}_+^n$ and $x, y \in \mathcal{D}$, implying that the form $\Lambda_{\Theta,\mathcal{G}}$ is $\mathcal{S}$-symmetric.

We say, as above, that Θ is of *positive type with respect to* $\mathcal{S}$, resp. a *moment form with respect to* $\mathcal{G}$, if the form (1.10′) is positive semi-definite , resp. a moment form.

When $\Lambda_{\Theta,\mathcal{G}}$ has a representing measure F, then F is also said to be a *representing measure* for Θ. In this case we have $\Theta_\alpha(x,y) = \int g_\alpha dF_{x,y}$ for all $\alpha \in \mathbb{Z}_+^n, x, y \in \mathcal{D}$.

Conversely, the equalities $\Theta_\alpha(x,y) = \int g_\alpha dF_{x,y}$ for all $\alpha \in \mathbb{Z}_+^n, x, y \in \mathcal{D}$, with $g_\alpha \in L^2(F_{x,x})$ for all $\alpha \in \mathbb{Z}_+^n, x \in \mathcal{D}$, imply that F is a representing measure for Θ.

We can apply, in particular, the above remarks to n-sequences $\Theta = (\Theta_\alpha)_{\alpha\in\mathbb{Z}_+^n}$ consisting of hermitian forms on $\mathcal{D}$. As involution on $\mathbb{Z}_+^n$ we take simply the identity, and we suppose that the basis $\mathcal{G}$ has the additional property $g_\alpha = \bar{g}_\alpha, g_{\alpha+\beta} = g_\alpha g_\beta$ for all $\alpha, \beta \in \mathbb{Z}_+^n$.

By setting $\Theta_{\alpha,\beta} = \Theta_{\alpha+\beta}, \alpha, \beta \in \mathbb{Z}_+^n$, we associate the sequence Θ with the hermitian form $\Lambda_{\Theta,\mathcal{G}}$ given by

$$\Lambda_{\Theta,\mathcal{G}}(\varphi,\psi) = \sum_{\alpha,\beta\in\mathbb{Z}_+^n} \Theta_{\alpha+\beta}(x_\alpha, y_\beta), \tag{1.10''}$$

where $\varphi = \sum_\alpha g_\alpha \otimes x_\alpha, \psi = \sum_\beta g_\beta \otimes y_\beta$, corresponding to (1.10′).

Of course, as before, the sequence Θ is *of positive type with respect to* $\mathcal{S}$, resp. *a moment sequence with respect to* $\mathcal{G}$, if the associated form $\Lambda_{\Theta,\mathcal{G}}$ is positive semi-definite, resp. a moment form.

When $\Lambda_{\Theta,\mathcal{G}}$ has a representing measure F, then F is also said to be a *representing measure* for Θ.

2) For a given function space $\mathcal{S}$, we denote (as in [Vas4]) by $\mathcal{S}^{(1)}$ the vector space spanned by all products of the form fg, with $f, g \in \mathcal{S}$. We clearly have $\mathcal{S} \subset \mathcal{S}^{(1)}$, and $\mathcal{S} = \mathcal{S}^{(1)}$ if $\mathcal{S}$ is an algebra.

Let $L: \mathcal{S}^{(1)} \to \mathcal{E}(\mathcal{D}, \mathcal{H})$ be a unital linear map which preserves the involution, that is, a linear application such that $L(1)$ is the identity on $\mathcal{D}$, and $\langle L(\bar{f})x, y\rangle = \langle x, L(f)y\rangle$, $f \in \mathcal{S}^{(1)}$, $x, y \in \mathcal{D}$.

In this case, the map L can also be associated with a hermitian form Λ_L on the tensor product $\mathcal{S} \otimes \mathcal{D}$, via the equality

$$\Lambda_L(\varphi,\psi) = \sum_{j,k} \langle L(f_j \bar{g}_k)x_j, y_k\rangle, \tag{1.11}$$

where $\varphi = \sum_j f_j \otimes x_j, \psi = \sum_k g_k \otimes y_k$ are arbitrary elements from $\mathcal{S} \otimes \mathcal{D}$.

As in the previous example, one says that the linear map L is of *positive type* if Λ_L is positive semi-definite.

The map L is said to be a *moment function* (this is an extension of the corresponding concept from [Vas4]) if the associated hermitian form Λ_L is a moment form. In this case we have, in particular, $\langle L(f)x, y\rangle = \int f\, dF_{x,y}$ for all $x, y \in \mathcal{D}$, where F is the representing measure of Λ_L (also called a *representing measure* for L). In addition, Λ_L is positive semi-definite, by Lemma 1.4.

If $\mathcal{D} = \mathcal{H}$, one should replace $\mathcal{E}(\mathcal{D}, \mathcal{H})$ by $\mathcal{L}(\mathcal{H})$. Note also that, if $\mathcal{D} = \mathcal{H} = \mathbb{C}$, then a linear map $L: \mathcal{S}^{(1)} \to \mathbb{C}$ is of positive type if and only if $L(|\, f\,|^2) \geq 0$ for all $f \in \mathcal{S}$.

3) Let $\mathcal{S}$ be a function space, consisting of Borel measurable functions on $\mathbb{R}^n$. Let also $F: \mathrm{Bor}(\mathbb{R}^n) \to \mathcal{L}(\mathcal{H})$ be a positive measure. We define the set $\mathcal{D} = \mathcal{D}_{F,\mathcal{S}}$ as the collection of all $x \in \mathcal{H}$ such that $f \in L^2(F_{x,x})$ for all $f \in \mathcal{S}$. Since

$$F_{x+y,x+y} \leq 2F_{x,x} + 2F_{y,y}, \qquad F_{\alpha x,\alpha x} =|\, \alpha\,|^2\, F_{x,x}$$

for all $x, y \in \mathcal{H}$, $\alpha \in \mathbb{C}$, we deduce that $\mathcal{D}$ is a linear space.

Suppose that $\mathcal{D}$ is dense in $\mathcal{H}$. Define the hermitian form Λ by (1.8). Then Λ is a moment form, and therefore, positive semi-definite, via Lemma 1.4.

REMARK 1.6. Let Λ be a moment form on $\mathcal{S} \otimes \mathcal{D}$, and let $F: \mathrm{Bor}(\mathbb{R}^n) \to \mathcal{L}(\mathcal{H})$ be a representing measure for Λ. Suppose that there exists a function $h \in \mathcal{S}$ which is nonnegative on the support of F, and such that $h\mathcal{S} \subset \mathcal{S}$. Then we have $\Lambda(h\varphi, \varphi) \geq 0$ for all $\varphi \in \mathcal{S} \otimes \mathcal{D}$. This assertion can be obtained on the lines of the proof of Lemma 1.4. If $\varphi = \sum_j \chi_{\sigma_j} \otimes x_j$ and h_0 is bounded and positive on the support of F, then, as for (1.9), $\Lambda_F(h_0\varphi, \varphi) = \sum_j \langle (\int_{\sigma_j} h_0 dF)x_j, x_j\rangle \geq 0$. If $\varphi = \sum_j f_j \otimes x_j \in \mathcal{S} \otimes \mathcal{D}$ is arbitrary, if $(h_m)_{m\geq 1}$ is a sequence of bounded Borel functions, with $0 \leq h_m \leq h$ on the support of F, pointwise convergent to h, and if $\varphi_m = \sum_j f_{j,m} \otimes x_j$ are constructed as in the proof of Lemma 1.4, then we have $\Lambda(h\varphi, \varphi) = \lim_{m\to\infty} \Lambda_F(h_m\varphi_m, \varphi_m) \geq 0$.

REMARK 1.7. Let $\mathcal{S}$ be a function space on $\mathbb{R}^n$, let Λ be a moment form on $\mathcal{S} \otimes \mathcal{D}$, and let F be a representing measure for Λ. Define $\mathcal{S}_F$ as the set of those Borel measurable functions f such that $f \in L^2(F_{x,x})$ for all $x \in \mathcal{D}$. Clearly, $\mathcal{S}_F$ is a function space which contains $\mathcal{S}$. Moreover, we can extend Λ to a moment form Λ_F on $\mathcal{S}_F \otimes \mathcal{D}$, given by the corresponding version of (1.7) (or (1.8)).

Let $\tau: \mathbb{R}^n \to \mathbb{R}^{n'}$ be a Borel measurable map and let $\mathcal{S}'$ be a function space on $\mathbb{R}^{n'}$ consisting of Borel measurable functions. Suppose that $f' \circ \tau \in \mathcal{S}_F$ for each $f' \in \mathcal{S}'$. We define on $\mathcal{S}' \otimes \mathcal{D}$ the hermitian form

$$\Lambda'(\varphi', \psi') = \Lambda_F(\varphi, \psi), \tag{1.12}$$

where $\varphi' = \sum_j f'_j \otimes x_j, \psi' = \sum_k g'_k \otimes y_k$ are arbitrary elements from $\mathcal{S}' \otimes \mathcal{D}$, and $\varphi = \sum_j f_j \otimes x_j, \psi = \sum_k g_k \otimes y_k$, with $f_j = f'_j \circ \tau, g_k = g'_k \circ \tau$ for all j, k. It is easily checked that Λ' is a moment form and that $F'(*) = F(\tau^{-1}(*))$ is a representing measure for Λ'.

REMARK 1.8. Let $\tau: \mathbb{R}^n \to \mathbb{R}^{n'}$ be a Borel measurable map and let $\mathcal{S}'$ be a function space on $\mathbb{R}^{n'}$. Set $\mathcal{S} = \{f = f' \circ \tau; f' \in \mathcal{S}'\}$, which is clearly a function space on $\mathbb{R}^n$.

Let Λ' be a hermitian form on $\mathcal{S}' \otimes \mathcal{D}$, and set $\mathcal{S}'_\tau = \{f' \in \mathcal{S}'; f' \circ \tau = 0\}$. Suppose that $\Lambda'(\varphi', \psi') = 0$ if either φ' or ψ' is in $\mathcal{S}'_\tau \otimes \mathcal{D}$. We set

$$\Lambda(\varphi, \psi) = \Lambda'(\varphi', \psi'), \tag{1.12'}$$

where $\varphi' = \sum_j f'_j \otimes x_j, \psi' = \sum_k g'_k \otimes y_k$ are arbitrary elements from $\mathcal{S}' \otimes \mathcal{D}$, and $\varphi = \sum_j f_j \otimes x_j, \psi = \sum_k g_k \otimes y_k$, with $f_j = f'_j \circ \tau, g_k = g'_k \circ \tau$ for all j, k. Formula (1.12′) makes sense and it defines a hermitian form Λ on $\mathcal{S} \otimes \mathcal{D}$. Indeed, the map $\sum_j f'_j \otimes x_j \to \sum_j (f'_j \circ \tau) \otimes x_j$ from $\mathcal{S}' \otimes \mathcal{D}$ into $\mathcal{S} \otimes \mathcal{D}$ is a surjection whose kernel is precisely $\mathcal{S}'_\tau \otimes \mathcal{D}$. In addition, if Λ' is positive semi-definite, then Λ is positive semi-definite too.

2. Operator moment problems in unbounded semi-algebraic sets

Let $\mathcal{P}_n = \mathcal{P}(\mathbb{R}^n)$ be the complex algebra of all polynomial functions on $\mathbb{R}^n$. Let also $\mathcal{S}$ be a function space on $\mathbb{R}^n$, which is a $\mathcal{P}_n$-module.

Let Λ be a positive semi-definite form on $\mathcal{S} \otimes \mathcal{D}$, where $\mathcal{D}$ and $\mathcal{H}$ are as in the previous section. Assume that Λ is $\mathcal{P}_n$-symmetric. It follows from Remark 1.1 that the linear spaces $\mathcal{S} \otimes \mathcal{D}$, $\mathcal{N}_\Lambda$, as well as the quotient $(\mathcal{S} \otimes \mathcal{D})/\mathcal{N}_\Lambda$, are all $\mathcal{P}_n$-modules.

Let $t = (t_1, \ldots, t_n)$ be the current variable of the space $\mathbb{R}^n$. We denote also by t_j the function $t \to t_j$ for all $j = 1, \ldots, n$. The function t_j induces a linear map on the space $(\mathcal{S} \otimes \mathcal{D})/\mathcal{N}_\Lambda$, by the formula

$$T_j(\varphi + \mathcal{N}_\Lambda) = t_j \varphi + \mathcal{N}_\Lambda, \qquad j = 1, \ldots, n. \tag{2.1}$$

Notice that each operator $T_j, j = 1, \ldots, n$, whose domain of definition $D(T_j)$ in the Hilbert space $\mathcal{H}_\Lambda$ is the (dense) subspace $(\mathcal{S} \otimes \mathcal{D})/\mathcal{N}_\Lambda$, is symmetric (via (1.5)).

For further development we need the following result (see [Vas3]).

THEOREM 2.1. *Let* $\theta_j(t) = (1+t_j^2)^{-1}, 1 \le j \le n$, $t = (t_1, \ldots, t_n) \in \mathbb{R}^n$, *and let* $\theta = (\theta_j)_{1\le j\le n}$. *Denote by* $\mathcal{R}_\theta$ *the complex algebra generated by* $\mathcal{P}_n$ *and by* $(\theta_j)_{1\le j\le n}$.

If $L: \mathcal{R}_\theta \to \mathbb{C}$ *is a linear map of positive type, then* L *has a uniquely determined representing measure.*

The proof of this result is obtained in [Vas3] (see especially Remark 2.4).

Let $\mathcal{R}_\theta$ be the complex algebra defined in Theorem 2.1. The next assertion is an operator version of Theorem 2.1.

THEOREM 2.2. *Let* Λ *be a positive semi-definite form on* $\mathcal{R}_\theta \otimes \mathcal{D}$, *which is unital and* $\mathcal{R}_\theta$*-symmetric. Then* Λ *is a moment form having a uniquely determined representing measure.*

If, moreover, $\Lambda(t_j\varphi, \varphi) \ge 0$ *for all* $\varphi \in \mathcal{R}_\theta \otimes \mathcal{D}$ *and* $j = 1, \ldots, n$, *then the support of the representing measure of* Λ *is concentrated in* $\mathbb{R}_+^n$.

Proof. We follow some ideas from the proof of [Vas3, Theorem 2.3] (see also [Vas3, Remark 2.7]).

Let $\mathcal{N} = \mathcal{N}_\Lambda$, be given by (1.2), for $\mathcal{S} = \mathcal{R}_\theta$, with Λ as stated. Let also $\mathcal{H}_\Lambda$ be the completion of $(\mathcal{R}_\theta \otimes \mathcal{D})/\mathcal{N}$.

We consider in $\mathcal{H}_\Lambda$ the operators $T_j, j = 1, \ldots, n$, given by (2.1). These operators are symmetric and densely defined, and $D(T_j) = (\mathcal{R}_\theta \otimes \mathcal{D})/\mathcal{N}$ for all j.

Note that if S is a symmetric densely defined operator in a Hilbert space $\mathcal{H}$, and if the sets $R(S \pm i)$ are dense in $\mathcal{H}$ (where $R(C)$ stands for the range of the linear operator C), then the closure of S is a self-adjoint operator, which is an easy consequence of the Cayley transform.

We shall show that T_j satisfies the conditions of the remark above for each j. Indeed, if $\varphi \in \mathcal{R}_\theta \otimes \mathcal{D}$ is arbitrary, then the elements $\psi_\pm = (t_j \mp i)\theta_j\varphi$ are solutions in $\mathcal{R}_\theta \otimes \mathcal{D}$ of the equations $(t_j \pm i)\psi_\pm = \varphi$. This implies the equalities $R(T_j \pm i) = D(T_j)$, showing that T_j is essentially self-adjoint.

Let A_j be the closure of T_j. We shall show that the operators $(i - A_1)^{-1}, \ldots, (i - A_n)^{-1}$ mutually commute. Indeed, the previous argument shows that the map $(i - T_j)^{-1}$ is well defined on $D = D(T_j)$, and leaves this space invariant, for all j. Moreover, the maps $(i - T_1)^{-1}, \ldots, (i - T_n)^{-1}$ mutually commute on D. Since A_j extends T_j, we clearly have $(i - A_j)^{-1}|D = (i - T_j)^{-1}$. Therefore, for all $j, k, j \ne k$,

$$(i - A_j)^{-1}(i - A_k)^{-1}\xi = (i - A_k)^{-1}(i - A_j)^{-1}\xi,$$

where $\xi \in D$ is arbitrary. Since $(i - A_1)^{-1}, \ldots, (i - A_n)^{-1}$ are bounded and D is dense, this implies that they mutually commute. In particular, the self-adjoint operators $A_1, \ldots, A_n$ have a joint spectral measure (see, for instance, [Vas1]).

If E is the joint spectral measure of $A_1, \ldots, A_n$, then $F(*) = PE(*)|\mathcal{H}$ is a representing measure for Λ, with P the orthogonal projection of $\mathcal{H}_\Lambda$ onto $\mathcal{H}$, where the latter space is regarded as a closed subspace of the former, via the identification given by (1.4).

To prove this assertion, let $r(T)$ be the linear map on D given by $r(T)(\varphi + \mathcal{N}) = r\varphi + \mathcal{N}$, for all $r \in \mathcal{R}_\theta, \varphi \in \mathcal{R}_\theta \otimes \mathcal{D}$. Then we have $r(A) \supset r(T)$. This is clear when r is a polynomial. We also have $\theta(A)^\beta \supset \theta(T)^\beta$, for an arbitrary multi-index β, which follows from the obvious relations $\theta(A)^{-\beta} \supset \theta(T)^{-\beta}$, and $\theta(A)^{-\beta}(\theta(A)^\beta - \theta(T)^\beta) = 0$. Using these remarks, we infer easily that $r(A) \supset r(T), r \in \mathcal{R}_\theta$. Therefore, with the identification given by (1.4), we have:

$$\infty > \Lambda(r \otimes x, r \otimes x) = \langle r(T)x, r(T)x \rangle = \langle r(A)x, r(A)x \rangle = \int |r|^2 \, dF_{x,x}$$

for all $r \in \mathcal{R}_\theta$ and $x \in \mathcal{D}$. Moreover, if $\varphi = \sum_j f_j \otimes x_j, \psi = \sum_k g_k \otimes y_k$ are arbitrary elements from $\mathcal{R}_\theta \otimes \mathcal{D}$, we can write that

$$\begin{aligned}\Lambda(\varphi, \psi) = \sum_{j,k} \Lambda(f_j \otimes x_j, g_k \otimes y_k) &= \sum_{j,k} \langle f_j(T)x_j, g_k(T)y_k \rangle \\ &= \sum_{j,k} \langle f_j(A)x_j, g_k(A)y_k \rangle = \sum_{j,k} \int f_j \bar{g}_k \, dF_{x_j,y_k},\end{aligned}$$

showing that F is a representing measure for Λ.

To prove the uniqueness of the representing measure of Λ, let $F^{(1)}, F^{(2)}$ be two such measures, and let $x \in \mathcal{D}$. Notice that

$$\Lambda_x(r) = \Lambda(r \otimes x, 1 \otimes x) = \int r \, dF^{(j)}_{x,x}, \qquad r \in \mathcal{R}_\theta, \; j = 1, 2,$$

defines a scalar-valued map of positive type. By virtue of Theorem 2.1, the representing measure of Λ_x has to be unique. Therefore, $F^{(1)}_{x,x} = F^{(2)}_{x,x}$. As $x \in \mathcal{D}$ is arbitrary and $\mathcal{D}$ is dense in $\mathcal{H}$, we infer $F^{(1)} = F^{(2)}$.

Finally, the property $\Lambda(t_j\varphi, \varphi) \geq 0$ for all $\varphi \in \mathcal{R}_\theta \otimes \mathcal{D}$ and $j = 1, \ldots, n$ implies that the operators $A_1, \ldots, A_n$ are all positive, and so their spectral measure E is concentrated in $\mathbb{R}^n_+$. This implies that the support of the representing measure of Λ is in $\mathbb{R}^n_+$, and completes the proof of the theorem.

□

REMARK 2.3. Let $\Theta = (\Theta_\alpha)_{\alpha \in \mathbb{Z}_+^n}$ be an n-sequence of hermitian forms on $\mathcal{D}$. Let also $\mathcal{G}_n = (t^\alpha)_{\alpha \in \mathbb{Z}_+^n}$ be the canonical basis of $\mathcal{P}_n$.

We associate Θ with the hermitian form $\Lambda_\Theta = \Lambda_{\Theta,\mathcal{G}_n}$ on $\mathcal{P}_n \otimes \mathcal{D}$ given by (1.10″).

Since the function space $\mathcal{P}_n$ and its canonical basis $\mathcal{G}_n$ will be fixed in this section (although n may vary), we say that the sequence Θ is of *positive type*, resp. *a moment sequence*, if it is of positive type with respect to $\mathcal{P}_n$, resp. a moment sequence with respect to $\mathcal{G}_n$ (see Example 1.5.1).

We can prove now the following operator version of [Vas3, Theorem 2.3 and Theorem 2.6]. We use the symbol e_j to designate the multi-index $(\epsilon_1, \dots, \epsilon_n)$, with $\epsilon_j = 1$ and $\epsilon_k = 0$ if $k \neq j$.

THEOREM 2.4. *Let $\Theta = (\Theta_\alpha)_{\alpha \in \mathbb{Z}_+^n}$ be an n-sequence consisting of sesquilinear forms on $\mathcal{D}$, with Θ_0 the restriction to $\mathcal{D}$ of the scalar product of $\mathcal{H}$.*

The n-sequence Θ is a moment sequence if and only if there exists a $2n$-sequence $\Omega = (\Omega_{(\alpha,\beta)})_{(\alpha,\beta) \in \mathbb{Z}_+^n \times \mathbb{Z}_+^n}$ consisting of sesquilinear forms on $\mathcal{D}$, which is of positive type, with the following properties:

(1) $\Theta_\alpha = \Omega_{(\alpha,0)}$ for all $\alpha \in \mathbb{Z}_+^n$,

(2) $\Omega_{(\alpha,\beta)} = \Omega_{(\alpha,\beta+e_j)} + \Omega_{(\alpha+2e_j,\beta+e_j)}$ for all $\alpha, \beta \in \mathbb{Z}_+^n, 1 \leq j \leq n$.

In this case, the n-sequence Θ has a uniquely determined representing measure in $\mathbb{R}^n$ if and only if the $2n$-sequence Ω is unique.

The n-sequence Θ has a representing measure whose support is concentrated in $\mathbb{R}_+^n$ if and only if there exists a $2n$-sequence Ω satisfying (1) and (2), and which, in addition, has the property $\Lambda_\Omega(t_j\varphi, \varphi) \geq 0$ for all $\varphi \in \mathcal{P}_{2n} \otimes \mathcal{D}$ and $j = 1, \dots, n$.

Proof. Let $\mathcal{R}_\theta$ be as in Theorem 2.1. Assume first that the n-sequence $\Theta = (\Theta_\alpha)_{\alpha \in \mathbb{Z}_+^n}$ has a representing measure F. As we clearly have $r \in L^2(F_{x,x})$ for all $r \in \mathcal{R}_\theta$ and $x \in \mathcal{D}$, we may define the hermitian forms

$$\Omega_{(\alpha,\beta)}(x,y) = \int t^\alpha \theta(t)^\beta \, dF_{x,y}(t), \qquad \alpha, \beta \in \mathbb{Z}_+^n,\ x, y \in \mathcal{D}. \tag{2.2}$$

Set $\Omega = (\Omega_{(\alpha,\beta)})_{(\alpha,\beta) \in \mathbb{Z}_+^n \times \mathbb{Z}_+^n}$. If $\tau(t) = (t, \theta(t)) \in \mathbb{R}^{2n}, t \in \mathbb{R}^n$, and if $F'(*) = F(\tau^{-1}(*))$, formula (2.2) shows that F' is a representing measure for the $2n$-sequence Ω. Consequently, Ω is of positive type, by Lemma 1.4 (see also Remark 1.7).

The sequence Ω obviously satisfies condition (1).

Next, since

$$\int (\theta_j(t)(1 + t_j^2) - 1) t^\alpha \theta(t)^\beta \, dF_{x,y}(t) = 0 \tag{2.3}$$

for all $\alpha, \beta \in \mathbb{Z}_+^n$, $j = 1, \ldots, n$ and $x, y \in \mathcal{D}$, we infer (2).

Conversely, assume that the $2n$-sequence $\Omega = (\Omega_{(\alpha,\beta)})_{\alpha,\beta \in \mathbb{Z}_+^n}$ exists. We define a map Λ of positive type on $\mathcal{R}_\theta \otimes \mathcal{D}$, via the equality

$$(2.4) \quad \Lambda(r_1 \otimes x_1, r_2 \otimes x_2) = \Lambda_\Omega(p_1 \otimes x_1, p_2 \otimes x_2), \qquad r_1, r_2 \in \mathcal{R}_\theta,\ x_1, x_2 \in \mathcal{D},$$

where $p_j \in \mathcal{P}_{2n}$ satisfies $r_j(t) = p_j(t, \theta(t))$, $t \in \mathbb{R}^n$, $j = 1, 2$ (with Λ_Ω as in Remark 2.3).

That Λ is correctly defined follows from Remark 1.8. Indeed, by virtue of [Vas3, Lemma 2.2], the algebra $\mathcal{R}_\theta$ is isomorphic to the quotient $\mathcal{P}_{2n}/\mathcal{I}_\sigma$, where $\mathcal{I}_\sigma$ is the ideal generated in $\mathcal{P}_{2n}$ by the polynomials $\sigma_j(t, s) = s_j(1 + t_j^2) - 1$, $j = 1, \ldots, n$. Note that (2) implies that $\Lambda_\Omega(\varphi, \psi) = 0$ if either φ or ψ is in $\mathcal{I}_\sigma \otimes \mathcal{D}$. Therefore, the form Λ is correctly defined and of positive type on $\mathcal{R}_\theta \otimes \mathcal{D}$, by Remark 1.8.

To apply Theorem 2.2 to Λ, we have only to observe that Λ is unital and $\mathcal{R}_\theta$-symmetric. This follows from (2.4), using the fact that the form Λ_Ω is unital and $\mathcal{P}_{2n}$-symmetric, as noticed in Example 1.5.1.

By virtue of Theorem 2.2, there exists a uniquely determined representing measure F for Λ. In particular

$$\Theta_\alpha(x, y) = \Omega_{(\alpha,0)}(x, y) = \Lambda_\Omega(t^\alpha \otimes x, 1 \otimes y) = \int t^\alpha \, dF_{x,y}(t), \qquad \alpha \in \mathbb{Z}_+^n,\ x, y \in \mathcal{D},$$

showing that Θ has a representing measure (see Example 1.5.1).

If Ω is uniquely determined, and if F', F'' are two representing measures for Θ, then we must have

$$\int t^\alpha \theta(t)^\beta \, dF'_{x,y}(t) = \int t^\alpha \theta(t)^\beta \, dF''_{x,y}(t)$$

by the uniqueness of Ω. Therefore, $\int r(t) dF'_{x,y}(t) = \int r(t) dF''_{x,y}(t)$ for all $r \in \mathcal{R}_\theta$, $x, y \in \mathcal{D}$. This implies that both F', F'' are representing measures for the same hermitian form on $\mathcal{R}_\theta \otimes \mathcal{D}$, and so $F' = F''$, by Theorem 2.2.

Conversely, if the representing measure F of Θ is unique, and if the sequences Ω', Ω'' satisfy (1), (2), then by (2.4) we have $\Omega'_{\alpha,\beta}(x, y) = \int t^\alpha \theta(t)^\beta dF_{x,y}(t) = \Omega''_{\alpha,\beta}(x, y)$ for all indices α, β and all vectors $x, y \in \mathcal{D}$.

To prove the last part of the theorem, assume that the n-sequence $\Theta = (\Theta_\alpha)_{\alpha \in \mathbb{Z}_+^n}$ has a representing measure F whose support is in $\mathbb{R}_+^n$. Define $\mathcal{S}_F$ as the set of all Borel measurable functions f such that $f \in L^2(F_{x,x})$ for all $x \in \mathcal{D}$. We clearly have $\mathcal{R}_\theta \subset \mathcal{S}_F$. Let Λ_F be the extension of Λ_Θ to $\mathcal{S}_F \otimes \mathcal{D}$ (see Remark 1.7). Fix $\varphi = \sum_k p_k \otimes x_k \in \mathcal{P}_{2n} \otimes \mathcal{D}$. Set $\tilde{\varphi} = \sum_k \tilde{p}_k \otimes x_k \in \mathcal{R}_\theta \otimes \mathcal{H}$, where $\tilde{p}_k(t) = p_k(t, \theta(t))$ for all k. If Ω is given by (2.2), we have:

$$\Lambda_\Omega(t_j \varphi, \varphi) = \Lambda_F(t_j \tilde{\varphi}, \tilde{\varphi}) \geq 0, \qquad j = 1, \ldots, n,$$

by (1.12) and Remark 1.6.

Conversely, assume the existence of a $2n$-sequence Ω of positive type, with the properties (1), (2), and such that $\Lambda_\Omega(t_j\varphi,\varphi)\geq 0$ for all $\varphi\in\mathcal{P}_{2n}\otimes\mathcal{D}$ and $j=1,\ldots,n$. If $\tilde{\varphi}=\sum_k\tilde{p}_k\otimes x_k\in\mathcal{R}_\theta\otimes\mathcal{H}$, is arbitrary, we find a $\varphi=\sum_k p_k\otimes x_k\in\mathcal{P}_{2n}\otimes\mathcal{H}$ such that $\tilde{p}_k(t)=p_k(t,\theta(t))$ for all k. With this notation, relation (2.4) shows that $\Lambda(t_j\tilde{\varphi},\tilde{\varphi})=\Lambda_\Omega(t_j\varphi,\varphi)\geq 0$, $j=1,\ldots,n$. The assertion follows via Theorem 2.2.

□

REMARK 2.5. Let $\Gamma=(\Gamma_\alpha)_{\alpha\in\mathbb{Z}_+^n}$ be an n-sequence of self-adjoint operators in $\mathcal{L}(\mathcal{H})$, with Γ_0 the identity on $\mathcal{H}$. The sequence Γ is associated with the hermitian form

$$\Lambda_\Gamma(\varphi,\psi)=\sum_{\alpha,\beta\in\mathbb{Z}_+^n}\langle\Gamma_{\alpha+\beta}x_\alpha,y_\beta\rangle,$$

where $\varphi=\sum_\alpha t^\alpha\otimes x_\alpha,\psi=\sum_\beta t^\beta\otimes y_\beta$ are arbitrary elements from $\mathcal{P}_n\otimes\mathcal{H}$, which corresponds to (1.10′).

As in Example 1.5.1, we say that Γ is of *positive type*, resp. a *moment sequence*, if Λ_Γ is of positive type with respect to the function space $\mathcal{P}_n$, resp. Λ_Γ is a moment sequence with respect to the the standard basis $(t^\alpha)_{\alpha\in\mathbb{Z}_+^n}$ of $\mathcal{P}_n$.

If Γ is a moment sequence, and if F is a representing measure for Γ, since $\mathcal{P}_n\subset\mathcal{I}_F^2$, we must have $\Gamma_\alpha=\int t^\alpha dF(t)$ for all $\alpha\in\mathbb{Z}_+^n$, via Remark 1.2. Conversely, if $\Gamma_\alpha=\int t^\alpha dF(t)$ for all $\alpha\in\mathbb{Z}_+^n$, with $\mathcal{P}_n\subset\mathcal{I}_F^2$, then Γ is a moment sequence.

COROLLARY 2.6. *Let $\Gamma=(\Gamma_\alpha)_{\alpha\in\mathbb{Z}_+^n}$ ($\Gamma_0=1_\mathcal{H}$) be an n-sequence of self-adjoint operators in $\mathcal{L}(\mathcal{H})$.*

The n-sequence Γ is a moment sequence if and only if there exists $2n$-sequence $\Delta=(\Delta_{(\alpha,\beta)})_{(\alpha,\beta)\in\mathbb{Z}_+^n\times\mathbb{Z}_+^n}$ consisting of self-adjoint operators in $\mathcal{L}(\mathcal{H})$, which is of positive type, with the following properties:

(1) *$\Gamma_\alpha=\Delta_{(\alpha,0)}$ for all $\alpha\in\mathbb{Z}_+^n$.*

(2) *$\Delta_{(\alpha,\beta)}=\Delta_{(\alpha,\beta+e_j)}+\Delta_{(\alpha+2e_j,\beta+e_j)}$ for all $\alpha,\beta\in\mathbb{Z}_+^n, 1\leq j\leq n$.*

In this case, the n-sequence Γ has a uniquely determined representing measure in $\mathbb{R}^n$ if and only if the $2n$-sequence Δ is unique.

The n-sequence Γ has a representing measure whose support is concentrated in $\mathbb{R}_+^n$ if and only if there exists a $2n$-sequence Δ satisfying (1) *and* (2), *and which, in addition, has the property that the $2n$-sequence $(\Delta_{(\alpha+e_j,\beta)})_{(\alpha,\beta)\in\mathbb{Z}_+^n\times\mathbb{Z}_+^n}$ is of positive type for all $j=1,\ldots,n$.*

Proof. We define the hermitian forms $\Theta_\alpha(x,y) = \langle \Gamma_\alpha x, y\rangle$ for all $\alpha \in \mathbb{Z}_+^n$ and $x, y \in \mathcal{H}$, and set $\Theta = (\Theta_\alpha)_{\alpha\in\mathbb{Z}_+^n}$.

If the sequence Γ has a representing measure F, since we clearly have $\mathcal{R}_\theta \subset \mathcal{I}_F^2$, we define the self-adjoint operators $\Delta_{(\alpha,\beta)}$ by the equation

$$\Delta_{(\alpha,\beta)} = \int t^\alpha \theta(t)^\beta \, dF, \qquad \alpha, \beta \in \mathbb{Z}_+^n,$$

via (1.6). The sequence $\Delta = (\Delta_{(\alpha,\beta)})_{(\alpha,\beta)\in\mathbb{Z}_+^n\times\mathbb{Z}_+^n}$ obviously has the properties (1) and (2) from the statement. If we put $\Omega_{(\alpha,\beta)}(x,y) = \langle \Delta_{(\alpha,\beta)}x, y\rangle$, $x, y \in \mathcal{H}$, then the sequence $\Omega = (\Omega_{(\alpha,\beta)})_{(\alpha,\beta)\in\mathbb{Z}_+^n\times\mathbb{Z}_+^n}$ is of positive type, by Theorem 2.4.

Conversely, if the $2n$-sequence Δ exists, we define the hermitian forms $\Omega_{(\alpha,\beta)}(x,y) = \langle \Delta_{(\alpha,\beta)}x, y\rangle$, $x, y \in \mathcal{H}$, and apply Theorem 2.4 to the sequences Θ and $\Omega = (\Omega_{(\alpha,\beta)})_{(\alpha,\beta)\in\mathbb{Z}_+^n\times\mathbb{Z}_+^n}$, which gives the desired conclusion.

The remaining assertions follow directly from Theorem 2.4. We only note that $\Lambda_\Omega(t_j\varphi,\varphi) \geq 0$ for all $\varphi \in \mathcal{P}_{2n} \otimes \mathcal{H}$ and $j = 1,\ldots,n$ if and only if $(\Delta_{(\alpha+e_j,\beta)})_{(\alpha,\beta)\in\mathbb{Z}_+^n\times\mathbb{Z}_+^n}$ is of positive type for all $j = 1,\ldots,n$. □

REMARK. Corollary 2.6 provides solutions to what may be called the Hamburger and Stieltjes moment problems in several variables, with operator data.

Following some ideas from [PuVa], we discuss in what follows a slightly different point of view, which reduces the number of parameters and gives a good control of the support of the representing measures. We start with the following known result (see [PuVa, Theorem 2.5]).

THEOREM 2.7. *Let* $\mathbf{p} = (p_1,\ldots,p_m)$ *be a given* m*-tuple of real polynomials from* $\mathcal{P}_n$*, and let*

$$\theta_{\mathbf{p}}(t) = (1 + t_1^2 + \cdots + t_n^2 + p_1(t)^2 + \cdots + p_m(t)^2)^{-1}, \qquad t \in \mathbb{R}^n.$$

Denote by $\mathcal{R}_{\theta_{\mathbf{p}}}$ *the complex algebra generated by* $\mathcal{P}_n$ *and* $\theta_{\mathbf{p}}$. *Let* Λ *be a linear map of positive type on* $\mathcal{R}_{\theta_{\mathbf{p}}}$ *such that* $\Lambda(p_k|r|^2) \geq 0, r \in \mathcal{R}_{\theta_{\mathbf{p}}}, k = 1,\ldots,m$. *Then* Λ *has a uniquely determined representing measure whose support is in the set* $\cap_{k=1}^m p_k^{-1}(\mathbb{R}_+)$.

As for Theorem 2.1, we can give an operator version of Theorem 2.7. We keep the notation from Theorem 2.7.

THEOREM 2.8. *Let Λ be a positive semi-definite form on $\mathcal{R}_{\theta_\mathbf{p}} \otimes \mathcal{D}$, which is unital and $\mathcal{R}_{\theta_\mathbf{p}}$-symmetric. Then Λ is a moment form having a uniquely determined representing measure.*

If, moreover, $\Lambda(p_k\varphi, \varphi) \geq 0$ for all $\varphi \in \mathcal{R}_{\theta_\mathbf{p}} \otimes \mathcal{D}$ and $k = 1, \dots, m$, then the support of the representing measure of Λ is concentrated in $\cap_{k=1}^m p_k^{-1}(\mathbb{R}_+)$.

Proof. We follow some ideas from [PuVa, Theorem 2.5].

Let Λ be as in the statement, let $\mathcal{N} = \mathcal{N}_\Lambda$ be given by (1.2) (with $\mathcal{S} = \mathcal{R}_{\theta_\mathbf{p}}$) and let $\mathcal{H}_\Lambda$ be the completion of the space $(\mathcal{R}_{\theta_\mathbf{p}} \otimes \mathcal{D})/\mathcal{N}$ (corresponding to (1.3)).

We consider in $\mathcal{H}_\Lambda$ the operators T_j, $j = 1, \dots, n$, given by (2.1), which are symmetric and densely defined on $(\mathcal{R}_{\theta_\mathbf{p}} \otimes \mathcal{D})/\mathcal{N}$ for all j. In addition, set

$$S_k(\varphi + \mathcal{N}) = p_k\varphi + \mathcal{N}, \qquad \varphi \in \mathcal{R}_{\theta_\mathbf{p}} \otimes \mathcal{D}, \qquad k = 1, \dots, m,$$

which are also densely defined and symmetric.

Let $B = T_1^2 + \dots + T_n^2 + S_1^2 + \dots + S_m^2$ be defined on $D(B) = (\mathcal{R}_{\theta_\mathbf{p}} \otimes \mathcal{D})/\mathcal{N}$. The domain $D(B)$ is clearly invariant under B. We want to show that B is essentially selfadjoint. To obtain this property we use the following.

REMARK. *Let A be a positive densely defined operator in $\mathcal{H}$. Suppose that $R(I + A)$ is dense in $\mathcal{H}$. Then the closure of A is self-adjoint.*

A proof of this remark can be found in [StZs], Lemma 9.5.

We next show that one can apply this Remark to B. Let $\omega = t_1^2 + \dots + t_n^2 + p_1^2 + \dots + p_m^2$. If $\varphi \in \mathcal{R}_{\theta\mathbf{p}} \otimes \mathcal{D}$, $\varphi = \sum_\ell r_\ell \otimes x_\ell$, we define $\varphi_j' = \sum_\ell t_j r_\ell \otimes x_\ell$, $j = 1, \dots, n$, and $\varphi_k'' = \sum_\ell p_k r_\ell \otimes x_\ell$, $k = 1, \dots, m$. We obviously have

$$\Lambda(\omega\varphi, \varphi) = \sum_{j=1}^n \Lambda(\varphi_j', \varphi_j') + \sum_{k=1}^m \Lambda(\varphi_k'', \varphi_k'') \geq 0,$$

showing that the operator B is positive.

If $\varphi \in \mathcal{R}_{\theta\mathbf{p}} \otimes \mathcal{D}$ is arbitrary, then the element $\psi(t) = \theta_\mathbf{p}(t)\varphi(t) \in \mathcal{R}_{\theta\mathbf{p}} \otimes \mathcal{D}$ satisfies the equation $(1 + \omega)\psi = \varphi$, whence we infer that the map $I + B$ is bijective on $D(B)$.

According to the Remark above, the operator B is essentially self-adjoint. By virtue of Nelson's criterion of self-adjointness [Nel, Corollary 9.2] (see also [EsVa, Theorem 3.2]), this implies that the operators $T_1, \dots, T_n, S_1, \dots, S_m$ are essentially selfadjoint, and their closures mutually commute. In particular, $A_1, \dots, A_n$, where $A_j = \bar{T}_j$ for all j, have a joint spectral measure (see [Vas1]).

If E is the joint spectral measure of $A_1, \ldots, A_n$, then $F(*) = PE(*)|\mathcal{H}$ is a representing measure for Λ, where P is the orthogonal projection of $\mathcal{H}_\Lambda$ onto $\mathcal{H}$ (via the identification given by (1.4)). Indeed, as in the proof of Theorem 2.2 above, if $r(T)$ is the linear map on $(\mathcal{R}_{\theta_\mathbf{p}} \otimes \mathcal{D})/\mathcal{N}$ given by $r(T)(\varphi+\mathcal{N}) = r\varphi+\mathcal{N}$, for all $r \in \mathcal{R}_{\theta_\mathbf{p}}, \varphi \in \mathcal{R}_{\theta_\mathbf{p}} \otimes \mathcal{D}$, then we have $r(A) \supset r(T)$, where $r(A)$ is given by the functional calculus of $A = (A_1, \ldots, A_n)$. Therefore, again as in the proof of Theorem 2.2, if $\varphi = \sum_j f_j \otimes x_j, \psi = \sum_k g_k \otimes y_k$ are arbitrary elements from $\mathcal{R}_{\theta_\mathbf{p}} \otimes \mathcal{D}$, with the identification given by (1.4), we have $\Lambda(\varphi, \psi) = \sum_{j,k} \int f_j \bar{g}_k \, dF_{x_j, y_k}$, showing that F is a representing measure for Λ.

The uniqueness of the representing measure of Λ also follows as in the proof of Theorem 2.2. The only modification is the use of Theorem 2.7 instead of that of Theorem 2.1.

We have only to show the assertion concerning the support of the representing measure. Note that $\bar{S}_k = p_k(A), k = 1, \ldots, m$, where $p_k(A)$ is given by the functional calculus of A. Indeed, as we have $S_k \subset p_k(A)$, and S_k is essentially selfadjoint, we must have $\bar{S}_k = p_k(A)$ for all k. The fact that $\Lambda(p_k\varphi, \varphi) \geq 0$ for all $\varphi \in \mathcal{R}_{\theta_\mathbf{p}} \otimes \mathcal{D}$ implies that the operator S_k is positive, and so $p_k(A)$ is positive for all k. In addition, the spectral measure E_k of $p_k(A)$ is given by $E_k(G) = E(p_k^{-1}(G))$ for all Borel sets $G \subset \mathbb{R}$. Since the spectral measure E_k must be concentrated in $\mathbb{R}_+$ for all k, it follows that the spectral measure E of A is concentrated in the set $\cap_{k=1}^m p_k^{-1}(\mathbb{R}_+)$, which implies that the representing measure of Λ itself is concentrated in the same set $\cap_{k=1}^m p_k^{-1}(\mathbb{R}_+)$.

This completes the proof of the theorem.

□

We now present an extension of [PuVa, Theorem 2.7] to sequences of hermitian forms, which provides a general solution to various operator moment problems in some not necessarily bounded semi-algebraic sets.

THEOREM 2.9. *Let $\Theta = (\Theta_\alpha)_{\alpha \in \mathbb{Z}_+^n}$ be an n-sequence consisting of hermitian forms on $\mathcal{D}$, with Θ_0 the restriction to $\mathcal{D}$ of the scalar product of $\mathcal{H}$. Let also $\mathbf{p} = (p_1, \ldots, p_m)$ be an m-tuple of real polynomials from $\mathcal{P}_n$, where $p_k(t) = \sum_{\xi \in I_k} a_{k\xi} t^\xi$, $k = 1, \ldots, m$, with $I_k \subset \mathbb{Z}_+^n$ finite for all k.*

The n-sequence Θ is a moment sequence and has a representing measure whose support is in the set $\cap_{k=1}^m p_k^{-1}(\mathbb{R}_+)$ if and only if there exists a $(n+1)$-sequence of positive type $\Omega = (\Omega_{(\alpha,\beta)})_{(\alpha,\beta) \in \mathbb{Z}_+^n \times \mathbb{Z}_+}$, consisting of hermitian forms on $\mathcal{D}$, with the following properties:

(1) $\Theta_\alpha = \Omega_{(\alpha,0)}$ *for all* $\alpha \in \mathbb{Z}_+^n$,

(2) $\Omega_{(\alpha,\beta)} = \Omega_{(\alpha,\beta+1)} + \sum_{j=1}^n \Omega_{(\alpha+2e_j,\beta+1)} + \sum_{k=1}^m \sum_{\xi,\eta \in I_k} a_{k\xi} a_{k\eta} \Omega_{(\alpha+\xi+\eta,\beta+1)}$ *for all* $\alpha \in \mathbb{Z}_+^n, \beta \in \mathbb{Z}_+$,

(3) $\Lambda_\Omega(p_k\varphi,\varphi)\ge 0$ *for all* $\varphi\in\mathcal{P}_{n+1}\otimes\mathcal{D}$ *and* $k=1,\dots,m$.

The n*-sequence* Θ *has a uniquely determined representing measure on* $\cap_{k=1}^m p_k^{-1}(\mathbb{R}_+)$ *if and only if the* $(n+1)$*-sequence* Ω *is unique.*

Proof. We follow the lines of the proof of Theorem 2.4 (see also [PuVa, Theorem 2.7]). Let $\mathcal{R}_{\theta_\mathbf{p}}$ be as in Theorem 2.7.

Assume first that the sequence $\Theta=(\Theta_\alpha)_{\alpha\in\mathbb{Z}_+^n}$ has a representing measure F whose support is in the set $\Sigma_\mathbf{p}=\cap_{k=1}^m p_k^{-1}(\mathbb{R}_+)$. Since $r\in L^2(F_{x,x})$ for all $r\in\mathcal{R}_{\theta_\mathbf{p}}$ and $x\in\mathcal{D}$, we may define the hermitian forms

$$\Omega_{(\alpha,\beta)}(x,y)=\int_{\Sigma_\mathbf{p}} t^\alpha\theta_\mathbf{p}(t)^\beta\,dF_{x,y}(t),\qquad \alpha\in\mathbb{Z}_+^n,\ \beta\in\mathbb{Z}_+,\ x,y\in\mathcal{D}. \tag{2.5}$$

Note that $\Omega=(\Omega_{(\alpha,\beta)})_{(\alpha,\beta)\in\mathbb{Z}_+^n\times\mathbb{Z}_+}$ is of positive type, by Lemma 1.4. Indeed, as in the proof of Theorem 2.4, if $\tau_\mathbf{p}(t)=(t,\theta_\mathbf{p}(t))$, $t\in\mathbb{R}^n$, then $F^\mathbf{p}(*)=F(\tau_\mathbf{p}^{-1}(*))$ is a representing measure for the $(n+1)$-sequences Ω, via (2.5).

Condition (1) is obviously fulfilled.

Next, since

$$\int_{\Sigma_\mathbf{p}}(\theta_\mathbf{p}(t)(1+t_1^2+\cdots+t_n^2+p_1(t)^2+\cdots+p_m(t)^2)-1)t^\alpha\theta_\mathbf{p}(t)^\beta\,dF_{x,y}(t)=0$$

for all $\alpha\in\mathbb{Z}_+^n,\beta\in\mathbb{Z}_+$ and $x,y\in\mathcal{D}$, we infer (2).

To prove (3), let $\varphi=\sum_\ell q_\ell\otimes x_\ell\in\mathcal{P}_{n+1}\otimes\mathcal{D}$ and let $\tilde\varphi=\sum_\ell\tilde q_\ell\otimes x_\ell\in\mathcal{R}_{\theta_\mathbf{p}}\otimes\mathcal{D}$, where $\tilde q_k(t)=q_k(t,\theta_\mathbf{p}(t))$, $t\in\mathbb{R}^n$. A direct calculation shows that $\Lambda_\Omega(p_k\varphi,\varphi)=\Lambda_F(p_k\tilde\varphi,\tilde\varphi)$, where Λ_F is as in Remark 1.7, and the latter is positive since p_k is positive on the support of F for all k, via Remark 1.6.

Conversely, assume that the $(n+1)$-sequence $\Omega=(\Omega_{(\alpha,\beta)})_{(\alpha,\beta)\in\mathbb{Z}_+^n\times\mathbb{Z}_+}$ exists. We define a form Λ on $\mathcal{R}_{\theta_\mathbf{p}}\otimes\mathcal{D}$ by the equality

$$\Lambda(r_1\otimes x_1,r_2\otimes x_2)=\Lambda_\Omega(q_1\otimes x_1,q_2\otimes x_2),\qquad r_1,r_2\in\mathcal{R}_{\theta_\mathbf{p}},\ x_1,x_2\in\mathcal{D}, \tag{2.6}$$

where $q_j\in\mathcal{P}_{n+1}$ satisfies $r_j(t)=q_j(t,\theta_\mathbf{p}(t))$, $t\in\mathbb{R}^n$, $j=1,2$ (with Λ_Ω as in Remark 2.3).

The form Λ is correctly defined. Indeed, the algebra $\mathcal{R}_{\theta_\mathbf{p}}$ is isomorphic to the quotient $\mathcal{P}_{n+1}/\mathcal{I}_\sigma$, where $\mathcal{I}_\sigma$ is the ideal generated in $\mathcal{P}_{n+1}$ by the polynomial $\sigma(t,s)=s(1+t_1^2+\cdots+t_n^2+p_1(t)^2+\cdots+p_m(t)^2)-1$, by virtue of [PuVa, Lemma 2.3]. As one has $\Lambda_\Omega(\varphi,\psi)=0$ if either φ or ψ is in $\mathcal{I}_\sigma\otimes\mathcal{D}$ by virtue of (2), the form Λ can be identified with the form induced by Λ_Ω on the quotient $(\mathcal{P}_{n+1}\otimes\mathcal{D})/(\mathcal{I}_\sigma\otimes\mathcal{D})$, and therefore, it is correctly defined and positive semi-definite on $\mathcal{R}_{\theta_\mathbf{p}}$ (see Remark 1.8).

Let $\tilde{\varphi} = \sum_\ell \tilde{r}_\ell \otimes x_\ell \in \mathcal{R}_{\theta_{\mathbf{p}}} \otimes \mathcal{D}$ be arbitrary. Then we can find $\varphi = \sum_\ell r_\ell \otimes x_\ell \in \mathcal{P}_{n+1} \otimes \mathcal{D}$ such that $\tilde{r}_\ell(t) = r_\ell(t, \theta_{\mathbf{p}}(t))$, $t \in \mathbb{R}^n$. Thus we have $\Lambda(p_k\tilde{\varphi}, \tilde{\varphi}) = \Lambda_\Omega(p_k\varphi, \varphi) \geq 0$ for all k, by condition (3) and (2.6).

By virtue of Theorem 2.8, there exists a uniquely determined representing measure F for Λ, whose support is in $\Sigma_{\mathbf{p}}$. In particular

$$\Theta_\alpha(x,y) = \Omega_{(\alpha,0)}(x,y) = \int_{\Sigma_{\mathbf{p}}} t^\alpha \, dF_{x,y}(t), \qquad \alpha \in \mathbb{Z}_+^n,\ x,y \in \mathcal{D},$$

i.e., Θ has a representing measure.

That the representing measure of Θ is unique if and only if the sequence Ω is uniquely determined follows as in the corresponding part of the proof of Theorem 2.4, where one should replace the use of Theorem 2.2 by that of Theorem 2.8. This completes the proof of the theorem.

□

COROLLARY 2.10. *Let* $\Gamma = (\Gamma_\alpha)_{\alpha \in \mathbb{Z}_+^n}$ $(\Gamma_0 = 1_{\mathcal{H}})$ *be an n-sequence of self-adjoint operators in* $\mathcal{L}(\mathcal{H})$. *Let also* $\mathbf{p} = (p_1, \ldots, p_m)$ *be an m-tuple of real polynomials from* $\mathcal{P}_n$, *where* $p_k(t) = \sum_{\xi \in I_k} a_{k\xi} t^\xi$, $k = 1, \ldots, m$, *with* $I_k \subset \mathbb{Z}_+^n$ *finite for all k.*

The n-sequence Γ is a moment sequence, and it has a representing measure whose support is in the set $\cap_{k=1}^m p_k^{-1}(\mathbb{R}_+)$, *if and only if there exists an $(n+1)$-sequence of positive type* $\Delta = (\Delta_{(\alpha,\beta)})_{(\alpha,\beta) \in \mathbb{Z}_+^n \times \mathbb{Z}_+}$ *in* $\mathcal{L}(\mathcal{H})$ *with the following properties*:

(1) $\Gamma_\alpha = \Delta_{(\alpha,0)}$ *for all* $\alpha \in \mathbb{Z}_+^n$,

(2) $\Delta_{(\alpha,\beta)} = \Delta_{(\alpha,\beta+1)} + \sum_{j=1}^n \Delta_{(\alpha+2e_j,\beta+1)} + \sum_{k=1}^m \sum_{\xi,\eta \in I_k} a_{k\xi} a_{k\eta} \Delta_{(\alpha+\xi+\eta,\beta+1)}$ *for all* $\alpha \in \mathbb{Z}_+^n, \beta \in \mathbb{Z}_+$,

(3) *the $(n+1)$-sequences* $(\sum_{\xi \in I_k} a_{k\xi} \Delta_{(\alpha+\xi,\beta)})_{(\alpha,\beta) \in \mathbb{Z}_+^n \times \mathbb{Z}_+}$ *are of positive type for all* $k = 1, \ldots, m$.

The n-sequence Γ has a uniquely determined representing measure on $\cap_{k=1}^m p_k^{-1}(\mathbb{R}_+)$ *if and only if the $(n+1)$-sequence Δ is unique.*

The result is a direct consequence of the previous theorem. As for Corollary 2.6, the only observation is that formula (2.5) leads, in this case, to an $(n+1)$-sequence of bounded self-adjoint operators, which is precisely the sequence Δ. The easy details are left to the reader.

REMARK. Corollary 2.10 provides, in particular, alternate solutions to the operator Hamburger and Stieltjes moment problems in several variables.

3. Subnormal multioperators

By *multioperator* in the Hilbert space $\mathcal{H}$ we shall mean an object of the form $T = (T_1, \ldots, T_n)$, where $T_j: D(T_j) \subset \mathcal{H} \to \mathcal{H}$ is a linear operator for each $j = 1, \ldots, n$.

Two operators $T_j: D(T_j) \subset \mathcal{H} \to \mathcal{H}, j = 1, 2$, are said to be *permutable* [IoVa] if

$$T_1 T_2 x = T_2 T_1 x, \qquad x \in D(T_1 T_2) \cap D(T_2 T_1).$$

A multioperator $T = (T_1, \ldots, T_n)$ in $\mathcal{H}$ is said to be *permutable* [IoVa] if T_j, T_k are permutable for all $j, k = 1, \ldots, n$.

According to [IoVa, Corollary 3.4], every multioperator consisting of commuting self-adjoint operators is permutable.

REMARK. Let $S = (S_1, \ldots, S_m)$ be permutable. Denote by Π_m the group of permutations of the set $\{1, \ldots, m\}$. If

$$D = \bigcap_{\pi \in \Pi_m} D(S_{\pi(1)} \cdots S_{\pi(m)}),$$

then one can easily see that

$$S_1 \cdots S_m x = S_{\pi(1)} \cdots S_{\pi(m)} x, \qquad x \in D.$$

Particularly, given a permutable multioperator $T = (T_1, \ldots, T_n)$ and a multi-index $\alpha = (\alpha_1, \ldots, \alpha_n) \in \mathbb{Z}_+^n$, we can define, unambiguously, the operator $T^\alpha = T_1^{\alpha_1} \cdots T_n^{\alpha_n}$ by setting $S_1 = \cdots = S_{\alpha_1} = T_1, S_{\alpha_1+1} = \cdots = S_{\alpha_1+\alpha_2} = T_2, S_{\alpha_1+\cdots\alpha_{n-1}+1} = \cdots = S_{\alpha_1+\cdots+\alpha_n} = T_n$ on the subspace

$$D^\alpha(T) = \bigcap_{\pi \in \Pi_m} D(S_{\pi(1)} \cdots S_{\pi(m)}),$$

with $m = |\alpha|$. We also set

$$D^\infty(T) = \bigcap_{\alpha \in \mathbb{Z}_+^n} D^\alpha(T).$$

The direct extension of the concept of bounded subnormal (multi)operator (see also [StSz1]) leads to the following

DEFINITION 3.1. Let $T = (T_1, \dots, T_n)$ be a multioperator in $\mathcal{H}$. We say that T is *subnormal* if there exists a Hilbert space $\mathcal{K} \supset \mathcal{H}$ and a multioperator $N = (N_1, \dots, N_n)$ in $\mathcal{K}$ consisting of commuting normal operators such that $T_j \subset N_j$ for all $j = 1, \dots, n$. In this case, N is said to be a *normal extension* of T.

We shall identify in the following the complex Euclidean space $\mathbb{C}^n$ with the real Euclidean space $\mathbb{R}^{2n}$ via the map

$$\mathbb{R}^{2n} \ni (t_1, \dots, t_n, s_1, \dots, s_n) \to (z_1, \dots, z_n) \in \mathbb{C}^n,$$

where $z_j = t_j + is_j$, $j = 1, \dots, n$. With this identification, the space $\mathcal{P}_{2n}$ is itself identified with the space of all finite sums of the form $\sum_{\alpha,\beta} c_{\alpha,\beta} \bar{z}^\alpha z^\beta$.

The function $z \to z_j$ will be denoted by z_j, $j = 1 \dots, n$.

REMARK 3.2. Suppose that $T = (T_1, \dots, T_n)$ is subnormal in $\mathcal{H}$ and let $N = (N_1, \dots, N_n)$ be a normal extension of T in $\mathcal{K}$. It follows easily from [IoVa, Corollary 3.4] that the multioperator N is permutable. This implies that T is also permutable. Then we have $D^\alpha(T) \subset D^\alpha(N)$ and $T^\alpha x = N^\alpha x$ for all $x \in D^\alpha(T)$ and $\alpha \in \mathbb{Z}_+^n$. In particular, $T^\alpha T^\beta = T^{\alpha+\beta}$ on $D^\infty(T)$ for all multi-indices α, β.

Let $\mathcal{D} \subset D^\infty(T)$ be a linear subspace dense in $\mathcal{H}$. Then the equation

$$\Lambda_T(\bar{z}^\alpha z^\beta \otimes x, \bar{z}^\xi z^\eta \otimes y) = \langle T^{\beta+\xi} x, T^{\alpha+\eta} y \rangle, \qquad \alpha, \beta, \xi, \eta \in \mathbb{Z}_+^n,\ x, y \in \mathcal{D},$$

(extended by linearity) defines a sesquilinear form on $\mathcal{P}_{2n} \otimes \mathcal{D}$, which corresponds to (1.10′). As a matter of fact, the form Λ_T is positive semi-definite. Indeed, if E is the spectral measure attached to N, because $\mathcal{P}_{2n} \subset L^2(E_{x,x})$ for each $x \in D^\infty(N)$, it follows that E is a representing measure for the corresponding form Λ_N on the space $\mathcal{P}_{2n} \otimes D^\infty(N)$. Therefore, Λ_N is positive semi-definite, by Lemma 1.4. Since Λ_T is the restriction of Λ_N to $\mathcal{P}_{2n} \otimes \mathcal{D}$, it follows that Λ_T must be positive semi-definite too.

REMARK 3.3. The identification of $\mathbb{R}^{2n+1}$ with $\mathbb{C}^n \times \mathbb{R}$ permits the identification of $\mathcal{P}_{2n+1}$ with the algebra generated by the family of (linearly independent) monomials

$$\mathcal{G} = \{\bar{z}^\alpha z^\beta u^k; z \in \mathbb{C}^n, u \in \mathbb{R}, \alpha, \beta \in \mathbb{Z}_+^n, k \in \mathbb{Z}_+\}.$$

The basis $\mathcal{G}$ is invariant under the involution

$$\mathbb{Z}_+^{2n+1} \ni (\alpha, \beta, k) \to (\beta, \alpha, k) \in \mathbb{Z}_+^{2n+1},$$

and under multiplication as well.

Let $\Omega = (\Omega_{(\alpha,\beta),k})_{\alpha,\beta\in\mathbb{Z}_+^n,k\in\mathbb{Z}_+}$ be a $(2n+1)$-sequence of sesquilinear forms on $\mathcal{D} \subset \mathcal{H}$ such that $\Omega_{(0,0),0}$ is the restriction of the scalar product to $\mathcal{D}$. Setting

$$\Lambda_\Omega(\bar{z}^{\alpha'} z^{\beta'} u^{k'} \otimes x', \bar{z}^{\alpha''} z^{\beta''} u^{k''} \otimes x'') = \Omega_{(\alpha'+\beta'',\alpha''+\beta'),k'+k''}(x', x'')$$

for all $\alpha', \alpha'', \beta'\beta'' \in \mathbb{Z}_+^n, k', k'' \in \mathbb{Z}_+, x', x'' \in \mathcal{D}$, we obtain, by extension, a sesquilinear form on $\mathcal{P}_{2n+1} \otimes \mathcal{D}$ (see also (1.10′)). Following the terminology from Example 1.5.1, we shall say that the sequence Ω is of *positive type* if the form Λ_Ω is positive semi-definite. Note that if Ω is of positive type, then

$$\Omega_{(\alpha,\beta),k}(x,y) = \overline{\Omega_{(\beta,\alpha),k}(y,x)}, \qquad \alpha,\beta \in \mathbb{Z}_+^n,\ k \in \mathbb{Z}_+,\ x,y \in \mathcal{D}, \tag{3.1}$$

via a standard argument of positive semi-definiteness.

THEOREM 3.4. *Let $T = (T_1, \ldots, T_n)$ be a multioperator in $\mathcal{H}$ and let $\mathcal{D} \subset D(T_1) \cap \cdots \cap D(T_n)$ be a linear subspace. Assume that $\mathcal{D}$ is dense in $\mathcal{H}$ and invariant under $T_1, \ldots, T_n$.*

There exist a Hilbert space $\mathcal{K} \supset \mathcal{H}$ and a multioperator $N = (N_1, \ldots, N_n)$ in $\mathcal{K}$ consisting of commuting normal operators such that $N_j \supset T_j|\mathcal{D}$ for all $j = 1, \ldots, n$ if and only if there exists a $(2n+1)$-sequence $\Omega = (\Omega_{(\alpha,\beta),k})_{\alpha,\beta\in\mathbb{Z}_+^n,k\in\mathbb{Z}_+}$ of sesquilinear forms on $\mathcal{D}$ with the following properties:

(1) *$\Omega_{(0,0),0}$ is the restriction of the scalar product to $\mathcal{D}$,*
(2) *Ω is of positive type,*
(3) *$\Omega_{(0,e_j),0}(x,y) = \langle T_j x, y\rangle$ for all $x, y \in \mathcal{D}$ and $j = 1, \ldots, n$,*
(4) *$\Omega_{(e_j,e_j),0}(x,x) = \|T_j x\|^2$, $j = 1, \ldots, n$, $x \in \mathcal{D}$,*
(5) *$\Omega_{(\alpha,\beta),k} = \Omega_{(\alpha,\beta),k+1} + \sum_{j=1}^n \Omega_{(\alpha+e_j,\beta+e_j),k+1}$ for all $\alpha, \beta \in \mathbb{Z}_+^n, k \in \mathbb{Z}_+$.*

In the affirmative case, if in addition the closure of $T_j|\mathcal{D}$ extends T_j for all $j = 1, \ldots, n$, then T is subnormal.

Proof. Assume first that there exists a multioperator N with the stated properties. Let also E be the spectral measure of N, which acts on the Borel subsets in $\mathbb{C}^n$. We define the maps

(3.2)
$$\Omega_{(\alpha,\beta),k}(x,y) = \Big\langle \int \bar{z}^\alpha z^\beta (1+\|z\|^2)^{-k}\, dE(z) x, y\Big\rangle, \quad \alpha,\beta \in \mathbb{Z}_+^n,\ k \in \mathbb{Z}_+,\ x,y \in \mathcal{D}.$$

We shall check that the sequence $\Omega = (\Omega_{(\alpha',\alpha''),k})_{\alpha',\alpha''\in\mathbb{Z}_+^n,k\in\mathbb{Z}_+}$ has properties (1)–(5).

Properties (1), (3) and (4) are easily verified.

Let us show that Ω is of positive type (see Remark 3.3). Indeed, if $\tau(z) = (z, (1+\|z\|^2)^{-1}), z \in \mathbb{C}^n$, then $F^\tau(*) = PE(\tau^{-1}(*))|\mathcal{H}$ is a representing measure for the form Λ_Ω, where P is the orthogonal projection of $\mathcal{K}$ onto $\mathcal{H}$, which follows from (3.2).

Property (5) is a consequence of the identity

$$[(1+\|z\|^2)(1+\|z\|^2)^{-1} - 1]\bar{z}^\alpha z^\beta (1+\|z\|^2)^{-k} = 0.$$

Conversely, if the $(2n+1)$-sequence Ω is given, and if Λ_Ω is the associated positive semi-definite form on $\mathcal{P}_{2n+1} \otimes \mathcal{D}$ (see Remark 3.3), we proceed as in the proof of Theorem 2.8 and Theorem 2.9.

The form Λ_Ω induces a positive semi-definite form, say Λ, on the space $\mathcal{R}_\theta \otimes \mathcal{D}$, where $\mathcal{R}_\theta$ is the algebra generated by $\mathcal{P}_{2n}$ and the rational function $\theta(z) = (1+\|z\|^2)^{-1}$. We find some operators $A_1, \ldots, A_n, B_1, \ldots, B_n$, extending, respectively, the multiplication by the independent variables $t_1, \ldots, t_n, s_1, \ldots, s_n$, which are commuting self-adjoint operators in the space $\mathcal{H}_\Lambda$. Therefore, the operators $N_j = A_j + iB_j$, $j = 1, \ldots, n$ are commuting normal operators. Set $N = (N_1, \ldots, N_n)$. We shall show that N is a normal extension of $(T_1|\mathcal{D}, \ldots, T_n|\mathcal{D})$. Notice that the multiplication by z_j extends $T_j|\mathcal{D}$ for all j. This follows from the fact that

$$\Lambda(z_j \otimes x - 1 \otimes T_j x, z_j \otimes x - 1 \otimes T_j x) = \Lambda_\Omega(z_j \otimes x - 1 \otimes T_j x, z_j \otimes x - 1 \otimes T_j x) = 0$$

via conditions (3), (4) and (3.1). Using the embedding (1.4), we infer easily the assertion.

Finally, if the closure of $T_j|\mathcal{D}$ extends T_j, then $T_j \subset N_j$ for all $j = 1, \ldots, n$, and so T is subnormal.

□

REMARKS 3.5. (1) As the referee noticed, the proof above needs the hypothesis $\mathcal{D}$ invariant under $T_1, \ldots, T_n$, omitted in an earlier version. He also remarked that our proof works without any commutativity hypothesis on $T_1, \ldots, T_n$, as in the bounded case (see [Sto2]).

The author was aware of some other suggestions of the referee's, as for instance the use of sequences of sesquilinear forms rather than that of sequences of linear operators on $\mathcal{D}$ (see the statements of Theorems 2.4 and 2.9).

(2) Although Theorem 3.4 is not an explicit characterization of the power subnormal multioperators (i.e., a characterization only in terms of the given multioperator), it leads to explicit characterizations, provided one can construct directly the sequence Ω. Such cases do exist, as will be shown in some work in progress.

(3) Our methods permit the control of the support of the spectral measure of a normal extension of a subnormal multioperator. As in Theorem 2.9, let $\mathbf{p} = (p_1, \dots, p_m)$ be an m-tuple of real polynomials from $\mathcal{P}_{2n}$, where $p_k(\bar{z}, z) = \sum_{\xi,\eta\in I_k} a_{k\xi\eta}\bar{z}^\xi z^\eta$, $k = 1, \dots, m$, with $I_k \subset \mathbb{Z}_+^n$ finite for all k. If we replace condition (5) from Theorem 3.4 by the stronger condition

$$\begin{aligned}\Omega_{(\alpha,\beta),\ell} =&\Omega_{(\alpha,\beta),\ell+1} + \sum_{j=1}^{n} \Omega_{(\alpha+e_j\beta+e_j),\ell+1}+\\ &+\sum_{k=1}^{m} \sum_{\xi,\eta,\lambda,\mu\in I_k} a_{k\xi\lambda}\bar{a}_{k\eta\mu}\Omega_{(\xi+\mu+\alpha,\eta+\lambda+\beta),\ell+1}\end{aligned}$$

for all $\alpha, \beta \in \mathbb{Z}_+^n, \ell \in \mathbb{Z}_+$, and add the condition (6) $\Lambda_\Omega(p_k\varphi, \varphi) \geq 0$ for all $\varphi \in \mathcal{P}_{2n+1}\otimes\mathcal{D}$ and $k = 1, \dots, m$, then we obtain that the support of a normal extension of T may be chosen to lie in $\cap_{k=1}^m p_k^{-1}(\mathbb{R}_+)$.

References

[AtPe] A. Athavale and S. Pedersen, Moment problems and subnormality, *J. Math. Anal. Appl.*, **146** (1990), 434–441.

[Ber] S. K. Berberian, *Notes on spectral theory,* Van Nostrand, Princeton, N. J., 1966.

[BCR] C. Berg, J. P. R. Christensen and P. Ressel, *Harmonic analysis on semigroups,* Springer-Verlag, New York – Berlin – Heidelberg – Tokyo, 1984.

[Bra] J. Bram, Subnormal operators, *Duke Math. J.*, **22** (1955), 75–94.

[DuSc] N. Dunford and J. T. Schwartz, *Linear Operators. Part* II, Interscience Publishers, New York – London – Sydney, 1963.

[EsVa] J. Eschmeier and F.-H. Vasilescu, On jointly essentially self-adjoint tuples of operators, *Acta Sci. Math.* (*Szeged*), **67** (2001), 375–388.

[Fri] J. Friedrich, Operator moment problems, *Math. Nachr.*, **151** (1991), 273–293.

[Fug] B. Fuglede, The multidimensional moment problem, *Expo. Math.*, **1** (1983), 47–65.

[Hal] P. R. Halmos, Normal dilations and extensions of operators, *Summa Brasiliensis Math.*, **2**:9 (1950), 125–134.

[IoVa] E. J. Ionaşcu and F.-H. Vasilescu, Joint spectral properties for permutable linear transformations, *J. Reine Angew. Math.*, **426** (1992), 23–45.

[Ito] T. Itô, On the commutative family of subnormal operators, *J. Fac. Sci. Hokkaido Univ.*, **14** (1958), 1–15.

[Mac] J. S. MacNerney, Hermitian moment sequences, *Trans. Amer. Math. Soc.*, **103** (1962), 45–81.

[Mas1] P. Masani, Quasi-isometric measures and their applications, *Bull. Amer. Math. Soc.*, **76** (1970), 427–528.

[Mas2] P. MASANI, Dilations as propagators of Hilbertian varieties, *SIAM J. Math. Anal.*, **9** (1978), 414–456.

[Nar] F. J. NARCOWICH, *R*-Operators II. On the Approximation of Certain Operator-Valued Analytic Functions and the Hermitian Moment Problem, *Indiana Univ. Math. J.*, **26** (1977), 483–513.

[Nel] E. NELSON, Analytic vectors, *Ann. Math.*, **70** (1959), 572–615.

[PuVa] M. PUTINAR and F.-H. VASILESCU, Solving moment problems by dimensional extension, *Ann. Math.*, **149** (1999), 1087–1107.

[Sch] K. SCHMÜDGEN, *Unbounded Operator Algebras and Representation Theory,* Operator Theory: Advances and Applications Vol. **37**, Birkhäuser Verlag, Basel – Boston – Berlin, 1990.

[Sti] W. F. STINESPRING, Positive functions on C^*-algebras, *Proc. Amer. Math. Soc.*, **6** (1955), 211–216.

[StZs] S. STRATILA and L. ZSIDO, *Lectures on von Neumann algebras,* Editura Academiei, Bucharest, 1979.

[Sto1] J. STOCHEL, Decomposition and disintegration of positive definite kernels on convex $*$-semigroups, *Ann. Polon. Math.*, **56** (1992), 243–294.

[Sto2] J. STOCHEL, Positive definiteness and commutativity of operators, *Proc. Amer. Math. Soc.*, **126** (1998), 431–440.

[StSz1] J. STOCHEL and F. H. SZAFRANIEC, On normal extensions of unbounded operators. I, *J. Operator Theory*, **14** (1985), 31–55; II, *Acta Sci. Math. (Szeged)*, **53** (1989), 153–177.

[StSz2] J. STOCHEL and F. H. SZAFRANIEC, The complex moment problem and subnormality: A polar decomposition approach, *J. Functional Analysis*, **159** (1998), 432–491.

[SzN1] B. SZ.-NAGY, A moment problem for self-adjoint operators, *Acta Math. Acad. Sci. Hungar.*, **3** (1952), 285–292.

[SzN2] B. SZ.-NAGY, *Extensions of linear transformations in Hilbert space which extend beyond this space,* Appendix to F. Riesz, B. Sz.-Nagy, "Functional Analysis", Ungar, New York, 1960.

[Vas1] F.-H. VASILESCU, *Analytic Functional Calculus and Spectral Decompositions*, Editura Academiei and D. Reidel Publ. Co., Bucharest and Dordrecht, 1982.

[Vas2] F.-H. VASILESCU, Moment problems for multi-sequences of operators, *J. Math. Anal. Appl.*, **219** (1998), 246–259.

[Vas3] F.-H. VASILESCU, *Hamburger and Stieltjes moment problems in several variables*, Preprint, 1998.

[Vas4] F.-H. VASILESCU, Operator theoretic characterizations of moment functions, 17*th Operator Theory Conference – Proceedings, The Theta Foundation,* 2000, 405–415.

F.-H. VASILESCU, Université des Sciences et Technologies de Lille, U.F.R. de Mathématiques Pures et Appliquées, U.M.R. du C.N.R.S. 8524, 59655 Villeneuve d'Ascq Cedex, France; *e-mail*: fhvasil@gat.univ-lille1.fr

Operator Theory: Advances and Applications, Vol. 127, 639–659

The argument principle and boundaries of analytic varieties

JOHN WERMER

Dedicated to the memory of Professor Béla Szőkefalvi-Nagy

Let M be a smooth, compact, oriented manifold in the space of n complex variables. When does there exist a complex-analytic subvariety of $\mathbb{C}^n \setminus M$ having M as its boundary in the sense of currents? R. Harvey and B. Lawson posed and solved this problem in 1975. Recently Herbert Alexander and the author gave a different solution based on Linking Numbers, in "Linking numbers and boundaries of varieties", *Ann. of Math.*, **151** (2000).

In the present paper we show, basing ourselves on this earlier work, how degrees of certain polynomial maps from M to the sphere in a suitable $\mathbb{C}^q$ can be used to give a necessary and sufficient condition, which generalizes the usual argument principle of analytic function theory.

1. Introduction

Let M be a compact, smooth, oriented manifold in $\mathbb{C}^n$. The problem in giving conditions on M in order that there exist a complex-analytic variety bounded by M was tackled in a fundamental paper by Reese Harvey and Blaine Lawson in 1975, [1]. Their answer involved the dimensions of complex-linear subspaces of the real tangent spaces of M.

Recently, in [2], Herbert Alexander and the author revisited the problem and gave conditions on M involving linking numbers of M and certain subvarieties of $\mathbb{C}^n \setminus M$.

In the present paper we show how, instead of linking numbers, *winding numbers* of certain polynomial maps of M on $\mathbb{C}^q$, where $\dim M = 2q - 1$, can be used to

AMS Subject Classification (1991): 32E, 32Q, 57, 32D15.

characterize the boundaries of analytic varieties. The relevant condition on the winding numbers generalize the usual argument principle. Our proofs make essential use of results in [2] and [1]. For terminology and results concerning currents we refer the reader to [1] and [3]. In the last Section we look at a special case.

We shall use the word "smooth" to mean "of class $\mathbb{C}^2$".

2. Winding numbers

Consider a smooth map $f\colon M \to N$ between two compact smooth oriented manifolds of dimension p. Let ω be a smooth closed p-form on N with $\int_N \omega = 1$. The *degree* of f is defined to be $\int_M f^*\omega$, (see Bott and Tu, [4], §5).

Let now M be a compact oriented manifold of dimension $2q-1$ contained in $\mathbb{C}^n$ and let $P = (P_1, \dots, P_q)$ be a map: $\mathbb{C}^n \to \mathbb{C}^q$, where each P_j is a polynomial on $\mathbb{C}^n$. Assume that (i) $P^{-1}(0)$ has complex dimension $n-q$, and (ii) $P^{-1}(0)$ is disjoint from M.

Let f be the map: $M \to S^{2q-1}$,where S^{2q-1} is the unit sphere in $\mathbb{C}^q$, which is given by: $z \to |P(z)|^{-1}P(z), z \in M$.

DEFINITION 2.1. The *winding number* of the map P on M, denoted $W(M,P)$, is the degree of f.

We denote by β_q the normalized Bochner–Martinelli kernel on $\mathbb{C}^q$, with singularity at 0. Here $\beta_1 = \frac{1}{2\pi i}\frac{dz}{z}$.

CLAIM 2.1. $W(M,P) = \int_M P^*\beta_q$.

Proof. Without loss of generality, there exists $k < 1$ with $|P| \leq k$ on M. Put $f_t(z) = tP(z) + (1-t)f(z)$ for $z \in M, 0 \leq t \leq 1$. For each t, $|f_t(z)| > |P(z)| > 0$, and so f maps M into $\mathbb{C}^q \setminus \{0\}$. We have $f_0 = f$ and $f_1 = P$ on M. So f and P are homotopic as maps: $M \to \mathbb{C}^q \setminus \{0\}$ and so

$$\int_M P^*\beta_q = W(M,P). \qquad \square$$

LEMMA 2.1. *Let $q = 1$. Then M is a finite union of disjoint oriented simple closed curves in $\mathbb{C}^n$. Put $\gamma = M$. A polynomial map $P: \mathbb{C}^n \to \mathbb{C}^q$, satisfying* (i) *and* (ii), *here consists of a single polynomial P with $P \neq 0$ on γ. Then*

$$W(\gamma, P) = \frac{1}{2\pi} \operatorname{vararg}_\gamma P.$$

Proof. Since $q = 1$, we have $\beta_1 = \frac{1}{2\pi i}\frac{dz}{z}$, and so

$$W(\gamma, P) = \int_\gamma P^*\beta_1 = \frac{1}{2\pi i}\int_\gamma \frac{dP}{P} = \frac{1}{2\pi} \operatorname{vararg}_\gamma P. \qquad \square$$

Let γ be as in Lemma 2.1. Suppose that there exists a finite Riemann surface Σ having γ as its boundary. Let P be a polynomial on $\mathbb{C}^n$ with $P \neq 0$ on γ. Then, by the argument principle on Σ, $\frac{1}{2\pi}\operatorname{vararg}_\gamma P$ is the number of zeros of P on Σ and so $\operatorname{vararg}_\gamma P \geq 0$. In view of Lemma 2.1 this tells us that a necessary condition on γ for γ to be the boundary of a finite Riemann surface is

$$W(M, P) \geq 0 \text{ for every polynomial } P \text{ on } \mathbb{C}^n \text{ with } P \neq 0 \text{ on } \gamma.$$

Consider now the case when M is a $(2q-1)$-dimensional, smooth, compact oriented manifold in $\mathbb{C}^n$ and P is a polynomial map: $\mathbb{C}^n \to \mathbb{C}^q$ satisfying (i) and (ii).

LEMMA 2.2. *Assume that there exists a complex-analytic submanifold V of $\mathbb{C}^n \setminus M$ such that $V \cup M$ is a compact manifold with boundary. Let P be a polynomial map: $\mathbb{C}^n \to \mathbb{C}^q$ satisfying* (i) *and* (ii). *Then $W(M, P) \geq 0$.*

NOTE. We tacitly use the orientation on V induced by its complex structure.

Proof. Without loss of generality 0 is not a critical value of the map $P|V: V \to \mathbb{C}$. Then $P^{-1}(0)$ intersects V transversally at a finite set of points $p_1, \ldots, p_s$

For each j, P is a one-one orientation preserving map from some compact neighborhood B_j of p_j in V to $\mathbb{C}^q$. Then, putting $X = bdB_j$,

$$\int_M P^*\beta_q = \sum_{j=1}^{s} \int_X P^*\beta_q = \sum_{j=1}^{s}(+1) = s.$$

So $W(M, P) \geq 0$. $\square$

DEFINITION 2.2. Let M be a compact, smooth, oriented submanifold of $\mathbb{C}^n$. We say that M *satisfies the winding condition* if for every polynomial map $P: \mathbb{C}^n \to \mathbb{C}^q$ which satisfies (i) and (ii) we have $W(M, P) \geq 0$.

Lemma 2.2 told us that the winding condition is necessary for the existence of a complex manifold V with M as its boundary. Our main theorem is the following partial converse:

THEOREM 2.3. *Let M be a compact, smooth, oriented submanifold of $\mathbb{C}^n$ with dimension $(2q-1)$, $q \geq 1$. Assume that M is connected. If M satisfies the winding condition, then there exists a complex-analytic subvariety V of $\mathbb{C}^n \setminus M$,* $\dim_{\mathbb{C}} V = q$ *such that the current of integration $[V]$ has finite mass and has the current of integration $[M]$ as its boundary, i.e.*

$$\int_M \sigma = \int_V d\sigma,$$

for every smooth $(2q-1)$-form σ on $\mathbb{C}^n$.

3. A formula for the winding number

If W is a linear subspace of $\mathbb{R}^N$ and $\beta_1, \ldots, \beta_k$ is an ordered basis of W, we call the wedge product $\beta_1 \wedge \cdots \wedge \beta_k$ in $\bigwedge^k(\mathbb{R}^N)$ *an orientation of* W and write $\mathrm{or}(W)$ for this element.

If W is an oriented subspace of $\mathbb{R}^N$ and A and B are oriented subspaces of W, we write

$$A \oplus B = W,$$

("$A \oplus B$ equals W as oriented vector spaces") if W is the direct sum of A and B, and $\mathrm{or}(W) = \mu\, \mathrm{or}(A) \wedge \mathrm{or}(B)$, for some positive scalar μ.

Let X be a smooth, oriented (2q)-dimensional manifold and let F be a smooth map: $X \to \mathbb{C}^n$. For $x \in X$, we write T_x for the tangent space to X at x. Then T_x is an oriented vector-space.

DEFINITION 3.1. Let A be an oriented smooth submanifold of an open set in $\mathbb{C}^n$ with dim A = 2n-2q. Fix x in X such that x is a non-critical point of F and assume that $y = F(x)$ belongs to A. Then F *is transversal to A at x* if

$$(dF)_x(T_x(X)) + T_y(A) = T_y(\mathbb{C}^n), \tag{2}$$

as vector spaces.

DEFINITION 3.2. We write

$$i_x(F, A) = +1$$

if (2) holds in the sense of oriented vector spaces, i.e.

$$(dF)_x(T_x(X) \oplus T_y(A) = T_y(\mathbb{C}^n).$$

We write $i_x(F, A) = -1$ if (2) holds and the two sides of (2) have opposite orientations.

THEOREM 3.1. *Fix a smooth compact oriented submanifold M of $\mathbb{C}^n$ with* $\dim M = 2q - 1$. *Let Σ be a $(2q)$-dimensional smooth oriented manifold with boundary and let F be a smooth map of Σ into $\mathbb{C}^n$ such that F, restricted to bd Σ, is an orientation-preserving diffeomorphism of bdΣ onto M. Let P be a polynomial map: $\mathbb{C}^n \to \mathbb{C}^q$ satisfying* (i) *and* (ii), *such that $A = P^{-1}(0)$ is non-singular and that 0 is not a critical value of $P \circ F: \Sigma \to \mathbb{C}^q$. Then $(P \circ F)^{-1}(0)$ consists of a finite number of non-critical points $x_1, \ldots, x_s$, and we have*

$$\int_M P^* \beta_q = \sum_{k=1}^{s} i_{x_k}(F, A). \tag{3}$$

LEMMA 3.2. *Let M, Σ, P, A be as in Theorem* 3.1. *Fix x in Σ and assume that* $i_x(F, A) = +1$. *Then $P \circ F: \Sigma \to \mathbb{C}^q$ is orientation-preserving at x.*

Proof. We denote the standard basis of $\mathbb{C}^n$ over $\mathbb{R}$ by $e_1 = (1, 0, \ldots, 0)$, $f_1 = (i, 0, \ldots, 0)$, etc. Then $\mathrm{or}(\mathbb{C}^n) = e_1 \wedge f_1 \wedge \ldots \wedge e_n \wedge f_n$. Let $\xi_1, \ldots, \xi_{2q}$ be a basis of $T_x = T_x(\Sigma)$ such that $\mathrm{or}(T_x) = \xi_1 \wedge \ldots \wedge \xi_{2q}$. Put $\xi'_j = (dF)_x(\xi_j)$, $j = 1, \ldots, 2q$, and denote by W the real vector-space spanned by the ξ'_j and put $\mathrm{or}(W) = \xi'_1 \wedge \ldots \xi'_{2q}$.

Put $y = F(x)$ and let $L = (dP)_y$. Then $L: \mathbb{C}^n \to \mathbb{C}^q$. Without loss of generality, the subspace $\{L = 0\} = \{z_1 = \ldots = z_q = 0\}$. Then

$$L(e_j) = L(f_j) = 0, \qquad j = q + 1, \ldots, n \tag{4}$$

and

$$\mathrm{or}\{L = 0\} = e_{q+1} \wedge f_{q+1} \wedge \ldots \wedge e_n \wedge f_n. \tag{5}$$

By construction of W, $(dF)_x$ is an orientation-preserving map: $T_x \to W$. Also

$$(d(P \circ F))_x = (dP)_y \circ (dF)_x = L \circ (dF)_x.$$

To prove our Lemma, it suffices to prove the following:

CLAIM 3.1. *The restriction of L to W is orientation-preserving as map: $W \to \mathbb{C}^q$.*

Proof. We must show that there exists $\lambda' > 0$ such that

$$L(\xi_1') \wedge \ldots \wedge L(\xi_{2q}') = \lambda' \operatorname{or}(\mathbb{C}^q) \tag{6}$$

Our hypothesis: $i_x(F, A) = +1$ yields some $\lambda > 0$ with

$$(\xi_1' \wedge \ldots \wedge \xi_{2q}') \wedge (e_{q+1} \wedge f_{q+1} \wedge \ldots \wedge e_n \wedge f_n) = \lambda \operatorname{or}(\mathbb{C}^n). \tag{7}$$

Now for suitable real constants a_j^ν, we have

$$\xi_\nu' = \sum_{j=1}^{n} (a_j^\nu e_j + b_j^\nu f_j), \; \nu = 1, \ldots, 2q.$$

Put

$$\eta_\nu = \sum_{j=1}^{q} (a_j^\nu e_j + b_j^\nu f_j), \; \nu = 1, \ldots, 2q.$$

$$\left(\bigwedge_{\nu=1}^{2q} \eta_\nu \right) \wedge (e_{q+1} \wedge f_{q+1} \wedge \ldots \wedge e_n \wedge f_n) = \lambda \operatorname{or}(\mathbb{C}^n). \tag{8}$$

$$\bigwedge_{\nu=1}^{2q} \eta_\nu = \bigwedge_{\nu=1}^{2q} \left[\sum_{j=1}^{q} (a_j^\nu e_j + b_j^\nu f_j) \right] = \kappa e_1 \wedge f_1 \wedge \ldots \wedge e_q \wedge f_q, \tag{9}$$

where κ is a real constant. Hence by (9), we have

$$(\kappa e_1 \wedge f_1 \wedge \ldots \wedge e_q \wedge f_q) \wedge (e_{q+1} \wedge f_{q+1} \wedge \ldots \wedge e_n \wedge f_n) = \lambda \operatorname{or}(\mathbb{C}^n). \tag{10}$$

So $\kappa = \lambda$ and so $\kappa > 0$. Now

$$\bigwedge_{\nu=1}^{2q} L(\xi_\nu') = \bigwedge_{\nu=1}^{2q} \left[\sum_{j=1}^{q} a_j^\nu L(e_j) + b_j^\nu L(f_j) \right], \tag{11}$$

where we used the fact that $L(e_j) = L(f_j) = 0$ for $j \geq q + 1$. Since the exterior algebra computations are parallel in (9) and (11), we get

$$\bigwedge_{\nu=1}^{2q} L(\xi_\nu') = \kappa L(e_1) \wedge L(f_1) \wedge \ldots \wedge L(e_q) \wedge L(f_q), \tag{12}$$

where κ is the same constant as in (9).

By hypothesis, $i_x(F, A) = +1$, and so L is one-one on W, and hence the vectors

$$L(\xi_1'), \ldots, L(\xi_{2q}')$$

are linearly independent over $\mathbb{R}$. It follows that the vectors

$$L(e_1), L(f_1), \ldots, L(e_q), L(f_q)$$

are linearly dependent over $\mathbb{R}$. Also, $L(f_\nu) = iL(e_\nu)$ for each ν, and so $L(e_1), \ldots, L(e_q)$ form a basis of $\mathbb{C}^q$ over $\mathbb{C}$. Hence

$$L(e_1) \wedge L(f_1) \wedge \ldots \wedge L(e_q) \wedge L(f_q) = \mu \, \mathrm{or}(\mathbb{C}^q)$$

for some $\mu > 0$. From (12) we get

$$\bigwedge_{\nu=1}^{2q} L(\xi_\nu') = (\kappa\mu) \, \mathrm{or}(\mathbb{C}^q).$$

Since $\kappa\mu > 0$, we have (6), and so the Claim is proved. Lemma 3.2 follows.

□

The same argument yields that $P \circ F$ is orientation-reversing at x if $i_x(F, A) = -1$. We proceed to the proof of Theorem 3.1.

Proof. Fix x in Σ. Suppose $x \in bd\Sigma$. Then $F(x) \in M$, and so $(P \circ F)(x) \neq 0$, since $P \neq 0$ on M, by hypothesis. So $(P \circ F)^{-1}(0) \subseteq \Sigma \setminus bd\Sigma$. Suppose that there exists an infinite sequence $\{x_\nu\}$ of points in $(P \circ F)^{-1}(0)$. Let x be an accumulation point of this sequence. Without loss of generality, $x_\nu \to x$ as $\nu \to \infty$. Then $(P \circ F)(x) = 0$. Since $(P \circ F)(x_\nu) = 0$ for all ν, and $x_\nu \to x$, x is a critical point of $P \circ F$. This contradicts our hypothesis that 0 is not a critical value of $P \circ F$.

So $(P \circ F)^{-1}(0)$ is a finite set of points: $x_1, \ldots, x_s$ contained in $\Sigma \setminus bd\Sigma$. For each $\nu, 1 \leq \nu \leq s$, we construct a compact neighborhood B_ν of x_ν in Σ such that the B_ν are mutually disjoint and such that $P \circ F$ maps B_ν diffeomorphically on a ball $(|\zeta| \leq r_\nu)$ in $\mathbb{C}^q$. Now we may parametrise M by the map $\zeta = F^{-1}(x), x \in bd\Sigma$. Then

$$\int_M P^*\beta_q = \int_{bd\Sigma} (P \circ F)^*\beta_q.$$

Also, by Stokes' theorem,

$$\int_{bd\Sigma} (P \circ F)^*\beta_q = \sum_{\nu=1}^{s} \int_{bdB_\nu} (P \circ F)^*\beta_q = \sum_{\nu=1}^{s} \int_{S_\nu} \beta_q$$

where S_ν is the sphere $|\zeta| = r_\nu$, oriented positively or negatively according as $P \circ F$ preserves or reverses orientation. In view of Lemma 3.2, then, we get

$$\int_{S_\nu} \beta_\nu = i_{x_\nu}(F, A),$$

and so (3) holds. Theorem 3.1 is proved. □

4. Slicing M by a complex hyperplane

We shall repeatedly use the following fact: if A and B are oriented vector spaces of even dimension, then

$$A \oplus B = B \oplus A.$$

Let T be an oriented vector space and let L be a linear map: $T \to \mathbb{C}$, which maps T onto $\mathbb{C}$. We wish to orient the subspace $\{L = 0\}$ of T.

CLAIM 4.1. *There exists a complementary subspace K to $\{L = 0\}$ in T, such that L maps K isomorphically onto $\mathbb{C}$.*

Proof. Choose vectors X and Y in T such that $L(X) = 1$ and $L(Y) = i$. Define K to be the span of X and Y. Fix a vector ξ in T. Then $L(\xi) = c \cdot 1 + d \cdot i, c, d \in \mathbb{R}$, so $L(\xi - (cX + dY)) = 0$. Hence $y = \xi - (cX + dY) \in \{L = 0\}$. Hence we have $\xi = y + cX + dY$, where $y \in \{L = 0\}$ and $cX + dY \in K$. It is clear that $K \cap \{L = 0\} = \{0\}$, and so K is a complement to $\{L = 0\}$ in T. By construction, L maps K isomorphically on $\mathbb{C}$. So the Claim holds. □

DEFINITION 4.1. We orient K by pulling back the orientation on $\mathbb{C}$ under L, and so $\text{or}(K) = X \wedge Y$. We then orient $\{L = 0\}$ so that $\{L = 0\} \oplus K = T$.

Let next Y be an oriented smooth manifold and let Φ be a smooth map: $Y \to \mathbb{C}$. We assume that each point of the set $\{\Phi = 0\}$ is a non-critical point of Φ. Then $\{\Phi = 0\}$ is a submanifold of Y. Also for each y in $\{\Phi = 0\}$, the zero-set of $(d\Phi)_y$ in T_y is the tangent space to $\{\Phi = 0\}$ at y, and also $(d\Phi)_y$ maps T_y onto $\mathbb{C}$.

DEFINITION 4.2. We orient the manifold $\{\Phi = 0\}$ by orienting each tangent space to $\{\Phi = 0\}$, by means of Definition 4.1, as the zero-set of the linear map $(d\Phi)_y$ from T_y to $\mathbb{C}$.

The objective of this Section is to prove that the winding condition is inherited under slicing by almost all complex hyperplanes, (Theorem 4.1, below). Let $M \subseteq \mathbb{C}^n$ be as in Theorem 2.3. We fix a complex-linear function ϕ on $\mathbb{C}^n$, and for each λ in $\mathbb{C}$ we put $H_\lambda = \{\phi = \lambda\}$, and put $Q_\lambda = M \cap H_\lambda$.

By Sard's theorem we know that a.e. λ in $\mathbb{C}$ is a non-critical value of the map $\phi: M \to \mathbb{C}$, and hence Q_λ is a smooth $(2q-3)$-dimensional compact manifold contained in H_λ. We fix such a λ. Since Q_λ is the zero set of the map $\phi - \lambda: M \to \mathbb{C}$, we may use Definition 4.2 to orient Q_λ as a submanifold of $\mathbb{C}^{n-1}$.

We form the manifold $M \times I$, where I is the interval $0 \le t \le 1$, and we orient this manifold so that $bd(M \times I) = M \times \{0\} - M \times \{1\}$. Since $M \subseteq \mathbb{C}^n$, the coordinate functions $z_1, \dots, z_n$ are defined on $M \times I$. We put

$$F(z,t) = (1-t)(z_1, \dots, z_n), (z,t) \in M \times I.$$

So F is a map: $M \times I \to \mathbb{C}^n$. We put

$$\Sigma_\lambda = \{(z,t) \in M \times I \mid \phi \circ F = \lambda\}.$$

THEOREM 4.1. *Assume that M satisfies the winding condition. Fix λ_o in $\mathbb{C}$ such that λ_o is a non-critical value of ϕ on M and also λ_o is a non-critical value of $\phi \circ F$ on $M \times I$. Then Q_{λ_o} satisfies the winding condition in $\mathbb{C}^{n-1}$.*

Without loss of generality, ϕ is the n'th coordinate function on $\mathbb{C}^n$. Then for all $\lambda \in \mathbb{C}^n$,

$$\Sigma_\lambda = \{(z,t) \mid (1-t)z_n = \lambda\} = \{(z,t) \mid F(z,t) \in H_\lambda\}.$$

We denote by F_λ the restriction of F to Σ_λ. Since the image of F_λ lies in H_λ, we may regard F_λ as a map of Σ_λ into $\mathbb{C}^{n-1}$.

Let $S = (S_1, \dots, S_{q-1})$ be a polynomial map: $\mathbb{C}^{n-1} \to \mathbb{C}^{q-1}$ such that (i) and (ii) hold for Q_λ. Define $P: \mathbb{C}^n \to \mathbb{C}^q$ by

$$P(z_1, \dots, z_{n-1}) = (S(z'), z_n - \lambda),$$

where $z' = (z_1, \dots, z_{n-1})$. By abuse of language, we shall somtimes regard $S^{-1}(0)$ as lying in $\mathbb{C}^n$, by identifying $S^{-1}(0)$ with $S^{-1}(0) \times \{\lambda\}$.

We have

$$\{x \in M \times I \mid F(x) \in S^{-1}(0)\} = \{x \in \Sigma_\lambda \mid F_\lambda(x) \in S^{-1}(0)\}. \tag{13}$$

LEMMA 4.2. *Fix $x \in \Sigma_\lambda$ such that $F_\lambda(x) \in S^{-1}(0)$. Assume that x is not a critical point of F_λ and x is not a critical point of F, and $F(x)$ is a regular point of $S^{-1}(0)$. Then*

$$i_x(F, S^{-1}(0)) = i_x(F_\lambda, S^{-1}(0)). \tag{14}$$

Proof. Since x is not a critical point of F, there is a neighborhood U of x in $M \times I$ which F maps diffeomorphically to a $(2q)$-dimensional piece of smooth manifold $F(U)$ in $\mathbb{C}^n$. Since $M \times I$ is oriented, $F(U)$ gets an orientation via F. We let Φ be the function $z_n - \lambda$ restricted to $F(U)$ and we orient the set $\{\Phi = 0\}$ as a zero set. Next, we give Σ_λ its orientation as the zero-set of $\Phi \circ F$. Then F maps $\Sigma_\lambda \cap U$ to $\{\Phi = 0\}$, and this is contained in $F(U)$. We denote by T the image of $T_x(M \times I)$ under $(dF)_x$. Then T is an oriented vector space contained in $\mathbb{C}^n$. Put $y = F(x)$. By choice of x, y is a regular point of $S^{-1}(0)$. We write A for the oriented tangent space to $S^{-1}(0)$ at y.

We first assume that $i_x(F, A) = +1$. This means that

$$T \oplus A = \mathbb{C}^n.$$

The tangent space to $\mathbb{C}^n$ is identified with $\mathbb{C}^n$.

We define L to be the map: $\mathbb{C}^n \to \mathbb{C}$ given by $L(z_1, \ldots, z_n) = z_n$, and we denote by H the zero set of L in $\mathbb{C}^n$. Then the oriented tangent space to $\{\Phi = 0\}$ is $T \cap H$. Since $T \oplus A = \mathbb{C}^n$, L maps T onto $\mathbb{C}$. This permits us to apply Claim 4.1 and Definition 4.1 above to the restriction of L to T, and we get vectors X, Y in T with $L(X) = 1$ and $L(Y) = i$. As before we define K to be the span of X and Y, and put $\mathrm{or}(K) = X \wedge Y$. We get

$$(T \cap H) \oplus K = T. \tag{15}$$

CLAIM 4.2. *$H \oplus K = \mathbb{C}^n$*

Proof of Claim. Since $L(X) = 1$, there exist real scalars c_j, d_j, such that $X = \sum_{j=1}^{n-1}(c_j e_j + d_j f_j) + e_n$, and similarly, $y = \sum_{j=1}^{n-1}(c'_j e_j + d'_j f_j) + f_n$. Then

$$\begin{aligned}\mathrm{or}(\mathbb{C}^n) &= e_1 \wedge f_1 \wedge \ldots \wedge e_n \wedge f_n = e_1 \wedge f_1 \wedge \ldots \wedge e_{n-1} \wedge f_{n-1} \wedge X \wedge Y \\ &= \mathrm{or}(H) \wedge \mathrm{or}(K) = \mathrm{or}(H \oplus K).\end{aligned}$$

So Claim 4.2 holds.

□

CLAIM 4.3. $(T \cap H) \oplus A = H$.

Proof of Claim. $T = (T \cap H) \oplus K$. Since $T \oplus A = \mathbb{C}^n$, we get $(T \cap H) \oplus K \oplus A = \mathbb{C}^n$. Also, there exists $\mu \in \mathbb{R}$, $\mu \neq 0$, such that $\operatorname{or}((T \cap H) \oplus A) = \mu \operatorname{or}(H)$. So, using Claim 4.2, we have

$$\operatorname{or}(\mathbb{C}^n) = \operatorname{or}((T \cap H) \oplus A \oplus K) = (\mu \operatorname{or}(H)) \wedge \operatorname{or}(K) = \mu \operatorname{or}(H \oplus K) = \mu \operatorname{or}(\mathbb{C}^n).$$

So $\mu > 0$. Therefore $(T \cap H) \oplus A = H$, and Claim 4.3 is proved.

Now $T \cap H$ is the tangent space to $\{\Phi = 0\}$ and $(dF_\lambda)_x$ is an orientation-preserving linear map of $T_x(\Sigma_\lambda)$ to the tangent space of $\{\Phi = 0\}$. So $i_x(F_\lambda, S^{-1}(0)) = +1$. So (14) holds in this case. If $i_x(F < A) = -1$, the corresponding argument gives (14). Lemma 4.2 is proved.

□

Fix now $\lambda \neq 0$ as in Theorem 4.1, and let $Q_\lambda, H_\lambda, \Sigma_\lambda, F_\lambda$ be defined as earlier. Then Σ_λ is non-singular. Note that

$$bd\Sigma_\lambda = \{(z,t) \in M \times I \mid z_n = \lambda\} = Q_\lambda \times \{0\}.$$

Let $S: \mathbb{C}^{n-1} \to \mathbb{C}^{q-1}$ be as above, and define $P(y', y_n) = (S(y'), y_n - \lambda)$, as earlier.

LEMMA 4.3. *Assume that* 0 *is not a critical value of* $P \circ F$ *and that* 0 *is not a critical value of* $S \circ F$. *Then*

$$\int_M P^* \beta_q = \int_{Q_\lambda} S^* \beta_{q-1}. \tag{16}$$

Proof. We have $bd(M \times I) = M \times \{0\} - M \times \{1\}$. For $(z,0) \in M \times \{0\}, F(z,0) = z$. For $(z,1) \in M \times \{1\}, F(z,1) = 0$. It follows that the proof of Theorem 3.1 applies with Σ replaced by $M \times I$, and yields

$$\int_M P^* \beta_q = \sum_{\nu=1}^{s} i_{x_\nu}(F, A), \tag{17}$$

where $\{x_1, \ldots, x_s\} = (P \circ F)^{-1}(0)$. Furthermore, $(P \circ F)^{-1}(0)$ coincides with $\{x \in \Sigma_\lambda \mid x \in (S \circ F)^{-1}(0)\}$. We now apply Theorem 3.1 with Σ replaced by Σ_λ and M replaced by Q_λ and F replaced by F_λ. By Lemma 4.2 we then get

$$\int_{Q_\lambda} S^* \beta_{q-1} = \sum_{\nu=1}^{s} i_{x_\nu}(F_\lambda, A) = \sum_{\nu=1}^{s} i_{x_\nu}(F, A). \tag{18}$$

It follows that the left hand side of (17) and (18) are equal, and we get (16).

□

Proof of Theorem 4.1. Fix λ_o as in Theorem 4.1. Then Q_{λ_0} is a smooth $(2q-3)$-manifold. Fix a polynomial map $S: \mathbb{C}^{n-1} \to \mathbb{C}^{q-1}$ which satisfies (i) and (ii) for Q_{λ_o}. It follows that for λ near λ_o, Q_λ is a smooth $(2q-3)$-manifold and S satisfies (i) and (ii) for Q_λ. Also, as $\lambda \to \lambda_o$, we have

$$(19) \qquad \int_{Q_\lambda} S^*\beta_{q-1} \to \int_{Q_{\lambda_o}} S^*\beta_{q-1}.$$

CLAIM 4.4. *There exists λ arbitrarily close to λ_o such that*

$$\int_{Q_\lambda} S^*\beta_{q-1} \geq 0.$$

Once this Claim is proved, (19) yields

$$W(Q_{\lambda_o}, S) = \int_{Q_{\lambda_o}} S^*\beta_{q-1} \geq 0.$$

This holds for every S which satisfies (i) and(ii) for Q_{λ_o}, so Q_{λ_o} satisfies the winding condition. The proof of Theorem 4.1 is complete. □

Proof of Claim 4.4. Define the polynomial P_o by

$$P_o(y', y_n) = (S(y'), y_n - \lambda_o).$$

For a.a. (a', ϵ) in $\mathbb{C}_{q-1} \times \mathbb{C}$, (a', ϵ) is a non-critical value for $P_o \circ F$. Hence for a.a. ϵ in $\mathbb{C}$, we have the following:

(*) (a', ϵ) is a non-critical value for $P_o \circ F$ for a.a. $a' \in \mathbb{C}^{q-1}$, as well as that $\lambda = \lambda_o + \epsilon$ is a non-critical value for $\phi|_M$, (where ϕ is as defined above), and also is a non-critical value for $\phi \circ F$. So

(**) Q_λ is a smooth $(2q-3)$-manifold, and

(***) Σ_λ is a smooth manifold.

Fix ϵ satisfying (*), (**), and (***). Then for a.a. a' in $\mathbb{C}^{q-1}$, a' is a non-critical value for $S \circ F_\lambda$ and (a', ϵ) is a non-critical value for $P_o \circ F$. Fix such an a' and define $S_{a'} = S - a'$. Define

$$P(y', y_n) = (S_{a'}(y'), y_n - \lambda) = (S(y'), y_n - \lambda_o) - (a', \epsilon).$$

So $P = P_o - (a', \epsilon)$. Also $S_{a'} = S - a'$. Therefore $P \circ F = P_o \circ F - (a', \epsilon)$, and $S_{a'} \circ F_\lambda = S \circ F_\lambda - a'$. By choice of a' and ϵ, then, 0 is not a critical value for $P \circ F$, and 0 is not a critical value for $S_{a'} \circ F_\lambda$. By Lemma 4.3,

$$\int_M P^*\beta_q = \int_{Q_\lambda} S_{a'}\beta_{q-1}.$$

Since M satisfies the winding condition, $\int_M P^*\beta_q \geq 0$, and so

$$\int_{Q_\lambda} S^*_{a'}\beta_{q-1} \geq 0. \tag{20}$$

Let $a' \to 0$. We get

$$\int_{Q_\lambda} S^*\beta_{q-1} \geq 0. \tag{21}$$

Since ϵ is arbitrarily close to 0, the Claim is proved, and so Theorem 4.1 follows. □

5. Holomorphic chains and maximal complexity

We next give some definitions and results from the Harvey–Lawson paper [1].

Let Ω be a domain in $\mathbb{C}^n$. A current T is a *holomorphic k-chain* in Ω if T admits a representation

$$T = \sum_j n_j[V_j], \tag{22}$$

where each V_j is a k-dimensional (complex-analytic)subvariety of Ω, $[V_j]$ is the current of integration over V_j, and n_j is an integer. T is *positive* if each $n_j > 0$.

Let W be a $(2p-1)$-dimensional real-linear subspace of $\mathbb{C}^n$. Let L be a complex-linear subspace contained in W. Then $\dim_{\mathbb{C}} L \leq p-1$.

DEFINITION 5.1. If W is as above, then W is *maximally complex* if it contains a complex-linear subspace of dimension $p-1$. Let M be a smooth submanifold of some open subset of $\mathbb{C}^n$. Denote, for each x in M, by T_x the real tangent space to M at x. M is called *maximally complex* if for each x, T_x is maximally complex.

We shall refer to the following results from [1] as Harvey–Lawson Theorem 1 and Harvey–Lawson Theorem 2.

THEOREM 5.1. *Let M be a smooth, compact, oriented $(2p-1)$-manifold in $\mathbb{C}^n$. Assume that $p > 1$ and that M is maximally complex. Then there exists a bounded hoholomorphic p-chain T in $\mathbb{C}^n \setminus M$ such that $bd(T) = M$. Furthermore, if M is connected, this holomorphic chain consists of a single term $\tau[V]$, where $\tau = +1$ or -1.*

NOTE. If we reverse the orientation of M, maximal complexity is unaffected, and this accounts for the need of a possible $\tau = -1$.

THEOREM 5.2. *Let γ be a smooth, compact, oriented 1-dimensional manifold in $\mathbb{C}^n$. Say that "γ satisfies the moment condition" if*

$$\int_\gamma \omega = 0 \tag{23}$$

for every holomorphic 1-form ω on $\mathbb{C}^n$. Then there exists a bounded holomorphic 1-chain in $\mathbb{C}^n \setminus \gamma$ whose boundary is $[\gamma]$

6. The case $q = 1$

When $q = 1$, M is a finite union of disjoint, smooth, oriented simple closed curves in $\mathbb{C}^n$. In other words, M is a smooth 1-cycle in $\mathbb{C}^n$.

THEOREM 6.1. *Let γ be a smooth 1-cycle in $\mathbb{C}^n$. Assume that γ is connected, and assume that γ satisfies the winding condition. Then there exists an analytic variety V in $\mathbb{C}^n \setminus \gamma$, $\dim_{\mathbb{C}} V = 1$, with $bd[V] = [\gamma]$.*

Proof. By hypothesis, γ is a simple closed smooth curve. We make use of three known results regarding the polynomial hull $\widehat{\gamma}$ of a smooth simple oriented closed curve γ in $\mathbb{C}^n$.

(*) Either $\widehat{\gamma} \setminus \gamma$ is empty or $\widehat{\gamma} \setminus \gamma$ is an analytic variety of complex dimension 1.

(**) If $\widehat{\gamma} = \gamma$, then the uniform closure $\Pi(\gamma)$ of polynomials, restricted to γ, equals the space $C(\gamma)$ of all continuous complex-valued functions on γ.

(***) If $V = \widehat{\gamma} \setminus \gamma$ is non-empty, then V has finite 2-dimensional Hausdorff measure, and either $bd[V] = [\gamma]$ or $bd[V] = -[\gamma]$.

For (*) and (**) see Stout, [6], Chapter 6. (***) is given in Lawrence, [5]. For (***) see also [1], Theorem 4.7.

Suppose that $\widehat{\gamma} = \gamma$. We choose a continuous function f on γ with $f \neq 0$ and the variation of the argument of f over $\gamma = -2\pi$. Put $\delta = \inf |f|$ over γ. By (**), we can find a polynomial Q on $\mathbb{C}^n$ such that $|Q - f| \leq \frac{1}{2}\delta$ on γ.

$$Q = f + (Q - f) = f\left(1 + \frac{Q - f}{f}\right) \tag{24}$$

on γ.

By choice of Q, $|\frac{Q-f}{f}| \leq \frac{1}{2}$ on γ, so $Q \neq 0$ on γ, and

$$\text{vararg}(Q) = \text{vararg}(f) = -2\pi, \tag{25}$$

where the variations of the argument are taken over γ.

By our hypothesis on γ, $W(\gamma, Q) \geq 0$, and so by Lemma 2.1, $\text{vararg}_\gamma Q \geq 0$, and so we have a contradiction. So $\widehat{\gamma} \neq \gamma$. By (*), then, $V = \widehat{\gamma} \backslash \gamma$ is an analytic variety, and by (***), either $bd[V] = [\gamma]$ or $bd[V] = -[\gamma]$. The second case is ruled out by the winding condition. So we are done.

□

We shall later on need the generalization of Theorem 6.1 to the case of a disconnected 1-cycle. This case is treated in [2], Theorem 3 together with Lemma 1.2. Because of our Lemma 2.1 above, the hypothesis that γ satisfies the winding condition is equivalent to the following: for every polynomial P on $\mathbb{C}^n$ with $P \neq 0$ on γ, vararg P over $\gamma \geq 0$. The results of [2] then yield:

THEOREM 6.2. *Let γ be a smooth 1-cycle in $\mathbb{C}^n$. Assume that γ satisfies the winding condition. Then there exists a positive holomorphic 1-chain T in $\mathbb{C}^n \setminus \gamma$, of bounded support and finite mass, such that $bd(T) = [\gamma]$.*

7. The case $q = 2$

THEOREM 7.1. *Let M be a smooth, oriented, compact 3-dimensional manifold in $\mathbb{C}^3$. Assume that M satisfies the winding condition. Then M is maximally complex.*

Proof. Fix a complex-linear function $\phi: \mathbb{C}^3 \to \mathbb{C}$. For each λ in $\mathbb{C}$ let H_λ be the complex hyperplane $\{\phi = \lambda\}$ and put $\gamma_\lambda = M \cap H_\lambda$. For a.a. λ, γ_λ is either empty or is a smooth 1-cycle. Since the winding condition holds for M by hypothesis, Theorem 4.1 gives that for a.a. λ, γ_λ satisfies the winding condition in $\mathbb{C}^2$. Fix λ and assume that $\gamma_\lambda \neq \emptyset$.

Theorem 6.2 then gives that there exists a positive holomorphic 1-chain T in $\mathbb{C}^2 \setminus \gamma_\lambda$ such that $bd(T) = [\gamma_\lambda]$. Finally, the proof of Theorem 1 on pp. 144 and 145 in [2] shows that the existence of such holomorphic 1-chains for a.e. λ implies that M is maximally complex.

□

THEOREM 7.2. *Let M be a smooth, oriented, compact 3-manifold in $\mathbb{C}^n$, $n > 3$. Assume that M satisfies the winding condition. Then M is maximally complex.*

Proof. We follow the method of proof of Theorem 5 in [2], adapted to the case of winding numbers.

Fix a point x in M. We claim that $T_x(M)$ is maximally complex . There exists a complex-linear map $\phi: \mathbb{C}^n \to \mathbb{C}^3$, such that x is a regular point of $\phi|M$ and $\phi^{-1}(\phi(x)) \cap M = \{x\}$ (see Harvey [3], proof of Lemma 3.5, p. 349). Let M_o denote the image $\phi(M)$. M_o is an oriented immersed 3-manifold with singularities in $\mathbb{C}^3$ such that integration over M_o gives a current $[M_o]$ such that $\phi_*([M]) = [M_o]$, (see Harvey, [3], p. 368).

Let H be a complex hyperplane in $\mathbb{C}^3$ such that

(a) $L = \phi^{-1}(H)$ is a complex hyperplane in $\mathbb{C}^n$ that intersects M in a smooth 1-cycle γ_1, and

(b) $H \cap M_o$ is a 1-cycle $\gamma = \phi(\gamma_1)$.

As M satisfies the winding condition, Theorem 4.1 shows that γ_1 satisfies the winding condition in L (with L viewed as a copy of $\mathbb{C}^{n-1}$). It follows by Theorem 6.2 that γ_1 bounds a holomorphic 1-chain V_1. Hence γ bounds the holomorphic 1-chain $V = \phi(V_1)$. As (a) and (b) hold for almost all hyperplanes H, we conclude as in the proof of Theorem 1 in [2], that

$$\int_{M_o} \alpha \, dz_1 \wedge dz_2 \wedge dz_3 = 0 \tag{26}$$

for all $\mathbb{C}^\infty$ functions α on $\mathbb{C}^3$. This implies that $T_y(M_o)$ is maximally complex at points y where M_o is smooth, since α can be chosen to have support in arbitrarily small neighborhoods of y. It follows that $T_x(M)$ is maximally complex, because x is a regular point of $\phi|M$. We conclude that M is maximally complex. So Theorem 7.2 is proved.

□

8. Proof of the Main Theorem (Theorem 2.3)

We shall need the following elementary fact, proved as Lemma 5.1 in [2].

LEMMA 8.1. *Let W be a real-linear subspace of $\mathbb{C}^n$, of odd real dimension $k > 3$, that is not maximally complex. Then there exists a complex-linear hyperplane H, through 0, such that $W \cap H$ is not maximally complex.*

THEOREM 8.2. *Let k be an odd integer ≥ 3. Let M be a smooth, compact, oriented k-manifold in $\mathbb{C}^n$. Assume that M satisfies the winding condition. Then M is maximally complex.*

Proof. We follow the proof of Theorem 5 in [2].

We shall argue by induction on k. By the previous Section, the result holds when $k = 3$. Fix $k > 3$ and let M be as in the statement of Theorem 8.2. By induction hypothesis, the assertion of the Theorem is true for dimensions $3, 5, \dots, k-2$.

Suppose now that M is not maximally complex. Chose a point p in M such that $T_p(M)$ is not maximally complex. By Lemma 8.1 there is a complex hyperplane H in $\mathbb{C}^n$ (through 0) such that $T_p \cap H$ is not maximally complex. Without loss of generality the translate $H_p = H + p$ is such that $Q = M \cap H_p$ is a smooth $(k-2)$-dimensional manifold. Then the tangent space to Q at p, which is $T_p \cap H$, is not maximally complex.

But now M satisfies the winding condition, so by Theorem 4.1 Q satisfies the winding condition in H_p, viewed as $\mathbb{C}^{n-1}$. By the induction hypothesis, then Q is a maximally complex manifold. This is a contradiction. So M is maximally complex. The induction is complete. So Theorem 8.2 is proved.

□

We are now ready to give the proof of the main theorem, Theorem 2.3.

Proof of Theorem 2.3. We are given M in $\mathbb{C}^n$ such that M is connected and satisfies the winding condition.

Assume first that $\dim M = 1$, so $q = 1$. Thus here M is a simple closed curve γ satisfying the winding condition. Theorem 6.1 supplies the desired variety V in $\mathbb{C}^n \setminus \gamma$ with boundary γ.

Assume next that $\dim M = 2q - 1 \geq 3$. Theorem 8.2 gives that M is maximally complex. The first Harvey–Lawson theorem, given in Section 5 above, supplies us with a variety V with $bd[V] = [M]$ or with a variety V with $-bd[V] = [M]$. We denote by M' the manifold obtained by reversing the orientation of M. In the second case, $bd[V] = [M']$.

Suppose the second case occurs. We fix a point x in $V$and choose a complex-linear map $L: \mathbb{C}^n \to \mathbb{C}^q$ with $L(x) = 0$ and such that L satisfies (i) and (ii) relative to M'. We choose a smooth function ψ on $\mathbb{C}^n$ such that ψ vanishes on a neighborhood of $L^{-1}(0)$ and ψ is identically 1 outside of some larger neighborhood of $L^{-1}(0)$. We have

$$\int_{M'} L^*\beta_q = \int_{M'} \psi(L^*\beta_q). \tag{27}$$

Applying Stokes' theorem as in the proof of Lemma 2.2 in Section 2, and using that L(x) = 0, we get that

$$\int_{M'} L^*\beta_q > 0, \tag{28}$$

and so we conclude that $W(M', L) > 0$. But M satisfies the winding condition by hypothesis, so $W(M, L) \geq 0$. Finally, $W(M', L) = -W(M, L)$. This is a contradiction, so the second case does not occur. It follows that there is a variety bounded by M, so the proof of Theorem 2.3 is finished.

□

9. The case of a graph in $\mathbb{C}^2$

We consider a simple closed curve in $\mathbb{C}^2$ which is the graph of a smooth complex-valued function f defined on the unit circle. Denote by Γ_f the graph of f in $\mathbb{C}^2$ over the unit circle Γ.

THEOREM 9.1. *Suppose that for every polynomial P on $\mathbb{C}^2$ with $P \neq 0$ on Γ_f we have* $\operatorname{vararg}_\Gamma P(z, f(z)) \geq 0$. *Then f admits an analytic extension F to the unit disk D.*

Proof. By hypothesis, Γ_f satisfies the winding condition in $\mathbb{C}^2$. By the Main Theorem, there exists an analytic variety Σ in $\mathbb{C}^2 \setminus \Gamma_f$ with boundary Γ_f. By Stokes' theorem on Σ, then, we have, (using z, w as complex coordinates in $\mathbb{C}^2$)

$$\int_{\Gamma_f} z^n w dz = 0, \qquad n \geq 0. \tag{29}$$

This is the same as to say

$$\int_{\Gamma} z^n f(z) dz = 0, \qquad n \geq 0, \tag{30}$$

and that is the classical condition for f to have an analytic extension to D. So we are done.

□

NOTE. E. L. Stout, in [7], has the stronger result that Theorem 9.1 holds if "f is smooth on Γ" is replaced by "f is continuous on Γ".

If f is smooth, the hypothesis of Theorem 9.1 can be weakened as follows:

THEOREM 9.2. *Fix a function f in $C^2(\Gamma)$. Assume that for every polynomial P on $\mathbb{C}^2$ having the form: $P(z, w) = w + Q(z)$, where Q is a polynomial on $\mathbb{C}$, if $P(z, f(z)) \neq 0$ for each $z \in \Gamma$, then* $\operatorname{vararg}_\Gamma P(z, f(z)) \geq 0$. *Then f admits an analytic extension to D.*

Proof. Since $f \in C^2$, we may write

(31) $$f(z) = \sum_{n=-\infty}^{\infty} c_n z^n, z \in \Gamma,$$

with $\sum_{-\infty}^{\infty} n|c_n| < \infty$. We put $A(z) = -\sum_{n=o}^{\infty} c_n z^n, z \in \Gamma$.

Then A belongs to the disk algebra, and we have

(32) $$f(z) + A(z) = \sum_{-\infty}^{-1} c_n z^n = \overline{\sum_{1}^{\infty} \overline{c_{-n}} z^n}, z \in \Gamma.$$

Put $B(z) = \sum_1^\infty \overline{c_{-n}} z^n, |z| \leq 1$. Then B belongs to the disk algebra and $B \in C^1(\Gamma)$, and $B(0) = 0$. Suppose that B is not identically 0. We have $f + A = \overline{B}$ on Γ. Since B is smooth on Γ, we can find a complex number ϵ such that $B - \epsilon \neq 0$ on Γ and $B - \epsilon$ has a zero near $z = 0$. By (32), $f + A = \overline{B}$ on Γ so

(33) $$f + A - \overline{\epsilon} = \overline{B - \epsilon}$$

on Γ, and so $f + (A - \overline{\epsilon}) \neq 0$ on Γ. Also $\mathrm{vararg}_\Gamma(\overline{B - \epsilon}) < 0$.We now approximate $A - \overline{\epsilon}$ uniformly on Γ by a polynomial Q so closely that $f + Q \neq 0$ on Γ and $\mathrm{vararg}_\Gamma(f + Q) < 0$. This contradicts our hypothesis. So B is identically 0, and so (32) gives that f admits an analytic extension to D. We are done.

□

NOTE 1. We do not know whether Theorem 9.2 remains true if f is merely assumed continuous on Γ.

Tom Banchoff and Lee Rudolph asked what happens if instead of allowing arbitrary polynomials $Q(z)$ in the hypothesis of Theorem 9.2, we allowed merely polynomials of degree 1. In other words, one may ask: Let f be a smooth function on Γ and suppose that for every polynomial $w + az + b$, a, b constants, which does not vanish on Γ_f, we have $\mathrm{vararg}_\Gamma(f(z) + az + b) \geq 0$. Does it follow that f admits an analytic extension to D ?

The following example shows that the answer is No.

EXAMPLE. Take $f(z) = \overline{z} + z^2$ on Γ. We claim that for every pair of constants a, b such that $f + az + b \neq 0$ on Γ, we have $\mathrm{vararg}_\Gamma(f(z) + az + b) \geq 0$.

Fix a, b and put $g(z) = f(z) + az + b$. Assume that $g \neq 0$ on Γ. Define $N(z) = (z^3 + 1 + az^2 + bz)z^{-1}$ for all z in $\mathbb{C}$. Then $N(z) = g(z)$ on Γ. So $\frac{1}{2\pi}\,\mathrm{vararg}_\Gamma\, g = \frac{1}{2\pi}\,\mathrm{vararg}_\Gamma\, N$ = number of zeros of N in D minus the number of poles of N in D. Since $g \neq 0$ on Γ, we have that

(34) $$z^3 + 1 + az^2 + bz = (z - \alpha)(z - \beta)(z - \gamma),$$

where $|\alpha| \neq 1, |\beta| \neq 1, |\gamma| \neq 1$.

Also $\alpha\beta\gamma = -1$. Hence for one of them, say α, we have $|\alpha| < 1$. Hence the number of zeros of N in D is at least 1. So $\operatorname{vararg}_\Gamma g \geq 0$. This proves the claim.

Evidently, f has no analytic extension to D. So the answer, indeed, is No.

10. Comments

The referee pointed out that analogs of our main theorem should hold when $\mathbb{C}^n$ is replaced by a Stein manifold. Let us look at this for the case $q = 1$.

CLAIM 10.1. *Fix a Stein manifold X and let γ be a simple closed smooth oriented curve in X. Assume that for every holomorphic function ψ on X with $\psi \neq 0$ on γ, we have* $\operatorname{vararg}_\gamma \psi \geq 0$. *Then there exists an analytic subvariety V of X with $bdV = \gamma$.*

Proof. As is well known, we may embed X as an analytic submanifold X_o of $\mathbb{C}^N$ for some N. Let γ_o be the image of γ under this embedding. Fix a polynomial P on $\mathbb{C}^N$ with $P \neq 0$ on γ_o and denote by ψ the restriction of P to X_o. Our hypothesis yields that

$$\operatorname{vararg}_{\gamma o} P = \operatorname{vararg}_{\gamma o} \psi \geq 0. \tag{35}$$

So γ_o satisfies the winding condition in $\mathbb{C}^N$. By Theorem 2.3 above, there exists an analytic variety V_o of $\mathbb{C}^N \setminus \gamma_o$ with $bdV = \gamma_o$. Then V_o is contained in the polynomially convex hull of γ_o in $\mathbb{C}^N$. Since X_o is an analytic submanifold of $\mathbb{C}^n$, it follows that V_o is contained in X_o. The preimage V of V_o in X, then, is an analytic variety in X with boundary γ. We are done.

□

ACKNOWLEDGMENTS. I am very much indebted to my colleagues Bruno Harris and Tom Goodwillie for their enlightenment and suggestions regarding topological questions connected with the present paper.

References

[1] R. HARVEY and B. LAWSON, On boundaries of complex analytic varieties. I, *Ann. Math.*, **102** (1975), 233–290.

[2] H. ALEXANDER and JOHN WERMER, Linking numbers and boundaries of varieties, *Ann. Math.*, **151** (2000), 125–150.

[3] R. Harvey, Holomorphic chains and their boundaries, *Proc. Sympos. Pure Math. XXX*, Part **1** (1977), 302–382.

[4] R. Bott and L. Tu, *Differential Forms in Algebraic Topology*, Springer Verlag, New York, 1982.

[5] M. Lawrence, Polynomial hulls of rectifiable curves, *Amer. J. Math.*, **117** (1995), 405–417.

[6] E. L. Stout, *The Theory of Uniform Algebras*, Bogden and Quigley, Inc., 1971.

[7] E. L. Stout, *Boundary values and mapping degree*, Preprint, 1999.

J. Wermer, Department of Mathematics, Brown University, Providence, R.I. 02912, USA; *e-mail*: wermer@math.brown.edu

Functional Analysis and Applications: Memorial Conference for Béla Szőkefalvi-Nagy

August 2–August 6, 1999, Szeged

Conference Program

Organizers gratefully acknowledge support from:

Hungarian Academy of Sciences

Mathematical Department of the Hungarian Academy of Sciences

Regional Committee of the Hungarian Academy of Sciences, Szeged

József Attila University, Szeged

Kossuth Lajos University, Debrecen

Juhász Gyula Foundation

Soros Foundation

Pro Szeged Foundation

City Goverment of Szeged

OTP Bank, Szeged

Universitas Foundation of "Kereskedelmi & Hitelbank Rt., Debrecen"

Hungarian NFS Research grant no. T 022920, S 31357

Satelite program of the
WORLD CONFERENCE ON SCIENCE
Budapest, June 26–July 1, 1999

August 2, Monday

10:00–10:20: Opening Ceremony

Chairman: **R. G. Douglas**

10:30–11:10: C. Foias, A weighted Commutant Lifting Theorem

11:20–12:00: C. Sadosky, From the Hilbert transform to the commutant lifting theory

12:10–12:50: J. Esterle, Translation invariant subspaces for weighted Hilbert spaces of sequences

Section A

Chairman: **V. Müller**

15:00–15:30: M. Bakonyi, On an extension problem for polynomials

15:35–16:05: M. Omladic, Approximating commuting operators

16:10–16:30: M. Mbekhta, Elements with generalized bounded conjugation orbits

16:35–16:55: M. Ptak, Spherical isometries are hyporeflexive

17:00–17:20: J. Stochel, An asymmetric Putnam–Fuglede theorem for unbounded operators

17:25–17:45: M. Zajac, Reflexivity of intertwining operators

17:50–18:10: M. Kosiek, Invariant subspaces for n-tuples of contractions

Section B

Chairman: **C. Sadosky**

15:00–15:30: I. Suciu, Power dilation, minimality and semi-invariant subspaces

15:35–16:05: B. Pavlov, Differential operators on graphs: spectral theory and scattering

16:10–16:30: S. Fedorov, On the boundedness of weighted Hilbert transform in the multiply connected domain

16:35–16:55: K. Watanabe, On the E_{-2}-construction: Scattering theory

17:00–17:20: L. Molnár, Local automorphisms of some quantum mechanical structures

17:25–17:45: D. Petz, Monotone inner products

17:50–18:10: S. Hassi, On parametrization and admissibility of generalized resolvents of symmetric operators and their intermediate extensions

20:00 Welcome Party

August 3, Tuesday

Chairman: **C. Foias**

9:30–10:10: **R. G. Douglas**, Quotient Hilbert modules and Sz.-Nagy–Foias model theory

10:20–11:00: **S. Naboko**, Szőkefalvi-Nagy–Foias functional model for dissipative operators and applications to the neutron physics

11:10–11:50: **V. Vasyunin**, Sz.-Nagy–Foias model in a coordinate-free setting

12:00–12:40: **D. Hadwin**, Completely rank-nonincreasing linear maps

Section A

Chairman: **L. Zsidó**

14:30–15:00: **B. Chevreau**, Some applications of dilation theory in the study of dual algebras

15:10–15:40: **J. A. R. Holbrook**, Schur norms and the multivariate von Neumann inequality

15:50–16:10: **C.-G. Ambrozie**, Model and dilation theory in several variables

16:25–16:45: **G. Cassier**, Similarity problems and inequalities of von Neumann type

16:55–17:15: **S. Kupin**, A sufficient condition for the similarity of a weak contraction to a normal operator

17:25–17:45: **M. Dostal**, From similarity to $\mathcal{U}+\mathcal{K}$-orbits—a block tri-diagonal decomposition technique

Section B

Chairman: **S. Naboko**

14:30–15:00: **G. Misra**, New homogeneous operators from old ones via Nagy–Foias theory

15:10–15:40: **D. Gaspar**, Intertwinings for bicontractions and von Neumann inequalities

15:50–16:10: **D. Kalyuzhniy**, On the extension of B. Sz.-Nagy's Dilation Theorem to linear pencils of operators

16:25–16:45: **S. Marcantognini**, On liftings of intertwining operators

16:55–17:15: **E. Abakumov**, On shift invariant subspaces in Banach sequence spaces

17:25–17:45: **E. A. Suciu**, Nagy–Foias diagram over $A(D^n)$

19:15 Visit to the grave site

August 4, Wednesday

Chairman: **F.-H. Vasilescu**

9:30–10:10: **H. Langer**, Singularly perturbed selfadjoint operators
10:20–11:00: **A. Atzmon**, Maximal and minimal invariant subspaces
11:10–11:50: **V. Müller**, Dvoretzky's theorem in operator theory
12:00–12:40: **J. Eschmeier**, Invariant subspaces for spherical contractions

Section A

Chairman: **V. Vasyunin**

14:30–15:00: **D. Z. Arov**, On the class of strongly regular J-inner matrix functions
15:10–15:40: **I. Valusescu**, The maximal function of a semispectral measure
15:50–16:10: **H. de Snoo**, Operator models associated with Kac subclasses of generalized Nevanlinna functions
16:25–16:45: **M. Möller**, On the essential spectrum of some operators from mathematical physics
16:55–17:15: **M. Langer**, Resonances of a λ-rational Sturm–Liouville problem
17:25–17:45: **A. Luger**, A factorization result for matrix Nevanlinna functions

Section B

Chairman: **F. H. Szafraniec**

14:30–15:00: **M. Uchiyama**, Operator monotone functions which are defined implicitly and operator inequalities
15:10–15:40: **T. Furuta**, Order preserving operator inequalities and related classes of operators around p-hyponormal and log-hyponormal operators
15:50–16:10: **M. Fujii**, Furuta inequality for chaotic order and its applications
16:25–16:45: **E. Kamei**, Parametrized Furuta inequality under the chaotic order
16:55–17:15: **P. Kurasov**, Krein's formula in perturbation theory
17:25–17:45: **Z. D. Arova**, On Livsic–Brodskii nodes with strongly regular J-inner characteristic functions in the Hardy class
17:50–18:10: **S. Ota**, On q-normal operators

August 5, Thursday

Chairman: **H. Langer**

9:30–10:10: **L. Zsidó**, Analytic extension of vector valued functions

10:20–11:00: **J. Wermer**, The Argument Principle and boundaries of analytic varieties

11:10–11:50: **A. Korányi**, Hardy spaces of harmonic functions in old and new settings

12:00–12:40: **C. J. K. Batty**, Extensions and stability of semigroups of operators

Section A

Chairman: **C. J. K. Batty**

14:30–15:00: **Q.-P. Vu**, The Lyapunov–Sylvester equations and the asymptotic behaviour of solutions of difference equations

15:10–15:40: **A. Athavale**, Completely hyperexpansive operator tuples

15:50–16:10: **J. Janas**, The spectral phase transition for a class of unbounded Jacobi matrices

16:25–16:45: **J. van Neerven**, On the stochastic abstract Cauchy problem in Banach spaces

16:55–17:15: **I. Cioranescu**, Representation formulas for cosine functions in UMD-spaces

17:25–17:45: **A. Biswas**, Weighted variant of the three chains completion theorem

17:50–18:10: **R. Drnovsek**, Common invariant subspaces for collections of operators

Section B

Chairman: **V. Peller**

14:30–15:00: **T. Nakazi**, One dimensional perturbations of Hankel operators

15:10–15:40: **P. Y. Wu**, Polygon and numerical ranges

15:50–16:10: **V. Kapustin**, Spectral projections of non-contractive operators and exposed points in H^1

16:25–16:45: **H. Woracek**, On de Branges spaces of entire functions

16:55–17:15: **A. Gheondea**, Contributions to the harmonic analysis on Krein spaces

17:25–17:45: **V. Strauss**, On a spectral decomposition of a π-semiunitary operator

17:50–18:10: **M. Sabac**, Localization of the Wielandt–Wintner Theorem

19:00 Reception

August 6, Friday

Chairman: **D. Hadwin**

9:30–10:10: **S. Stratila**, The Commutation Theorem for tensor products of operator algebras over subalgebras

10:20–11:00: **H. Bercovici**, Continuous combinatorics in Jordan model theory and eigenvalue inequalities

11:10–11:50: **V. Peller**, Unitary interpolations of matrix functions

12:00–12:40: **F.-H. Vasilescu**, Operator moment problems in unbounded sets

Section A

Chairman: **M. Omladic**

14:30–15:00: **F. H. Szafraniec**, Subnormality in the quantum harmonic oscillator

15:10–15:40: **H. Helson**, The Smirnov class of a class of groups

15:50–16:10: **Z. Sebestyén**, Normalizable operators

16:25–16:45: **B. Le Gac**, A Beppo Levi type theorem for bundle convergence in noncommutative L^2-spaces

16:55–17:15: **F. Móricz**, Orthogonal series and SLLN in noncommutative L^2-spaces

Section B

Chairman: **P. Y. Wu**

14:30–15:00: **L. A. Fialkow**, The quartic complex moment problem

15:10–15:40: **J. Zemanek**, Powers and resolvents

15:50–16:10: **P. Semrl**, Preserving invertibility

16:25–16:45: **G. Hofmann**, On GNS representations on indefinite inner product spaces and applications to QFT

16:55–17:15: **L. Stachó**, Contractive projections from the viewpoint of holomorphy

List of Participants

Abakumov, Evgeny	Paris, France
Ambrozie, Calin-Grigore	Bucharest, Romania
Arov, Damir	Odessa, Ukraine
Arova, Zoya	Odessa, Ukraine
Athavale, Ameer	Pune, India
Atzmon, Aharon	Tel Aviv, Israel
Bakonyi, Mihály	Atlanta, Georgia, U.S.A.
Batty, Charles J. K.	Oxford, England
Bercovici, Hari	Bloomington, Indiana, U.S.A.
Biswas, Animikh	Bloomington, Indiana, U.S.A.
Bognár, János	Budapest, Hungary
Cassier, Gilles	Lyon, France
Chevreau, Bernard	Bordeaux, France
Cioranescu, Ioana	San Juan, Puerto Rico
Clark, Douglas N.	Athens, Georgia, U.S.A.
Craciunescu, Aurelian	Timisoara, Romania
De Snoo, Henk	Groningen, The Netherlands
Dostal, Michal	Praha, Czech Republic
Douglas, Ronald G.	College Station, Texas, U.S.A.
Drnovsek, Roman	Ljubljana, Slovenia
Durszt, Endre	Szeged, Hungary
Eschmeier, Jörg	Saarbrücken, Germany
Esterle, Jean	Bordeaux, France
Fedorov, Sergei	Auckland, New Zealand
Fialkow, Lawrence A.	New Paltz, New York, U.S.A.
Foias, Ciprian	Bloomington, Indiana, U.S.A.
Fujii, Masatoshi	Osaka, Japan
Furuta, Takayuki	Tokyo, Japan
Gaspar, Dumitru	Timisoara, Romania
Gaspar, Pastorel	Timisoara, Romania
Gheondea, Aurelian	Bucharest, Romania
Győry, Máté	Debrecen, Hungary
Hadwin, Don	Durham, New Hampshire, U.S.A.
Hassi, Seppo	Helsinki, Finland
Helson, Henry	Berkeley, California, U.S.A.

Hofmann, Gerald	Leipzig, Germany
Holbrook, John	Guelph, Ontario, Canada
Janas, Jan	Krakow, Poland
Kalyuzhniy, Dmitriy	Odessa, Ukraine
Kamei, Eizaburo	Maebashi, Japan
Kapustin, Vladimir	St.Petersburg, Russia
Kérchy, László	Szeged, Hungary
Korányi, Ádám	Bronx, New York, U.S.A.
Kosiek, Marek	Krakow, Poland
Kupin, Stanislav	Bordeaux, France – Kharkov, Ukraine
Kurasov, Pavel	Stockholm, Sweden
Langer, Heinz	Wien, Austria
Langer, Matthias	Wien, Austria
Le Gac, Barthélemy	Marseille, France
Luger, Annemarie	Wien, Austria
Luig, Klaus	Leipzig, Germany
Marcantognini, Stefania	Caracas, Venezuela
Mbekhta, Mostafa	Lille, France
Misra, Gadadhar	Bangalore, India
Molnár, Lajos	Debrecen, Hungary
Móricz, Ferenc	Szeged, Hungary
Möller, Manfred	Witwatersrand, South Africa
Müller, Vladimir	Praha, Czech Republic
Naboko, Sergey	St.Petersburg, Russia
Nakazi, Takahiko	Sapporo, Japan
Neerven, Jan van	Delft, The Netherlands
Omladic, Matjaz	Ljubljana, Slovenia
Ota, Schoichi	Fukuoka, Japan
Pavlov, Boris	Auckland, New Zealand
Peller, Vladimir	Manhattan, Kansas, U.S.A.
Petz, Dénes	Budapest, Hungary
Ptak, Marek	Krakow, Poland
Sabac, Mihai	Bucharest, Romania
Sadosky, Cora	Washington, DC, U.S.A.
Sebestyén, Zoltán	Budapest, Hungary
Semrl, Peter	Ljubljana, Slovenia
Stachó, László	Szeged, Hungary
Stochel, Jan	Krakow, Poland
Stratila, Serban	Bucharest, Romania

Strauss, Vladimir	Caracas, Venezuela
Suciu, Elena Alina	Bucharest, Romania
Suciu, Ion	Bucharest, Romania
Suciu, Nicolae	Timisoara, Romania
Szafraniec, Franciszek H.	Krakow, Poland
Uchiyama, Mitsuru	Fukuoka, Japan
Valusescu, Ilie	Bucharest, Romania
Vasilescu, Florian-Horia	Lille, France
Vasyunin, Vasily	St.Petersburg, Russia
Vu, Quoc-Phong	Athens, Ohio, U.S.A.
Watanabe, Kazuo	Tokyo, Japan
Wermer, John	Providence, Rhode Island, U.S.A.
Woracek, Harald	Wien, Austria
Wu, Pei Yuan	Hsinchu, Taiwan
Zajac, Michal	Bratislava, Slovak Republic
Zemanek, Jaroslav	Warsaw, Poland
Zsidó, László	Rome, Italy

MIX
Papier aus verantwortungsvollen Quellen
Paper from responsible sources
FSC® C105338

If you have any concerns about our products,
you can contact us on
ProductSafety@springernature.com

In case Publisher is established outside the EU,
the EU authorized representative is:
Springer Nature Customer Service Center GmbH
Europaplatz 3, 69115 Heidelberg, Germany

Printed by Libri Plureos GmbH
in Hamburg, Germany